Atlas of the Ultraviolet Sky

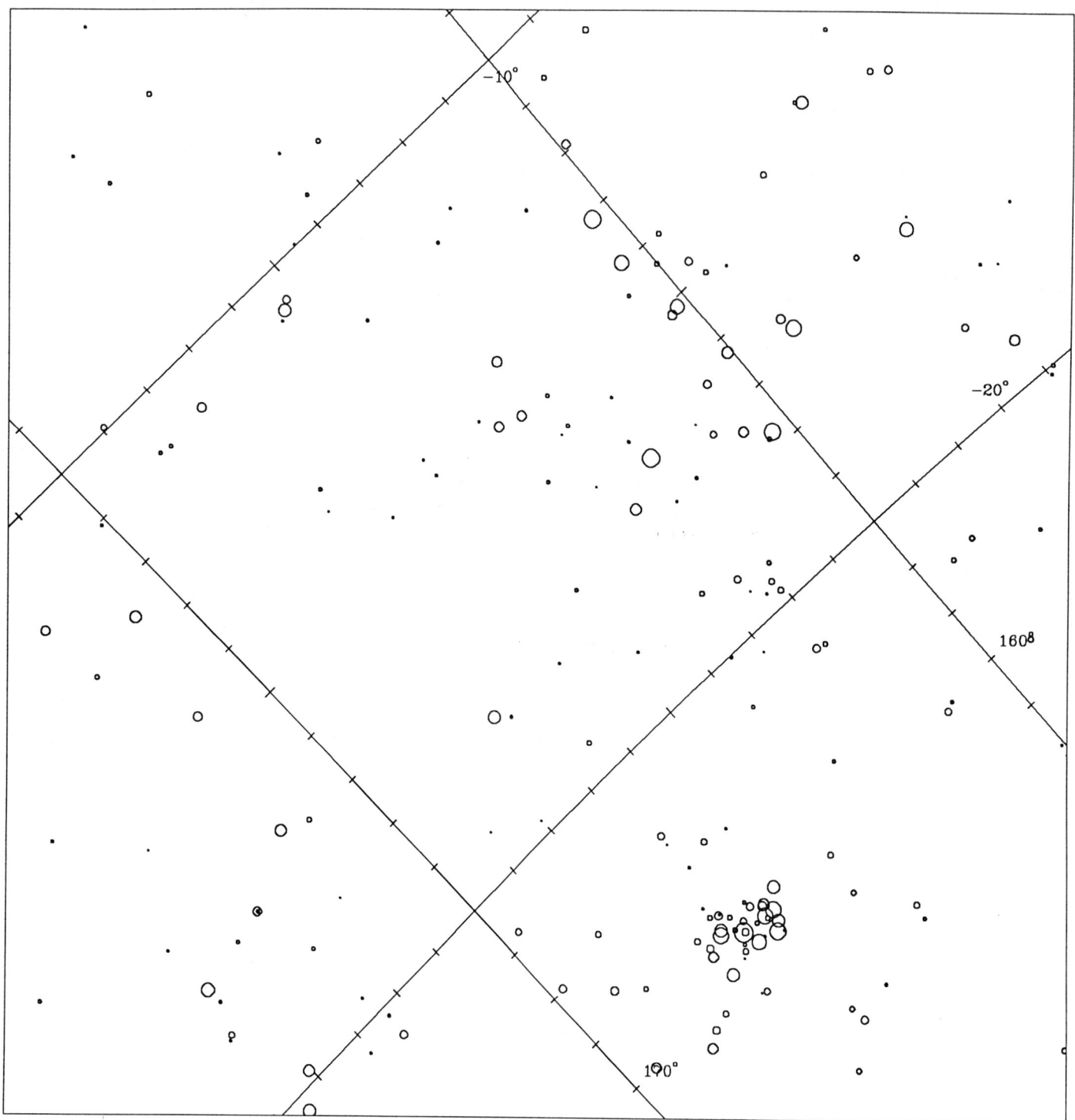

Atlas of the Ultraviolet Sky

Richard C. Henry,
Wayne B. Landsman,
Jayant Murthy,
Peter D. Tennyson,
James B. Wofford, and
Robert Wilson

The Johns Hopkins University Press
Baltimore and London

© 1988 The Johns Hopkins University Press
All rights reserved
Printed in the United States of America

The Johns Hopkins University Press, 701 West 40th Street, Baltimore,
Maryland 21211
The Johns Hopkins Press Ltd., London

The paper used in this publication meets the minimum requirements of
American National Standard for Information Sciences—Permanence of
Paper for Printed Library Materials, ANSI Z39.48-1984.

Library of Congress Cataloging-in-Publication Data

Atlas of the ultraviolet sky.

   Includes indexes.
   1. Stars—Atlases.   2. Ultraviolet astronomy—Atlases.
I. Henry, Richard C.
QB65.A86   1988   522'.68   88-45403
ISBN 0-8018-3738-3 (alk. paper)

Contents

Acknowledgments vi

Introduction 1

Index to Celestial Coordinates 21

Plates 27

Index to Galactic Coordinates 453

Acknowledgments

The ultraviolet stellar data on which the atlas is based were obtained by the European Space Research Organization. This work was partially supported by Infrared Astronomy Satellite (IRAS) contract 957653 and by NASA grant 5-619 to the Johns Hopkins University. The maps were generated on computers provided by NASA, in support of the Hopkins Ultraviolet Telescope (HUT). This work was also supported, in its initial phases, by a small Strategic Defense Initiative (SDI) subcontract from the Applied Physics Laboratory of the Johns Hopkins University, for which we thank Dr. Donald J. Williams and Dr. Ching-I. Meng. We thank E. Dorrit Hoffleit for permission to use the *Yale Bright Star Catalog* data.

Introduction

This book is a collection of 212 pairs of maps, together portraying the entire sky at both visible and ultraviolet wavelengths.

If the atlas is opened to any page, the map on the left shows an 18° by 18° portion of the sky, with the positions of the stars indicated by symbols whose size reflects the brightness of that star as seen by the human eye. All stars visible to the human eye are shown, as well as somewhat fainter stars; also, the names of the brightest stars are given. This visible-brightness map is intended only to orient the user.

The second map in the pair, the one on the right-hand page, shows the identical area of the sky, but the stars are now indicated by symbols whose size reflects the star's brightness in the ultraviolet part of the spectrum, in particular at 1565 Å. These maps go much fainter than do the visible, orientation maps, and the relative sizes of symbols between maps are not intended for comparison.

What is meant by "the ultraviolet"? Astronomers usually characterize electromagnetic radiation by its wavelength in Ångstrom units, where $1 \text{ Å} = 10^{-8}$ centimeters. Using these units, the human eye is sensitive to radiation in the approximate wavelength range 4000 Å to 6500 Å. Astronomers making precise measurements of the brightness of stars often use, in conjunction with a photoelectric photometer, a set of three filters, called U (peak transmission at 3600 Å), B (4200 Å), and V (5400 Å). The symbols U, B, and V are read "ultraviolet," "blue," and "visible," respectively, so, for this particular purpose, "ultraviolet" means "the wavelength range 3200 Å to 3900 Å," where indeed the human eye has little or no sensitivity.

The word "ultraviolet" is also used, in astronomy and elsewhere, for radiation of much shorter wavelengths. The transition from ultraviolet to X-rays occurs at perhaps 100 Å, or even at shorter wavelengths. So in fact "ultraviolet" means electromagnetic radiation in the approximate wavelength range 100 Å to 4000 Å.

However, the plane of our galaxy, where the Sun and Earth are located, is not a vacuum. It contains, between the scattered stars, vast quantities of interstellar gas (principally atomic hydrogen and helium), at extremely low density. Ultraviolet radiation having a wavelength shorter than 912 Å is capable of photoionizing such gas, and, in this interaction, the radiation itself vanishes. This process is extremely effective. As a result, only a very small number of relatively nearby stars (above all, the Sun) can be detected at all, at wavelengths shorter than 912 Å. For this reason, the wavelength range 912 Å to 4000 Å might be termed the "astronomical ultraviolet." Thus, the atlas presents the brightnesses of the stars at 1565 Å, a wavelength in the astronomical ultraviolet very far removed from visible wavelengths.

Even in the astronomical ultraviolet, the interstellar medium is not completely transparent, because of the existence, in addition to the interstellar gas, of interstellar dust. These dust particles, of uncertain composition, are known to be comparable in size to wavelengths of visible and ultraviolet light, and they tend to absorb such radiation. The effectiveness of the absorption increases as one moves, in the visible, to bluer wavelengths and, because of this, the visible light of distant stars is said to be "reddened." In the ultraviolet the extinction by dust is stronger still, reaching a peak at 2200 Å and then generally decreasing somewhat at still shorter wavelengths. So in the astronomical ultraviolet, extinction by dust is more effective than it is in the visible.

An important factor determining the apparent brightness of a star is, of course, its distance. Apart from distance itself, and the concomitant effects of extinction (that is, absorption of light, and also scattering of light) by interstellar dust, stars differ from one another principally in luminosity and temperature.

Luminosity differences do not produce large differences between the ultraviolet and visible-light appearances of stars. Indeed, two

Introduction

stars of identical temperature differing only in luminosity, and located at the same distance, will appear exactly the same as two stars of identical temperature and identical luminosity at appropriate different distances, and (apart from effects of interstellar extinction) differences in distance alone cannot produce any difference in the relative brightness between the visible and the ultraviolet.

We are left with temperature as the parameter that determines, most significantly by far, the relative brightnesses of stars in the ultraviolet, compared with their visible-light brightnesses. By temperature, we do not mean the interior temperature (which can be hundreds of millions of degrees), but rather the so-called photospheric temperature, the temperature of the outer layer of the star from which electromagnetic radiation actually succeeds in escaping from the star into interstellar space. For the Sun, that temperature is about 6000°K. The photospheric temperatures of other stars range from 2400°K to 40,000°K. A star with a photospheric temperature of 40,000°K will radiate predominately in the shortest ultraviolet wavelengths, whereas a star as cool as the Sun will emit only a negligible portion of its radiation in the astronomical ultraviolet.

Let us see the dramatic effect of temperature, by consulting the atlas.

Every astronomer knows that the brightest star in the prominent constellation Orion is the red star Betelgeuse, and also that Orion is located on the celestial equator between 5 and 6 hours of right ascension. Well, at least all *amateur* astronomers know that! Since it is the RA and Dec that we know, we consult the Index to Celestial Coordinates, which immediately precedes the atlas maps (had it been galactic coordinates that we had happened to know, we would have consulted the Index to Galactic Coordinates, which follows the maps). The index indicates that Orion appears on plates 100 and 101. Turning to those plates, we find Betelgeuse (α Ori) near the top of the visible-light version of plate 101. Farther down on that same map, we identify the three stars of Orion's belt, as well as his sword.

Now compare the ultraviolet map. In the ultraviolet, the three stars of the belt are easily picked out, and the sword region is prominent, but Betelgeuse cannot be found. The use of a transparent plastic overlay allows us to check that Betelgeuse is entirely absent in the ultraviolet.

Astronomers classify stars, according to their temperature, by means of their spectral class:

Spectral Class	Temperature
O	40,000
B	20,000
A	9,000
F	6,500
G	6,000
K	4,000
M	3,000

Only the hottest stars, spectral classes O, B, and A, emit sufficient radiation at 1565 Å to appear in the atlas. In contrast, stars of *all* spectral types appear in the visible-light maps of the atlas.

The Ultraviolet Data

In March 1972, from the U.S. Western Test Range, California, a satellite of the European Space Research Organization called TD-1 was launched. Among the experiments on board was an ultraviolet astronomy experiment, called S2/68, which was provided by scientists from Belgium and the United Kingdom. That experiment is described by A. Boksenberg, R. G. Evans, R. G. Fowler, I. S. K. Gardner, L. Houziaux, C. M. Humphries, C. Jamar, D. Macau,

D. Malaise, A. Monfils, K. Nandy, G. I. Thompson, R. Wilson, and H. Wroe (1973).

The data from that experiment have been used in a large number of important astronomical and astrophysical investigations, which have been published in the scientific literature. A small TD-1 map of the entire sky at 1565 Å has been published by Gondhalekar, Phillips, and Wilson (1980), and the great body of stellar ultraviolet data from the experiment has been published in three important catalogs (Jamar et al. 1976; Macau-Hercot et al. 1978; and Thompson et al. 1978).

In the course of our research in ultraviolet astrophysics, we have again and again found it extremely useful to use the magnetic tape version of these TD-1 data to create maps of limited regions of the ultraviolet sky. It occurred to us that systematic generation and publication of an atlas representing the entire sky in the ultraviolet would be useful, and might even find some audience beyond narrow specialists in the field. Although ultraviolet maps of limited regions have been published, notably by Page, Carruthers, and Heckathorn (1982), we felt that a systematic atlas, to a uniform limiting brightness, on a good scale, might be of value.

To construct the atlas, we used the complete, unpublished TD-1 survey catalog of 58,012 stars. A subset of this catalog, containing the 31,215 stars having a signal to noise greater than 10 in any passband, has been published by Thompson et al. (1978). The 1565 Å data have a bandpass of 330 Å.

For several reasons, the TD-1 sensitivity limit was not uniform across the sky. This nonuniformity was partially due to its polar orbit, which meant that the number of useable scans (and signal to noise) increased with a star's ecliptic latitude. There were also some variations in the backgrounds, which at 1565 Å were mainly due to scattered geocoronal Lyman Alpha (characteristic radiation of hydrogen gas). We desired a sensitivity limit that would ensure completeness over the entire sky. A study of the distribution of 1 sigma errors in the 1565 Å flux (for those stars with negligible photon statistics) shows that 99 percent of the observations had errors less than 6×10^{-13} ergs cm^{-2} s^{-1} Å^{-1}. Therefore, the sensitivity limit for this atlas was chosen to be 1.0×10^{-12} ergs cm^{-2} s^{-1} Å^{-1}.

A more serious problem with using the TD-1 data for a sky atlas occurred whenever two or more stars appeared simultaneously within the 17' × 11.9' spectrophotometer entrance slot, causing their spectra to overlap. No fluxes were recorded whenever stars were too closely blended to give useful results. Thus, approximately 10 percent of the stars in the atlas did not have fluxes recorded by TD-1. For these stars it was necessary to compute ultraviolet fluxes theoretically, based on the star's visible magnitude, spectral type, luminosity, and color excess. This information was converted to an ultraviolet flux using the table of intrinsic ultraviolet colors published by Carnochan (1982), and an average interstellar reddening curve. The procedure is discussed in more detail in Landsman (1984), and it is quite similar to the method described by Gondhalekar, Phillips, and Wilson (1980). To keep the star maps easy to read, we did not indicate whether the ultraviolet flux shown for a particular star was theoretically computed, or actually observed by TD-1. However, any star observed by TD-1, with a flux greater than 5×10^{-12} ergs cm^{-2} s^{-1} Å^{-1}, should be found in the catalog of Thompson et al. (1978). That includes 10,518 of the 25,314 stars in this atlas.

The Ultraviolet Sky

An overview of the ultraviolet sky was obtained, using a photometer on board a rocket, by Henry, Swandic, Shulman, and Fritz (1977). Their observed map of the sky is shown as figure 1(a), while figure 1(b) shows the ultraviolet brightness that was predicted using the method of Henry (1977). Good general agreement is seen.

The maps in figure 1 are Aitoff equal-area projections of the entire sky. They are in celestial coordinates, the same coordinate system that is used in the atlas. The north celestial pole is at the top of the map, and 19^h of right ascension is at each edge of the map. The darker the shading, the brighter the ultraviolet emission (at 1530 Å) at that location.

A glance at figure 1 shows that in the ultraviolet, as we know to be the case in the visible, the Milky Way is prominent. Superimposed on the RA, Dec maps of figure 1 is a coarse grid of galactic coordinates. The Milky Way (darkest shading) crudely follows the galactic equator, whereas the north and south galactic poles are quite dark (that is, white in the figure).

A closer examination of figure 1 shows that the Milky Way is tipped some 19° with respect to the galactic equator (which is defined with reference to much more distant reaches of the galaxy). A coordinate system more appropriate to the *local* region of the galaxy is the so-called Gould coordinates, and a predicted ultraviolet-brightness map of the sky in these coordinates is given in figure 2. This is the physically most meaningful coordinate system in which to view the ultraviolet sky. Also, in figure 2 (unlike figure 1), the shading is not saturated on the darkest regions. Thus figure 2 presents a true linear predicted ultraviolet picture of the sky.

The appearance of the ultraviolet sky, as seen in figure 2, is dramatic indeed. Not only are the ultraviolet-emitting stars tightly confined to the Gould equator, they are extremely unevenly distributed in Gould longitude. The very bright (heavily shaded) region near the center of the map is Orion. To the left, the southern Milky Way (Puppis, Carina, Norma, Scutum) is very bright, while to the right, the northern Milky Way (Auriga, Cassiopeia, Cygnus) is relatively very dark. Indeed, some parts of the Gould equator are almost as dark as high-latitude regions far from the galactic plane.

Powerful experimental verification of the great concentration of

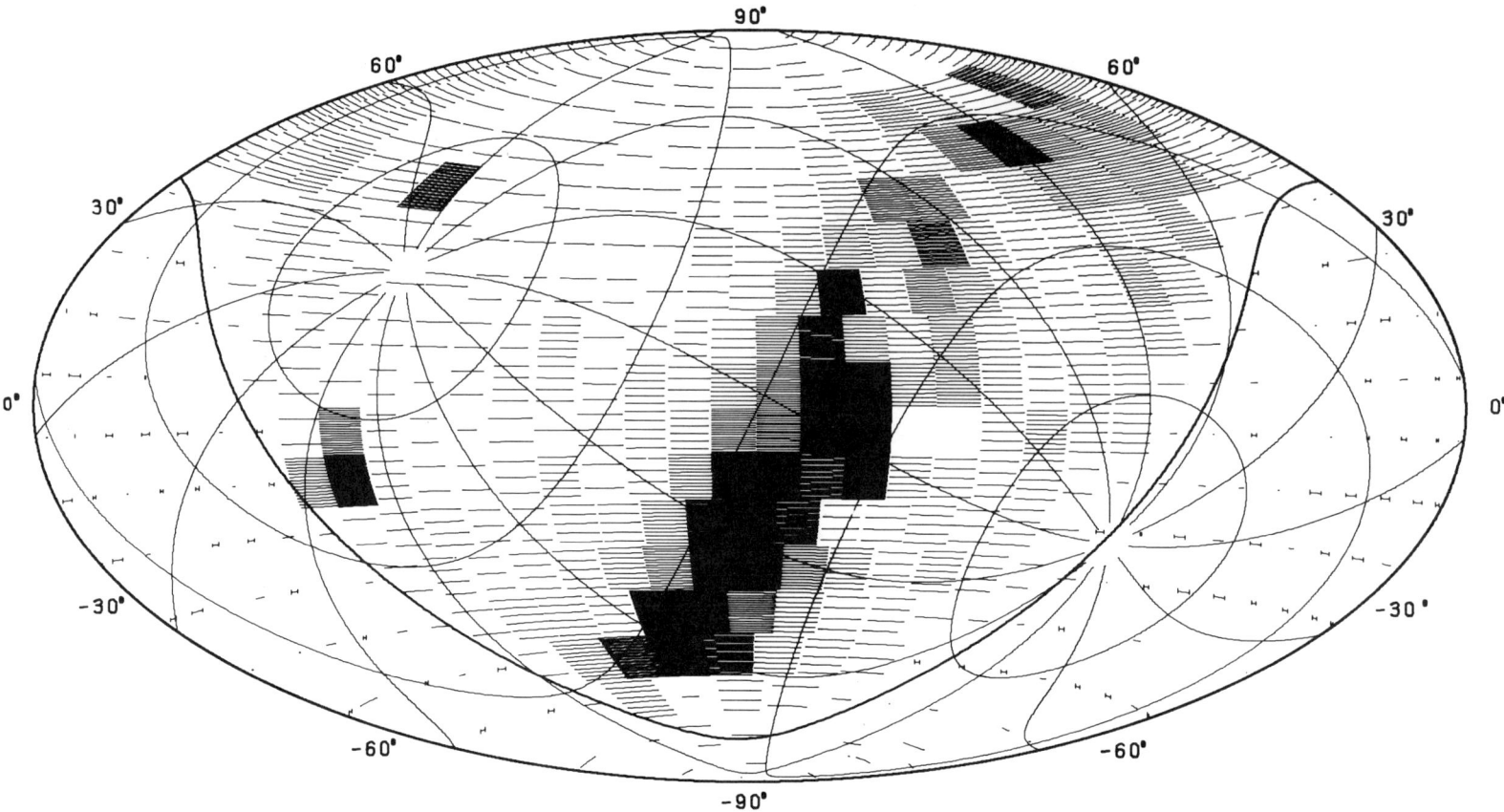

Figure 1(a). This is an experimentally measured map, at 1530 Å wavelength, of about half the sky, obtained using a photometer carried above the Earth's atmosphere on a rocket. The sky is plotted in celestial coordinates (Right Ascension and Declination), with the North Celestial Pole at the top of the map. The RAs of the side edges of the map are 19 hours (285°). The darkest areas are the regions that are brightest at 1530 Å. A coarse grid of galactic coordinates is superimposed. The two isolated very bright stars in the general vicinity of the North Galactic Pole are η UMa at a Declination of 49°, and Spica (α Vir) at a Declination of −10°. (A much more sensitive instrument aboard the TD-1 satellite was used to obtain the brightnesses of individual stars that were used in constructing the maps that form the main body of the atlas.)

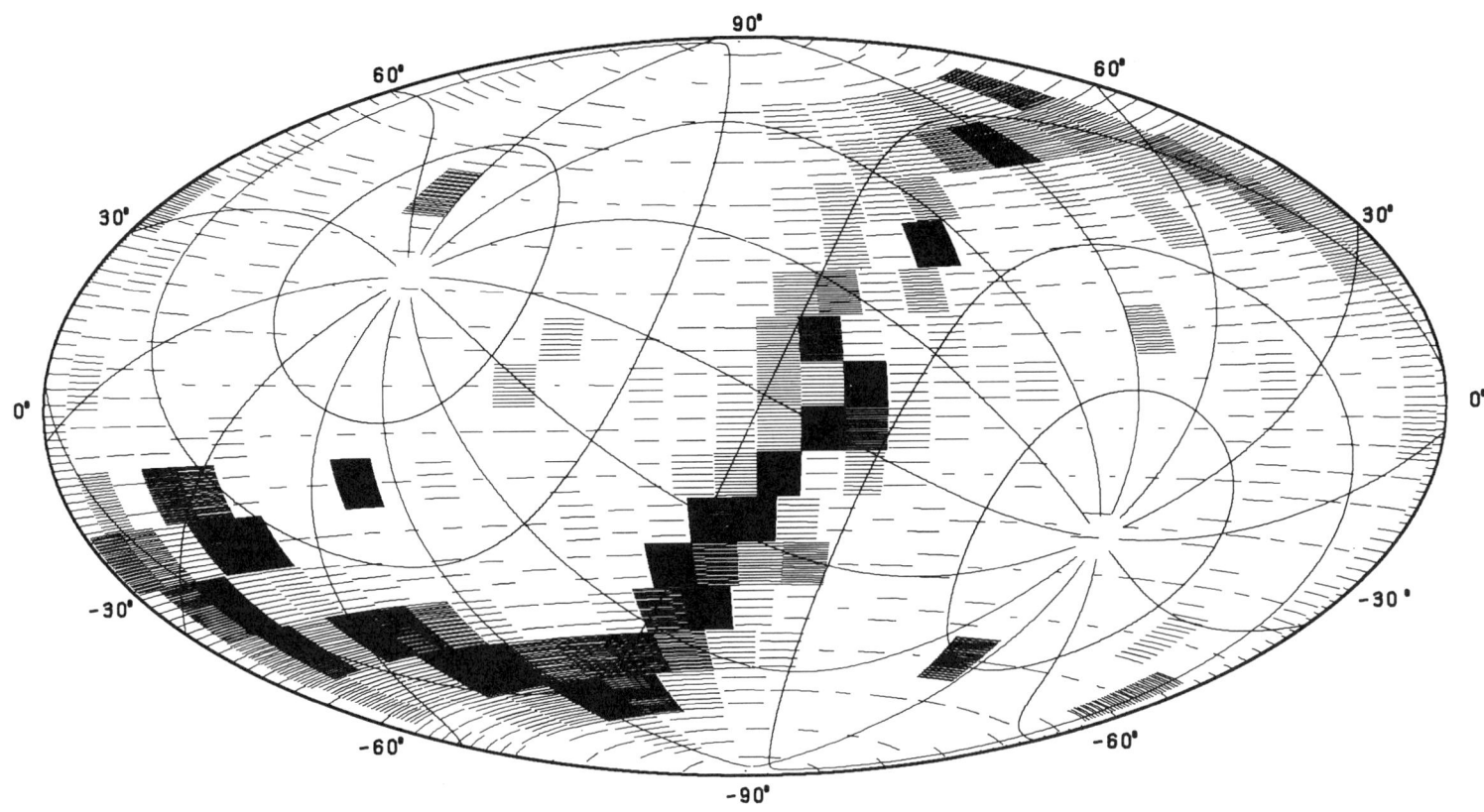

Figure 1(b). For comparison with the observed ultraviolet picture of the sky (shown in part *a* of this figure), we provide a computer-generated prediction of sky brightnesses, which is based on each star's visible brightness and spectral class. Good general agreement is seen in regions that were surveyed by the rocket-borne photometer, and predictions are seen for the regions of the sky that could not be seen by the photometer (because the Earth was in the way). The bright isolated star at galactic latitude $-59°$ is Achernar (α Eri).

ultraviolet radiation to parts of the Gould plane has been provided by an experiment by William G. Fastie (1973). An Ebert spectrometer was carried aboard the *Apollo 17* spacecraft to the moon, and later, on transearth coast, observations were made with the spectrometer of the distribution of ultraviolet light over the sky. During the first astronaut sleep period on transearth coast, the data shown in figure 3 were obtained. Ultraviolet brightness, on a linear intensity scale, is plotted versus time. The spacecraft rolled with a period of about 20 minutes, so the same regions of the sky were scanned repeatedly. The intensities shown are for various ultraviolet wavelengths (various solid and dashed lines), for the first of the many rotations. Notice the extremely high intensities observed at the two crossings of the Gould plane, in Orion, and in the southern Milky Way. Notice the single isolated star, η UMa, observed at 12 minutes. Notice the object observed at 16 minutes, and particularly notice that its brightness at different ultraviolet wavelengths varies dramatically, which is not true for η UMa or for the two galactic plane crossings. This object is nothing but the planet Earth, observed from deep cislunar space. Finally, notice the almost perfect blackness of the ultraviolet sky, far from the plane crossings; that is, before and after the observation of η UMa. Even between the two plane crossings, in the 0 to 4 minute period, when the look direction was less than 10° from the Gould plane, the sky is relatively dark.

The sky has an interestingly different appearance in the ultraviolet, compared with the visible. If you were located outside the Earth's atmosphere, and if your eyes were suddenly to lose their visible-light sensitivity but gain equal 1565 Å sensitivity, the appearance of the sky would change dramatically. Late-type (cooler) stars would vanish, while early-type (hot) stars would increase greatly in brightness, and early-type stars too faint to be seen in the visible would be readily apparent. That, plus thousands of still fainter stars, is what appears in the atlas.

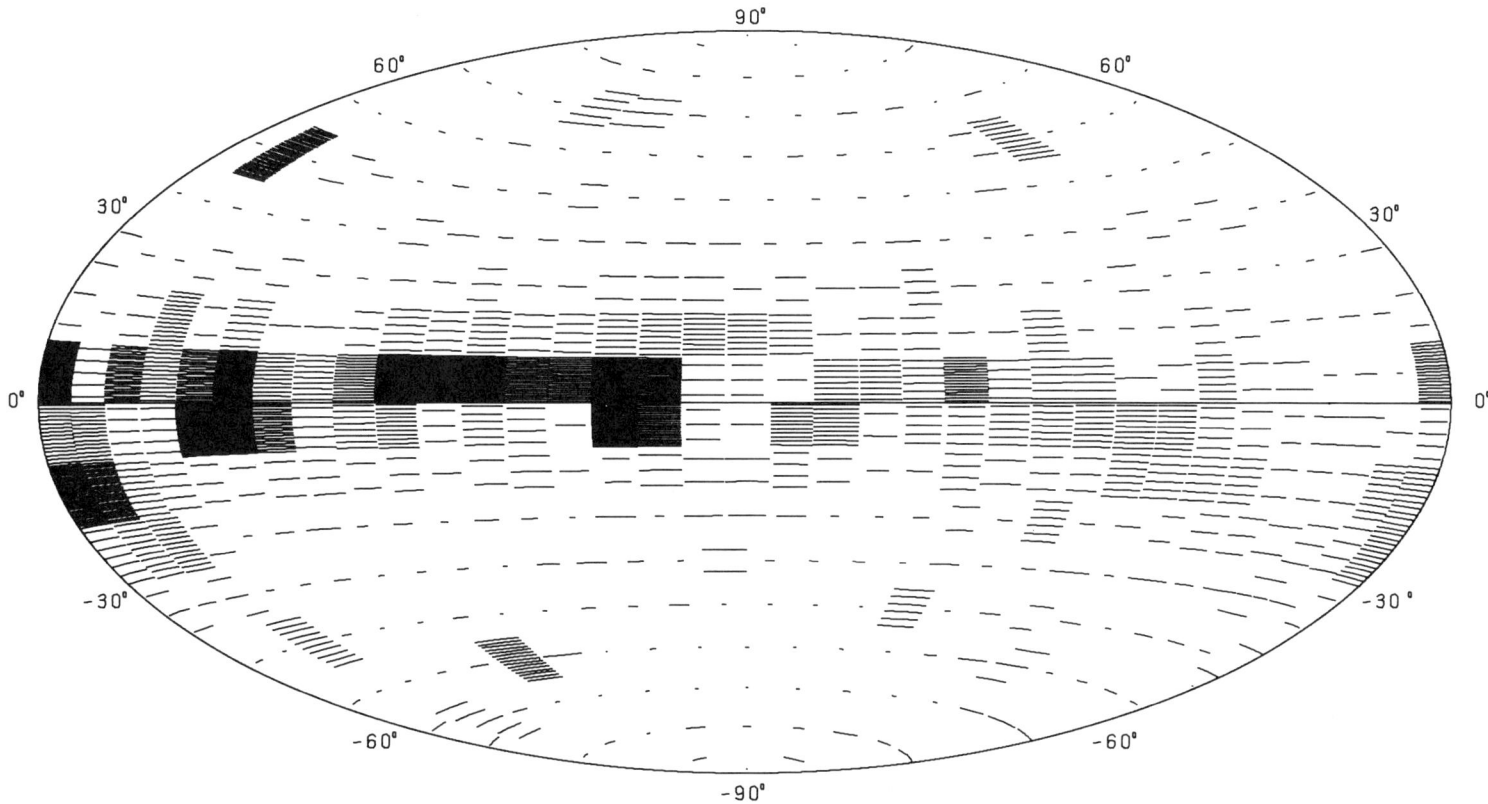

Figure 2. A computer-generated prediction of sky brightness at 1482 Å, based on each star's visible brightness and spectral class. The entire sky is shown, with the Gould belt forming the equator of this particular map. The Gould belt is a rather local portion of the Milky Way galaxy which happens to be tipped about 19° from the rest of the galactic plane. The figure indicates that ultraviolet light from the comparatively local stars of the Gould belt heavily dominates the appearance of the sky in the ultraviolet. That this is not just an artifact of the brightness limit of the star catalog that was used in constructing the map can be seen by examining the *observed* brightnesses that are seen in figure 1(a).

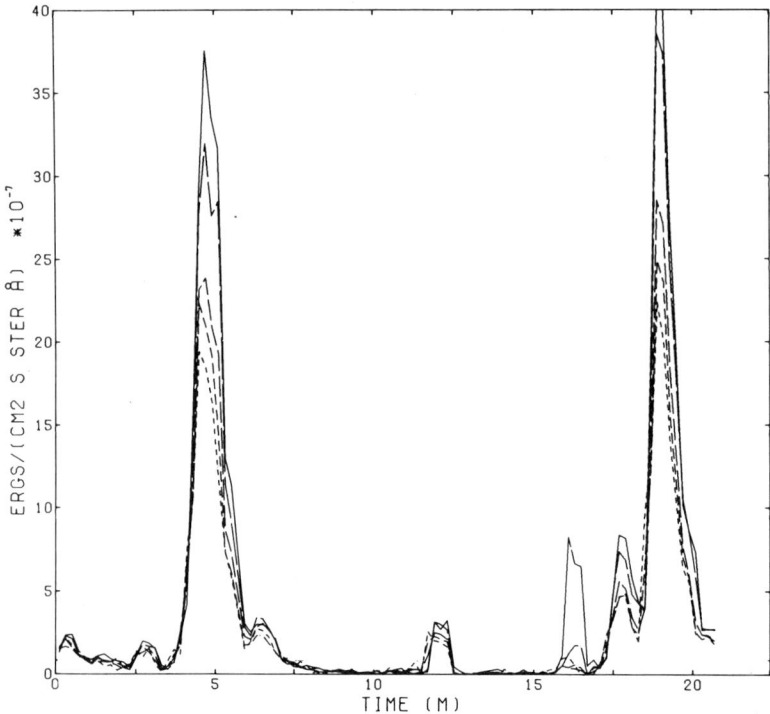

Figure 3. William G. Fastie had a far-ultraviolet spectrometer aboard the *Apollo 17* mission to the moon. The spectrometer was used to obtain this distribution of ultraviolet brightness around a large circle on the sky. The two crossings of the Gould belt are extremely prominent, as are the Earth (at 16 minutes) and η UMa (at 12 minutes).

Introduction

The Atlas

Coordinate Systems

All of the maps are in celestial coordinates (that is, RA and Dec), epoch 1950. They are 18° by 18° in size, and the centers of the maps have been chosen so that the maps cover the entire sky, with some degree of overlap for convenience.

Some users of the atlas will find our particular choice of centers for the maps especially convenient—we refer to scientists who also use data from the Infrared Astronomy Satellite (IRAS). The IRAS satellite mapped the entire sky in four wavelength bands: 120,000 Å; 250,000 Å; 600,000 Å; and 1 million Ångstrom units. These are all wavelengths that are much *longer* than the approximately 6500 Å limit of sensitivity of the human eye. For such long wavelengths, for convenience, astronomers usually use micrometers (called "microns" in the older literature) in place of Ångstroms as a unit of measurement; 1 micrometer = 1μ = 10^4Å = 10^{-4}cm, so the IRAS maps of the sky are at wavelengths of 12μ, 25μ, 60μ, and 100μ. These infrared maps are available, for use on large computers, from NASA, in 16° by 16° regions that are centered on the same 212 locations that we have chosen for the atlas.

Our visible-light maps have superimposed a coarse grid of RA and Dec coordinates, for location of stars. In the case of the ultraviolet maps, we felt that it would be useful to impose a coarse grid of galactic coordinates, so that the user is aware of the position of the field with respect to galactic coordinates, which are the more meaningful coordinates from a physical point of view.

The equations used to project RA and Dec onto a flat, two-dimensional coordinate system are as follows:

$$x = -3600 \frac{180}{\pi} \frac{\cot(d)\sin(a-a0)}{\sin(d0)+\cot(d)\cos(d0)\cos(a-a0)} \quad (1)$$

$$y = 3600 \frac{180}{\pi} \frac{\cos(d0)-\cot(d)\sin(d0)\cos(a-a0)}{\sin(d0)+\cot(d)\cos(d0)\cos(a-a0)}, \quad (2)$$

where $a0$ and $d0$ are, respectively, the RA and Dec of the center of the field in degrees, and a and d are, respectively, the RA and Dec of the point of interest. The two-dimensional coordinates, x and y, are in units of arc-seconds.

The equations to perform the reverse of the above transformation (that is, to go from x to y to RA and Dec, a and d, respectively, in units of degrees) are the following:

$$x' = -\frac{\pi}{180} \frac{x}{3600}$$

$$y' = \frac{\pi}{180} \frac{y}{3600}$$

$$a = \frac{180}{\pi} atan\,2(x'\,sec(d0), 1-y'\,tan(d0)) + a0 \quad (3)$$

$$d = \frac{180}{\pi} atan\,2((y'+tan(d0))\cos(a-a0), 1-y'\,tan(d0)), \quad (4)$$

where $a0$, $d0$, a, d, x, and y are defined as before, and $atan2(x,y)$ returns the arc-tangent of x/y.

Introduction

Brightness

The equations used for determining the diameters of the circles used in plotting the stars for the *Yale Bright Star Catalogue* and the TD1 ultraviolet maps are:

$$d = 0.2 + 2\frac{6.0 - mag}{3.0} \quad (5)$$

and

$$d = 0.2 + 4\frac{\log_{10}(flux)}{5.0}, \quad (6)$$

respectively, where d is the diameter of the circle in units of mm, *mag*, is the magnitude of the star, and *flux* is the flux of the star in units of 10^{-12} ergs cm^{-2} s^{-1} Å^{-1}. Stars with a visible magnitude greater than 6.0 are plotted as 0.2 mm diameter dots, as are the faintest (10^{-12} ergs cm^{-2} s^{-1} Å^{-1}) ultraviolet stars. These are the smallest dots that are plotted. The 1565 Å flux of a star can be transformed to a magnitude by using the absolute calibration of Hayes and Latham (1975):

$$m(1565) = -2.5\log_{10} flux(1565) - 21.175. \quad (7)$$

Finding the Correct Plate

The user may consult the Index to Celestial Coordinates or the Index to Galactic Coordinates, or compute the plate number from the RA and Dec using the following FORTRAN program:

```fortran
            program irasfields
c
            integer plate
            real*8 ra,dec
c
            write(6,10)
10          format(1x,'RA(hours):   ',$)
            read(5,*) ra
            write(6,20)
20          format(1x,'DEC(deg):   ',$)
            read(5,*) dec
c
            ra = ra * 15.
c
            if (dec .gt. 82.5) then
                    plate = 1
            else if (dec .gt. 67.5) then
                    plate = 2 + int(0.5 + ra / 36.)
            else if (dec .gt. 52.5) then
                    plate = 12 + int(0.5 + ra / 24.)
            else if (dec .gt. 37.5) then
                    plate = 27 + int(0.5 + ra / 18.)
            else if (dec .gt. 22.5) then
                    plate = 47 + int(0.5 + ra / 15.)
            else if (dec .gt. 7.5) then
                    plate = 71 + int(0.5 + ra / 15.)
            else if (dec .gt. -7.5) then
                    plate = 95 + int(0.5 + ra / 15.)
            else if (dec .gt. -22.5) then
                    plate = 119 + int(0.5 + ra / 15.)
            else if (dec .gt. -37.5) then
                    plate = 143 + int(0.5 + ra / 15.)
            else if (dec .gt. -52.5) then
                    plate = 167 + int(0.5 + ra / 18.)
            else if (dec .gt. -67.5) then
                    plate = 187 + int(0.5 + ra / 24.)
            else if (dec .gt. -82.5) then
                    plate = 202 + int(0.5 + ra / 36.)
            else
                    plate = 212
            endif
c
            write(6,30) plate
30          format(1x,'Plate:   ',i3)
            end
```

References

Boksenberg, A., Evans, R. G., Fowler, R. G., Gardner, I.S.K., Houziaux, L., Humphries, C. M., Jamar, C., Macau, D., Malaise, D., Monfils, A., Nandy, K., Thompson, G. I., Wilson, R., and Wroe, H. 1973, *Mon. Not. R. Astro. Soc.*, **163**, 291.

Carnochan, D. J. 1982, *Mon. Not. R. Astro. Soc.*, **201**, 1139.

Fastie, W. G. 1973, *Moon*, **7**, 49.

Gondhalekar, P. M., Phillips, A. P., and Wilson, R. 1980, *Astron. Astrophys.*, **85**, 272.

Hayes, D. S., and Latham, D. W. 1975, *Ap. J.*, **197**, 595.

Henry, R. C. 1977, *Ap. J. Suppl.*, **33**, 451.

Henry, R. C., Swandic, J. R., Shulman, S. D., and Fritz, G. 1977, *Ap. J.*, **212**, 707.

Jamar, C., Macau-Hercot, D., Monfils, A., Thompson, G. I., Houziaux, L., and Wilson, R. 1976, Ultraviolet Bright Star Spectrophotometric Catalogue, ESA SR-27.

Landsman, W. B. 1984, PhD Thesis, Johns Hopkins University.

Macau-Hercot, D., Jamar, C., Monfils, A., Thompson, G. I., Houziaux, L., and Wilson, R. 1978, Supplement to the Ultraviolet Bright Star Spectrophotometric Catalogue, ESA SR-28.

Page, T., Carruthers, G. R., and Heckathorn, H. M. 1982, Revised S201 Catalog of Far-Ultraviolet Objects, Naval Research Laboratory Report 8487.

Thompson, G. I., Nandy, K., Jamar, C., Monfils, A., Houziaux, L., Carnochan, D. J., and Wilson, R. 1978, Catalogue of Stellar Ultraviolet Fluxes, SRC.

Galactic Coordinates on RA and DEC

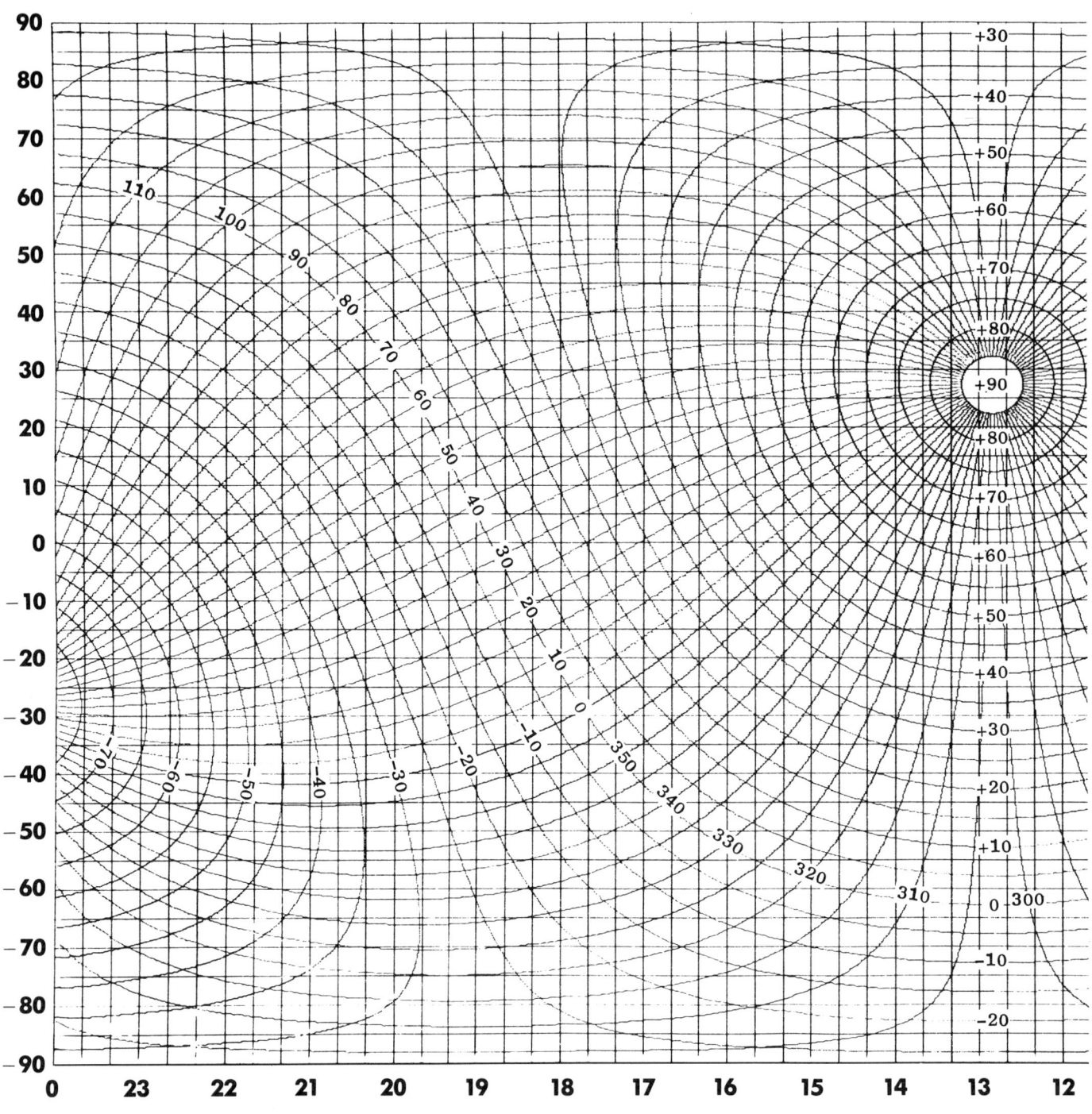

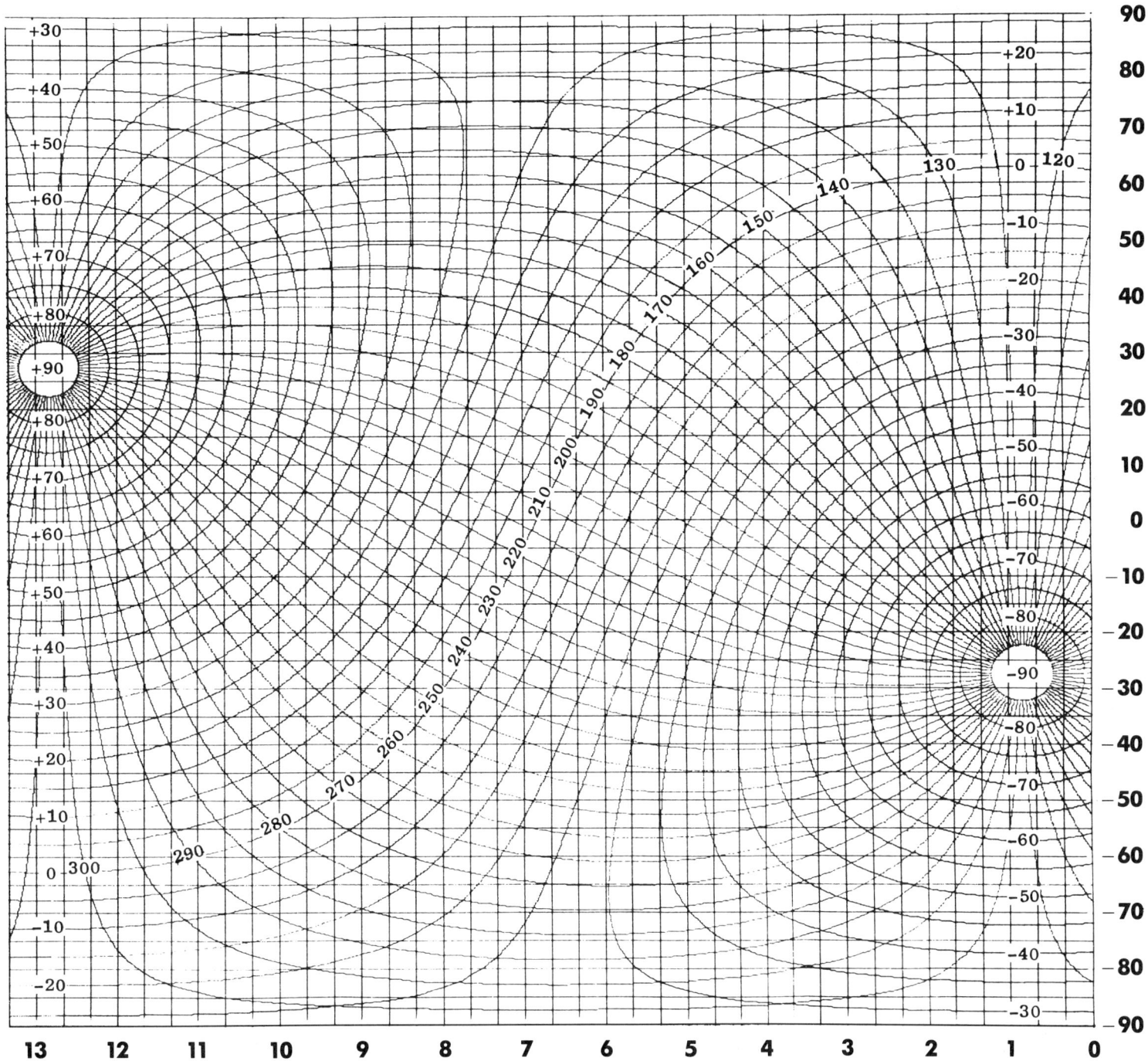

Index to Celestial Coordinates

PLATE	R.A. h m s	DEC. d m s	LONG. degrees	LAT. degrees
PL001	0 0 0.0	90 0 0.0	123.0	27.4
PL002	0 0 0.0	75 0 0.0	119.8	12.7
PL003	2 24 0.0	75 0 0.0	129.2	13.5
PL004	4 48 0.0	75 0 0.0	136.7	19.2
PL005	7 12 0.0	75 0 0.0	139.9	27.9
PL006	9 36 0.0	75 0 0.0	136.9	36.7
PL007	12 0 0.0	75 0 0.0	127.2	42.0
PL008	14 24 0.0	75 0 0.0	115.1	40.9
PL009	16 48 0.0	75 0 0.0	107.3	34.1
PL010	19 12 0.0	75 0 0.0	106.5	24.9
PL011	21 36 0.0	75 0 0.0	111.4	16.9
PL012	0 0 0.0	60 0 0.0	116.9	-2.0
PL013	1 36 0.0	60 0 0.0	128.8	-2.1
PL014	3 12 0.0	60 0 0.0	140.0	2.2
PL015	4 48 0.0	60 0 0.0	149.0	10.1
PL016	6 24 0.0	60 0 0.0	155.0	20.5
PL017	8 0 0.0	60 0 0.0	157.3	32.3
PL018	9 36 0.0	60 0 0.0	154.2	44.0
PL019	11 12 0.0	60 0 0.0	143.2	53.4
PL020	12 48 0.0	60 0 0.0	123.2	57.4
PL021	14 24 0.0	60 0 0.0	103.2	53.6
PL022	16 0 0.0	60 0 0.0	91.9	44.2
PL023	17 36 0.0	60 0 0.0	88.7	32.5
PL024	19 12 0.0	60 0 0.0	90.9	20.7
PL025	20 48 0.0	60 0 0.0	96.8	10.3
PL026	22 24 0.0	60 0 0.0	105.8	2.3
PL027	0 0 0.0	45 0 0.0	114.0	-16.7
PL028	1 12 0.0	45 0 0.0	127.3	-17.4
PL029	2 24 0.0	45 0 0.0	140.1	-14.4
PL030	3 36 0.0	45 0 0.0	151.4	-8.2
PL031	4 48 0.0	45 0 0.0	160.7	0.5
PL032	6 0 0.0	45 0 0.0	167.8	11.1
PL033	7 12 0.0	45 0 0.0	172.8	22.8
PL034	8 24 0.0	45 0 0.0	175.5	35.3
PL035	9 36 0.0	45 0 0.0	175.1	48.0
PL036	10 48 0.0	45 0 0.0	168.8	60.2
PL037	12 0 0.0	45 0 0.0	148.8	69.9
PL038	13 12 0.0	45 0 0.0	109.9	71.8
PL039	14 24 0.0	45 0 0.0	82.2	64.2
PL040	15 36 0.0	45 0 0.0	72.3	52.5
PL041	16 48 0.0	45 0 0.0	70.2	39.9
PL042	18 0 0.0	45 0 0.0	72.0	27.3
PL043	19 12 0.0	45 0 0.0	76.2	15.2
PL044	20 24 0.0	45 0 0.0	82.5	4.2
PL045	21 36 0.0	45 0 0.0	91.0	-5.3
PL046	22 48 0.0	45 0 0.0	101.6	-12.5
PL047	0 0 0.0	30 0 0.0	110.6	-31.4
PL048	1 0 0.0	30 0 0.0	125.8	-32.5
PL049	2 0 0.0	30 0 0.0	140.8	-30.1
PL050	3 0 0.0	30 0 0.0	154.0	-24.6

Index to Celestial Coordinates

PLATE	R.A. h	m	s	DEC. d	m	s	LONG. degrees	LAT. degrees
PL051	4	0	0.0	30	0	0.0	165.0	-16.7
PL052	5	0	0.0	30	0	0.0	173.9	-7.0
PL053	6	0	0.0	30	0	0.0	181.0	3.8
PL054	7	0	0.0	30	0	0.0	186.8	15.5
PL055	8	0	0.0	30	0	0.0	191.6	27.7
PL056	9	0	0.0	30	0	0.0	195.6	40.2
PL057	10	0	0.0	30	0	0.0	198.7	53.1
PL058	11	0	0.0	30	0	0.0	200.3	66.0
PL059	12	0	0.0	30	0	0.0	196.5	79.0
PL060	13	0	0.0	30	0	0.0	80.8	86.5
PL061	14	0	0.0	30	0	0.0	46.7	74.2
PL062	15	0	0.0	30	0	0.0	46.1	61.3
PL063	16	0	0.0	30	0	0.0	48.3	48.3
PL064	17	0	0.0	30	0	0.0	51.8	35.6
PL065	18	0	0.0	30	0	0.0	56.0	23.2
PL066	19	0	0.0	30	0	0.0	61.2	11.1
PL067	20	0	0.0	30	0	0.0	67.4	-0.2
PL068	21	0	0.0	30	0	0.0	75.2	-10.7
PL069	22	0	0.0	30	0	0.0	84.8	-19.8
PL070	23	0	0.0	30	0	0.0	96.6	-27.0
PL071	0	0	0.0	15	0	0.0	105.9	-46.0
PL072	1	0	0.0	15	0	0.0	126.9	-47.5
PL073	2	0	0.0	15	0	0.0	147.3	-44.2
PL074	3	0	0.0	15	0	0.0	163.9	-37.0
PL075	4	0	0.0	15	0	0.0	176.5	-27.2
PL076	5	0	0.0	15	0	0.0	186.2	-15.9
PL077	6	0	0.0	15	0	0.0	194.1	-3.6
PL078	7	0	0.0	15	0	0.0	200.8	9.2
PL079	8	0	0.0	15	0	0.0	207.1	22.4
PL080	9	0	0.0	15	0	0.0	213.8	35.7
PL081	10	0	0.0	15	0	0.0	221.7	48.9
PL082	11	0	0.0	15	0	0.0	233.5	61.8
PL083	12	0	0.0	15	0	0.0	258.0	73.2
PL084	13	0	0.0	15	0	0.0	315.2	77.3
PL085	14	0	0.0	15	0	0.0	359.7	69.4
PL086	15	0	0.0	15	0	0.0	17.6	57.2
PL087	16	0	0.0	15	0	0.0	27.5	44.1
PL088	17	0	0.0	15	0	0.0	34.8	30.8
PL089	18	0	0.0	15	0	0.0	41.2	17.5
PL090	19	0	0.0	15	0	0.0	47.6	4.5
PL091	20	0	0.0	15	0	0.0	54.7	-8.2
PL092	21	0	0.0	15	0	0.0	63.1	-20.2
PL093	22	0	0.0	15	0	0.0	73.7	-31.0
PL094	23	0	0.0	15	0	0.0	87.7	-40.0
PL095	0	0	0.0	0	0	0.0	97.7	-60.2
PL096	1	0	0.0	0	0	0.0	129.0	-62.5
PL097	2	0	0.0	0	0	0.0	157.8	-57.7
PL098	3	0	0.0	0	0	0.0	177.4	-48.3
PL099	4	0	0.0	0	0	0.0	190.3	-36.7
PL100	5	0	0.0	0	0	0.0	199.7	-24.0

Index to Celestial Coordinates　　　　　　　　　　　　　　　　　　23

PLATE	R.A. h	m	s	DEC. d	m	s	LONG. degrees	LAT. degrees
PL101	6	0	0.0	0	0	0.0	207.3	-10.9
PL102	7	0	0.0	0	0	0.0	214.3	2.4
PL103	8	0	0.0	0	0	0.0	221.4	15.7
PL104	9	0	0.0	0	0	0.0	229.5	28.7
PL105	10	0	0.0	0	0	0.0	239.9	41.1
PL106	11	0	0.0	0	0	0.0	254.8	52.1
PL107	12	0	0.0	0	0	0.0	277.7	60.2
PL108	13	0	0.0	0	0	0.0	309.0	62.5
PL109	14	0	0.0	0	0	0.0	337.8	57.7
PL110	15	0	0.0	0	0	0.0	357.4	48.3
PL111	16	0	0.0	0	0	0.0	10.3	36.7
PL112	17	0	0.0	0	0	0.0	19.7	24.0
PL113	18	0	0.0	0	0	0.0	27.3	10.9
PL114	19	0	0.0	0	0	0.0	34.3	-2.4
PL115	20	0	0.0	0	0	0.0	41.4	-15.7
PL116	21	0	0.0	0	0	0.0	49.5	-28.7
PL117	22	0	0.0	0	0	0.0	59.9	-41.1
PL118	23	0	0.0	0	0	0.0	74.8	-52.1
PL119	0	0	0.0	-15	0	0.0	78.0	-73.2
PL120	1	0	0.0	-15	0	0.0	135.2	-77.3
PL121	2	0	0.0	-15	0	0.0	179.7	-69.4
PL122	3	0	0.0	-15	0	0.0	197.6	-57.2
PL123	4	0	0.0	-15	0	0.0	207.5	-44.1
PL124	5	0	0.0	-15	0	0.0	214.8	-30.8
PL125	6	0	0.0	-15	0	0.0	221.2	-17.5
PL126	7	0	0.0	-15	0	0.0	227.6	-4.5
PL127	8	0	0.0	-15	0	0.0	234.7	8.2
PL128	9	0	0.0	-15	0	0.0	243.1	20.2
PL129	10	0	0.0	-15	0	0.0	253.7	31.0
PL130	11	0	0.0	-15	0	0.0	267.7	40.0
PL131	12	0	0.0	-15	0	0.0	285.9	46.0
PL132	13	0	0.0	-15	0	0.0	306.9	47.5
PL133	14	0	0.0	-15	0	0.0	327.3	44.2
PL134	15	0	0.0	-15	0	0.0	343.9	37.0
PL135	16	0	0.0	-15	0	0.0	356.5	27.2
PL136	17	0	0.0	-15	0	0.0	6.2	15.9
PL137	18	0	0.0	-15	0	0.0	14.1	3.6
PL138	19	0	0.0	-15	0	0.0	20.8	-9.2
PL139	20	0	0.0	-15	0	0.0	27.1	-22.4
PL140	21	0	0.0	-15	0	0.0	33.8	-35.7
PL141	22	0	0.0	-15	0	0.0	41.7	-48.9
PL142	23	0	0.0	-15	0	0.0	53.5	-61.8
PL143	0	0	0.0	-30	0	0.0	16.5	-79.0
PL144	1	0	0.0	-30	0	0.0	260.8	-86.5
PL145	2	0	0.0	-30	0	0.0	226.7	-74.2
PL146	3	0	0.0	-30	0	0.0	226.1	-61.3
PL147	4	0	0.0	-30	0	0.0	228.3	-48.3
PL148	5	0	0.0	-30	0	0.0	231.8	-35.6
PL149	6	0	0.0	-30	0	0.0	236.0	-23.2
PL150	7	0	0.0	-30	0	0.0	241.2	-11.1

Index to Celestial Coordinates

PLATE	R.A. h	m	s	DEC. d	m	s	LONG. degrees	LAT. degrees
PL151	8	0	0.0	-30	0	0.0	247.4	0.2
PL152	9	0	0.0	-30	0	0.0	255.2	10.7
PL153	10	0	0.0	-30	0	0.0	264.8	19.8
PL154	11	0	0.0	-30	0	0.0	276.6	27.0
PL155	12	0	0.0	-30	0	0.0	290.6	31.4
PL156	13	0	0.0	-30	0	0.0	305.8	32.5
PL157	14	0	0.0	-30	0	0.0	320.8	30.1
PL158	15	0	0.0	-30	0	0.0	334.0	24.6
PL159	16	0	0.0	-30	0	0.0	345.0	16.7
PL160	17	0	0.0	-30	0	0.0	353.9	7.0
PL161	18	0	0.0	-30	0	0.0	1.0	-3.8
PL162	19	0	0.0	-30	0	0.0	6.8	-15.5
PL163	20	0	0.0	-30	0	0.0	11.6	-27.7
PL164	21	0	0.0	-30	0	0.0	15.6	-40.2
PL165	22	0	0.0	-30	0	0.0	18.7	-53.1
PL166	23	0	0.0	-30	0	0.0	20.3	-66.0
PL167	0	0	0.0	-45	0	0.0	328.8	-69.9
PL168	1	12	0.0	-45	0	0.0	289.9	-71.8
PL169	2	24	0.0	-45	0	0.0	262.2	-64.2
PL170	3	36	0.0	-45	0	0.0	252.3	-52.5
PL171	4	48	0.0	-45	0	0.0	250.2	-39.9
PL172	6	0	0.0	-45	0	0.0	252.0	-27.3
PL173	7	12	0.0	-45	0	0.0	256.2	-15.2
PL174	8	24	0.0	-45	0	0.0	262.5	-4.2
PL175	9	36	0.0	-45	0	0.0	271.0	5.3
PL176	10	48	0.0	-45	0	0.0	281.6	12.5
PL177	12	0	0.0	-45	0	0.0	294.0	16.7
PL178	13	12	0.0	-45	0	0.0	307.3	17.4
PL179	14	24	0.0	-45	0	0.0	320.1	14.4
PL180	15	36	0.0	-45	0	0.0	331.4	8.2
PL181	16	48	0.0	-45	0	0.0	340.7	-0.5
PL182	18	0	0.0	-45	0	0.0	347.8	-11.1
PL183	19	12	0.0	-45	0	0.0	352.8	-22.8
PL184	20	24	0.0	-45	0	0.0	355.5	-35.3
PL185	21	36	0.0	-45	0	0.0	355.1	-48.0
PL186	22	48	0.0	-45	0	0.0	348.8	-60.2
PL187	0	0	0.0	-60	0	0.0	314.0	-56.3
PL188	1	36	0.0	-60	0	0.0	292.4	-56.4
PL189	3	12	0.0	-60	0	0.0	276.4	-49.4
PL190	4	48	0.0	-60	0	0.0	269.5	-38.5
PL191	6	24	0.0	-60	0	0.0	269.2	-26.5
PL192	8	0	0.0	-60	0	0.0	273.4	-15.3
PL193	9	36	0.0	-60	0	0.0	281.0	-5.9
PL194	11	12	0.0	-60	0	0.0	291.1	0.4
PL195	12	48	0.0	-60	0	0.0	302.9	2.6
PL196	14	24	0.0	-60	0	0.0	314.6	0.4
PL197	16	0	0.0	-60	0	0.0	324.8	-5.7
PL198	17	36	0.0	-60	0	0.0	332.5	-15.0
PL199	19	12	0.0	-60	0	0.0	336.7	-26.3
PL200	20	48	0.0	-60	0	0.0	336.5	-38.2

Index to Celestial Coordinates

PLATE	R.A. h	m	s	DEC. d	m	s	LONG. degrees	LAT. degrees
PL201	22	24	0.0	-60	0	0.0	329.9	-49.2
PL202	0	0	0.0	-75	0	0.0	307.2	-42.0
PL203	2	24	0.0	-75	0	0.0	295.1	-40.9
PL204	4	48	0.0	-75	0	0.0	287.3	-34.1
PL205	7	12	0.0	-75	0	0.0	286.5	-24.9
PL206	9	36	0.0	-75	0	0.0	291.4	-16.9
PL207	12	0	0.0	-75	0	0.0	299.8	-12.7
PL208	14	24	0.0	-75	0	0.0	309.2	-13.5
PL209	16	48	0.0	-75	0	0.0	316.7	-19.2
PL210	19	12	0.0	-75	0	0.0	319.9	-27.9
PL211	21	36	0.0	-75	0	0.0	316.9	-36.7
PL212	0	0	0.0	-90	0	0.0	303.0	-27.4

Plates

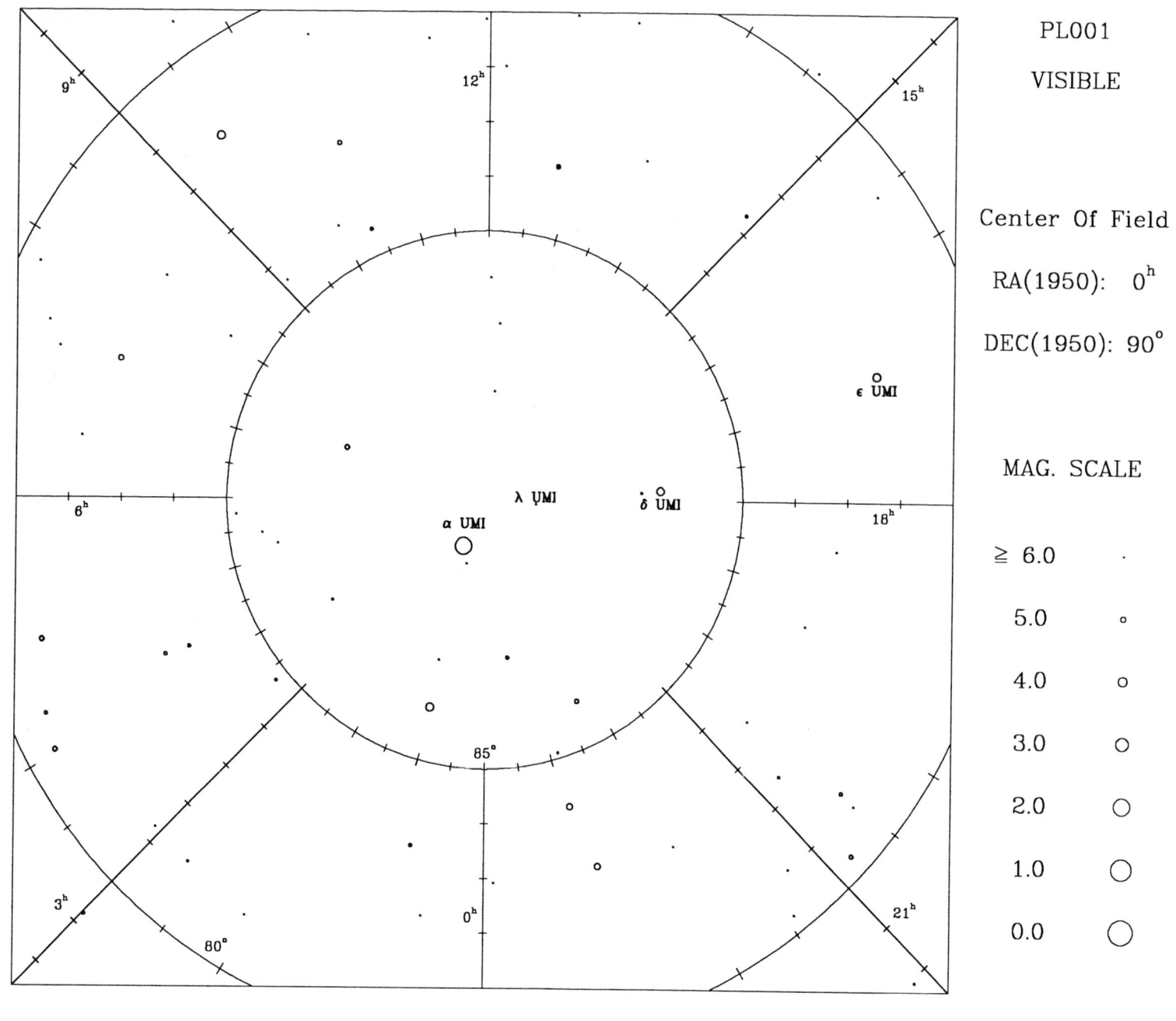

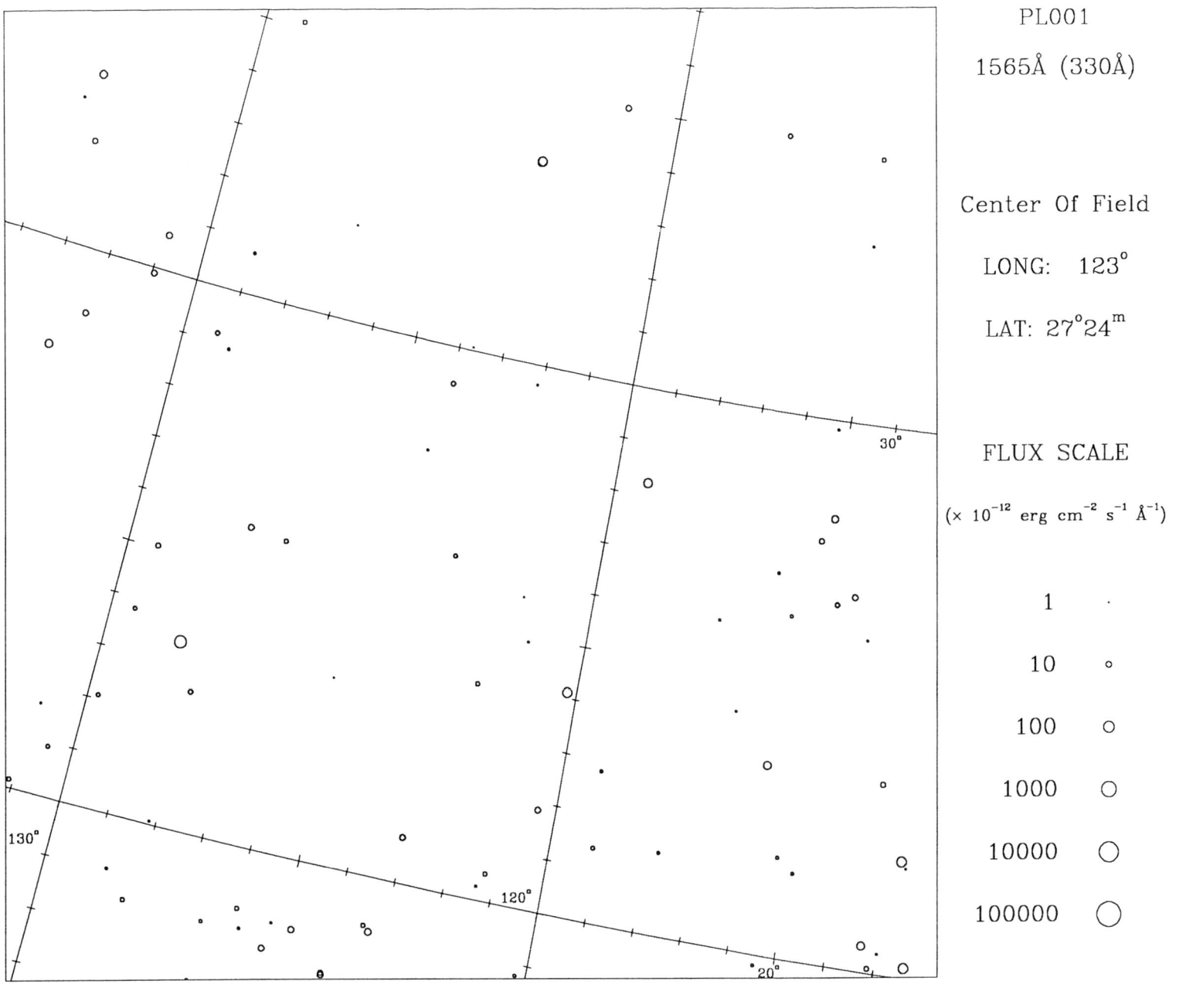

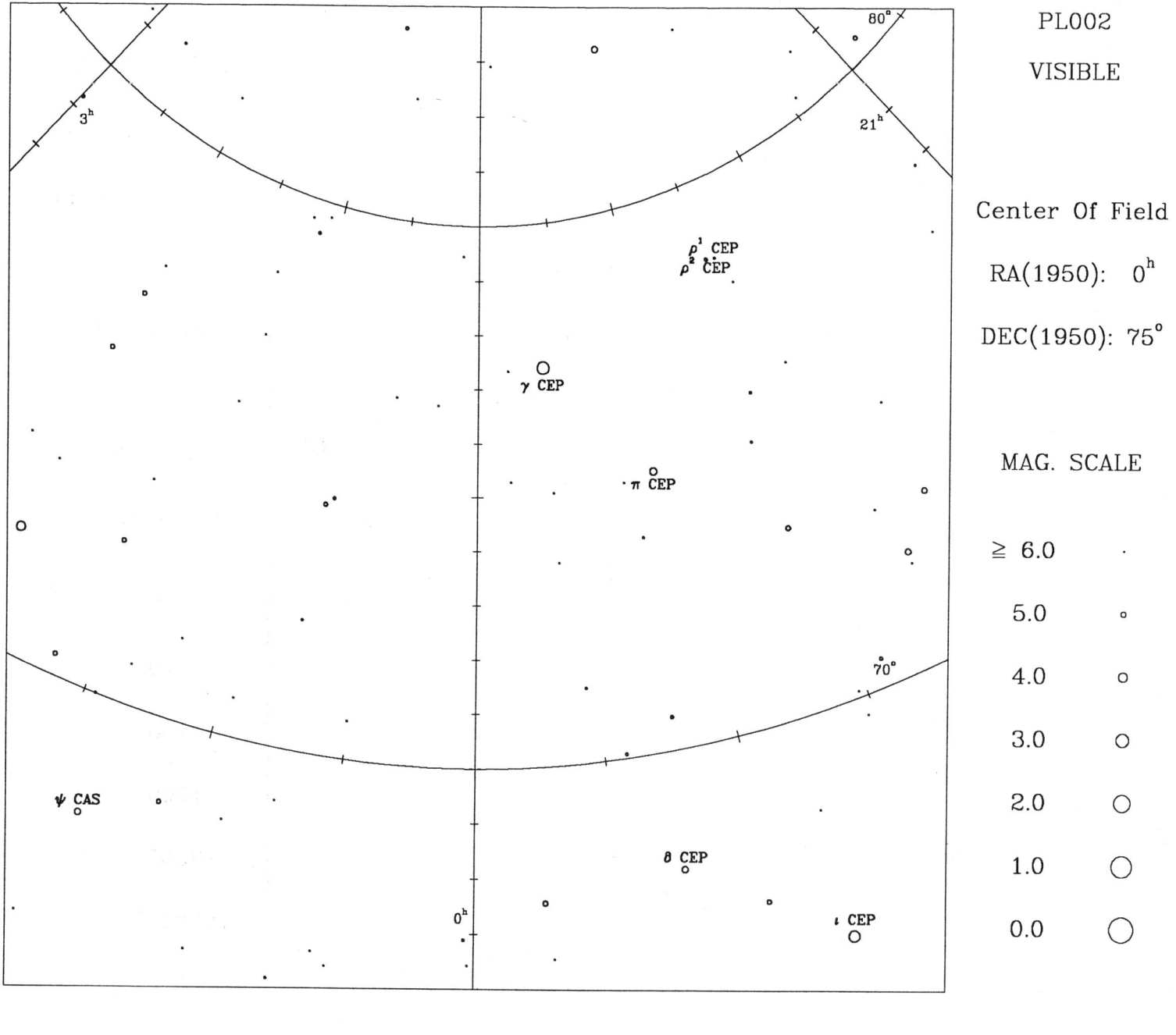

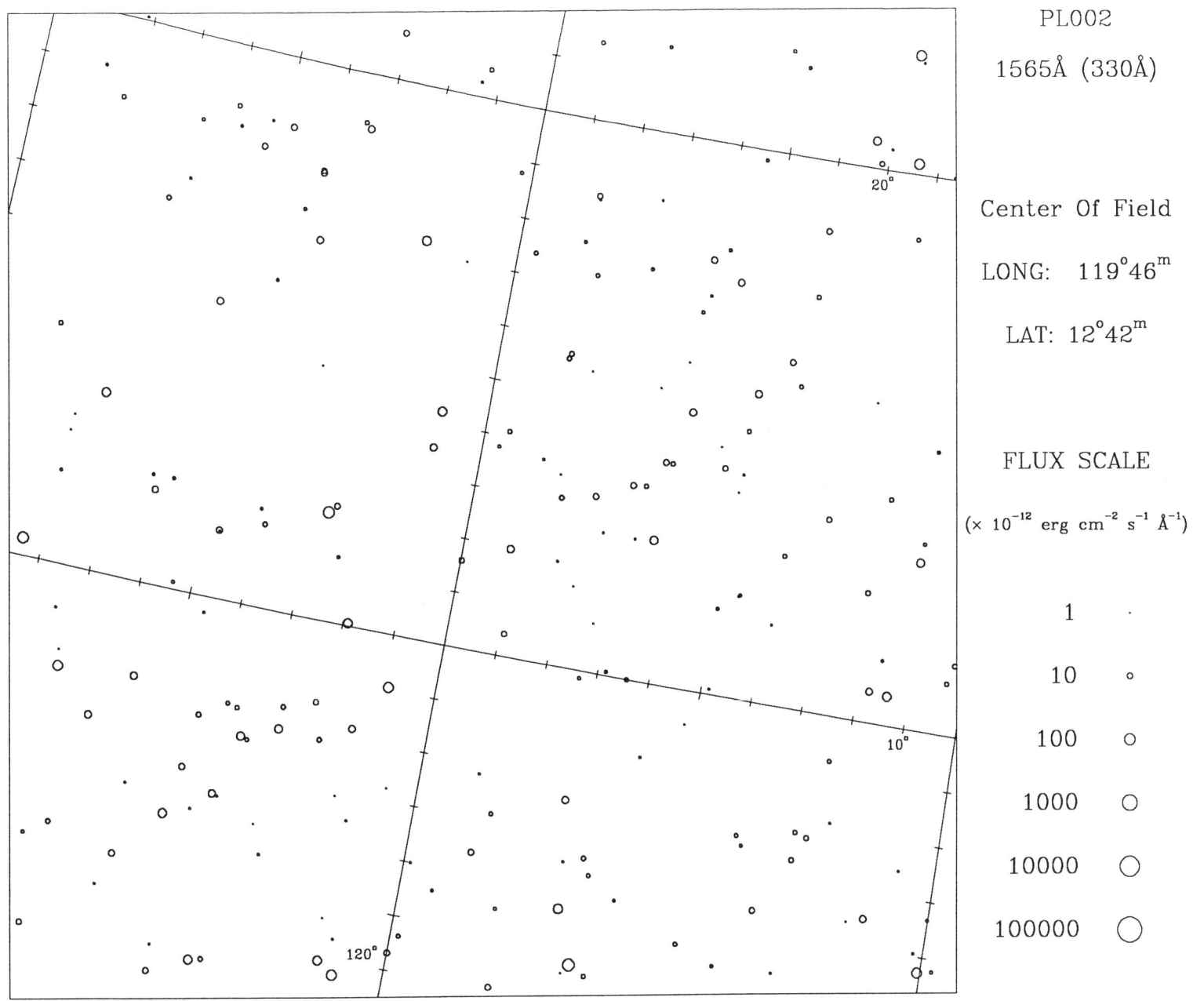

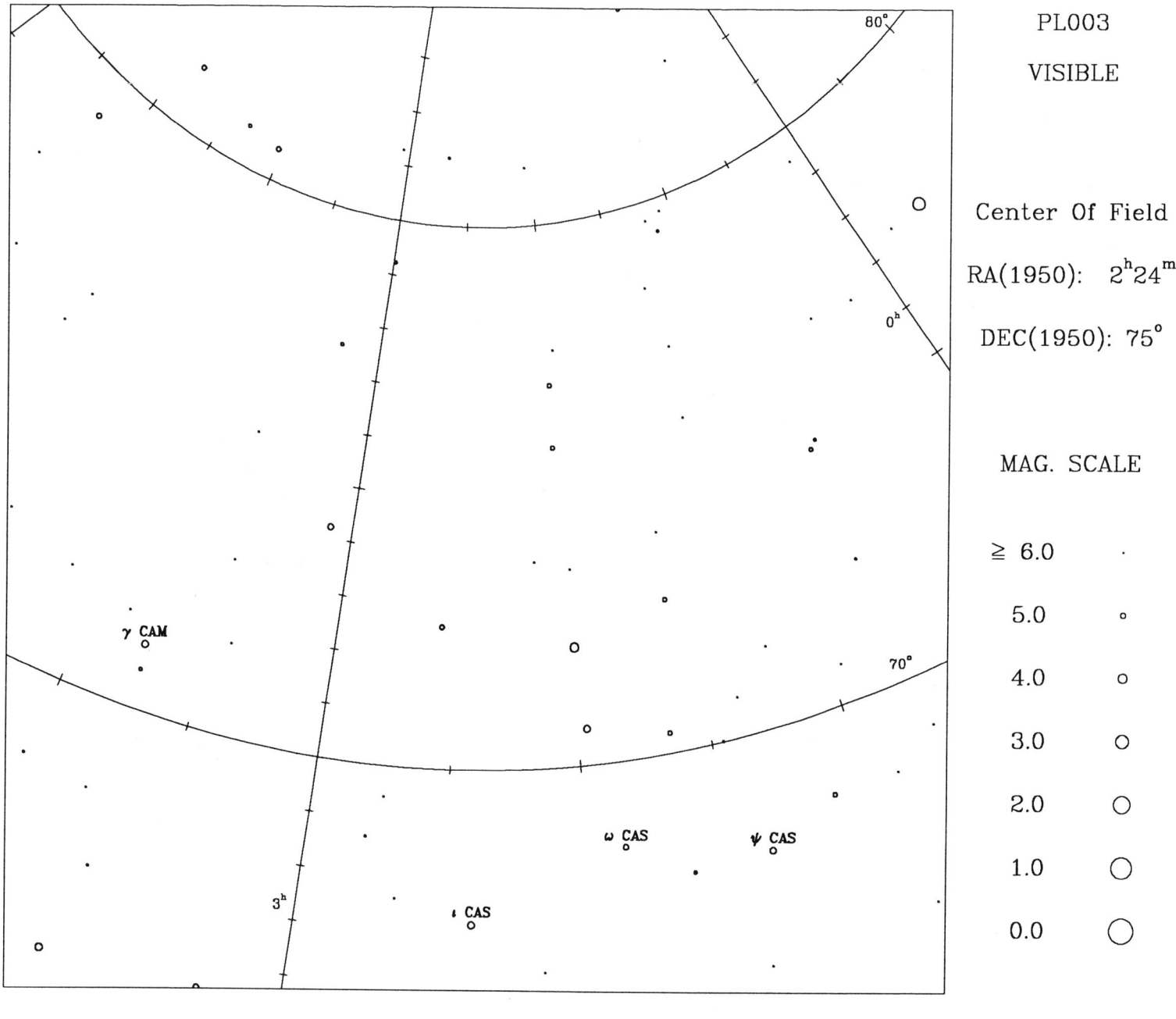

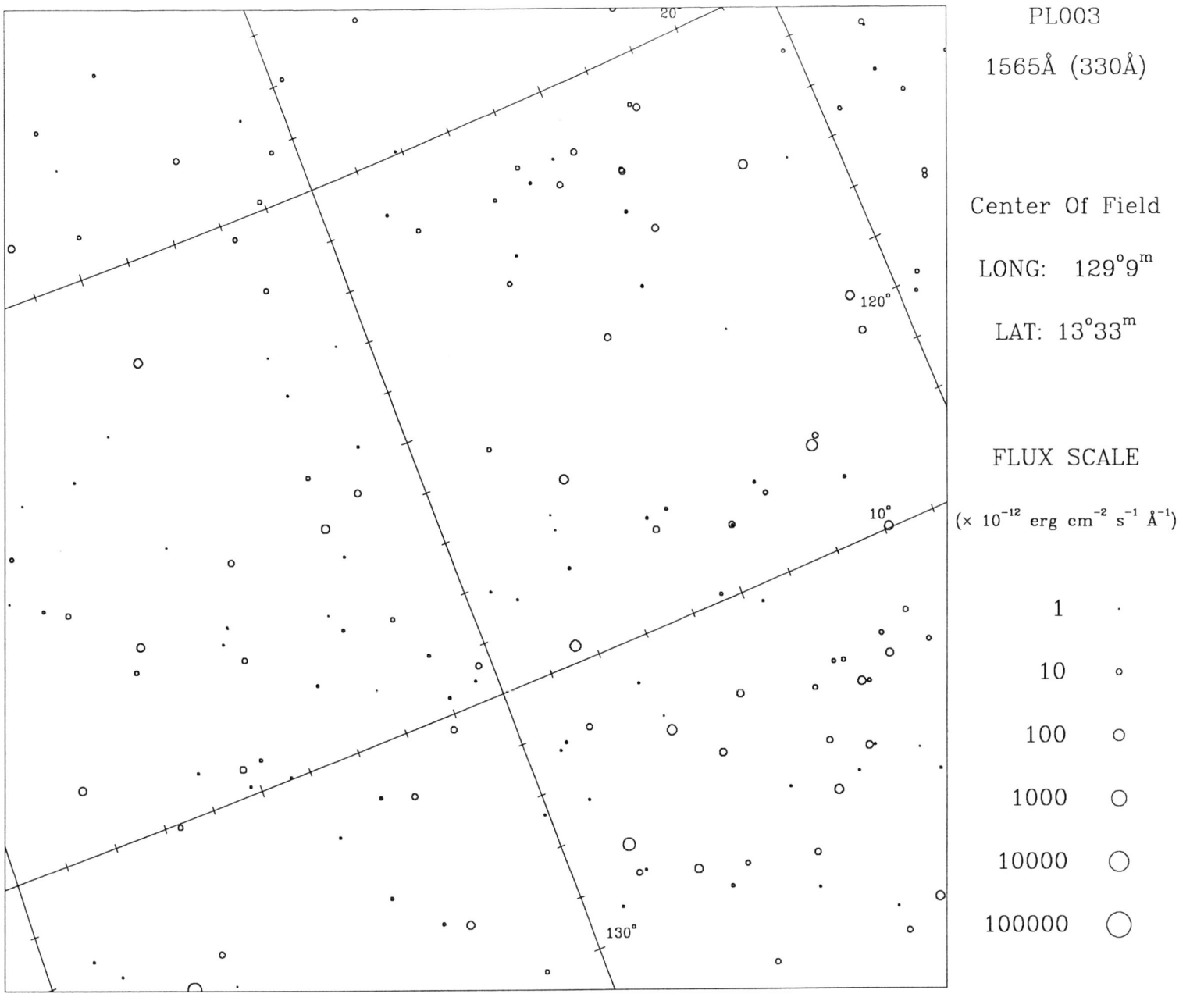

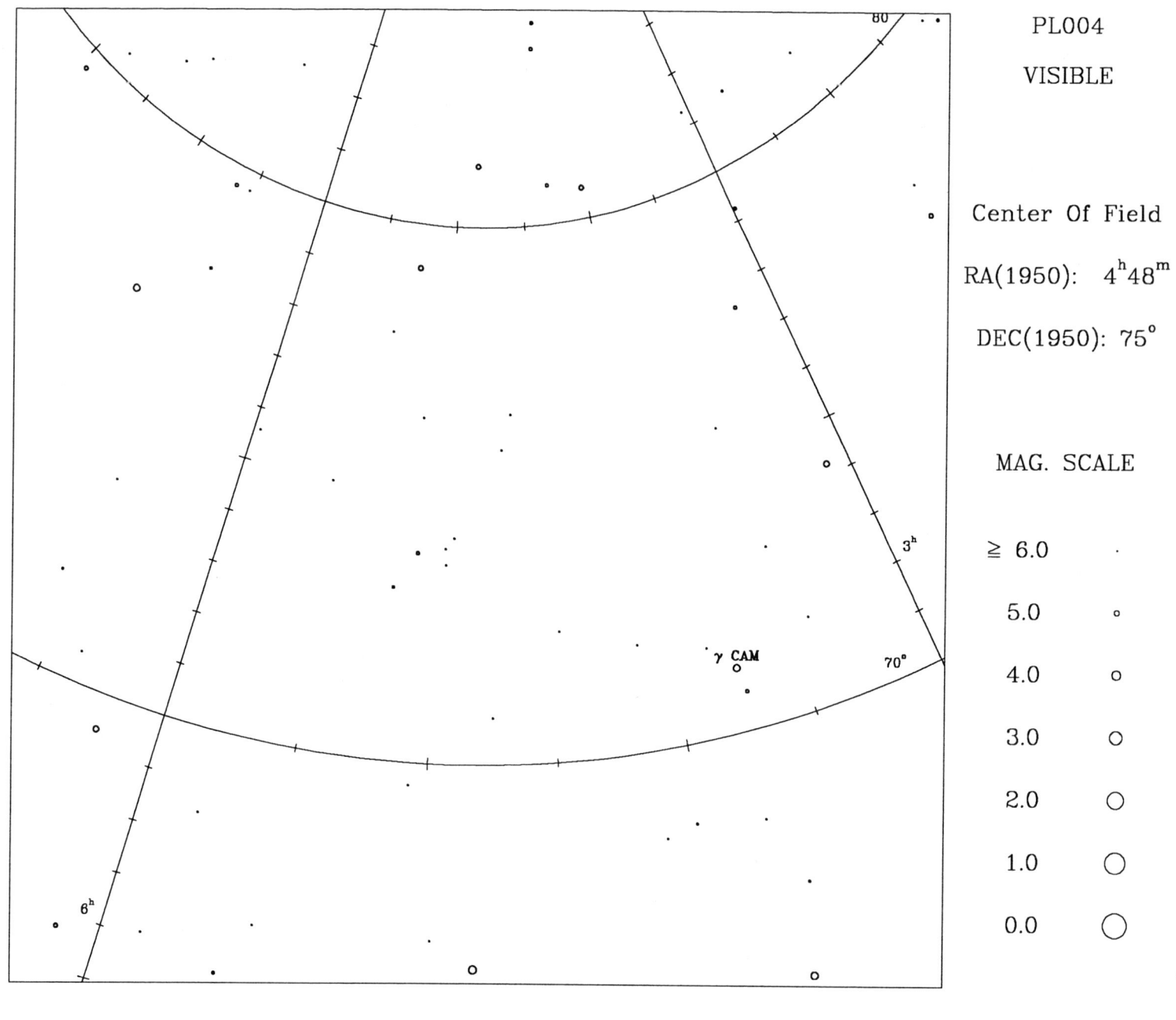

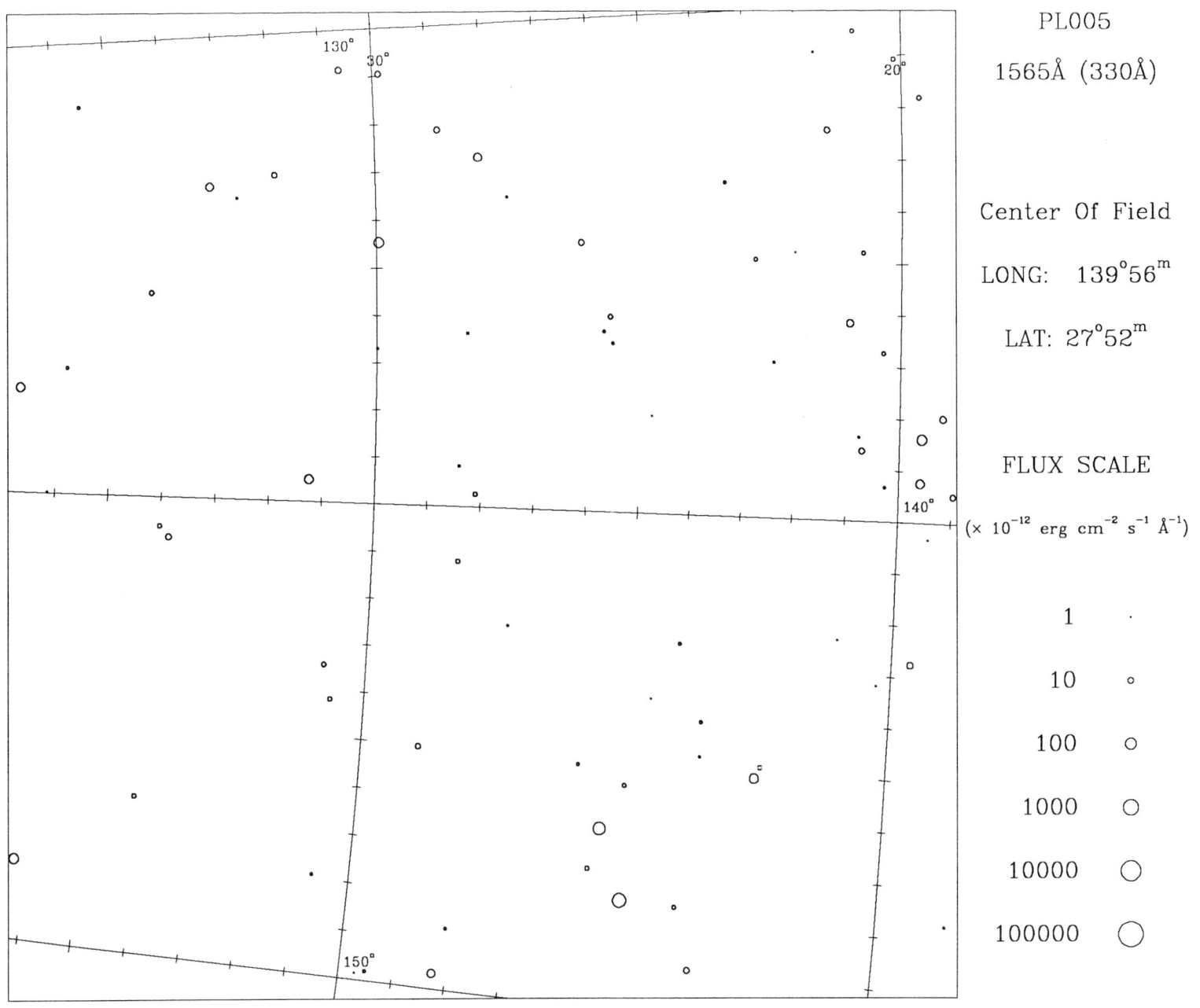

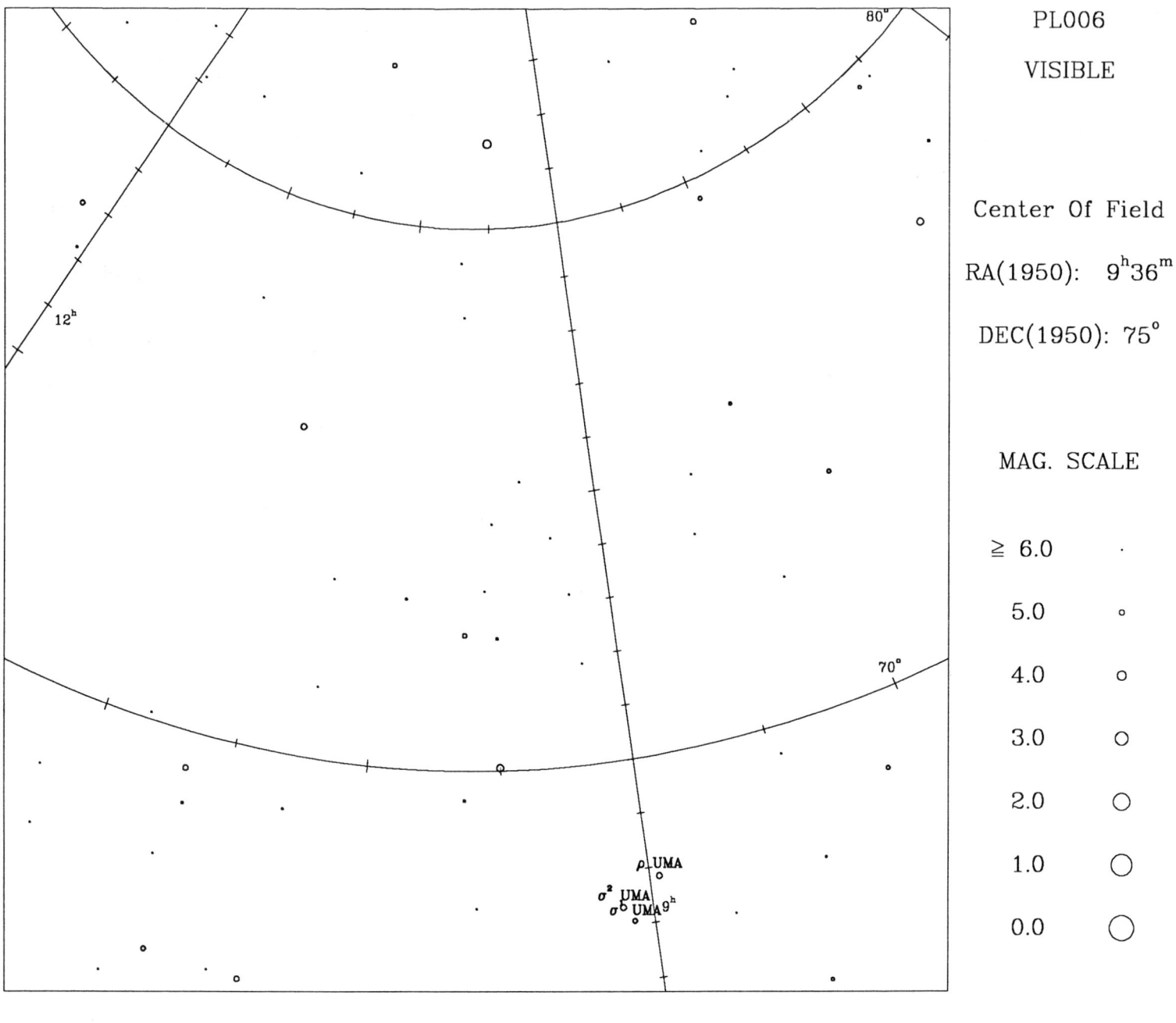

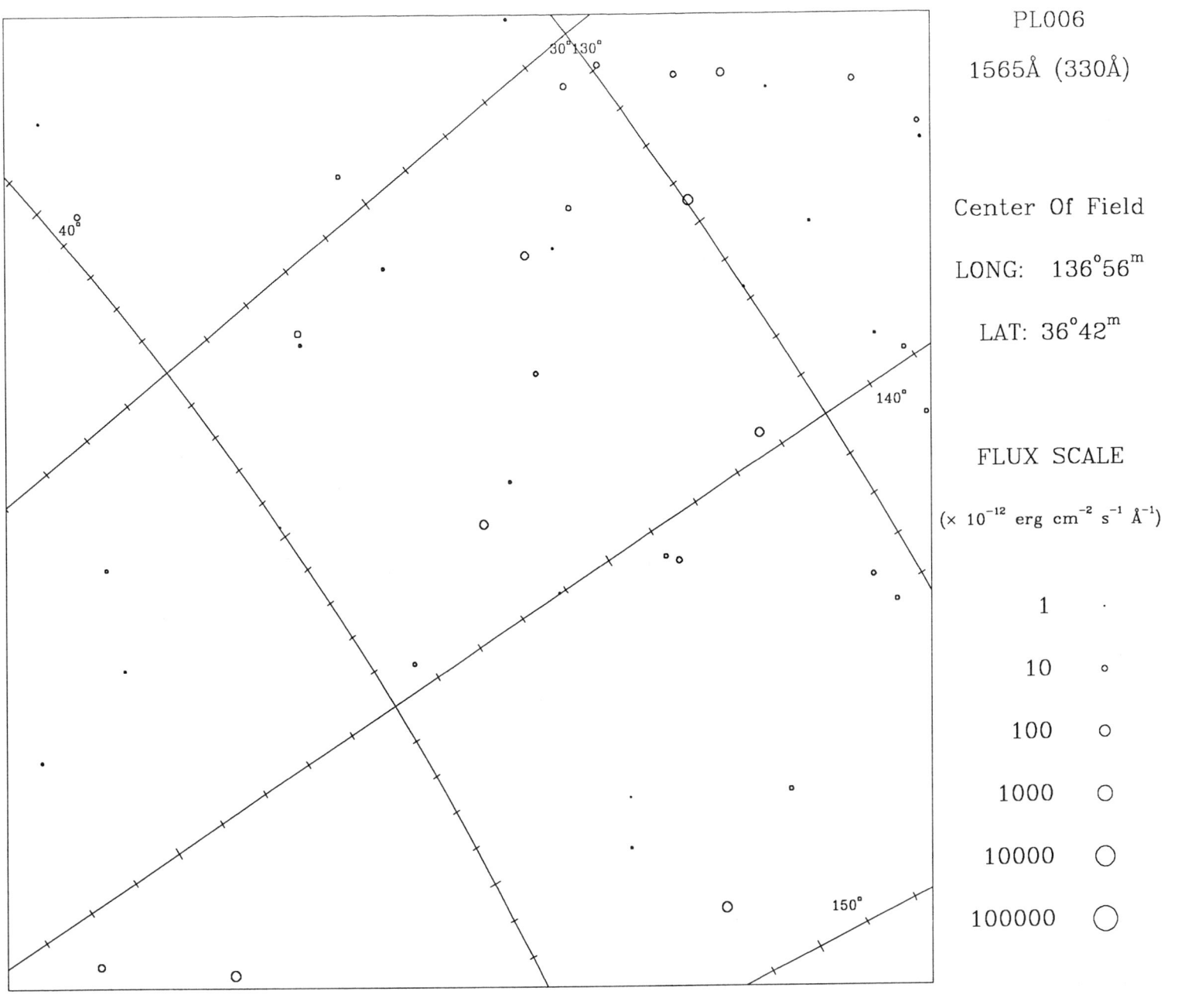

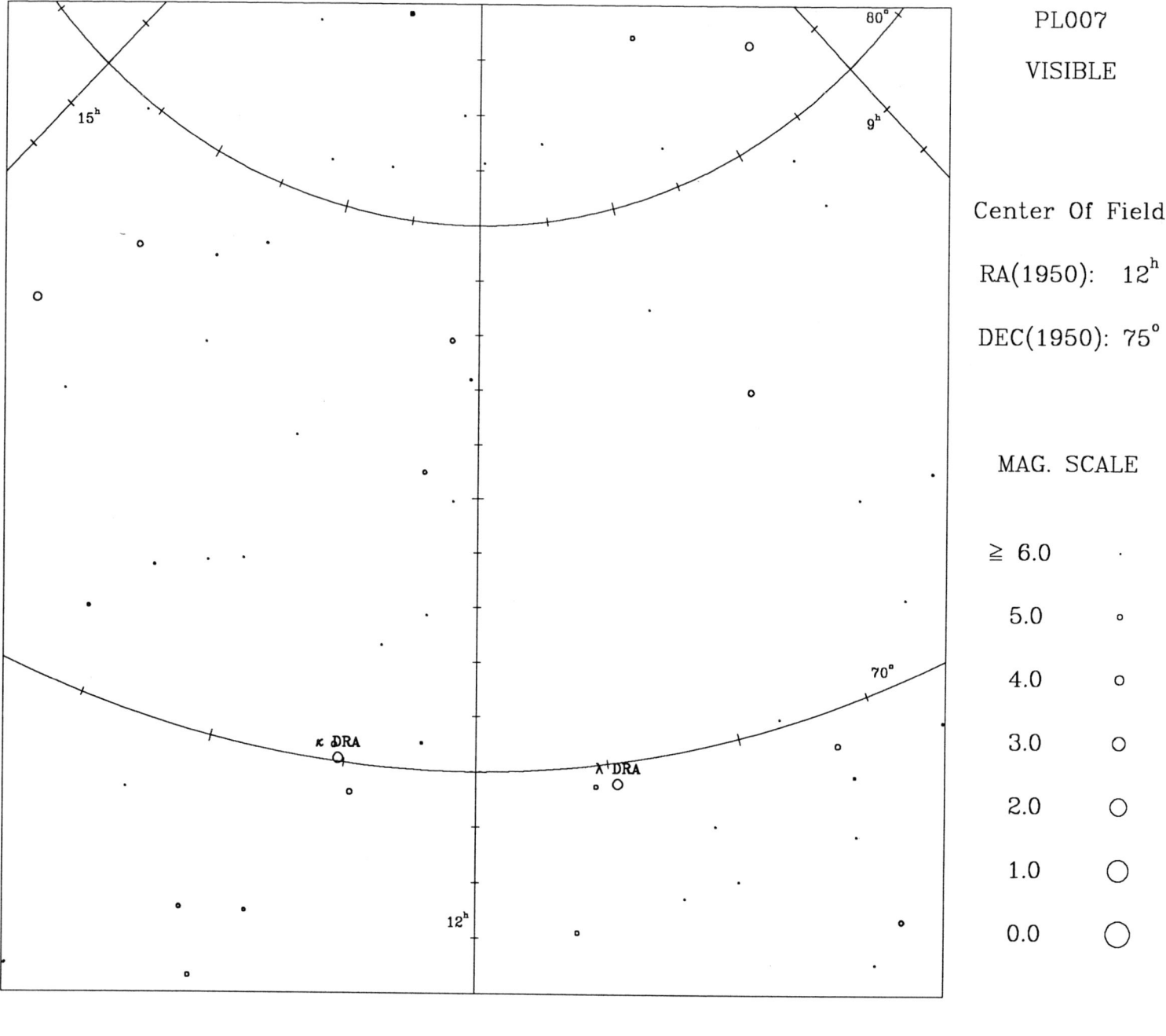

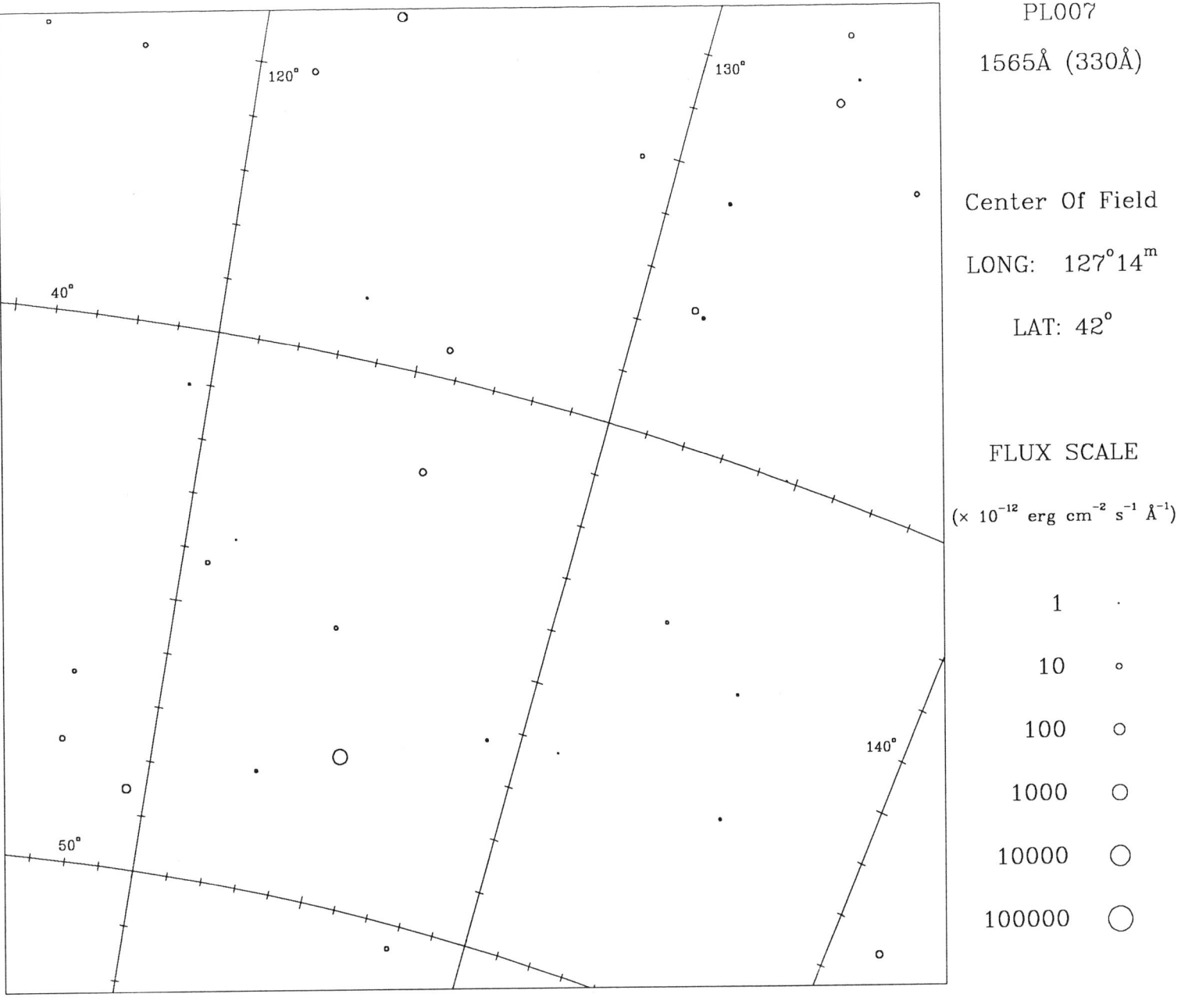

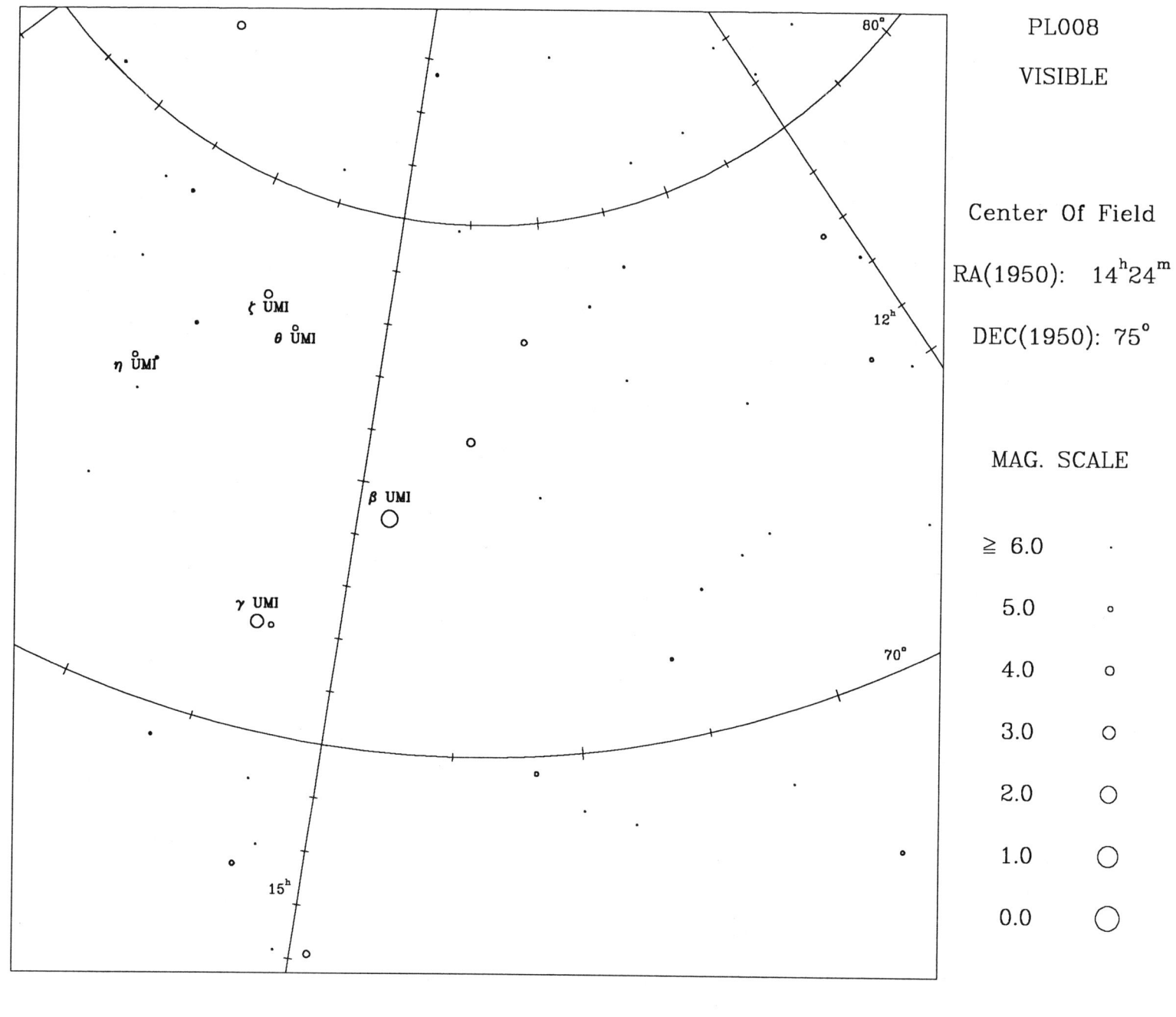

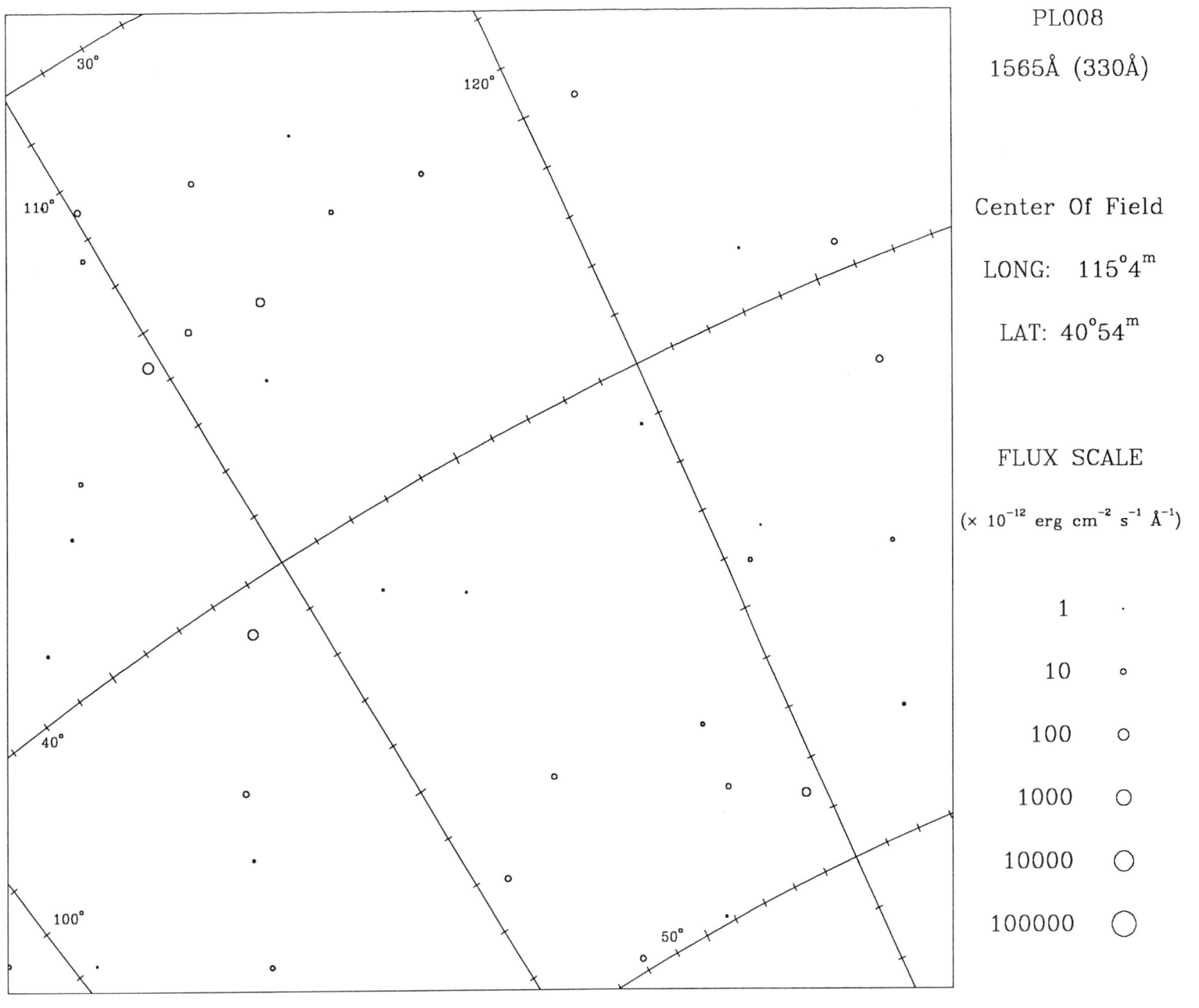

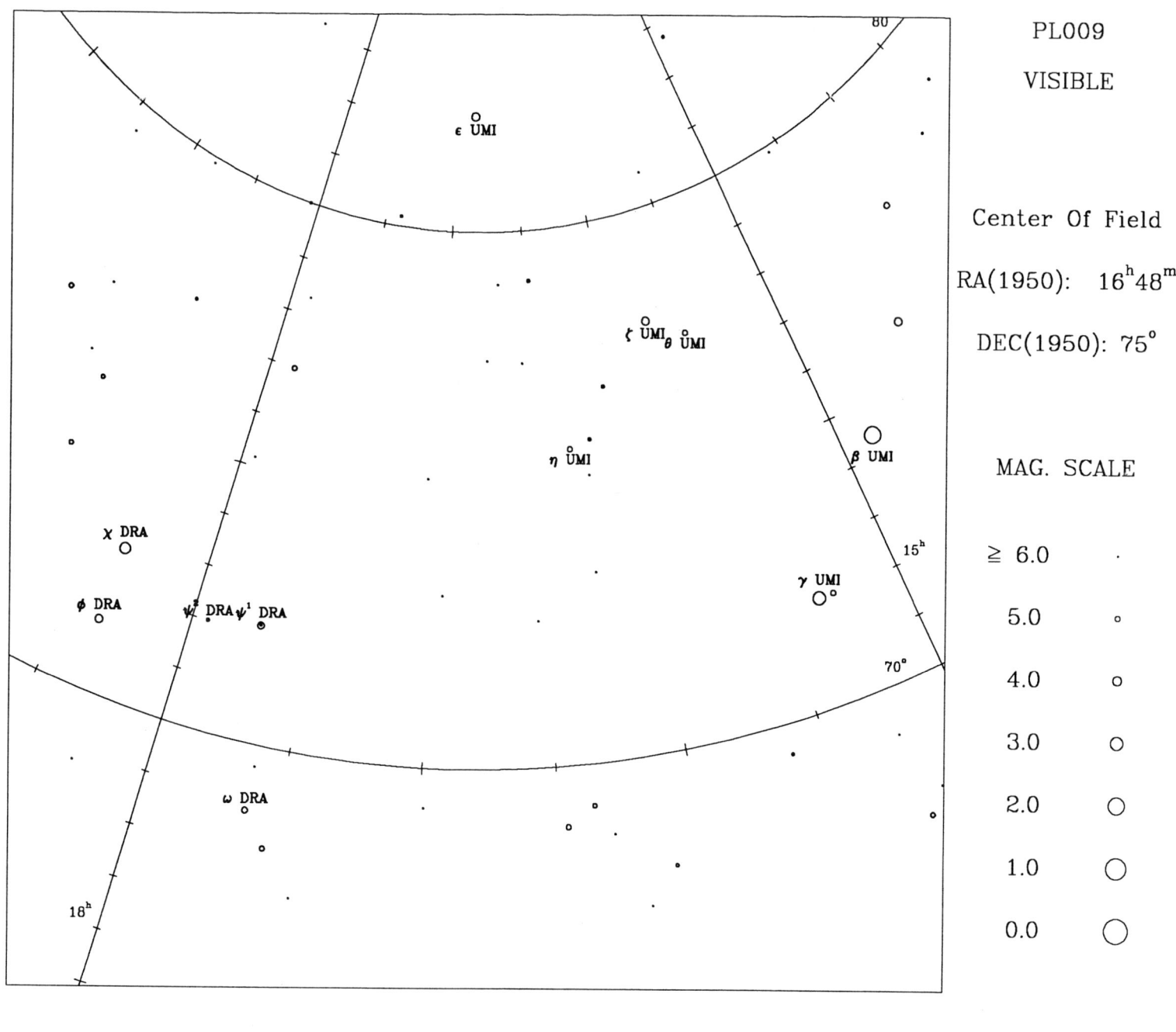

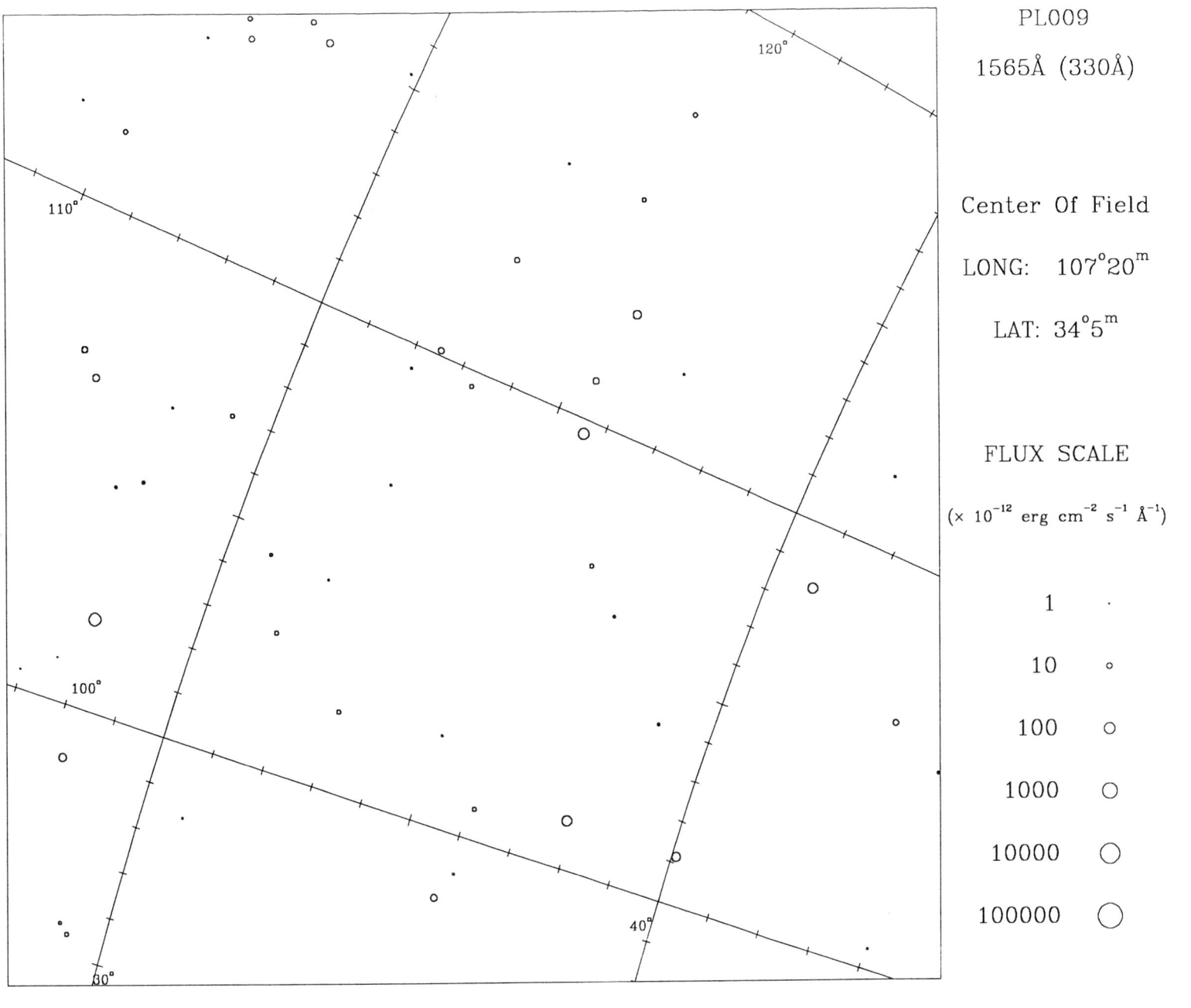

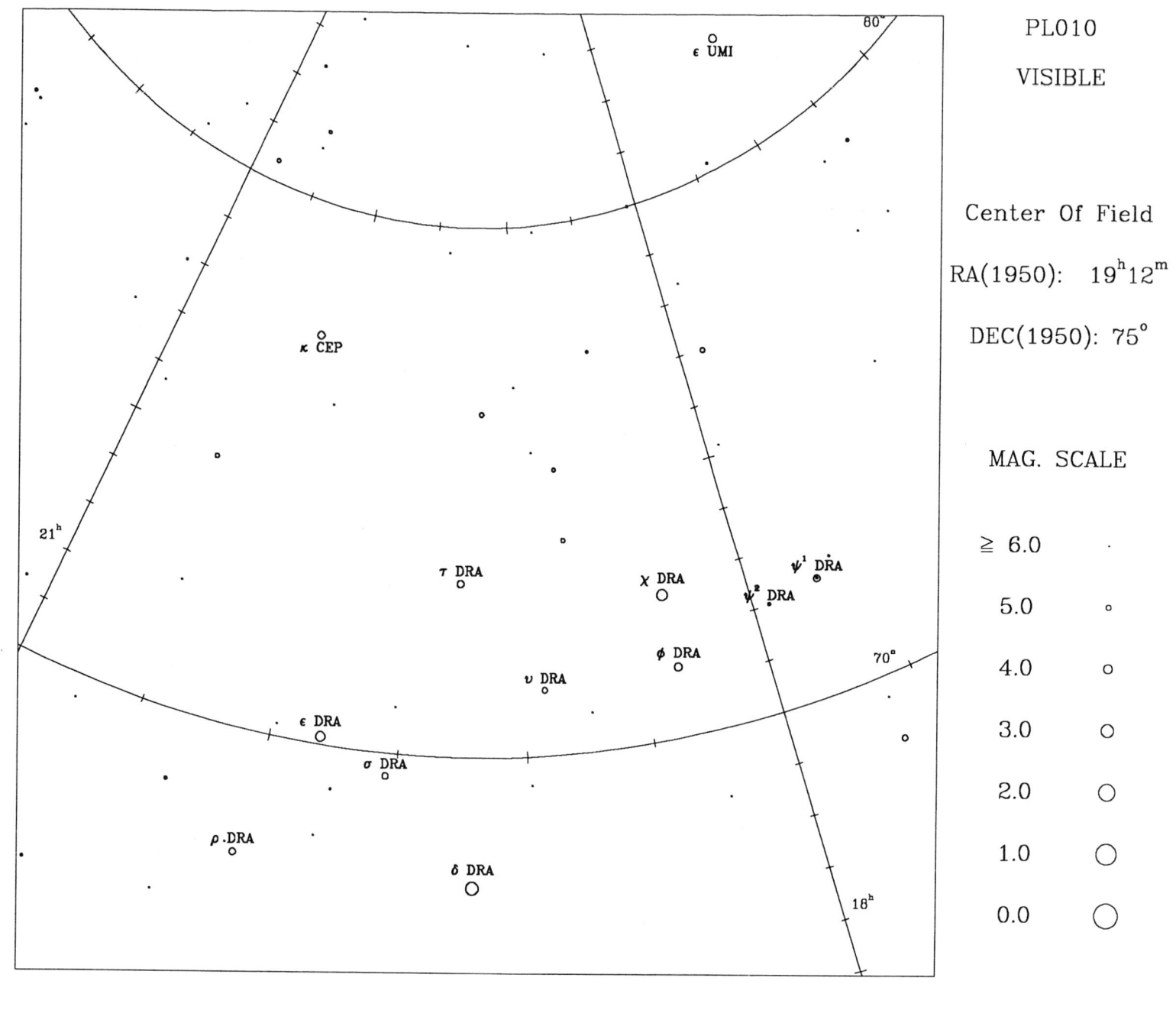

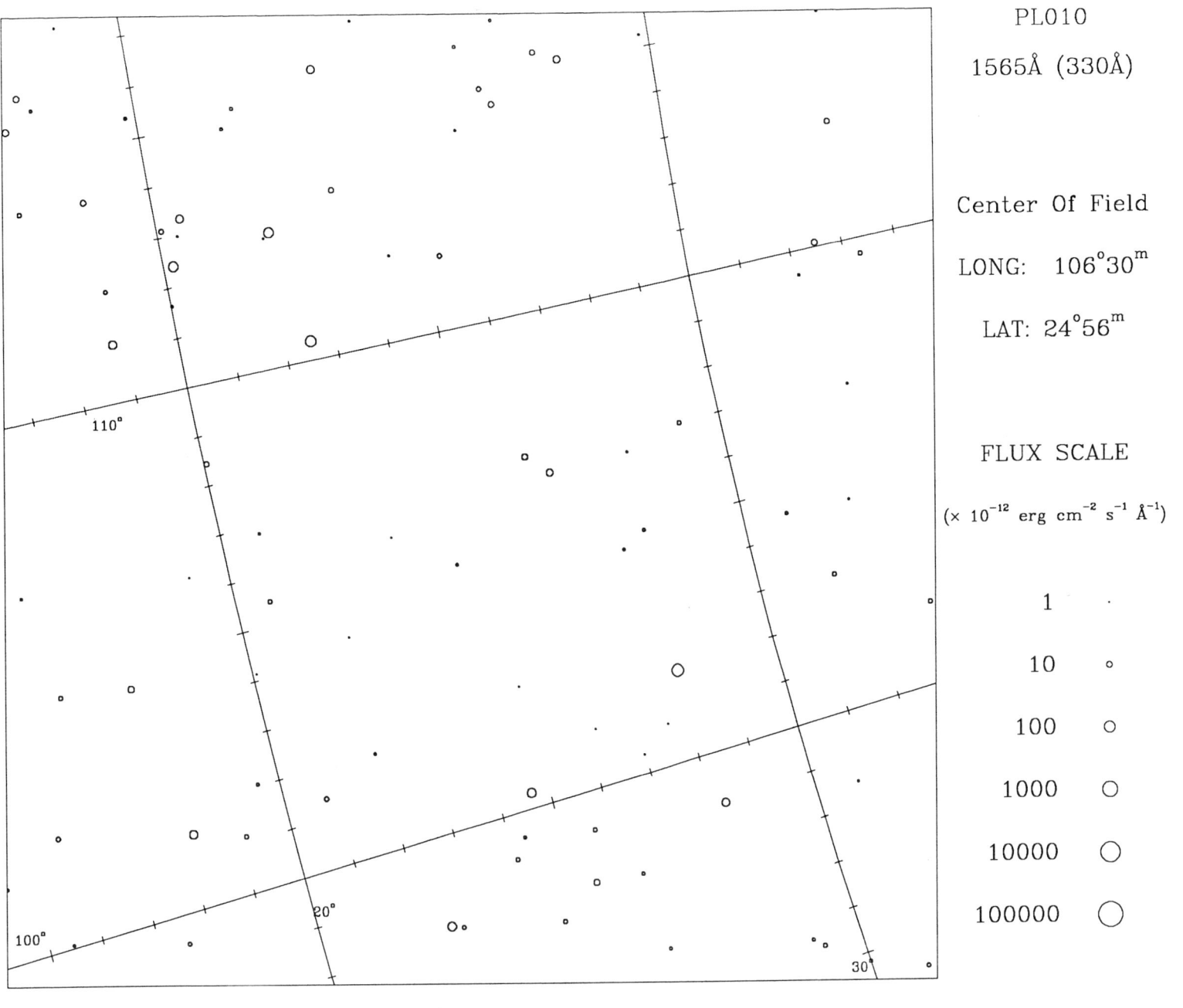

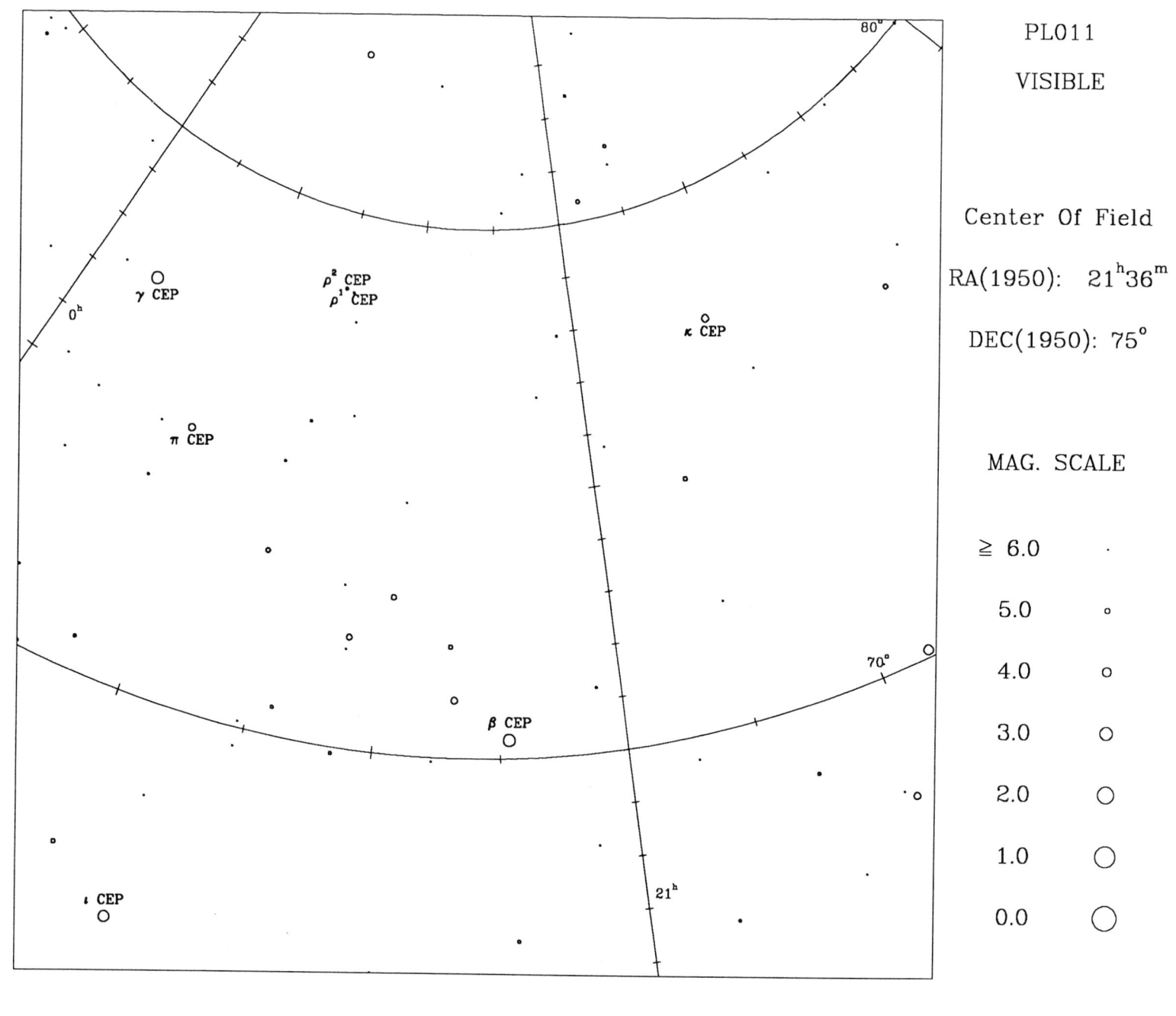

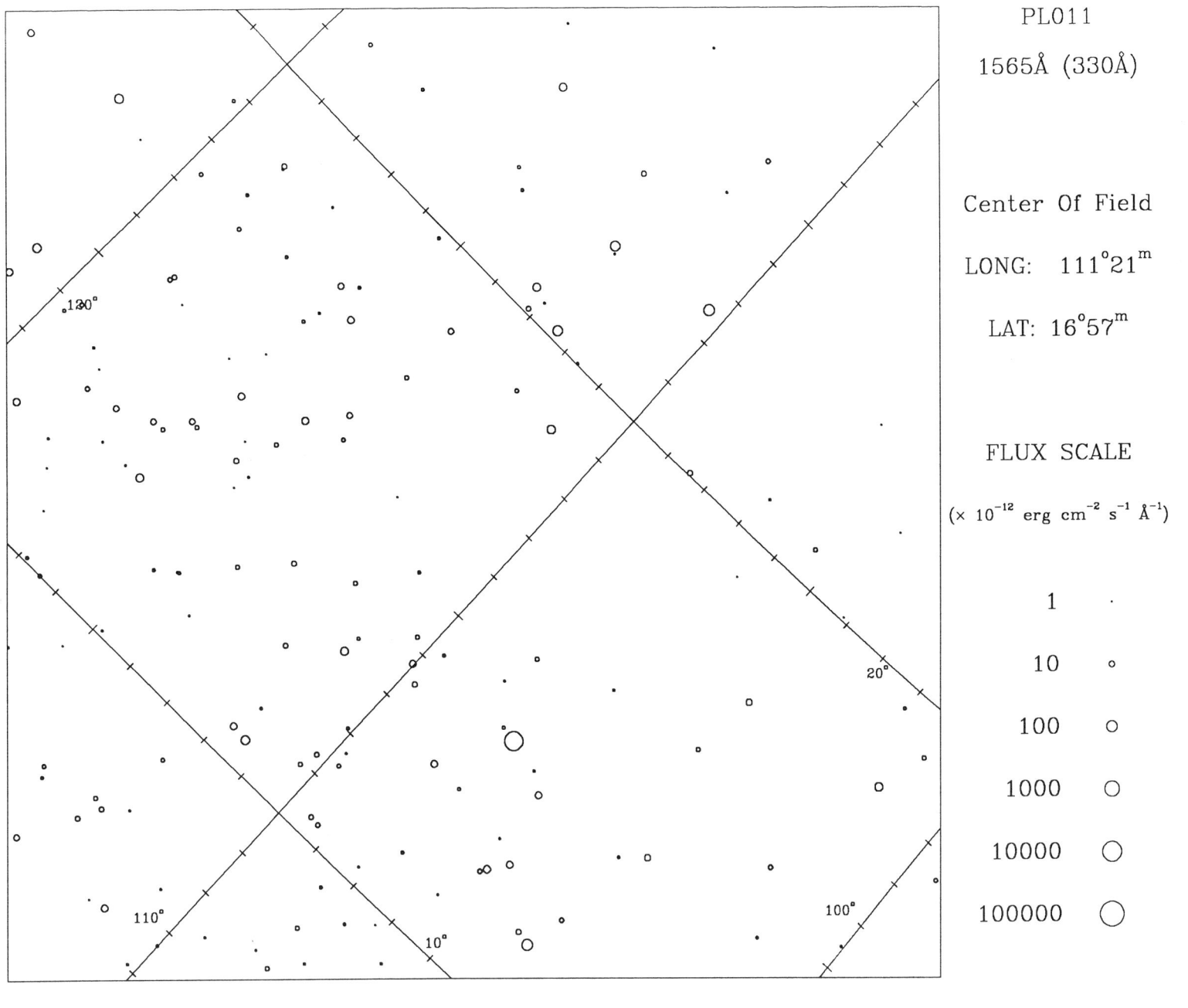

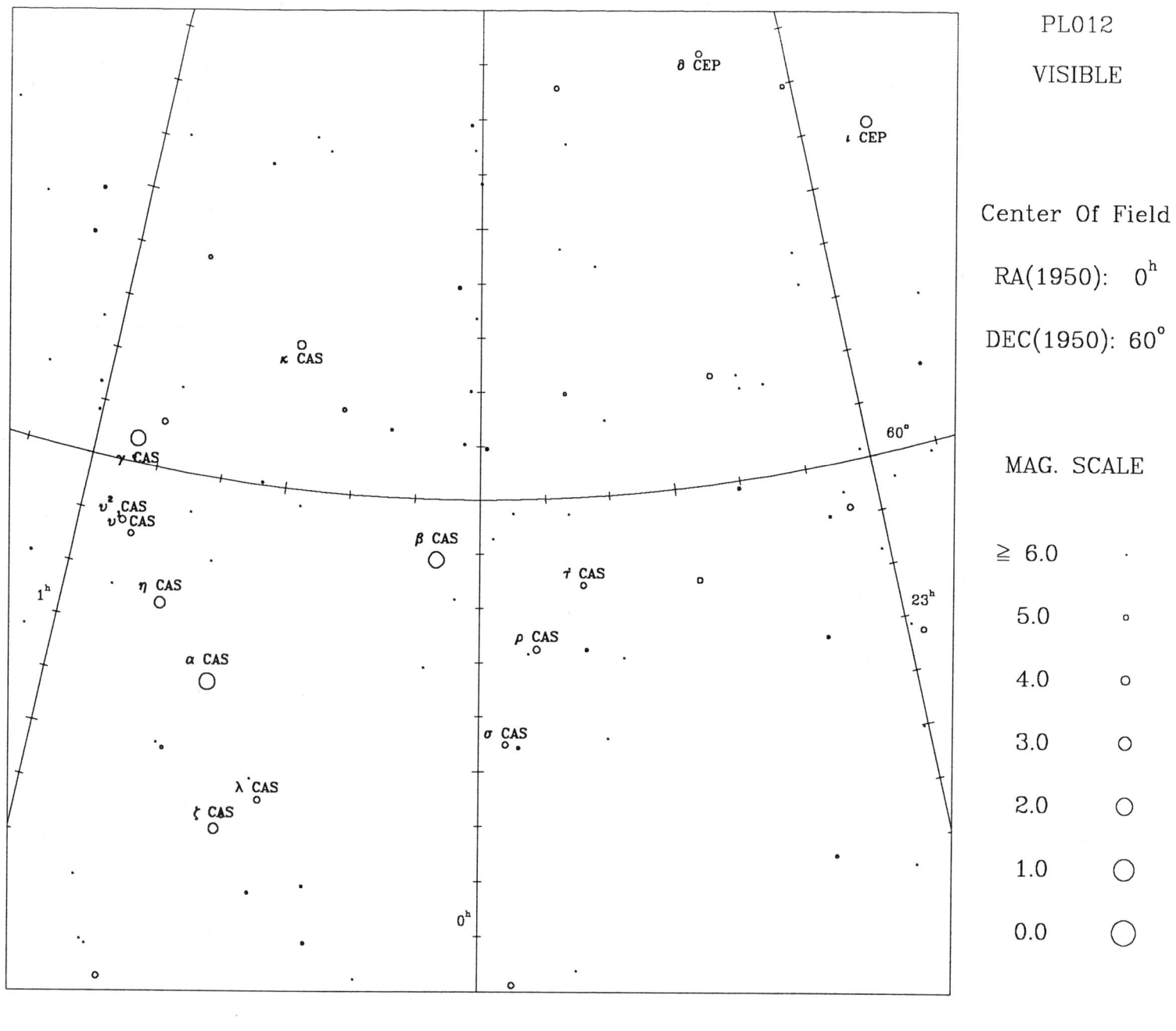

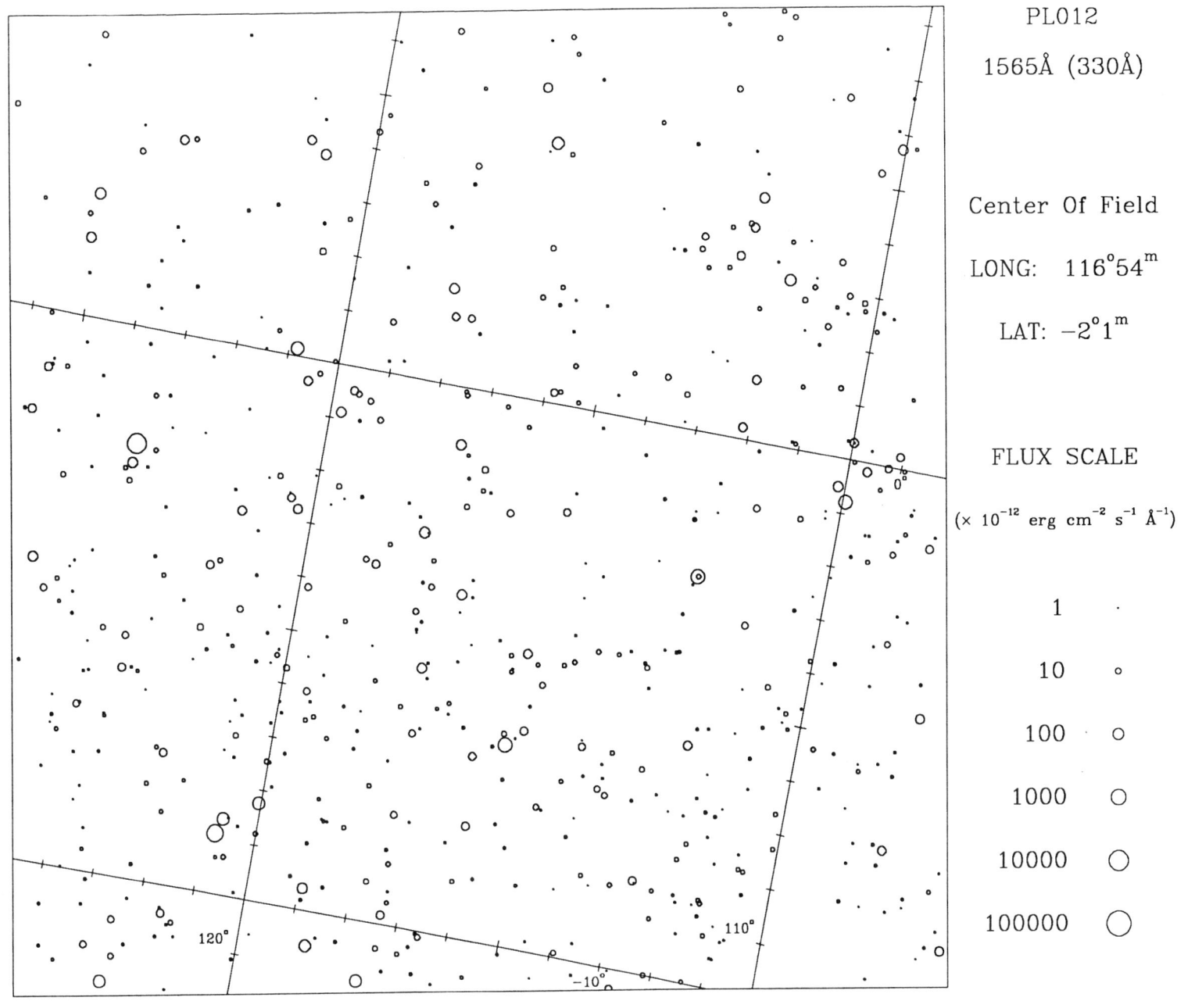

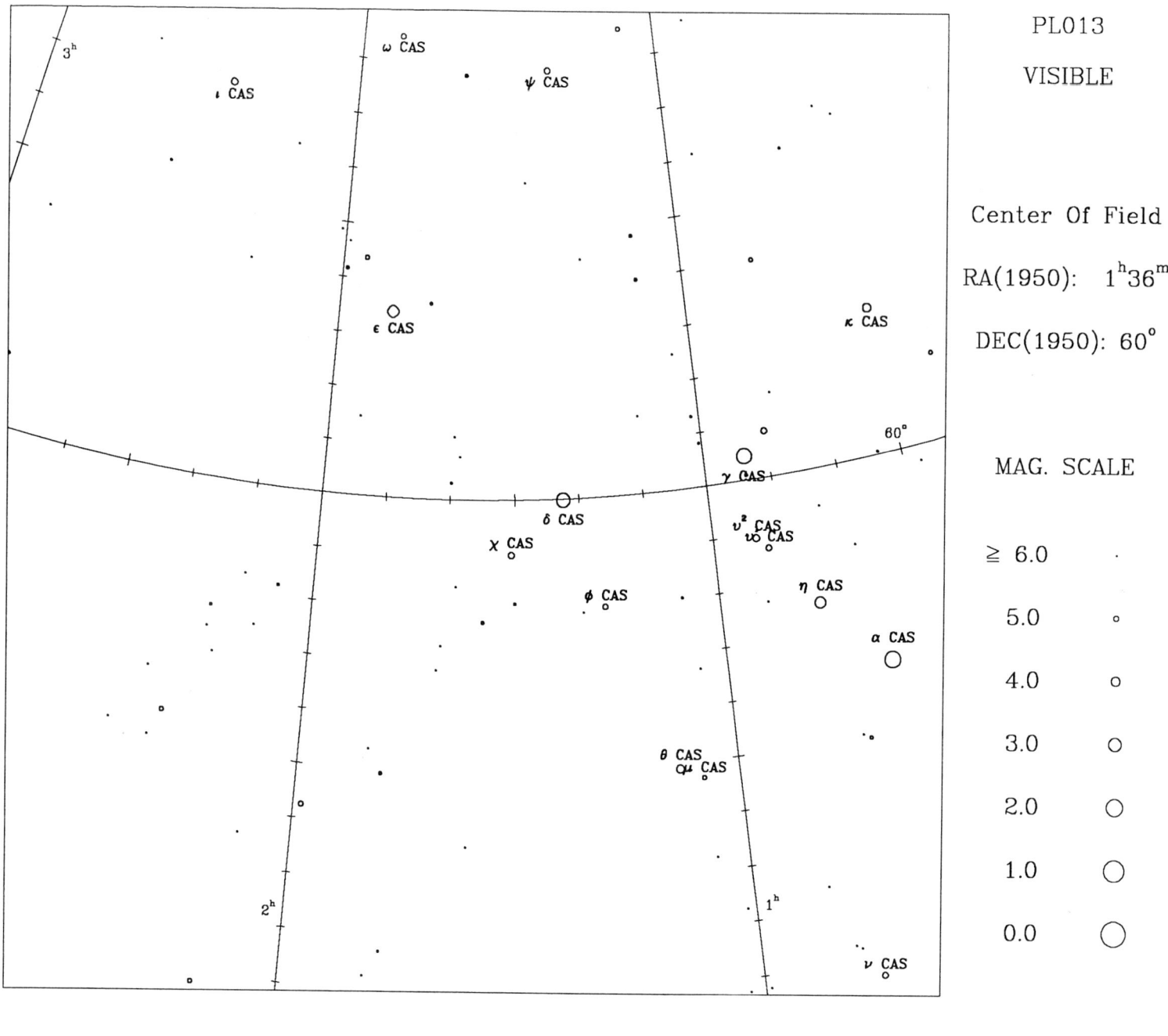

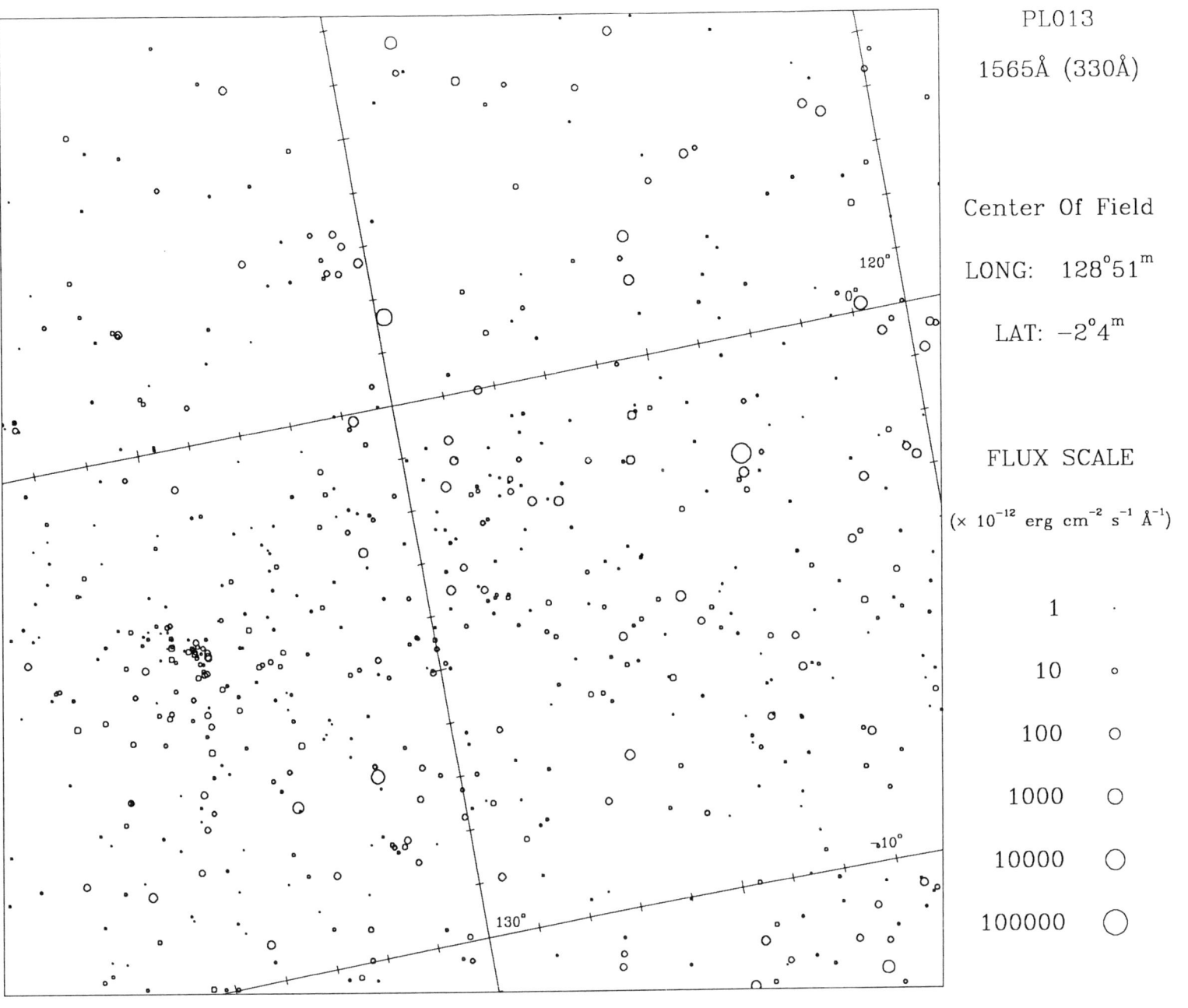

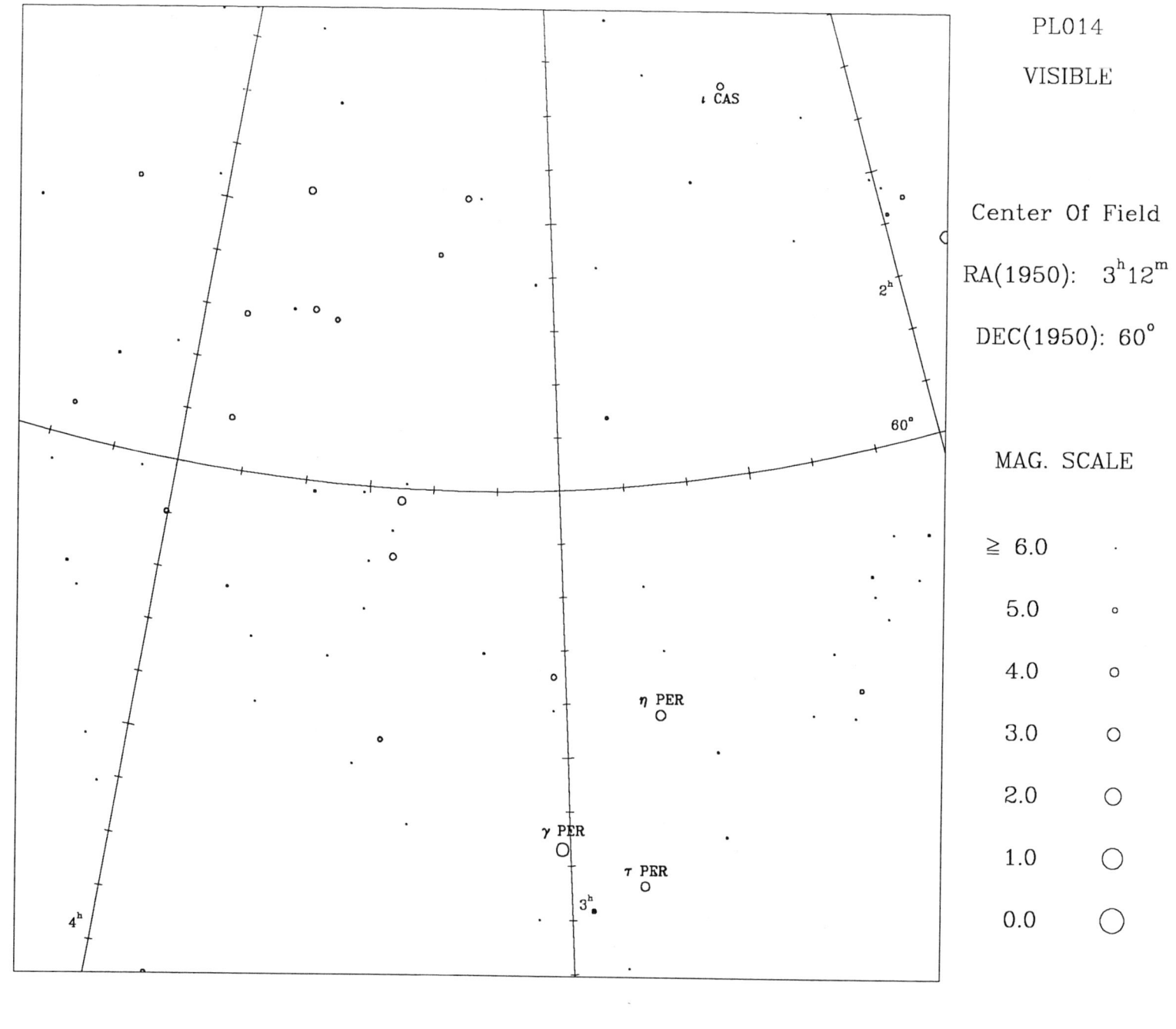

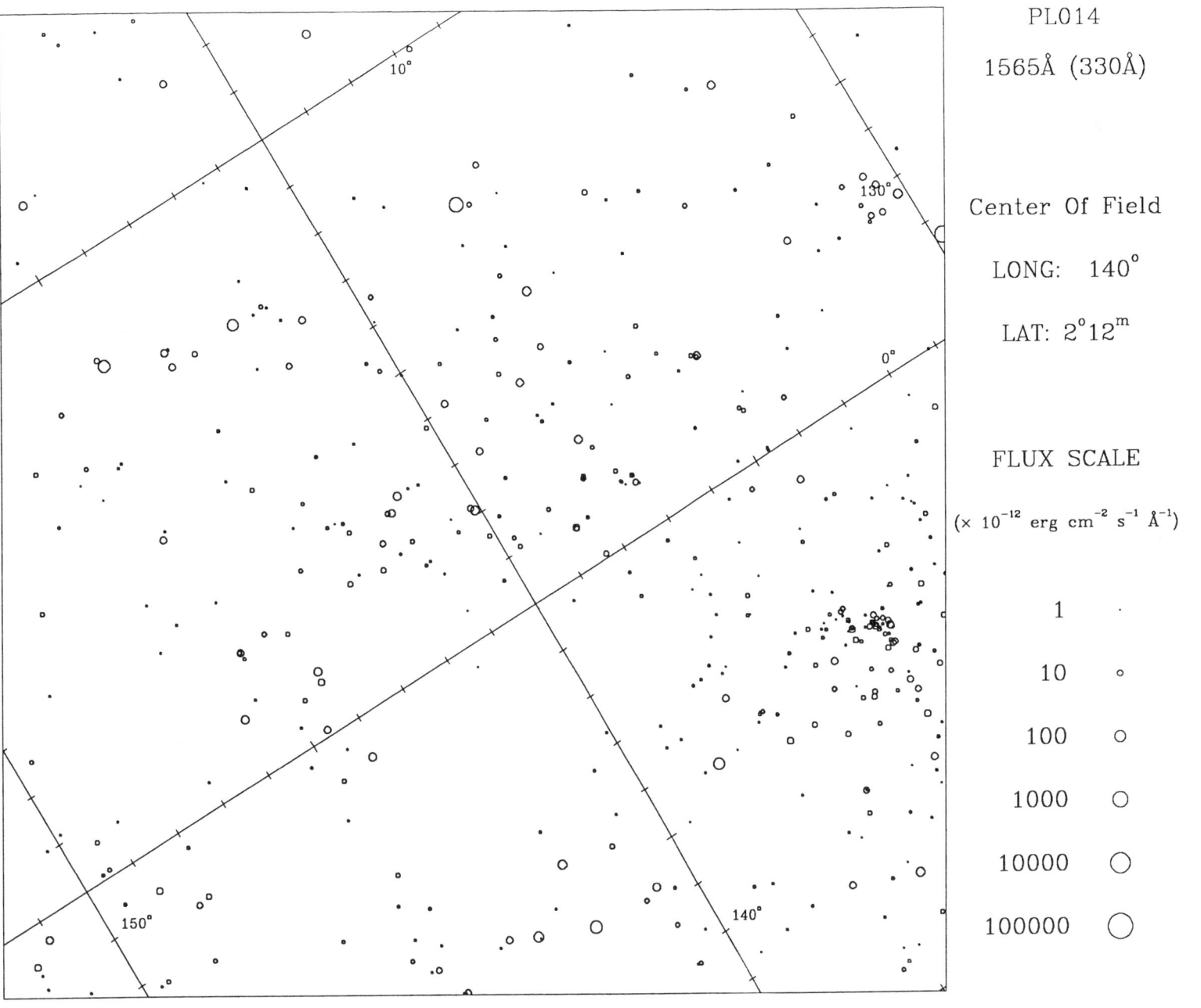

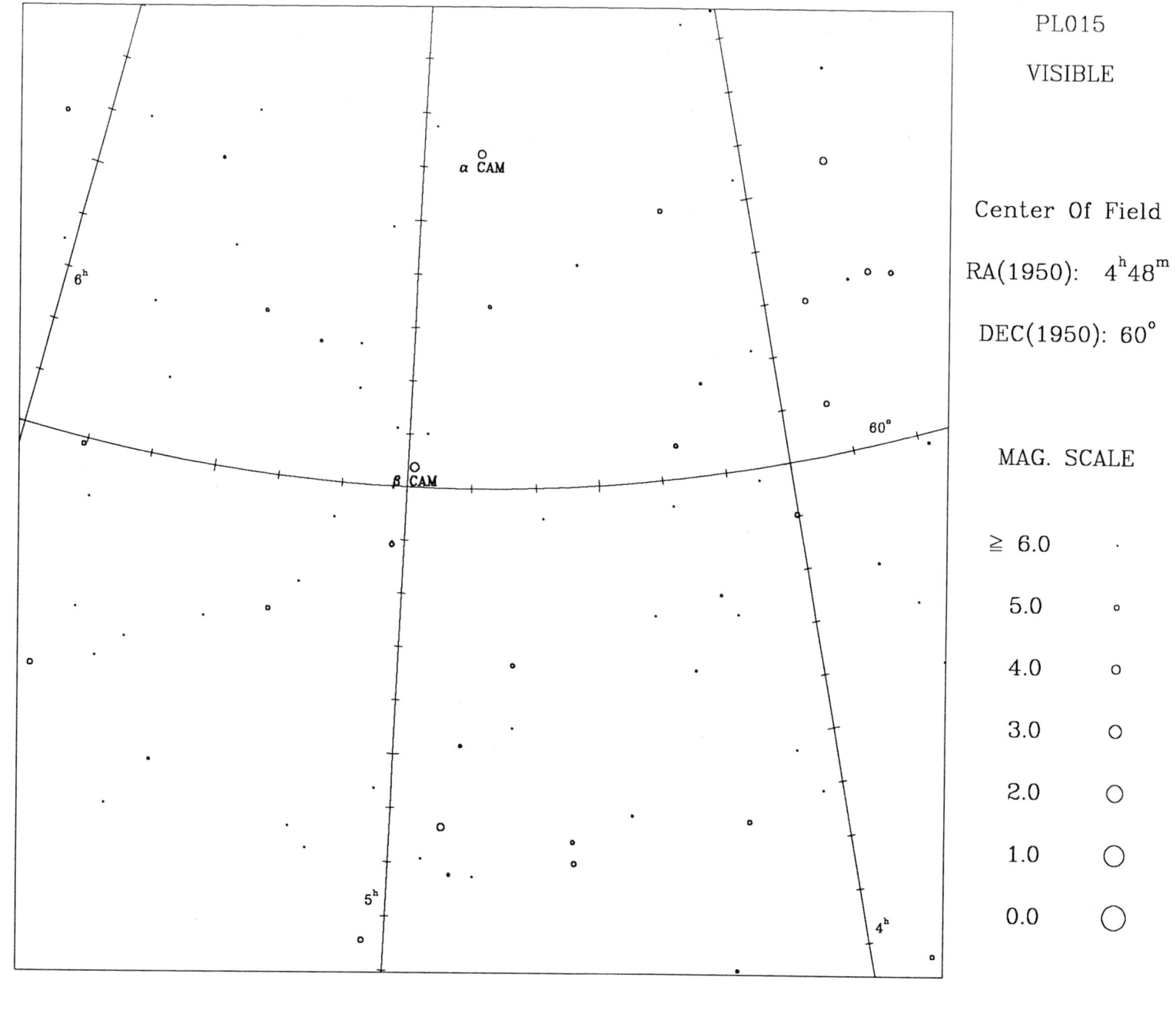

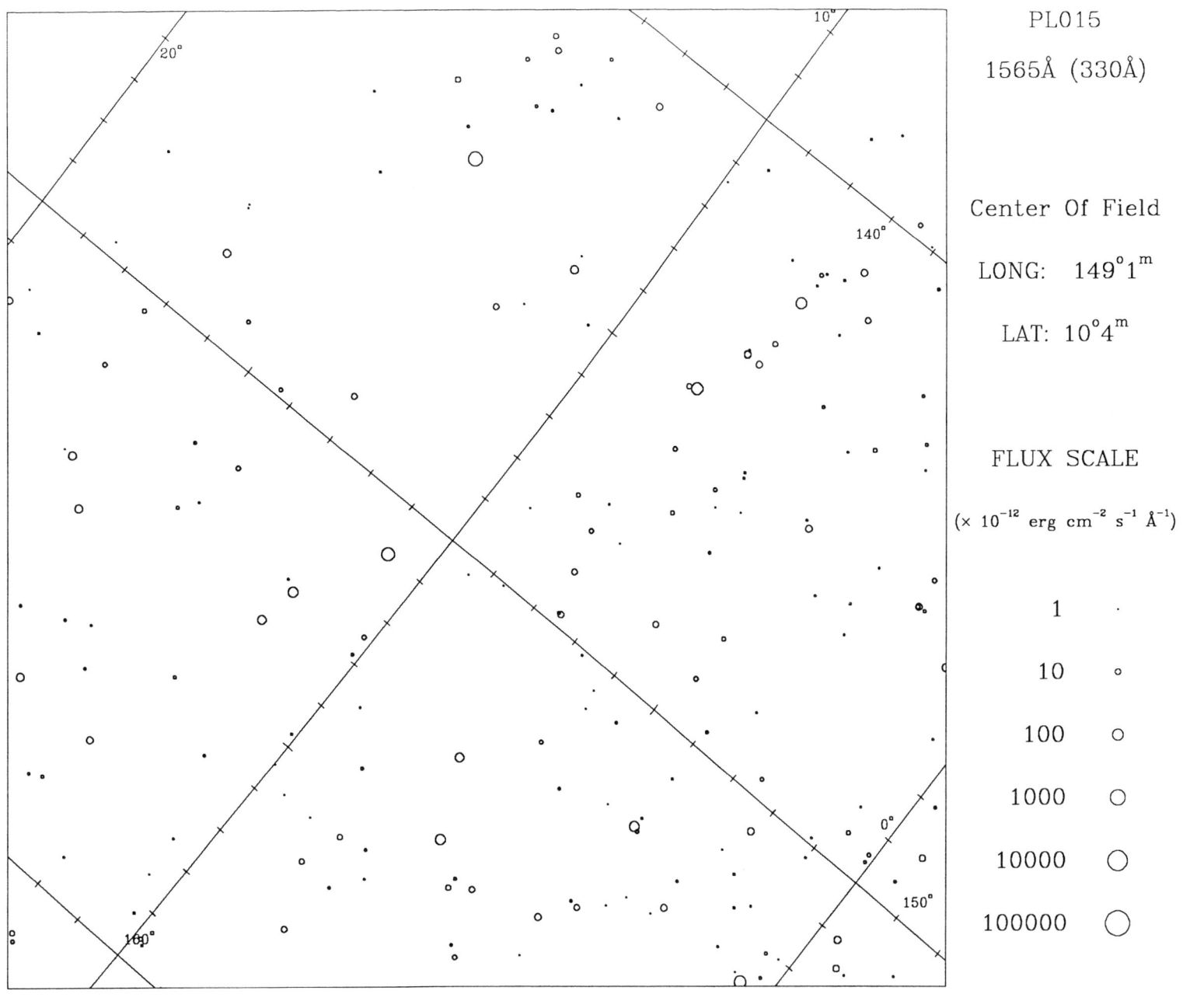

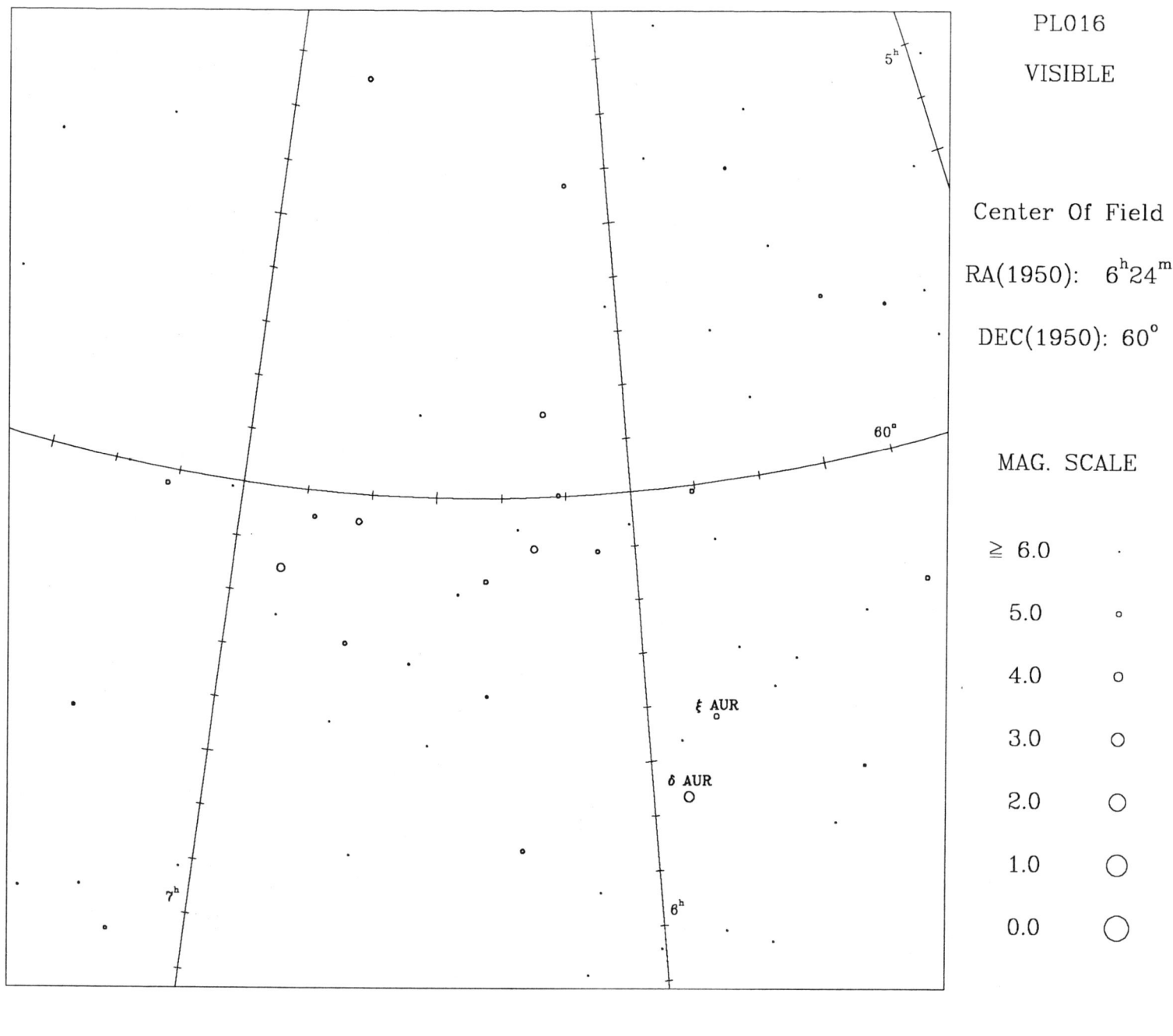

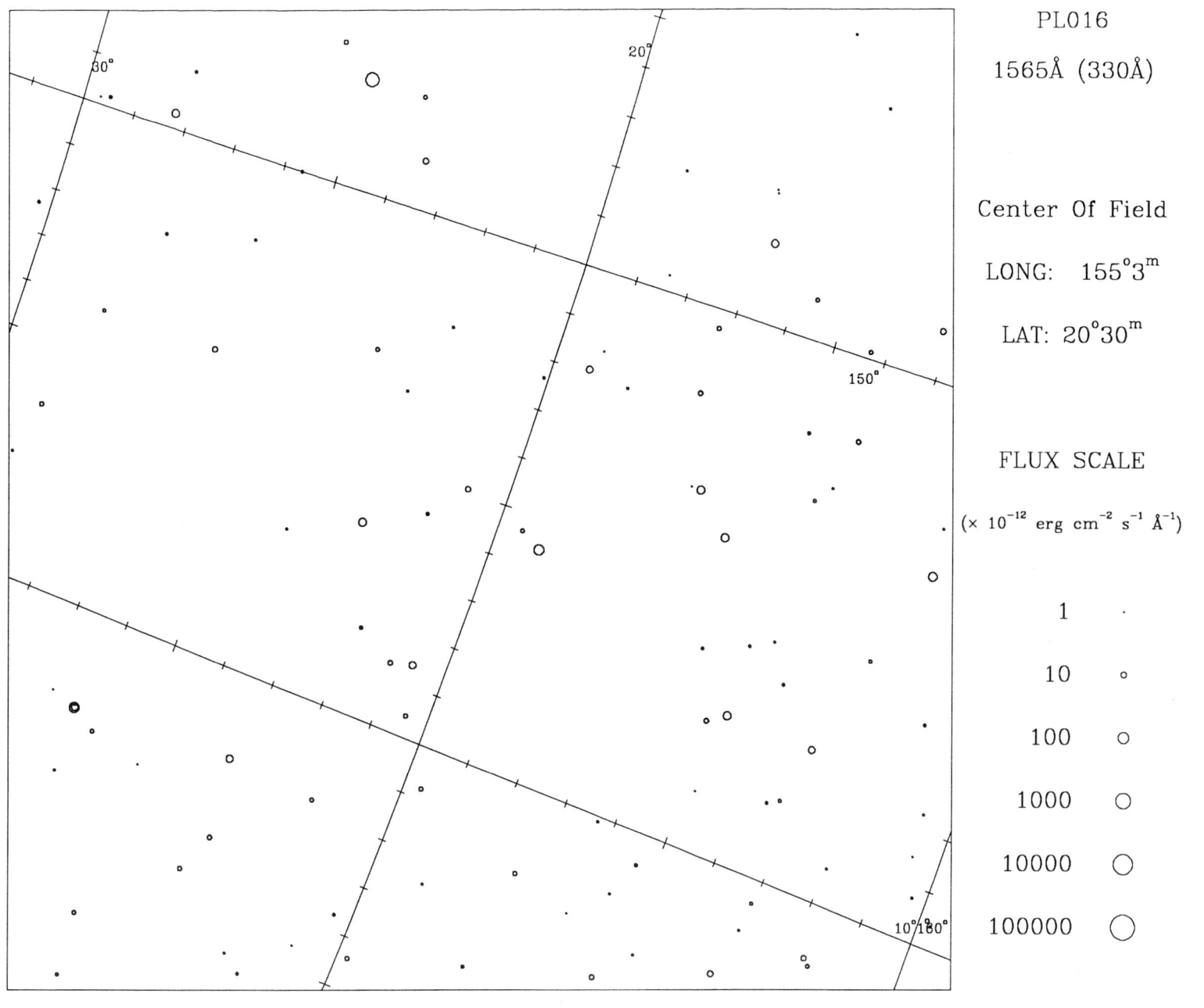

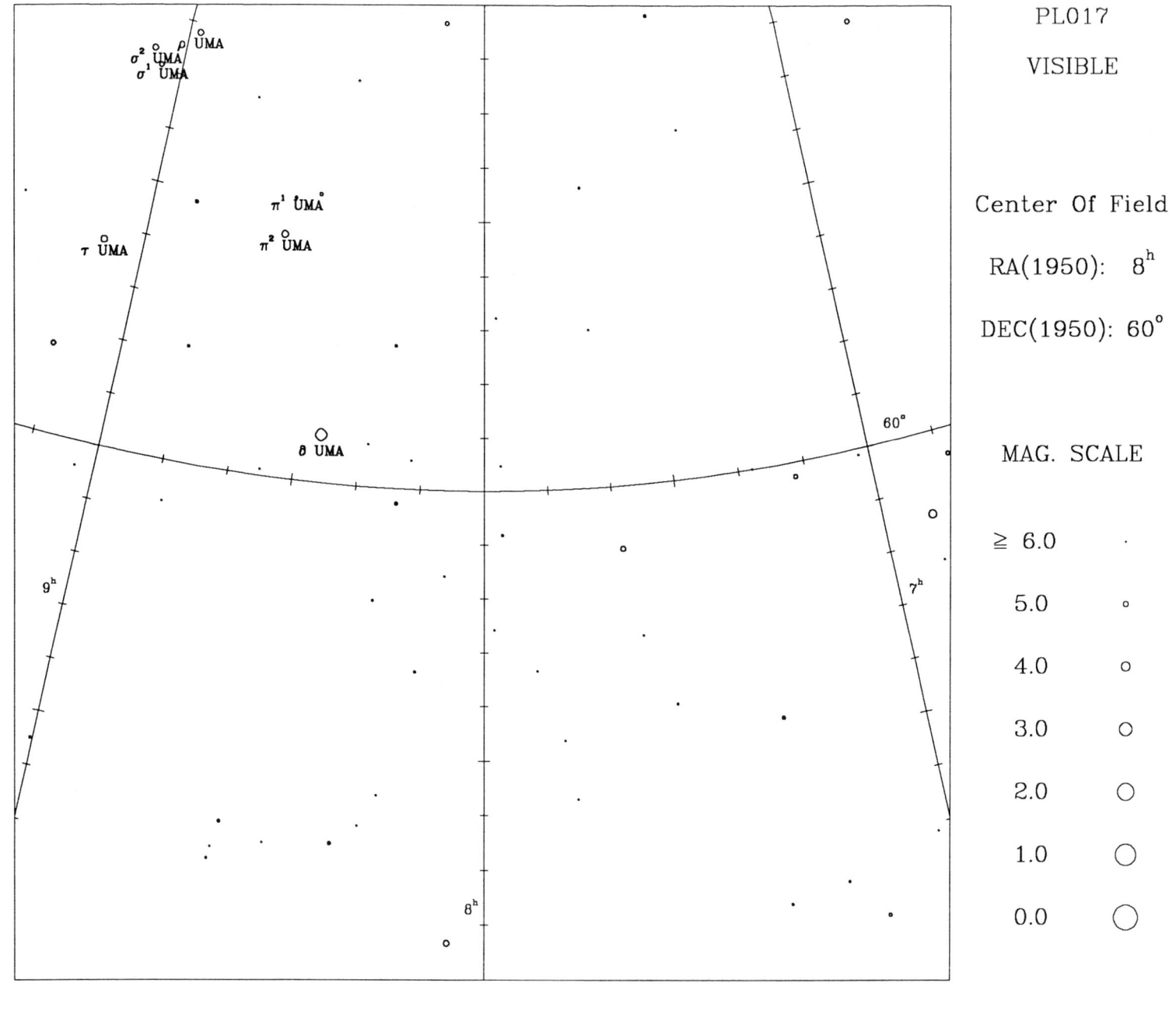

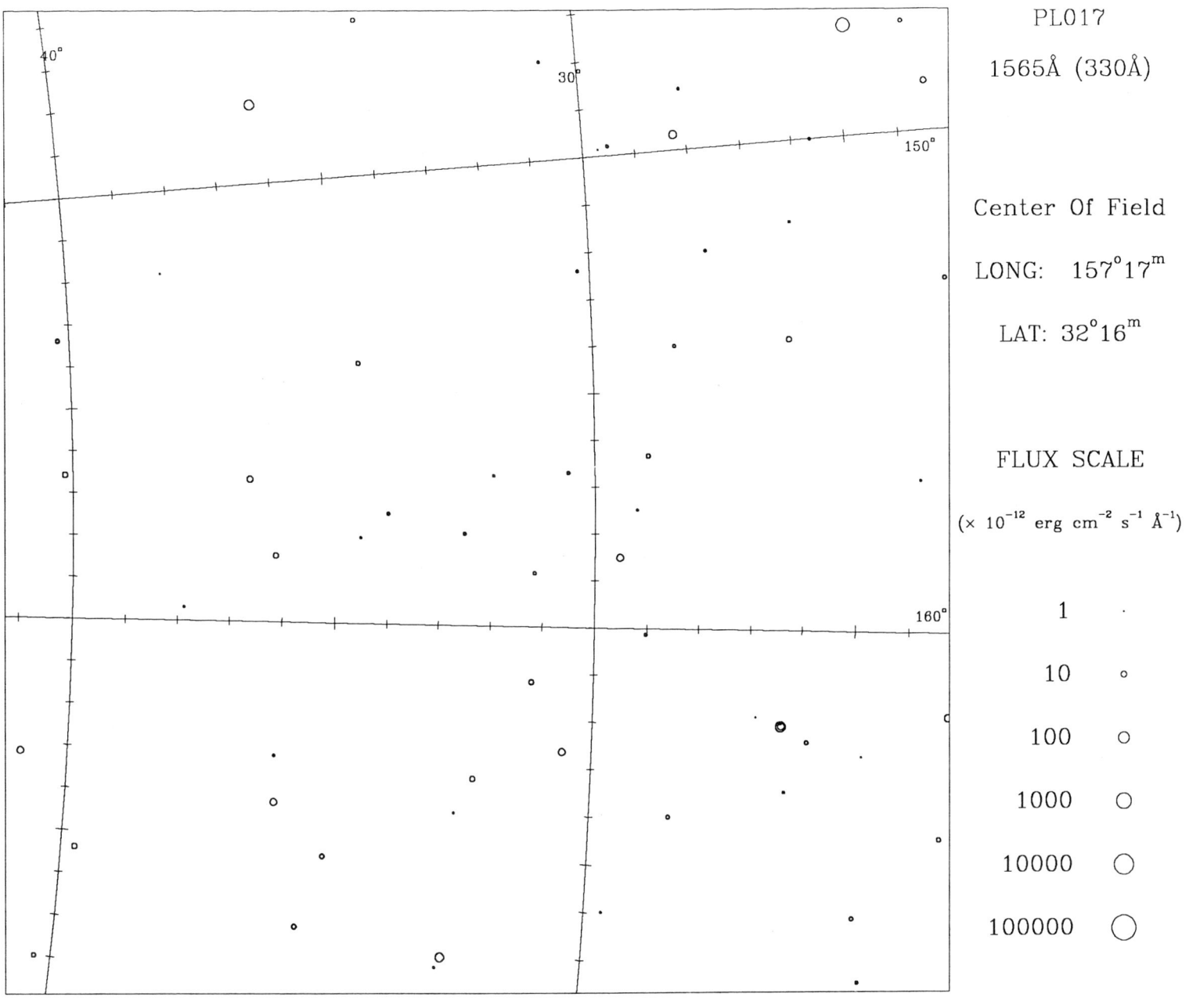

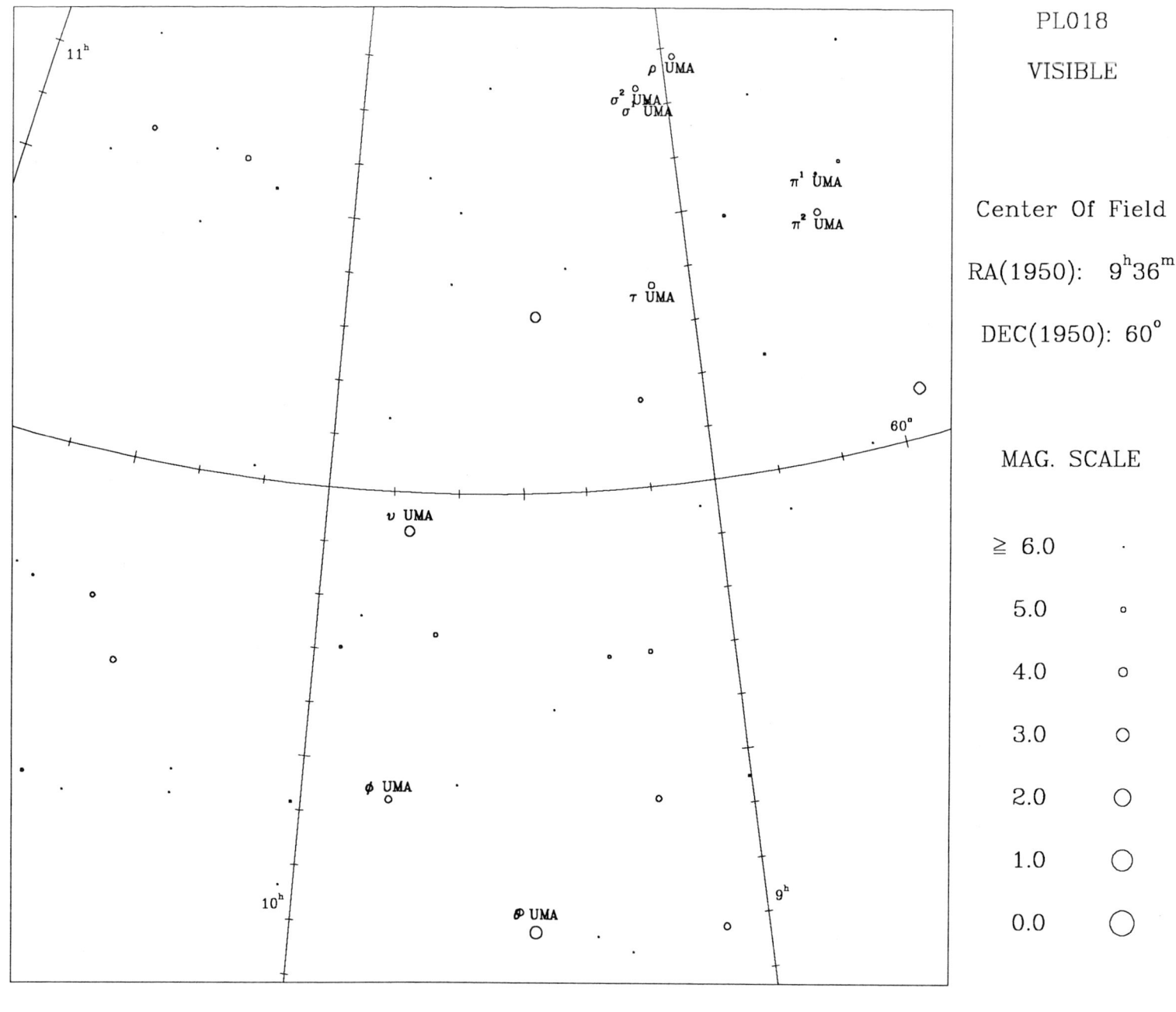

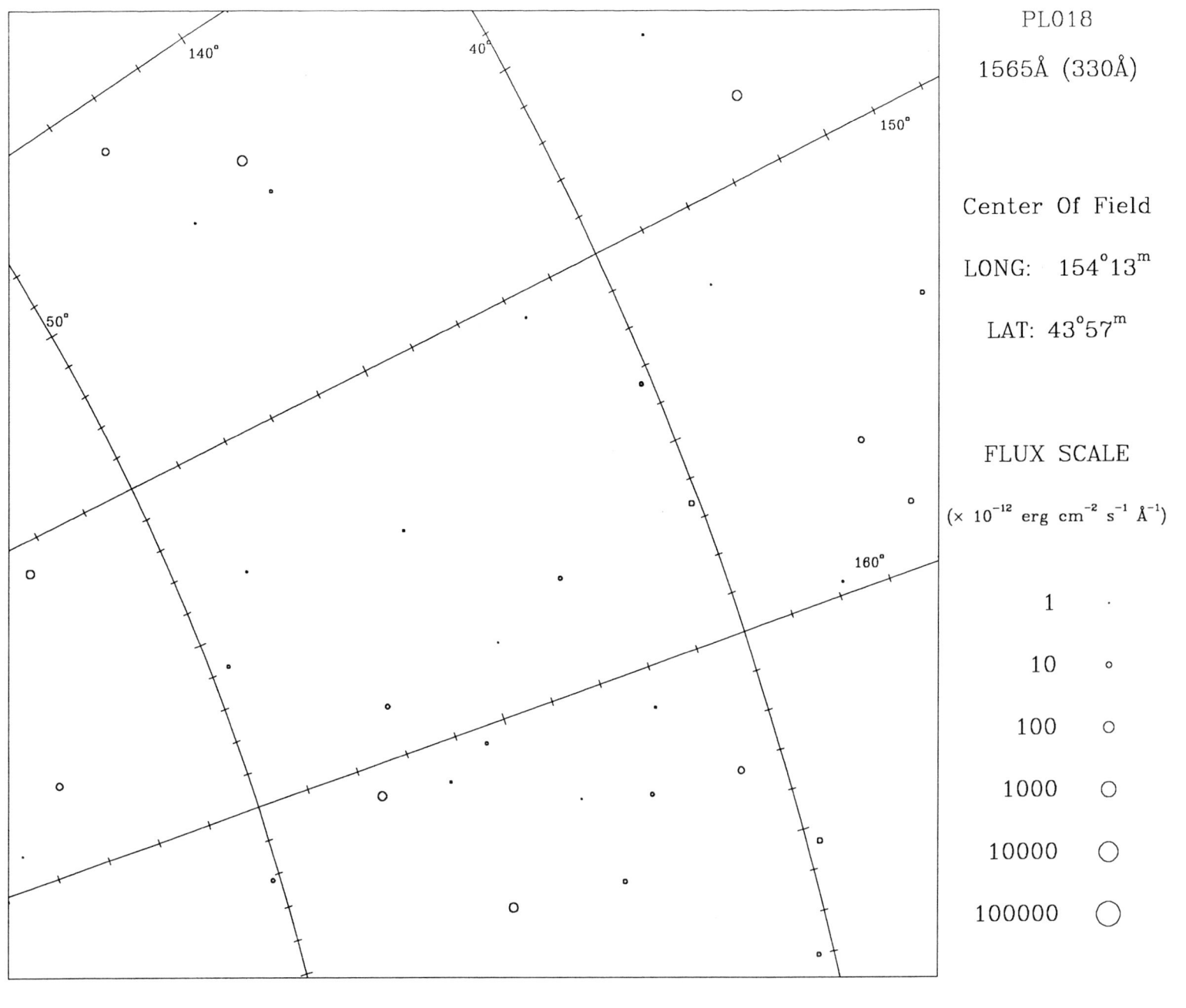

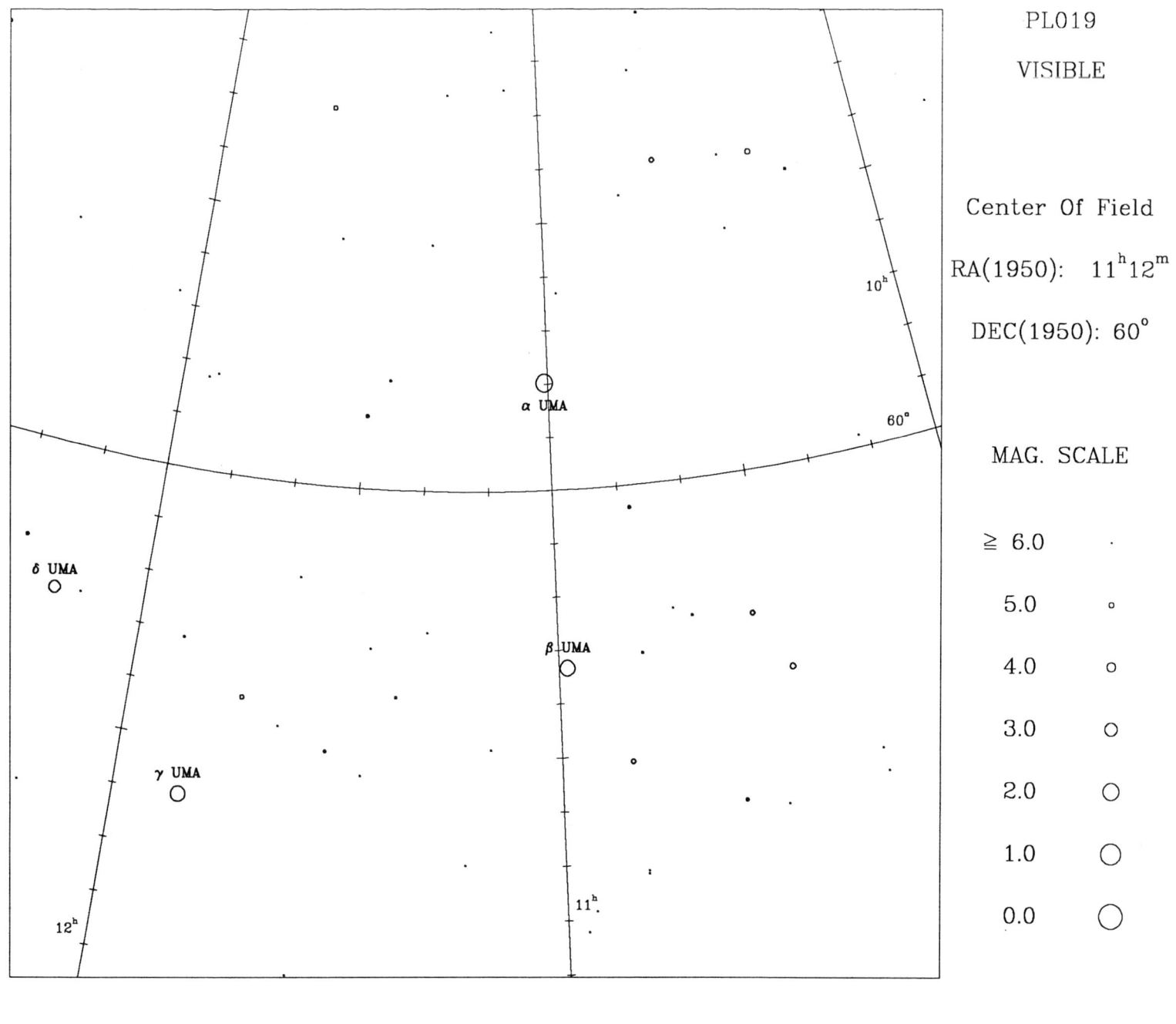

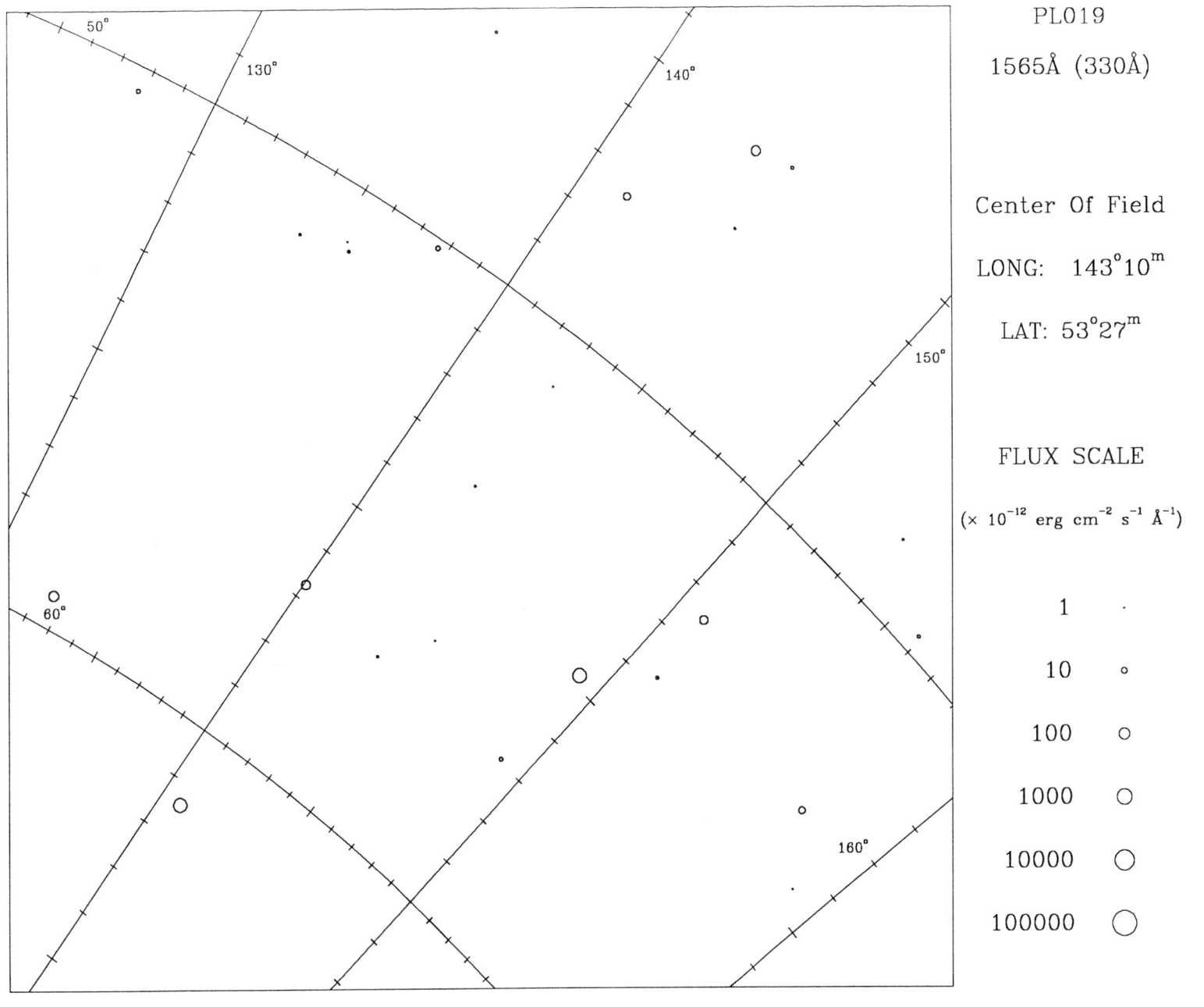

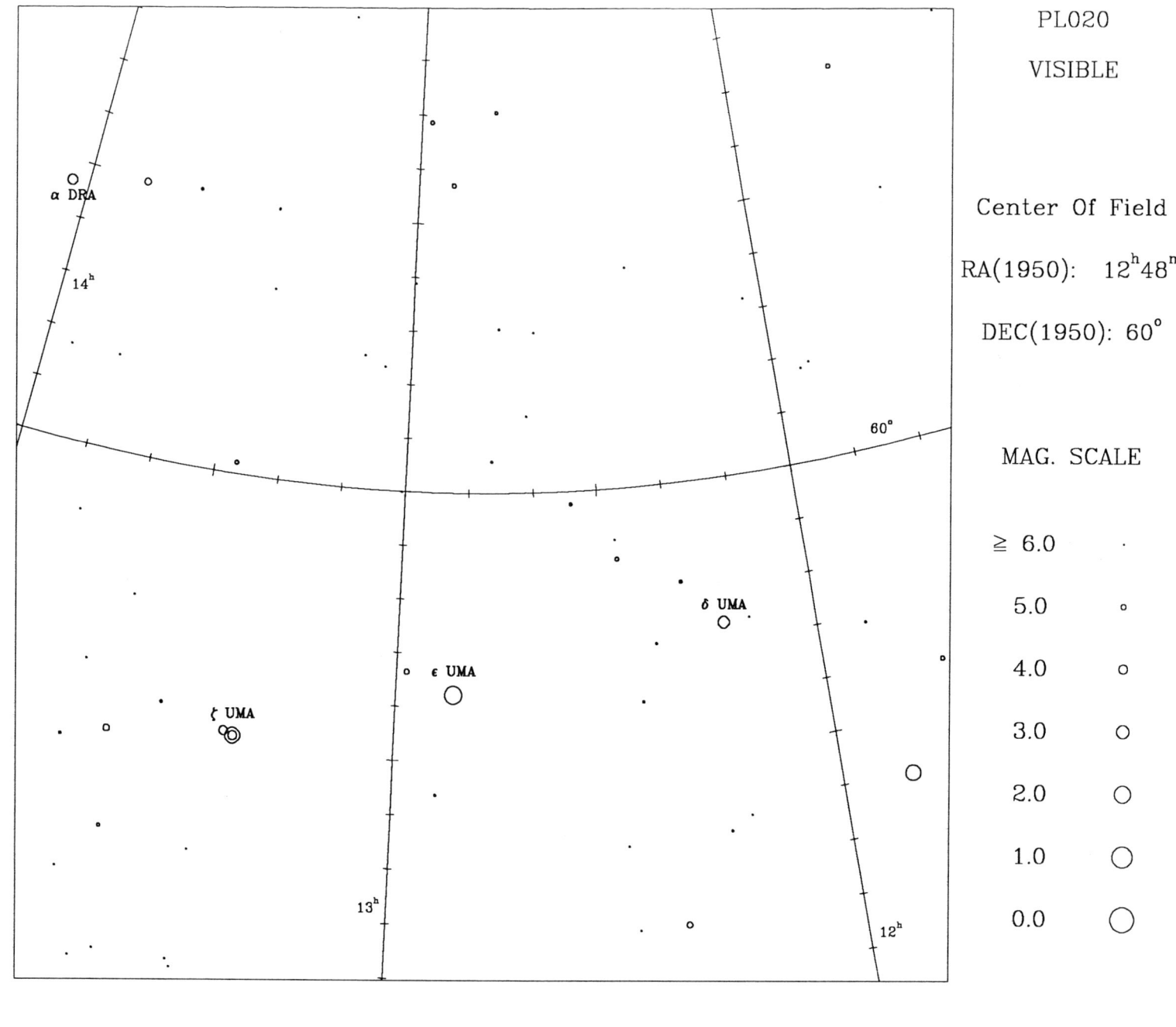

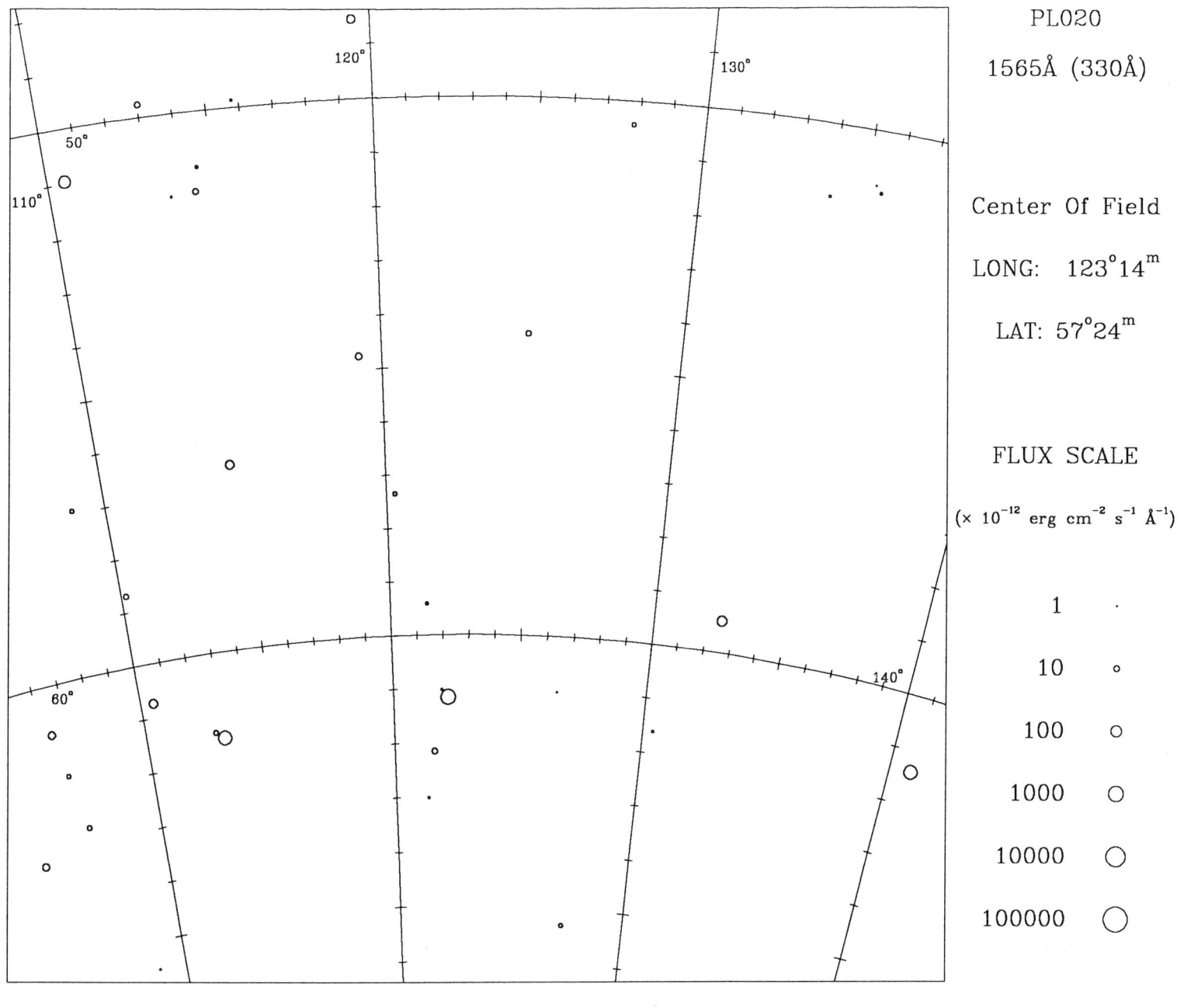

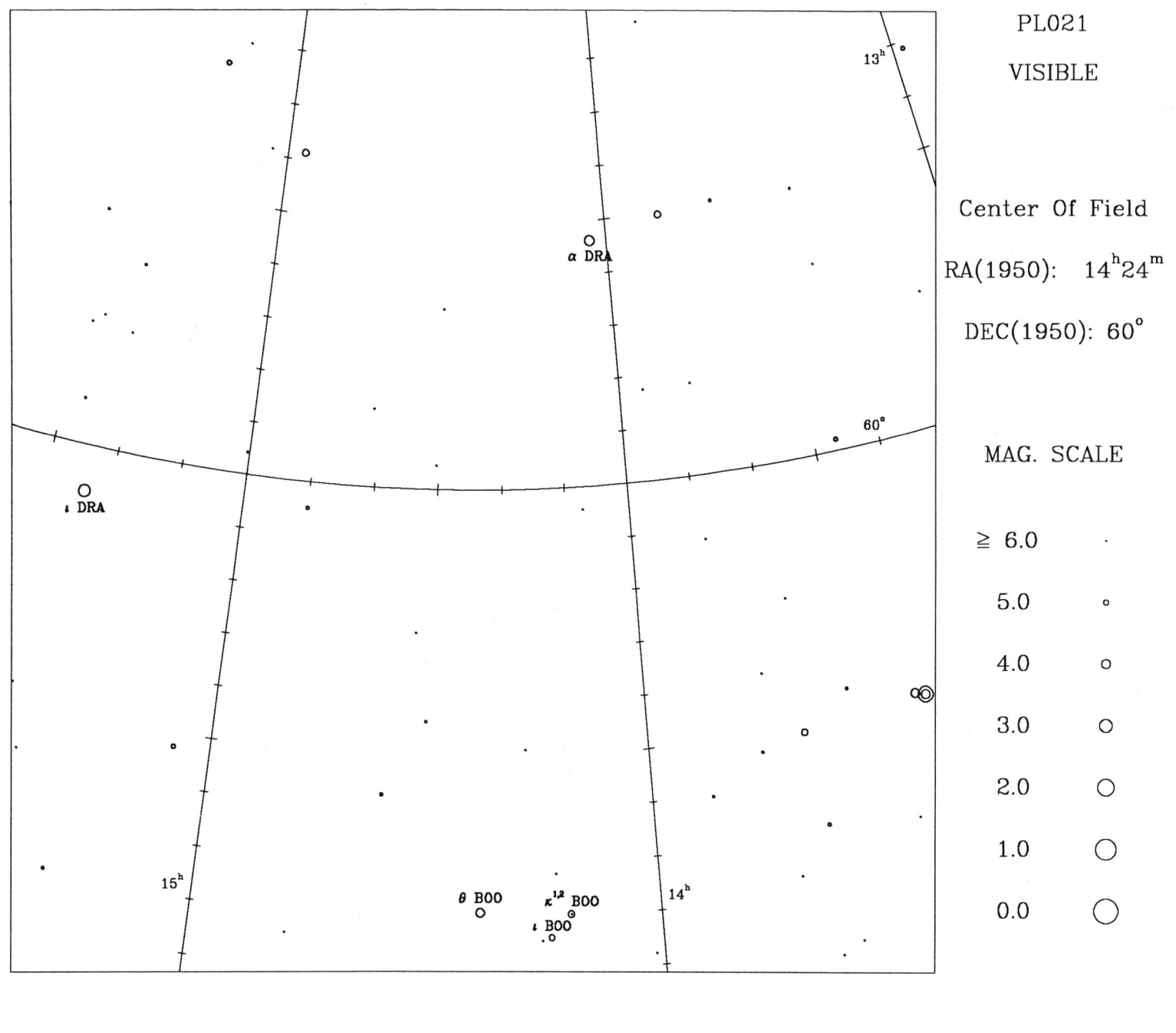

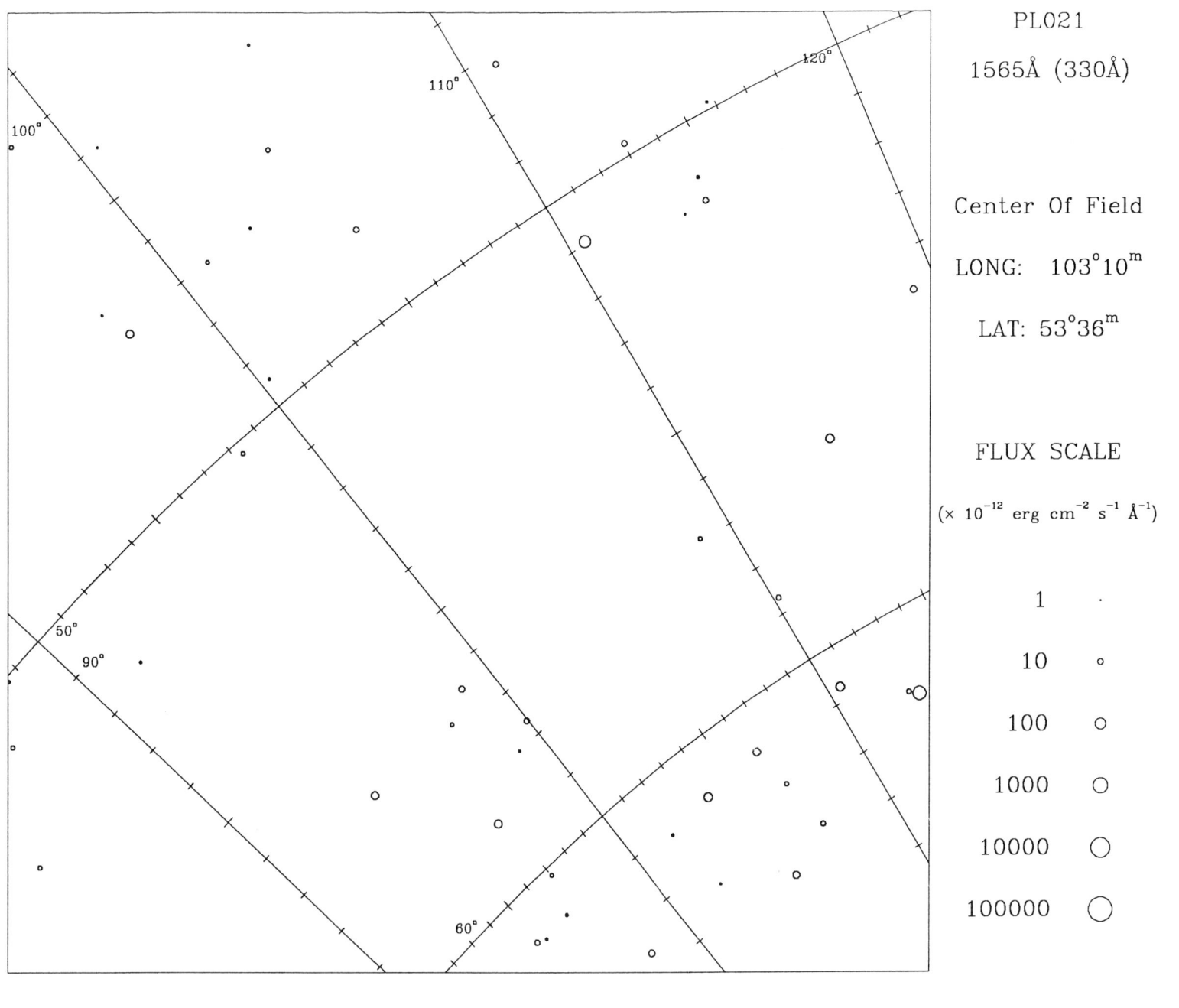

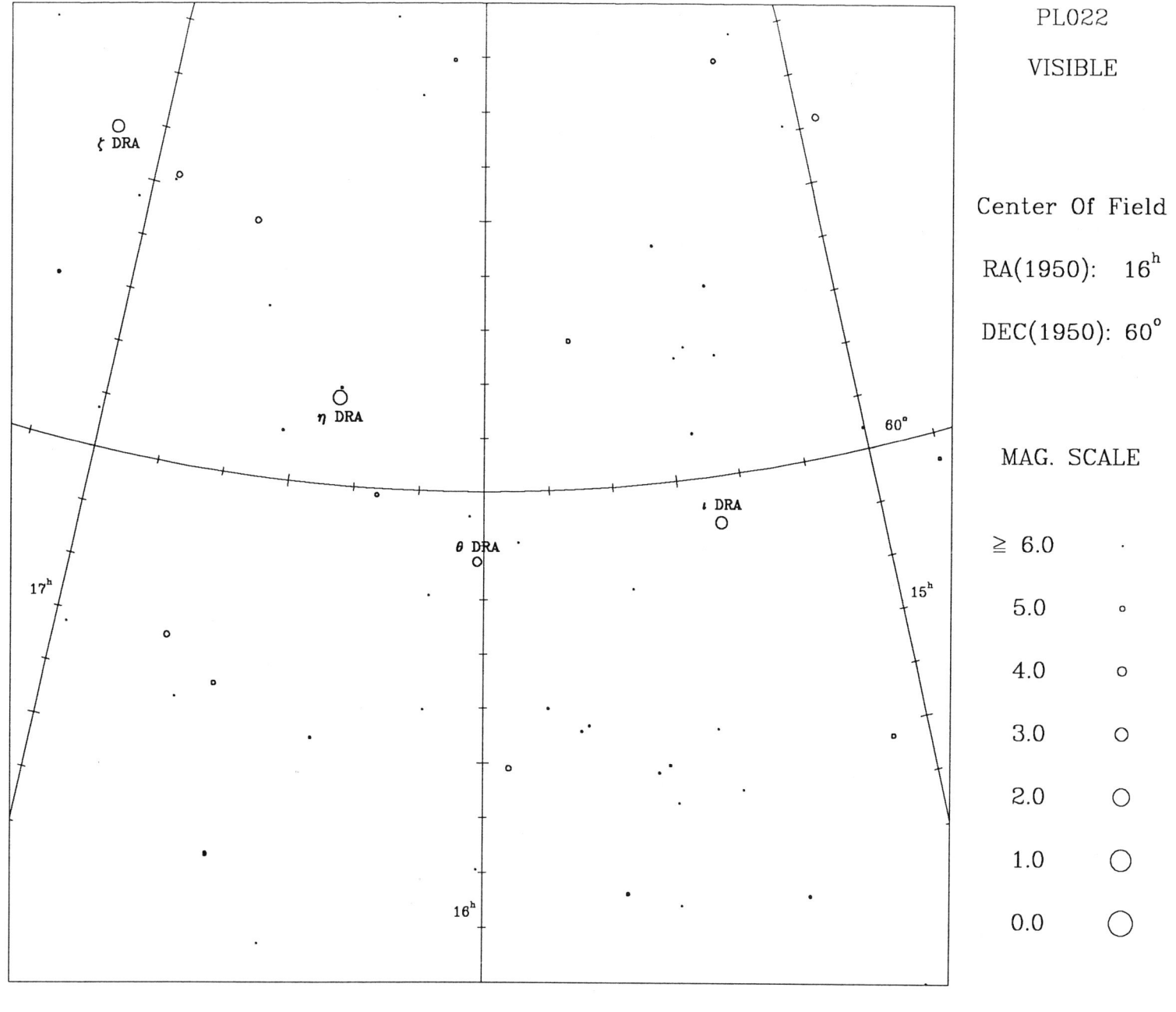

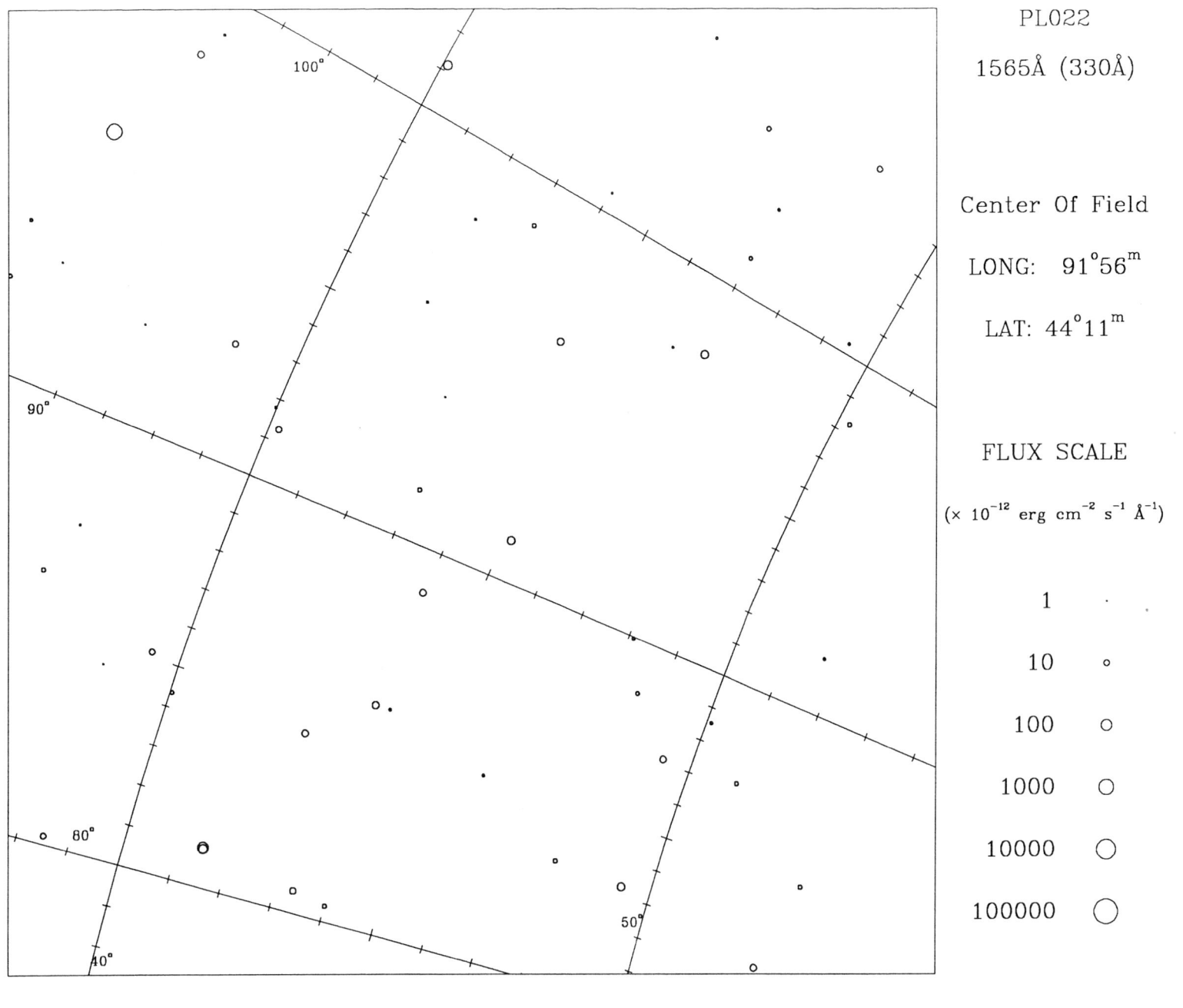

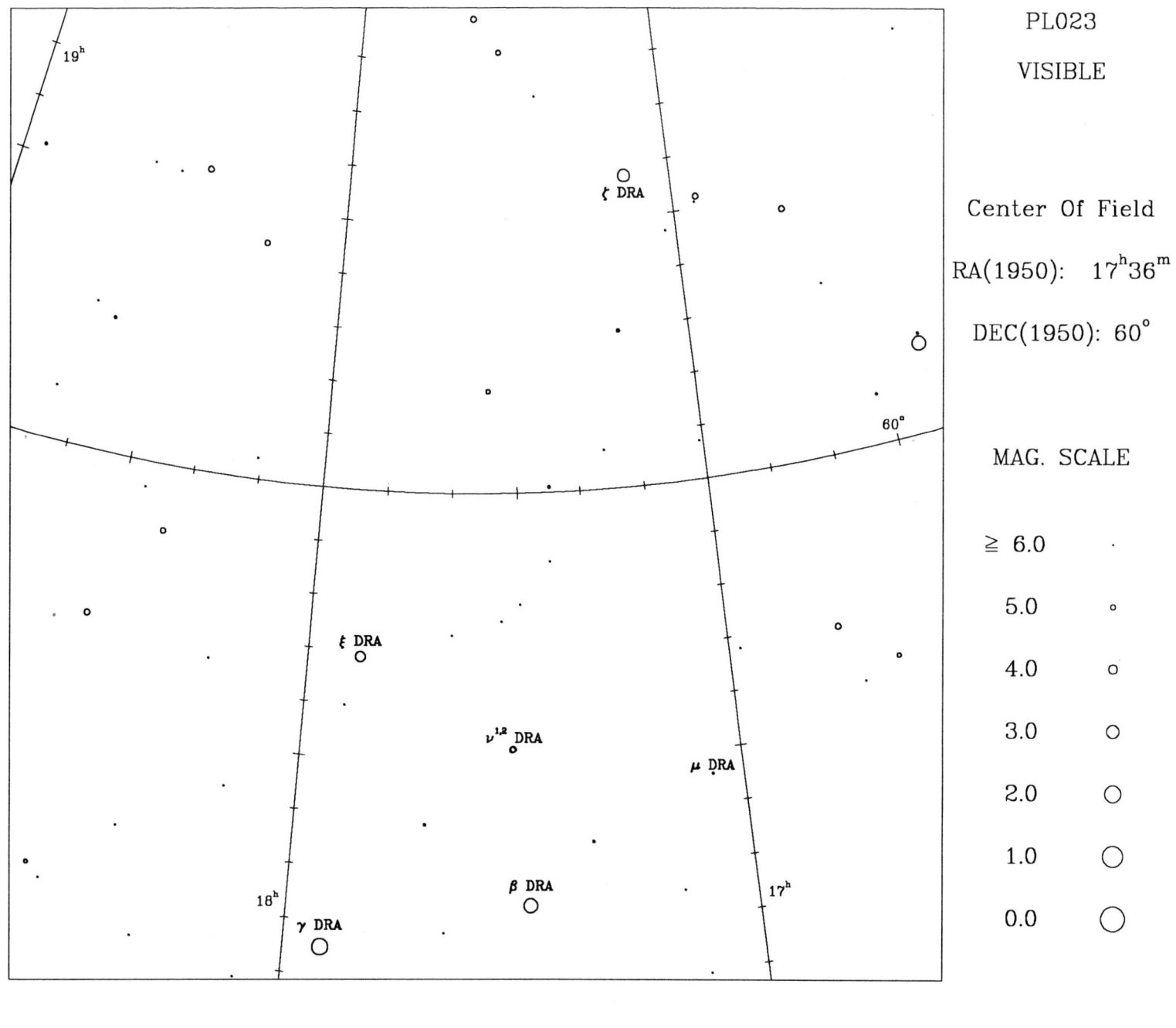

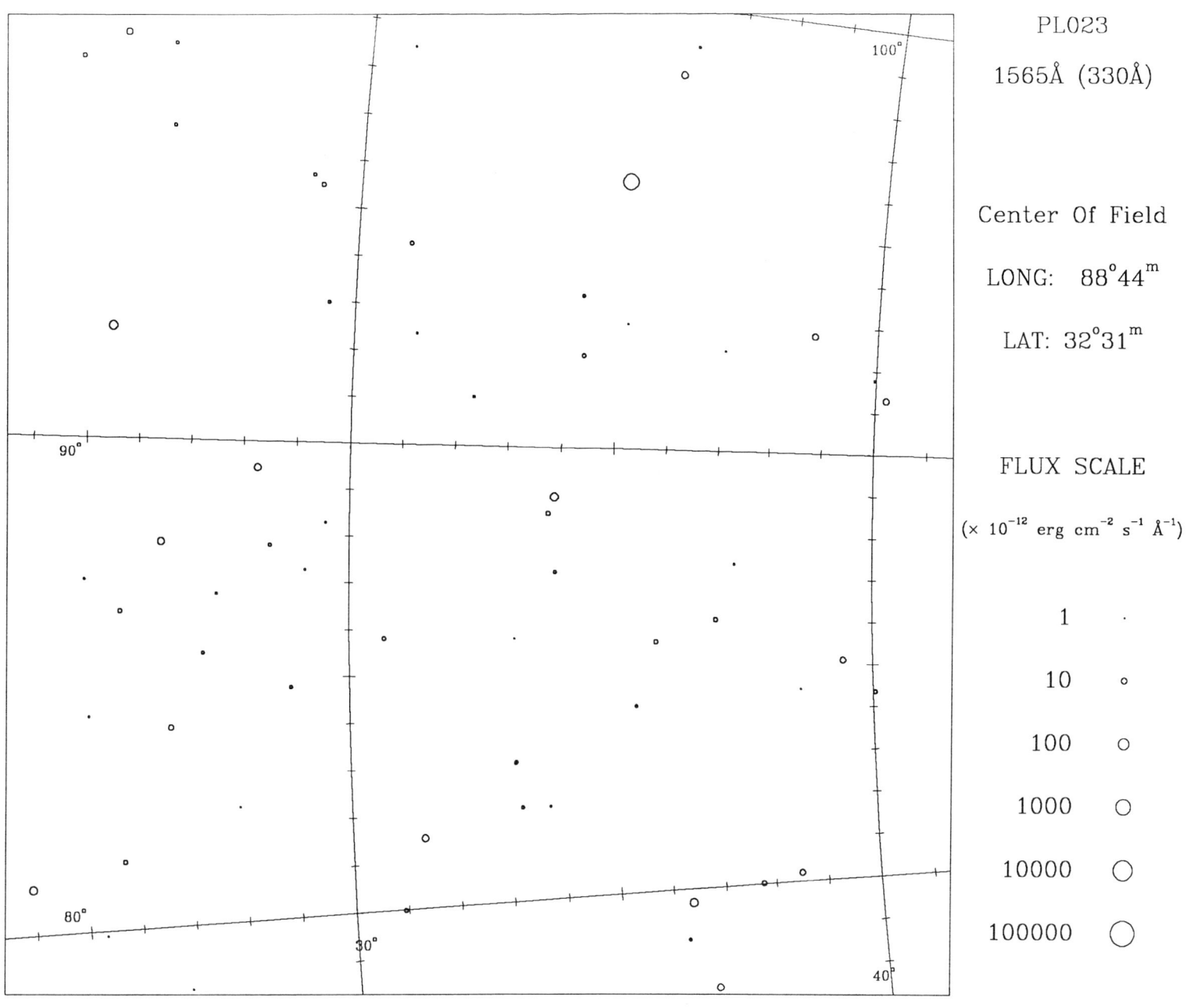

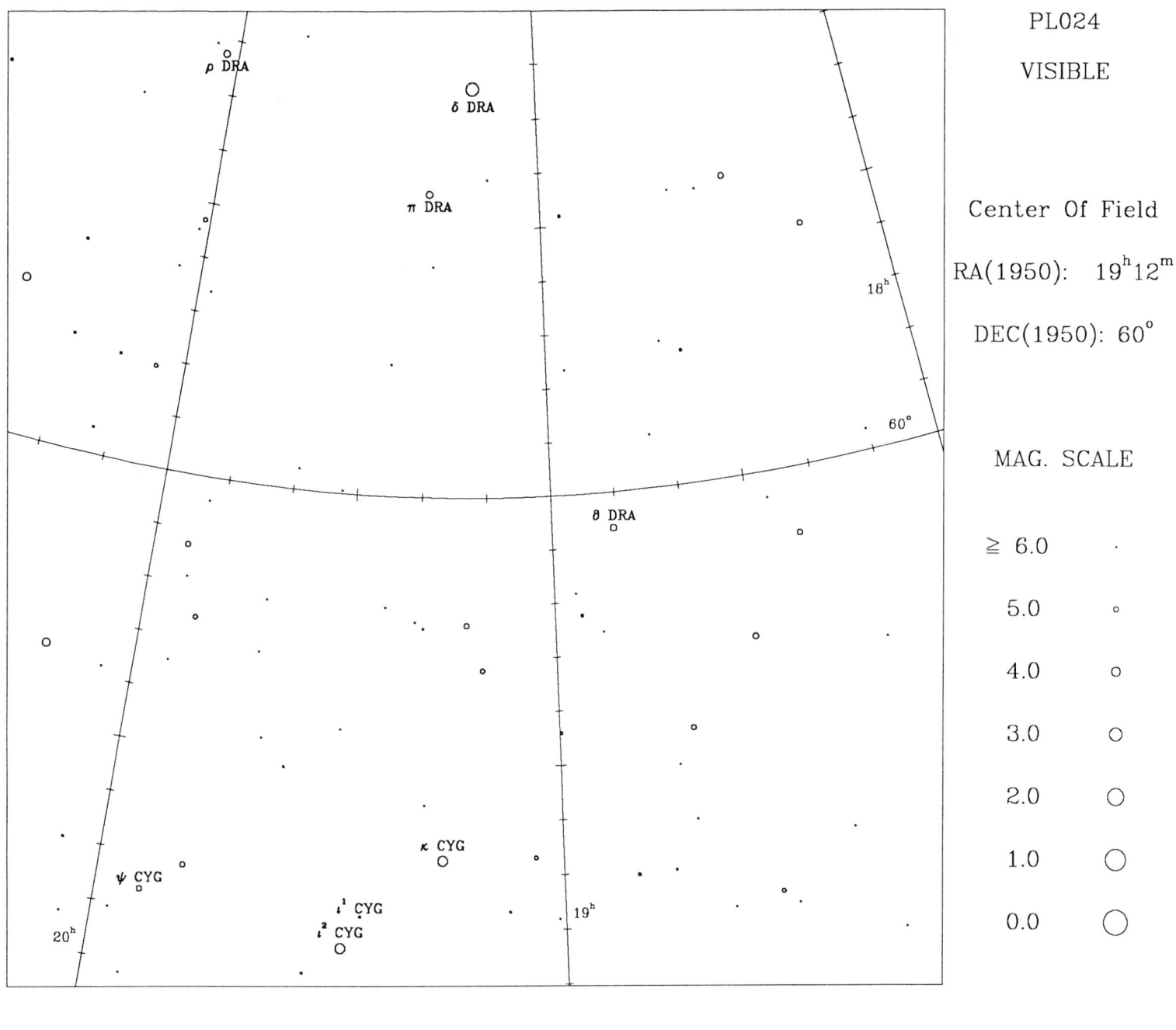

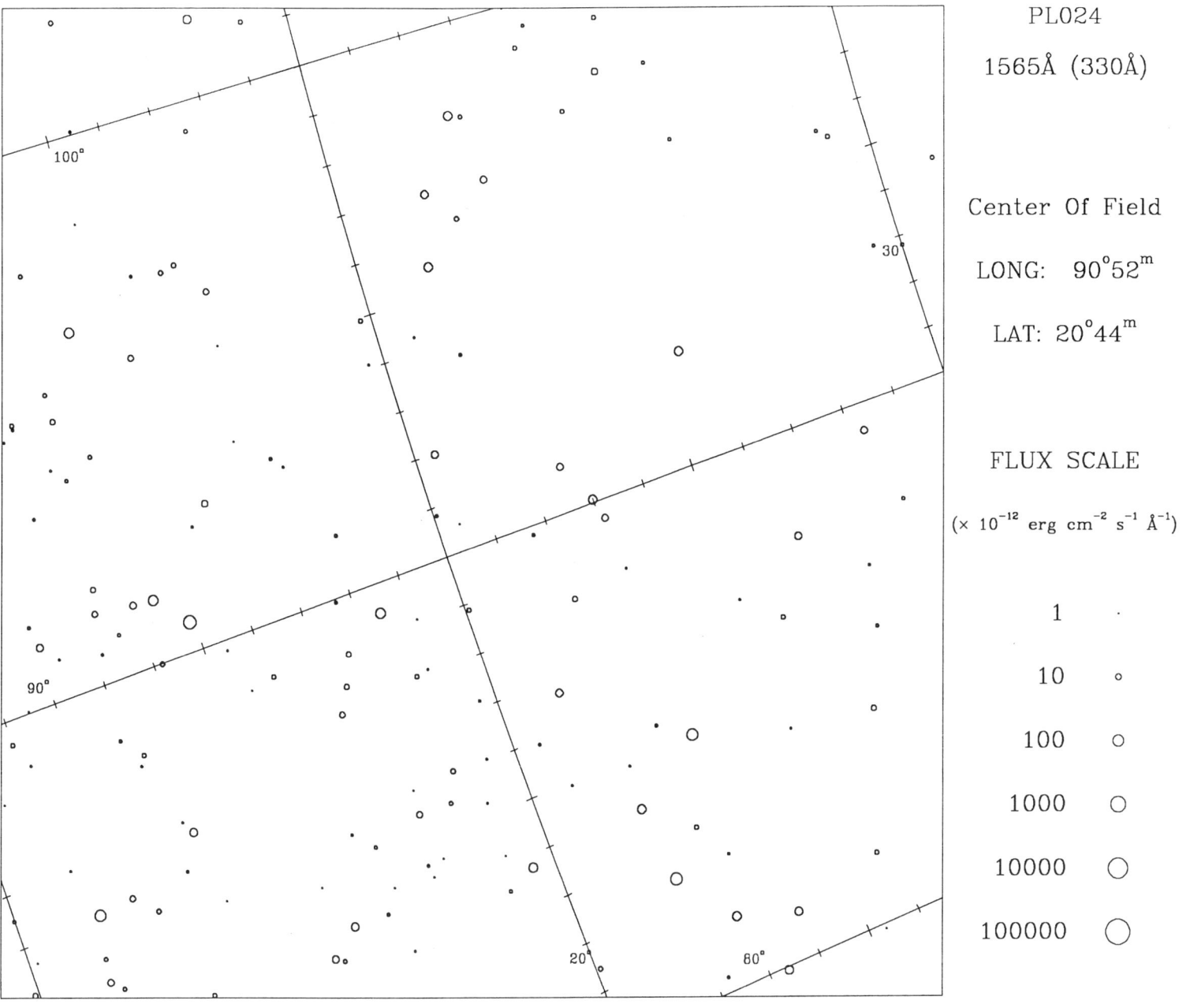

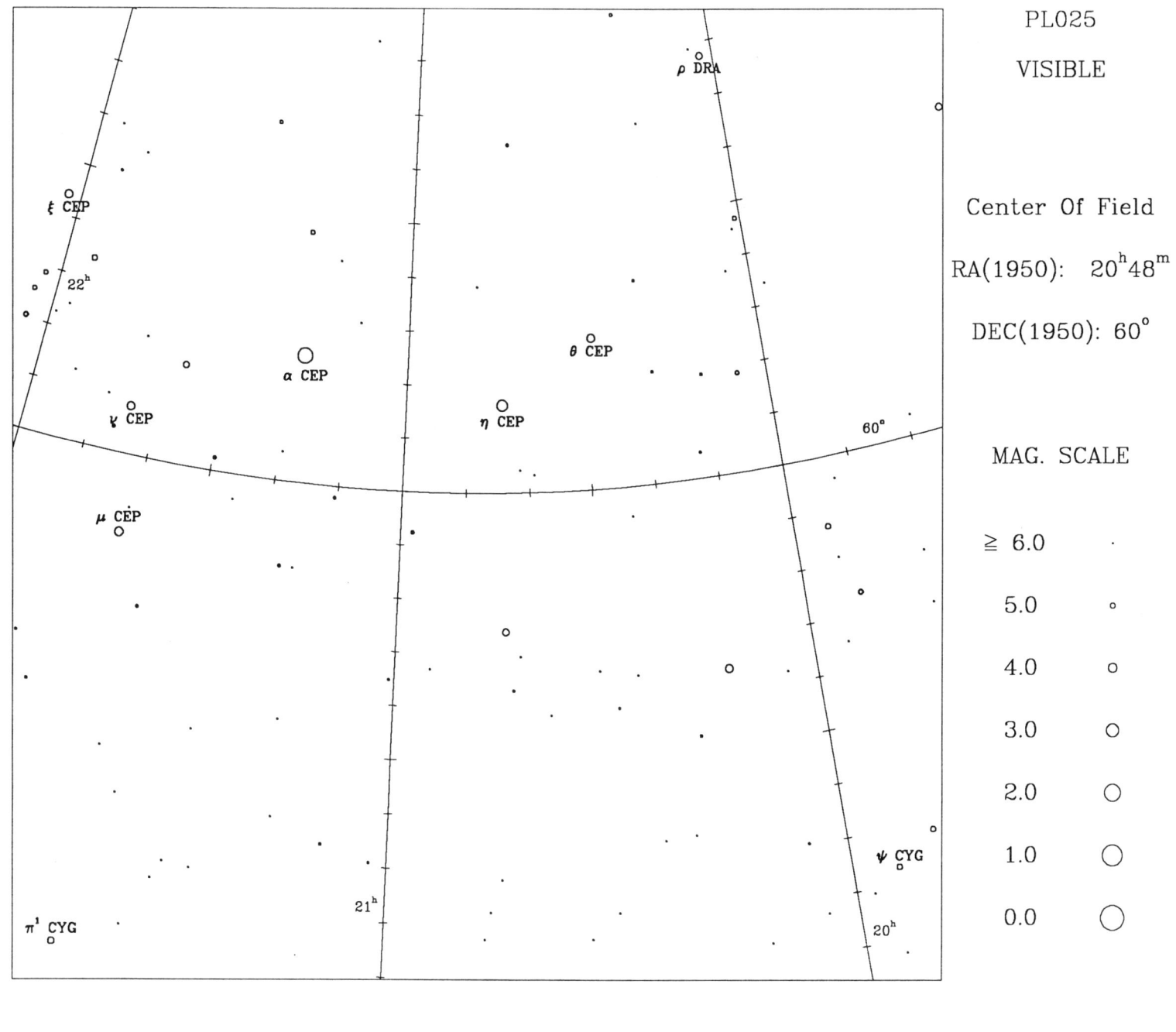

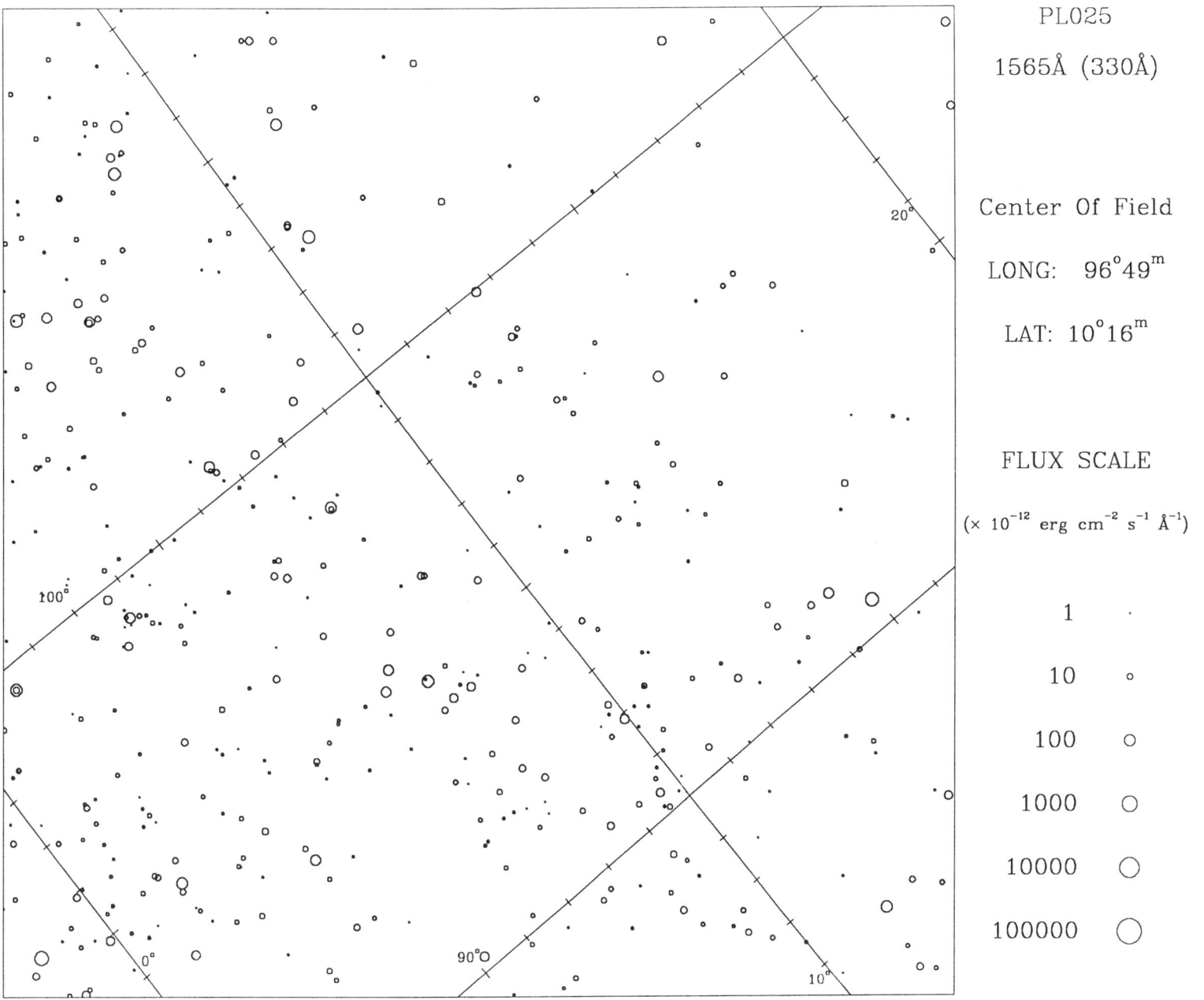

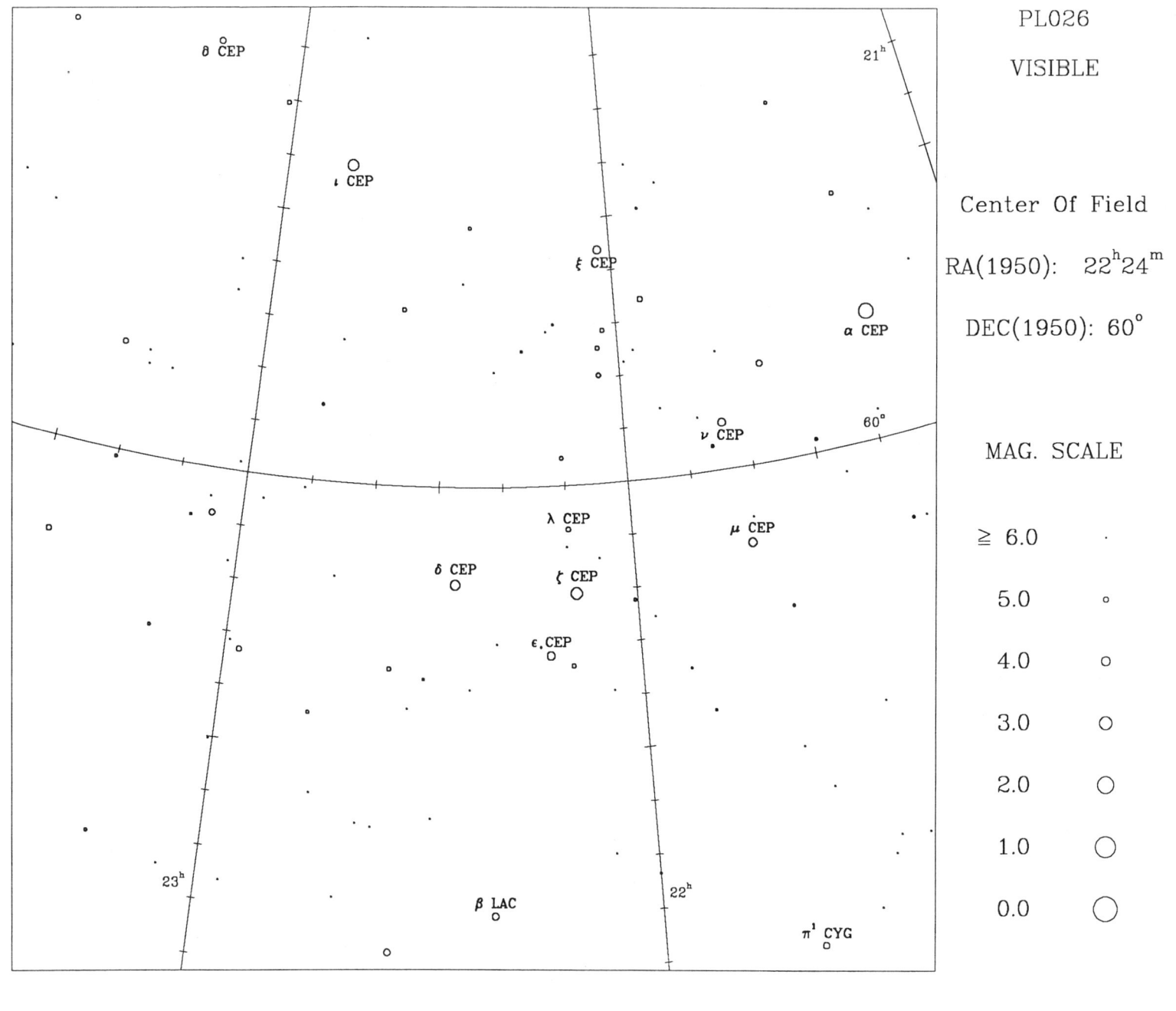

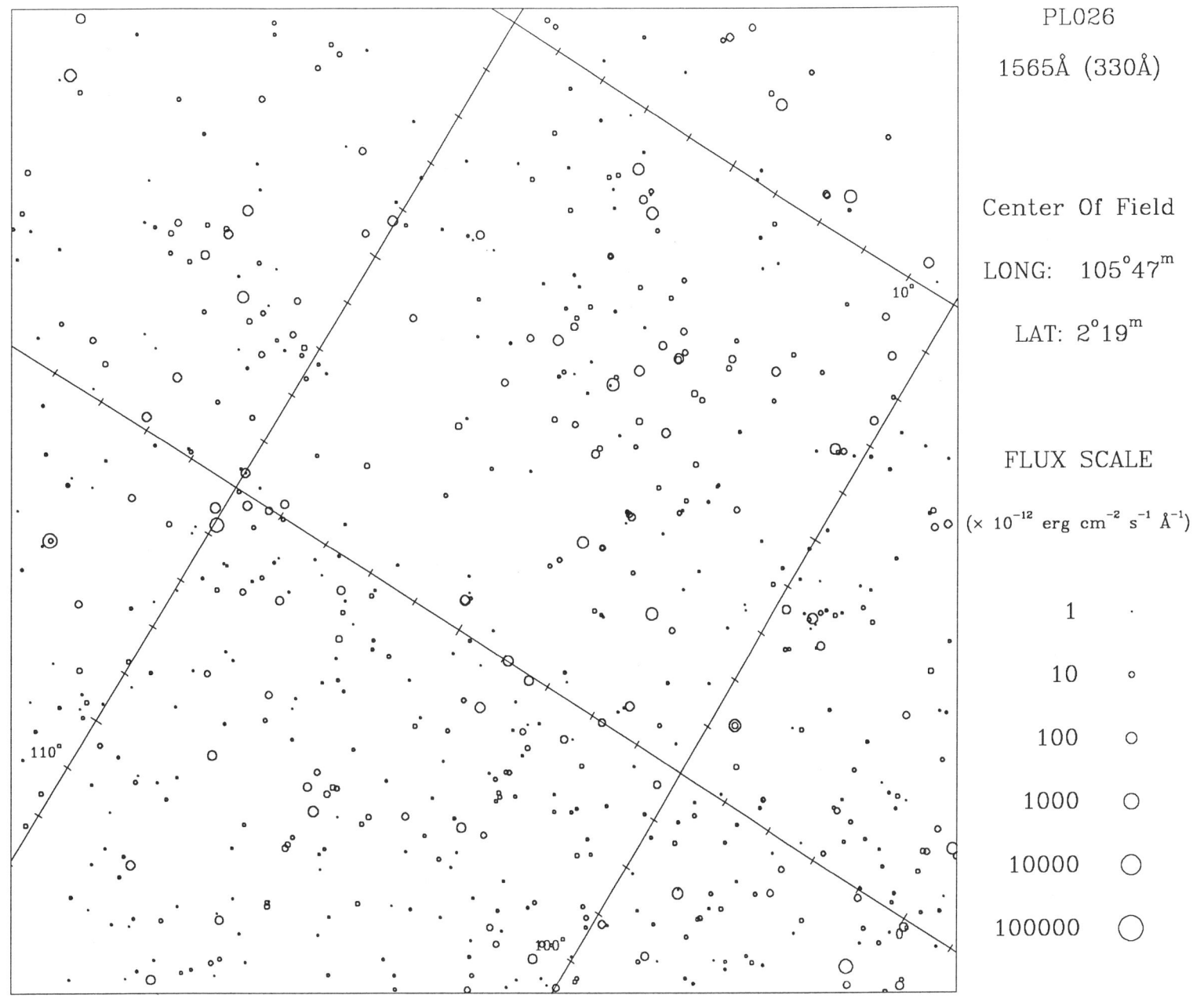

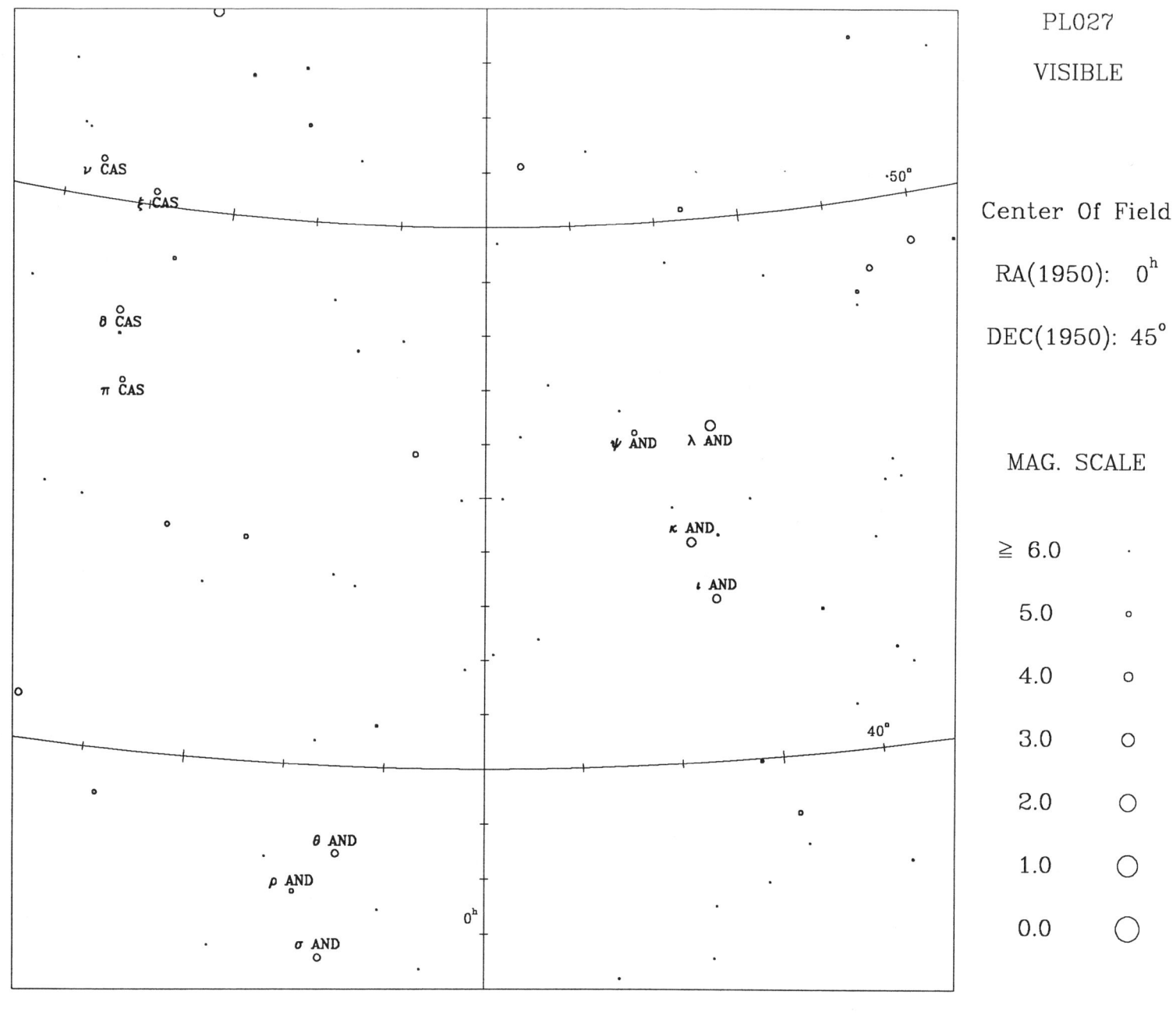

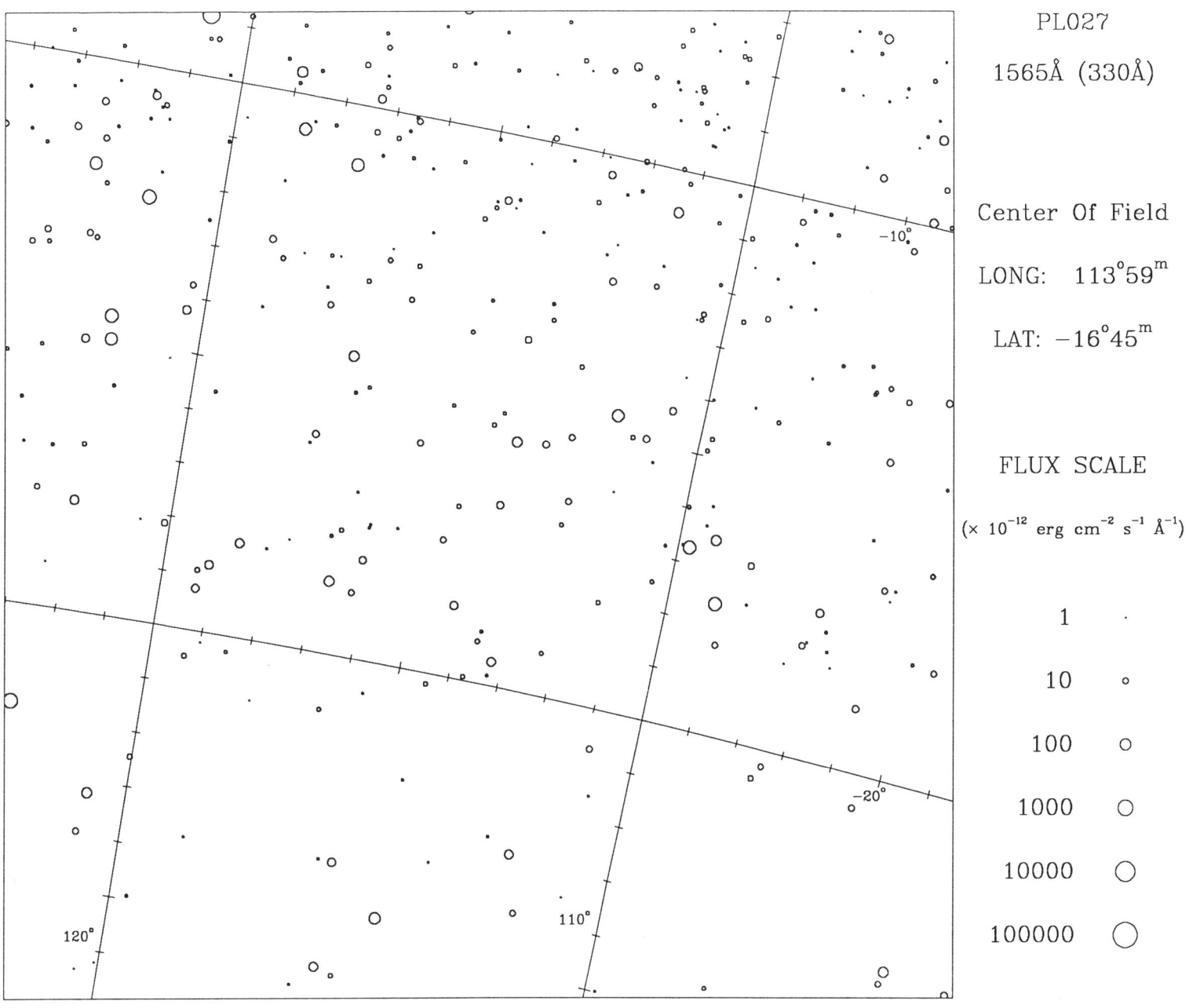

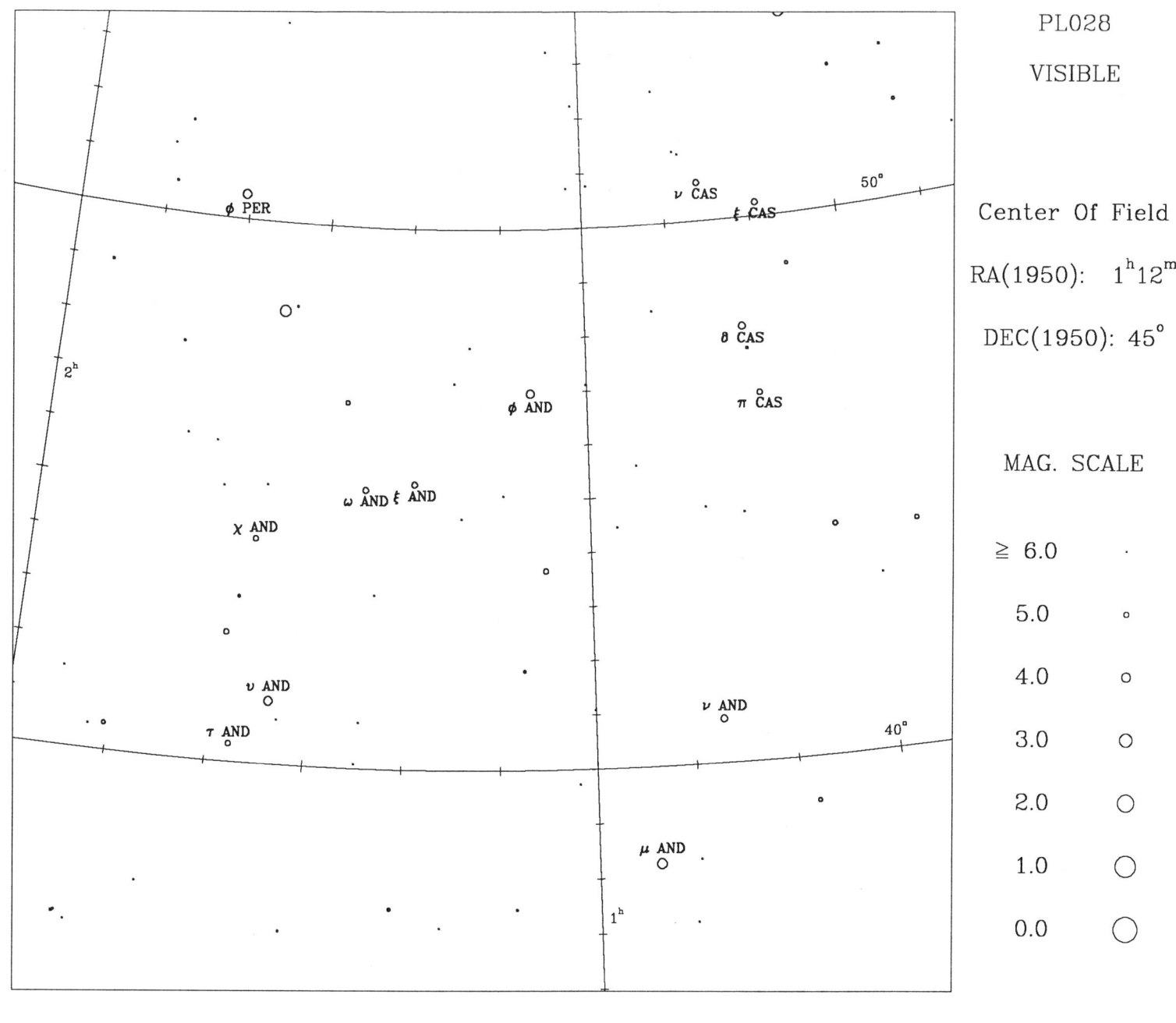

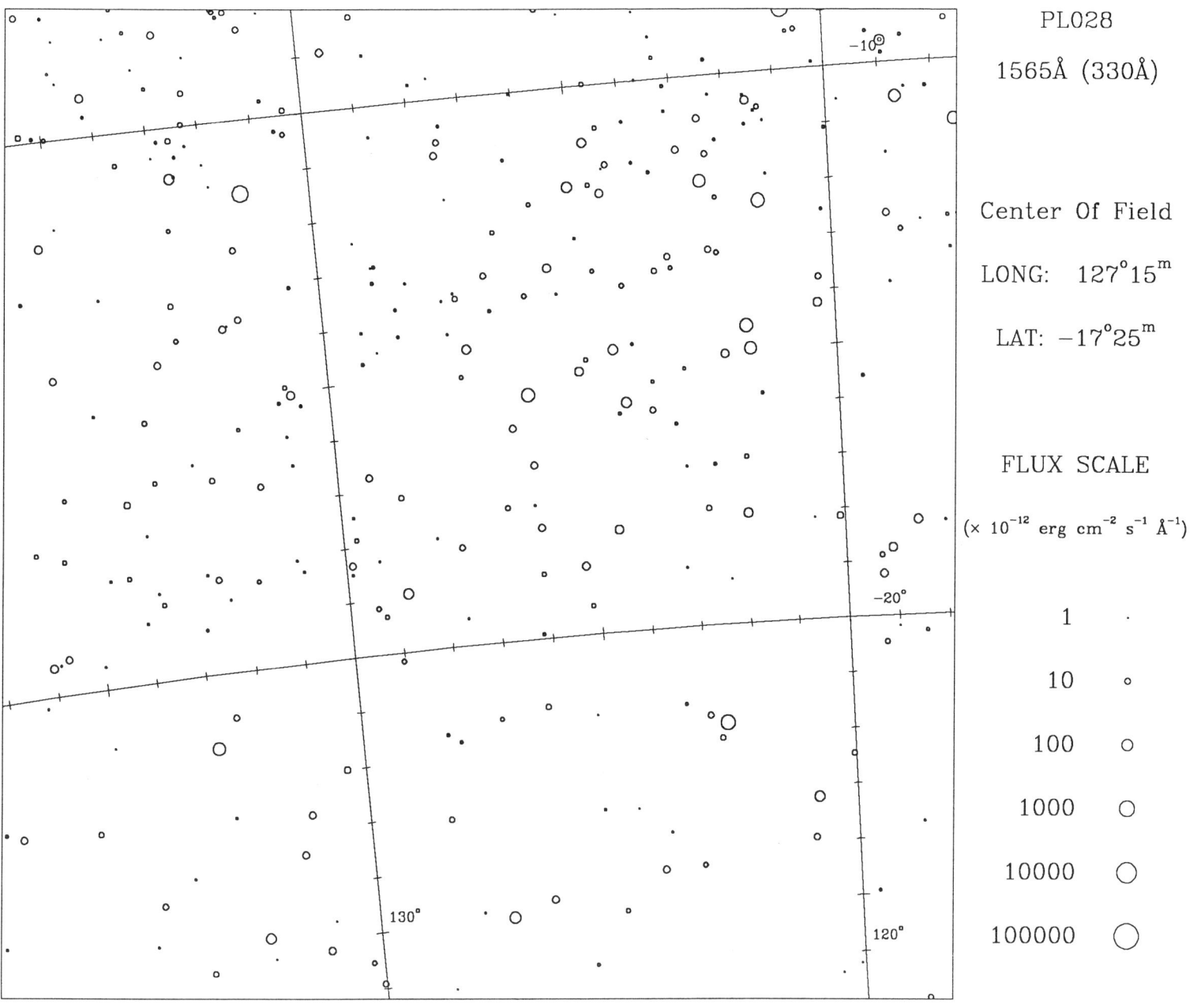

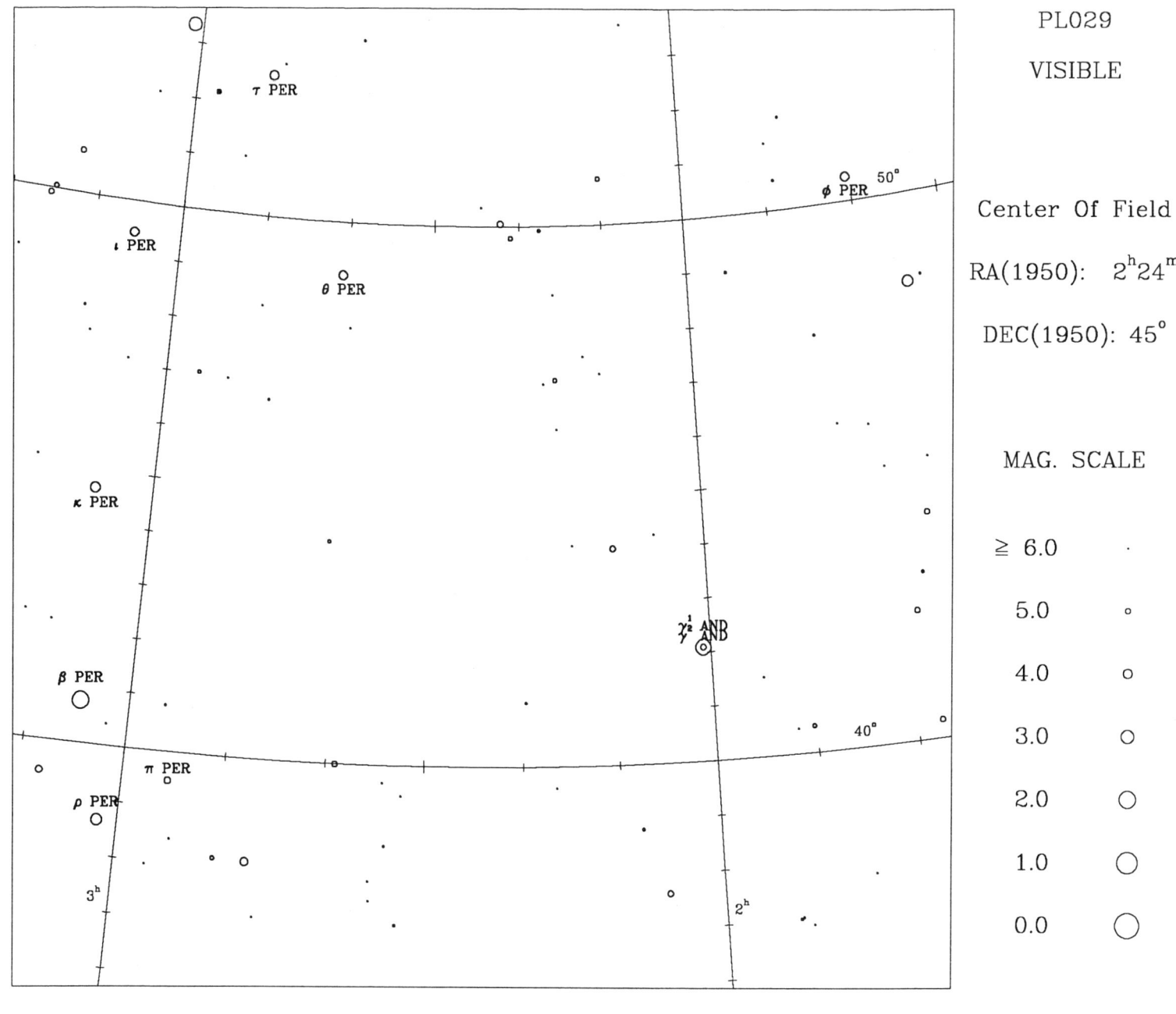

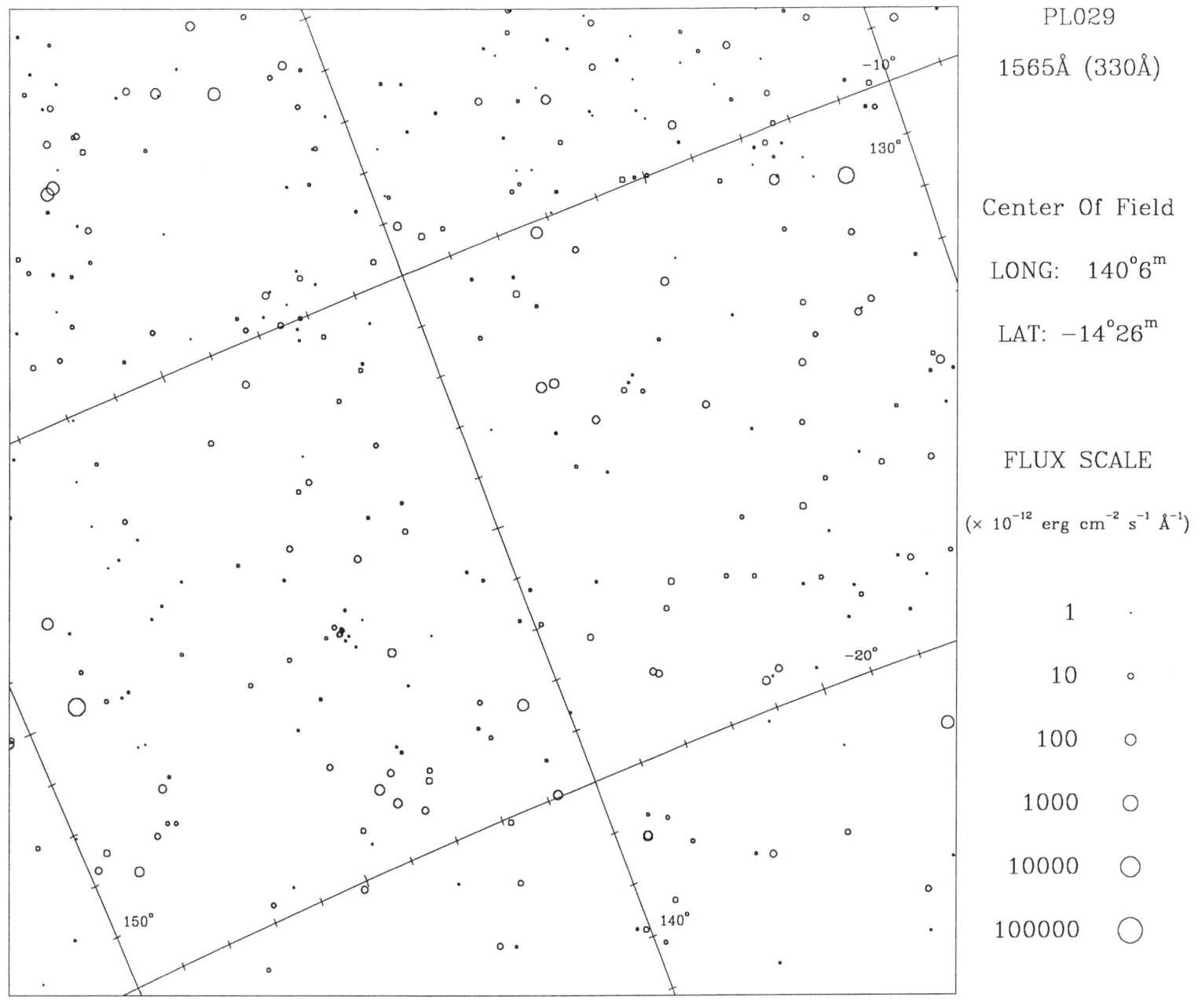

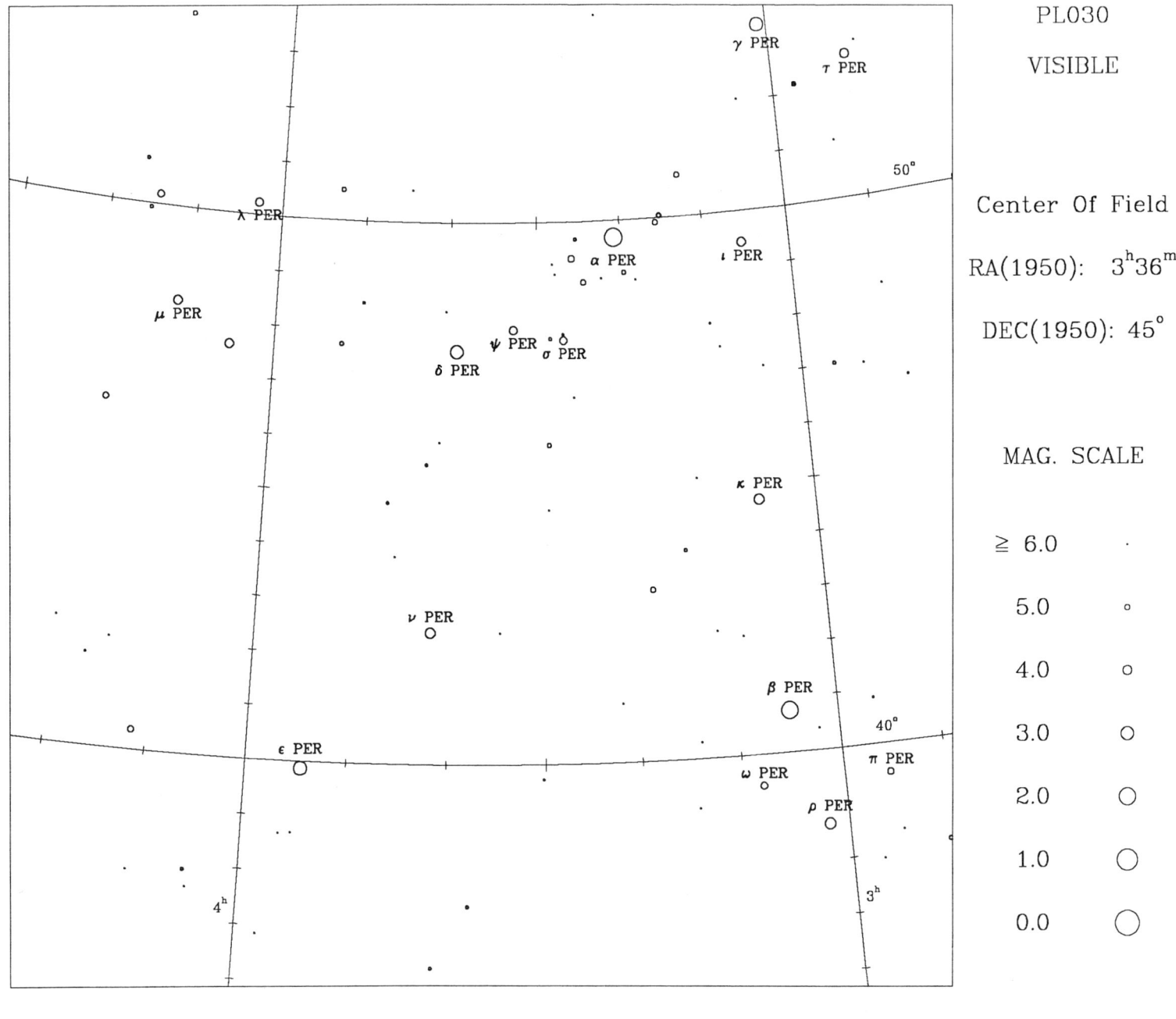

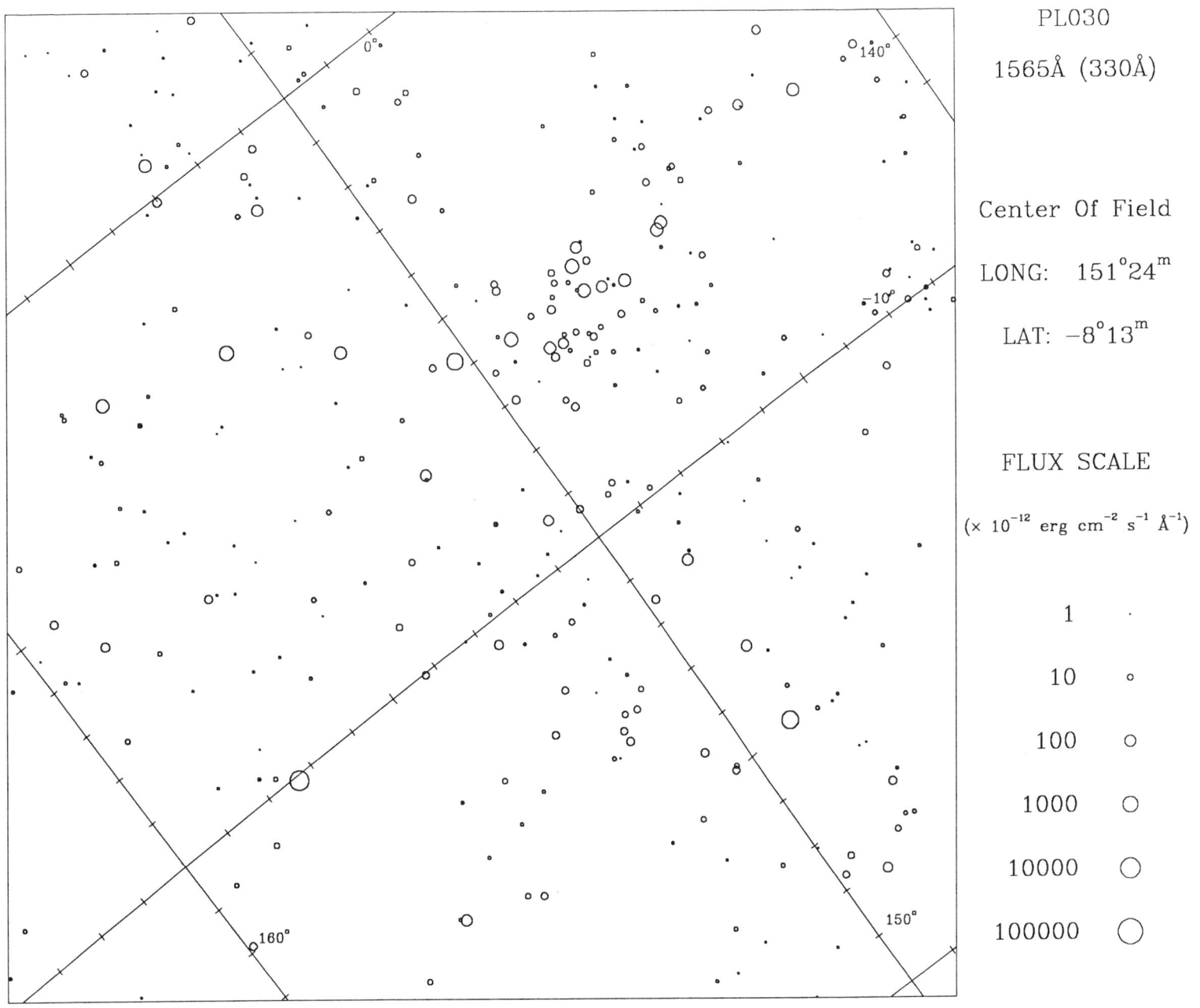

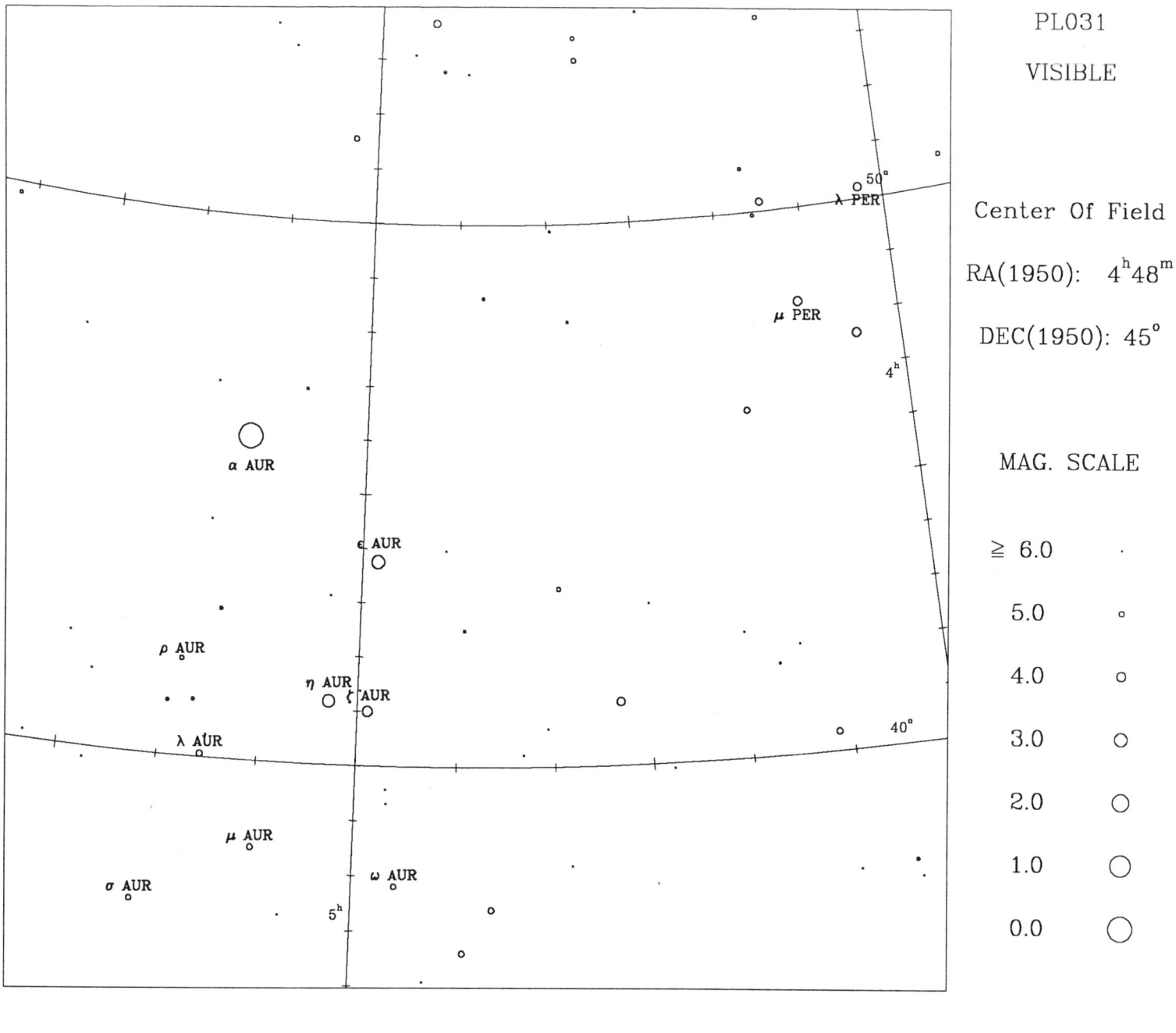

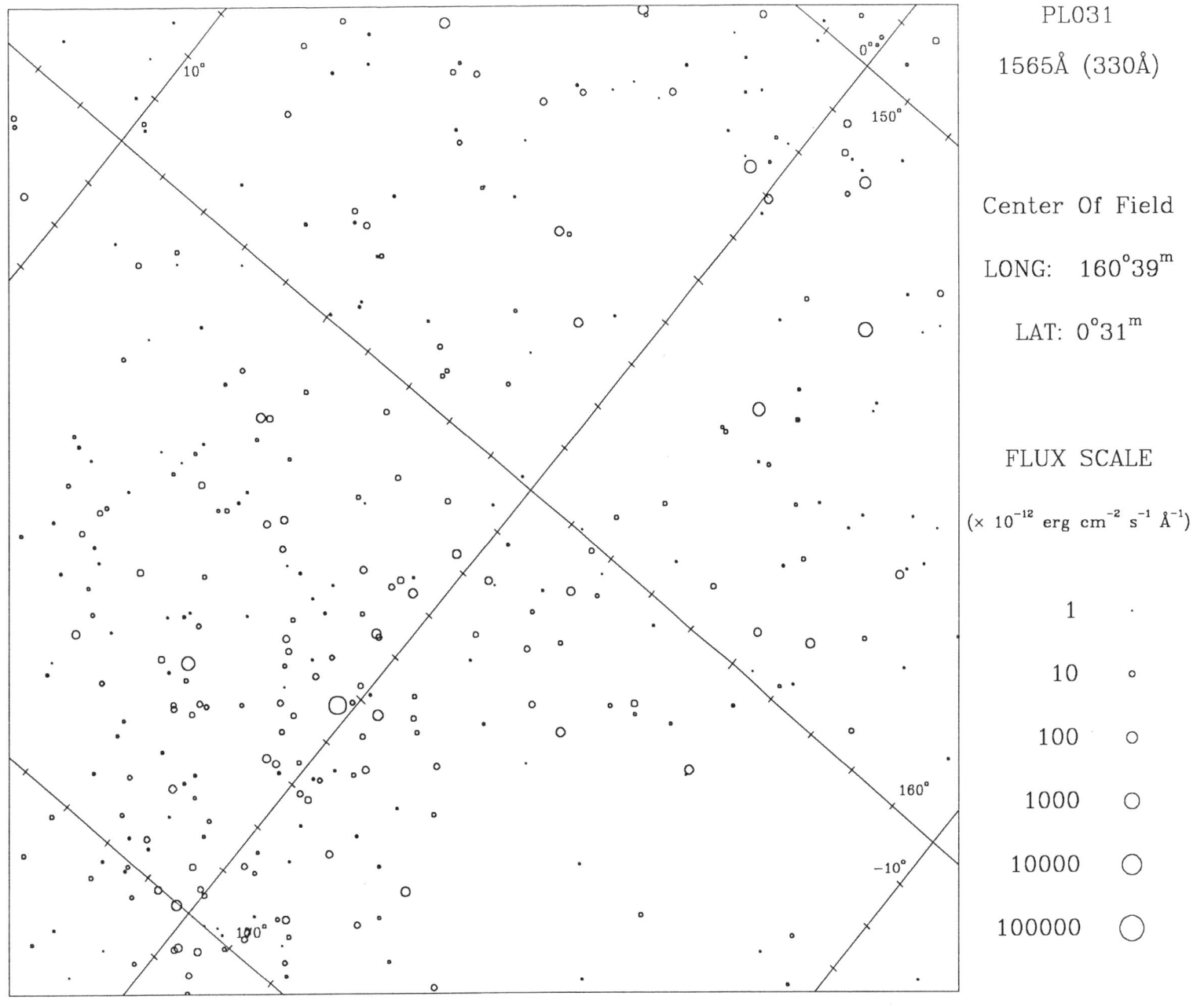

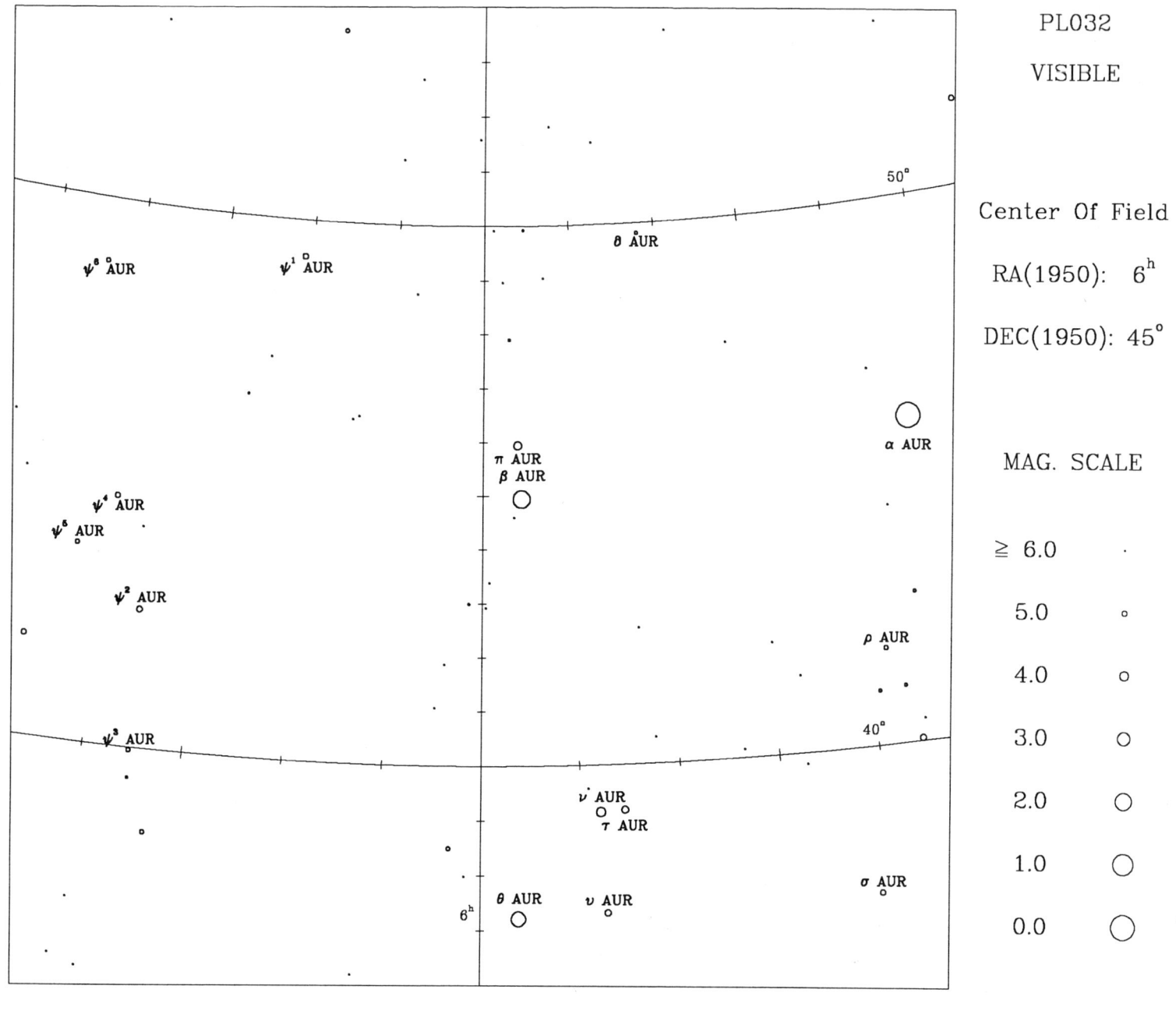

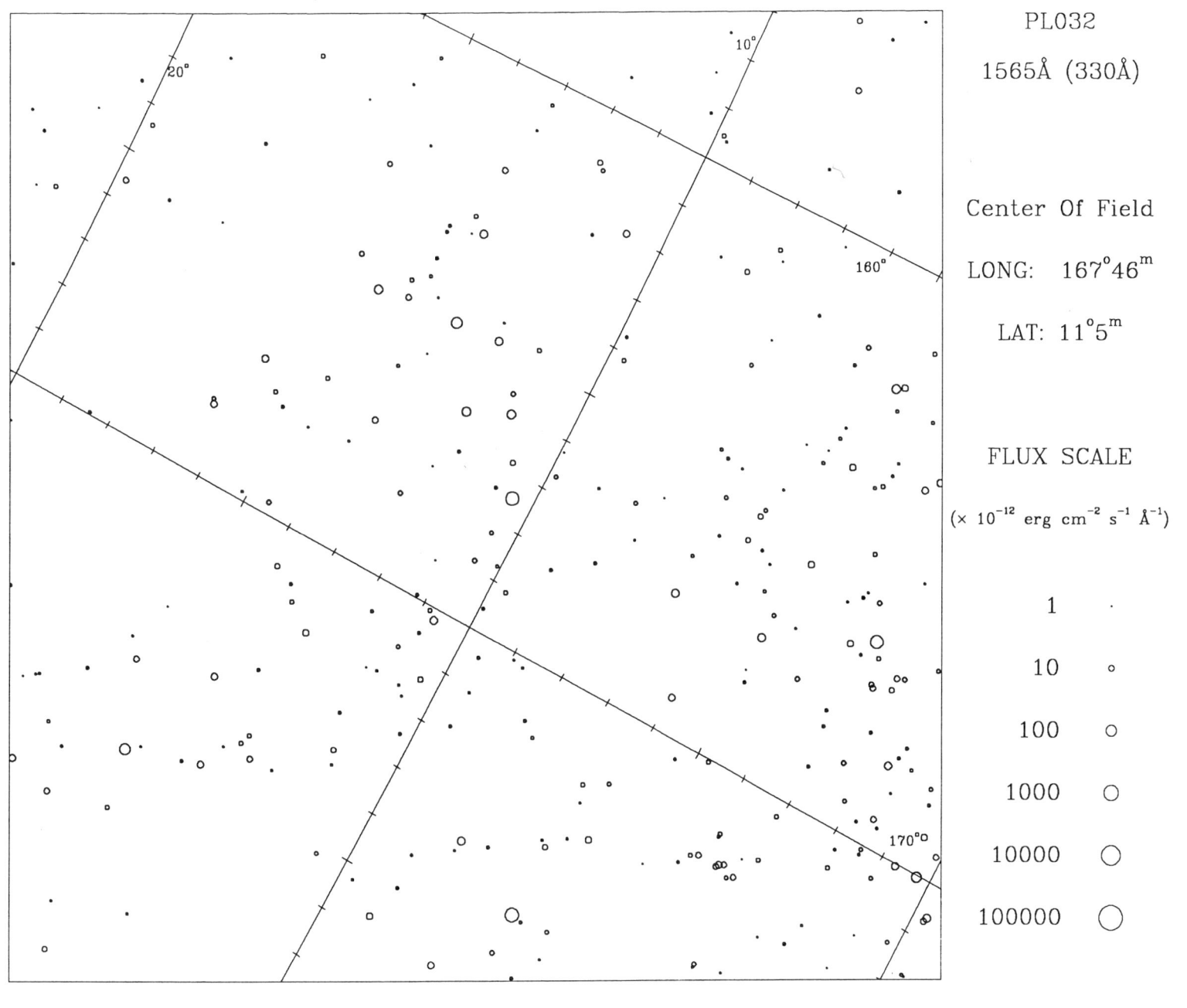

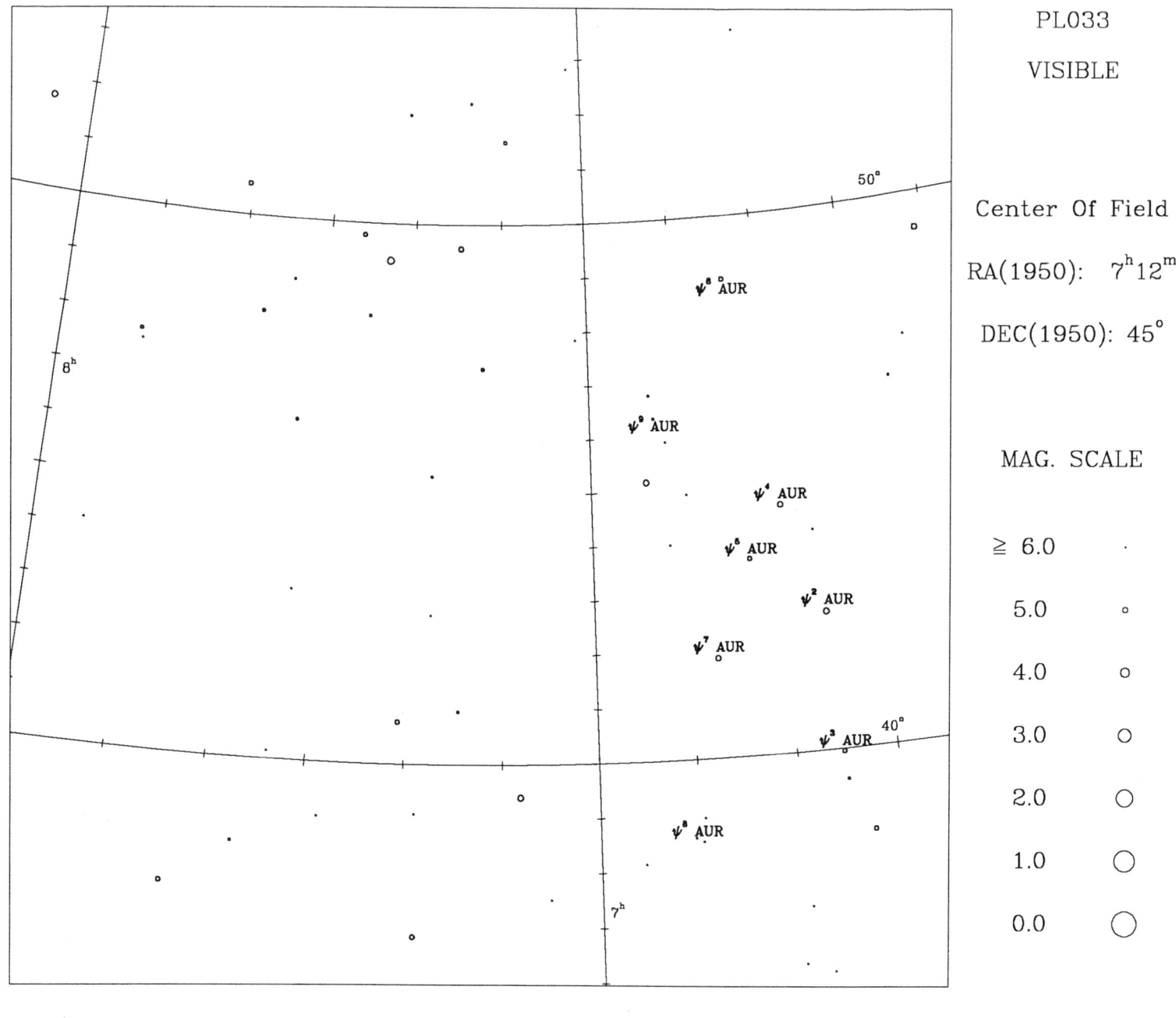

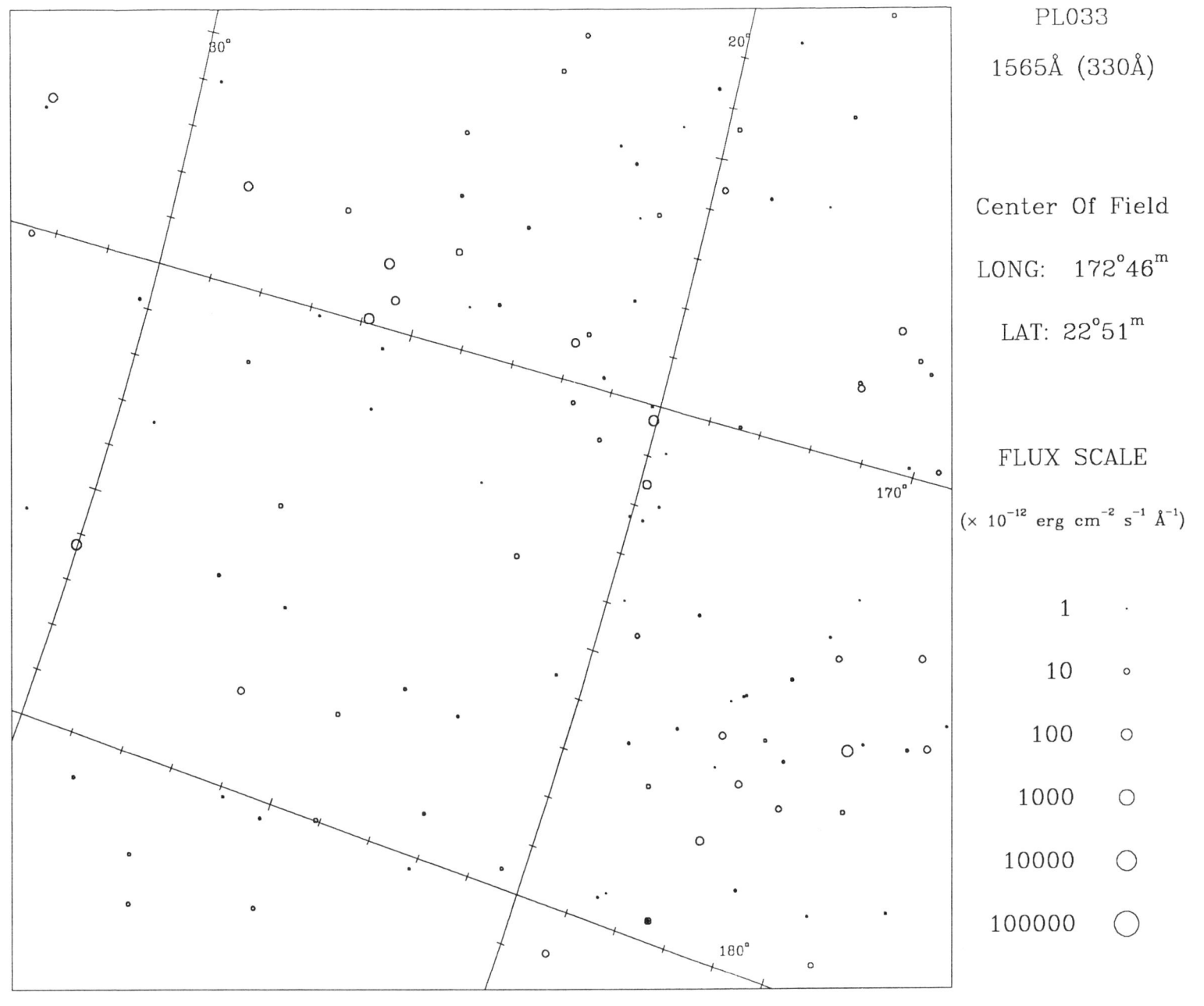

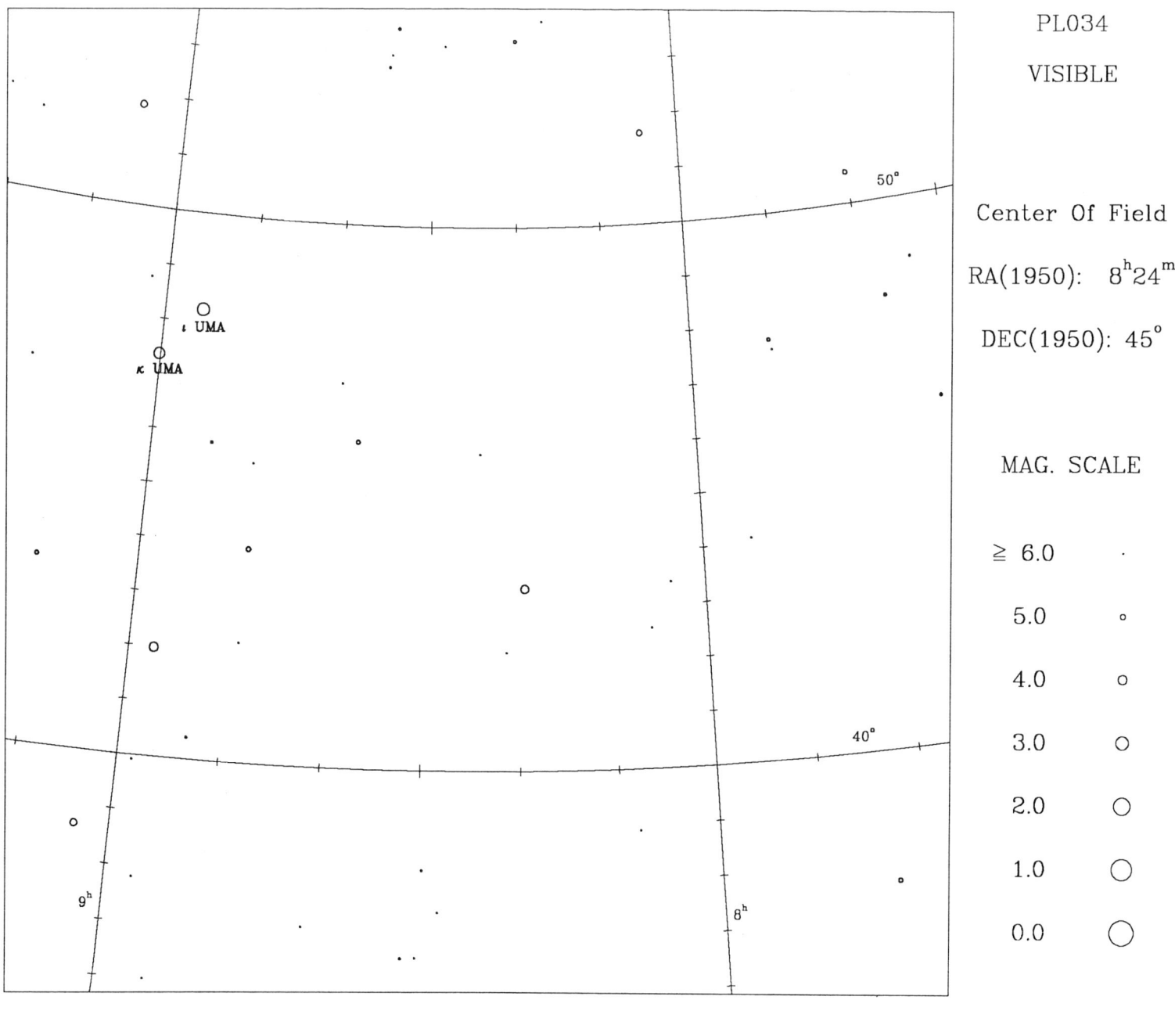

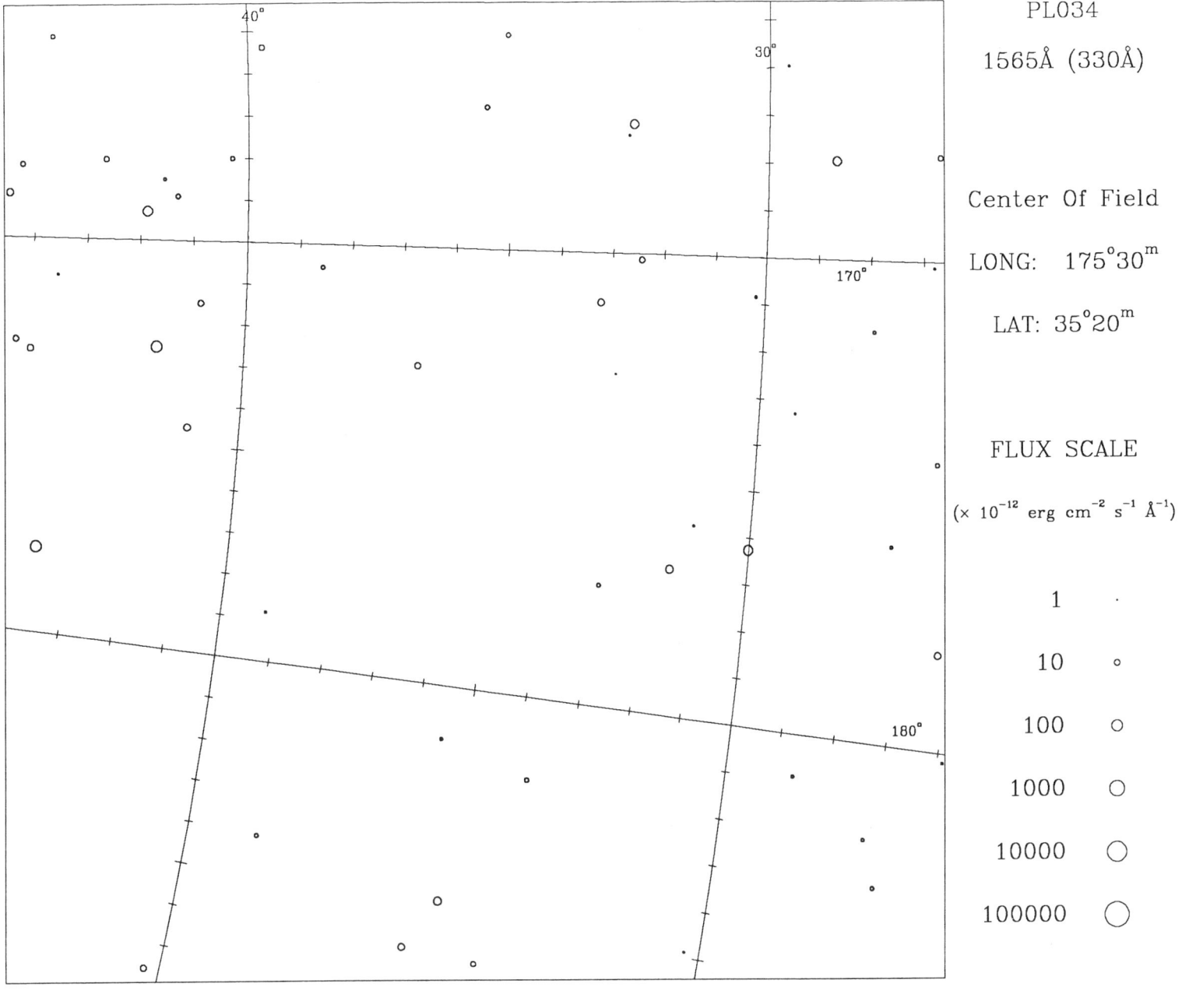

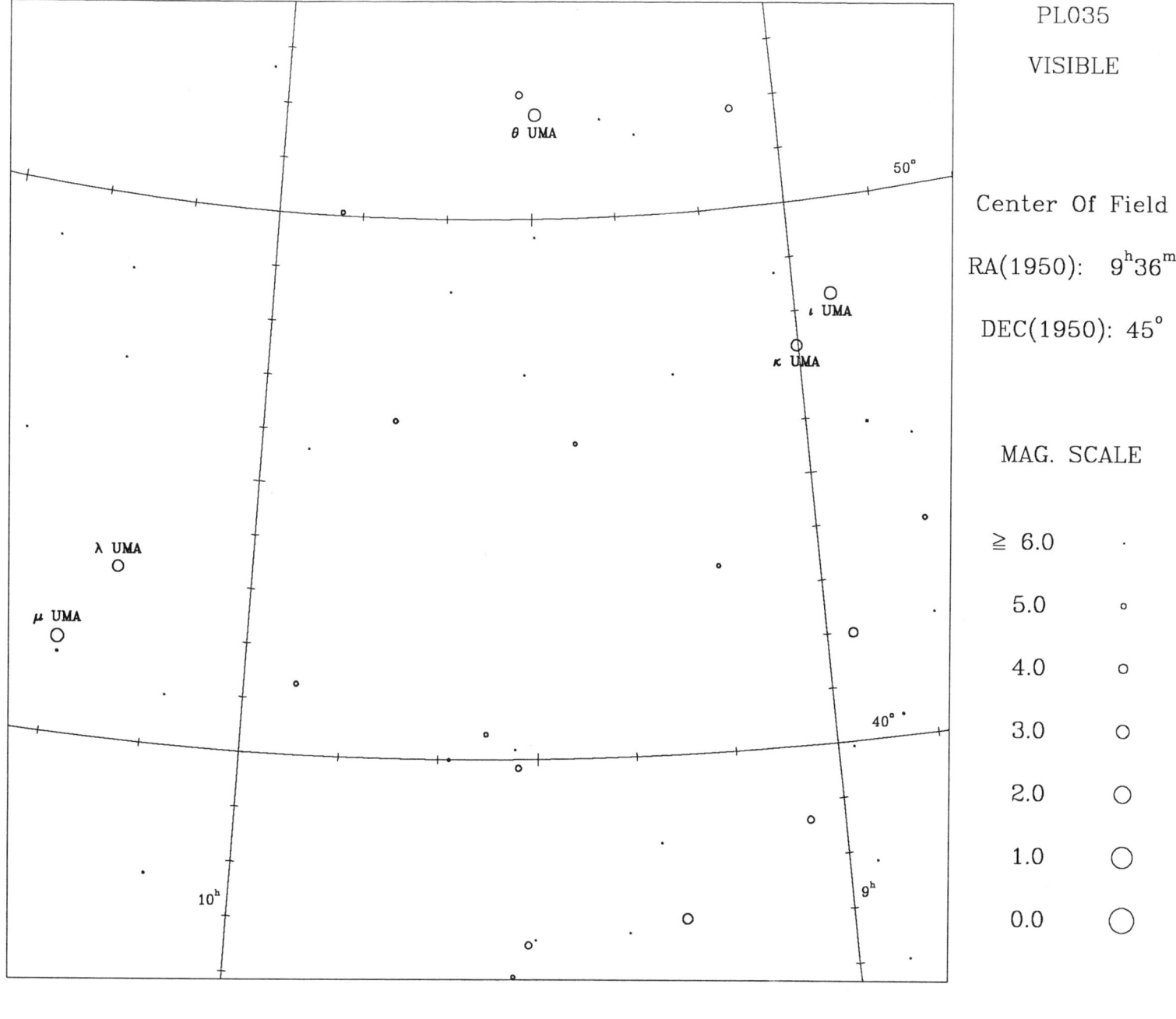

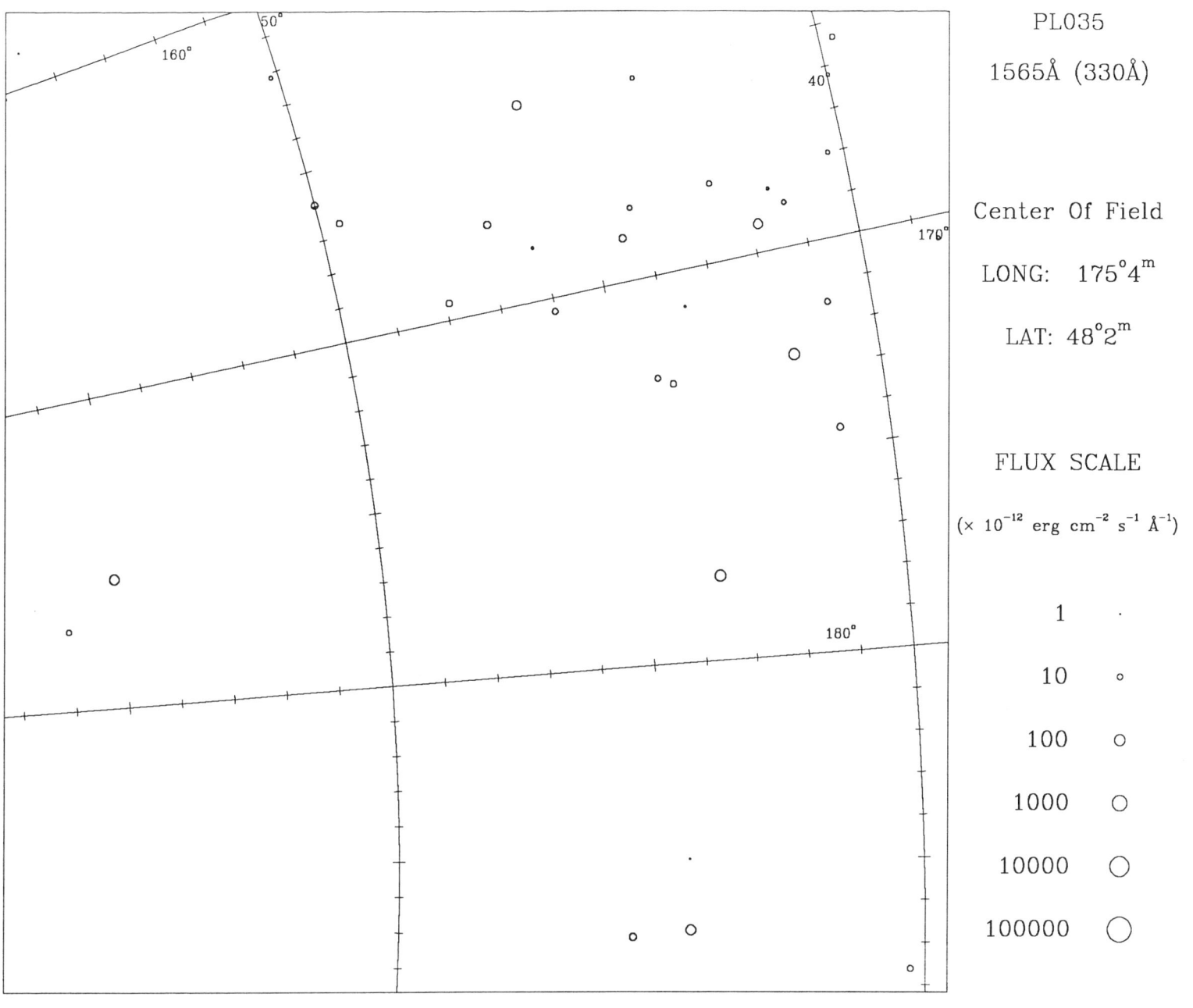

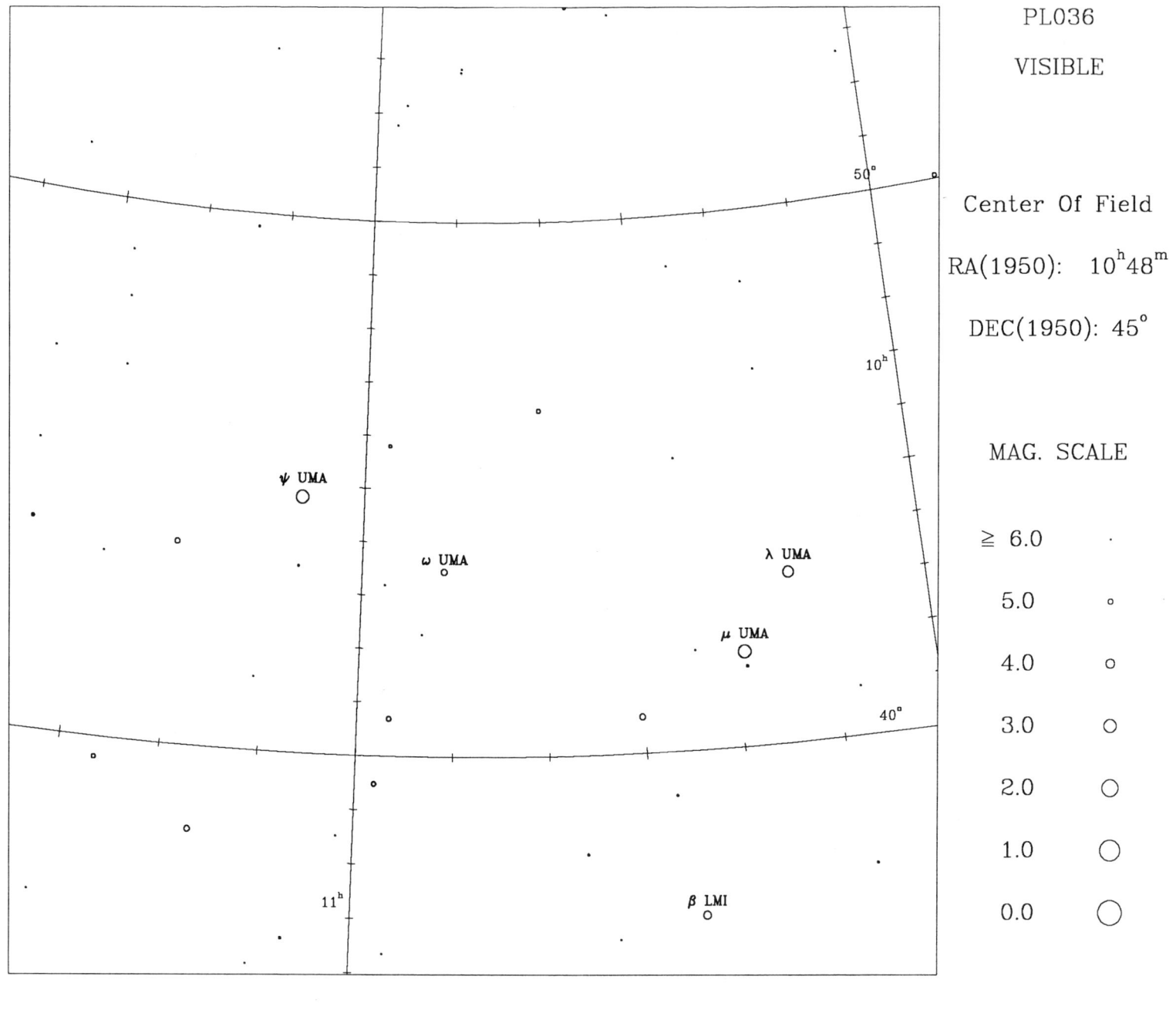

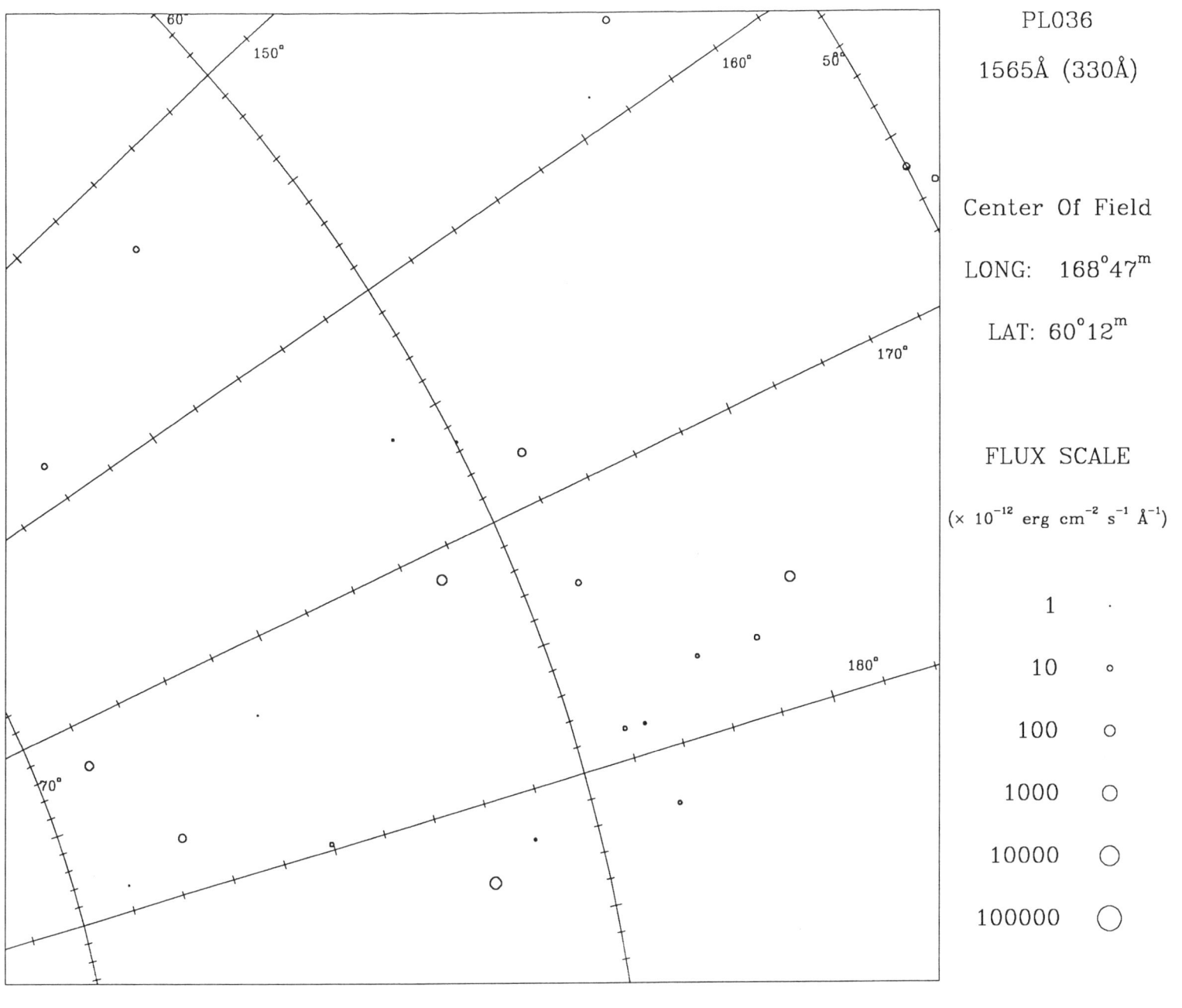

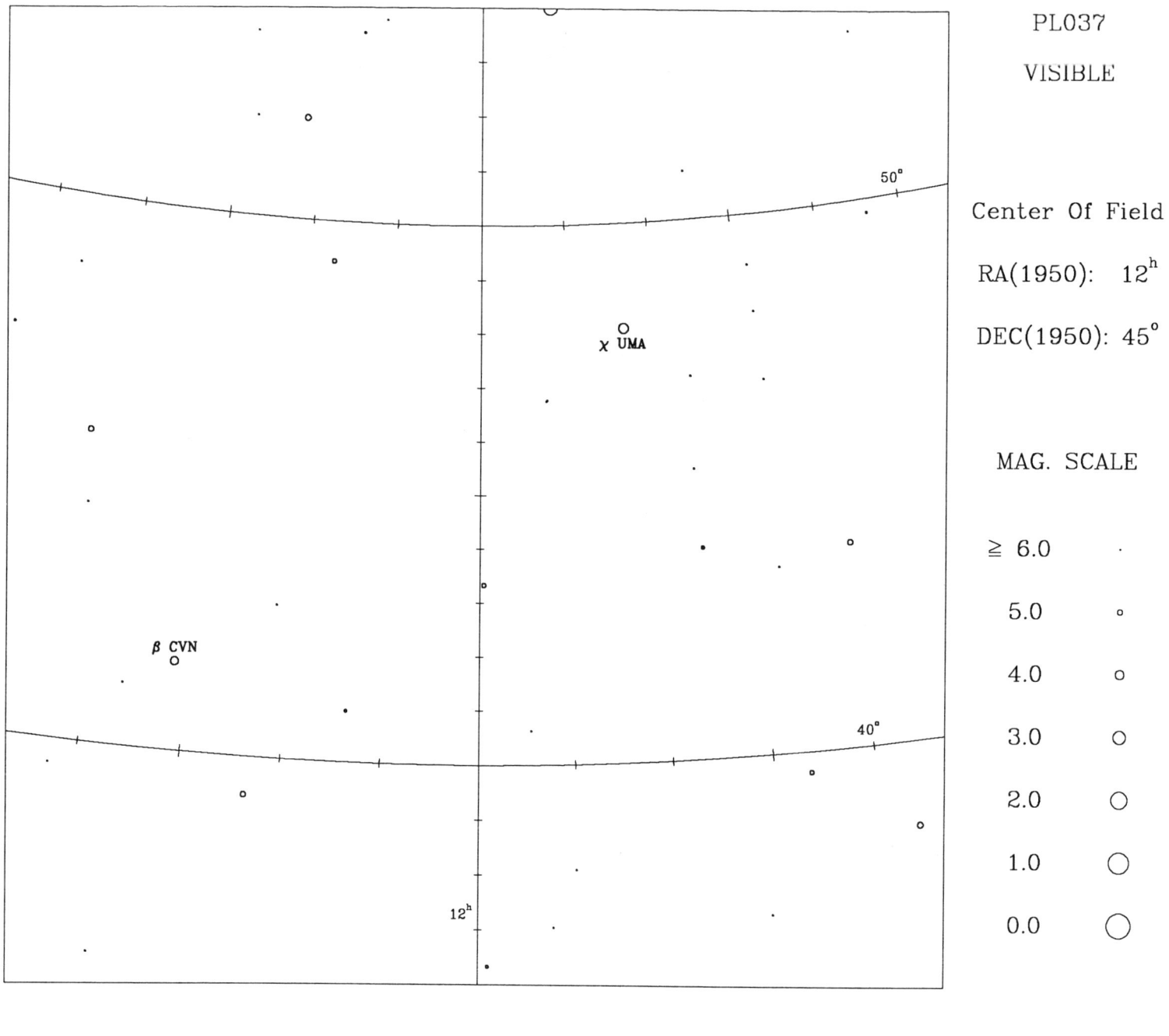

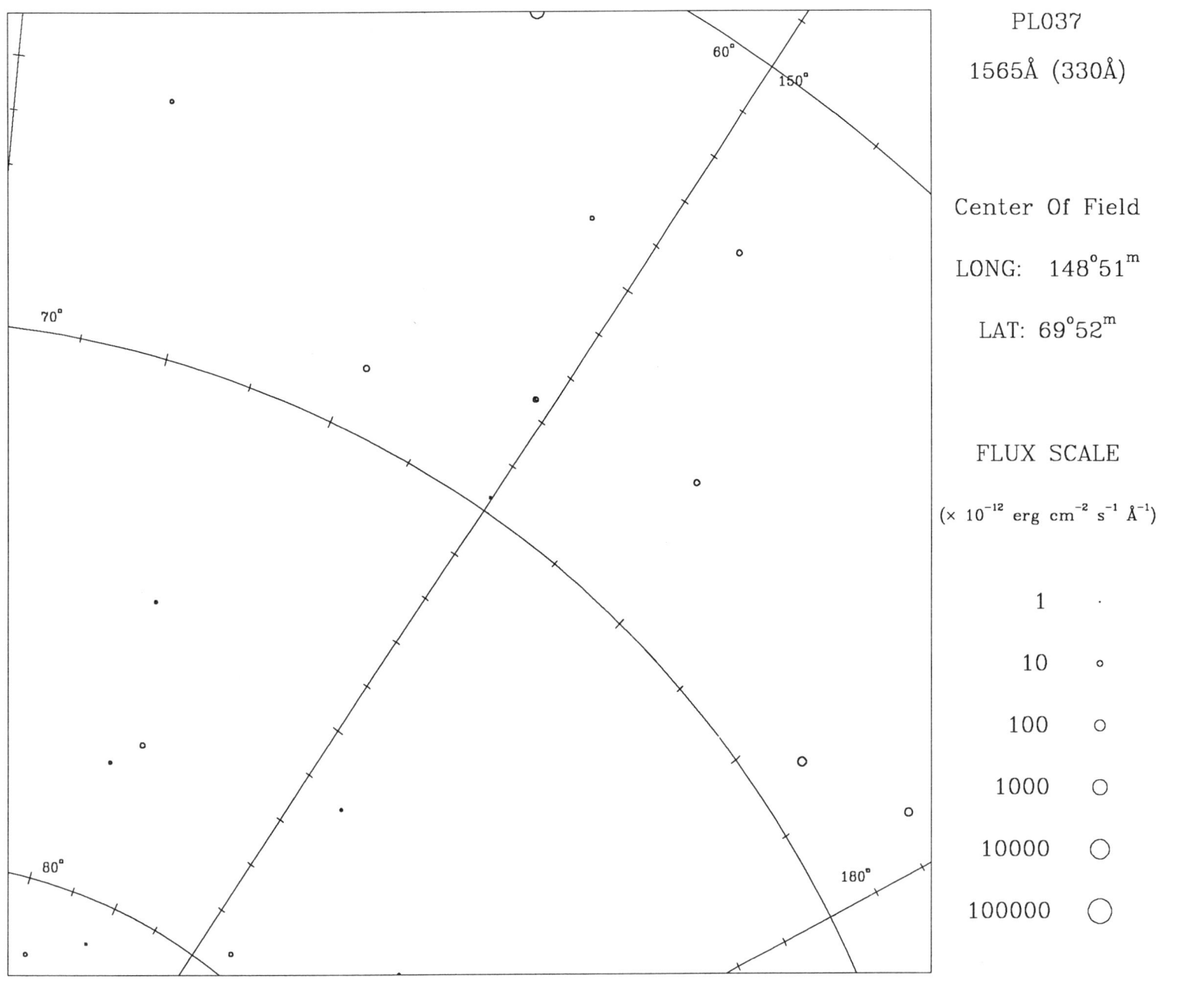

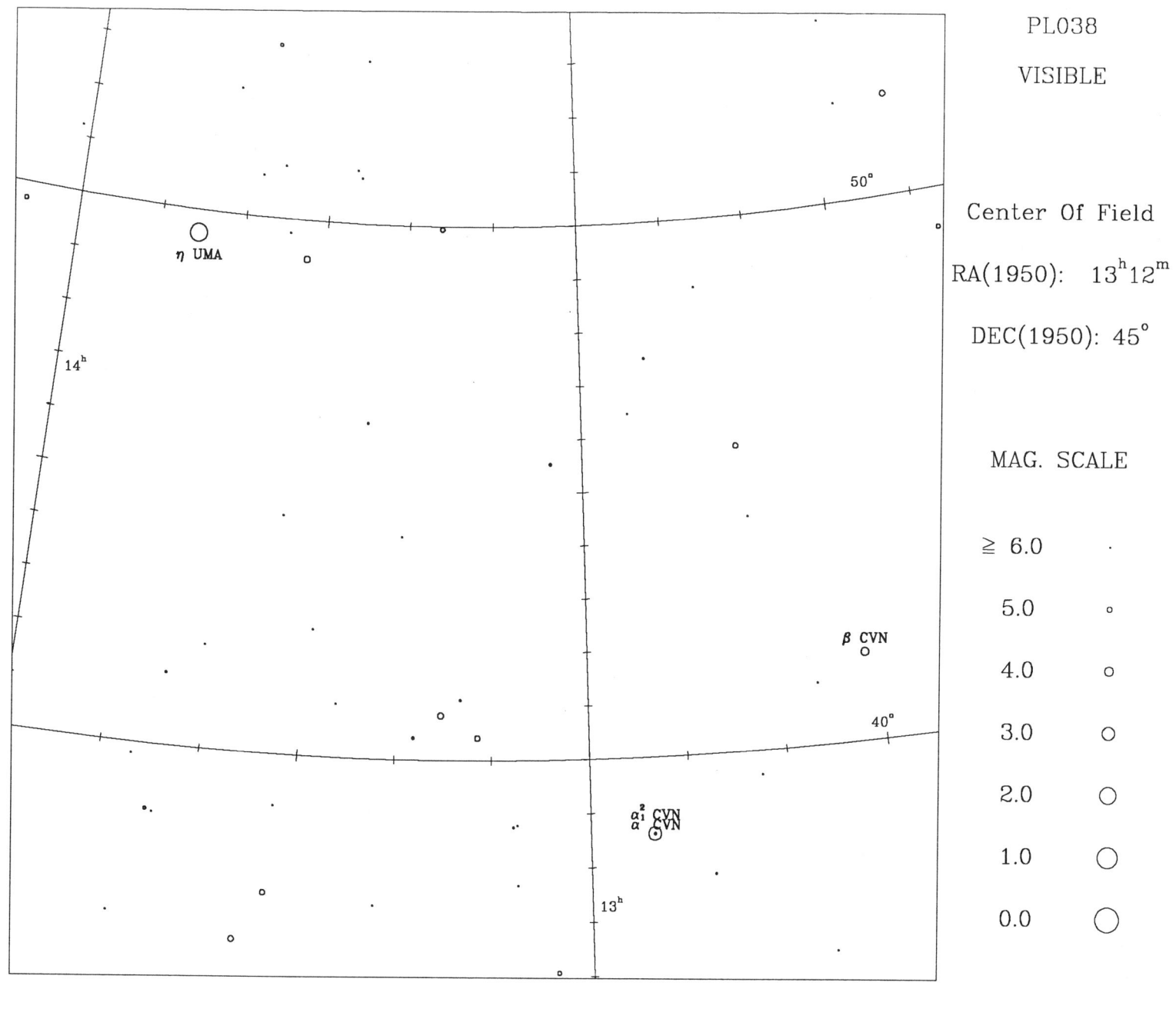

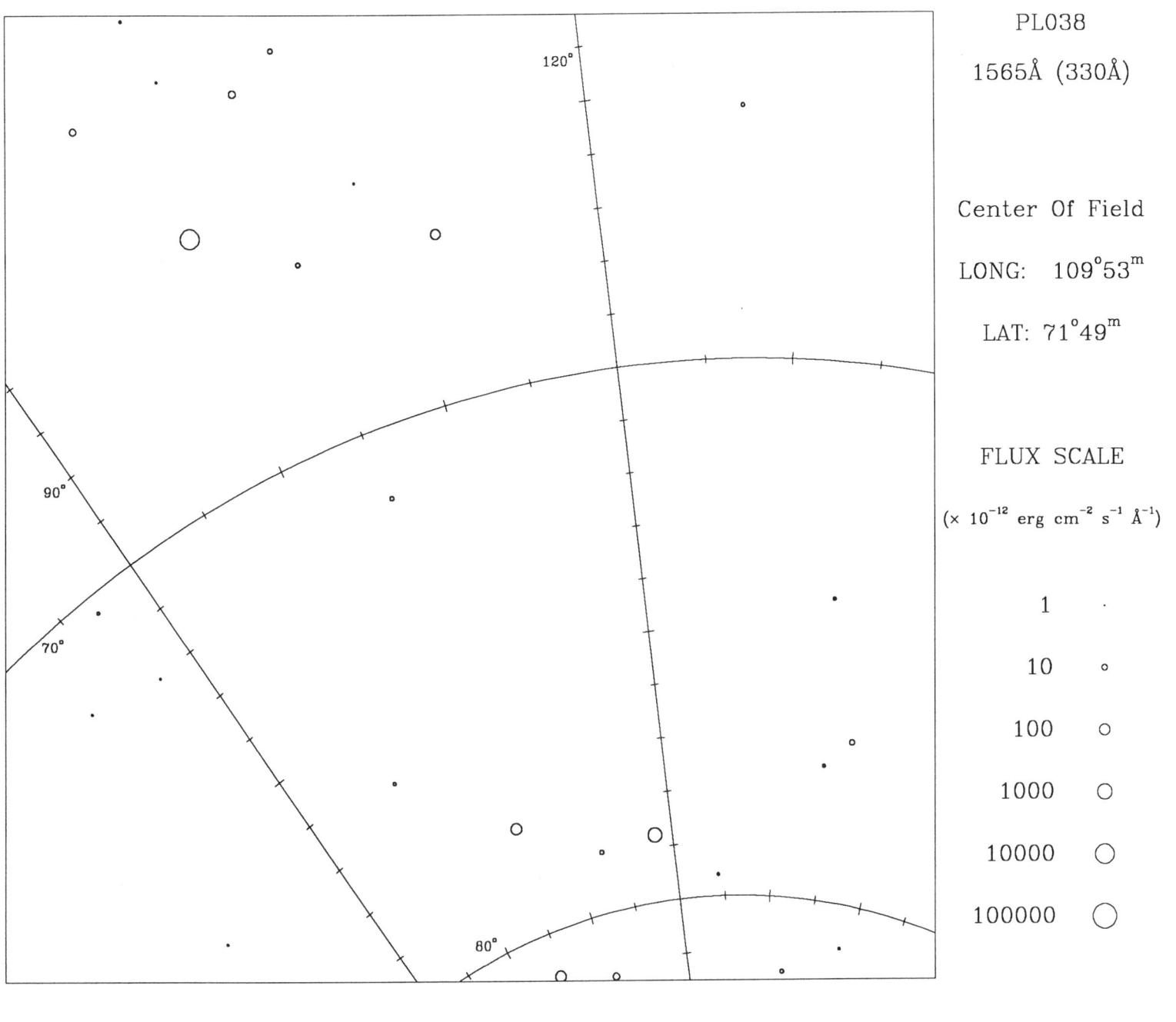

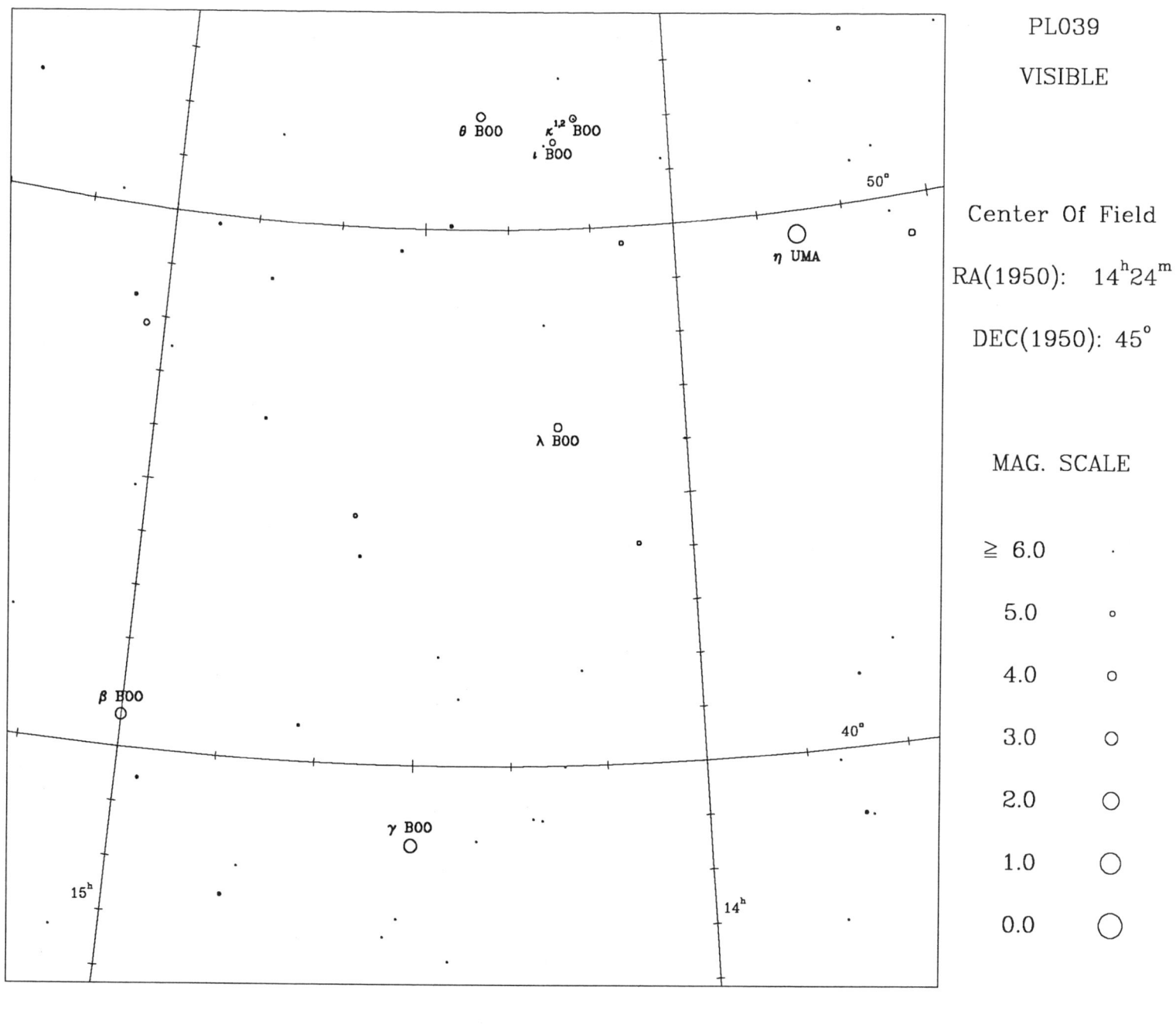

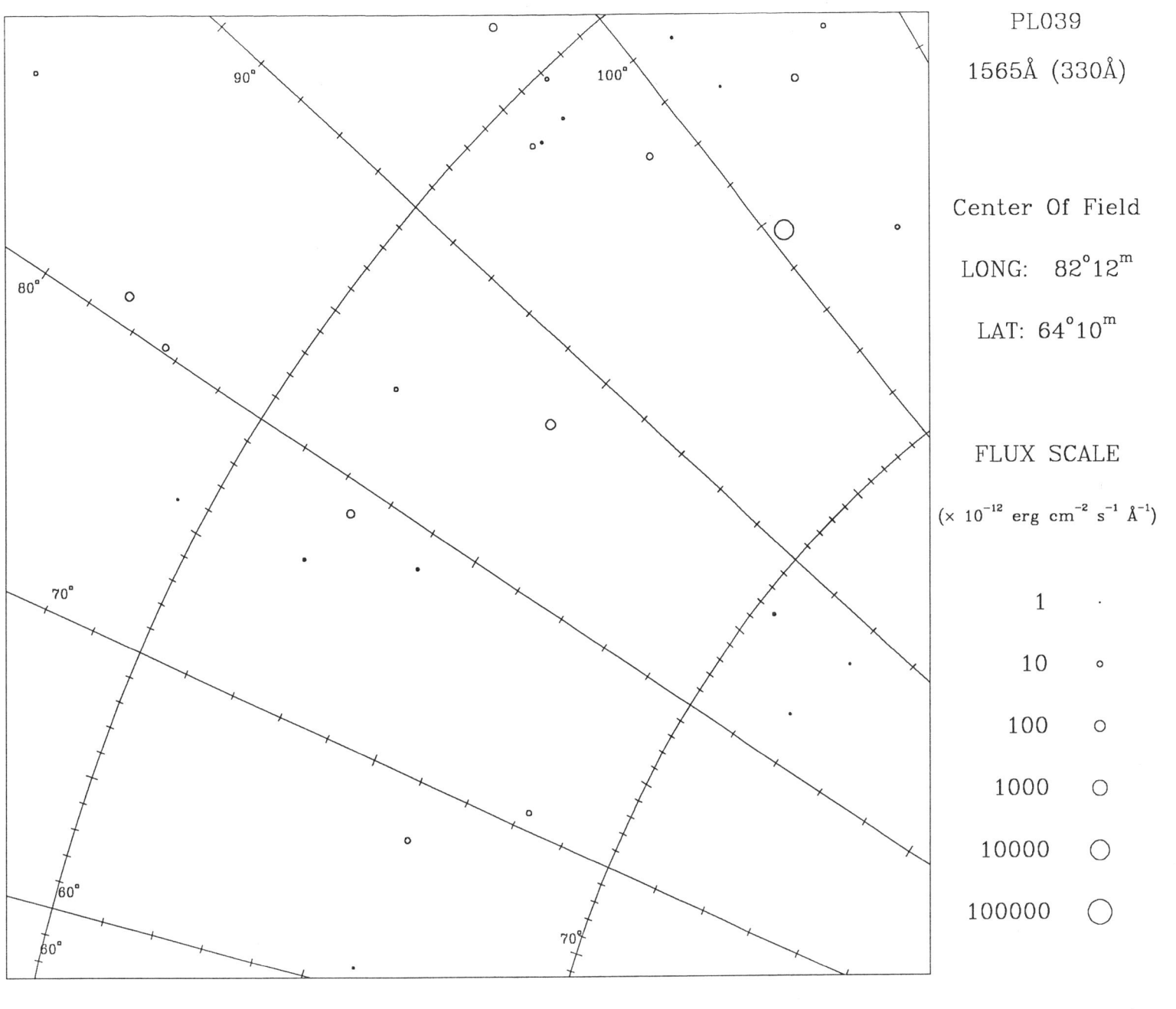

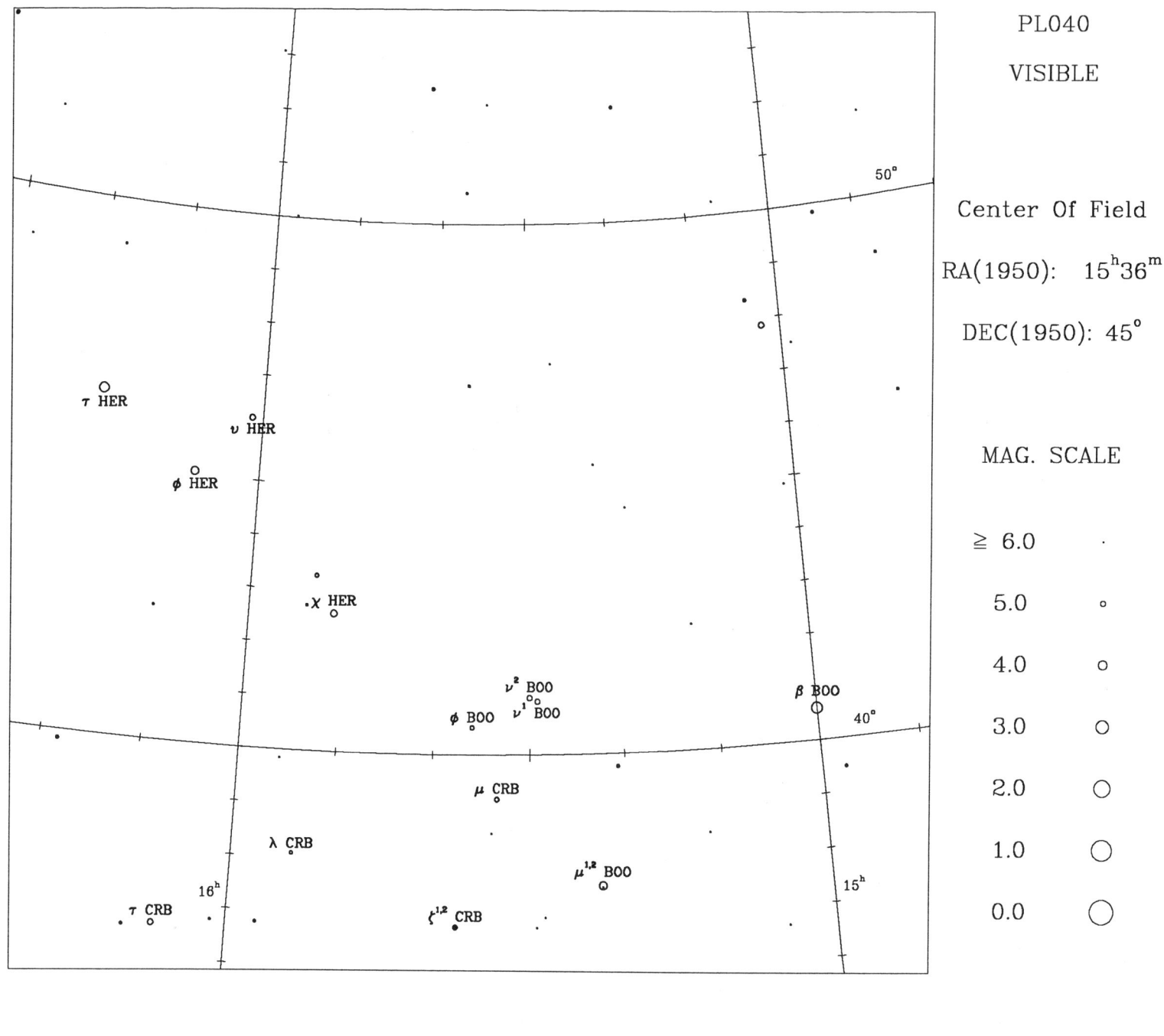

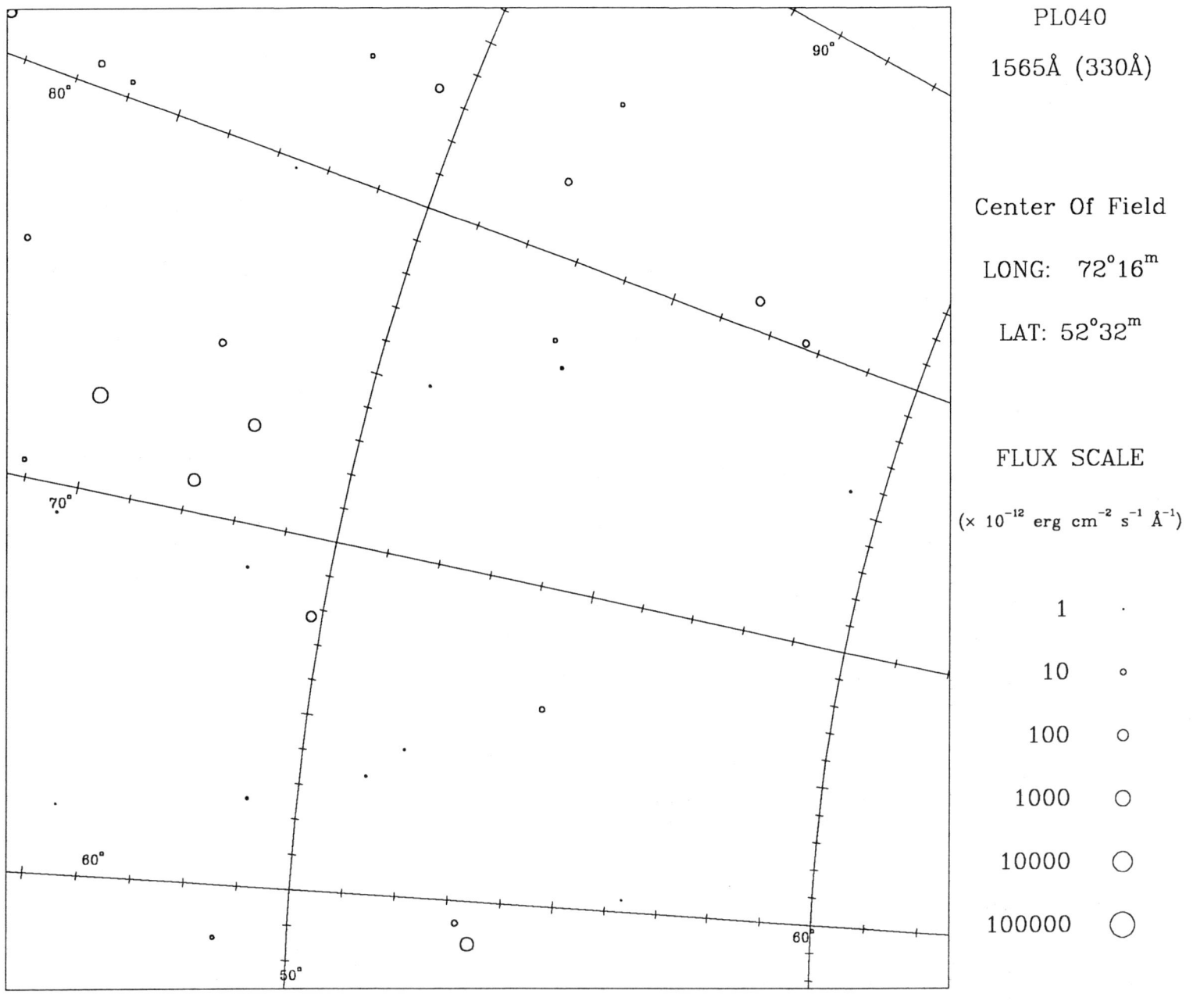

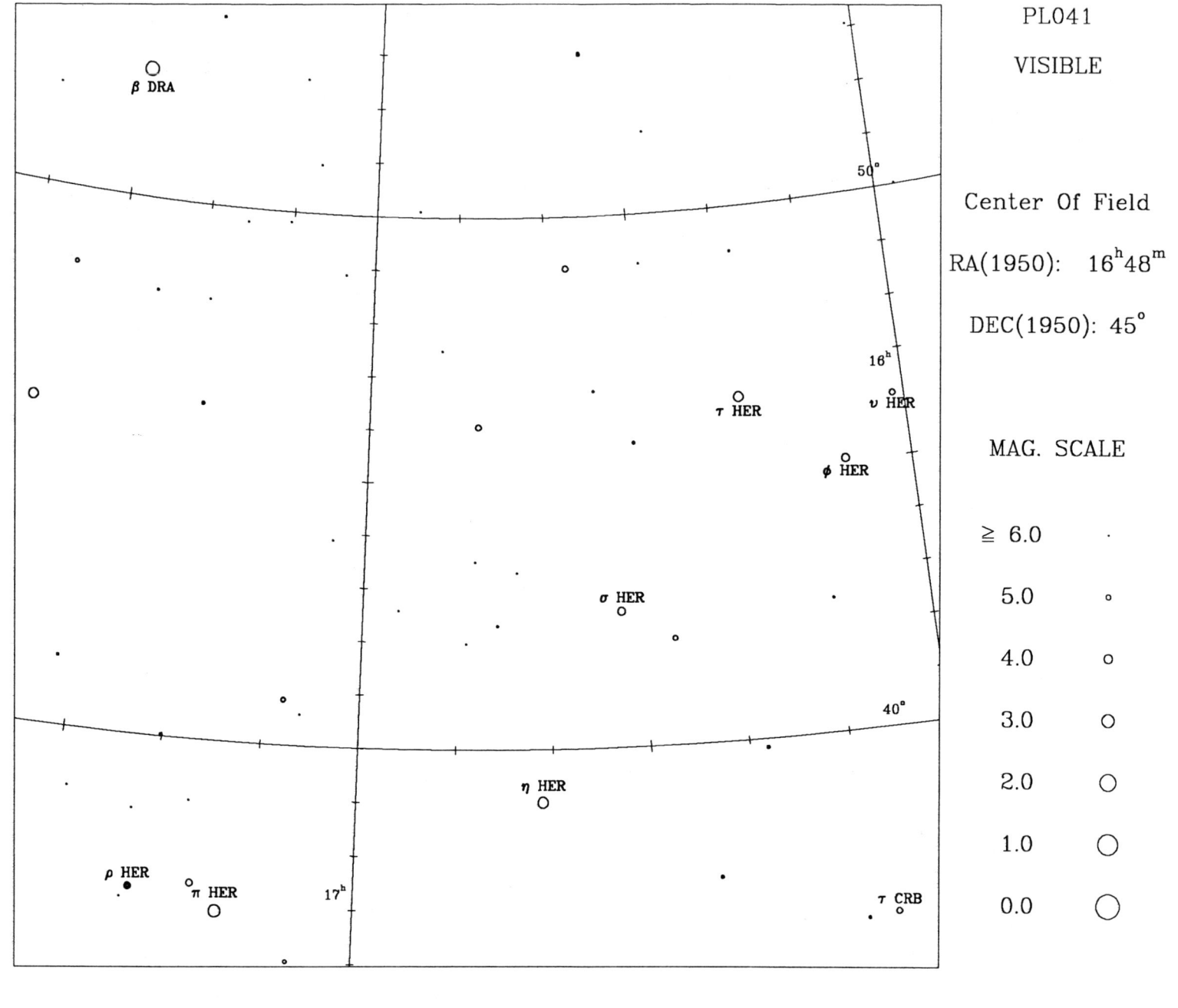

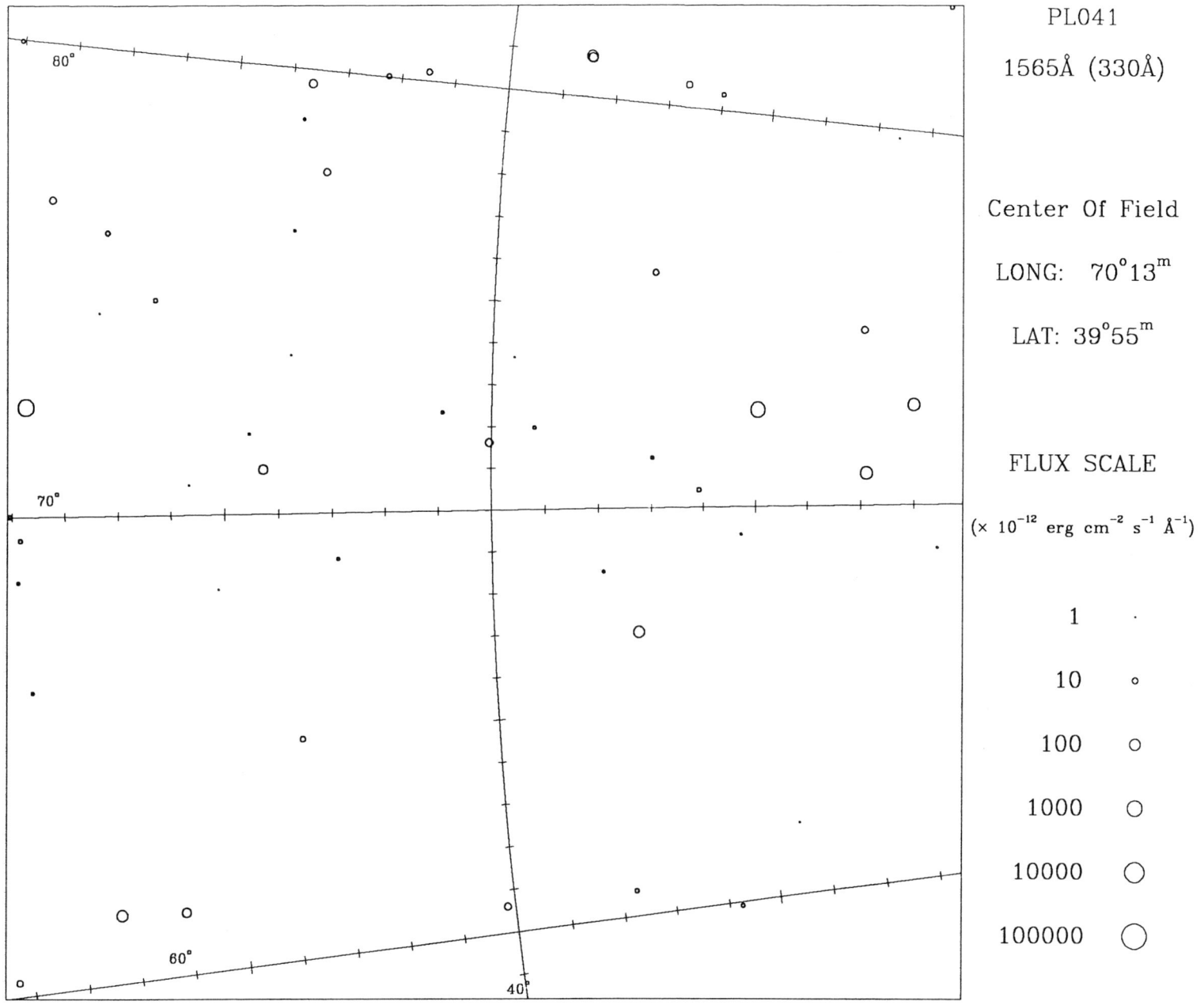

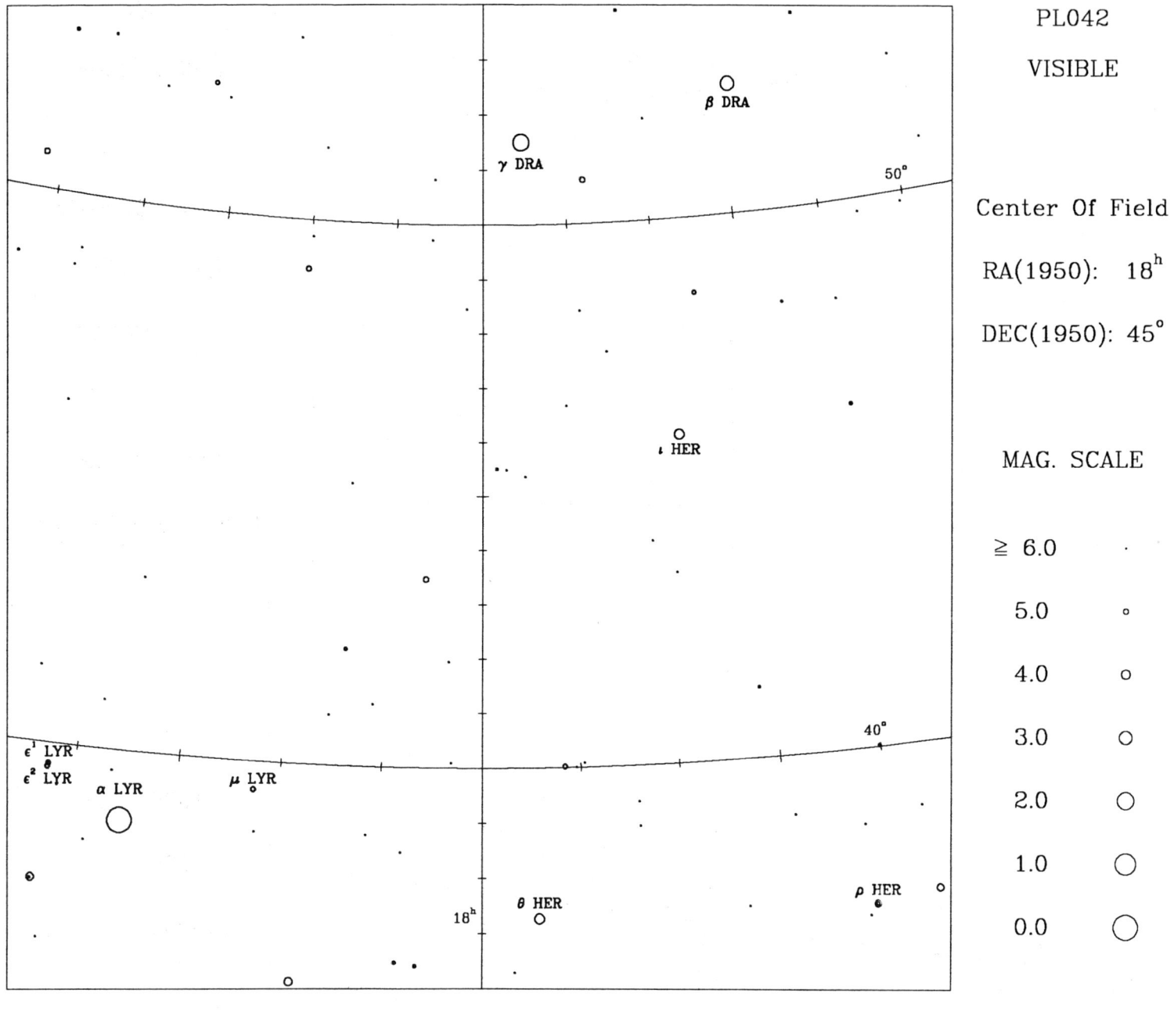

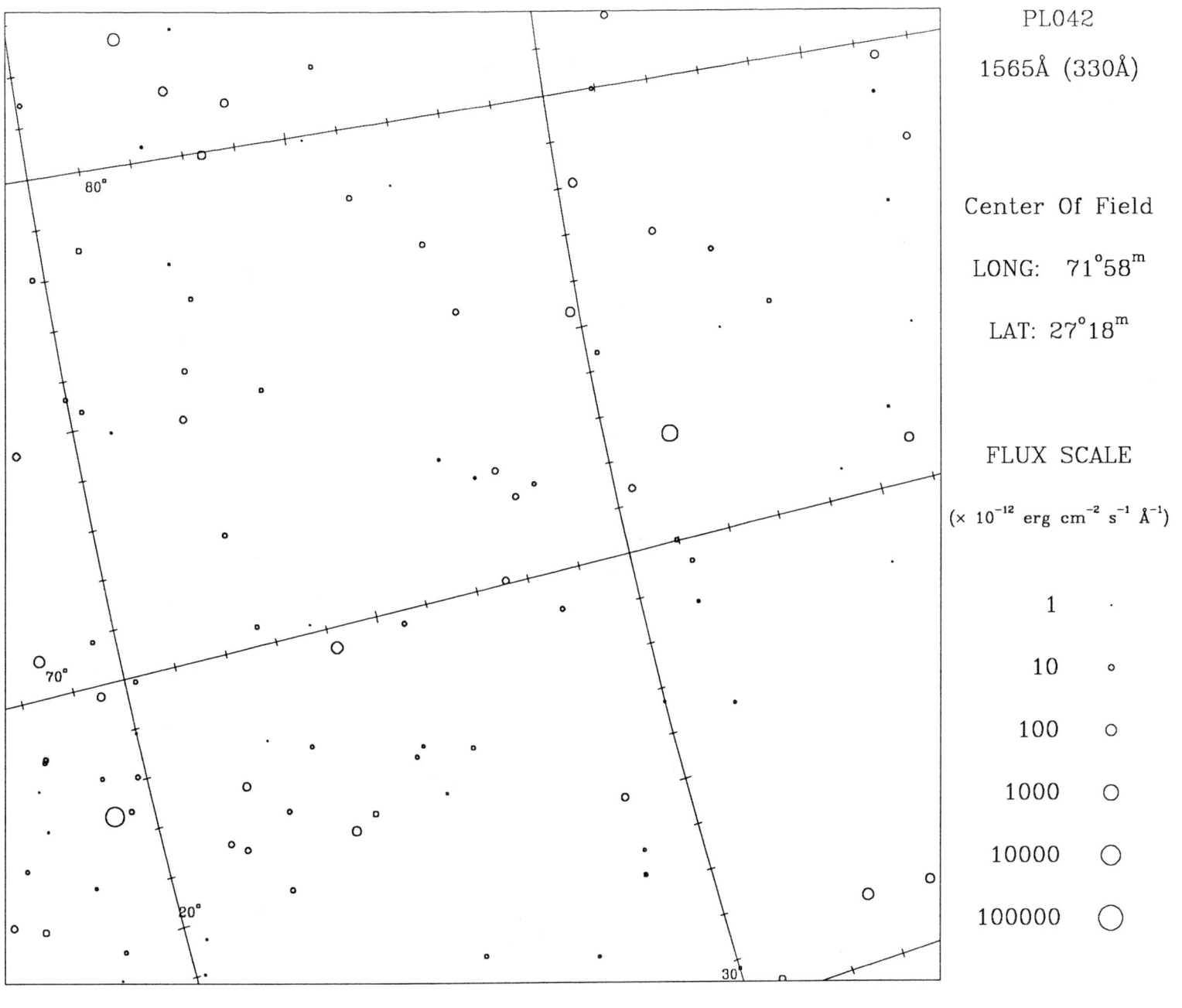

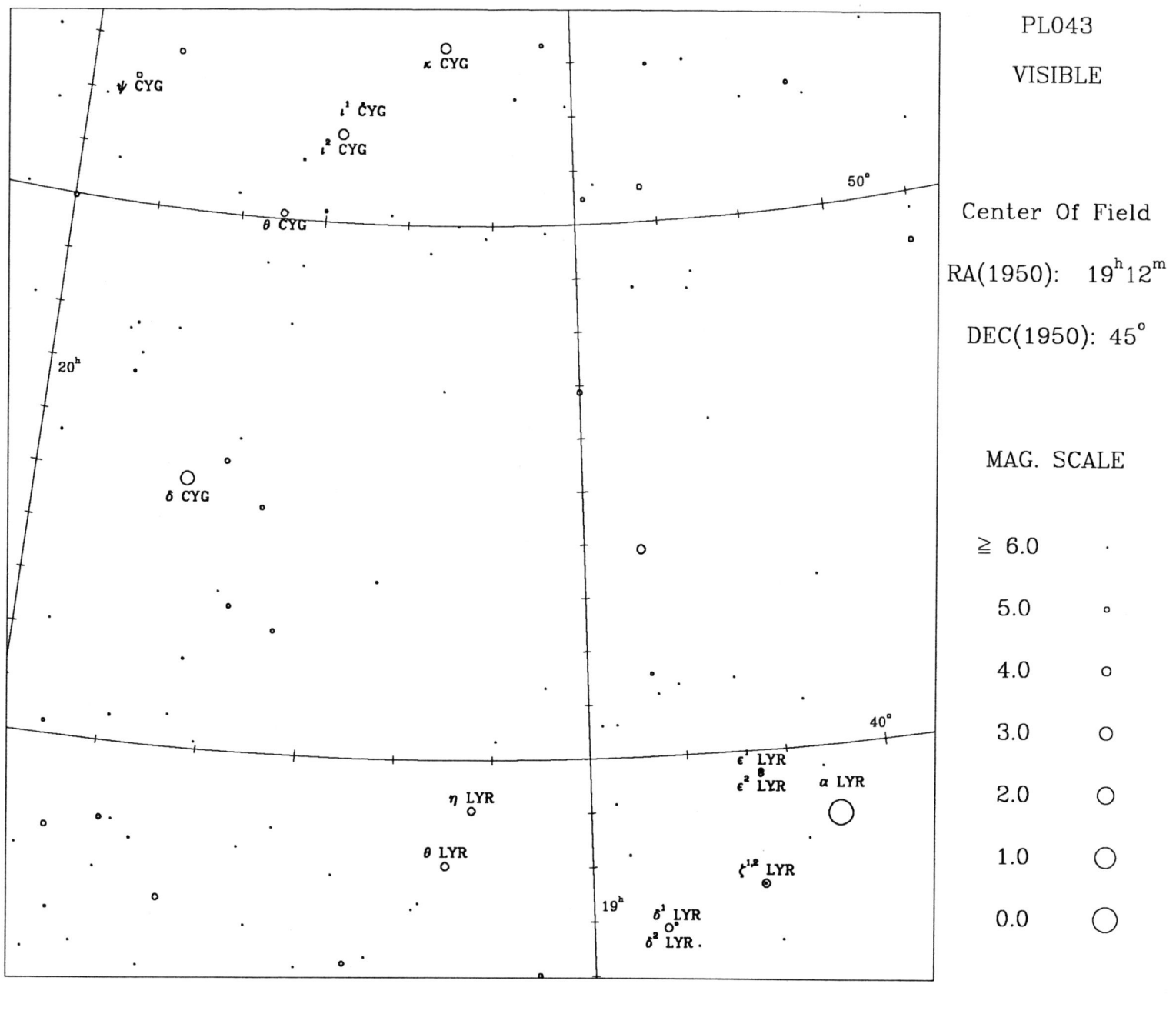

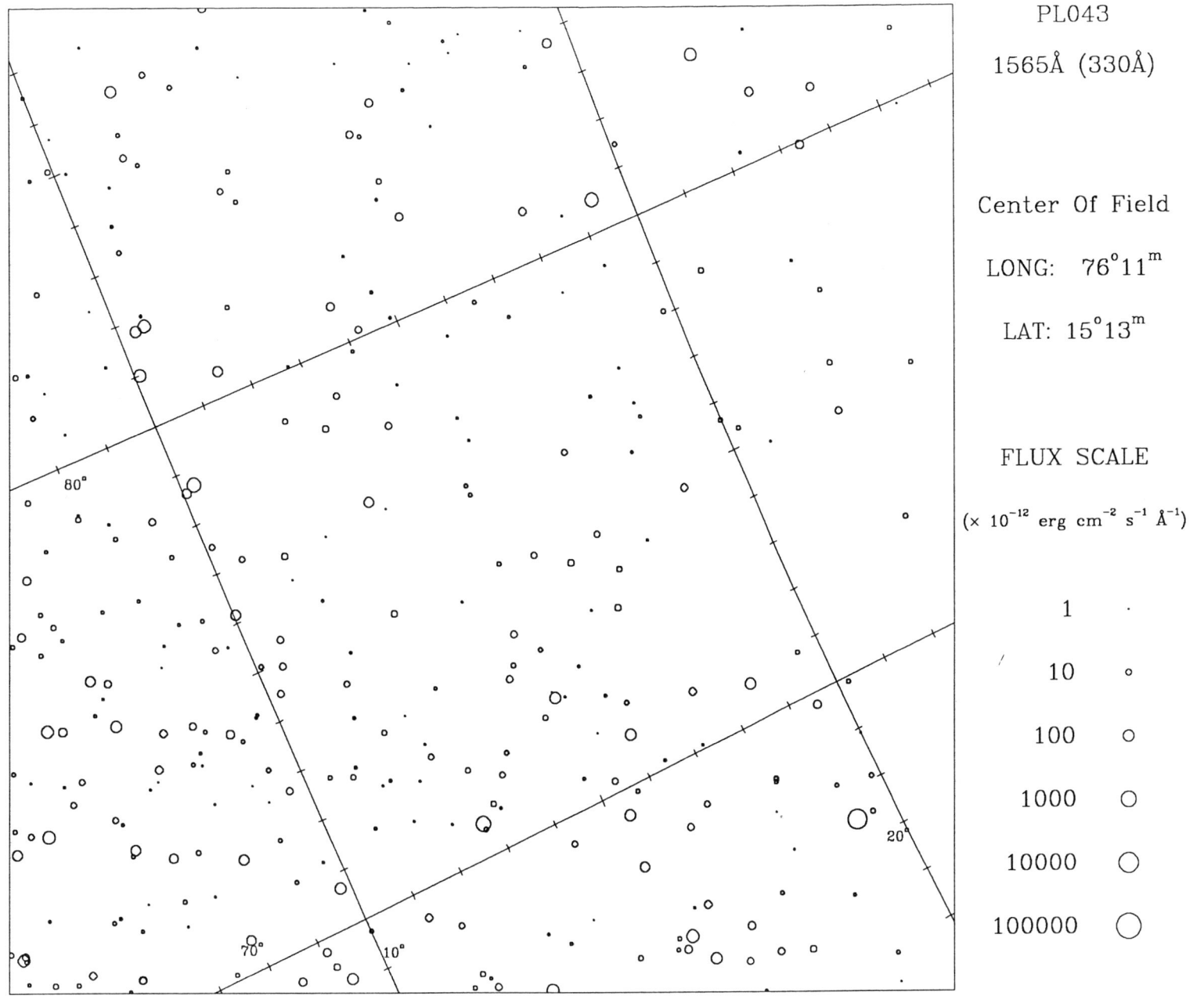

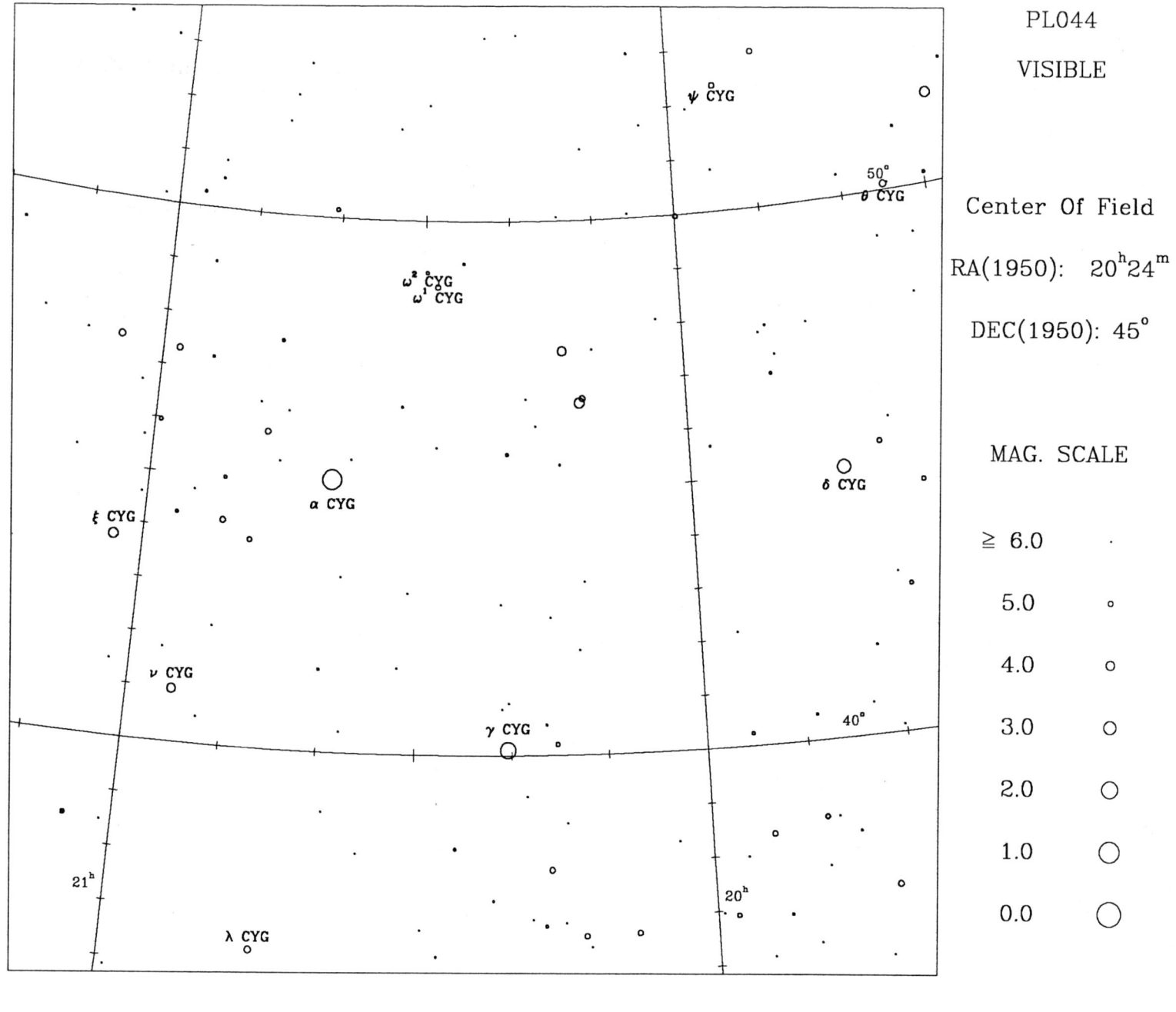

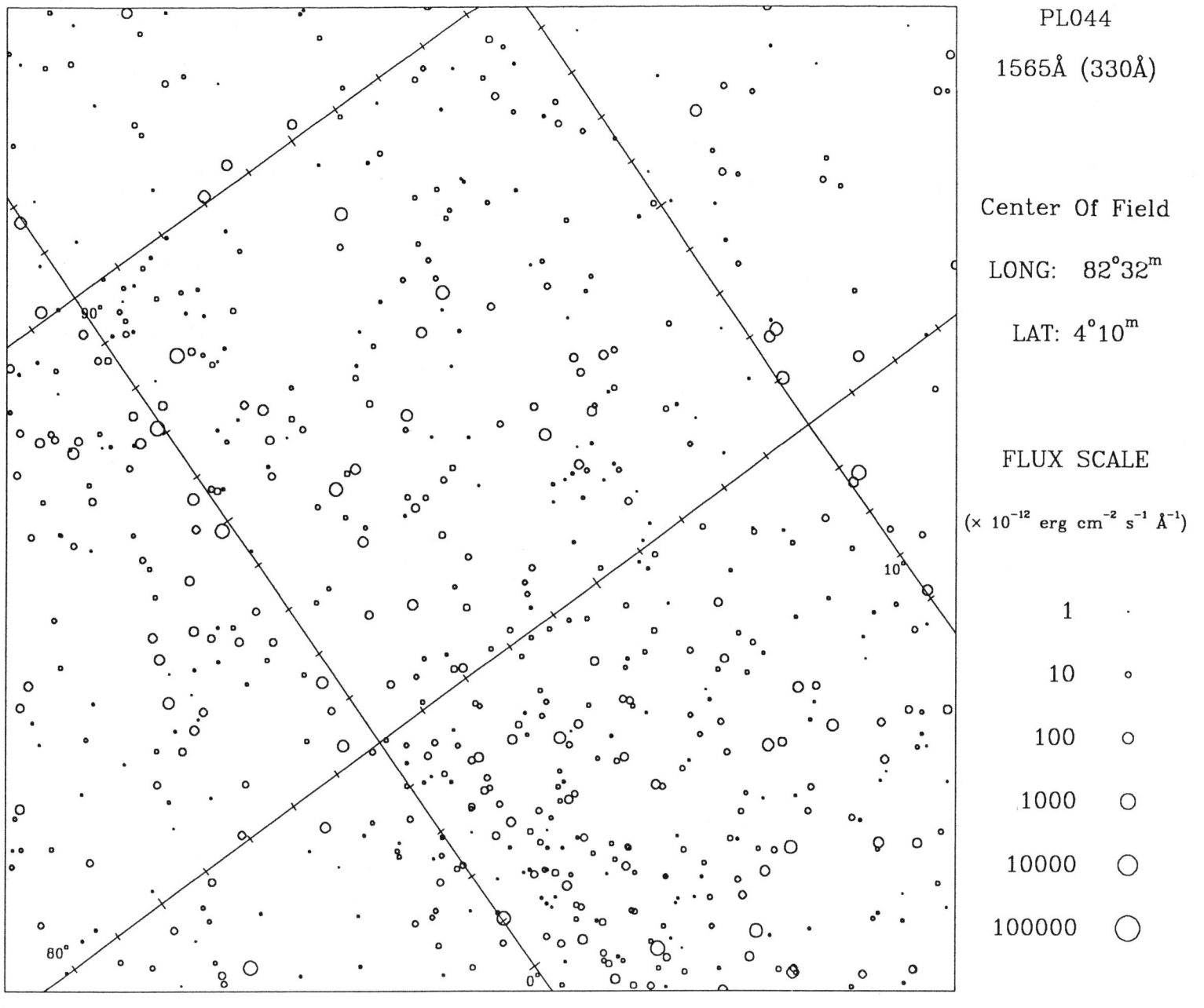

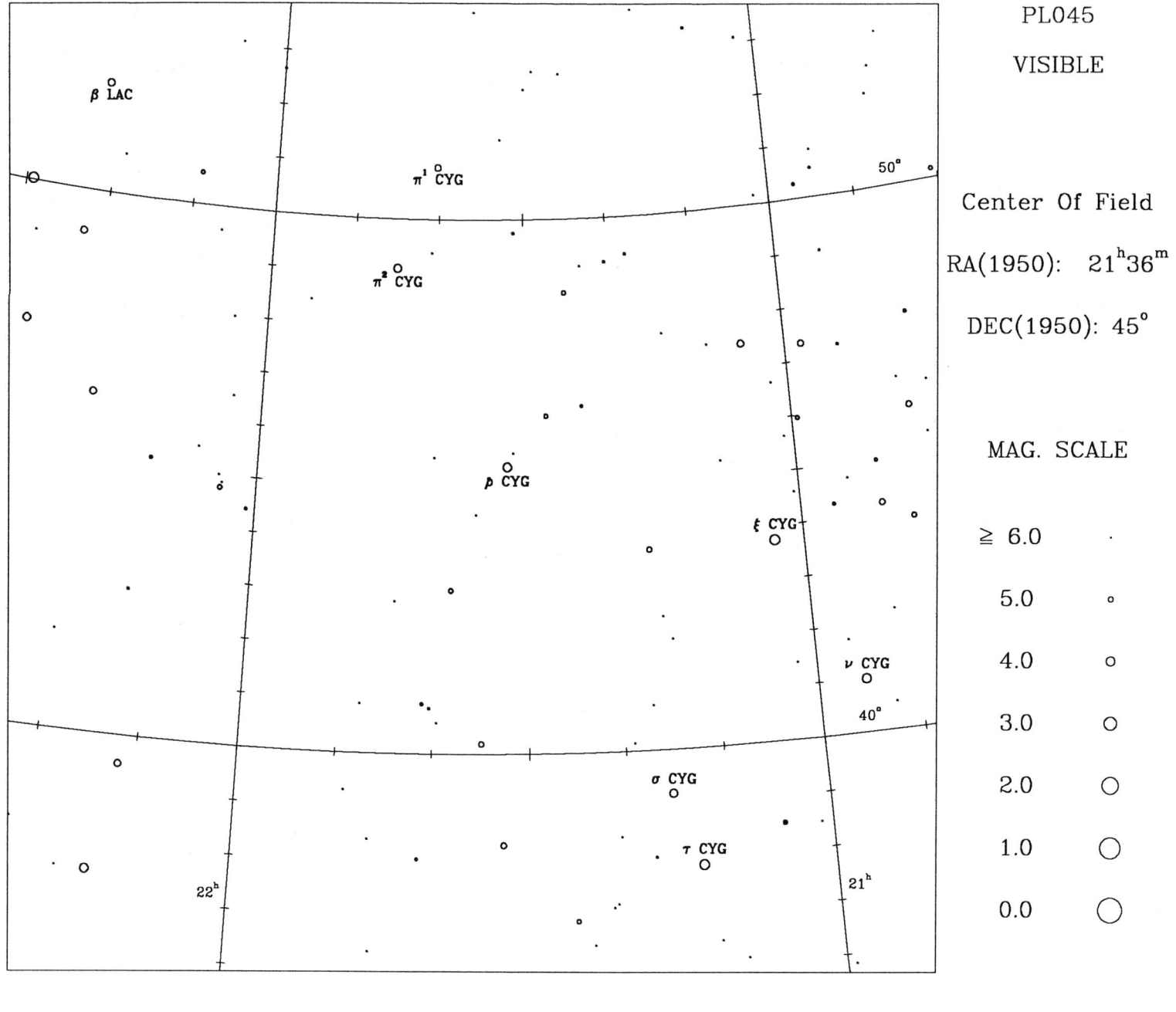

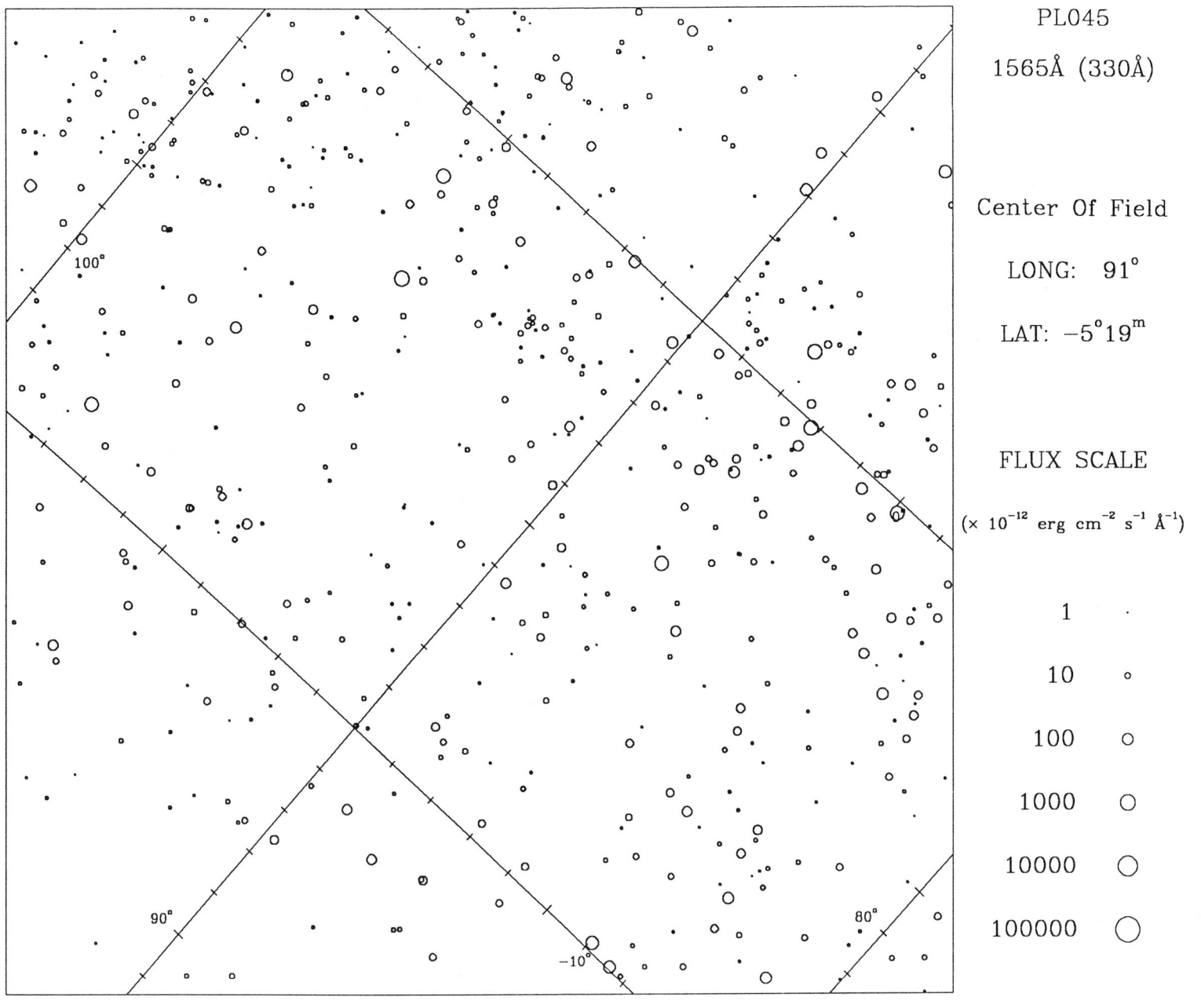

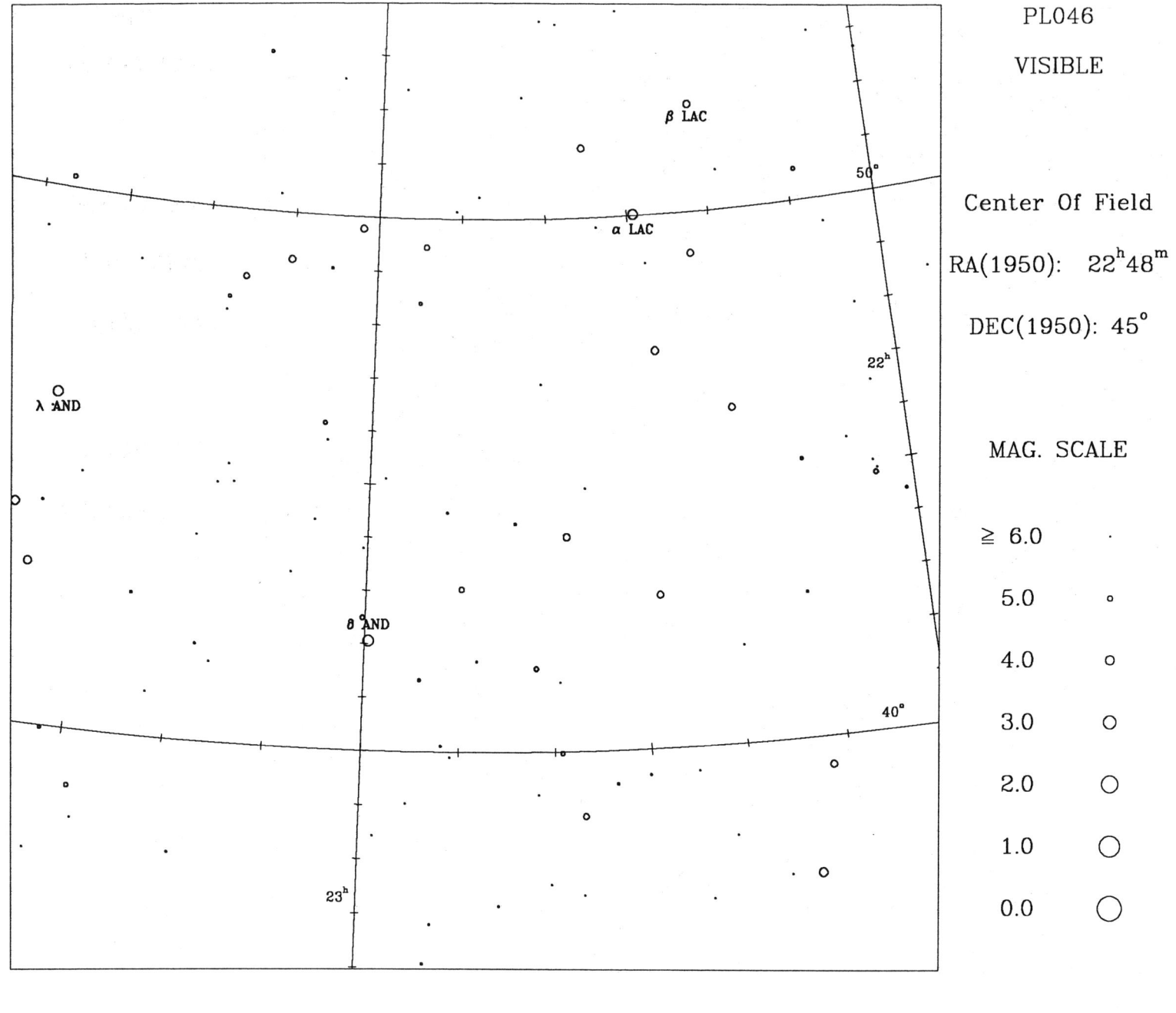

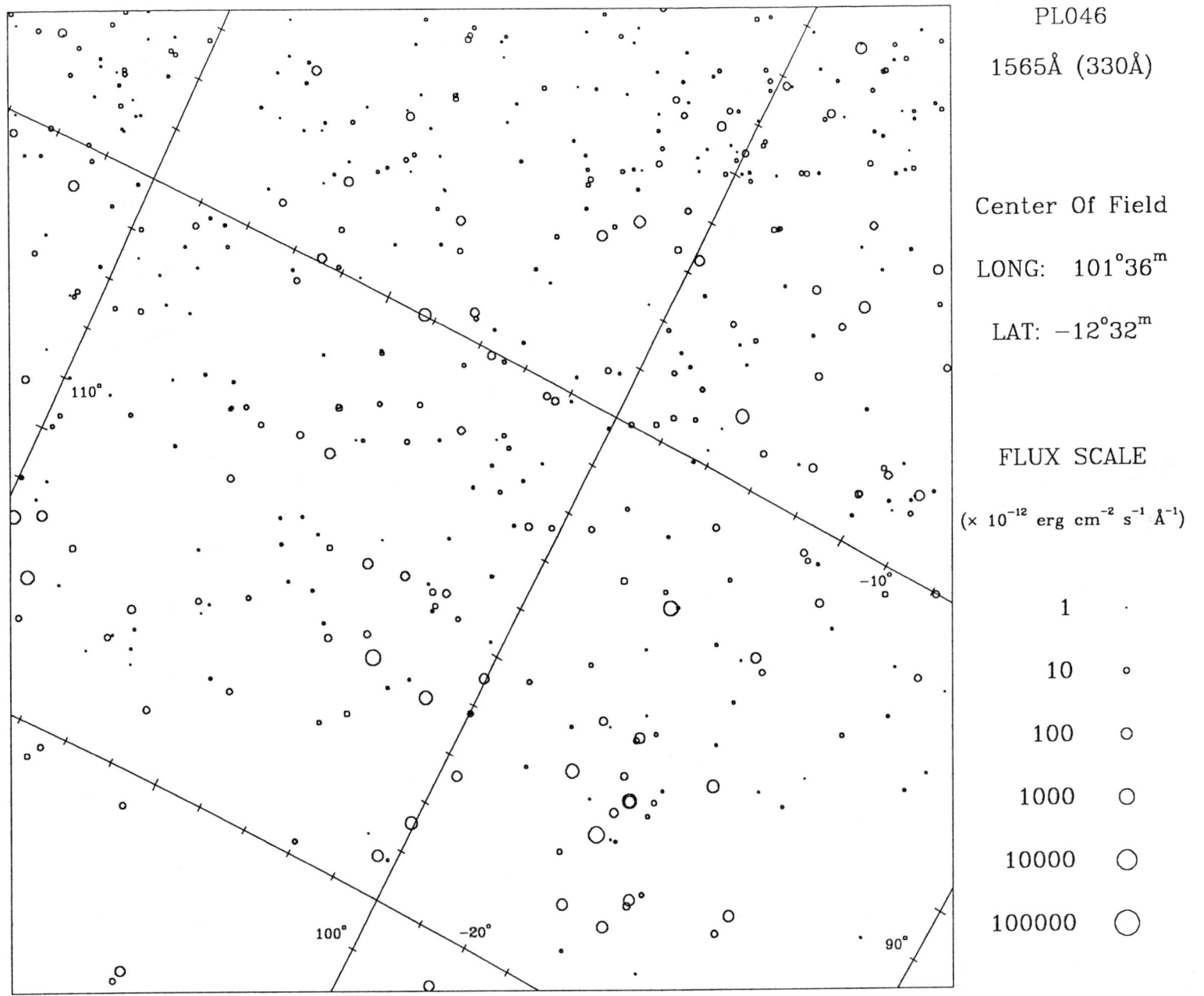

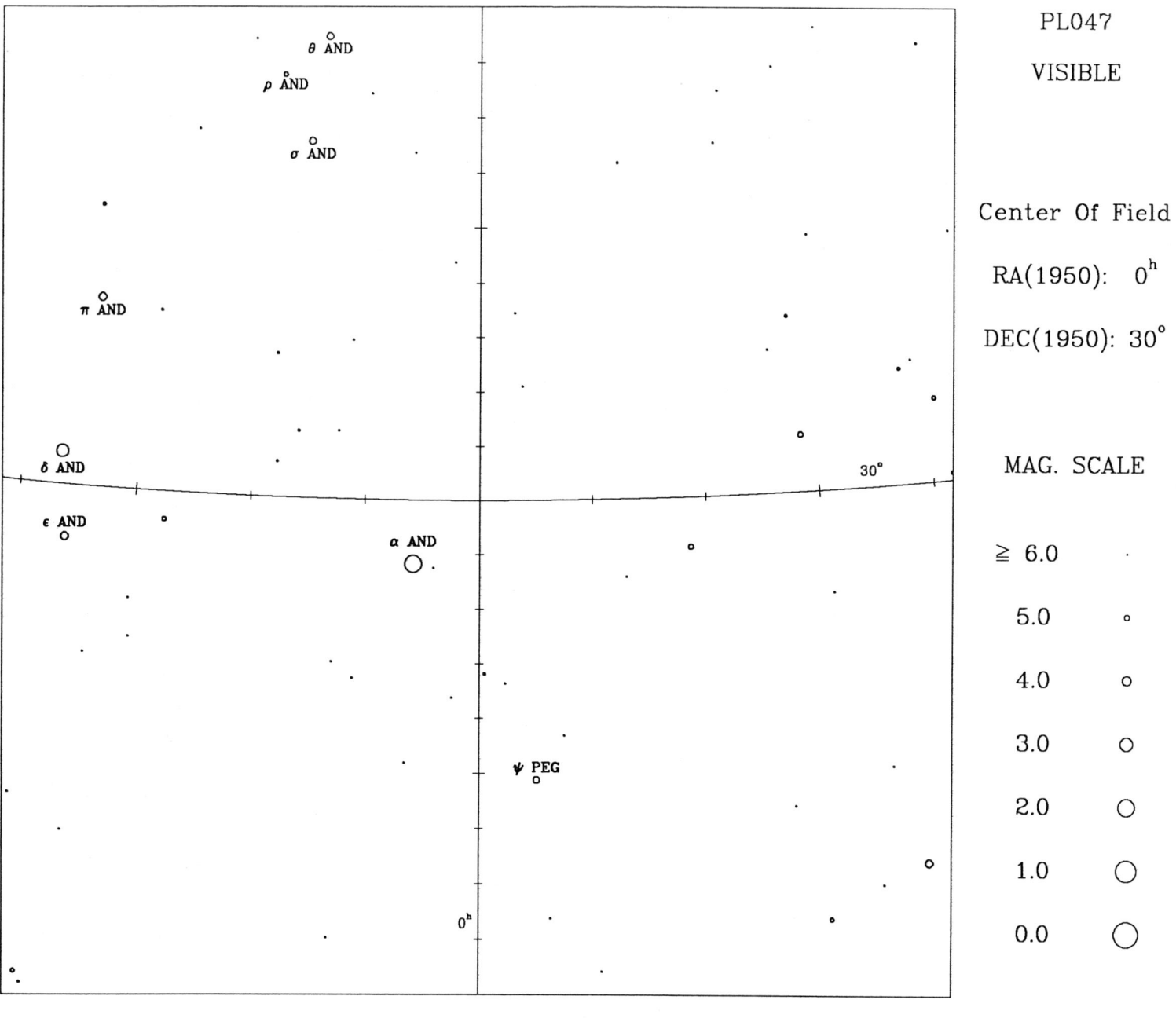

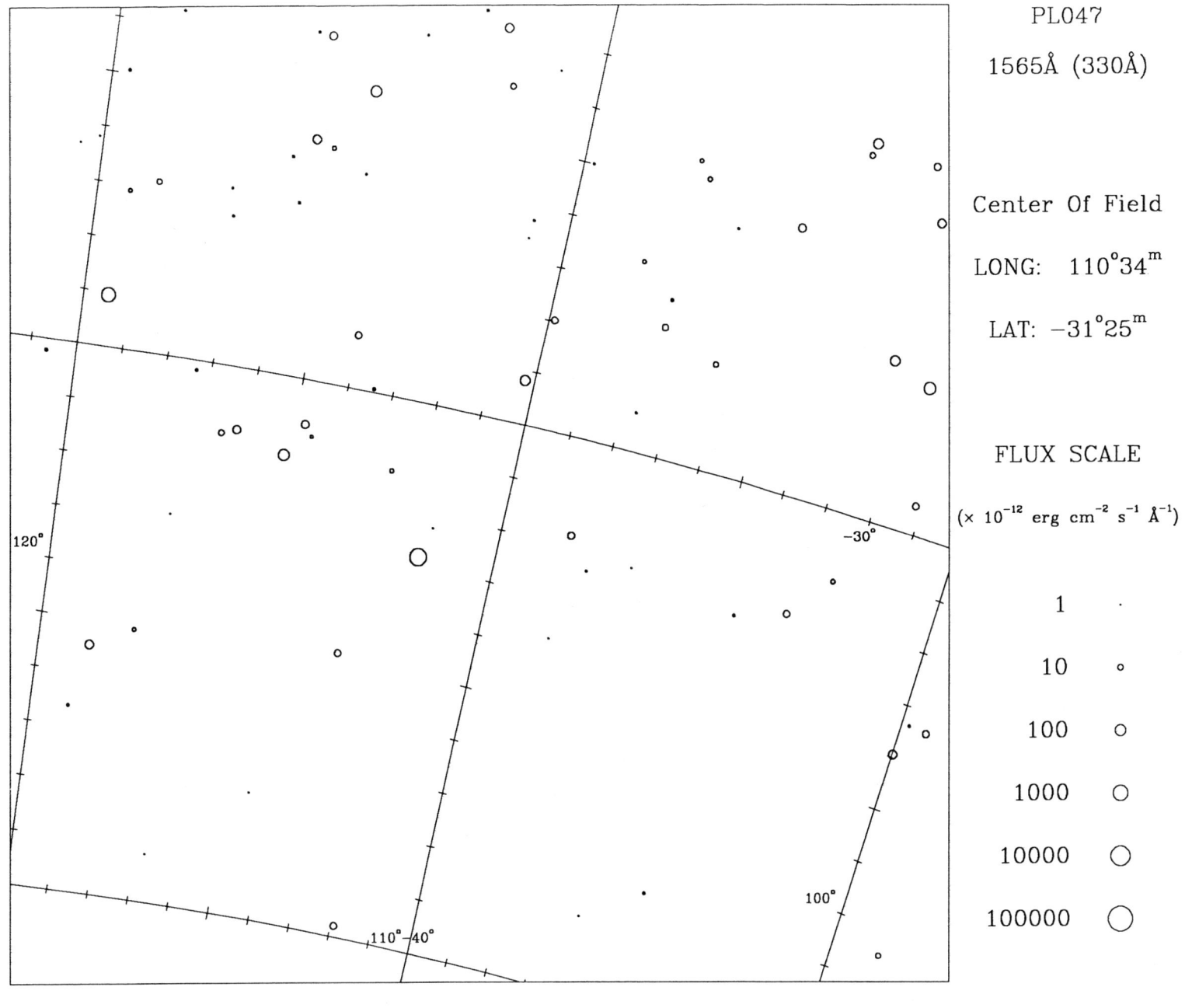

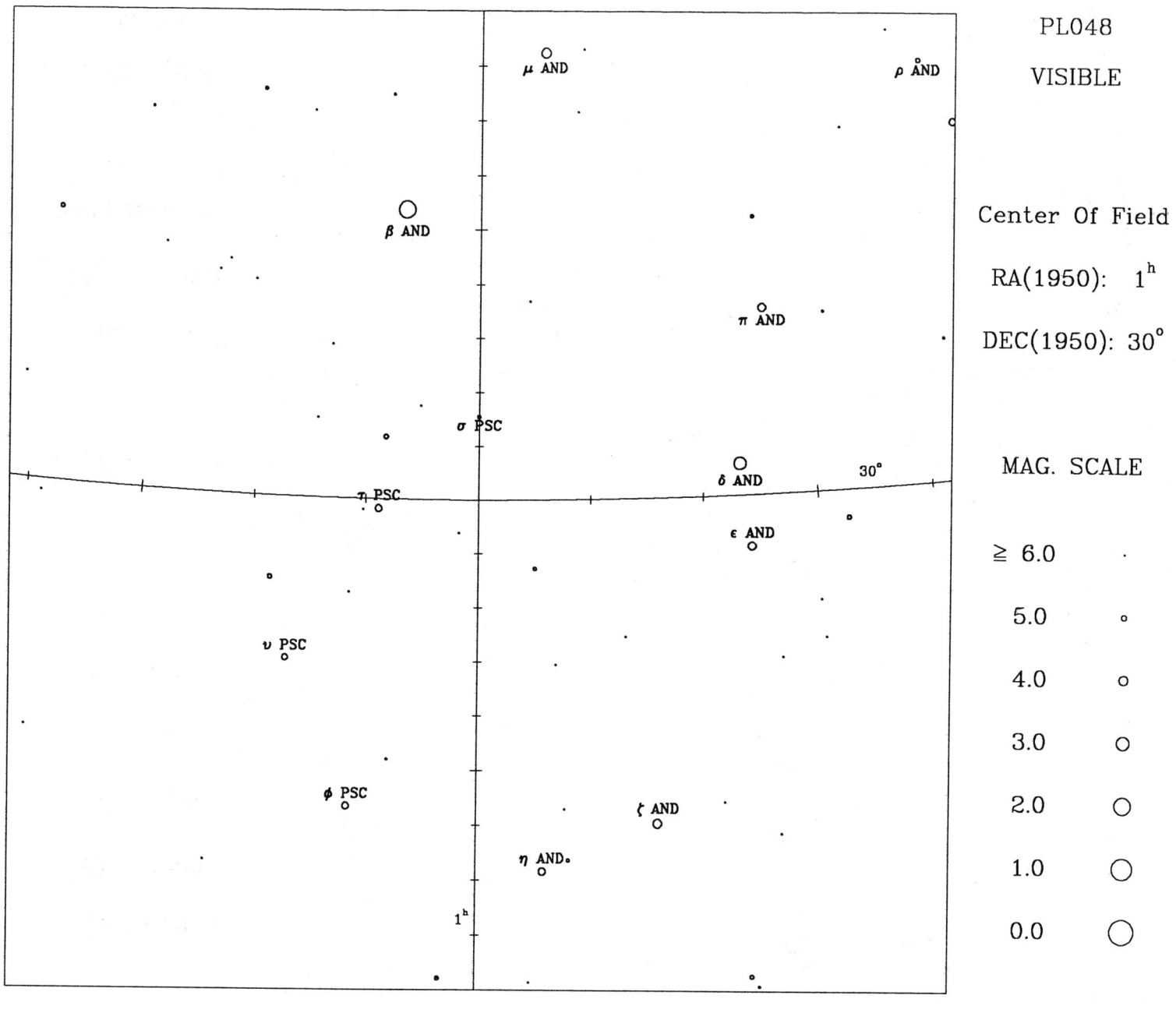

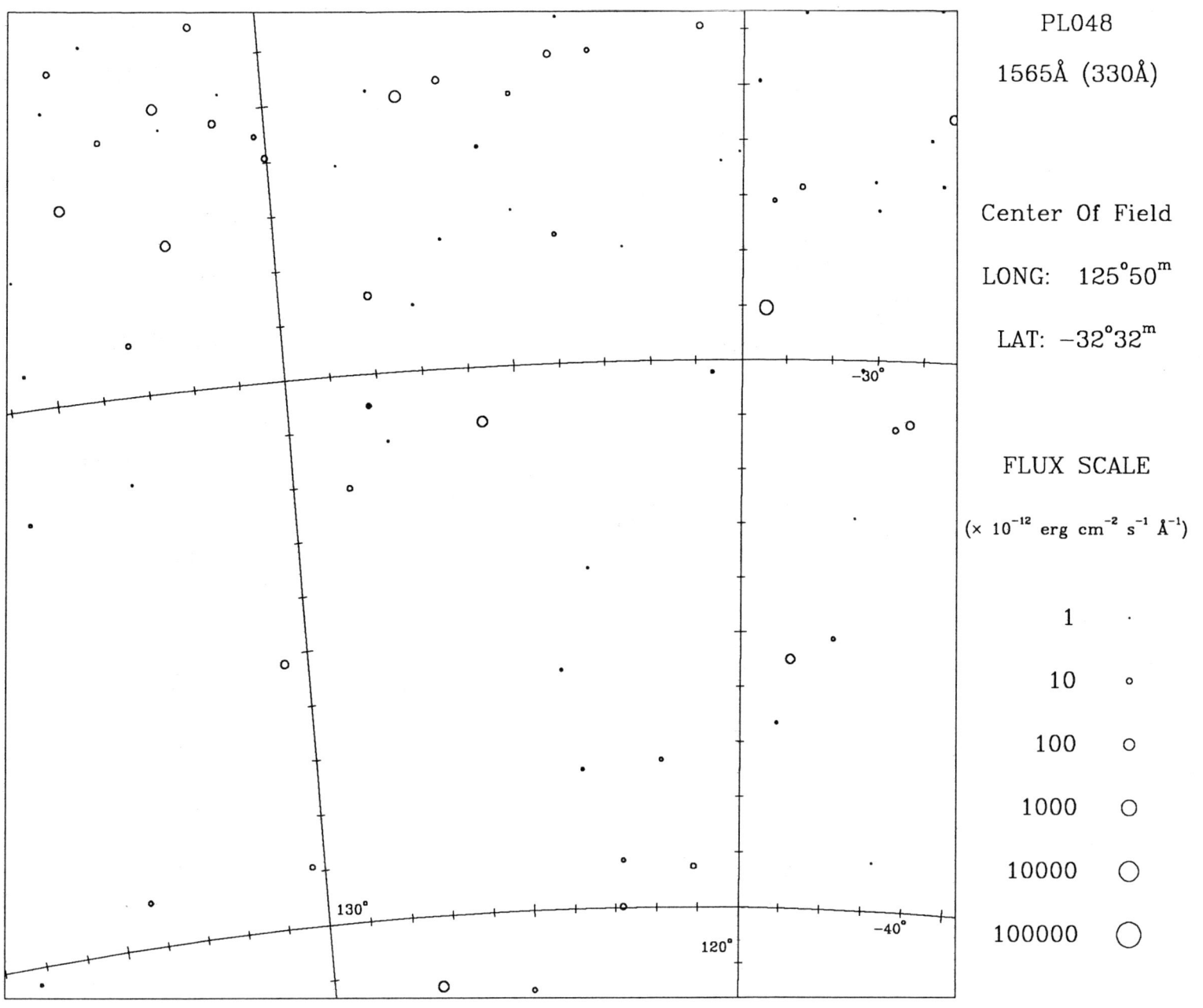

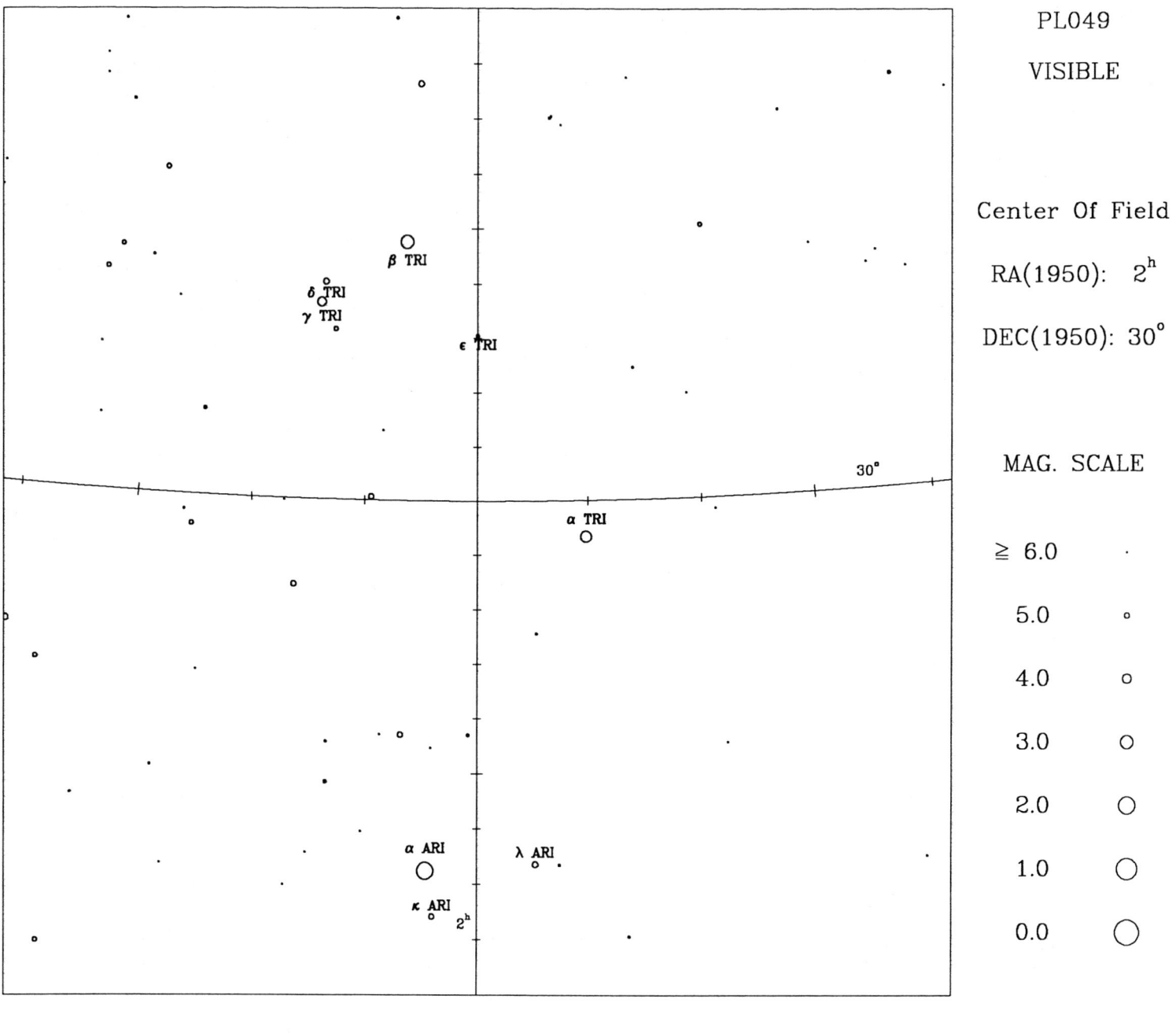

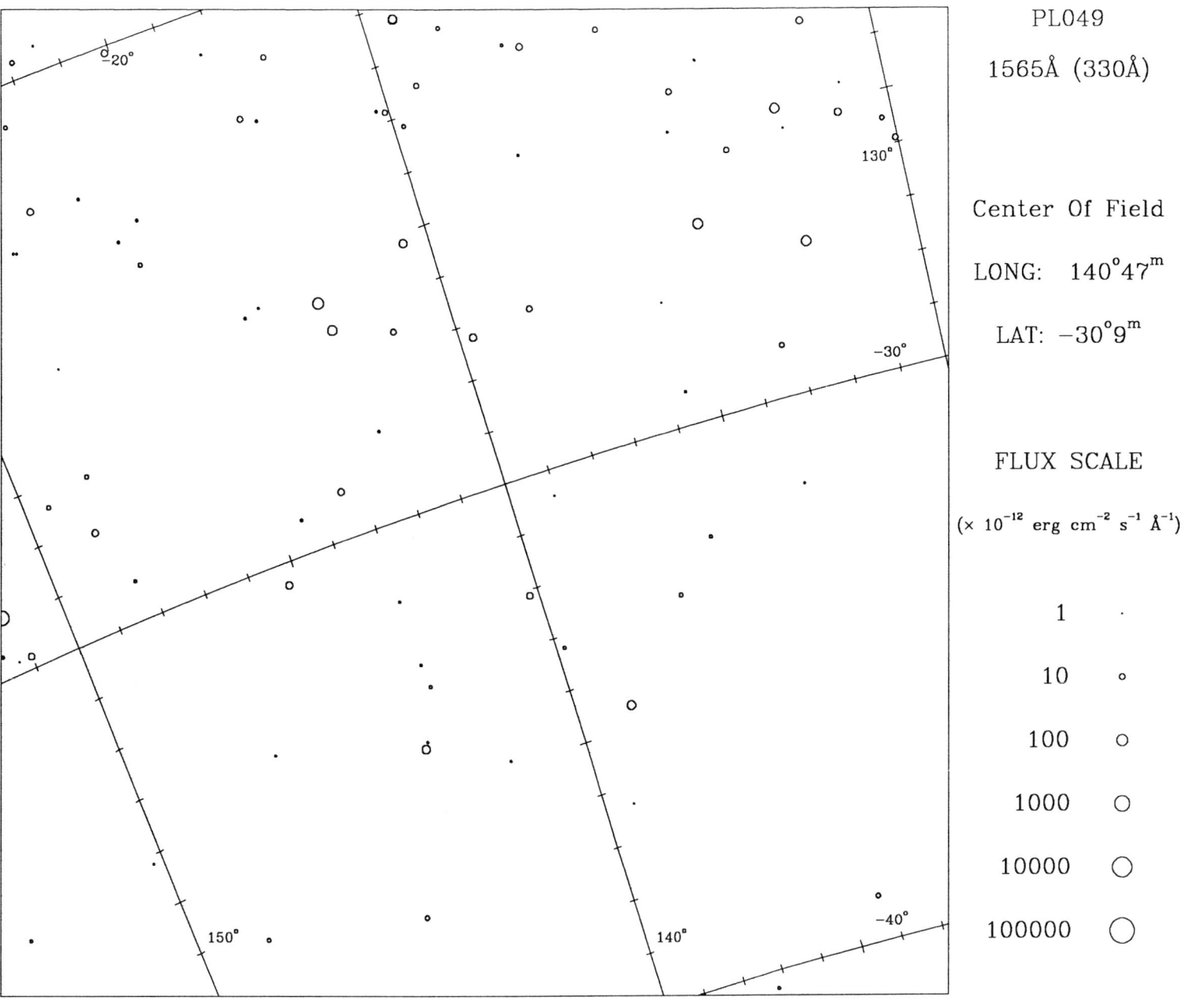

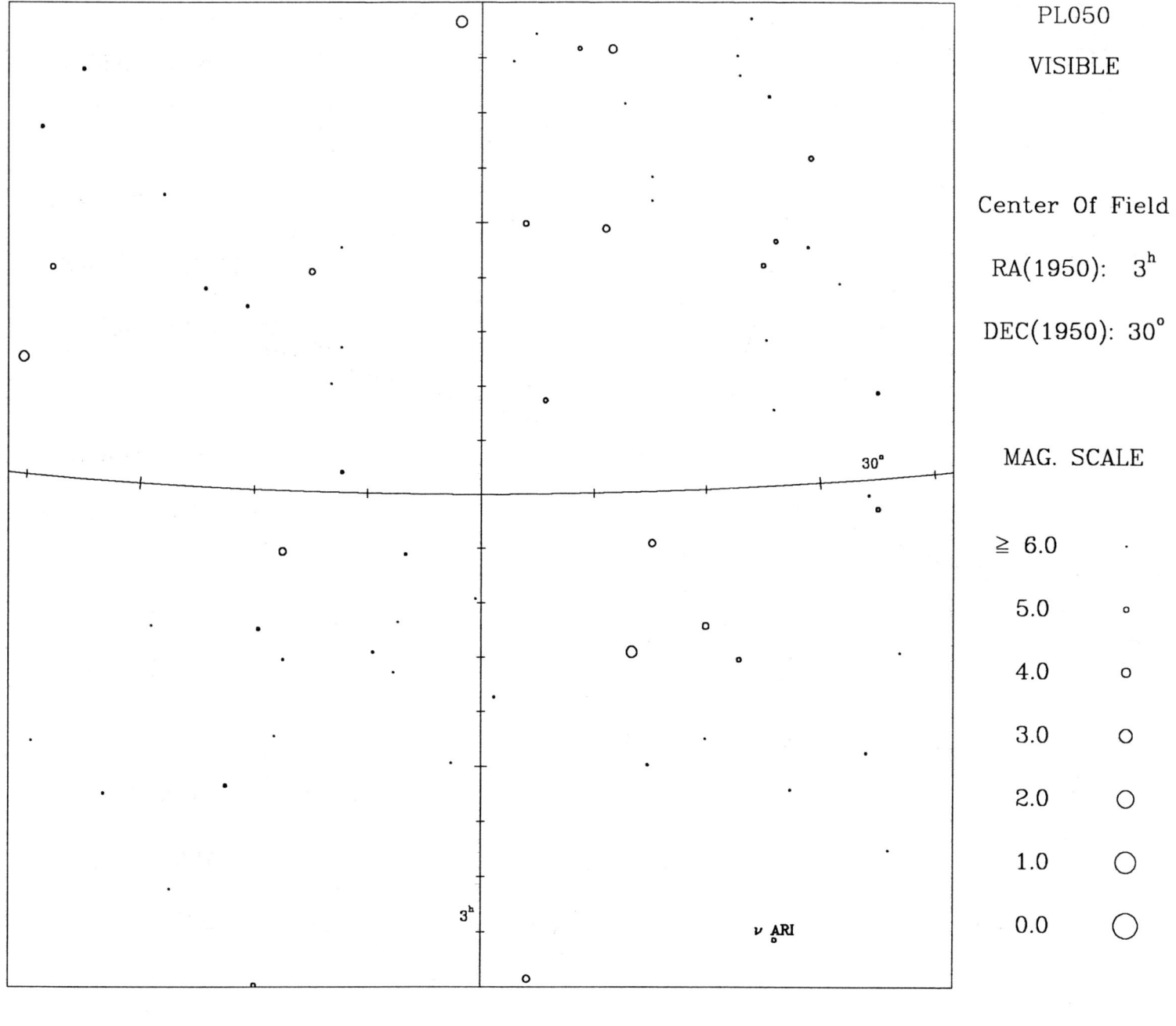

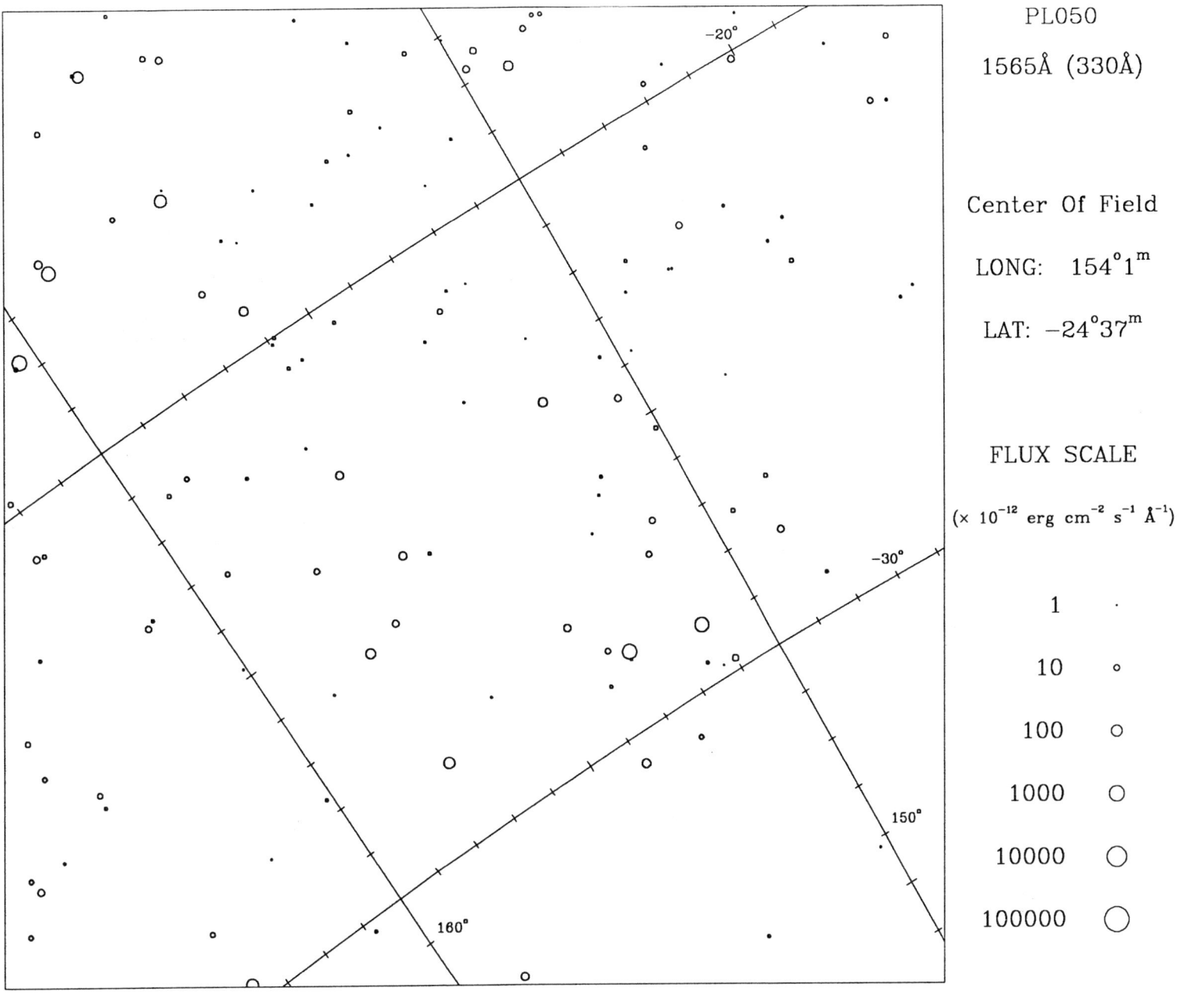

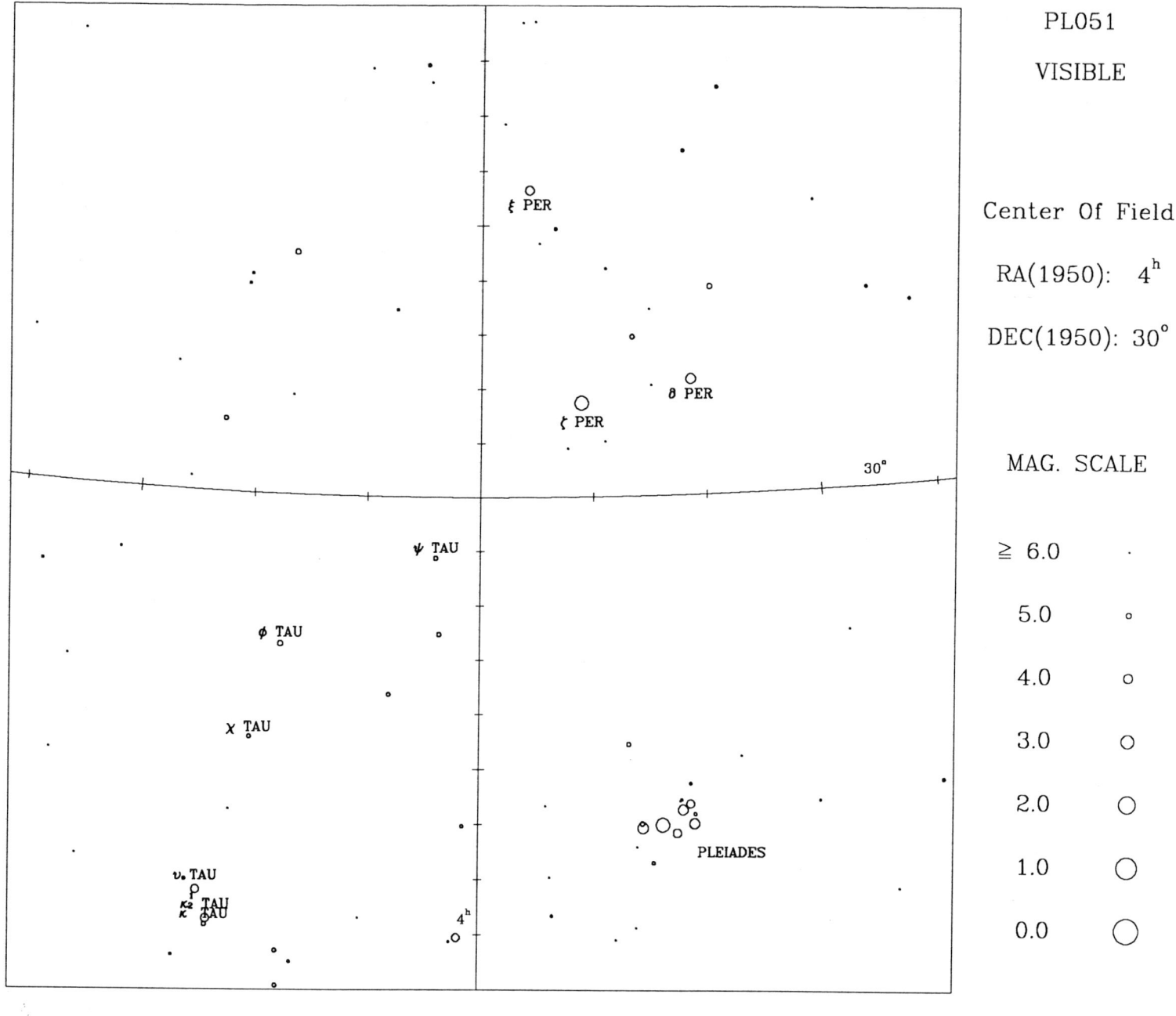

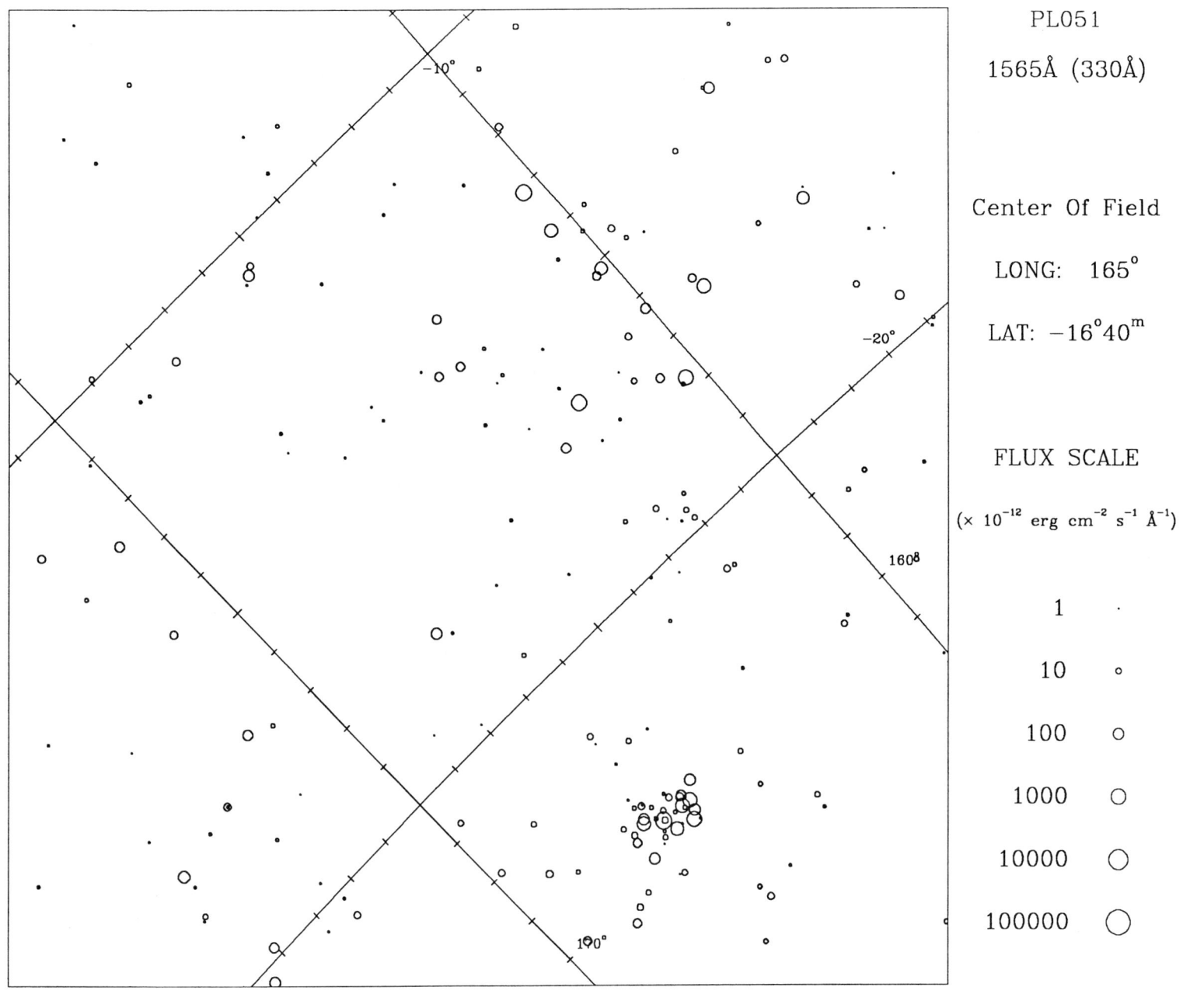

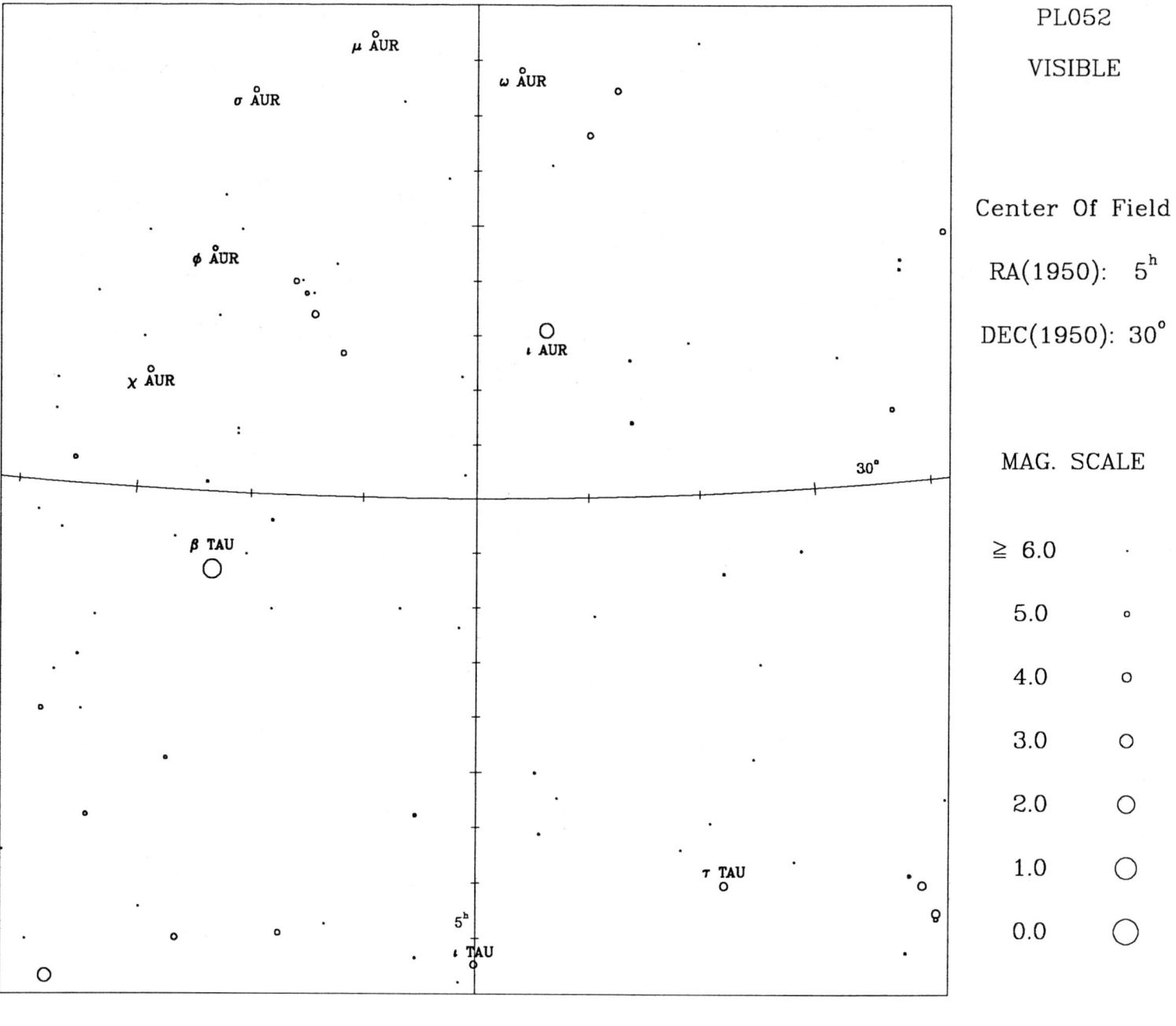

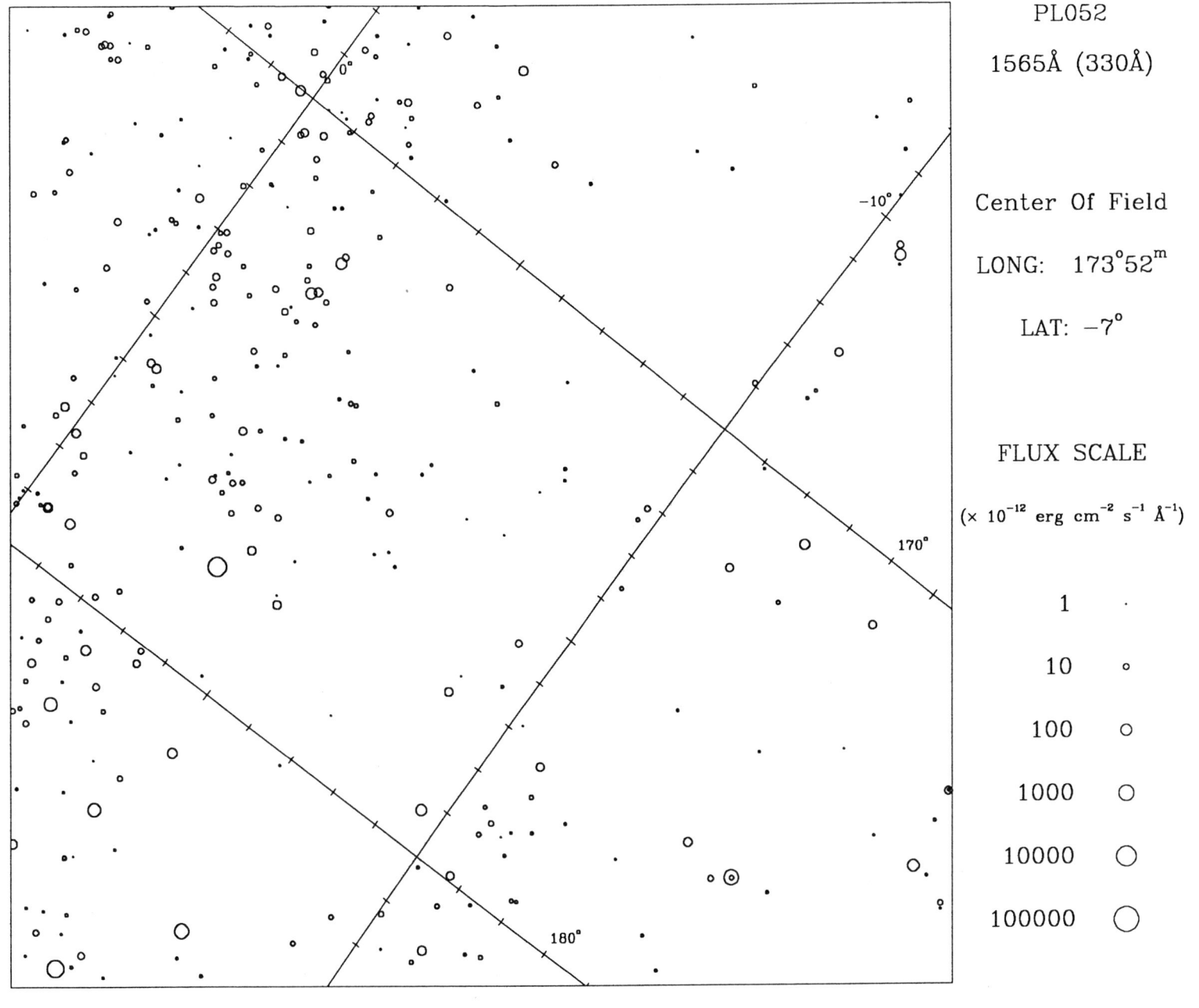

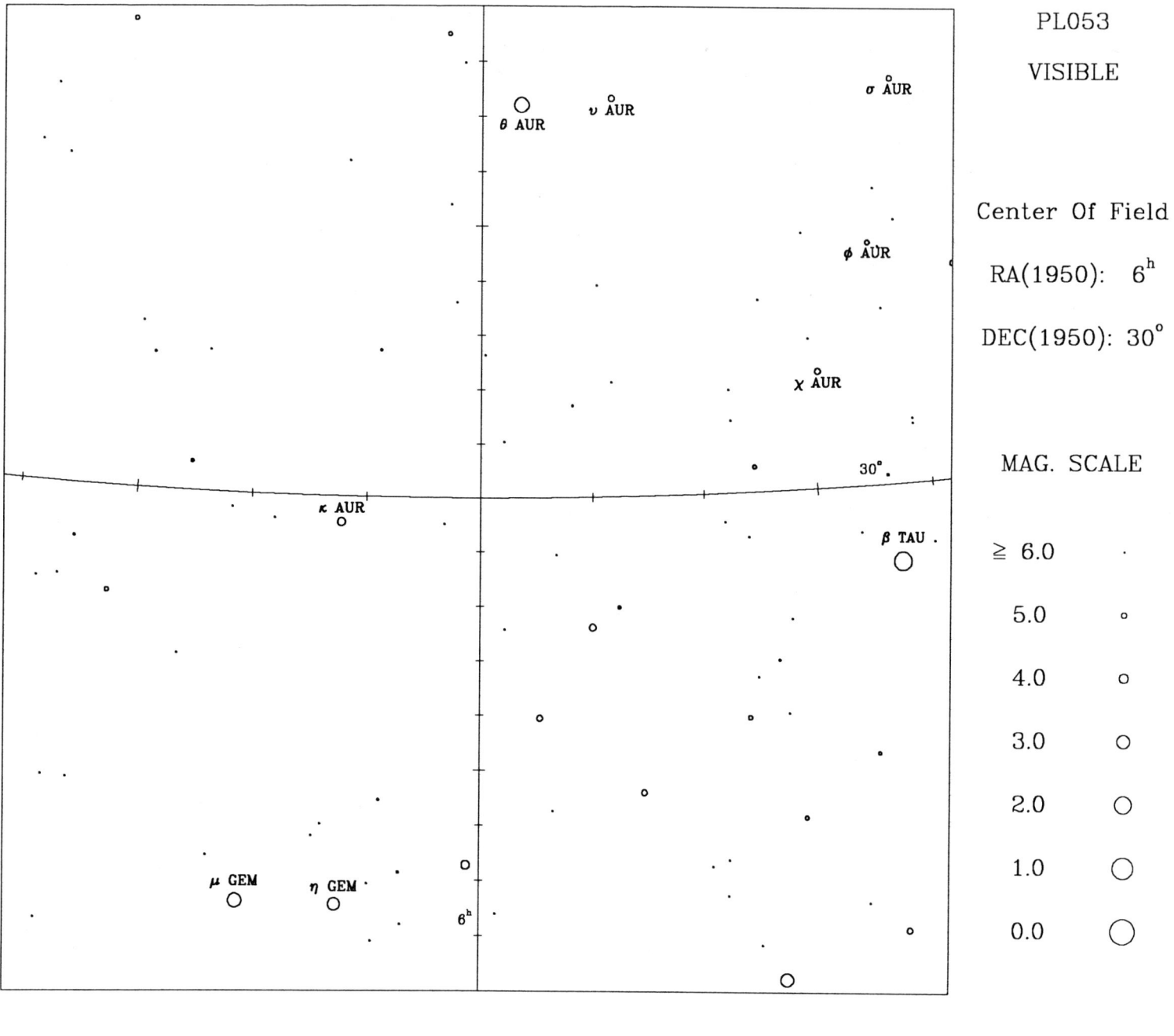

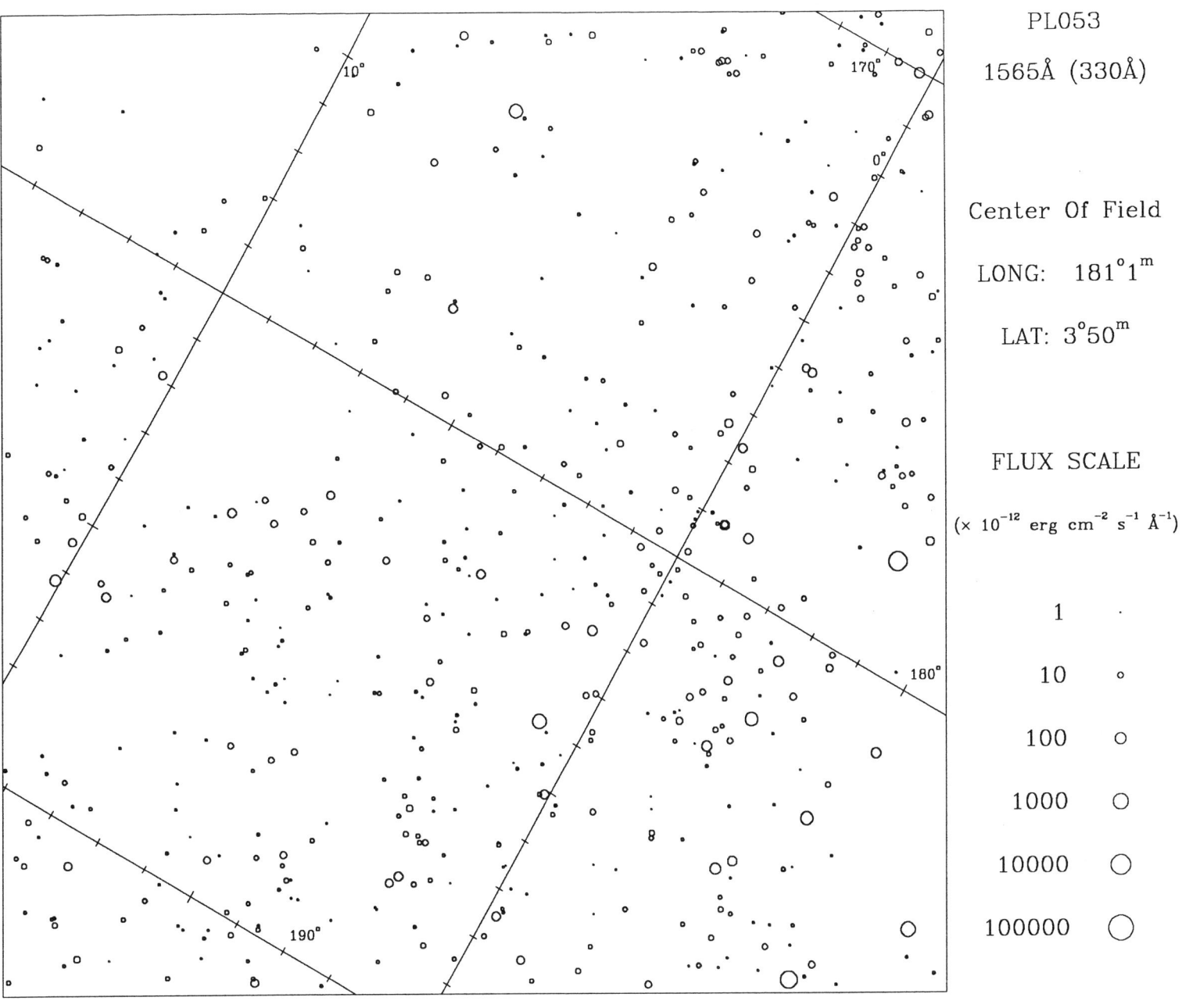

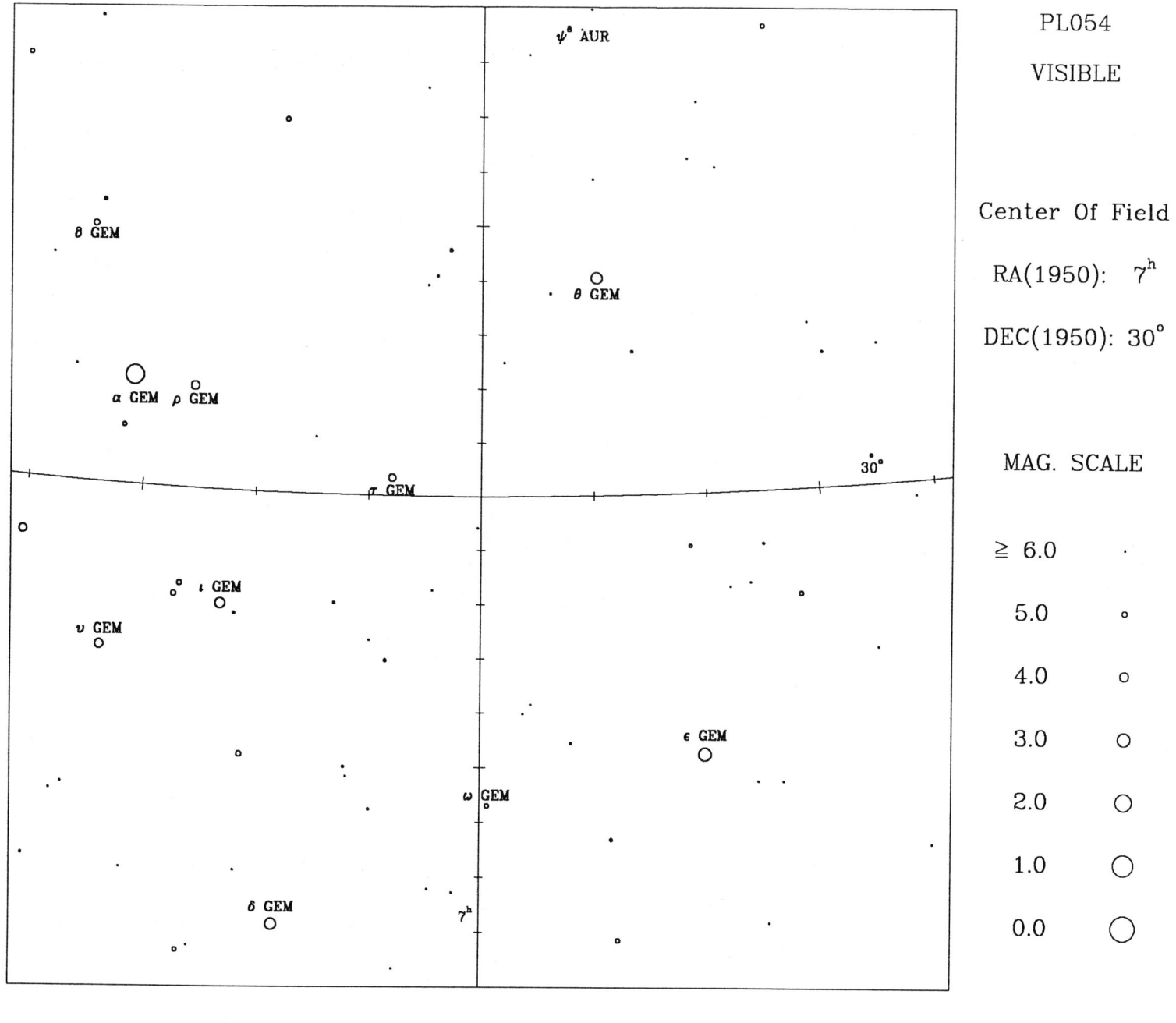

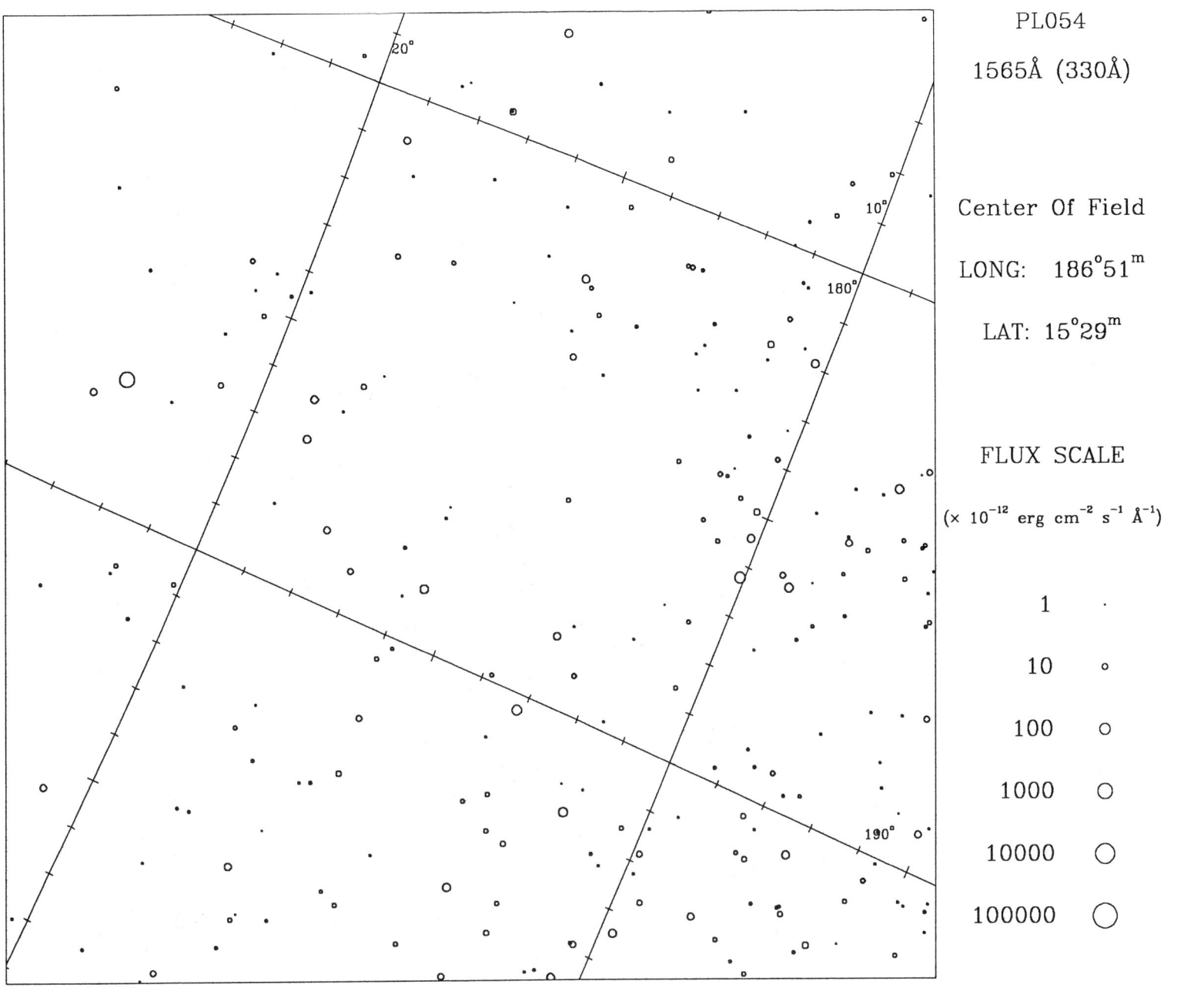

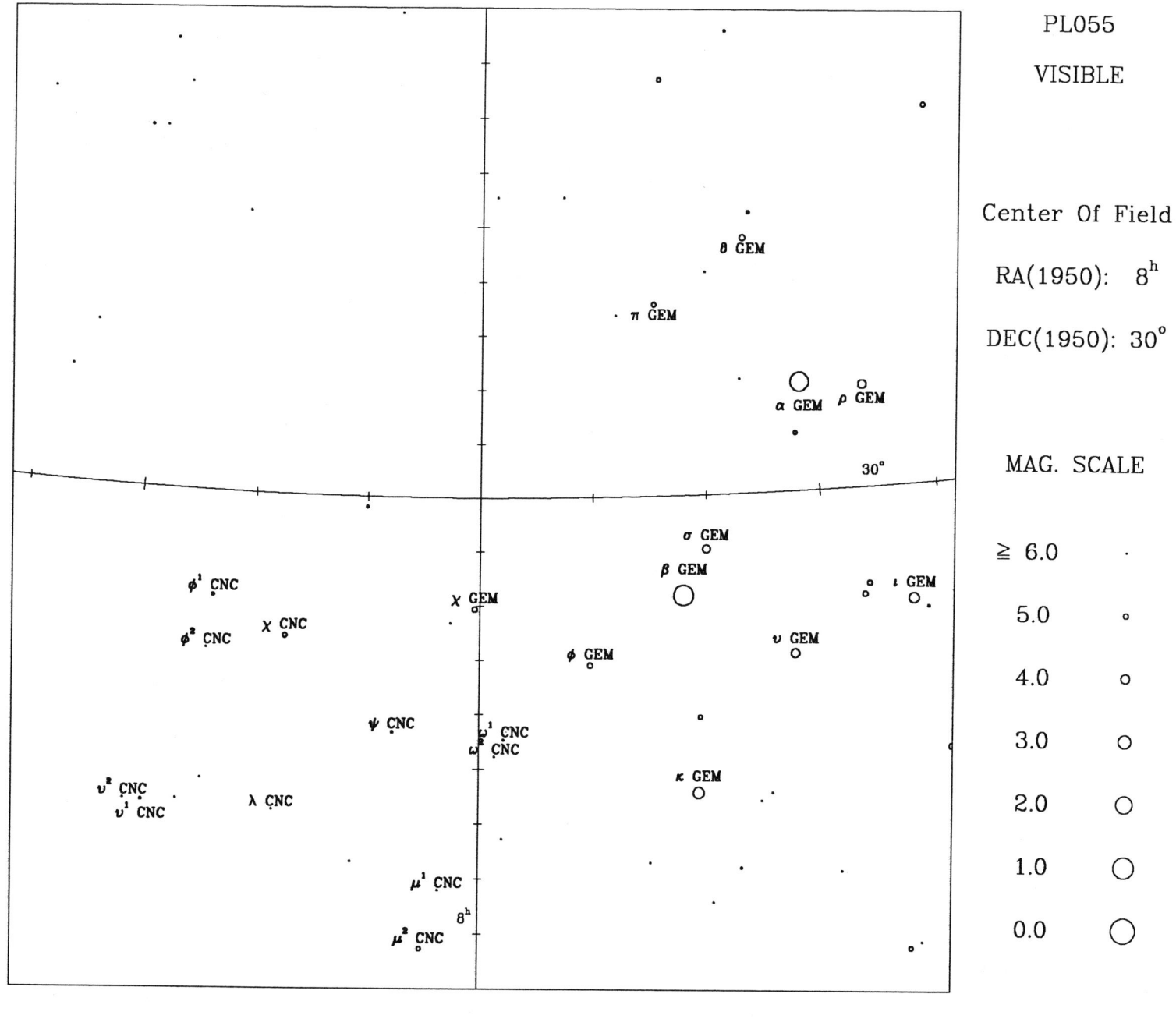

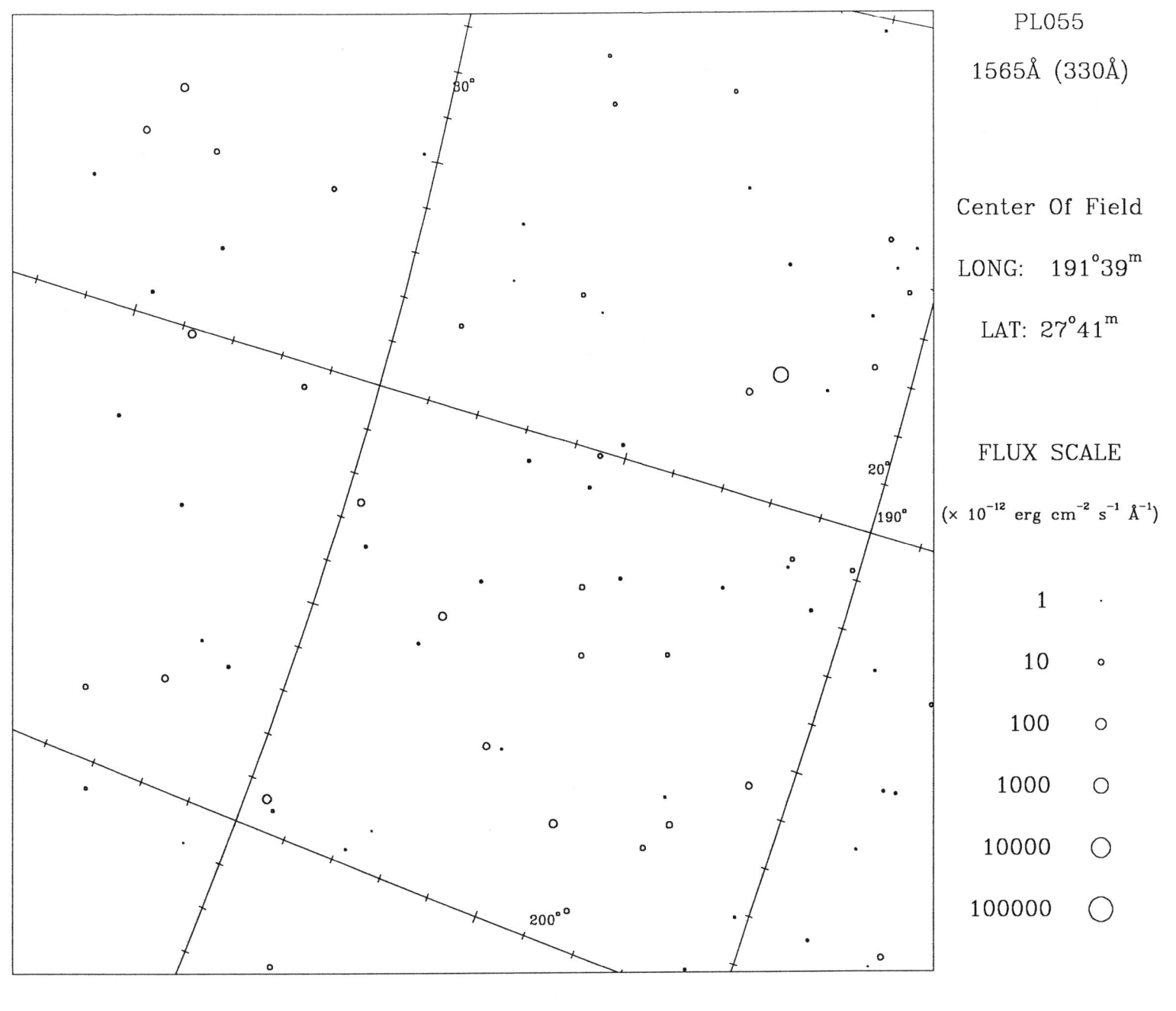

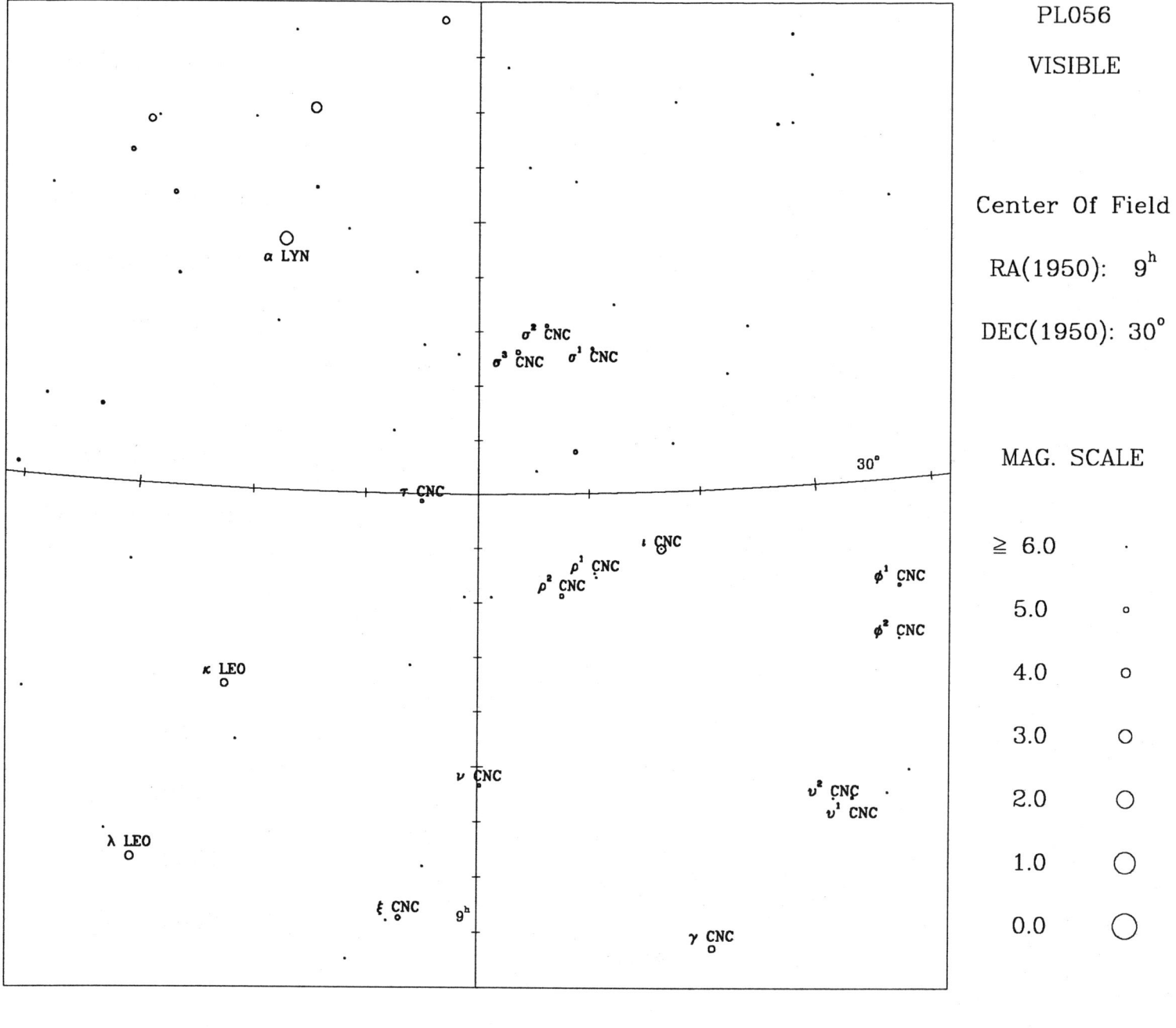

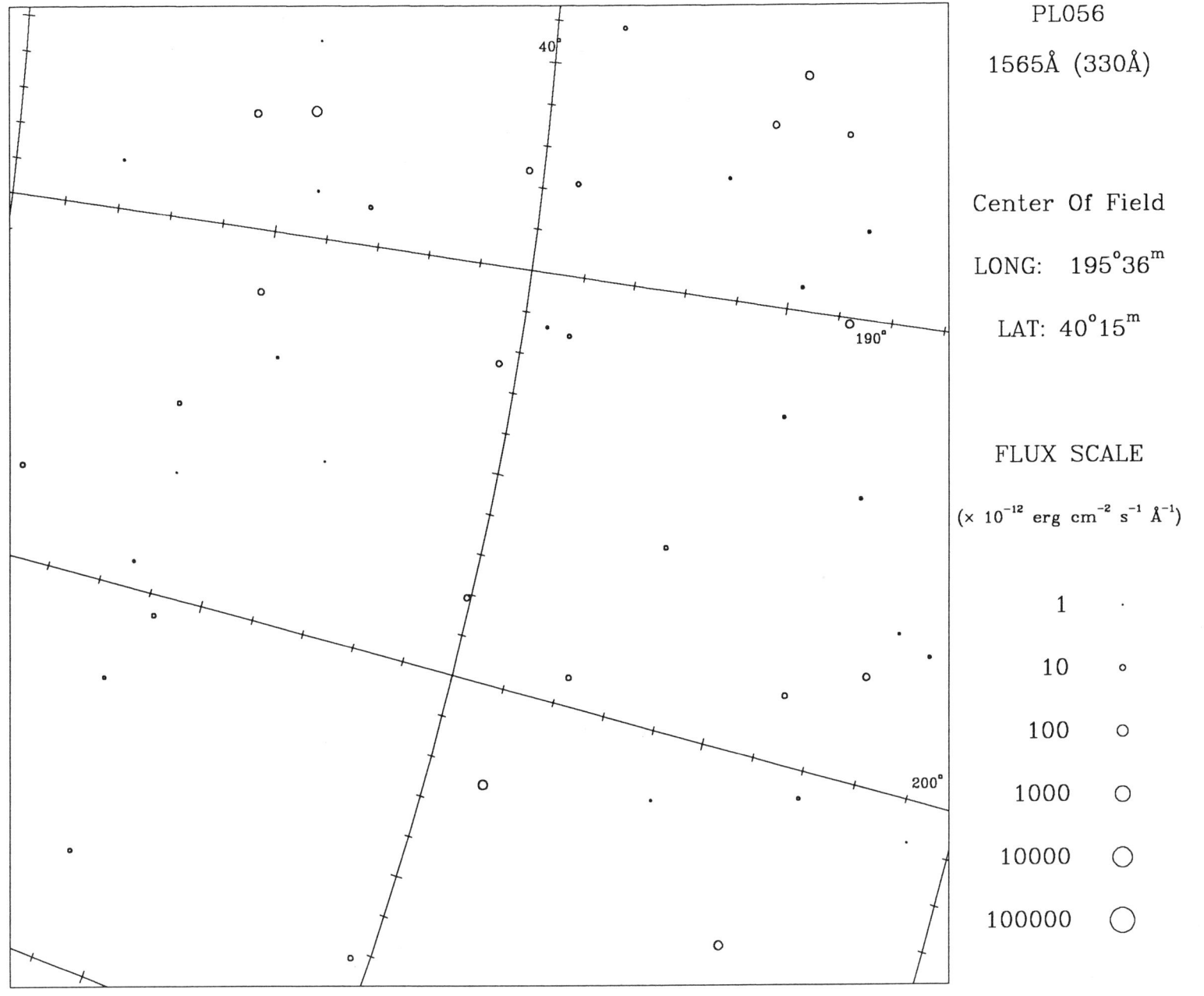

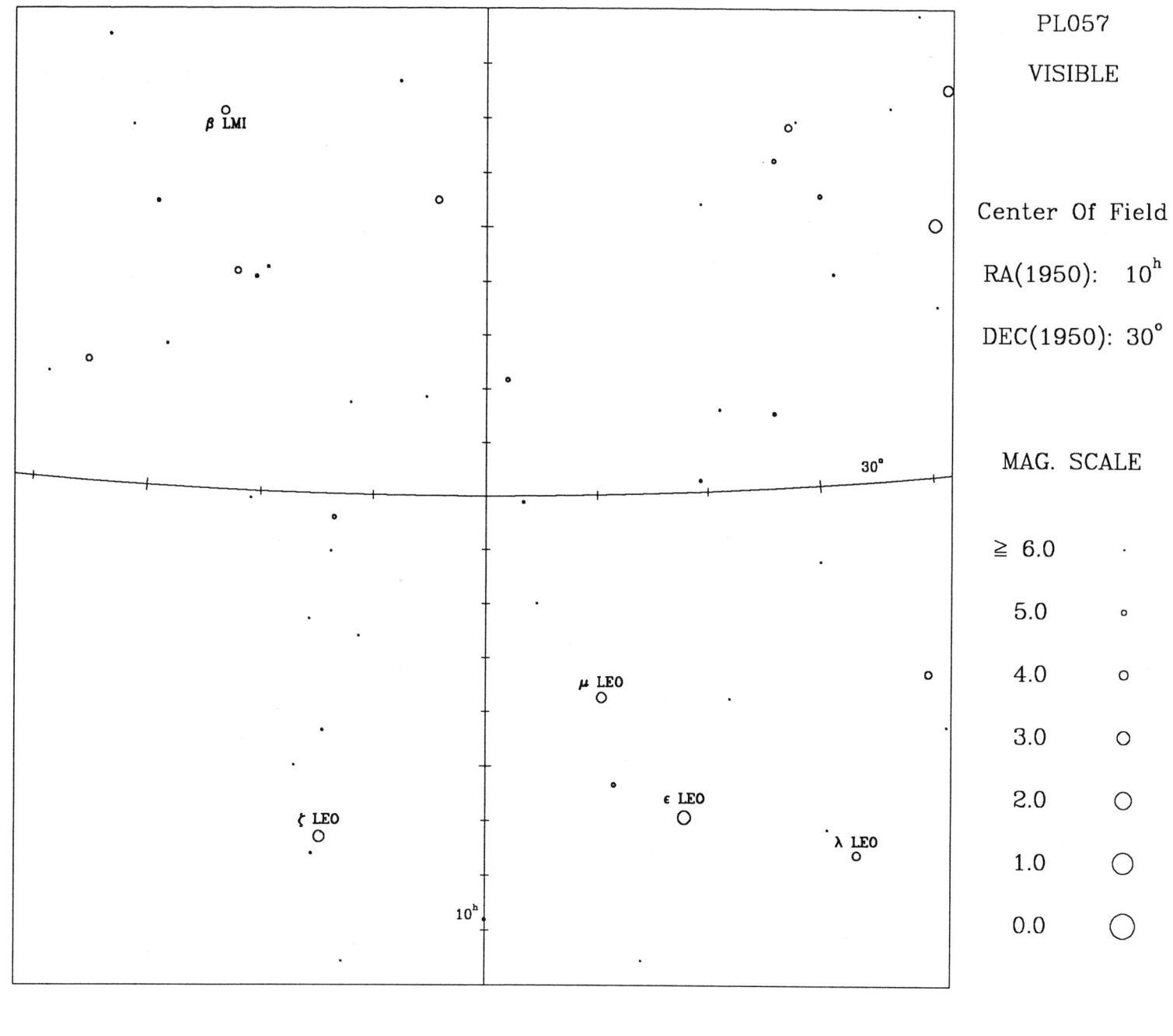

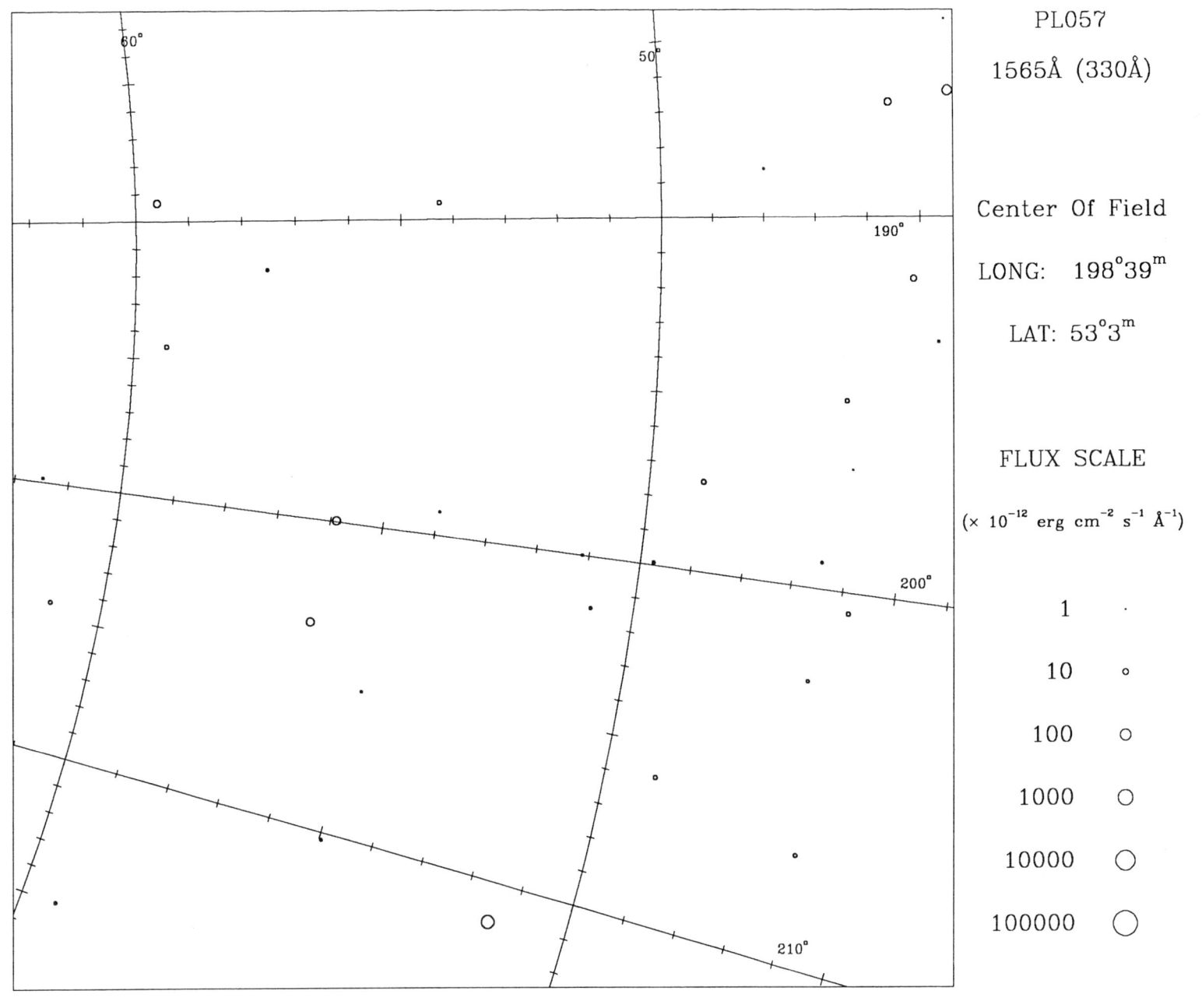

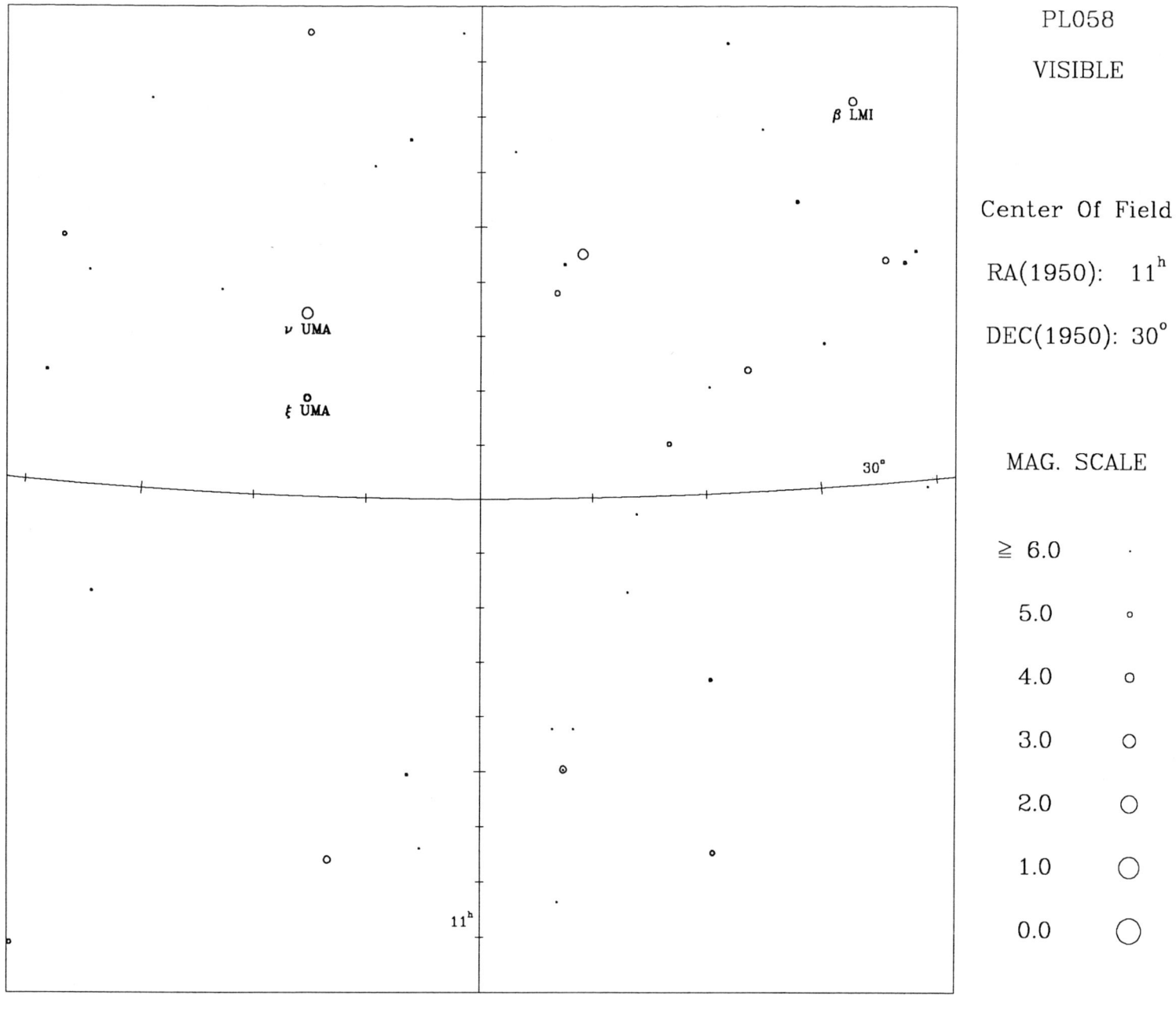

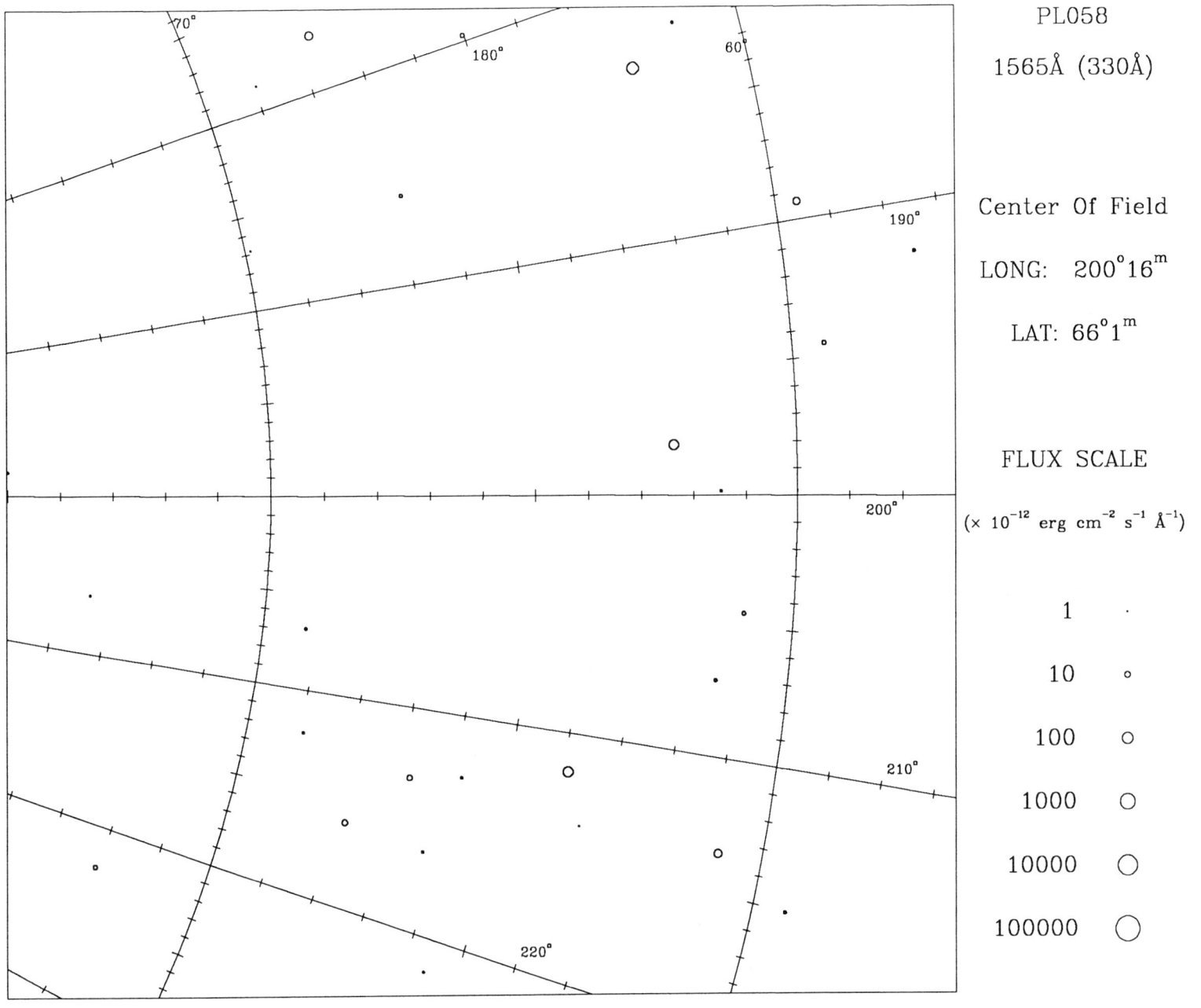

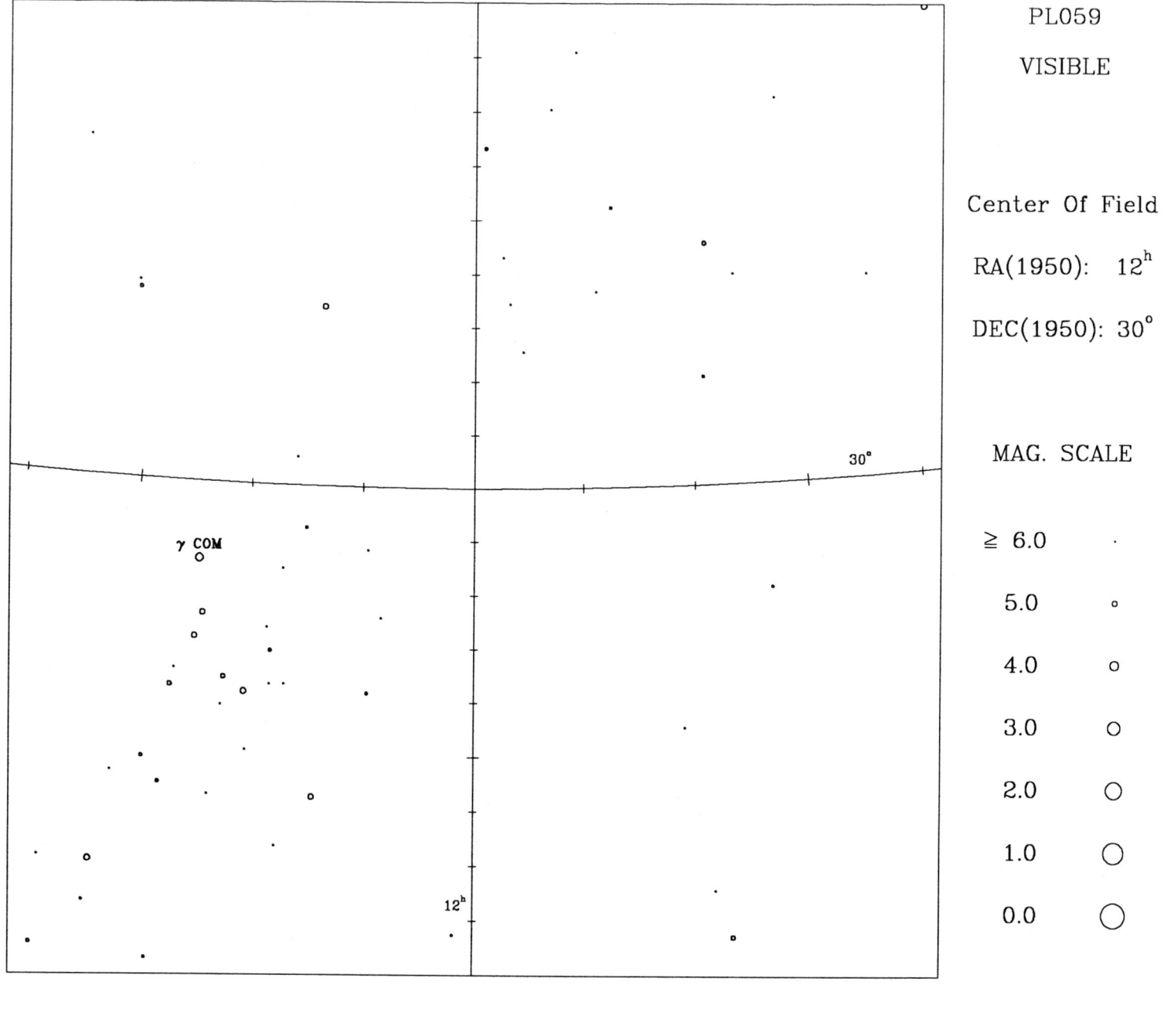

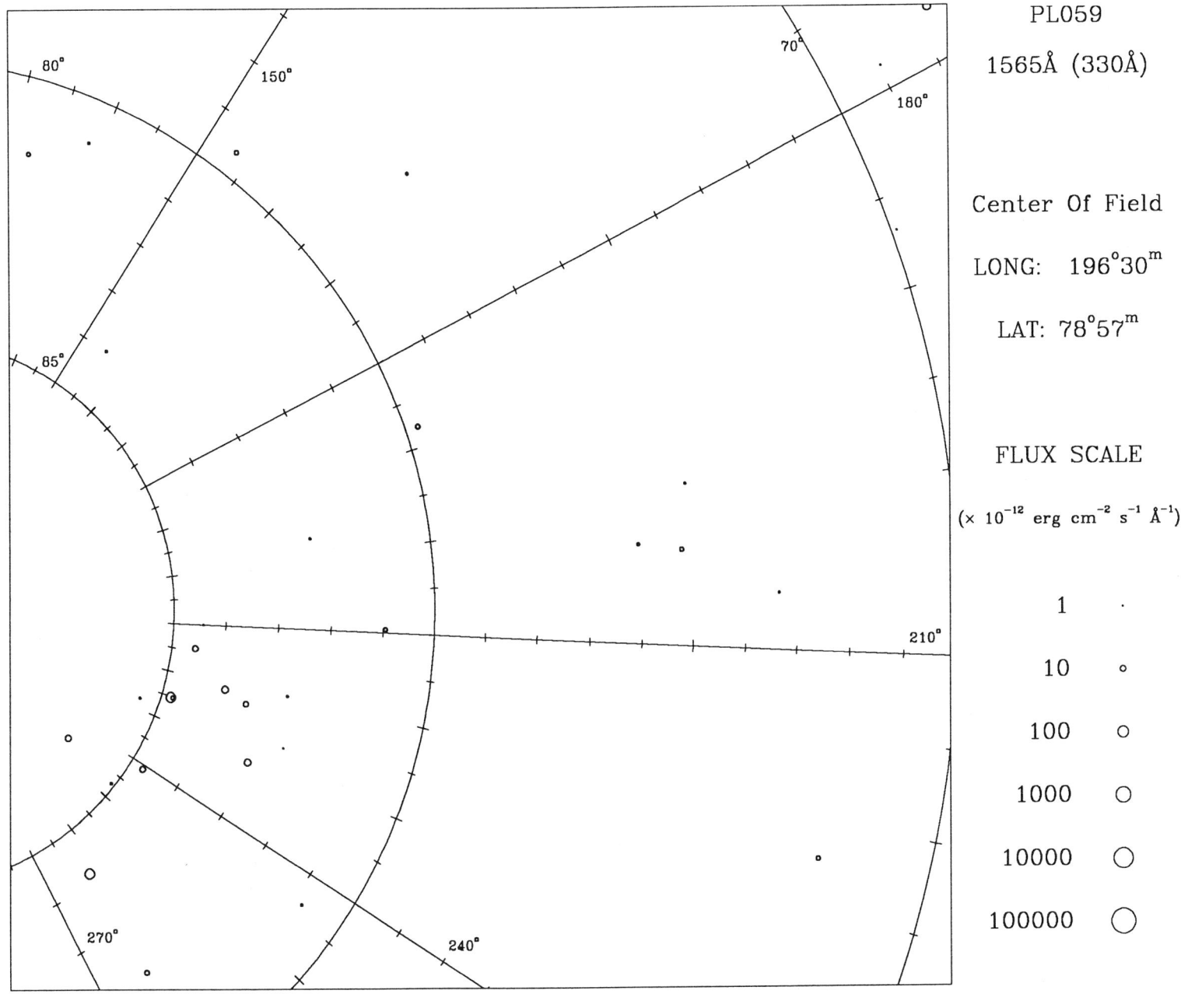

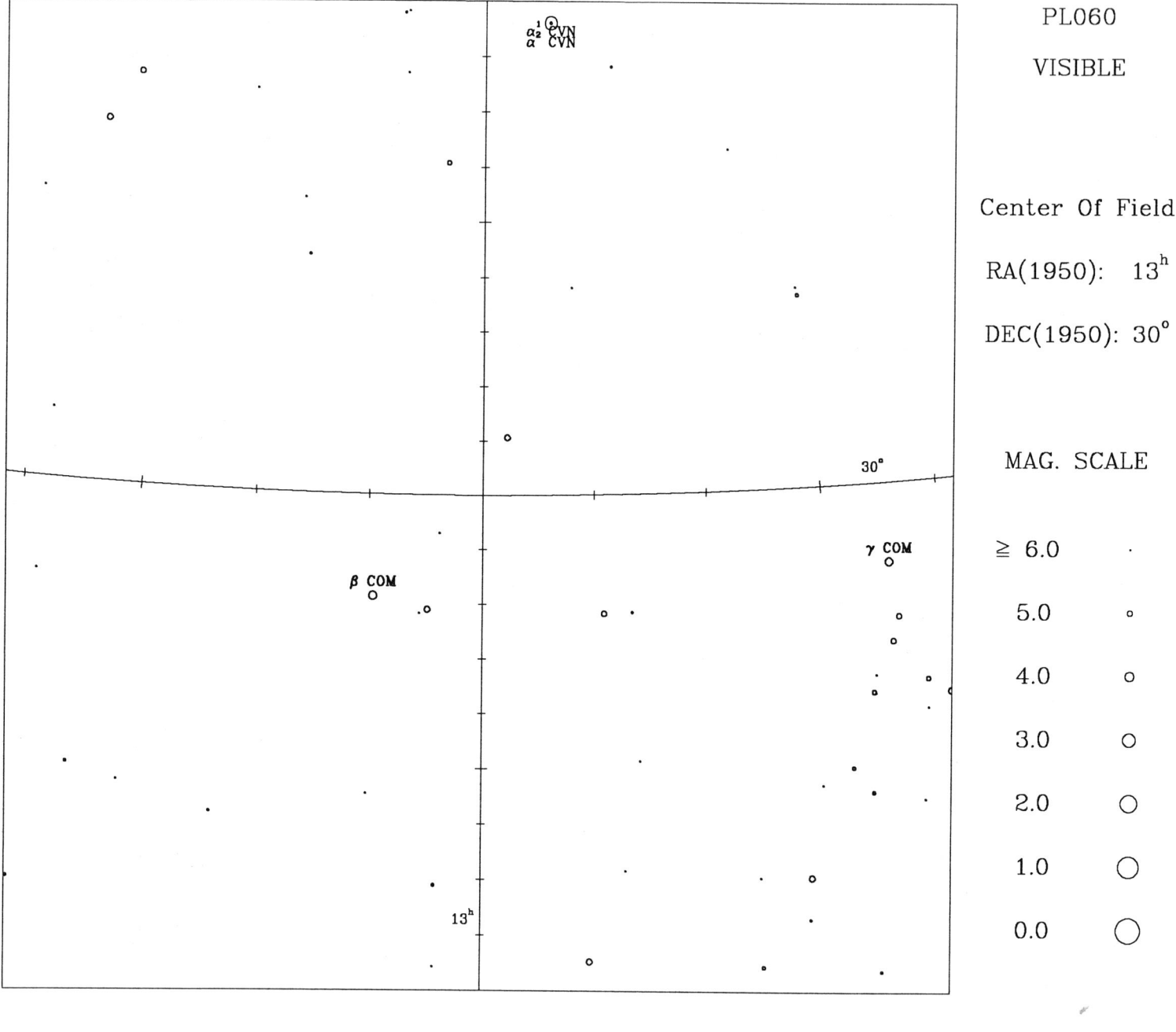

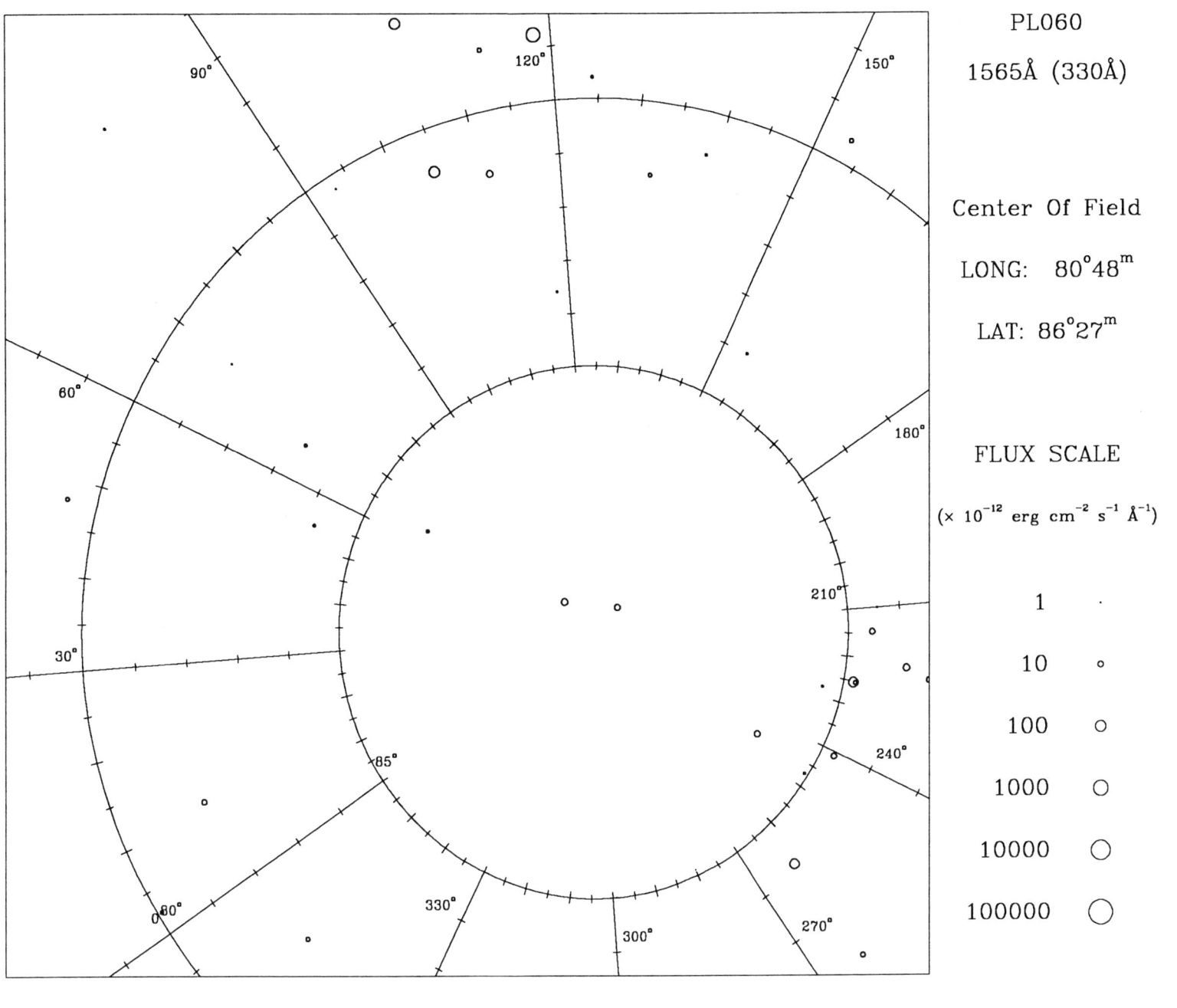

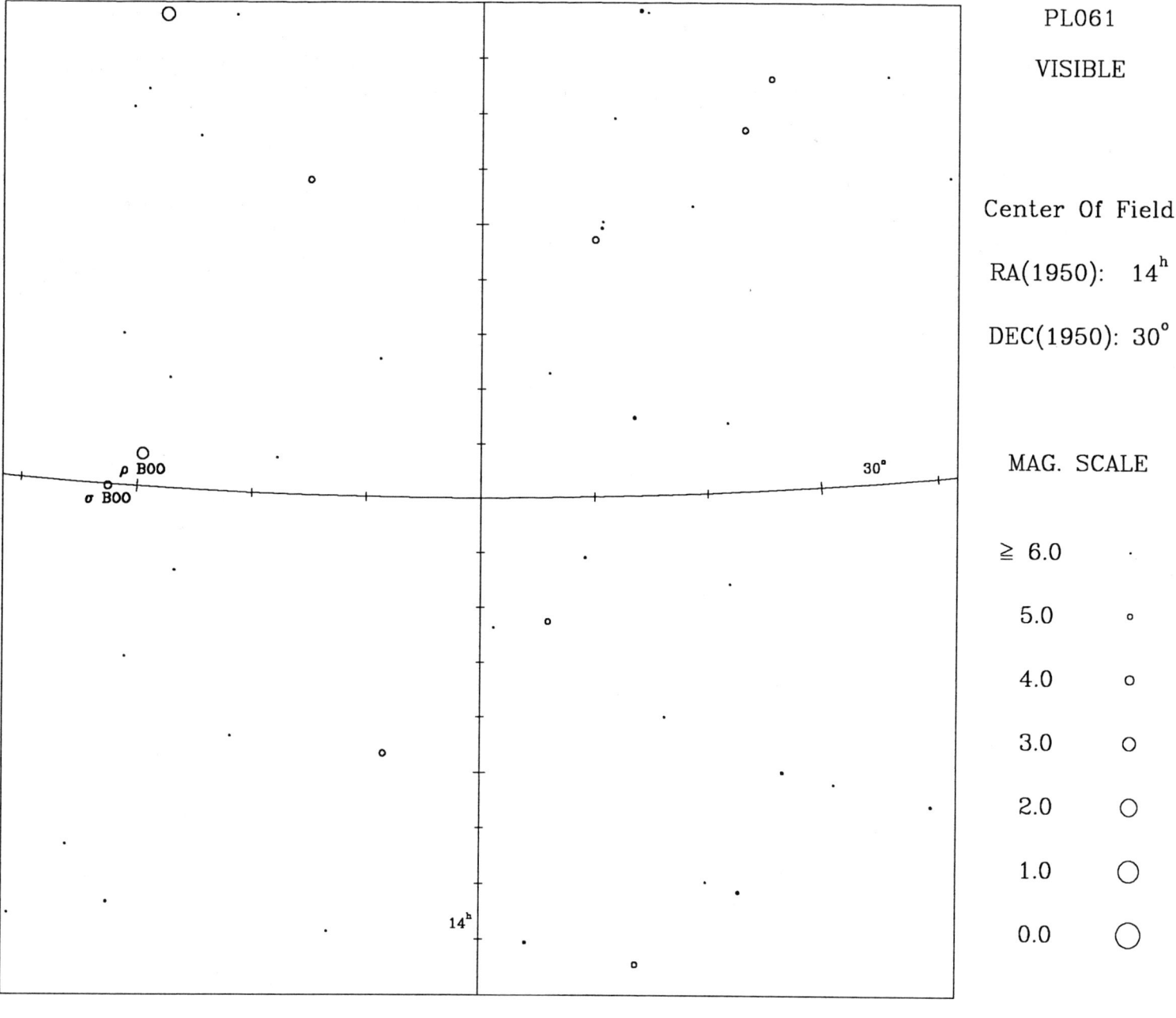

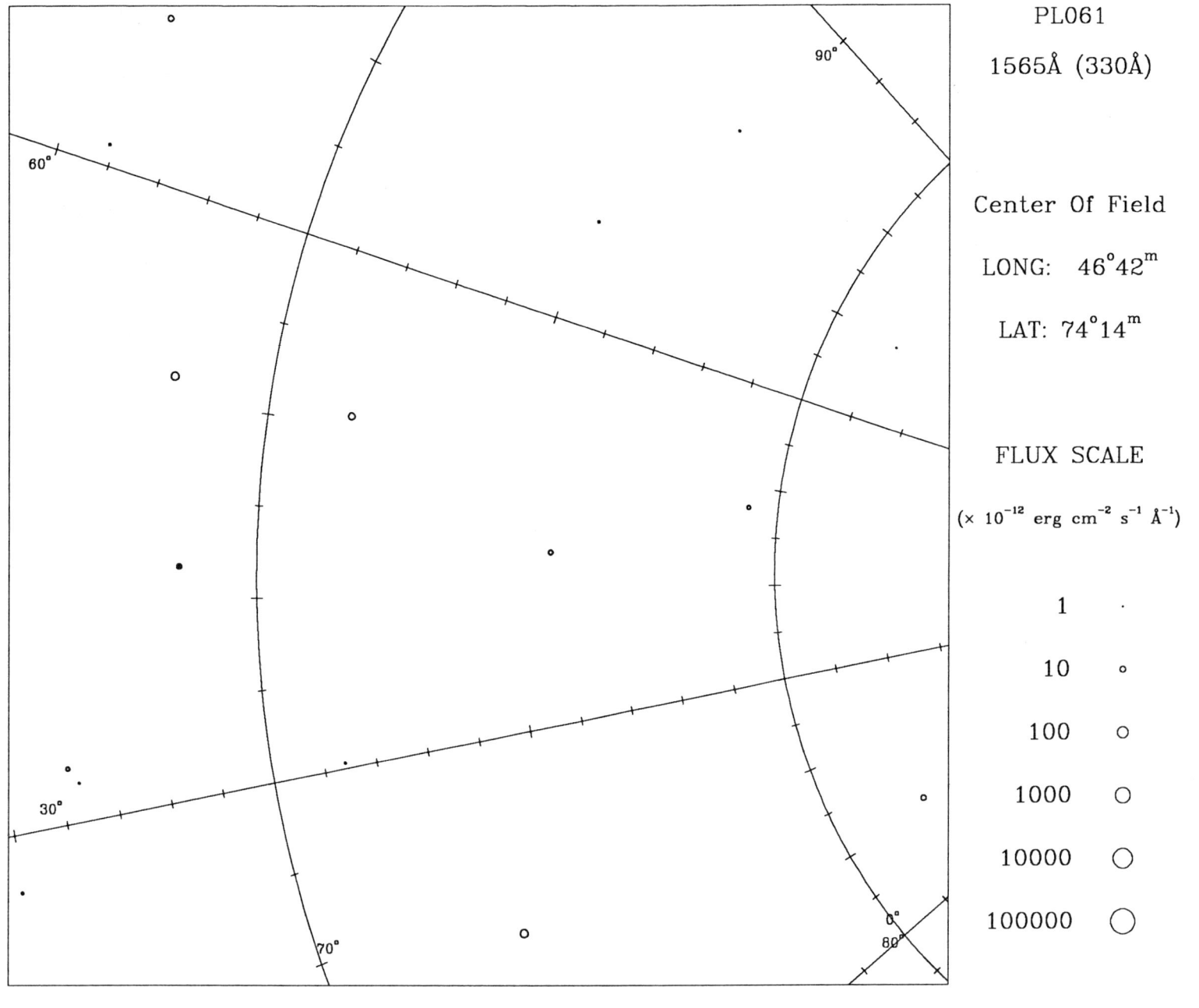

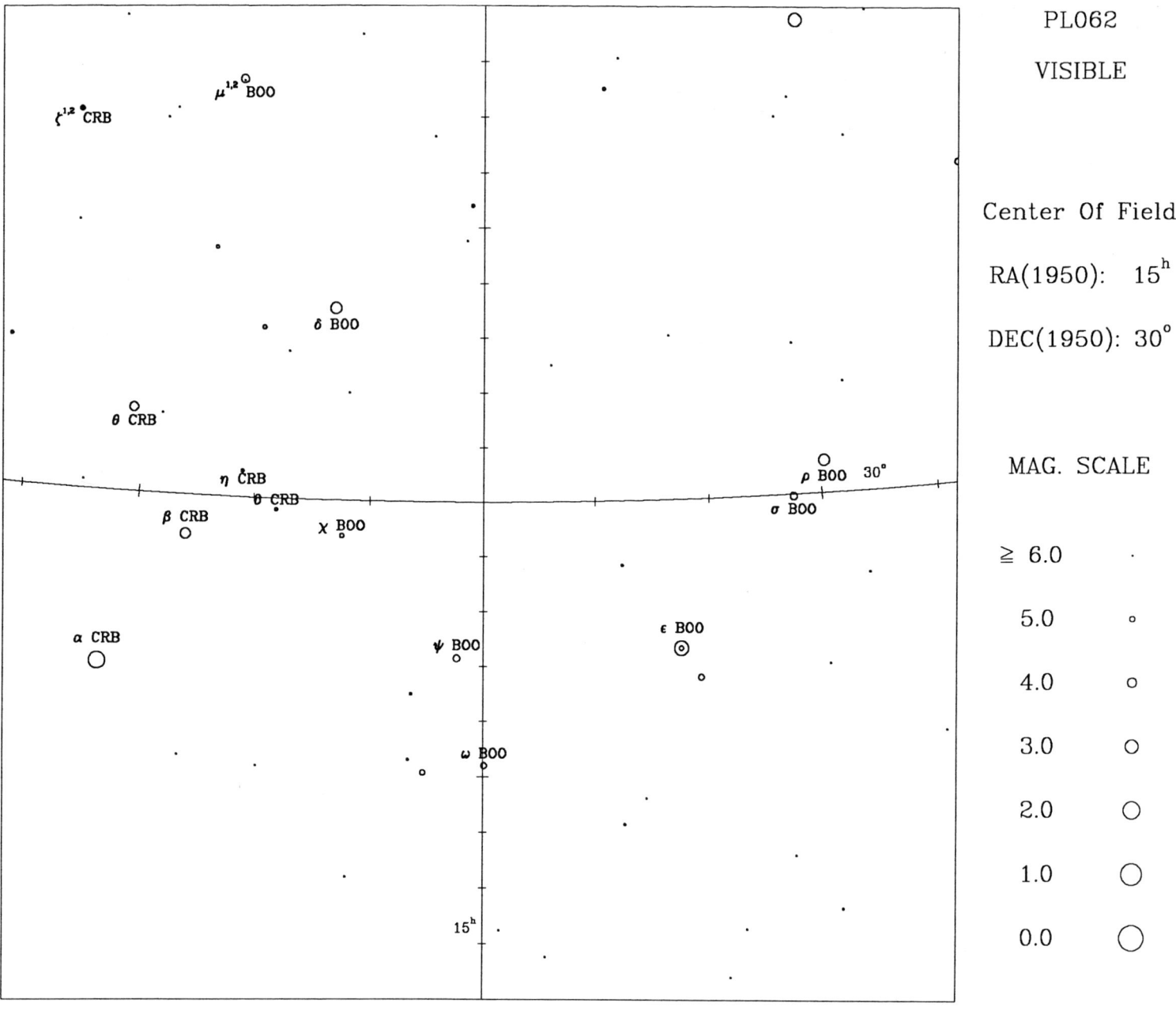

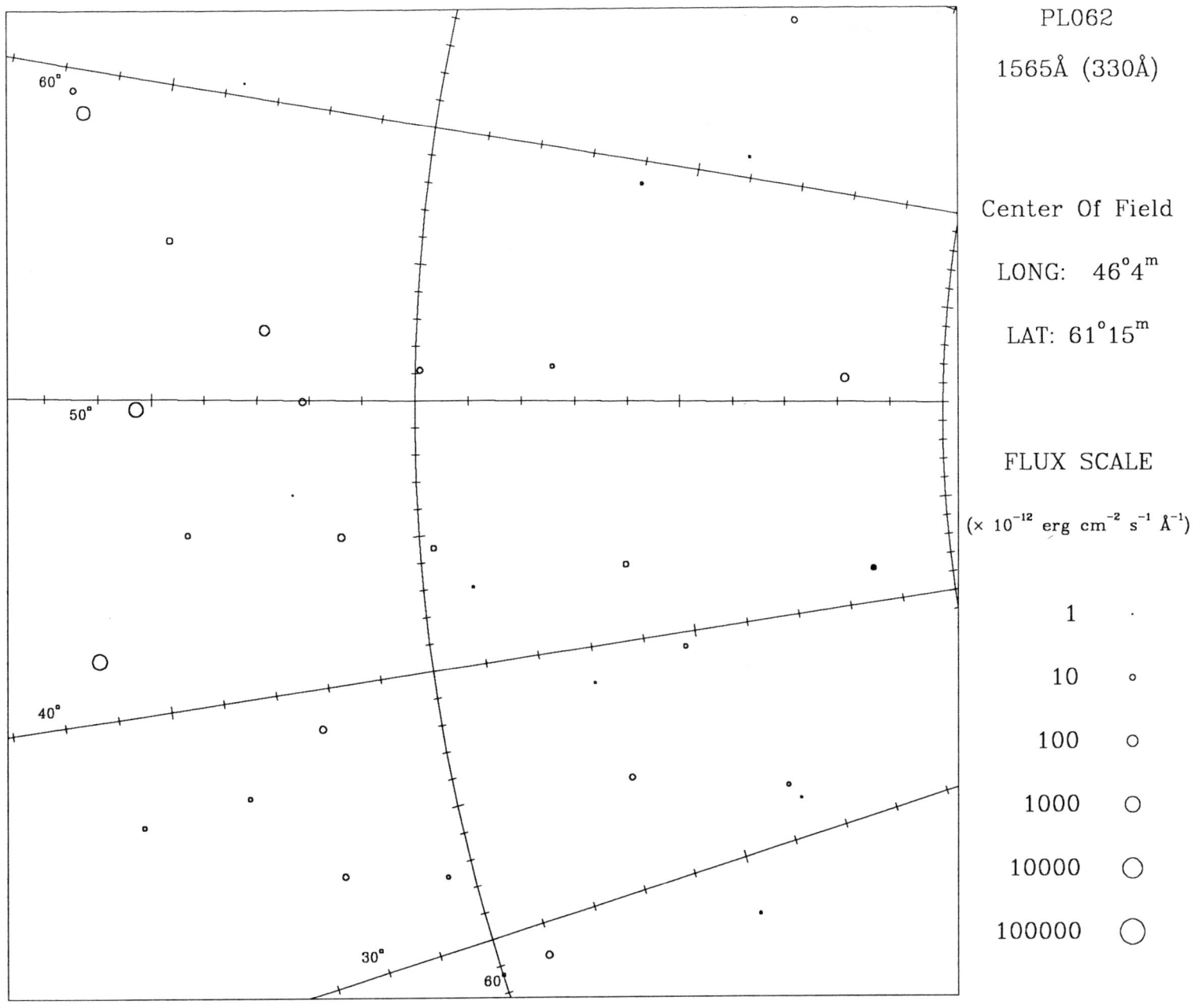

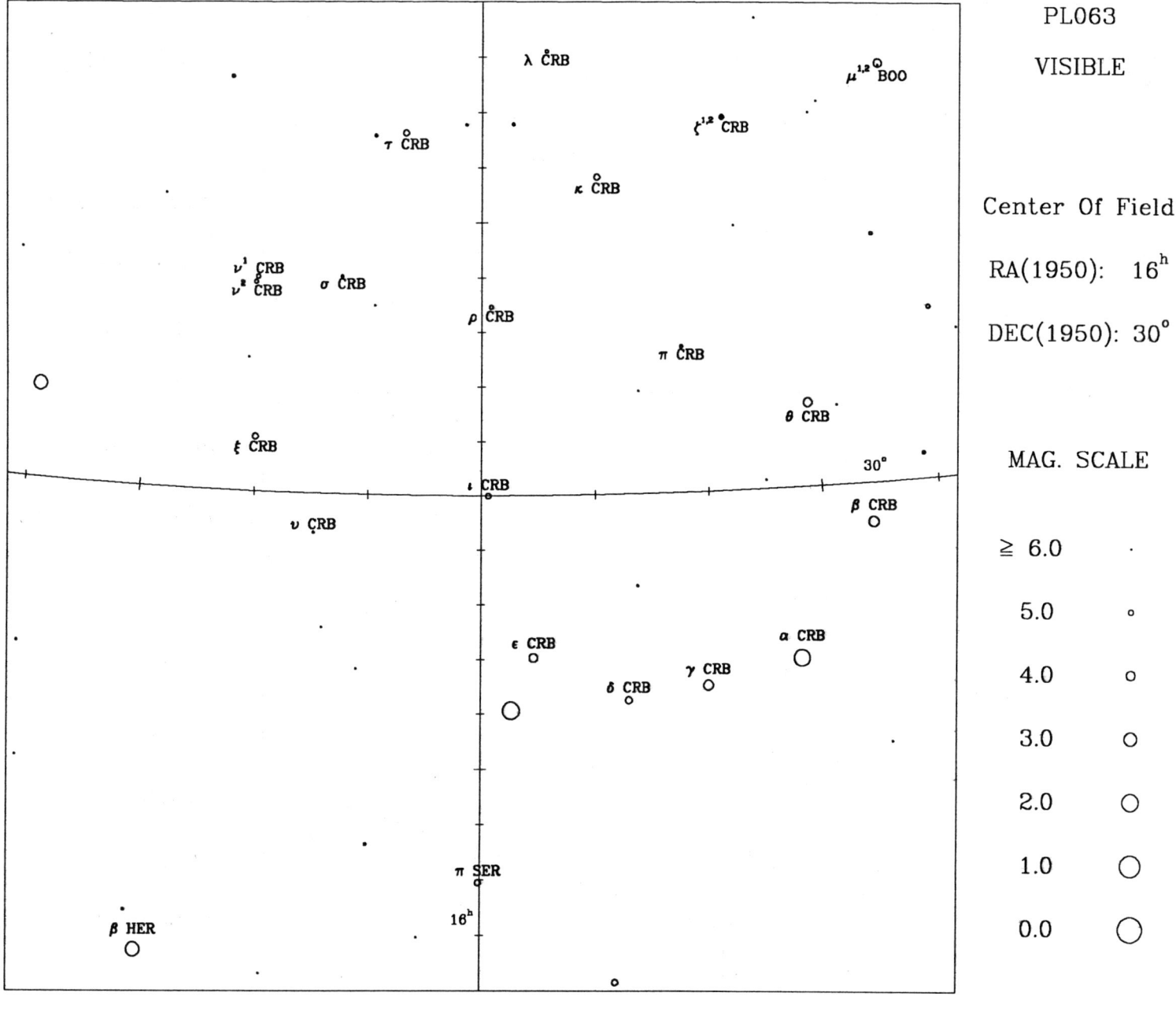

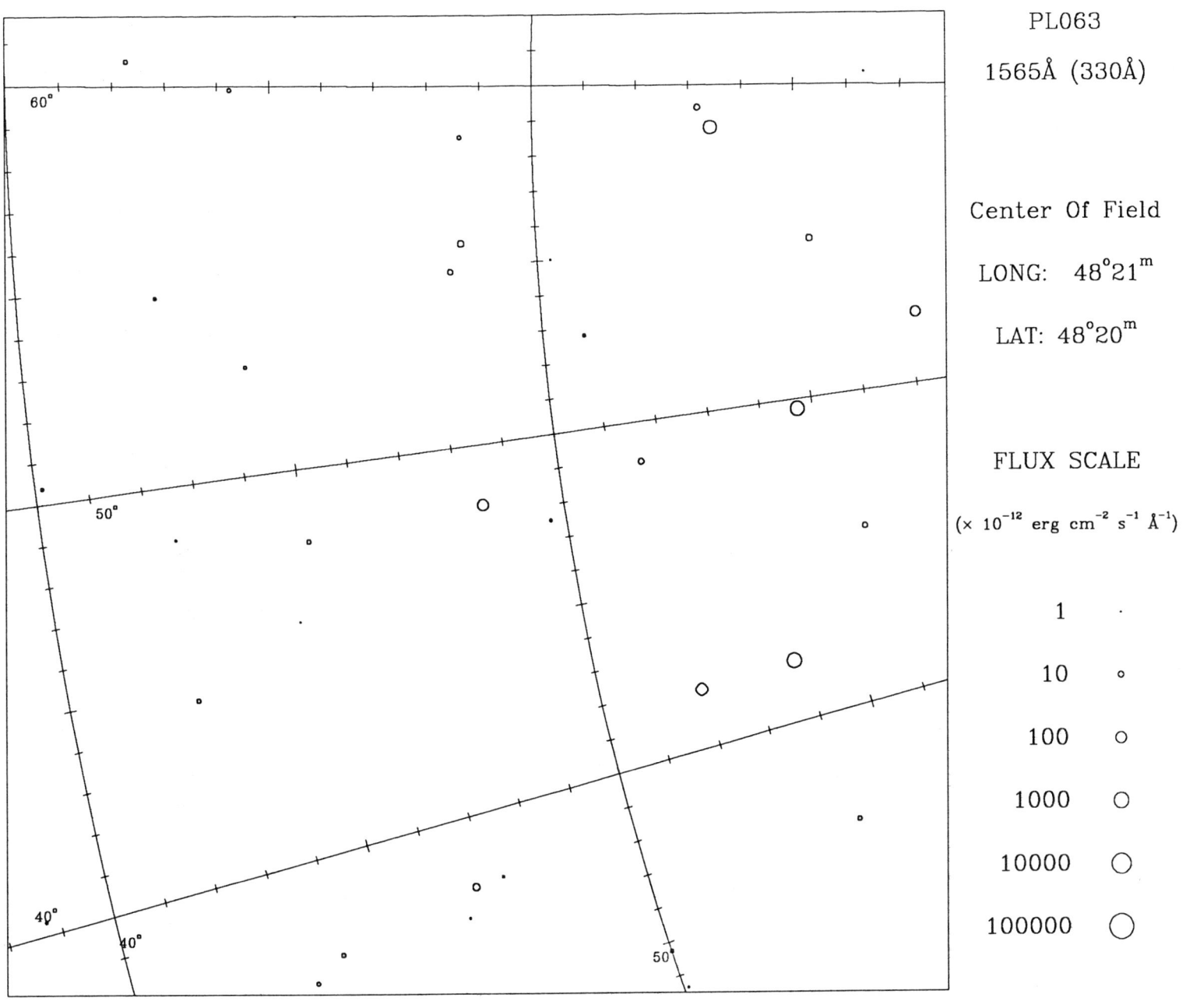

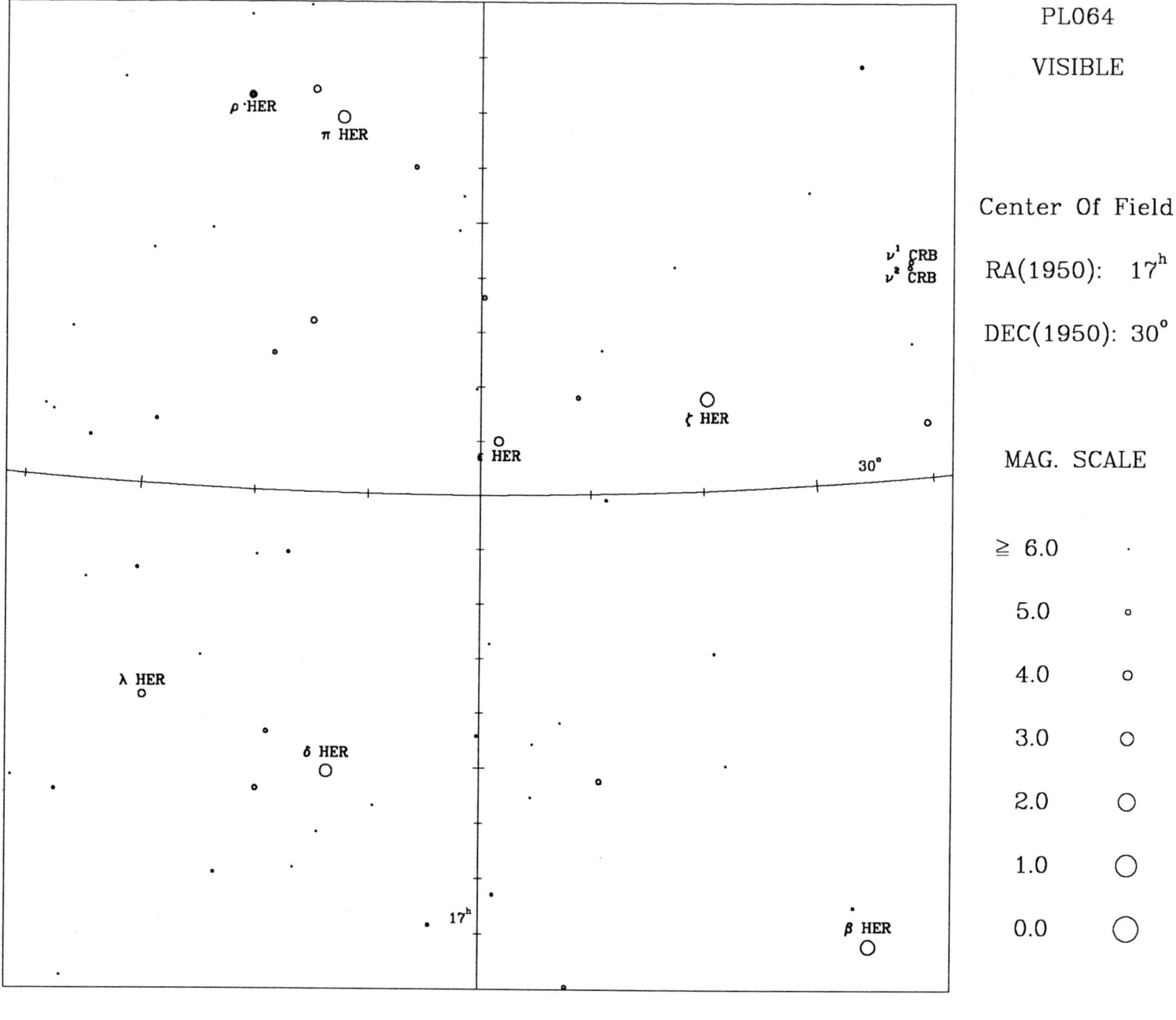

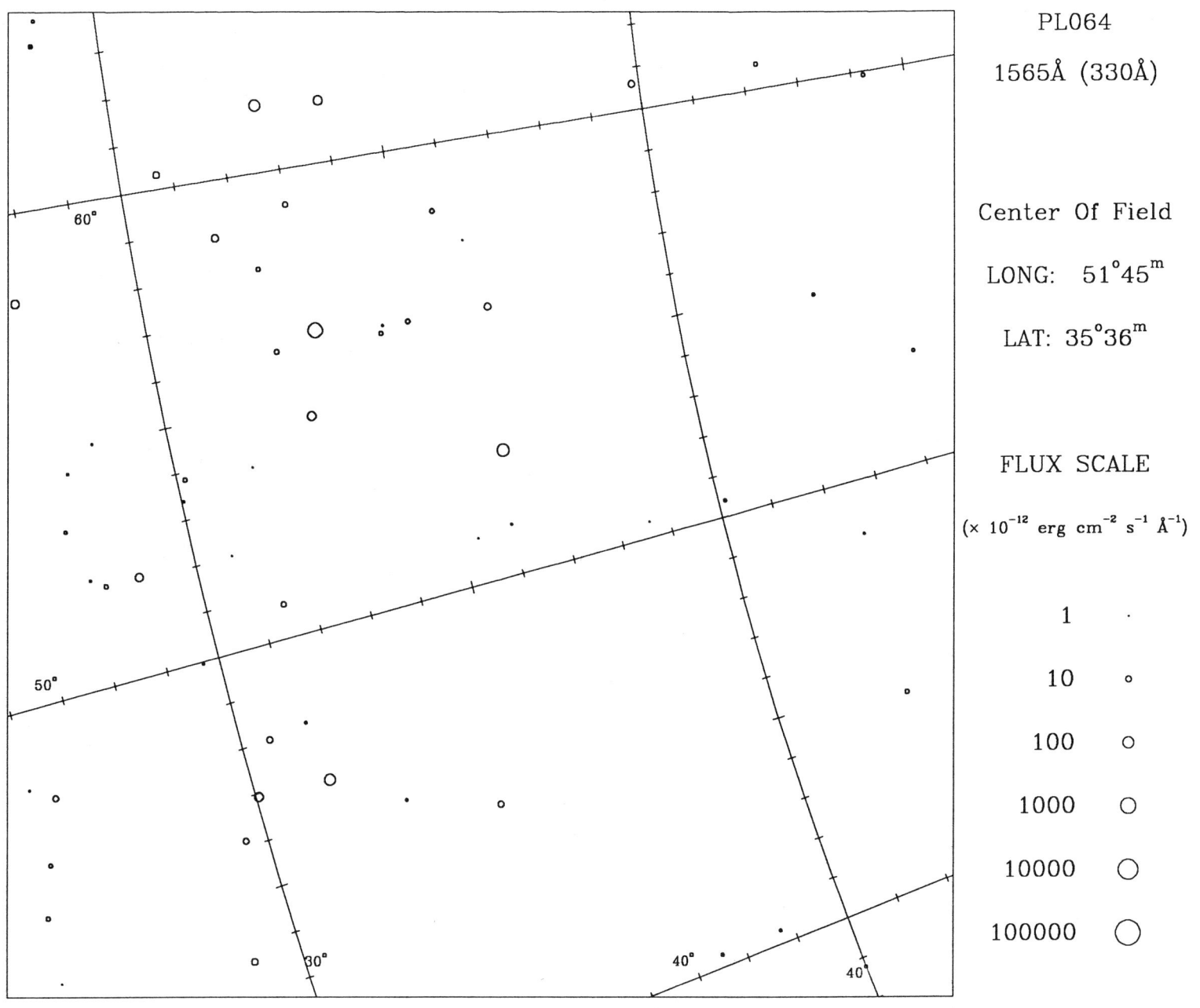

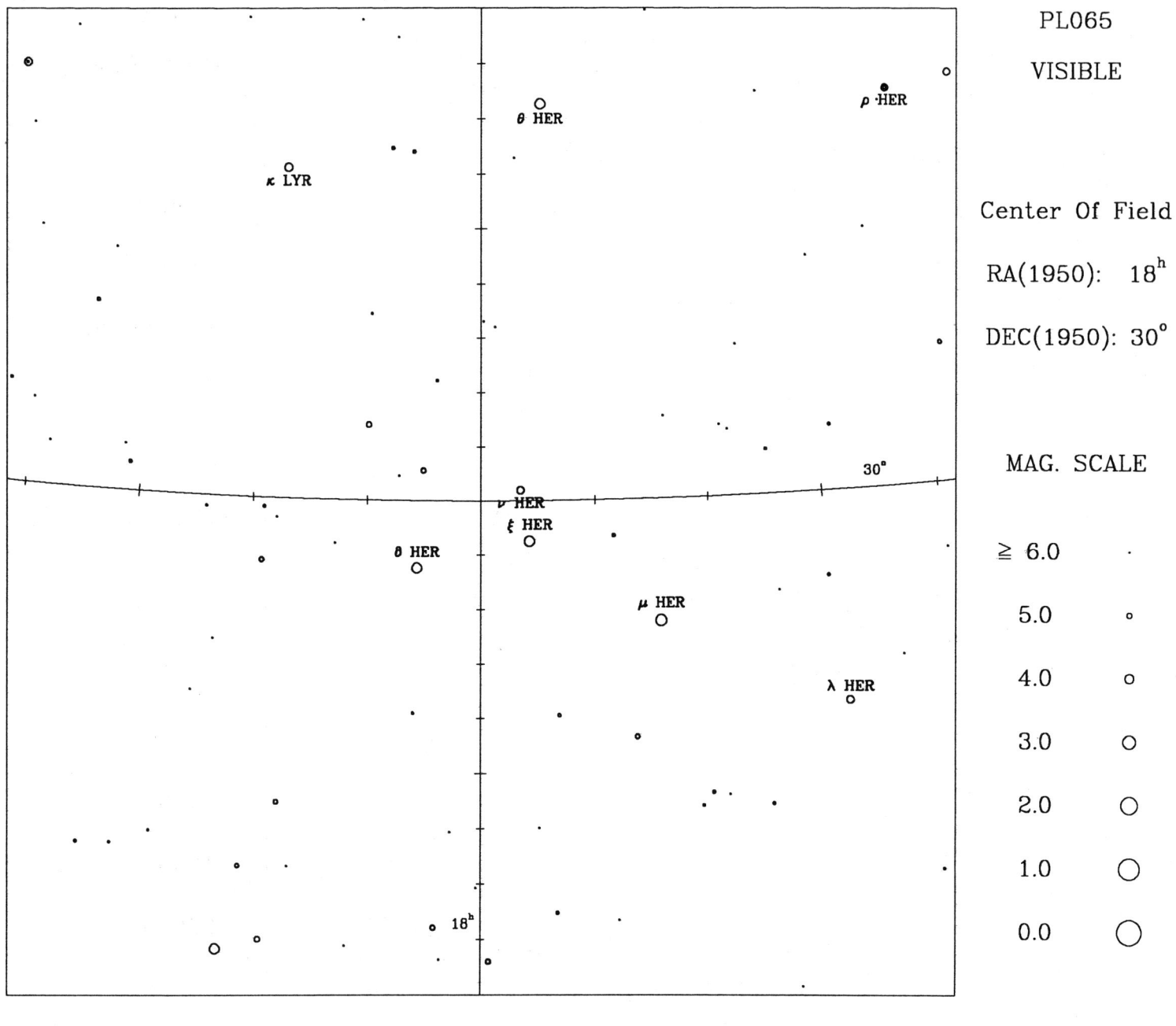

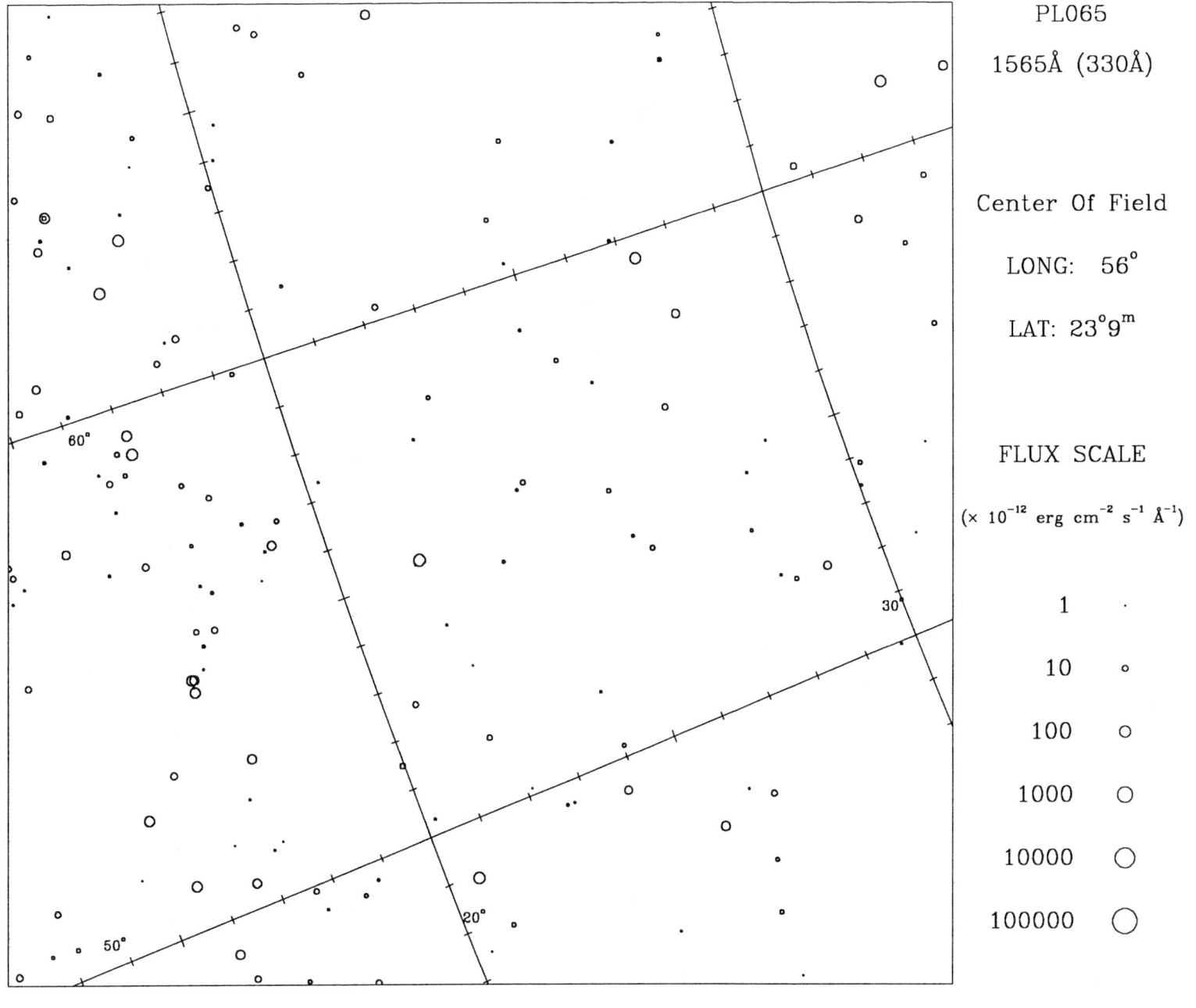

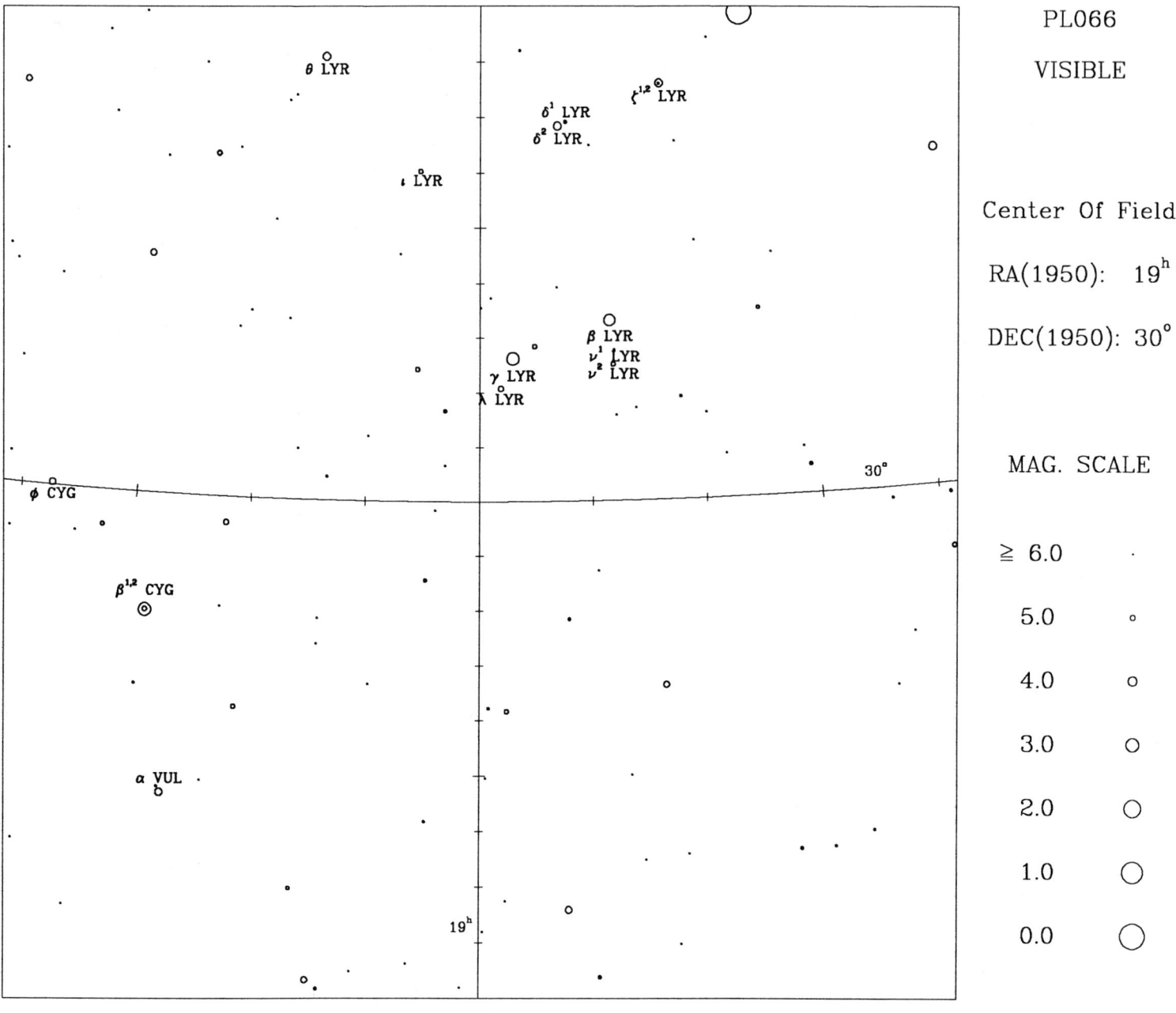

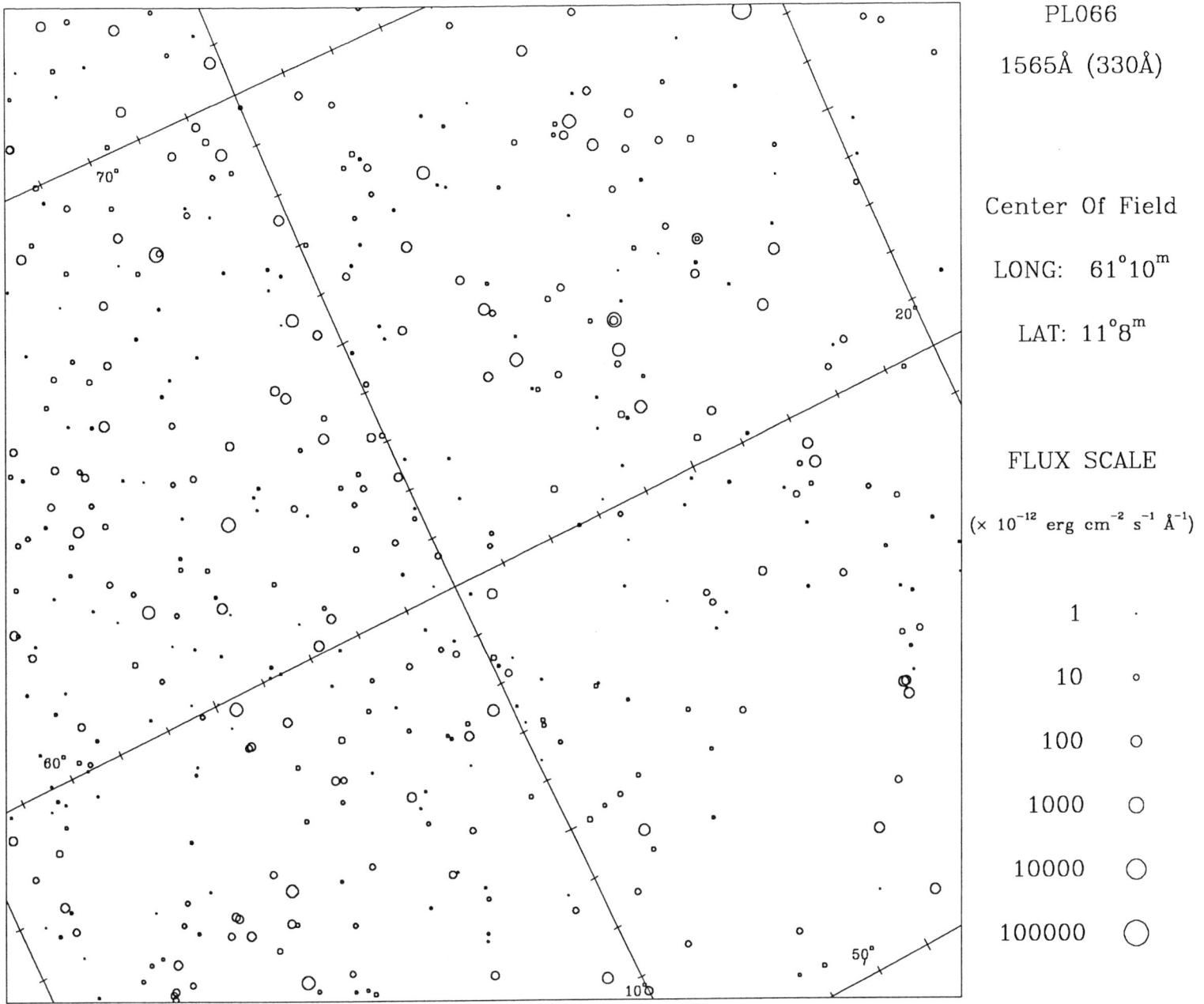

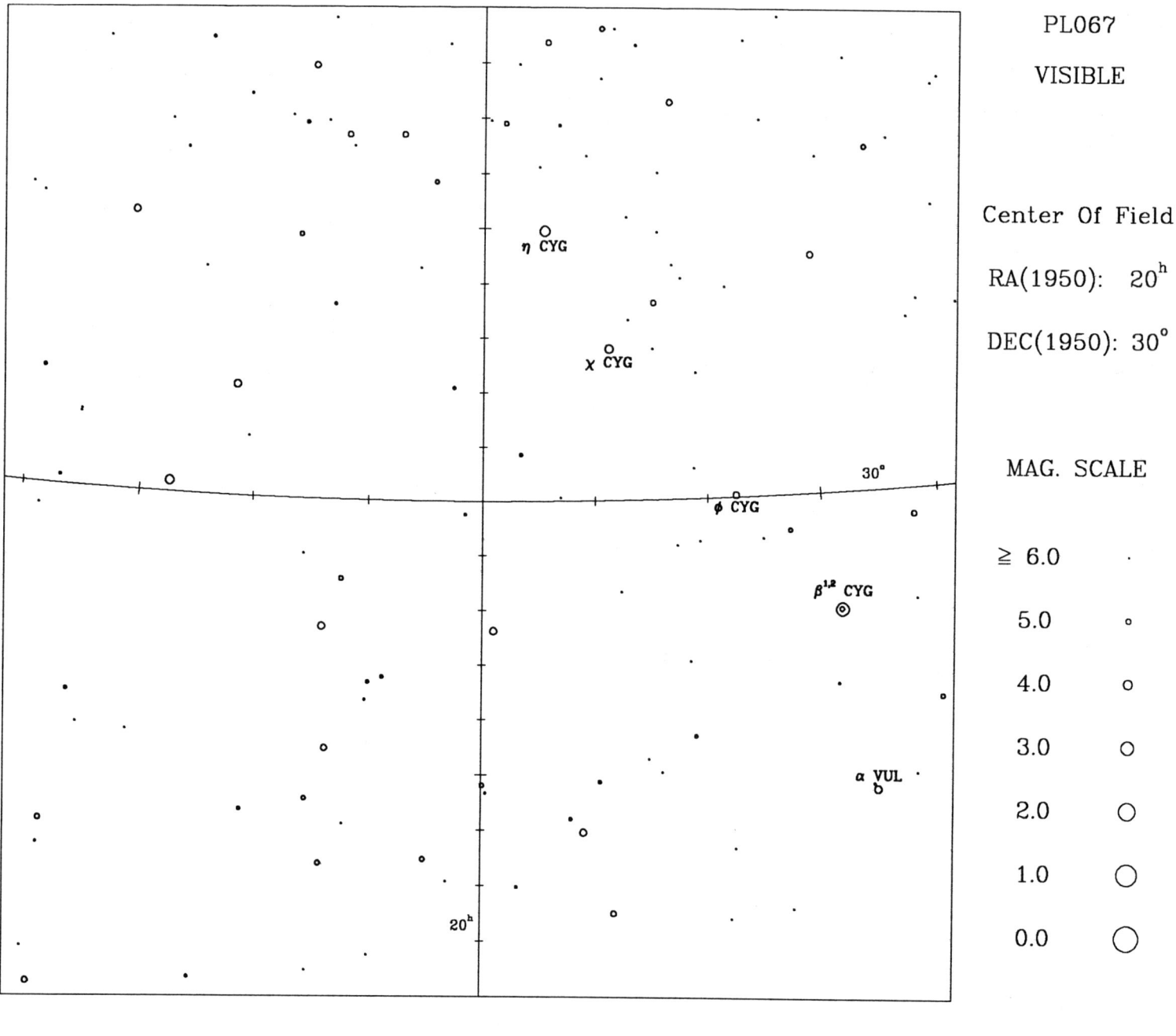

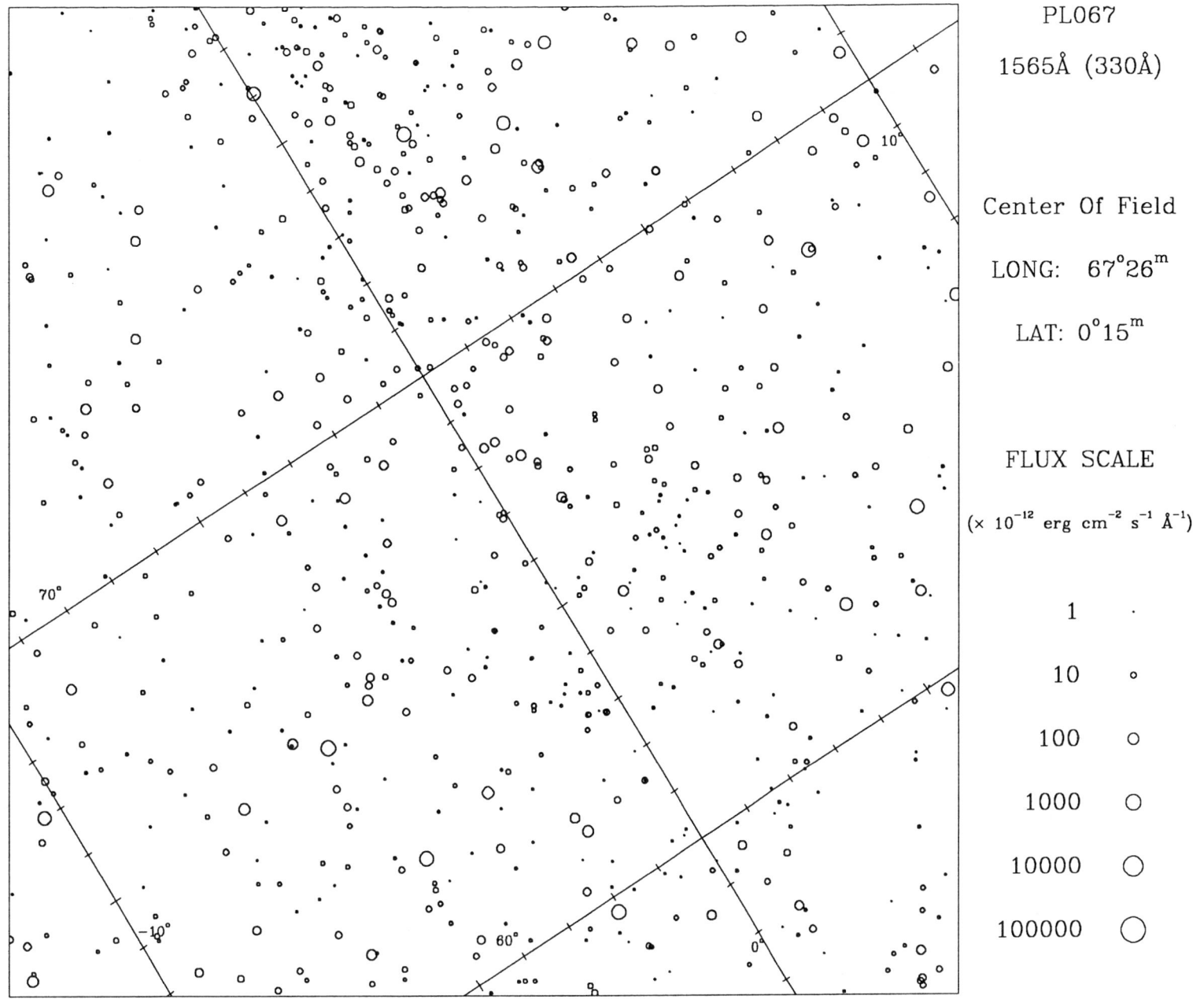

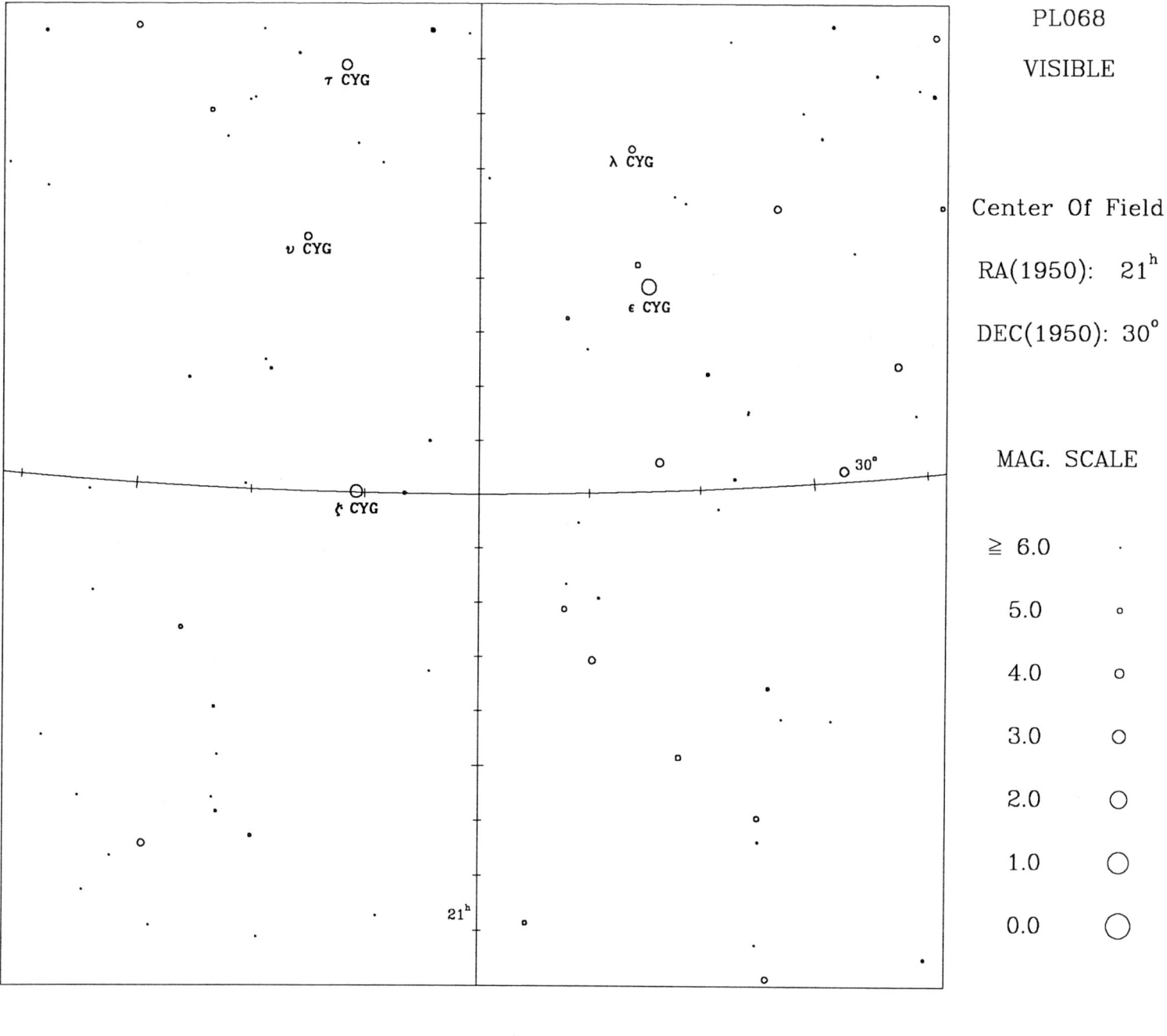

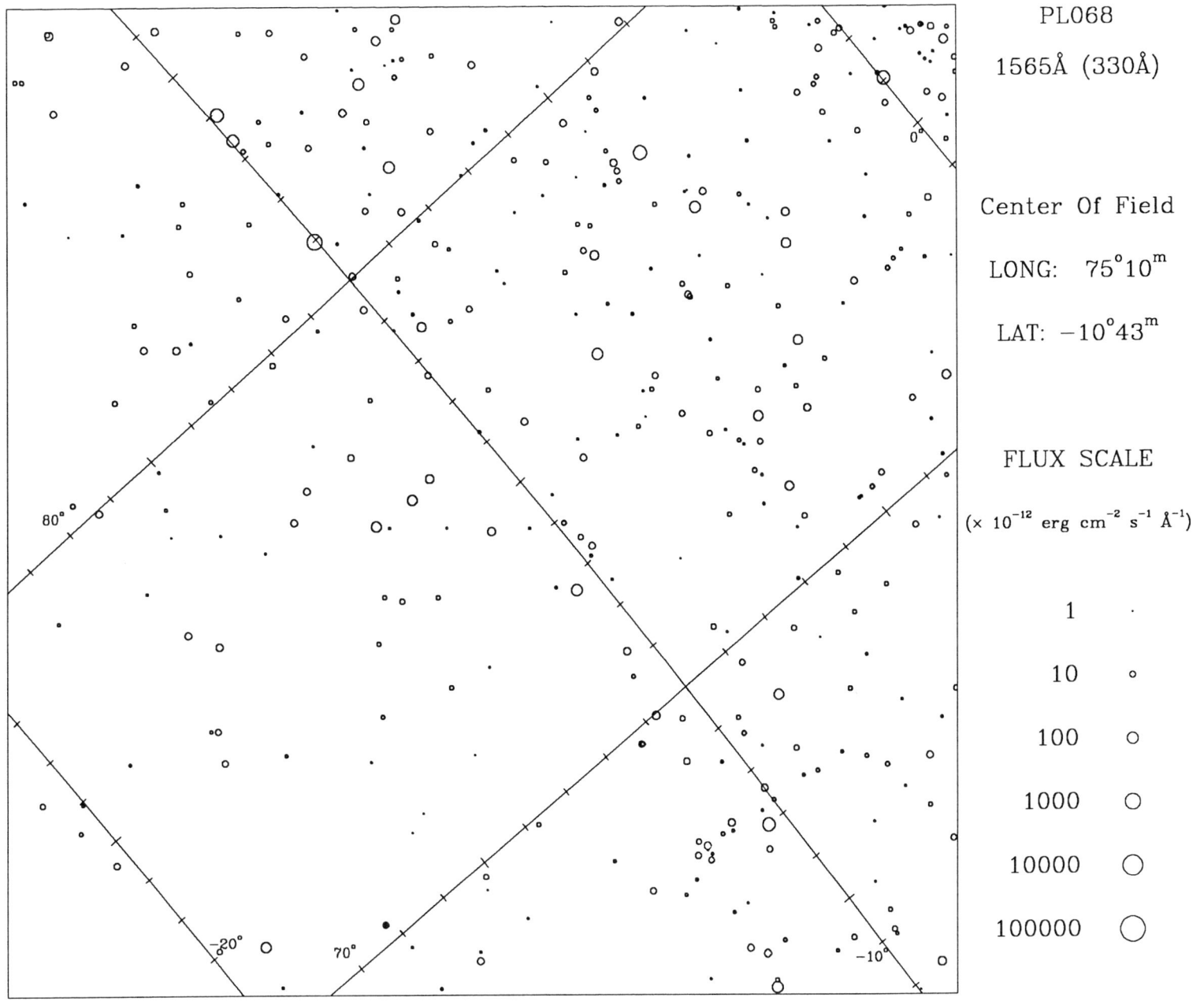

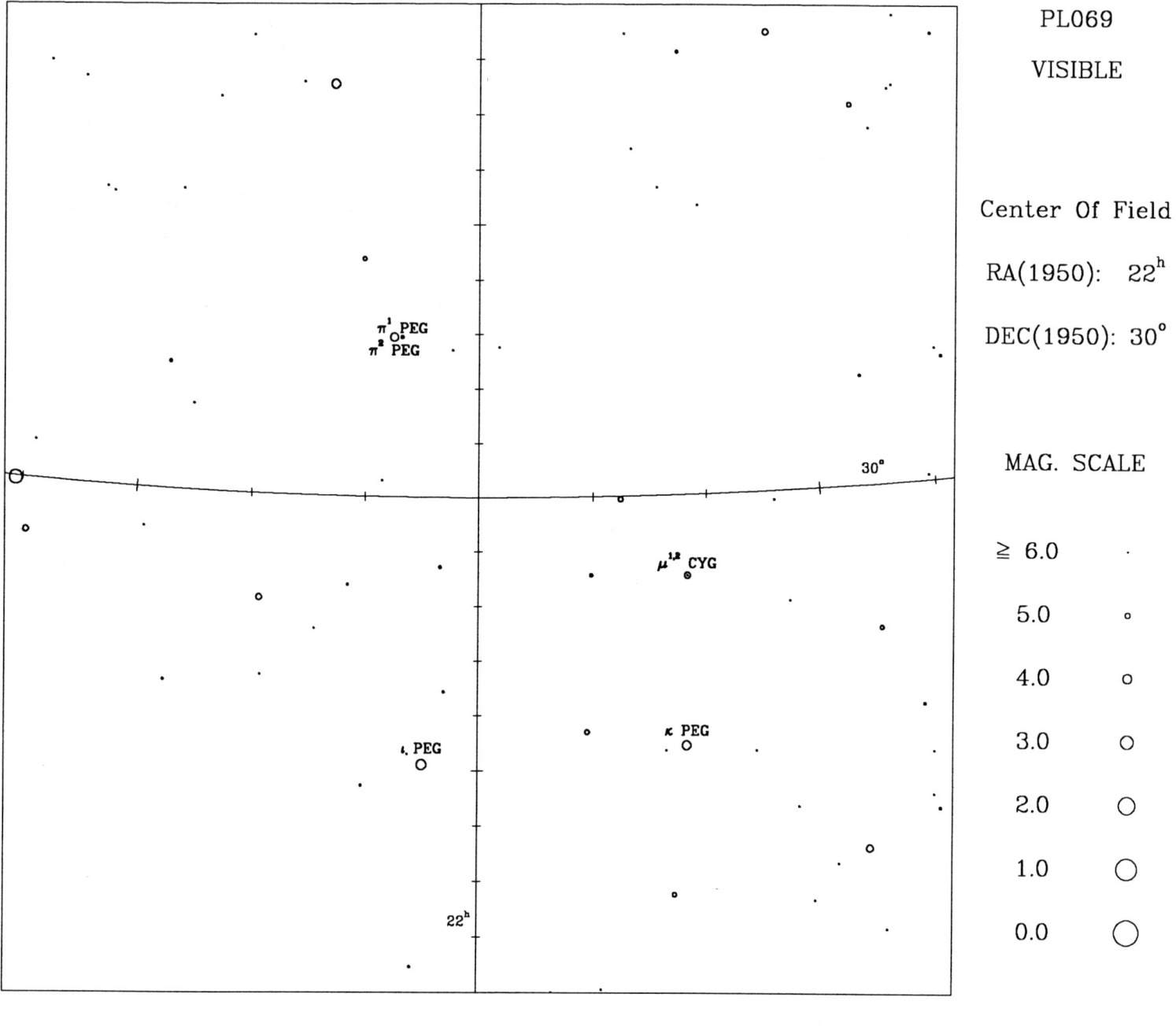

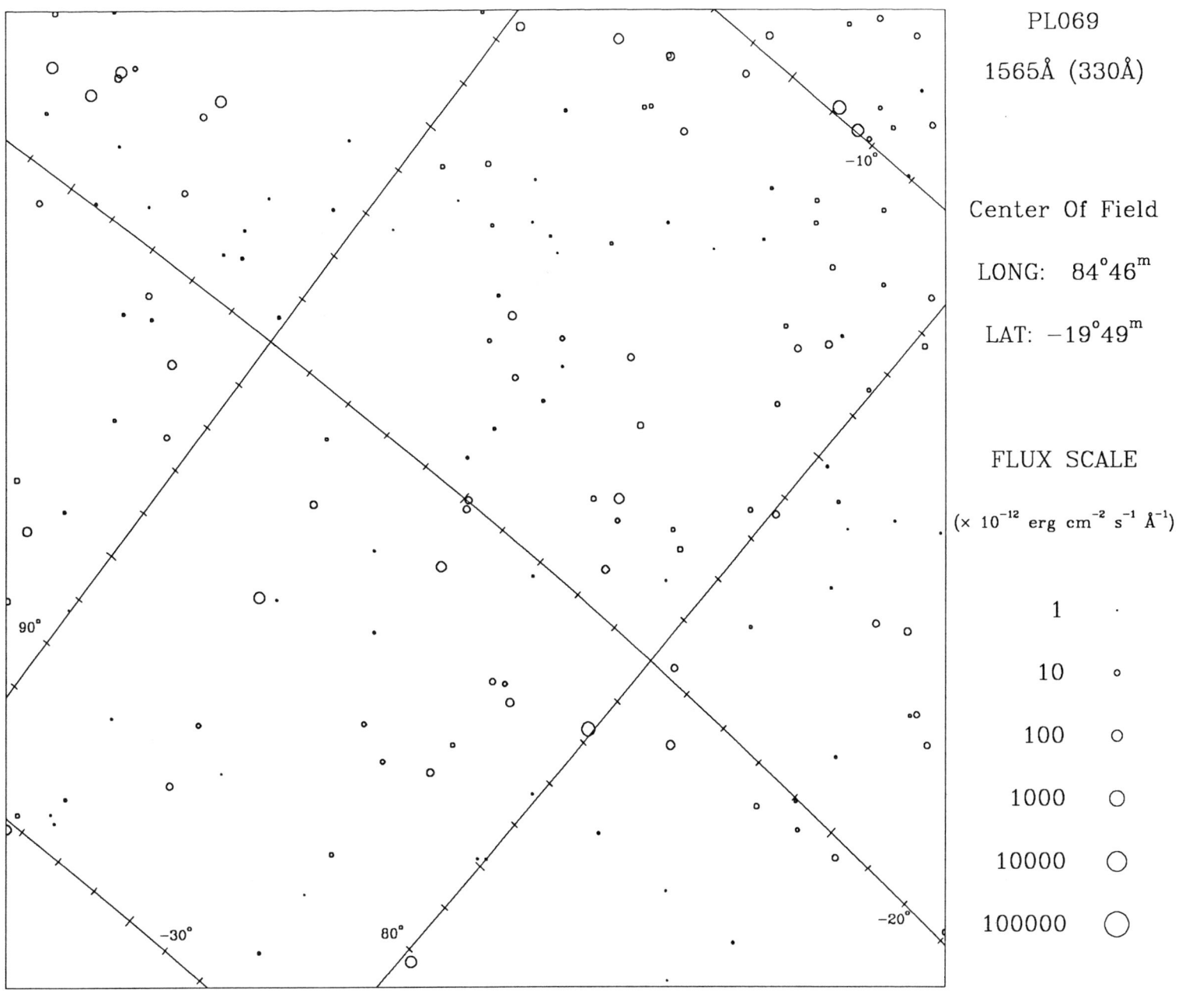

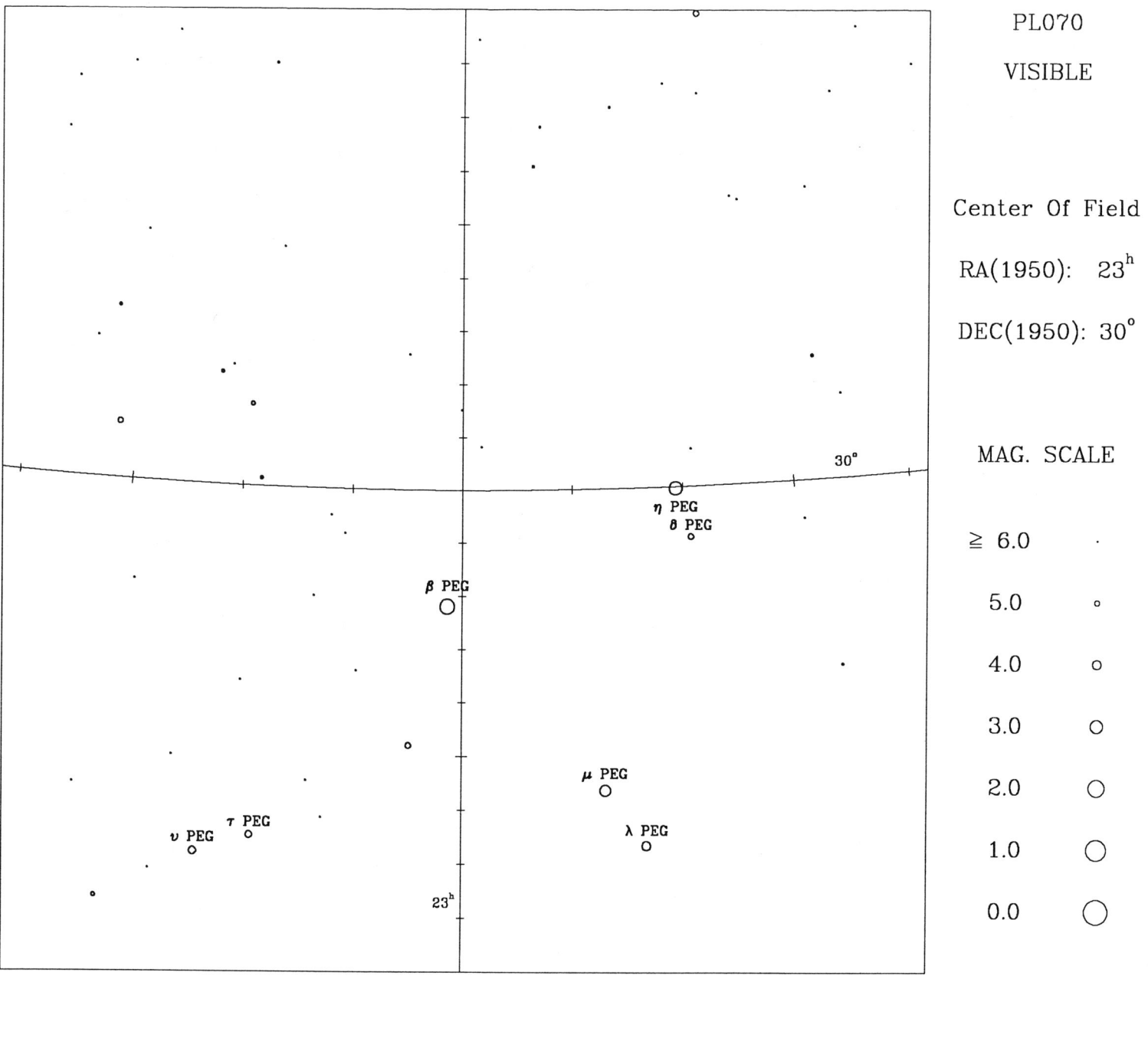

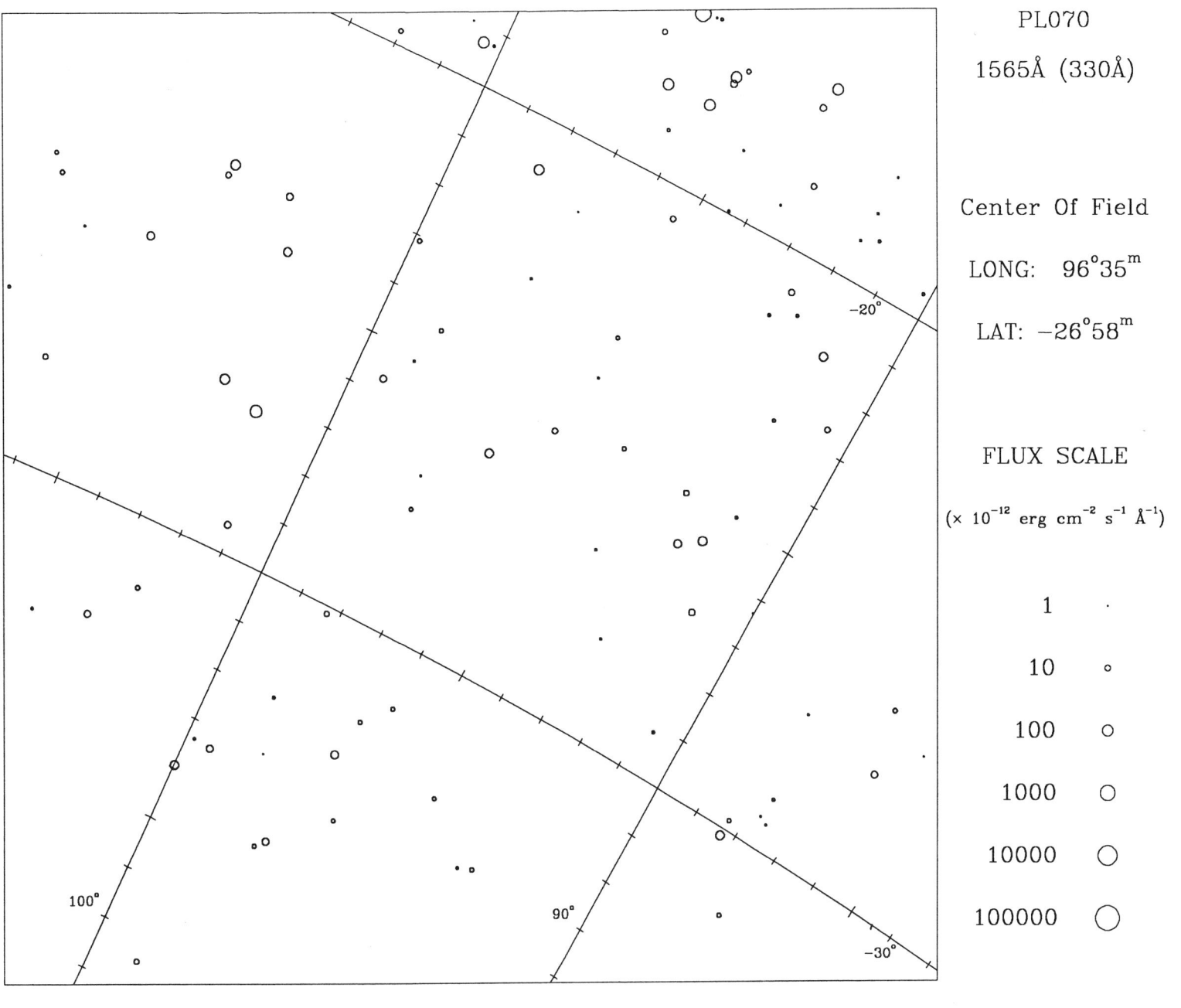

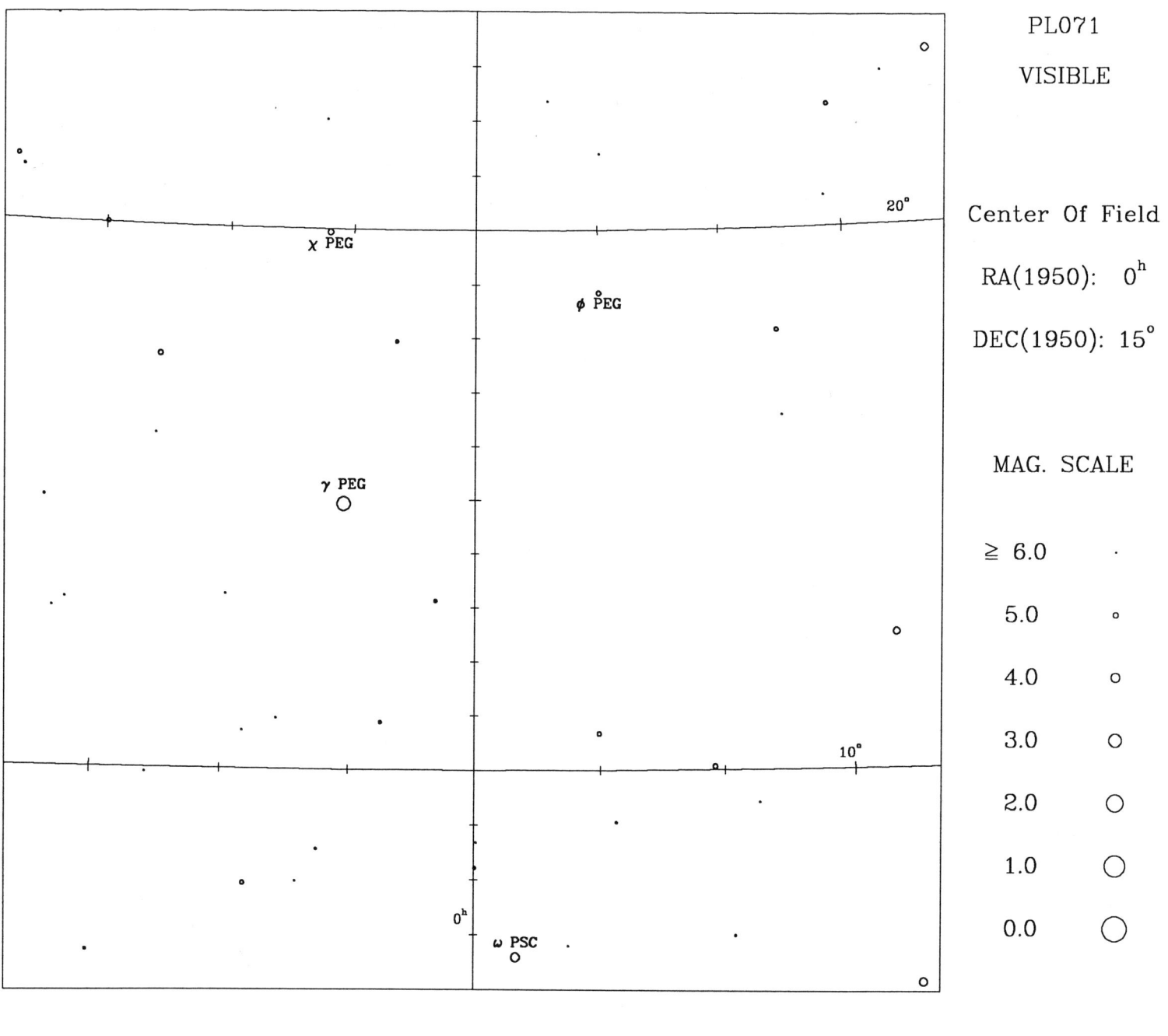

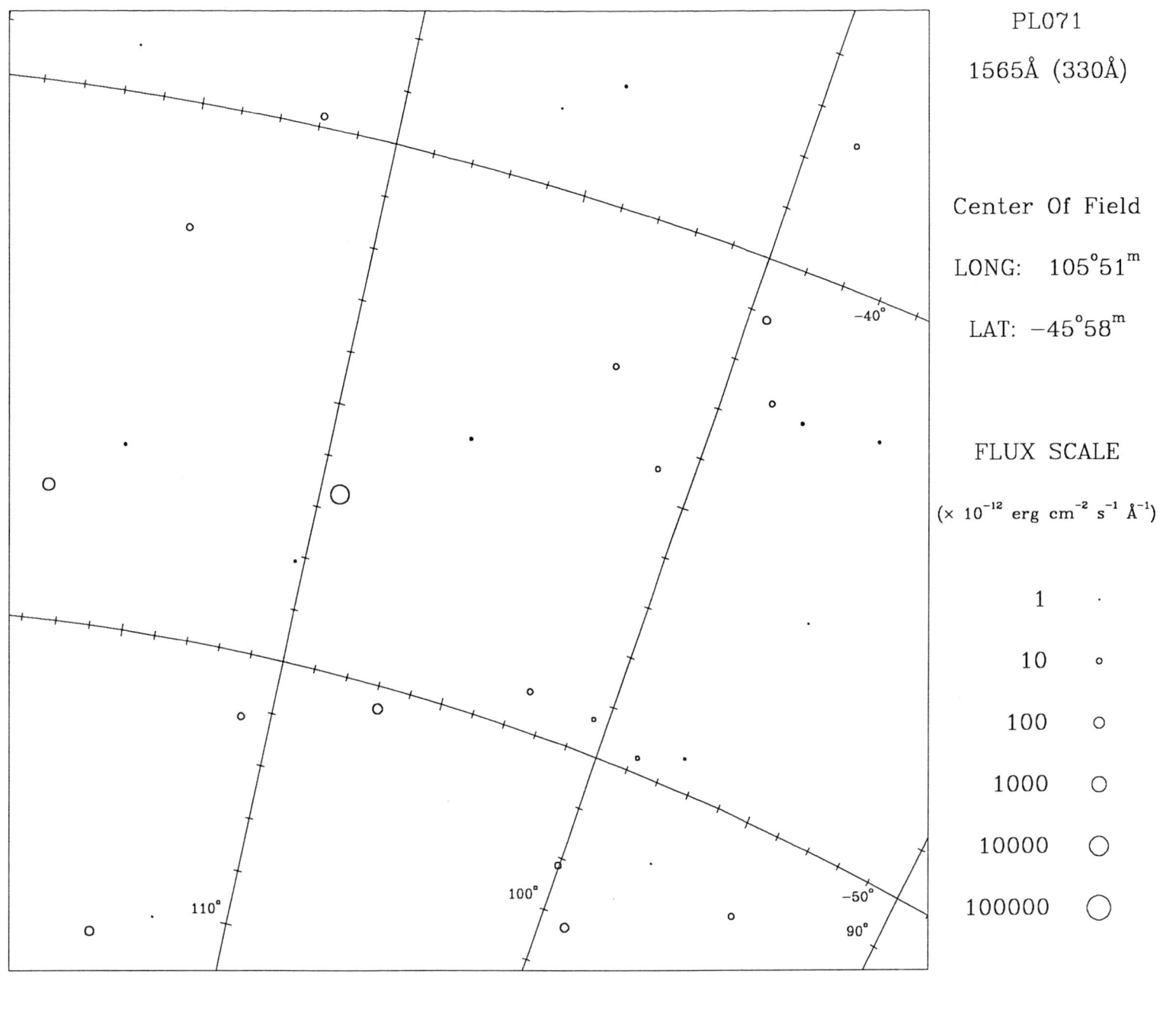

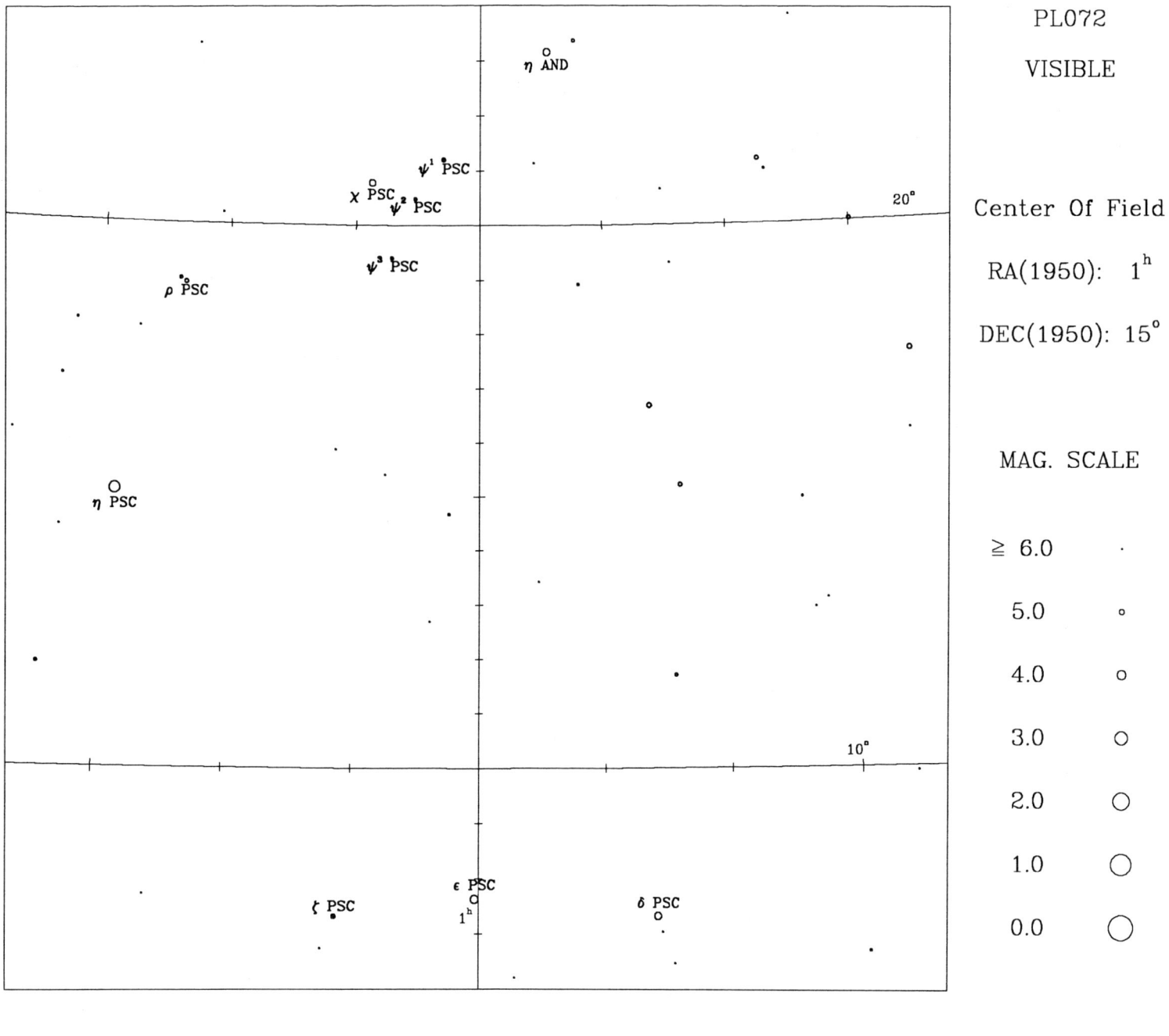

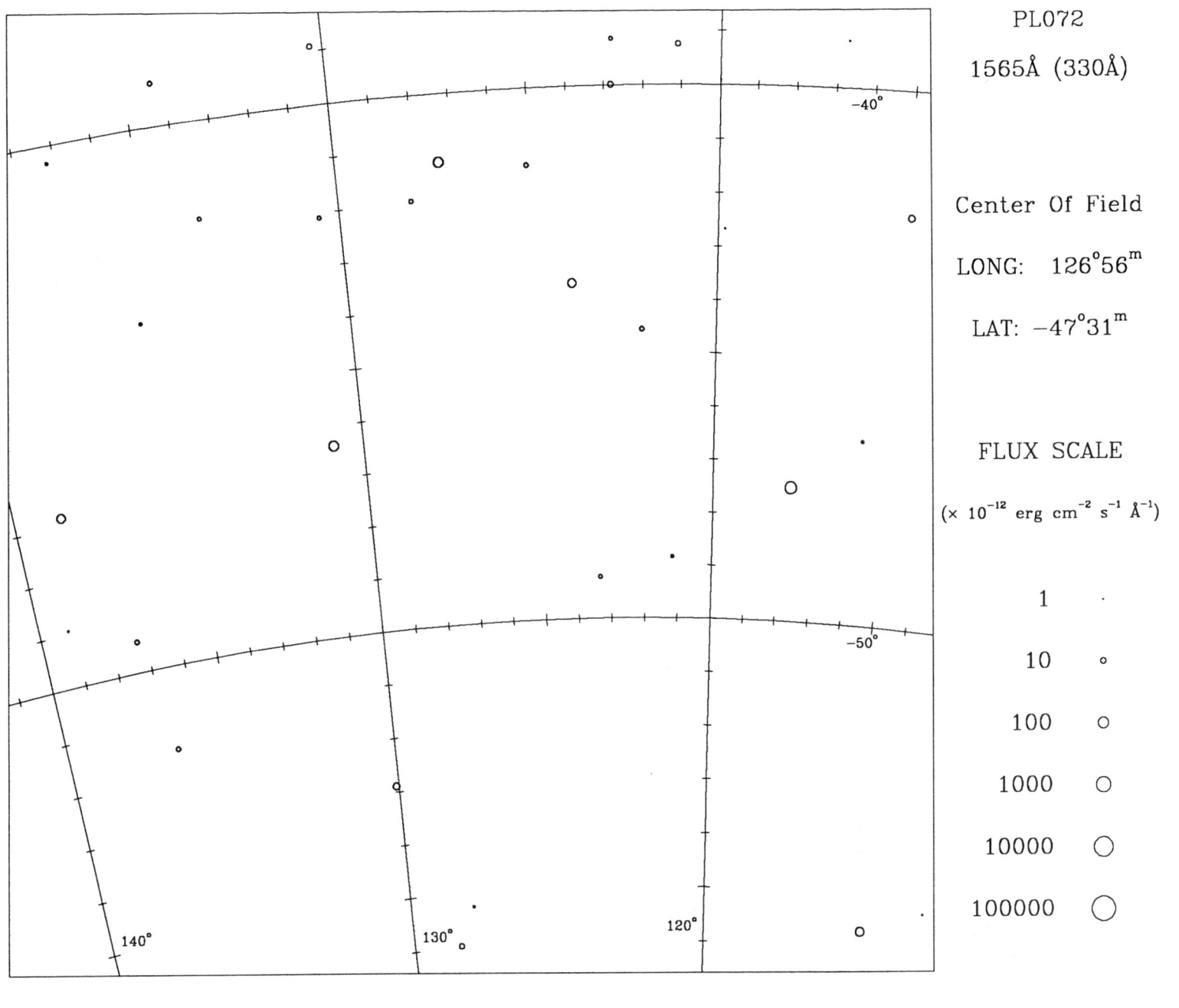

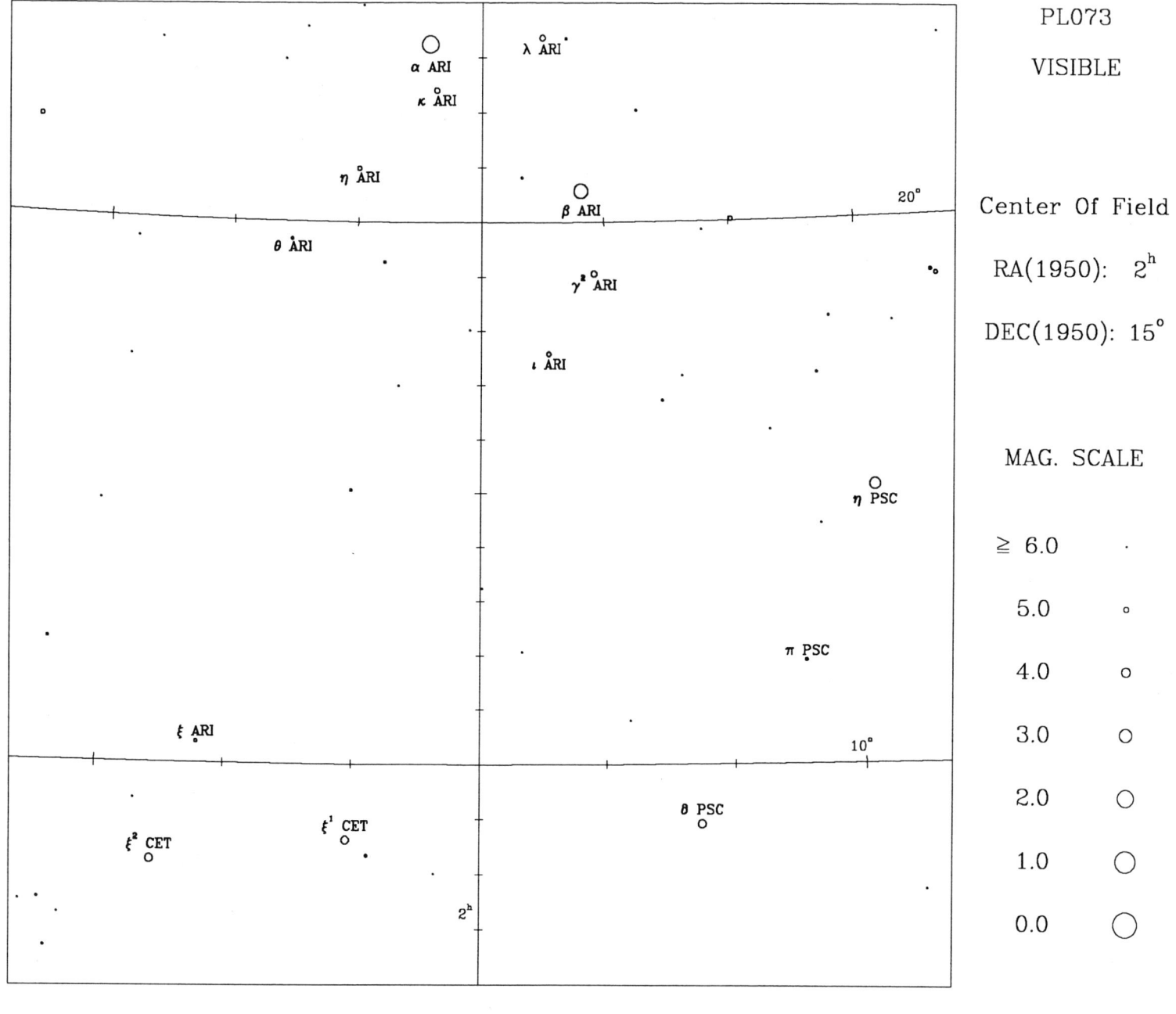

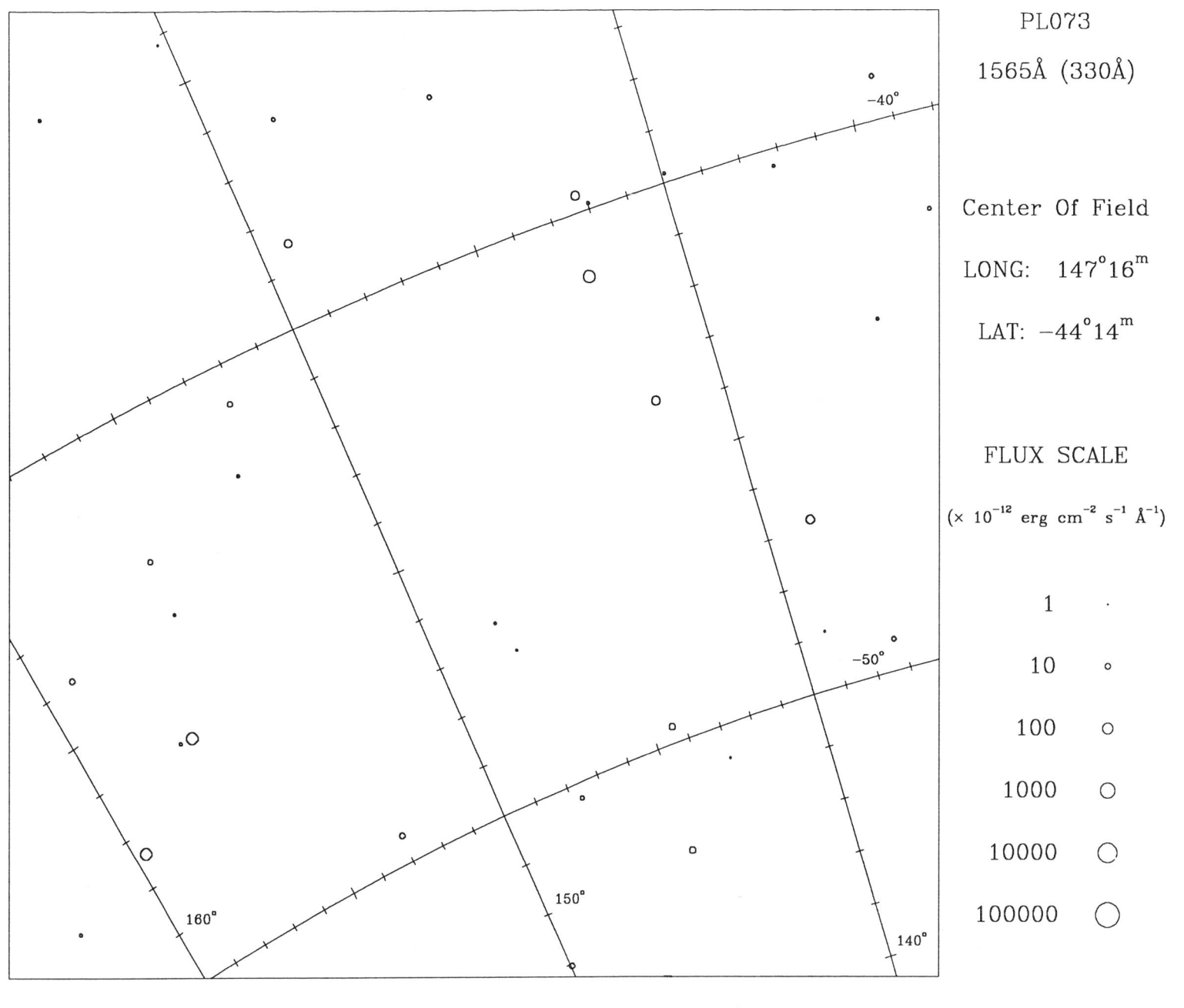

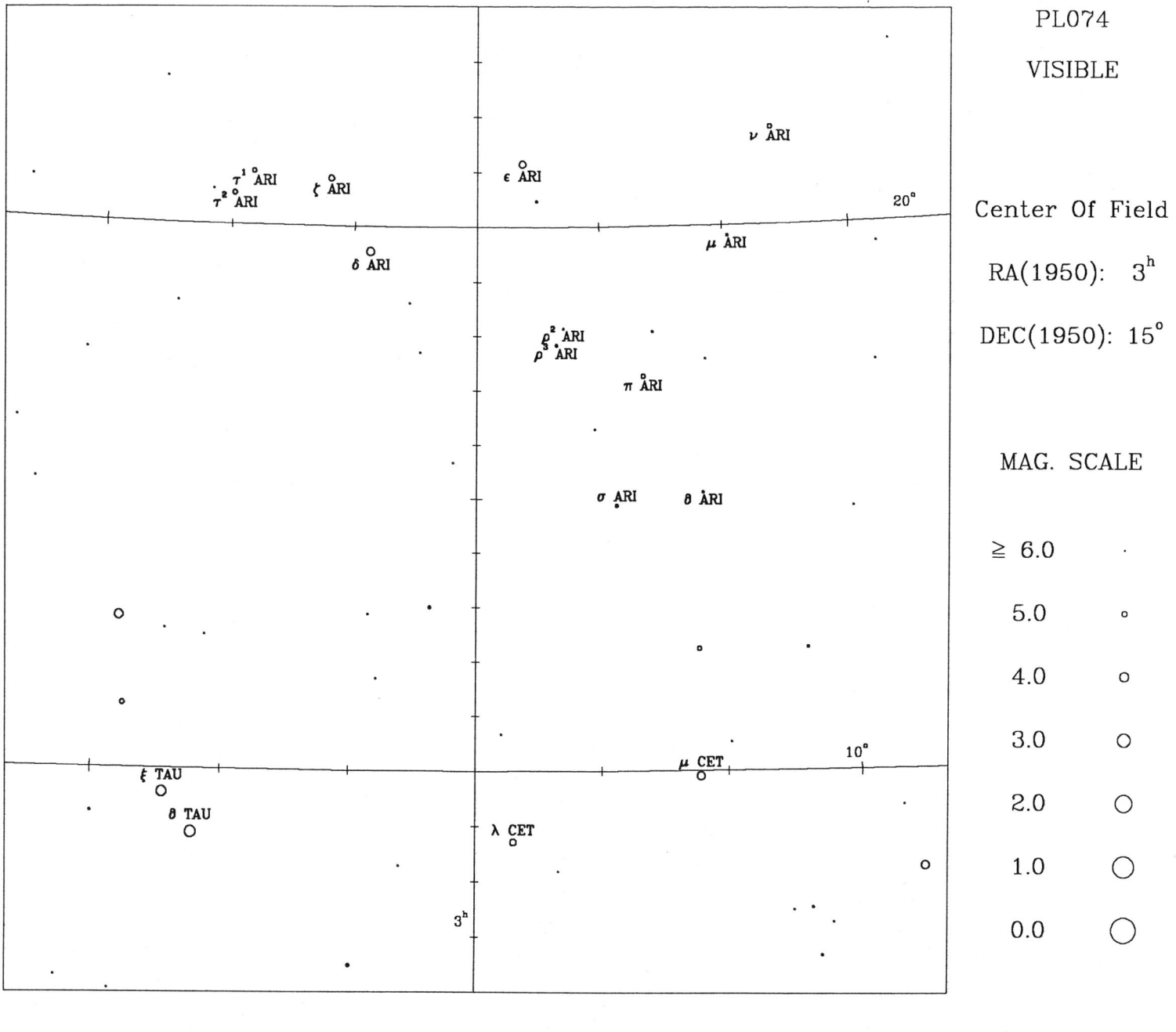

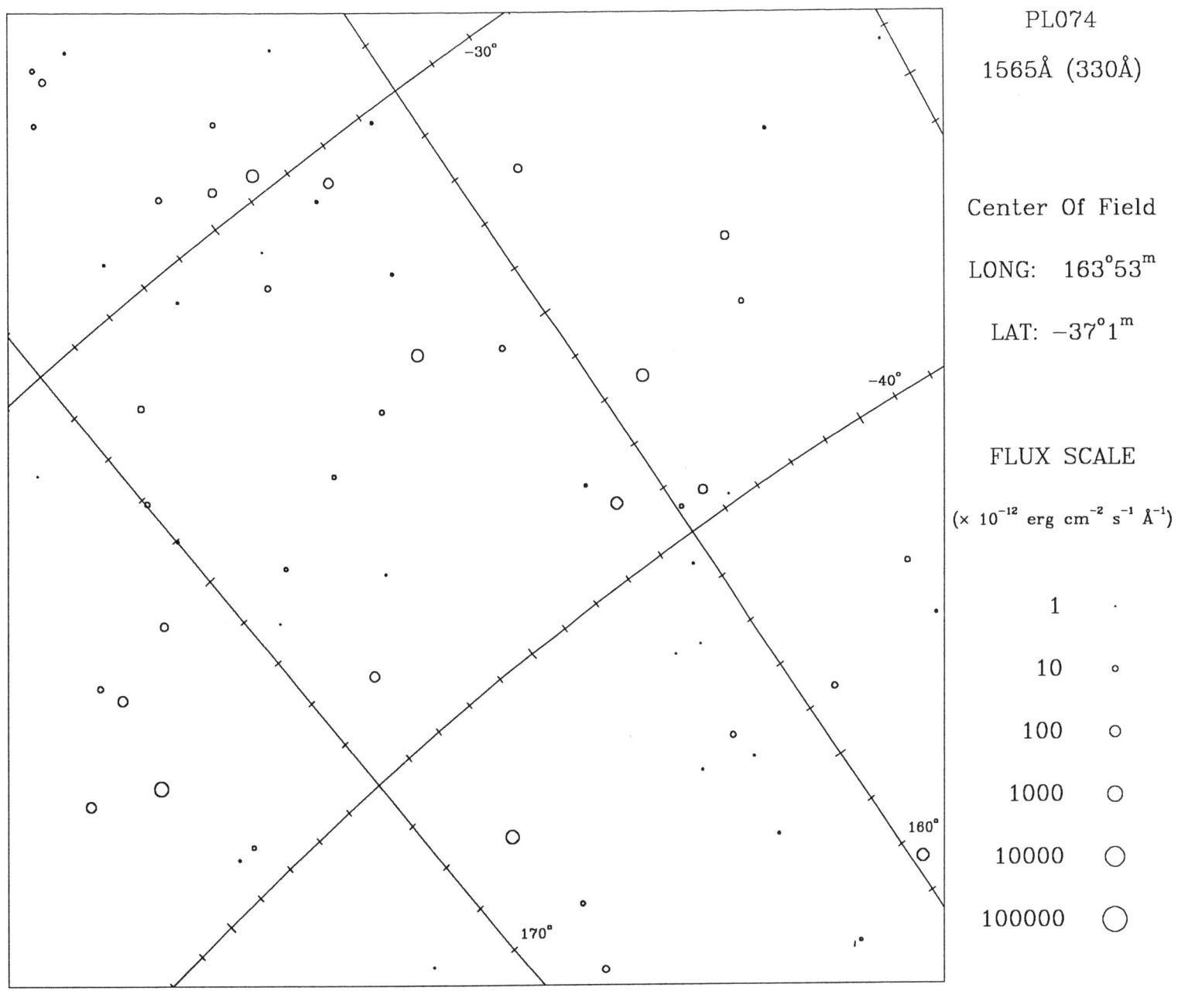

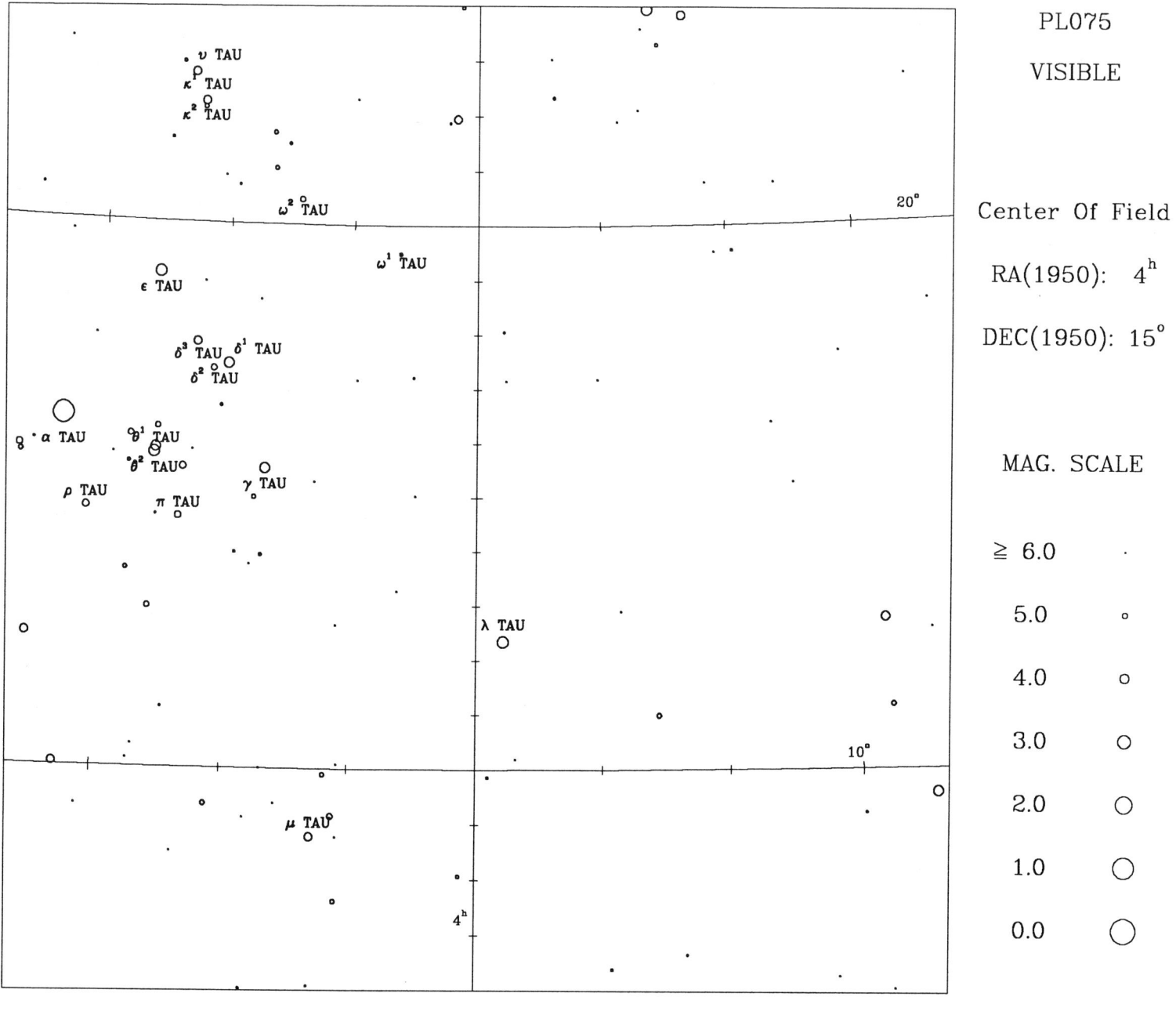

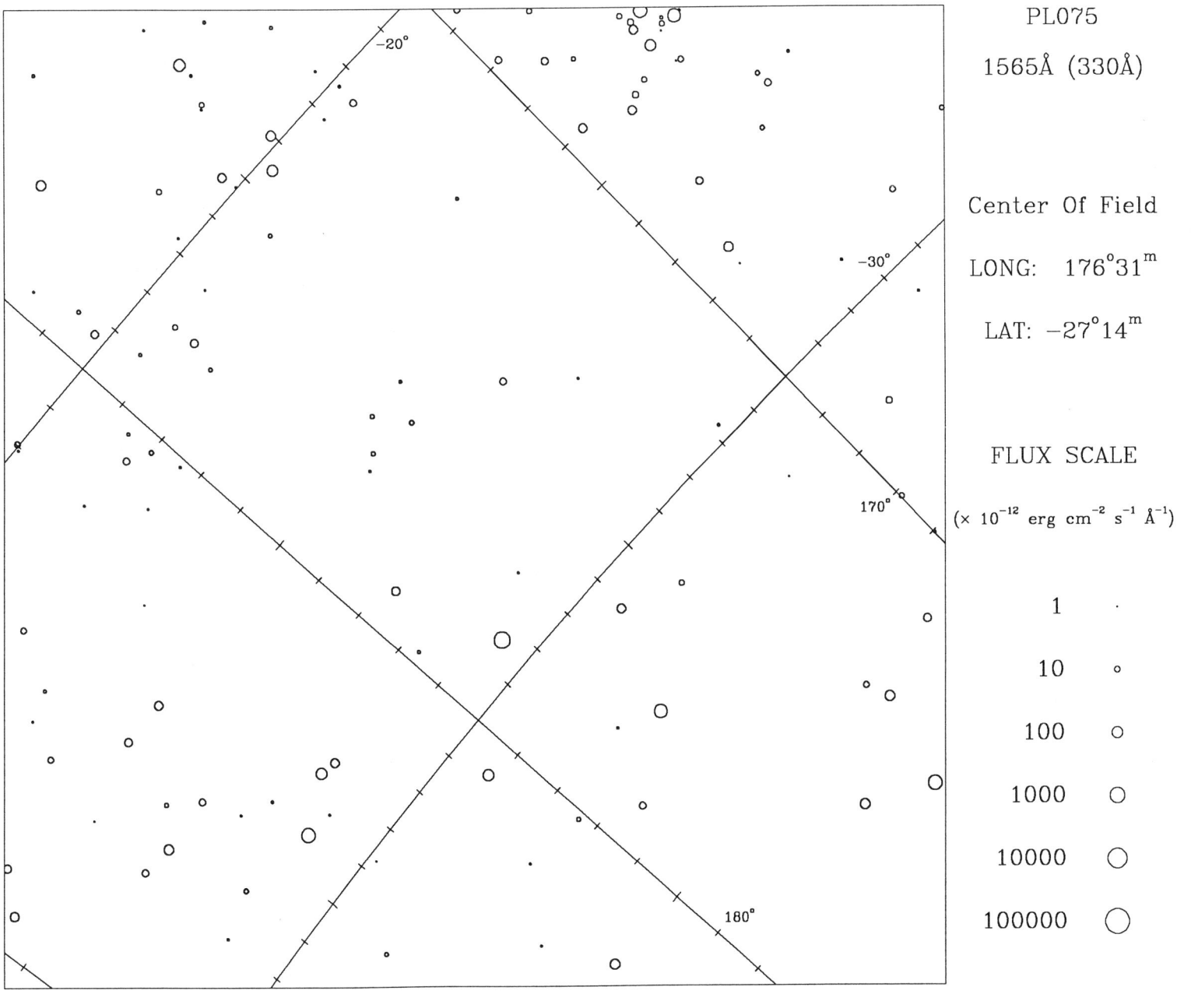

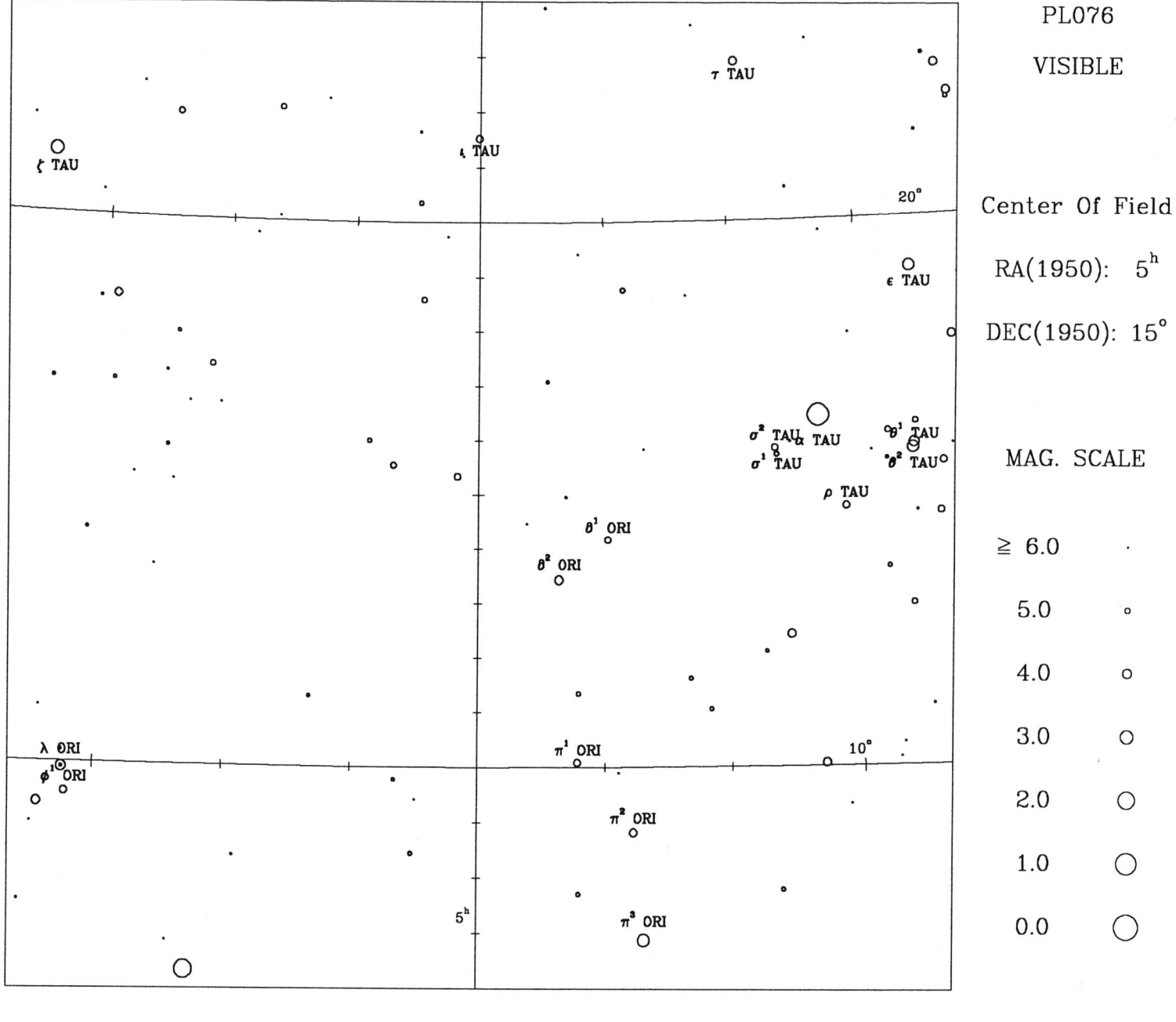

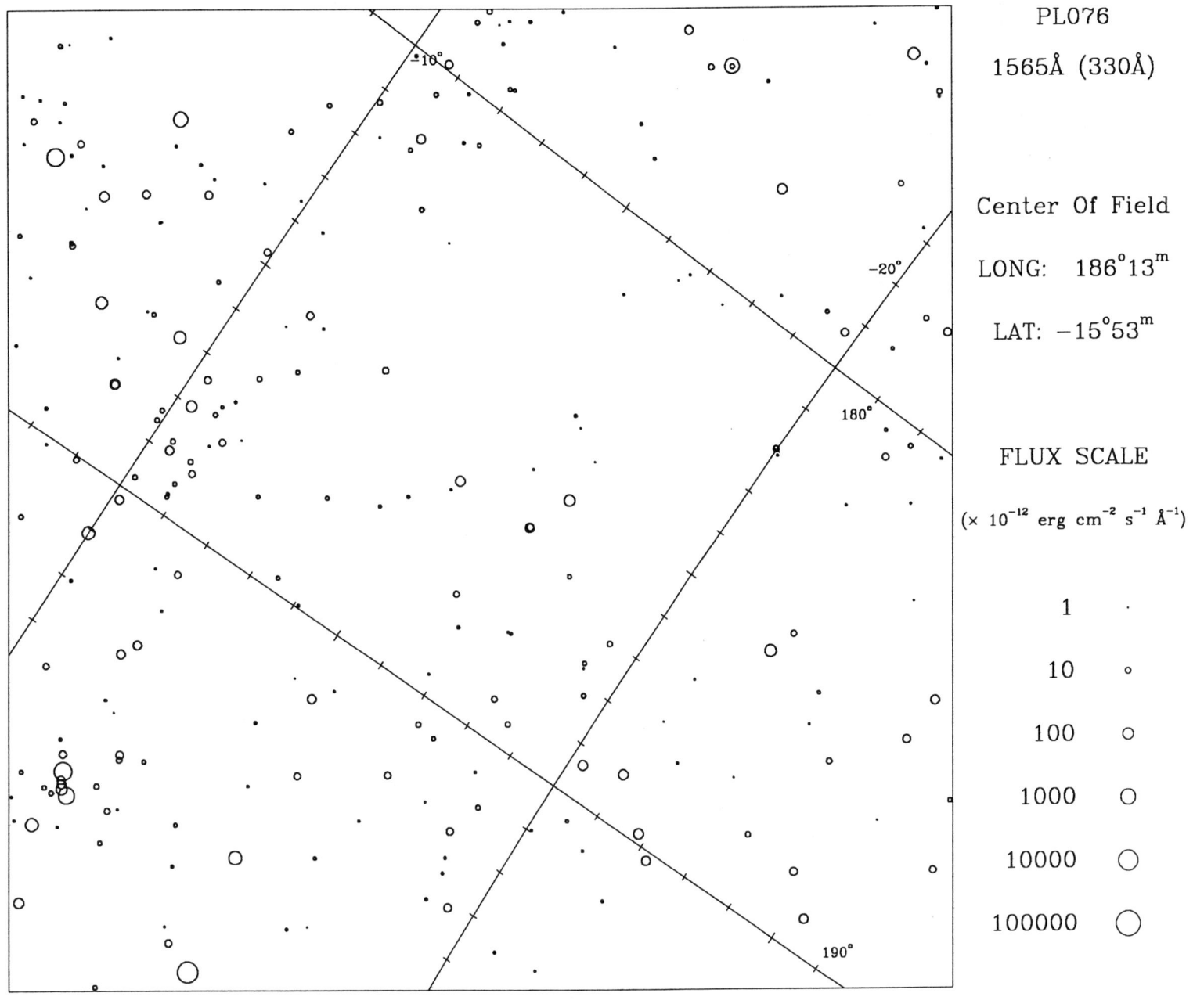

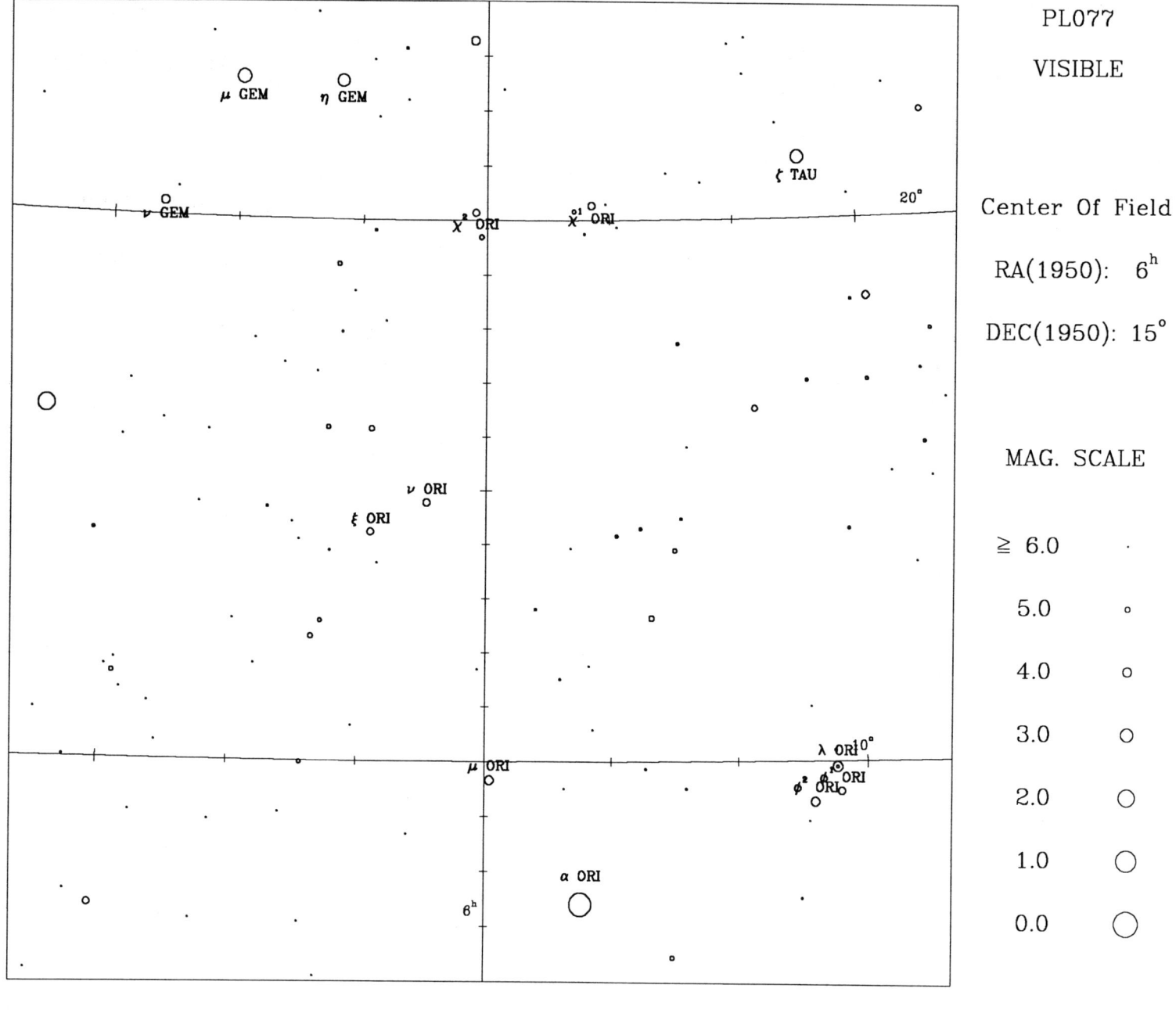

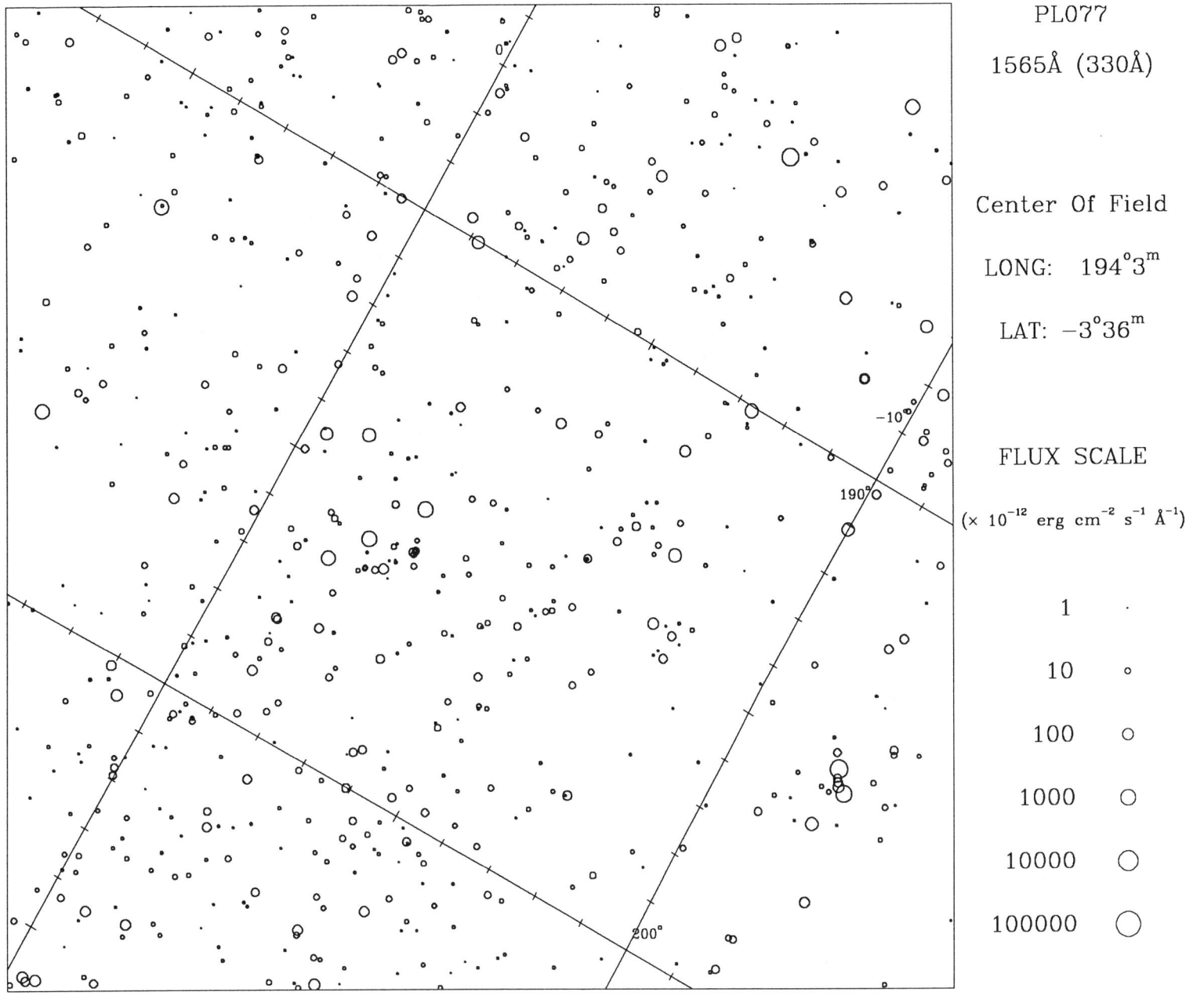

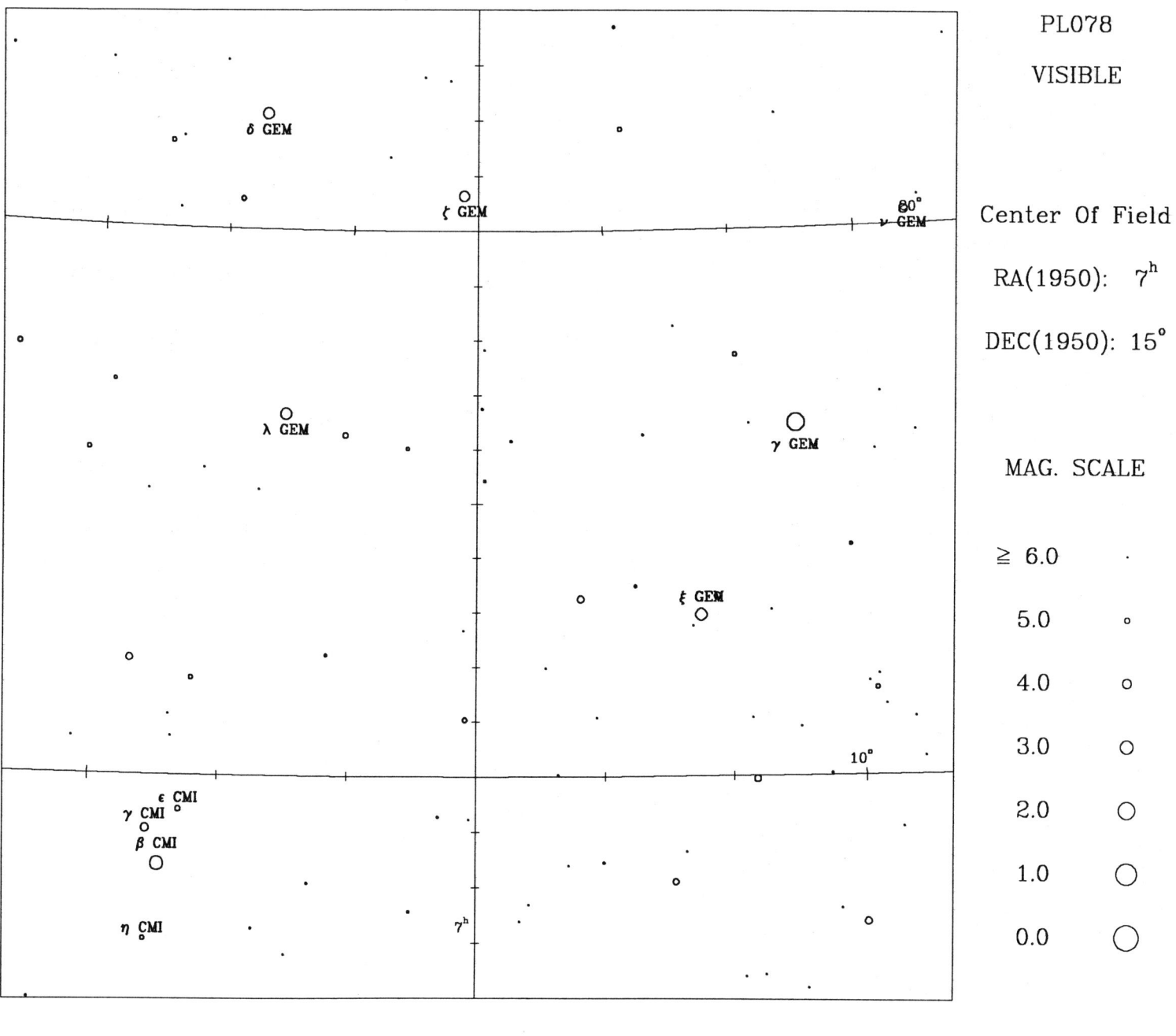

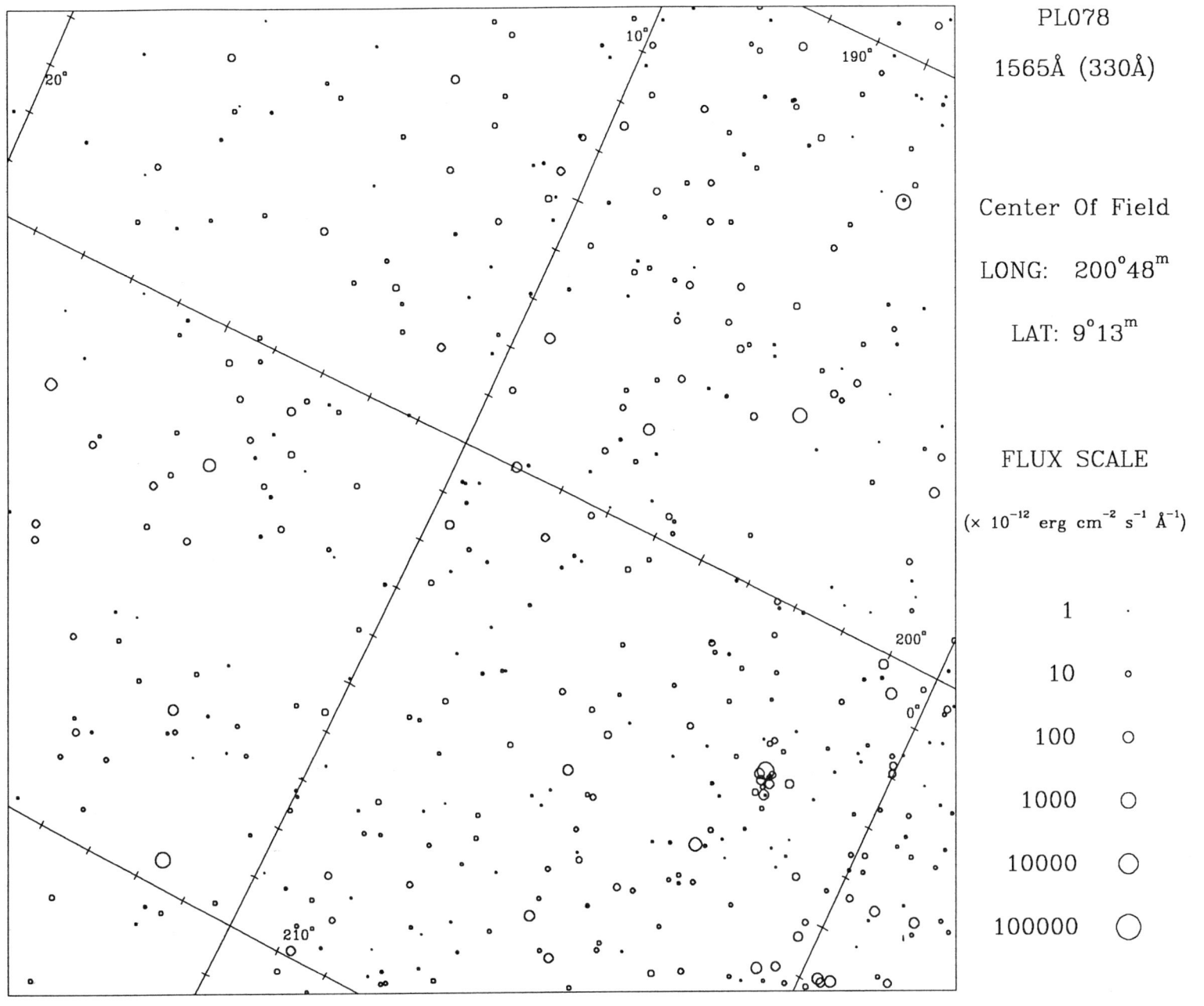

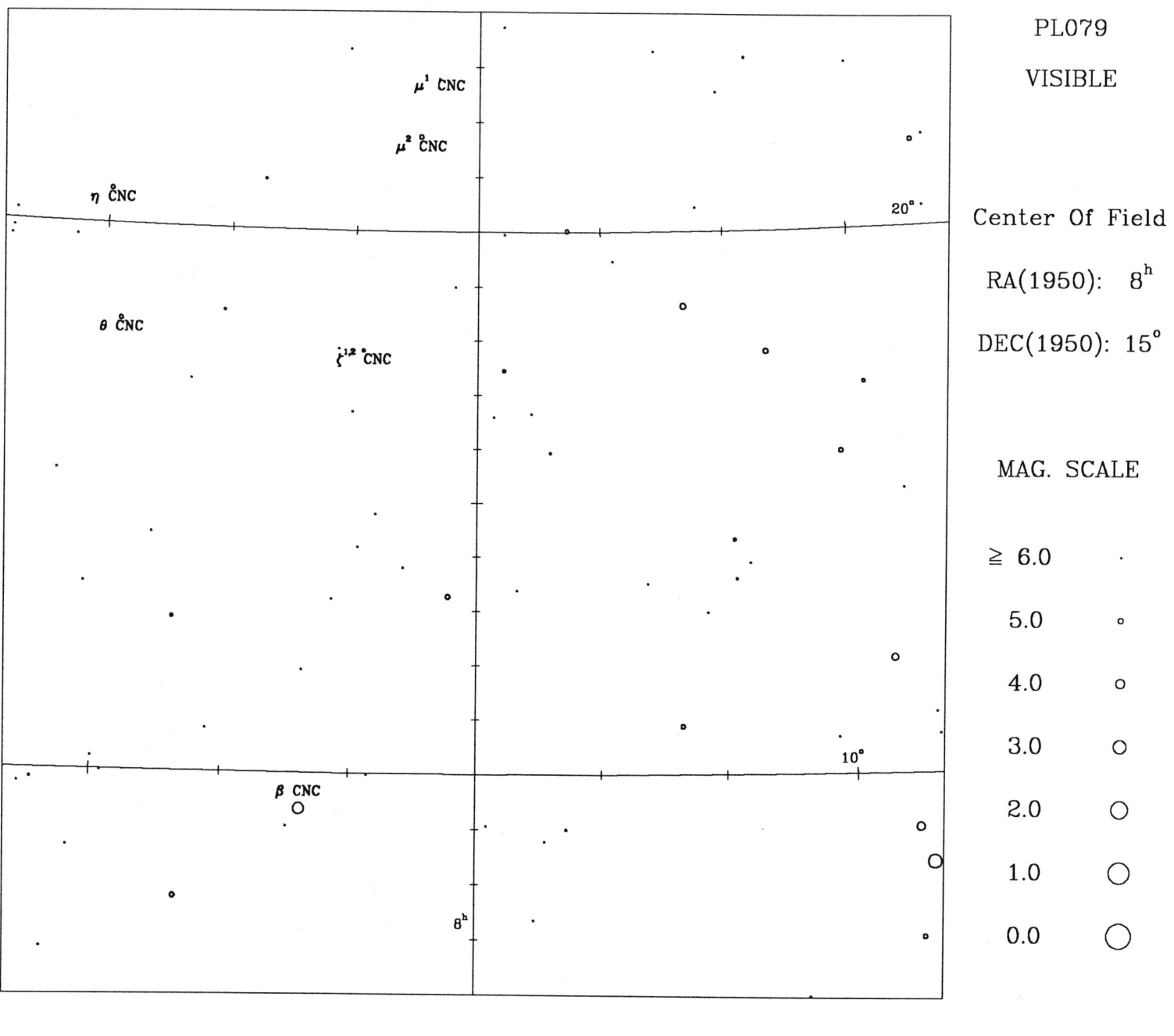

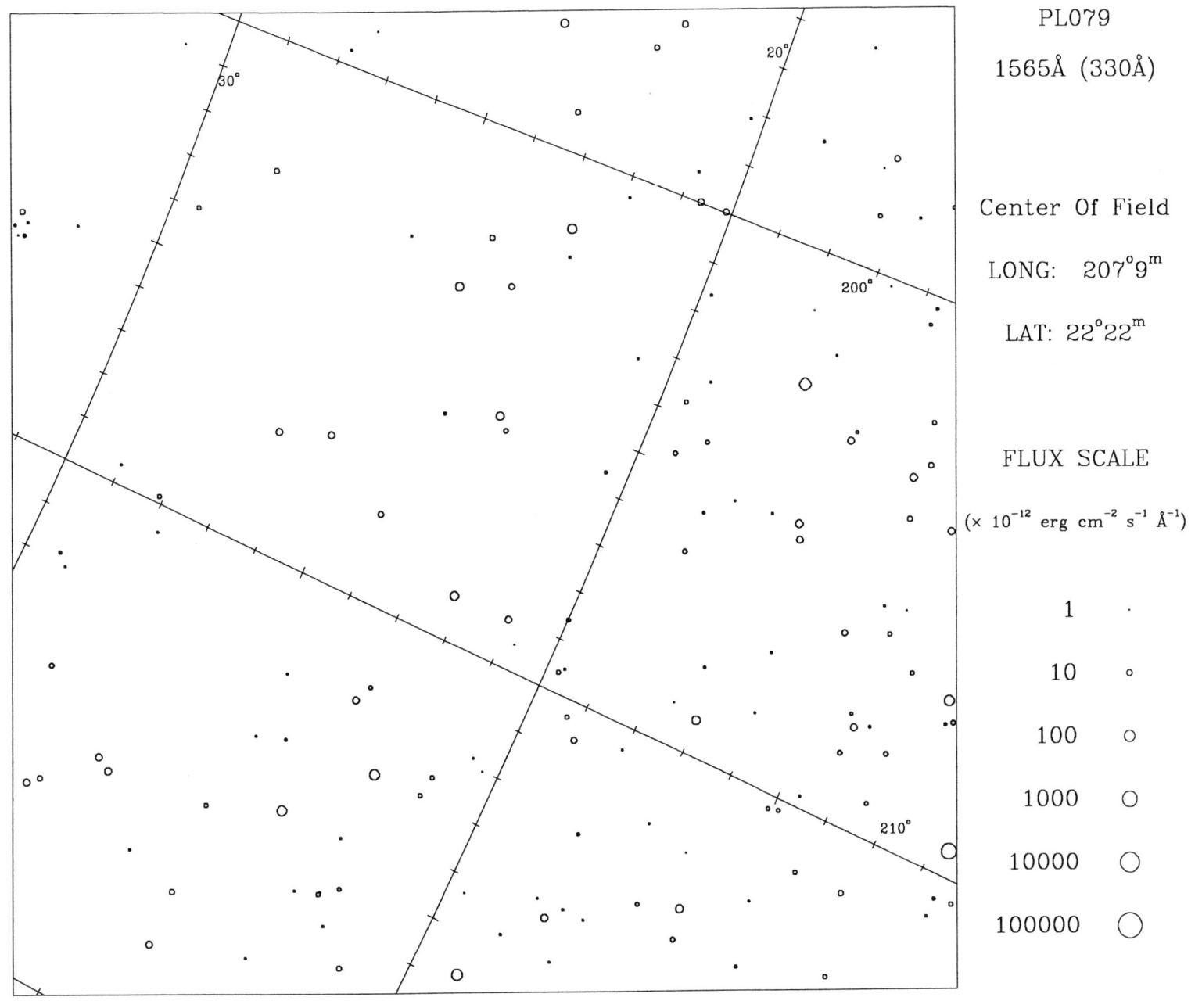

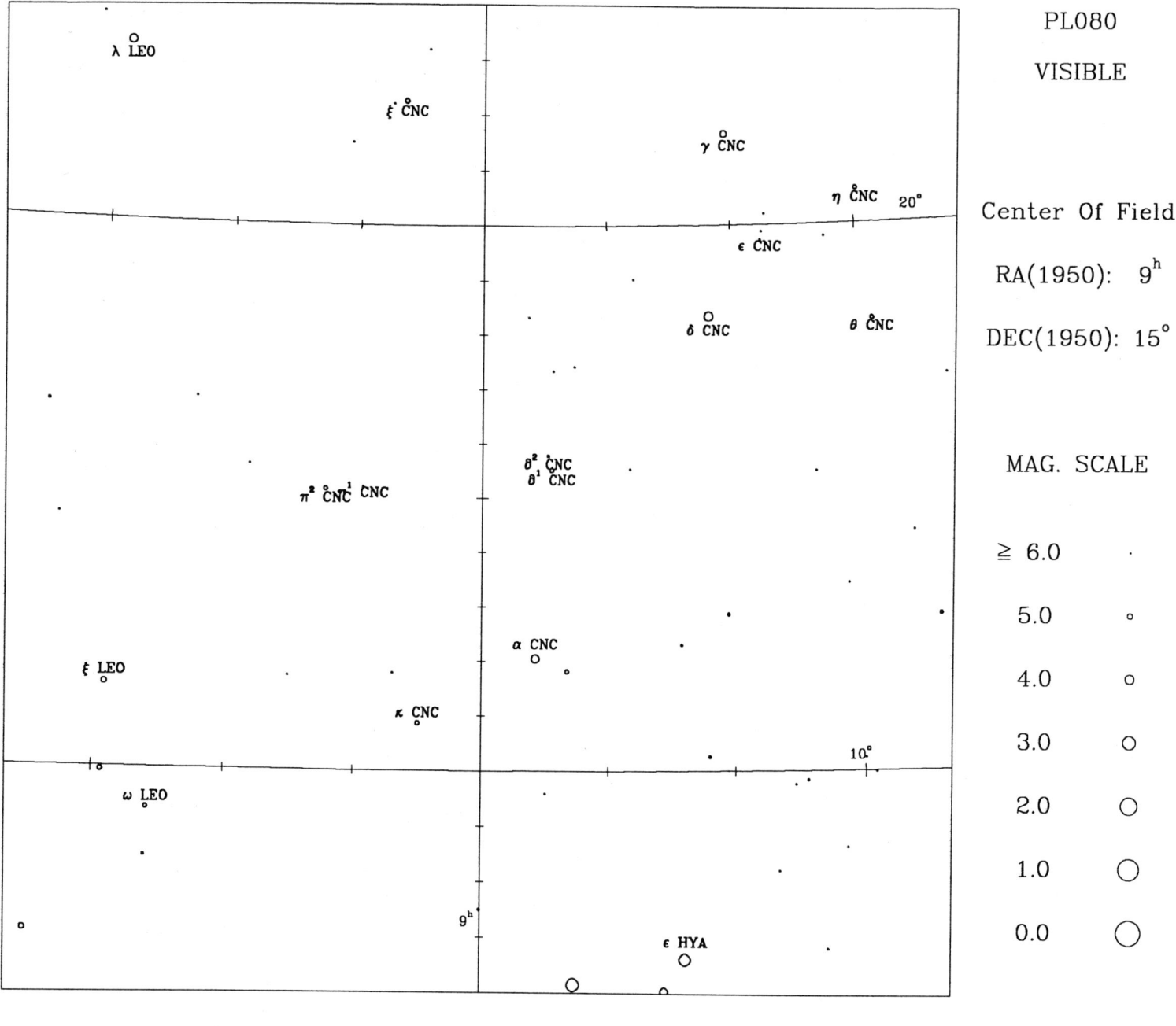

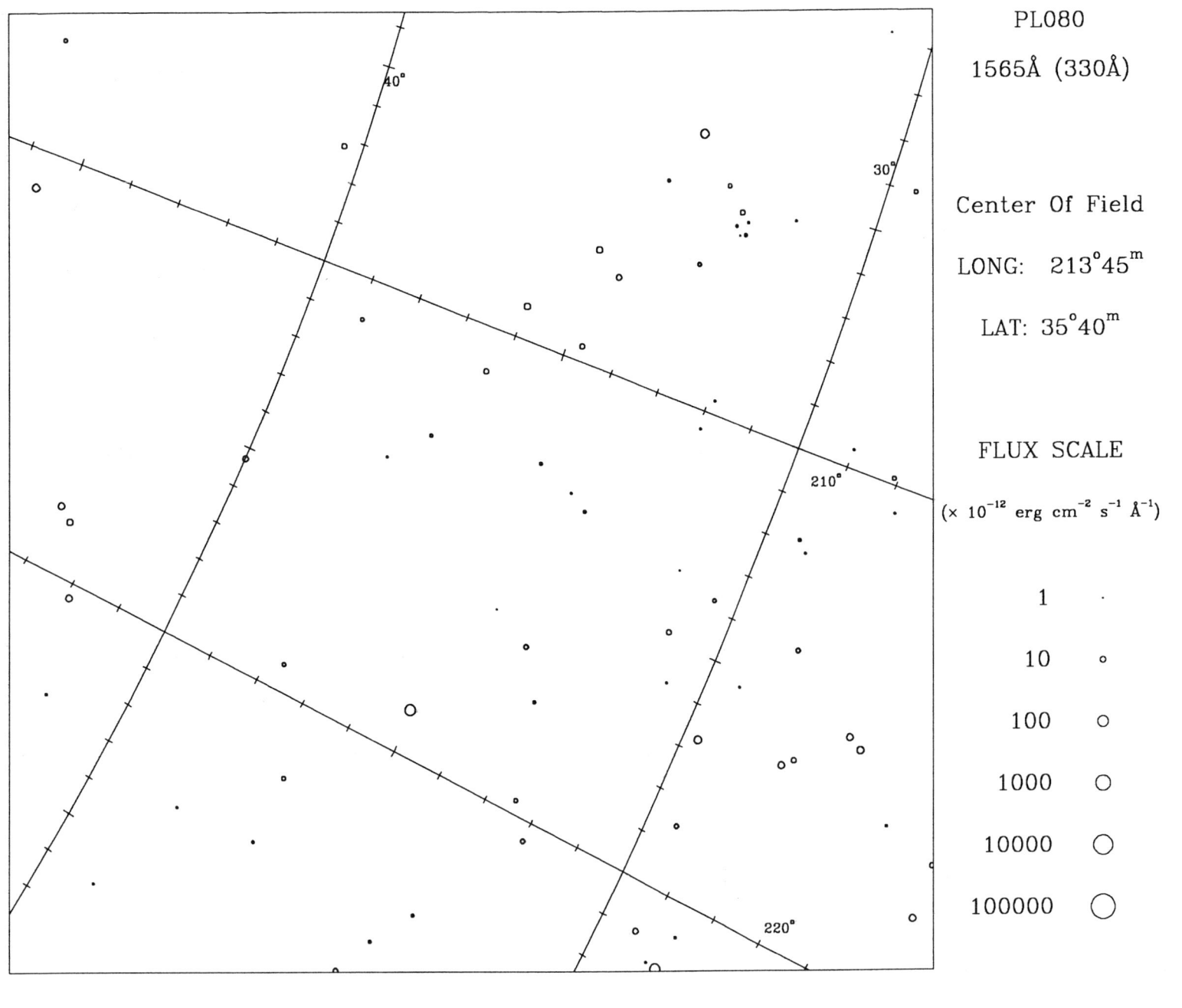

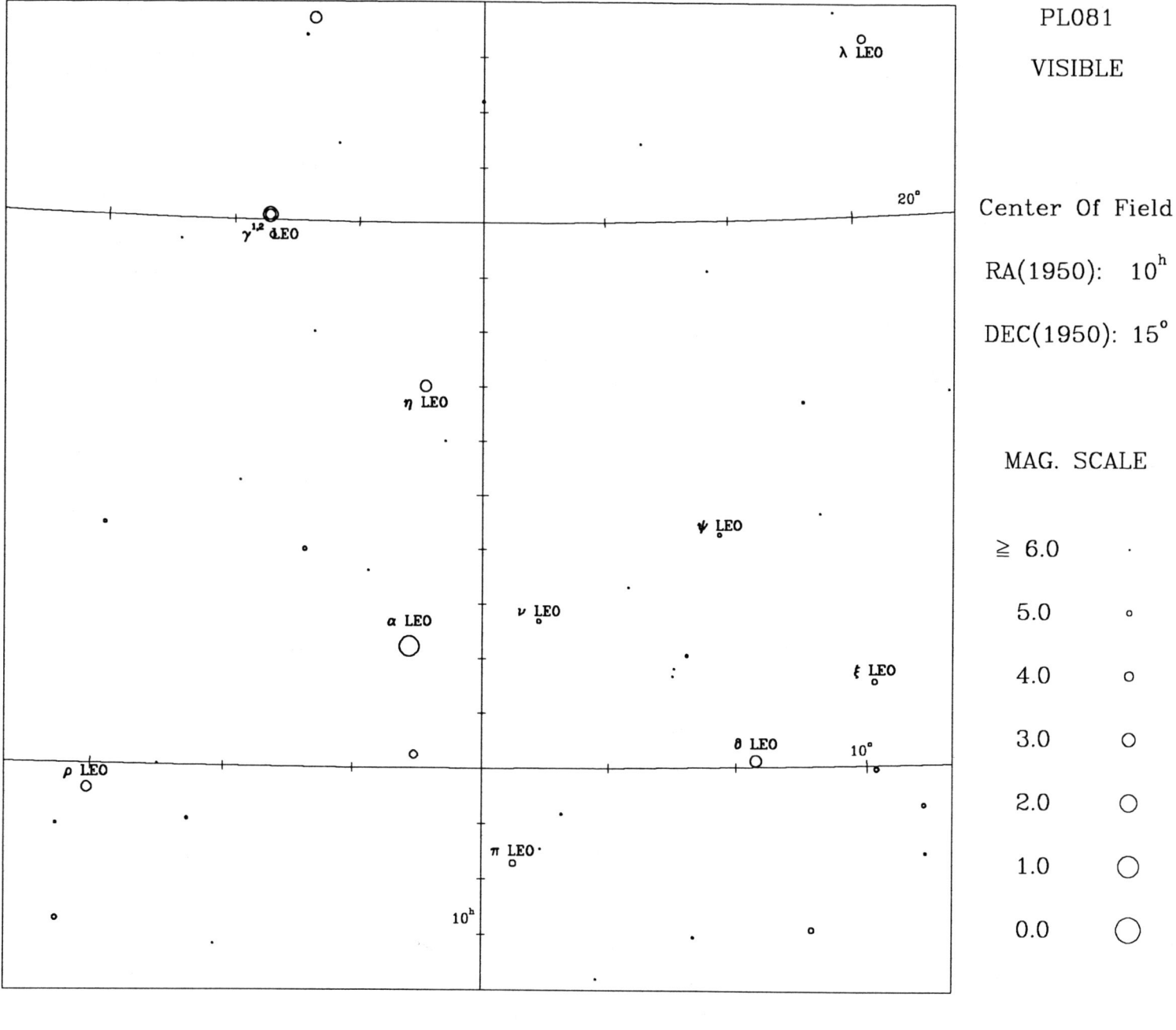

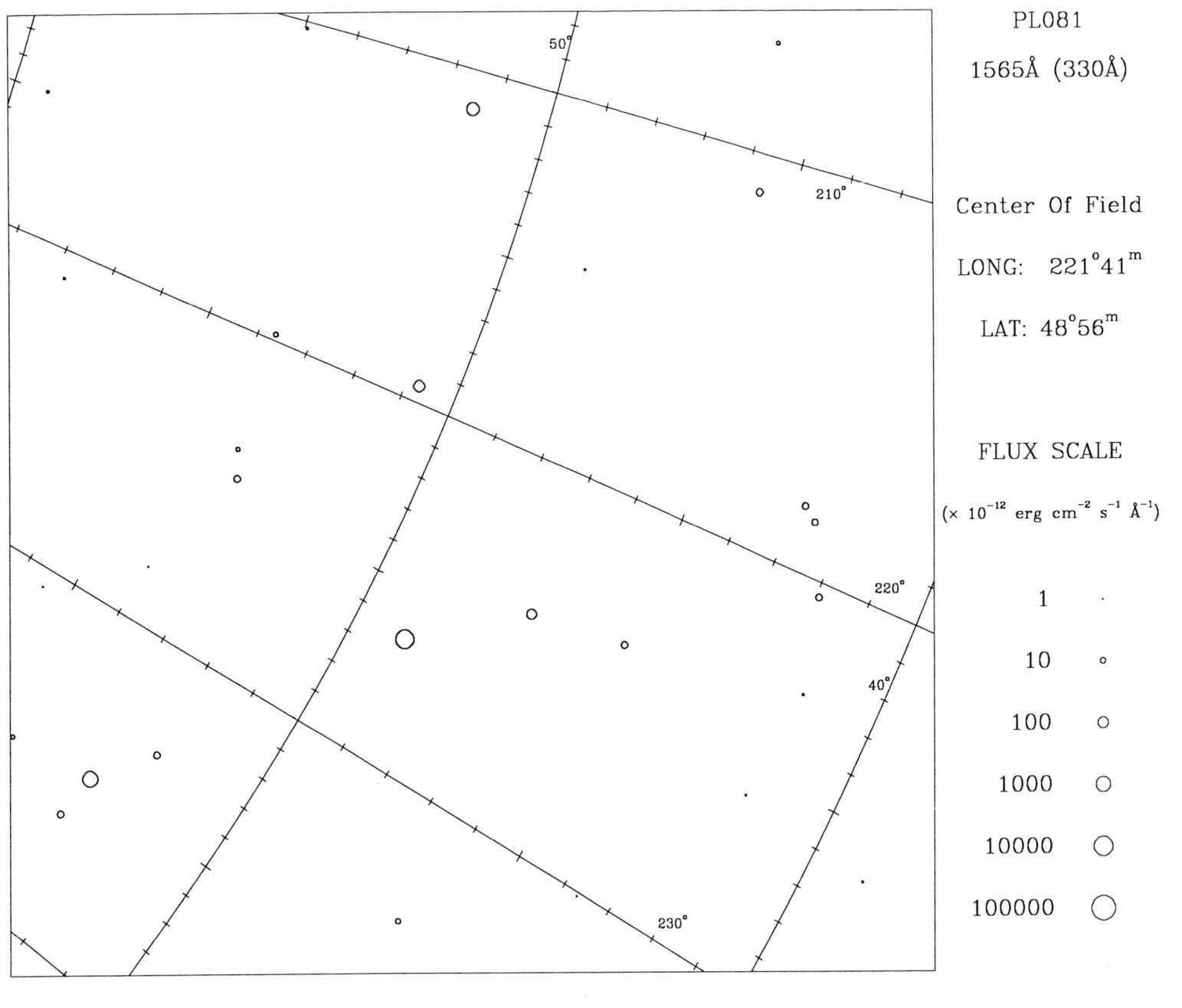

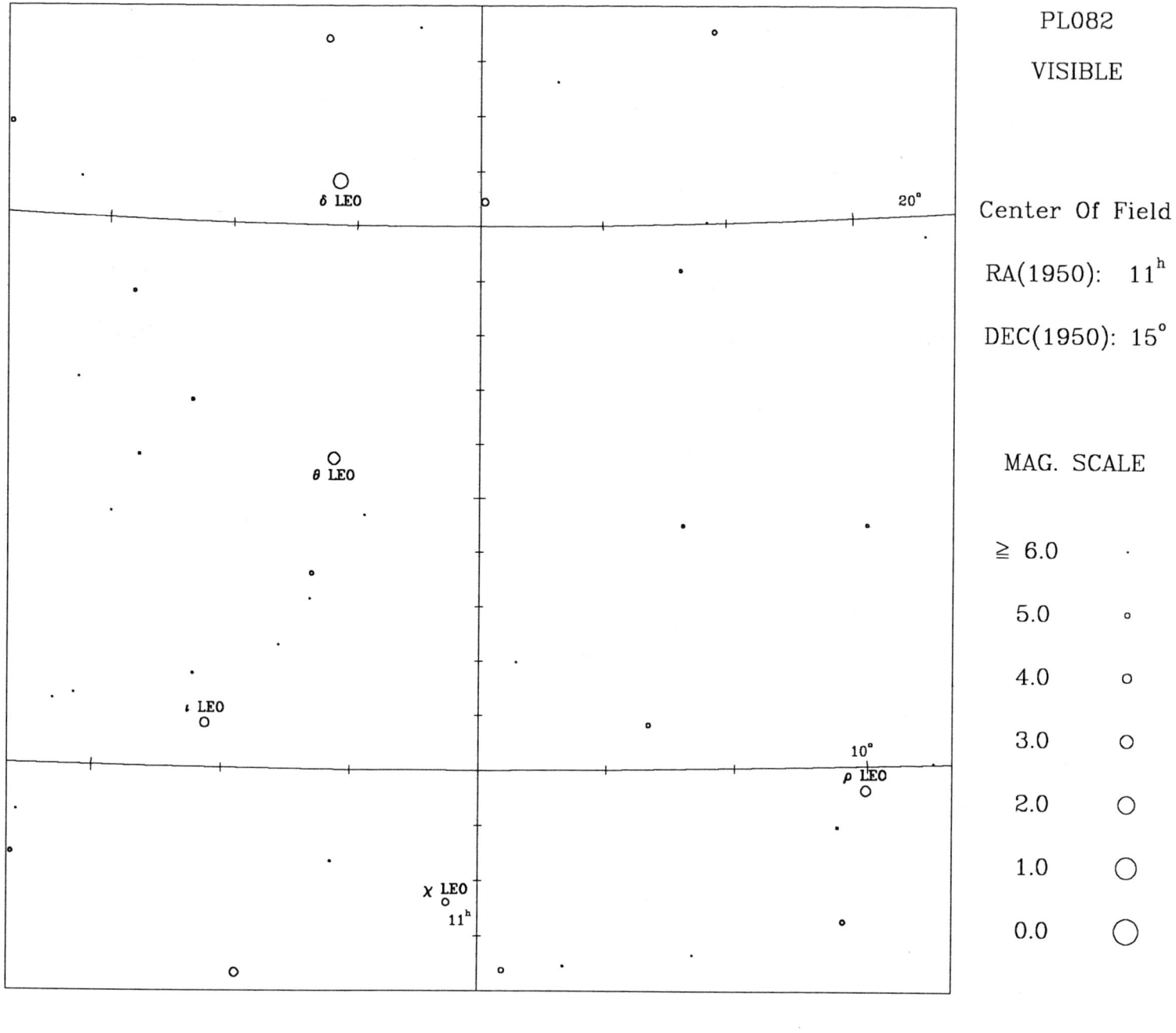

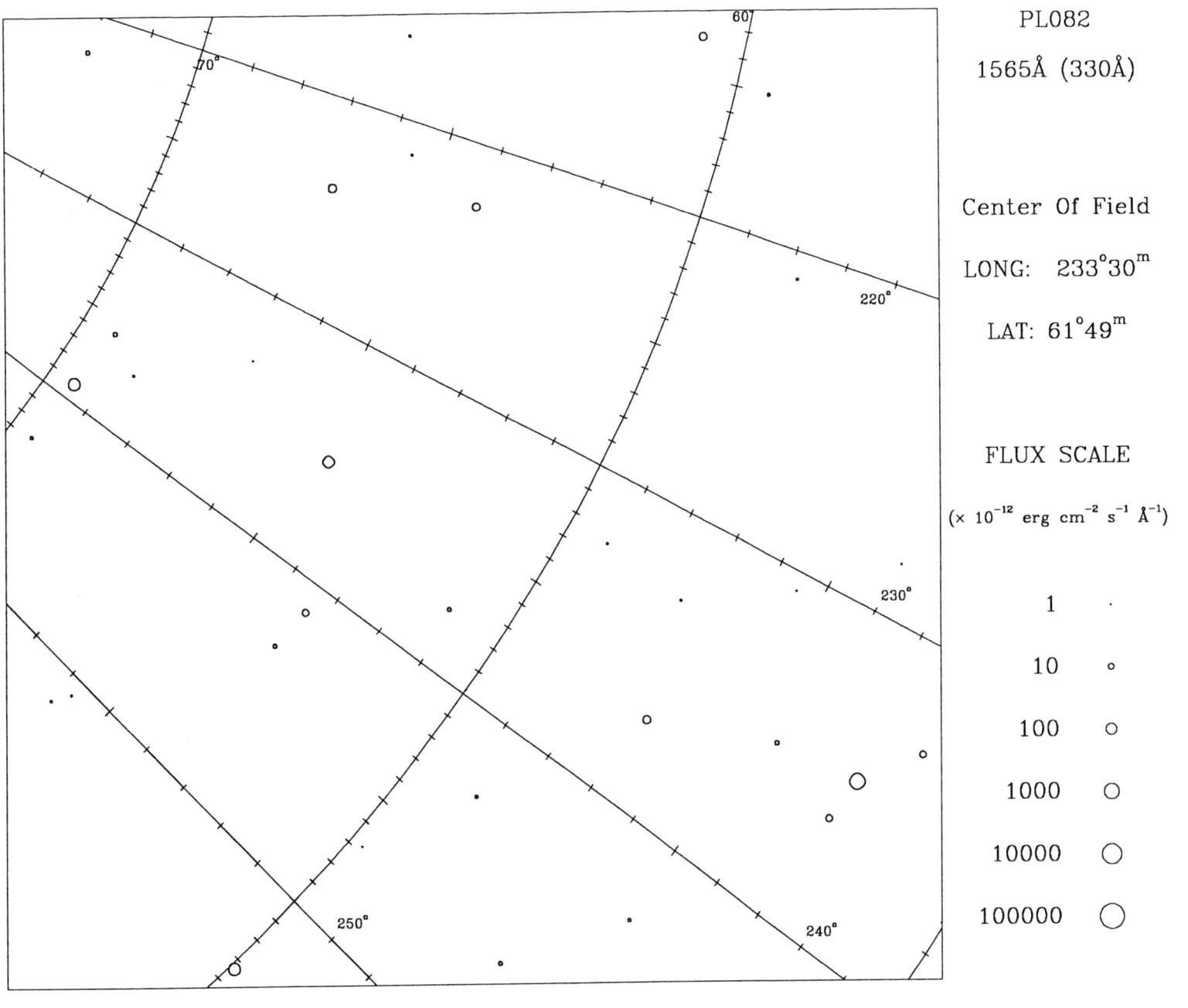

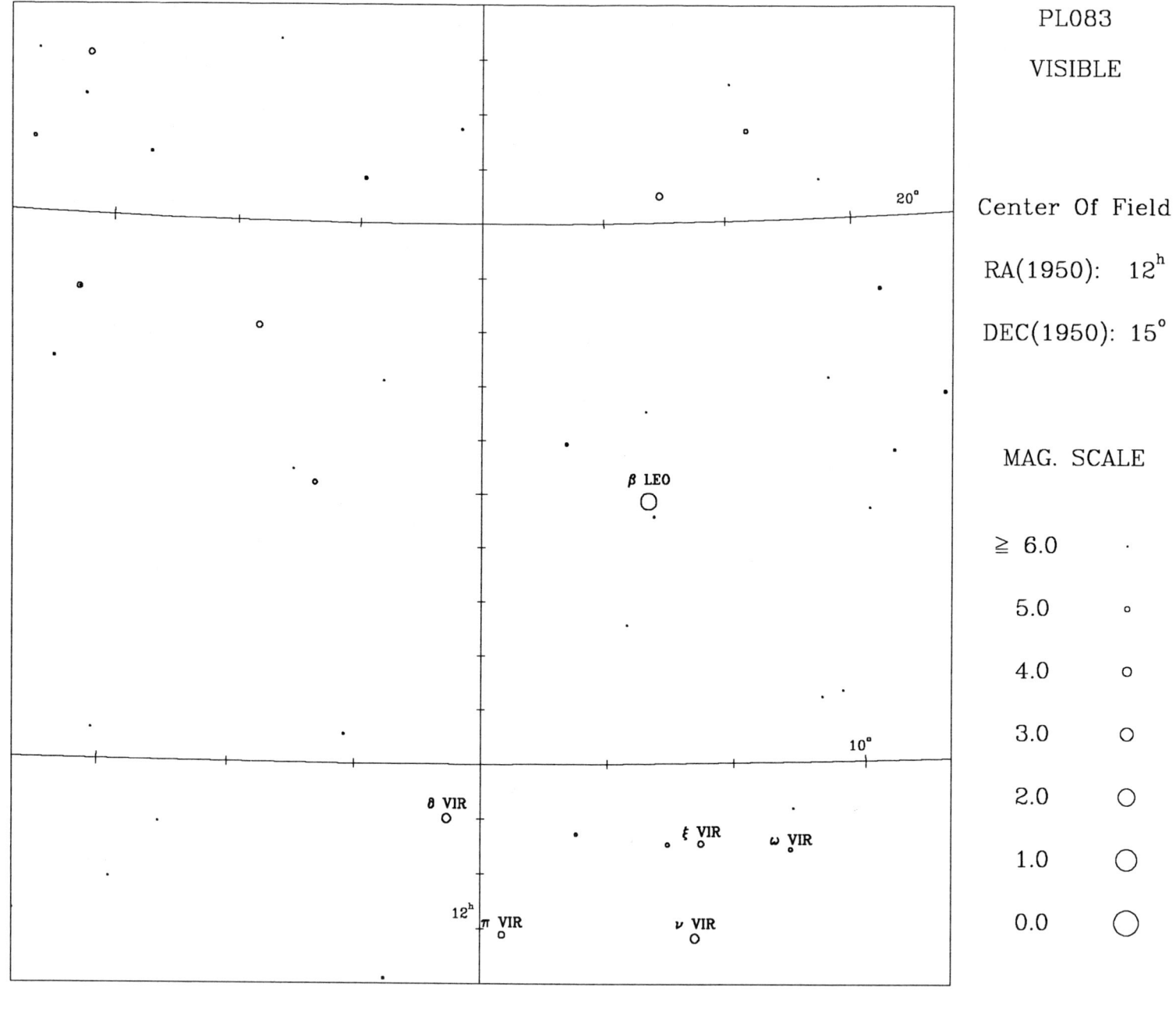

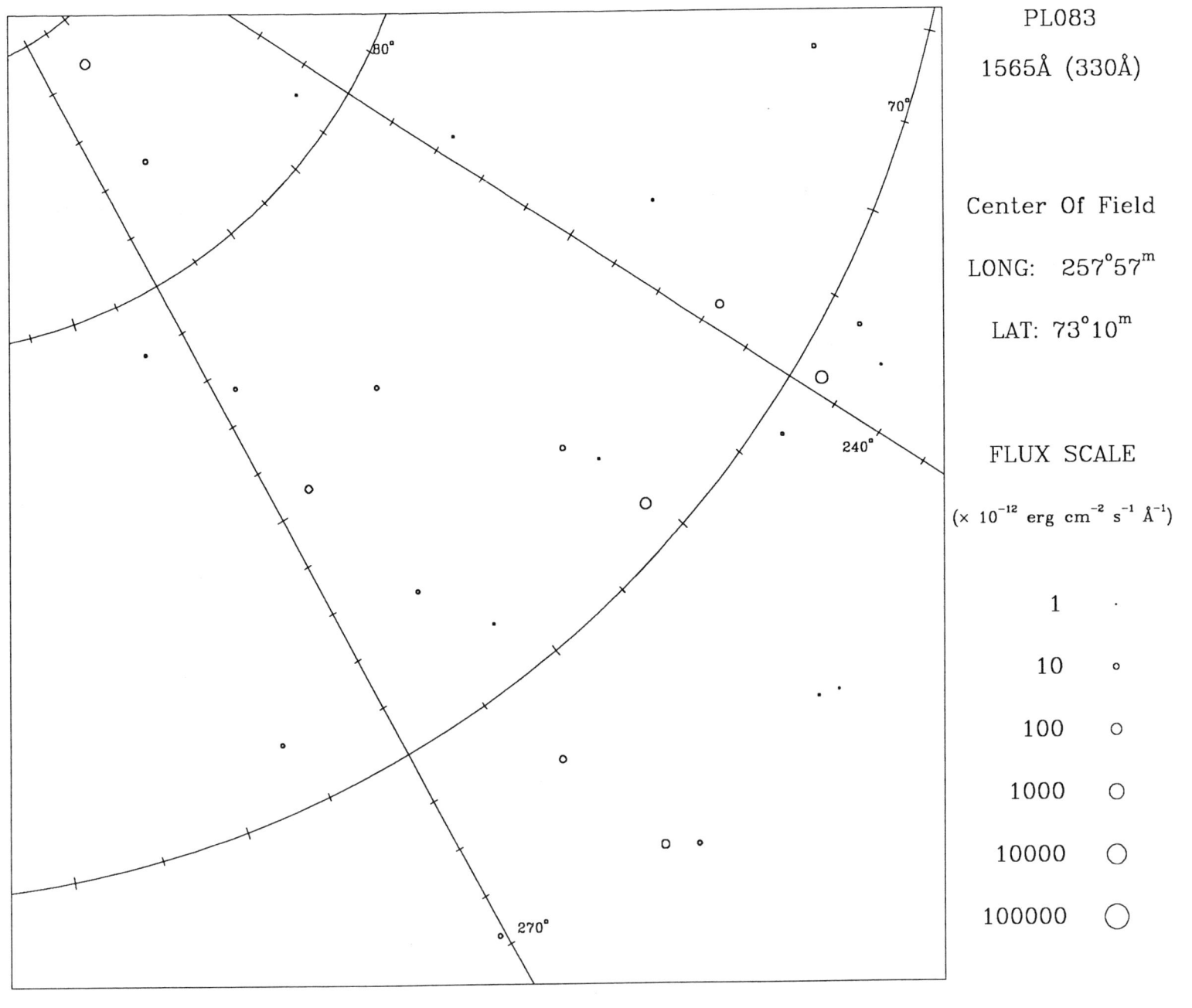

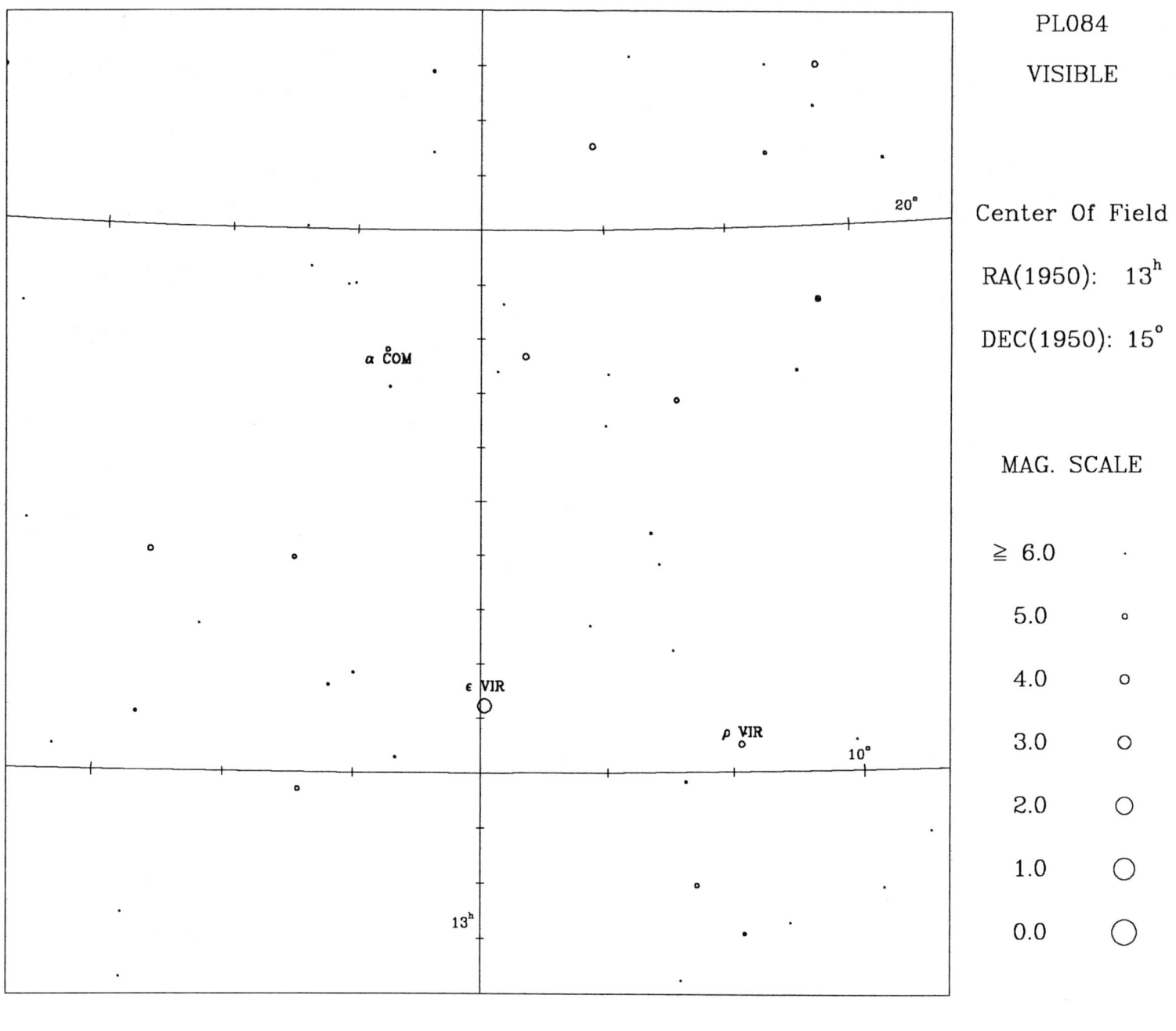

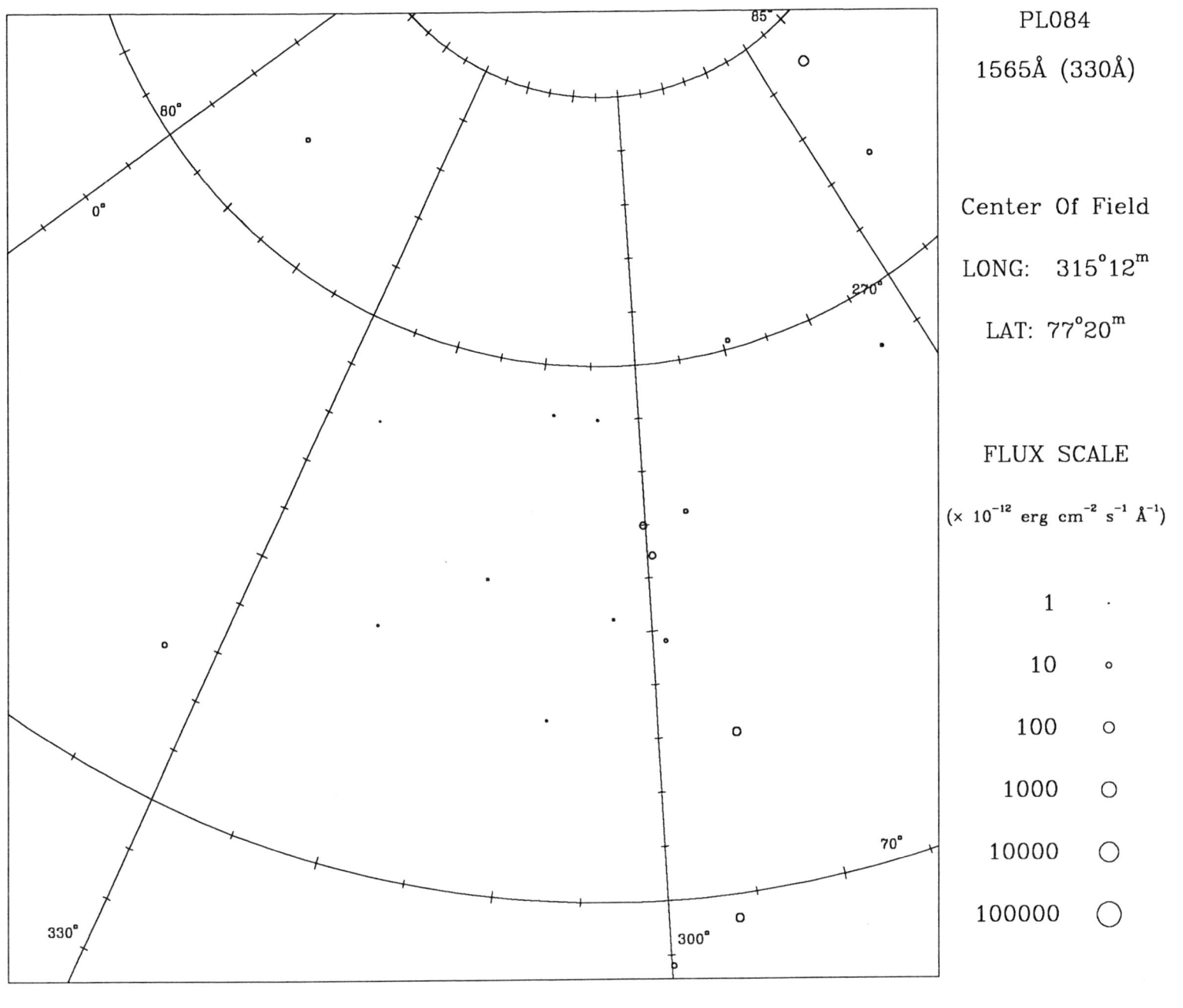

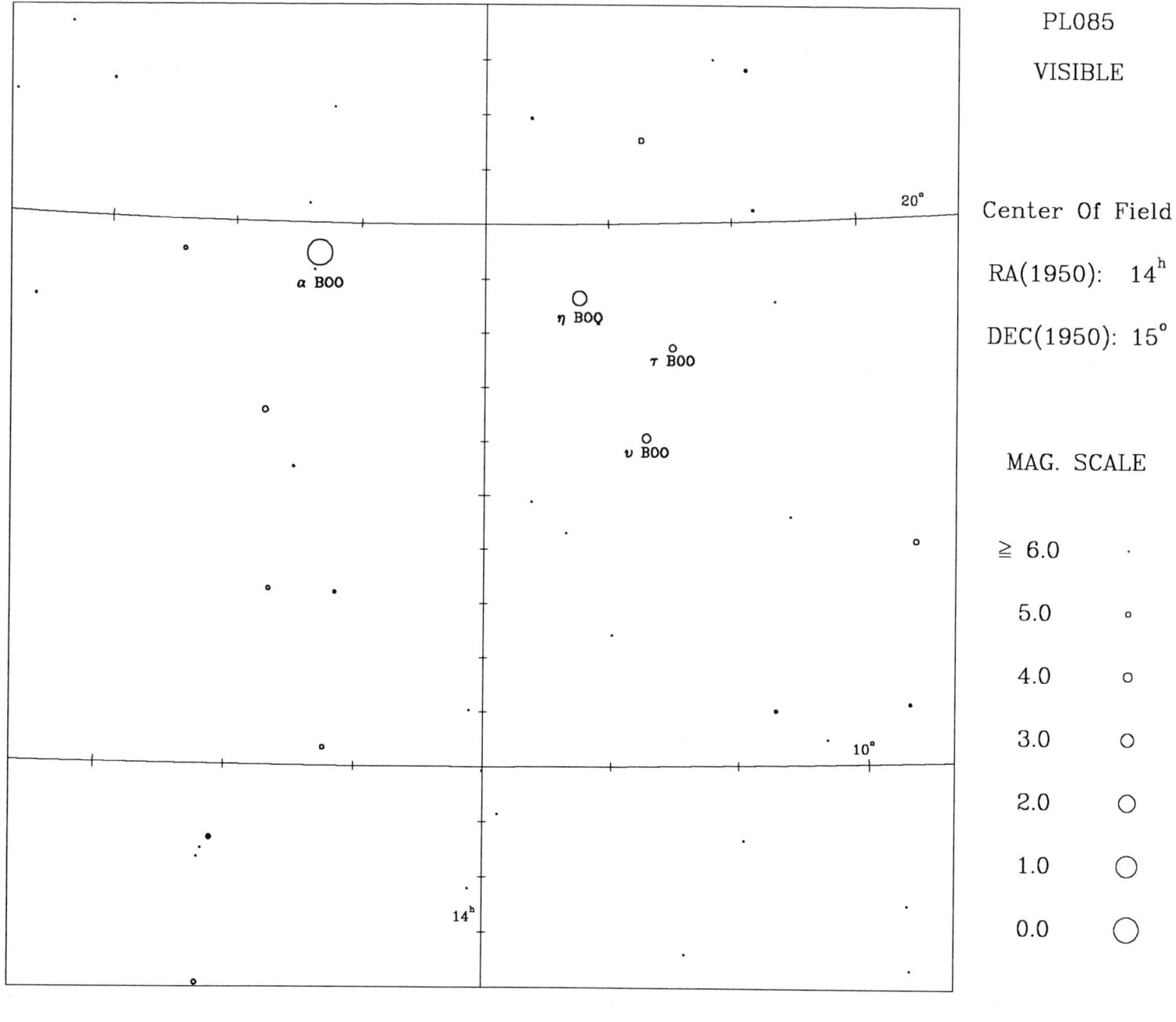

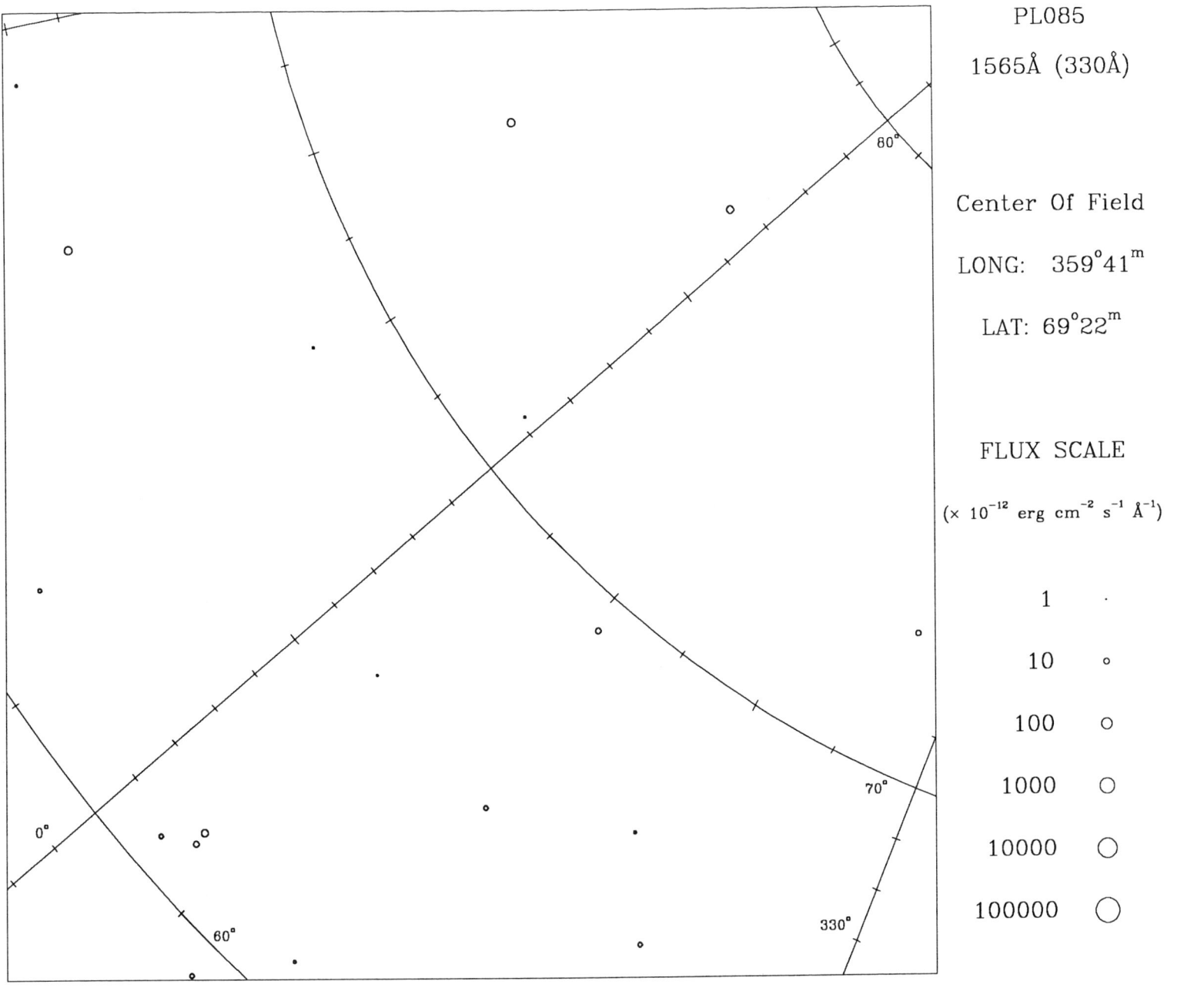

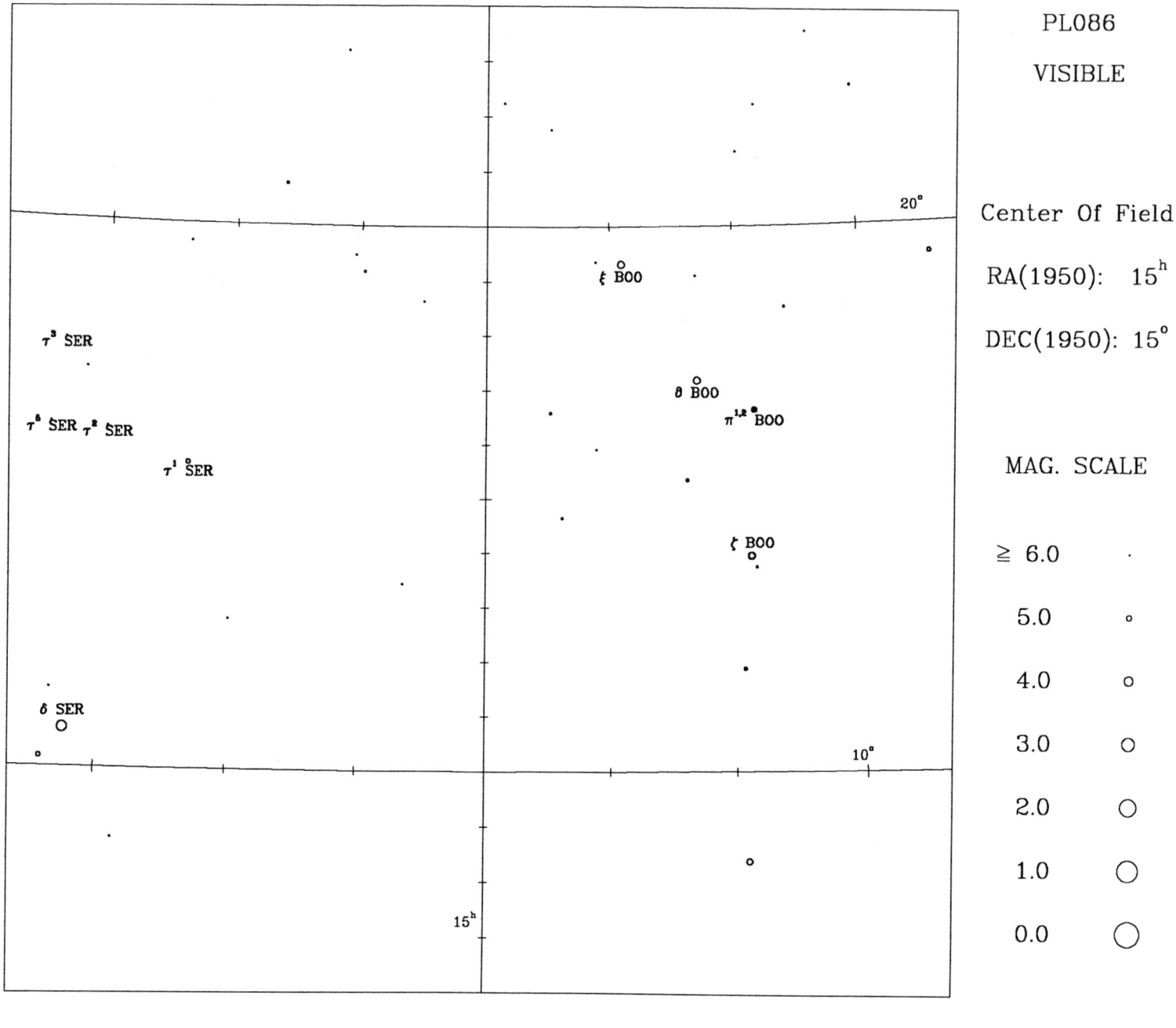

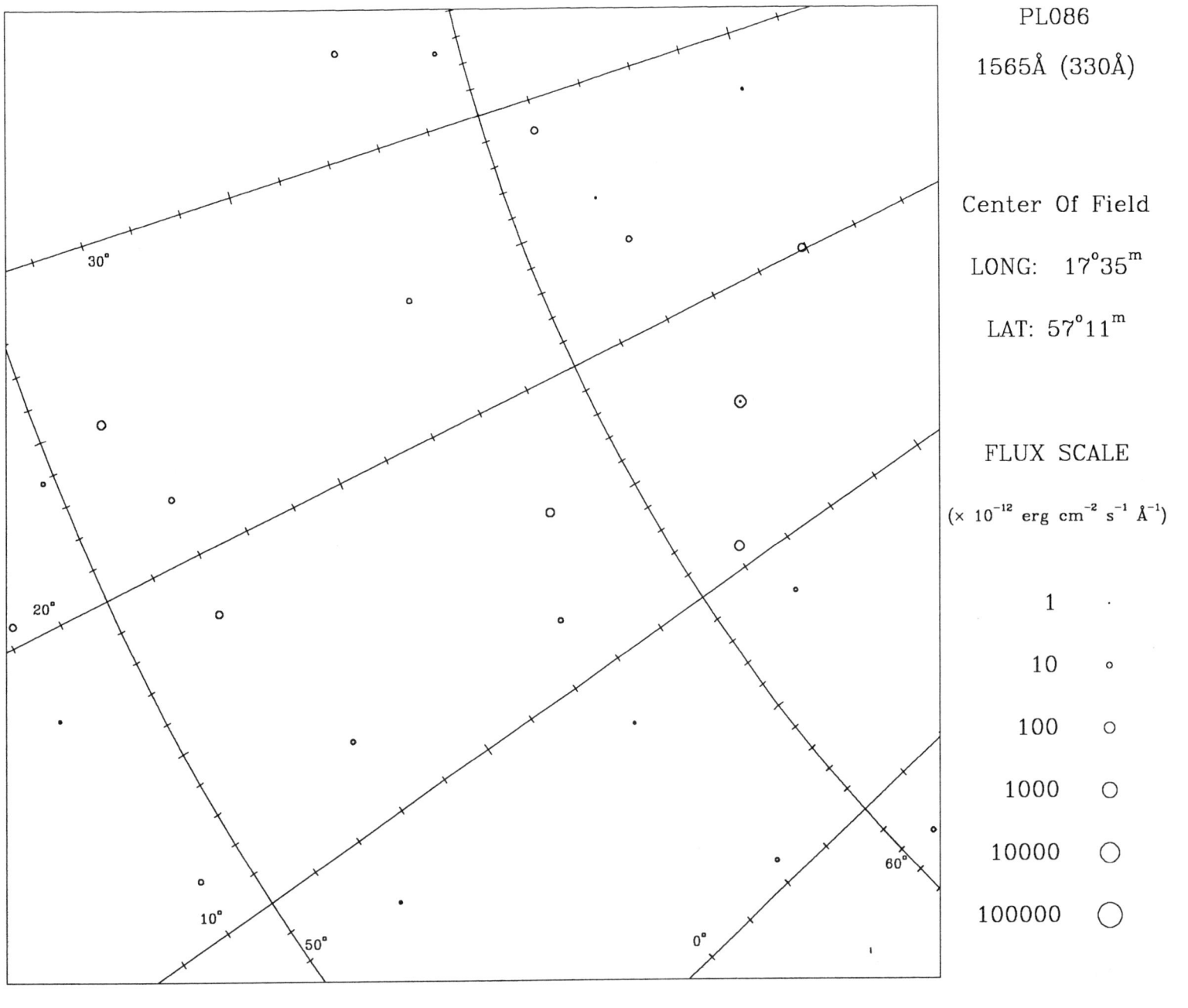

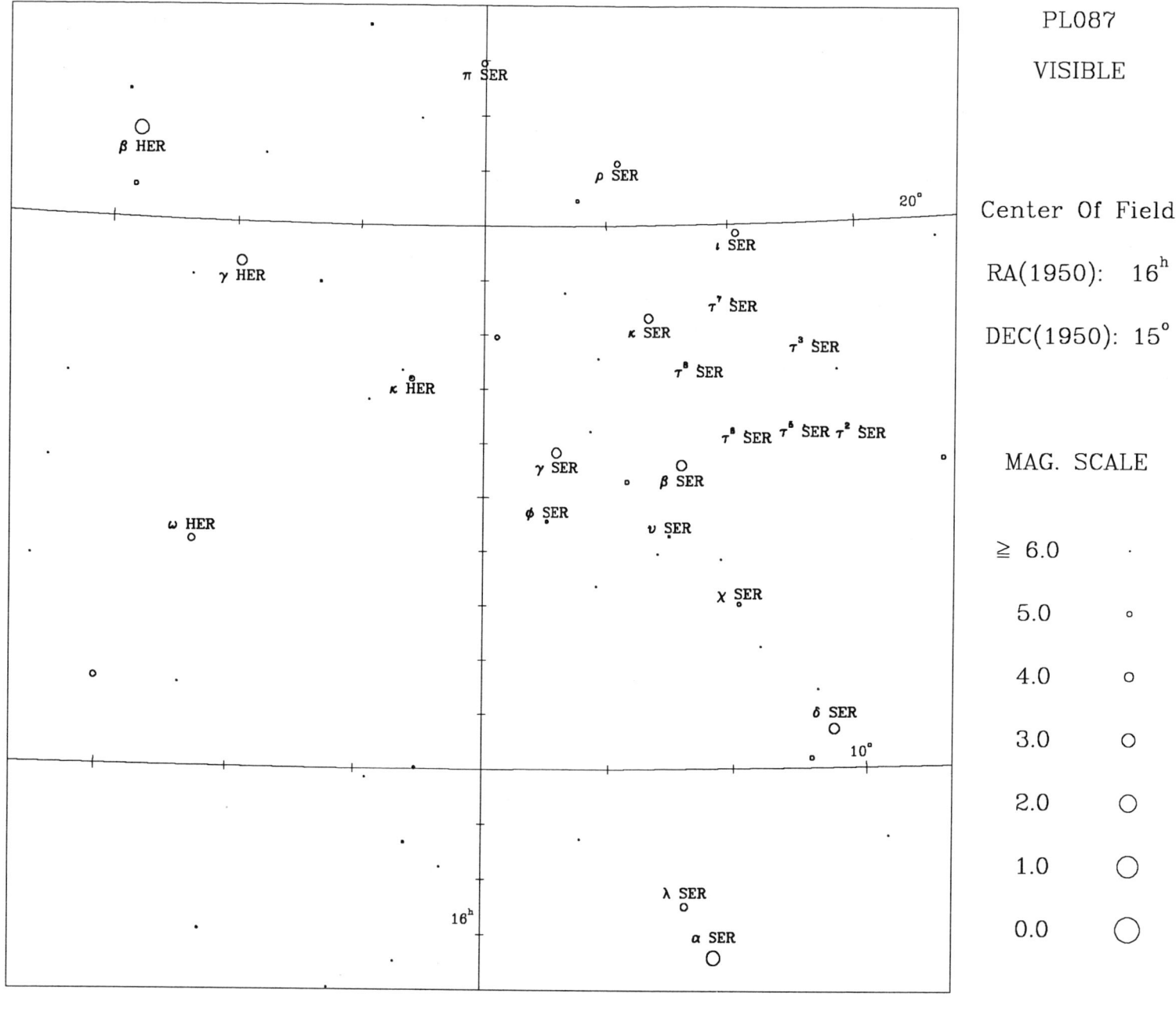

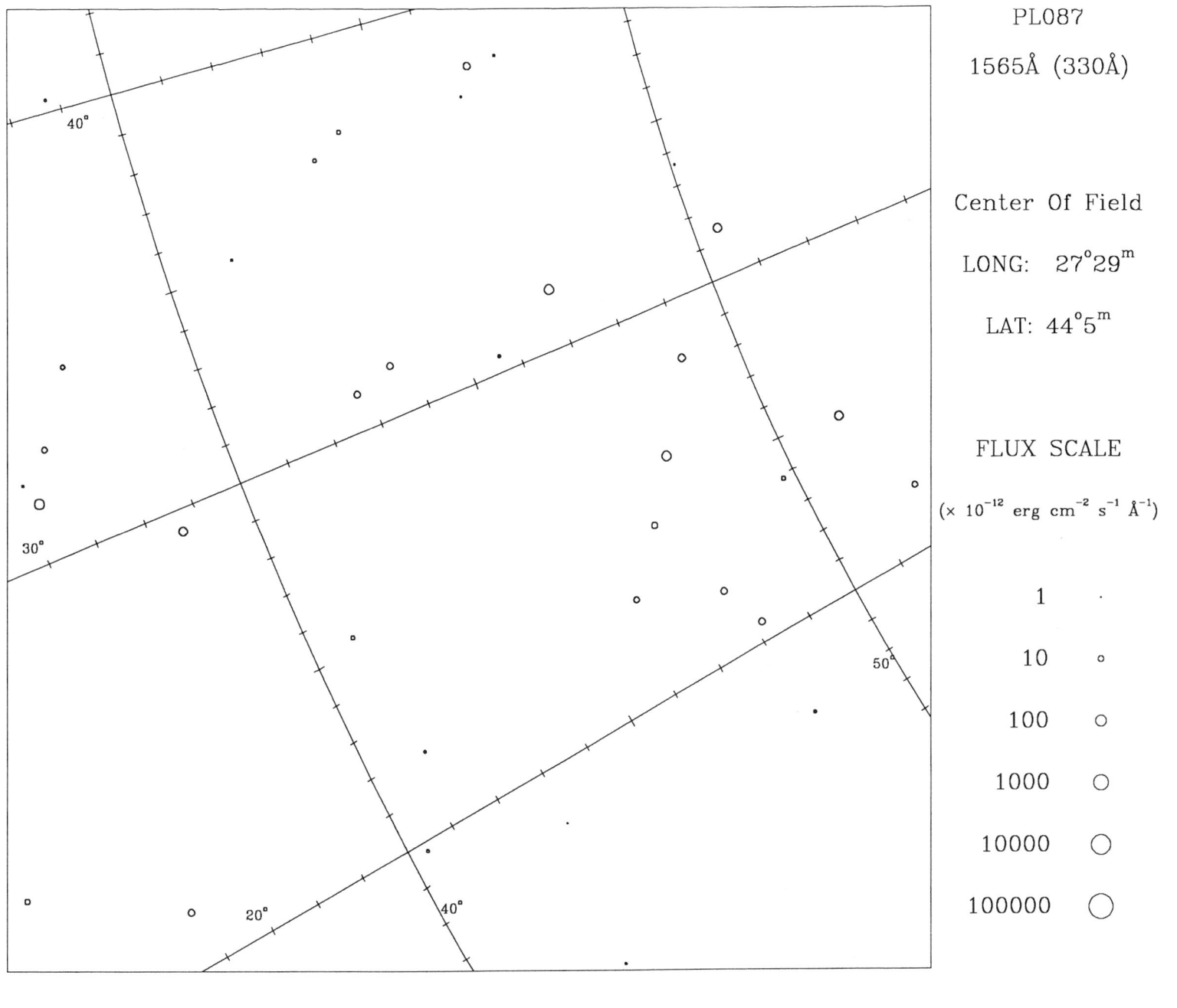

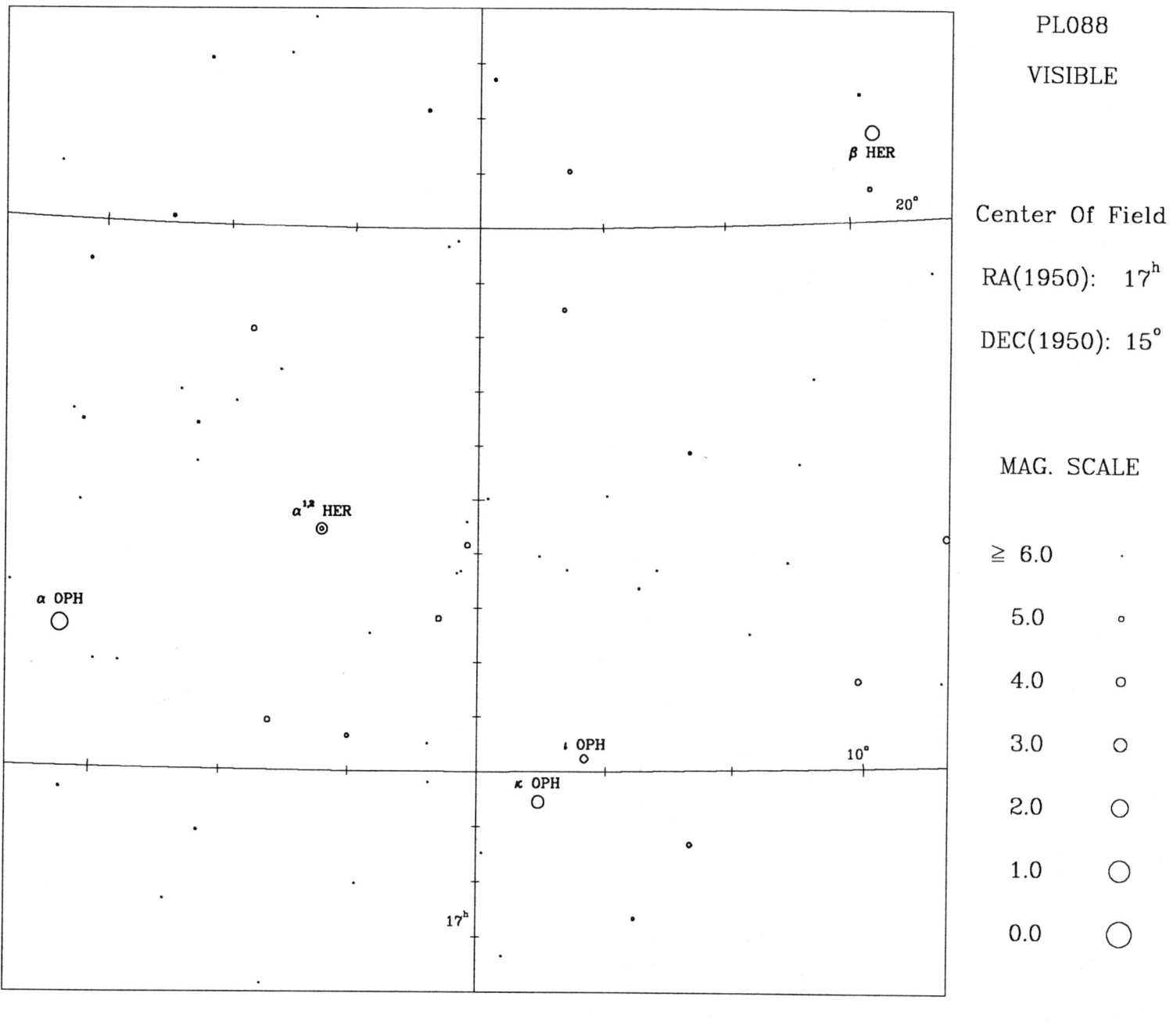

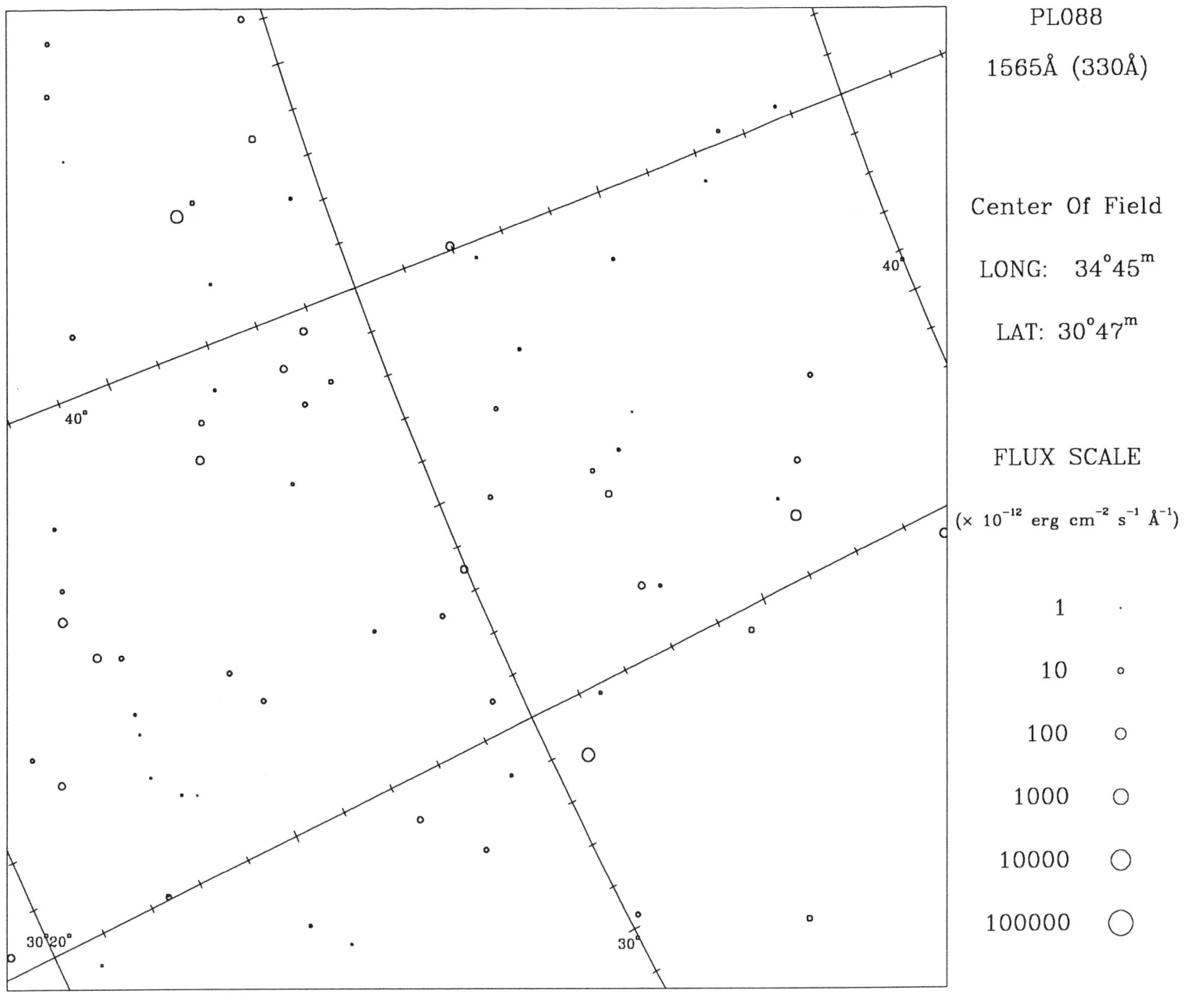

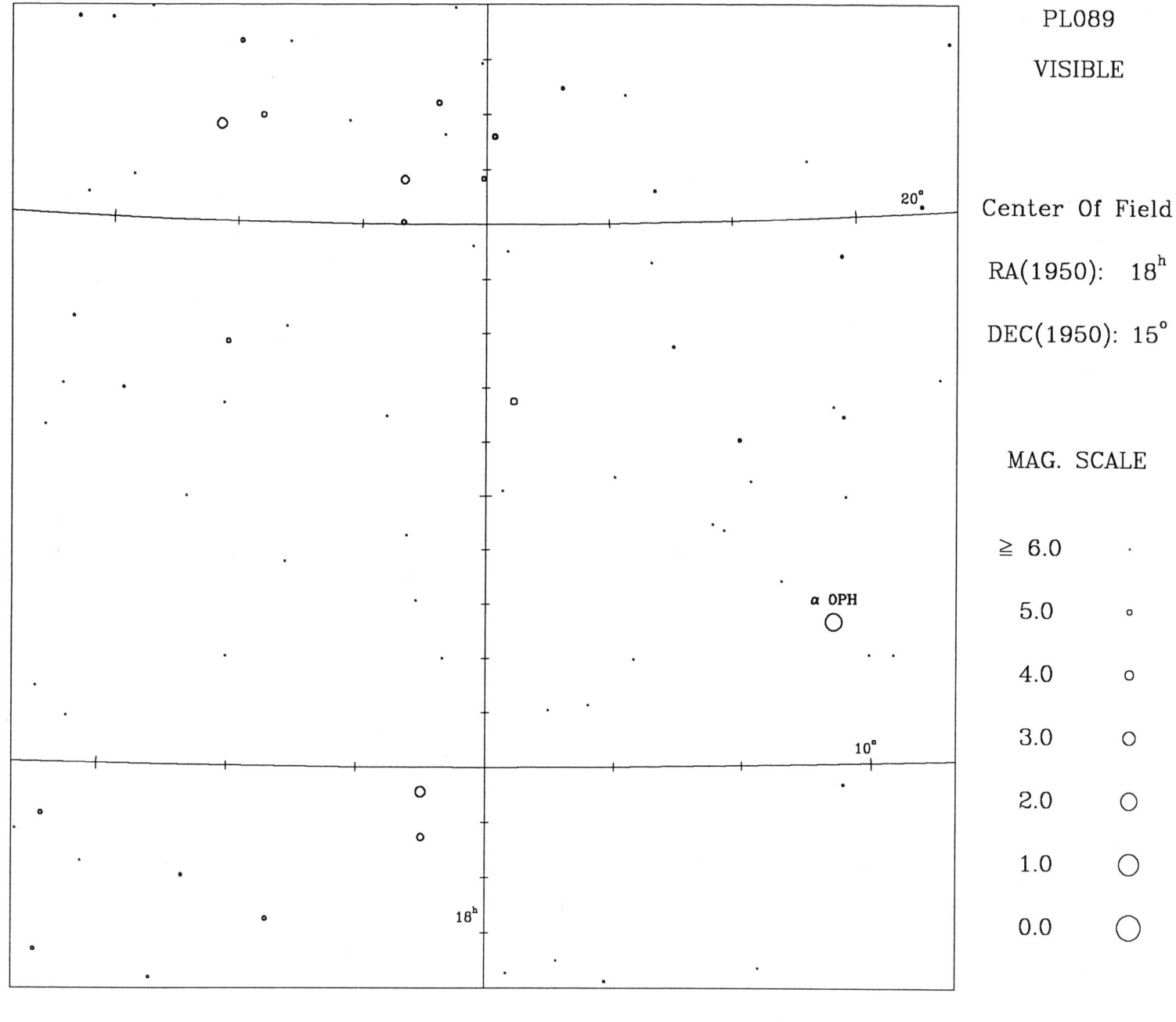

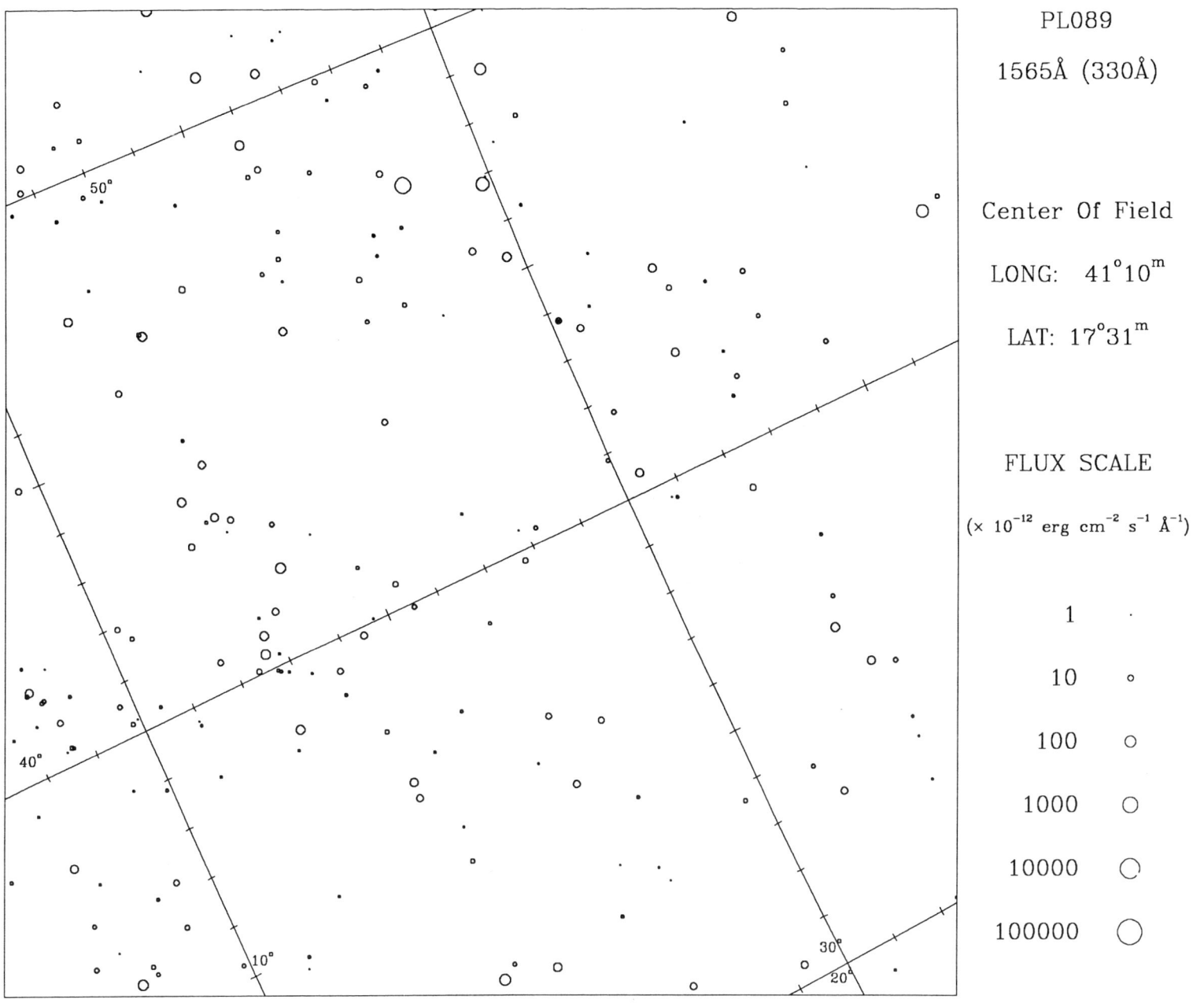

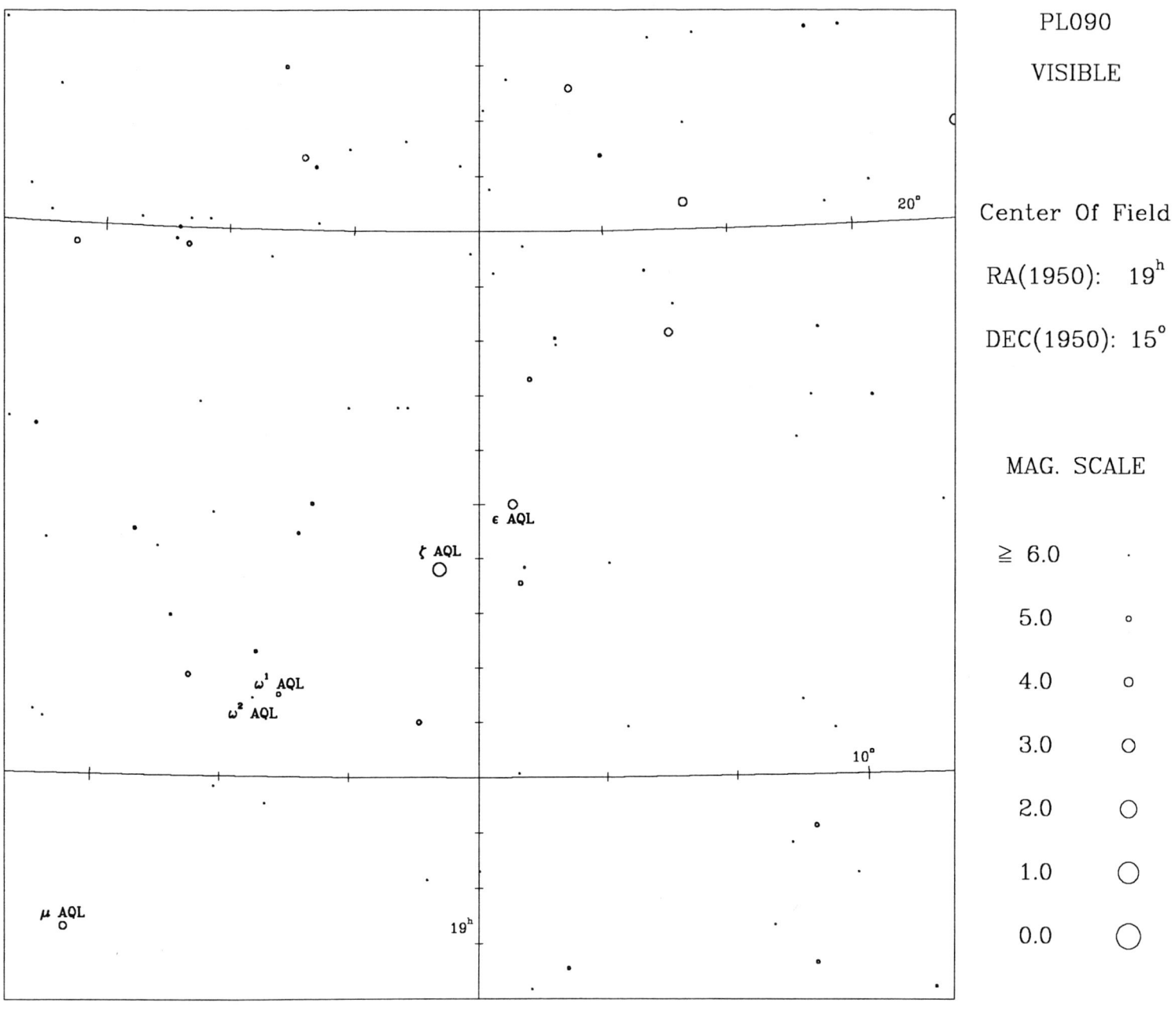

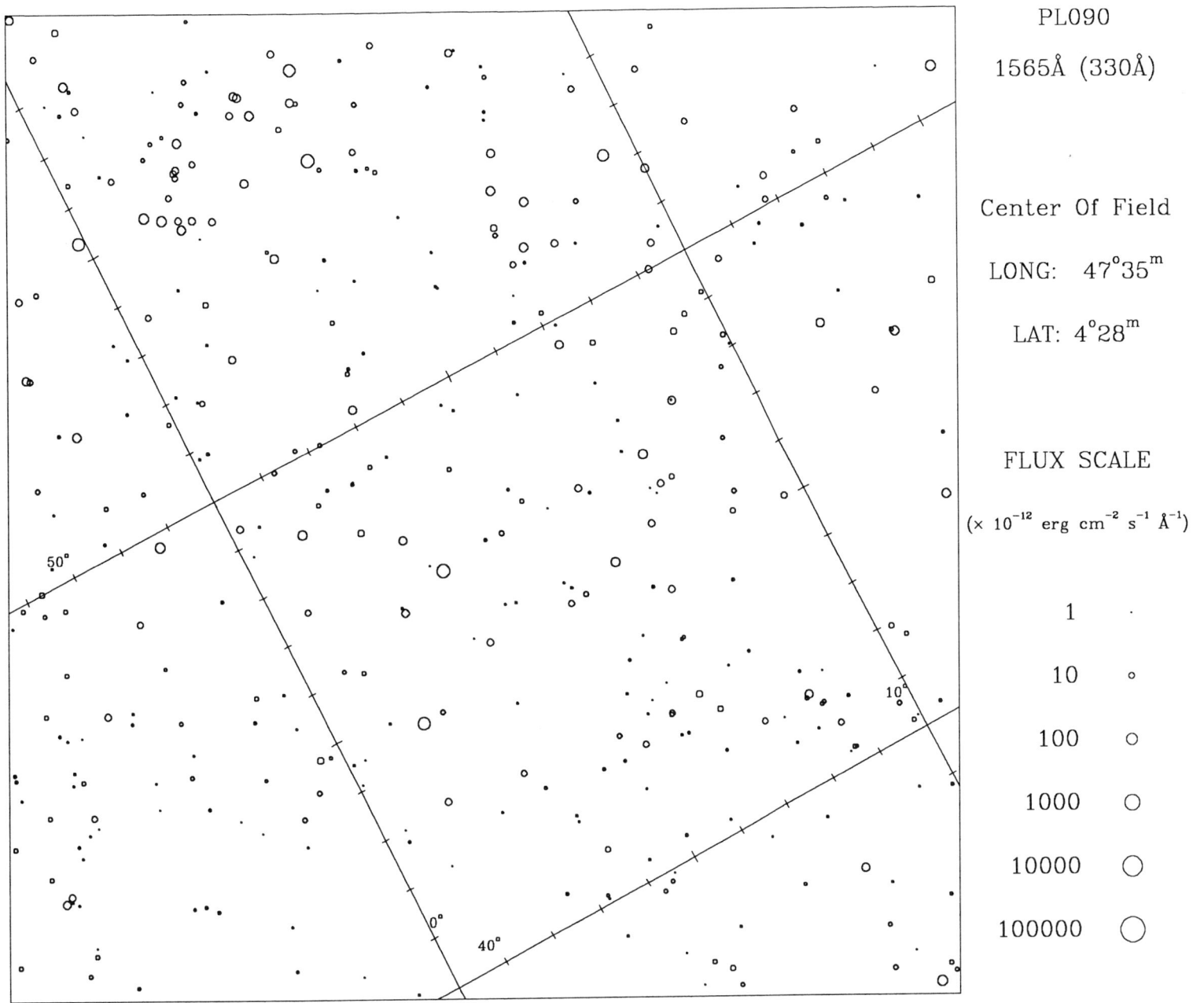

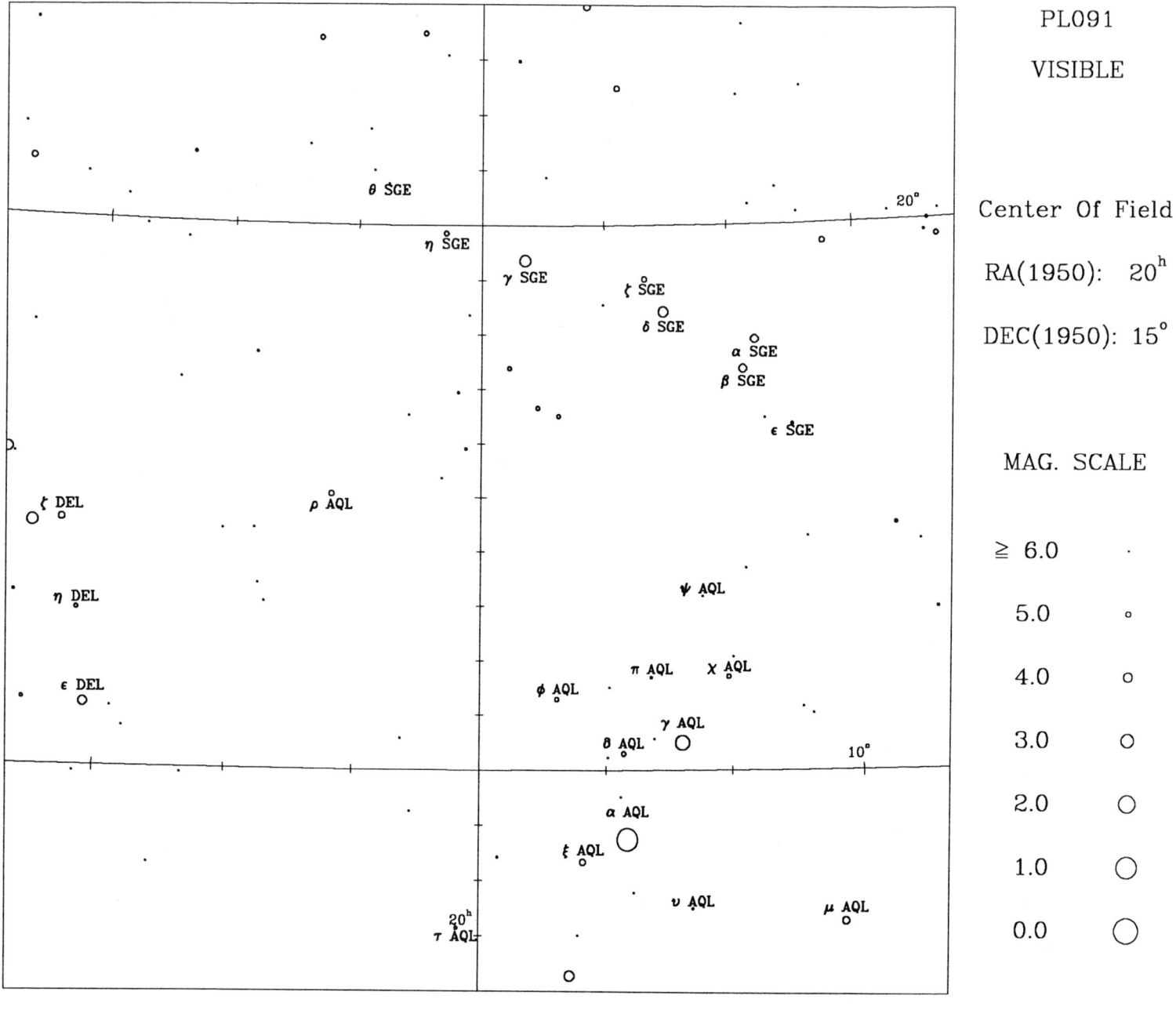

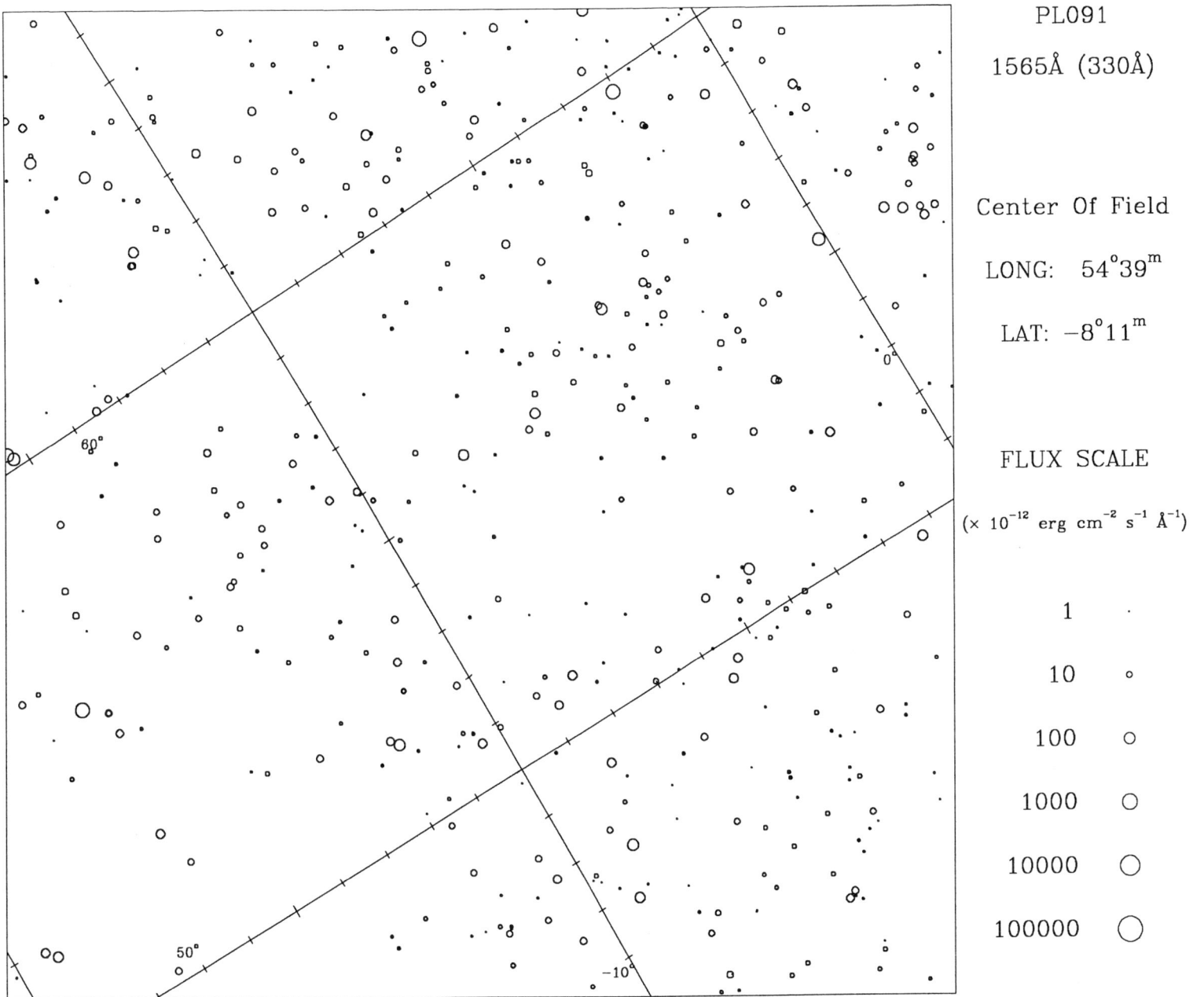

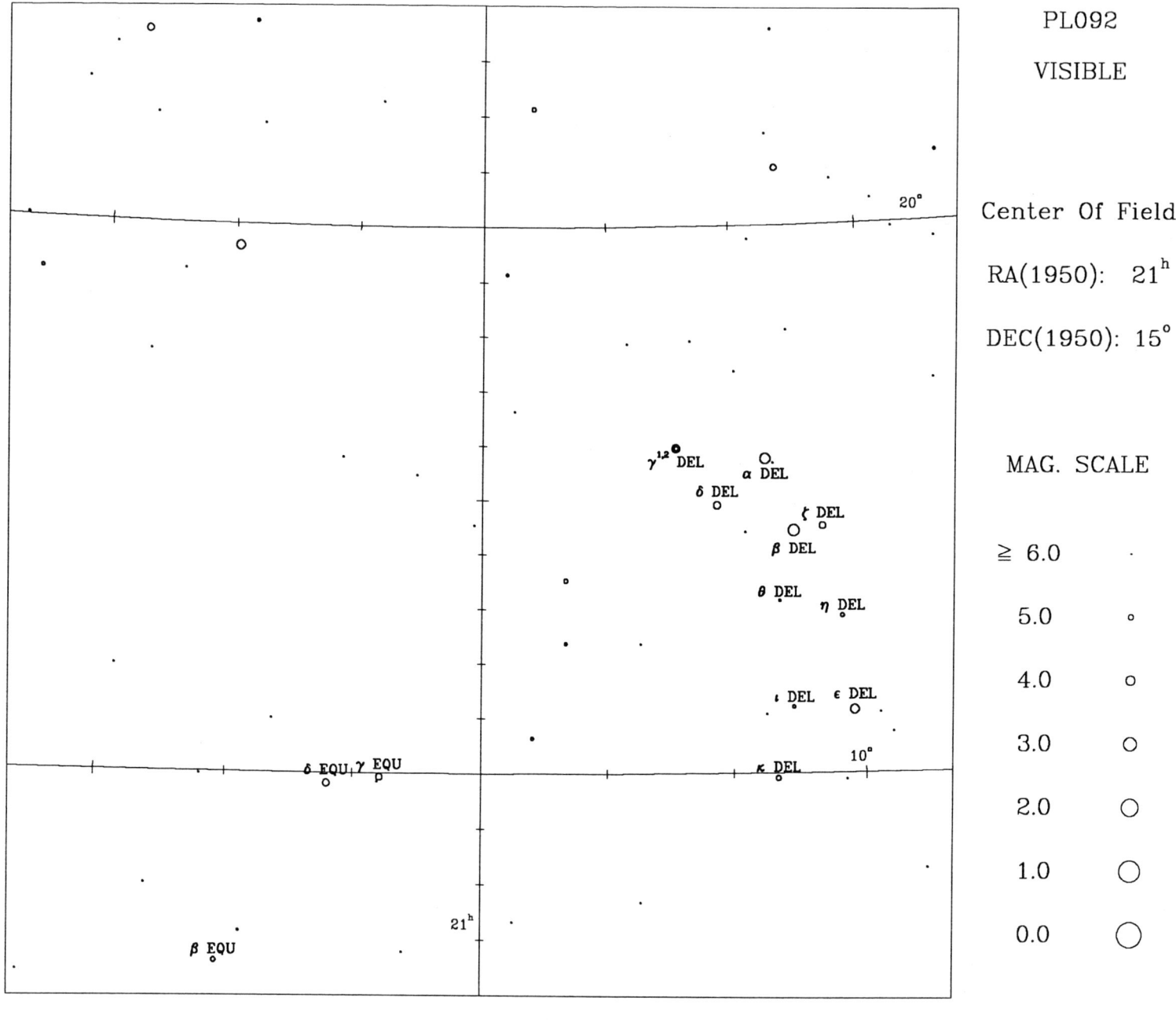

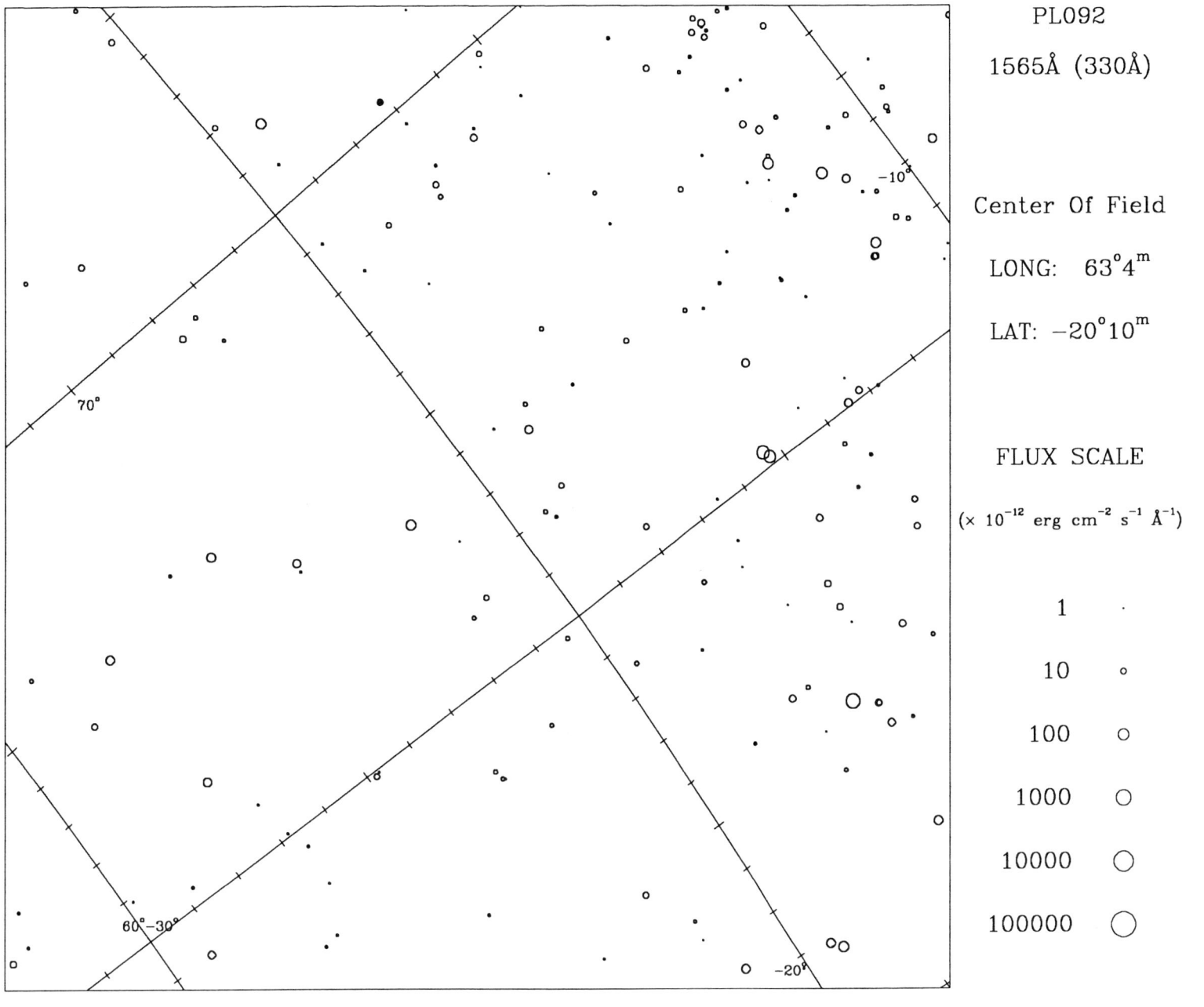

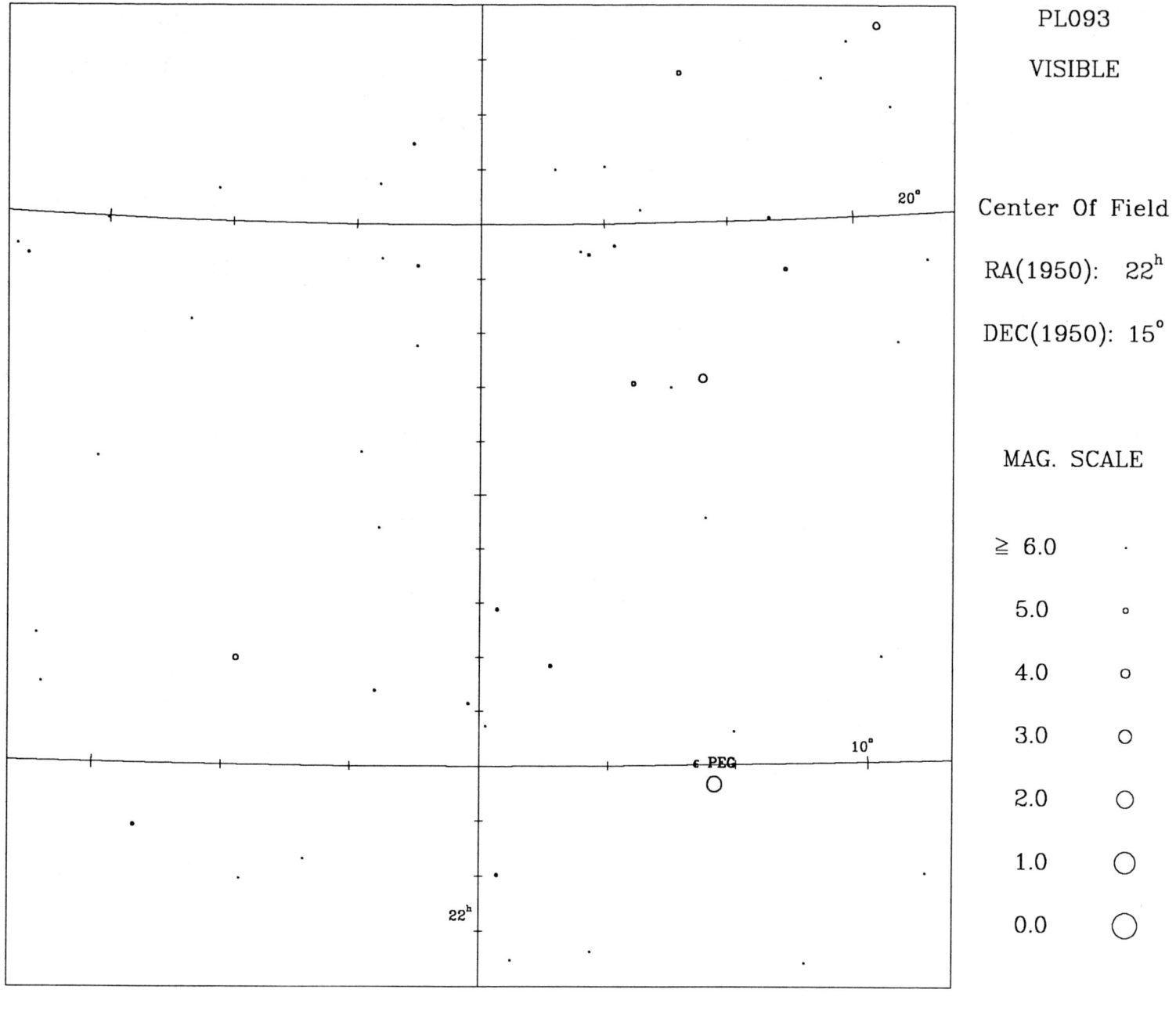

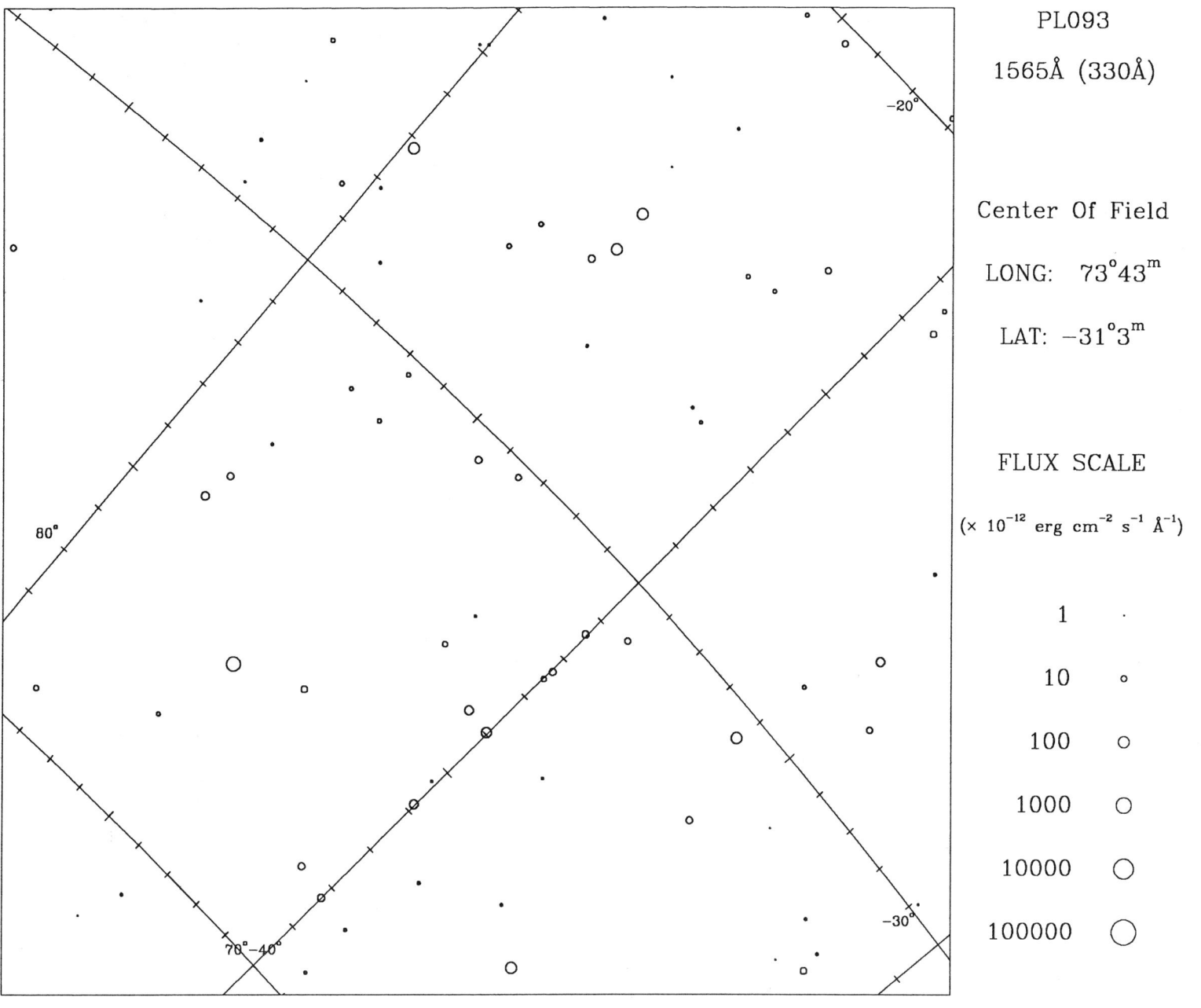

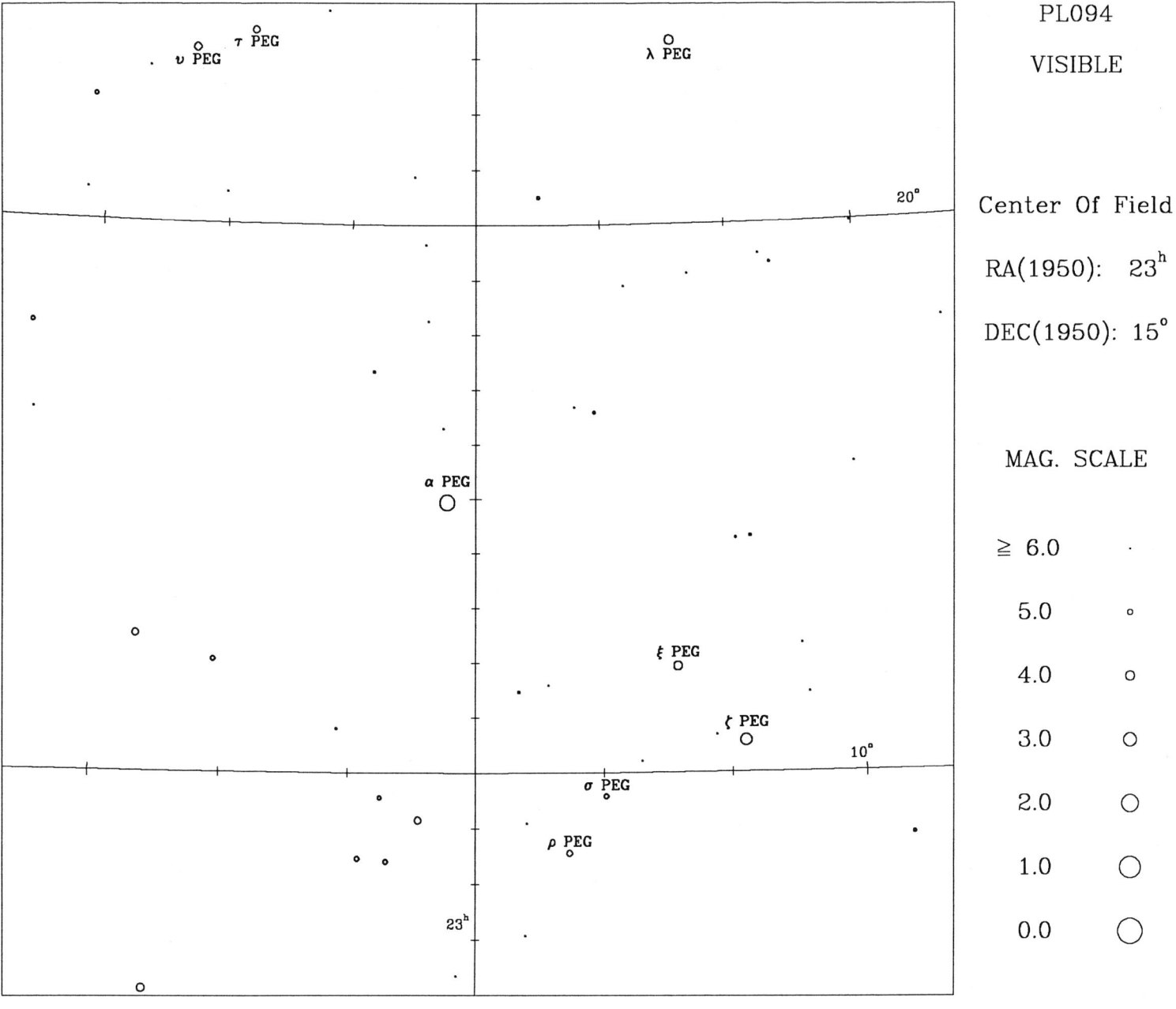

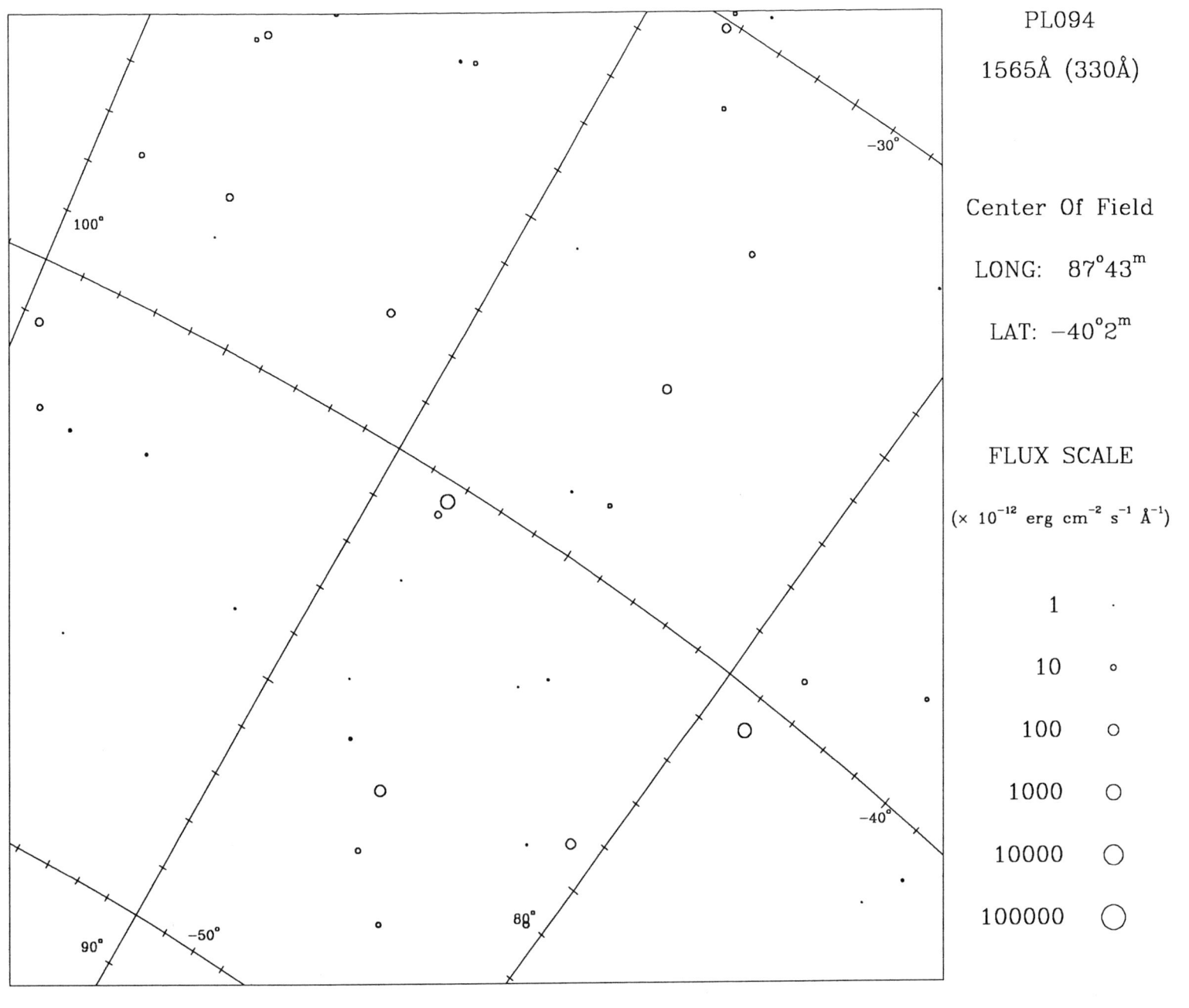

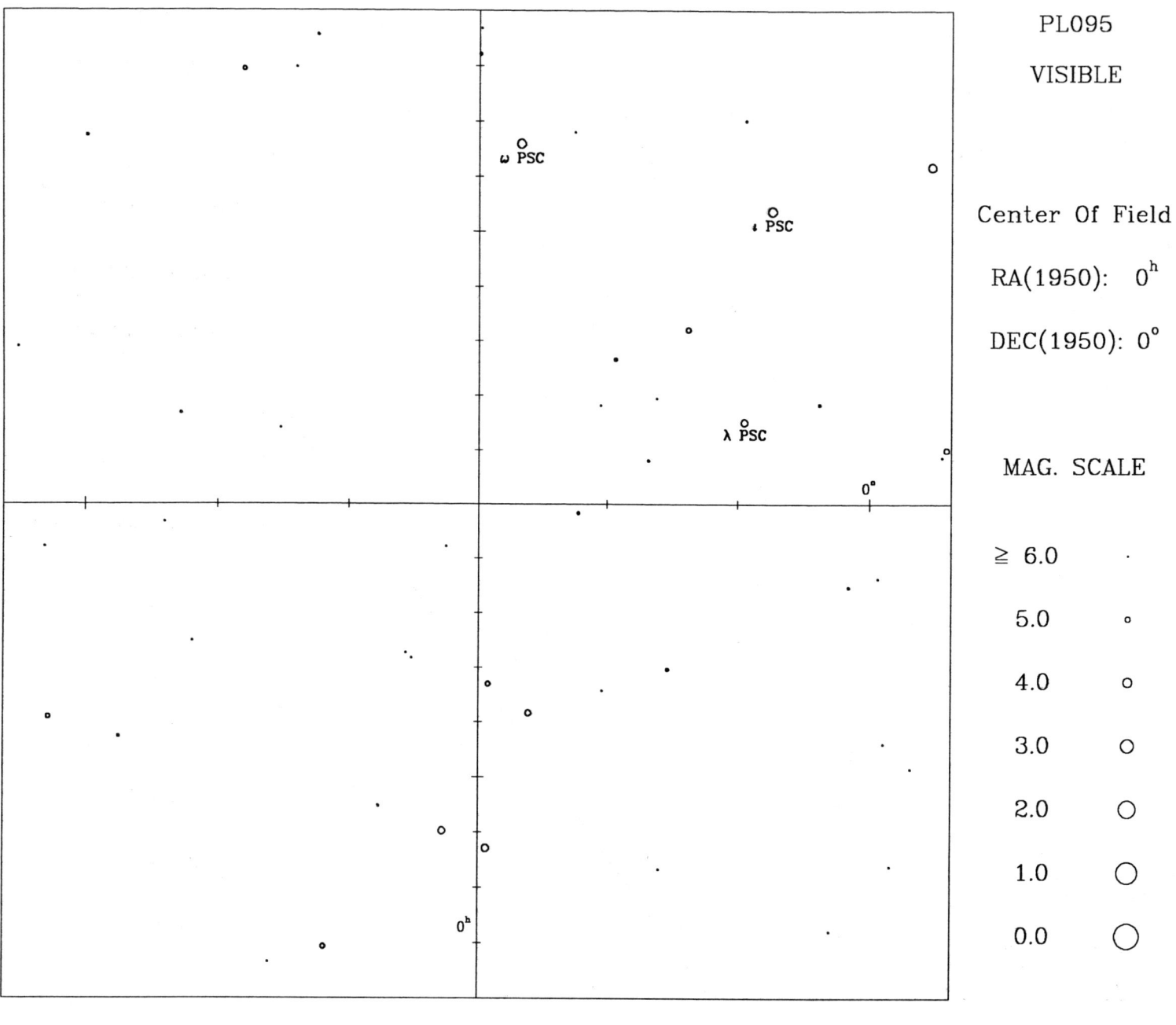

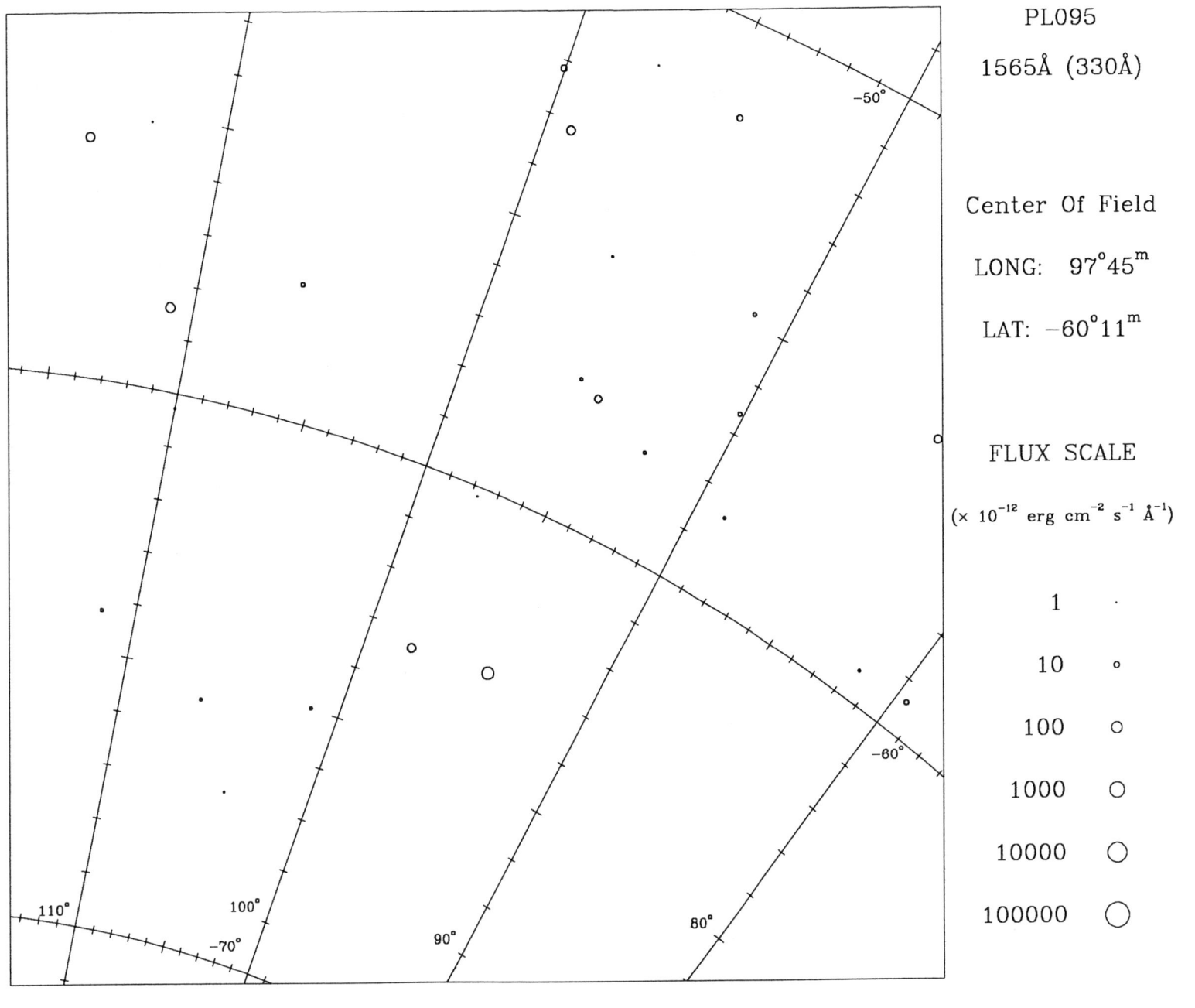

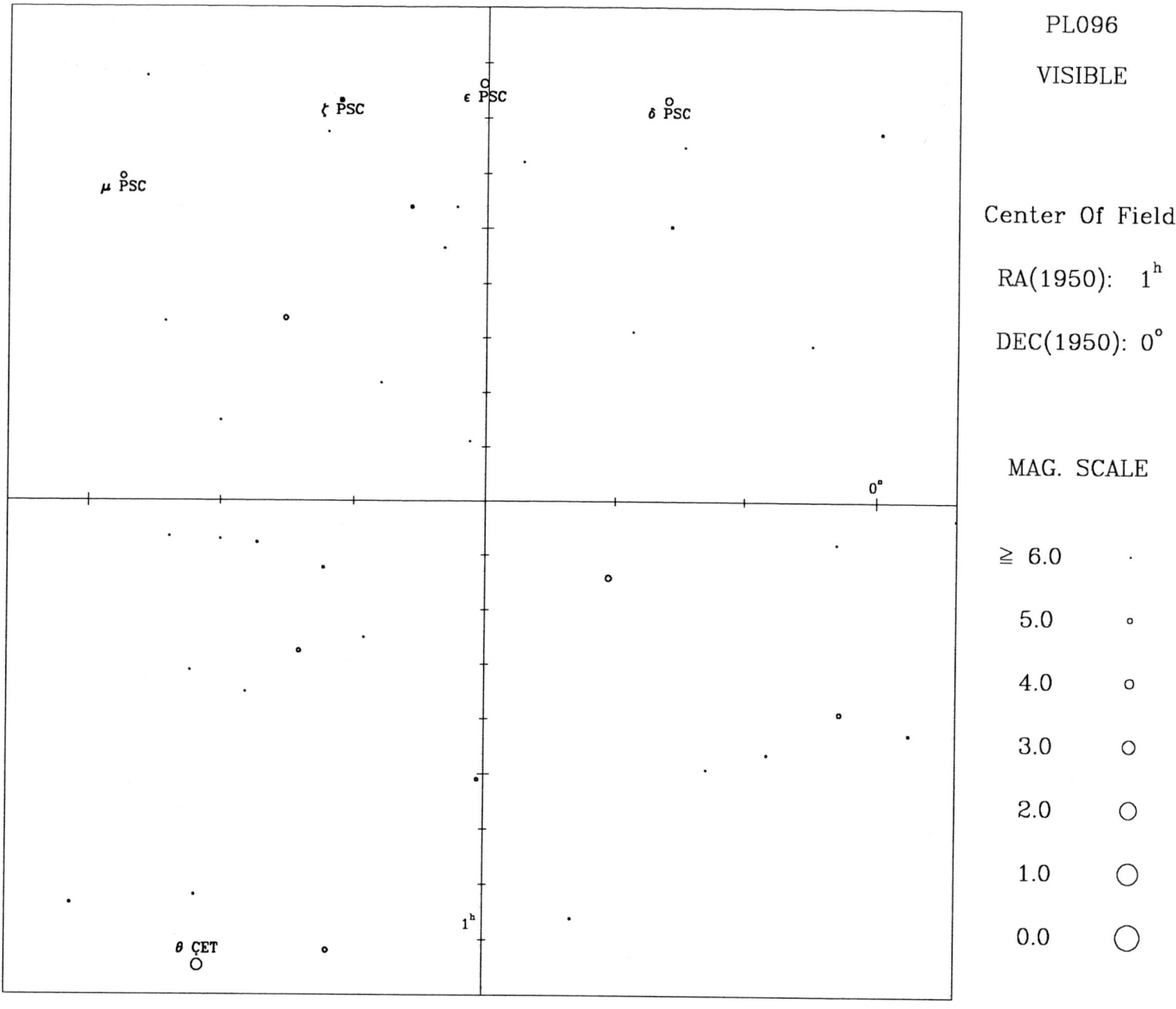

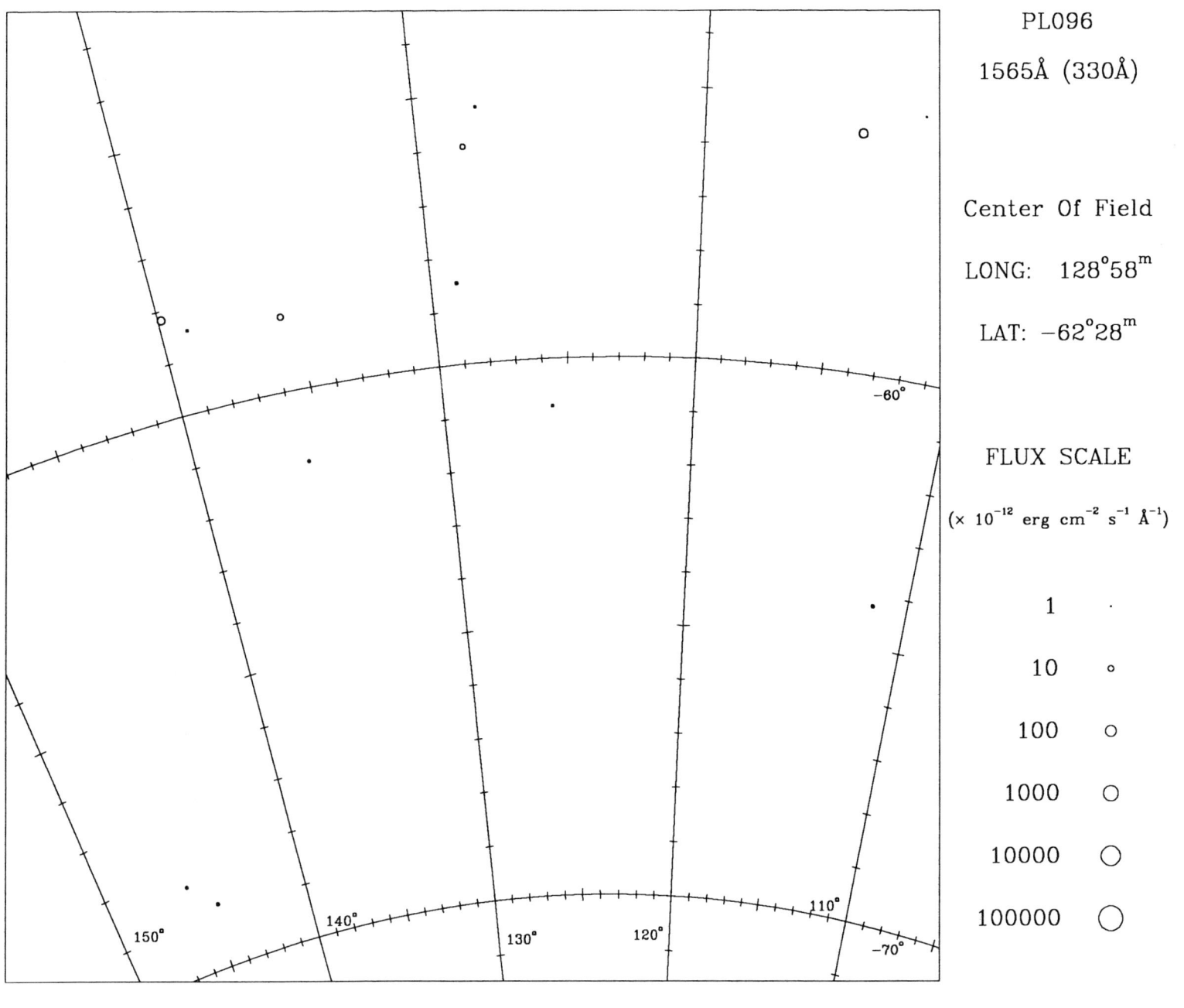

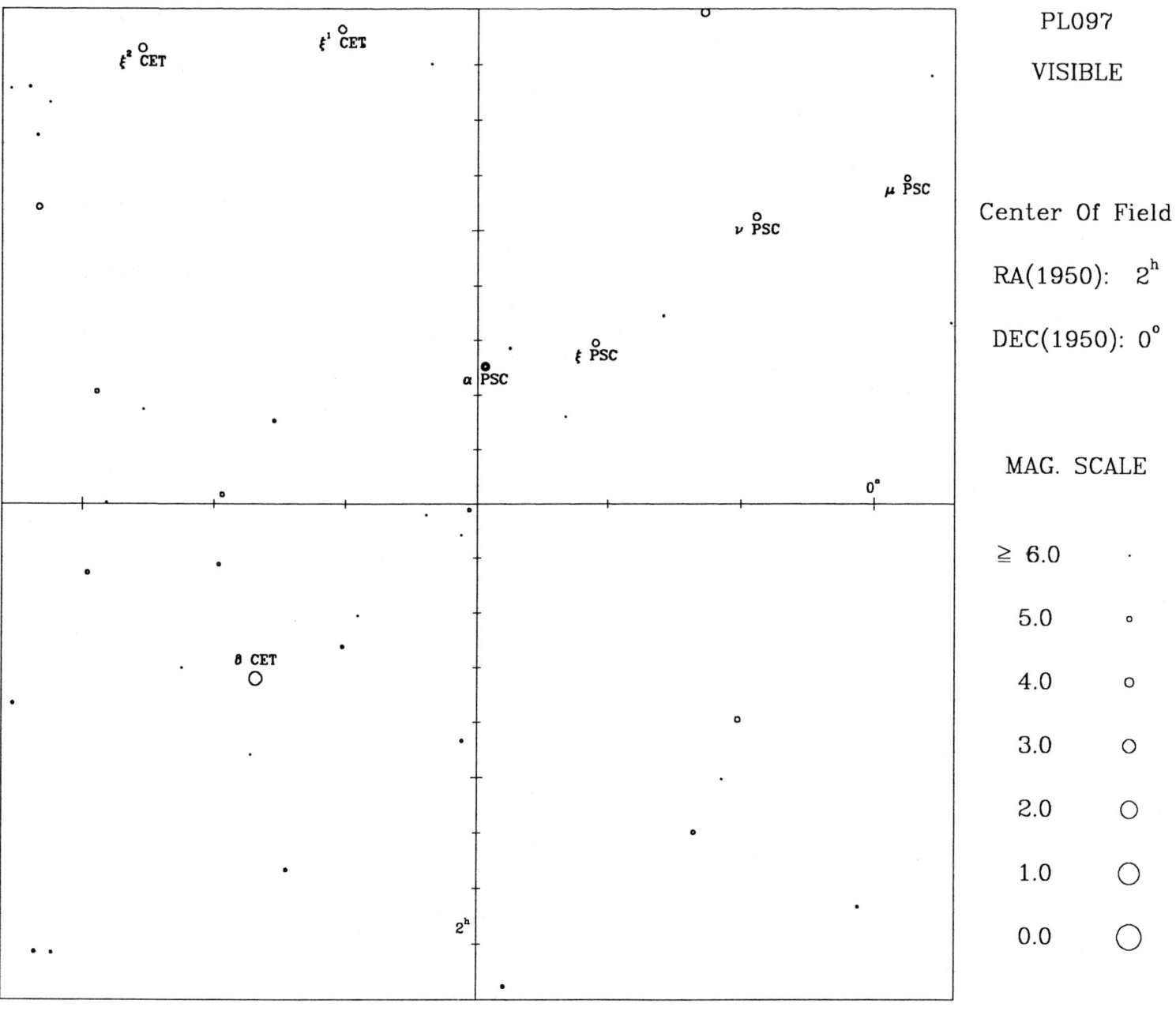

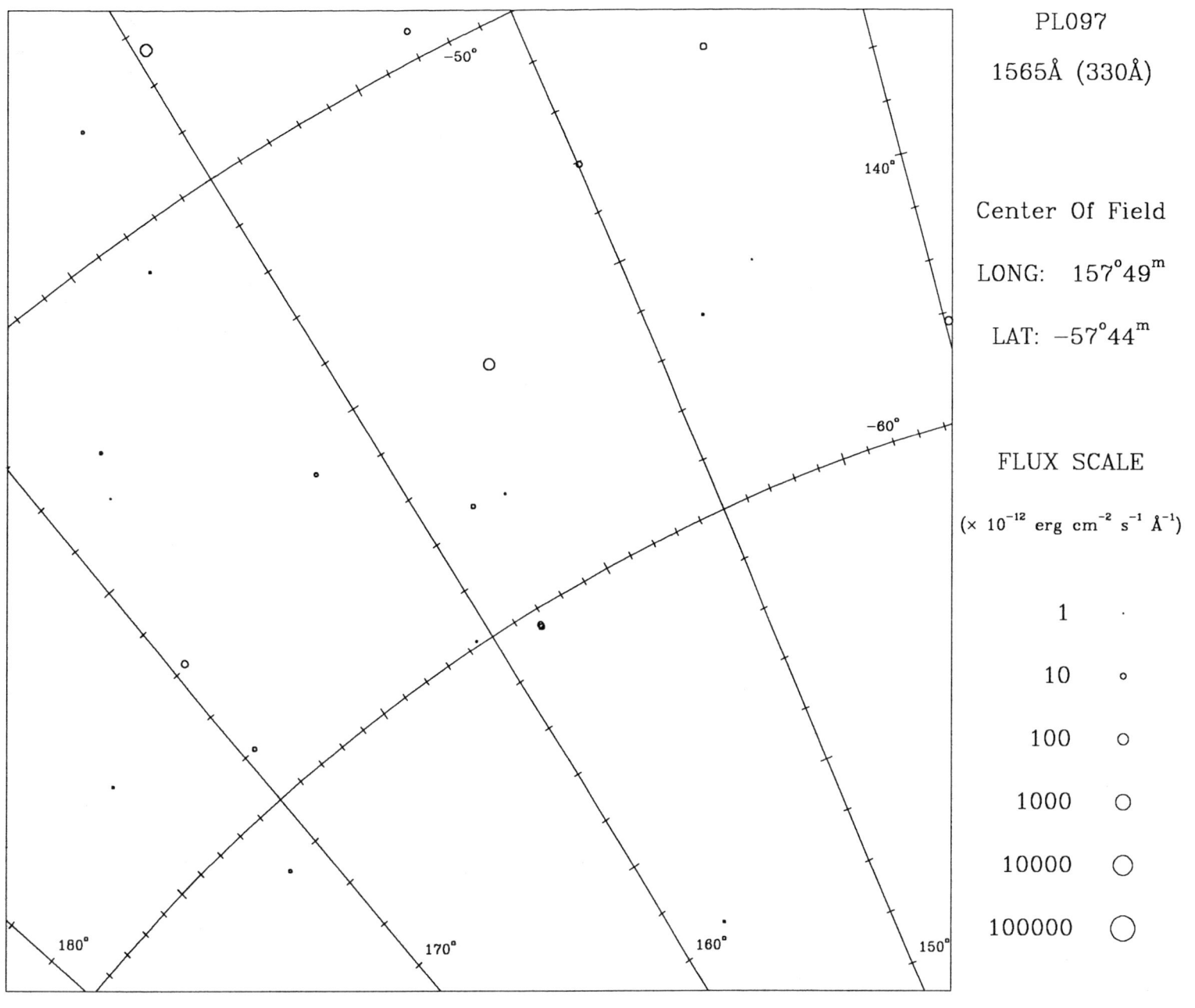

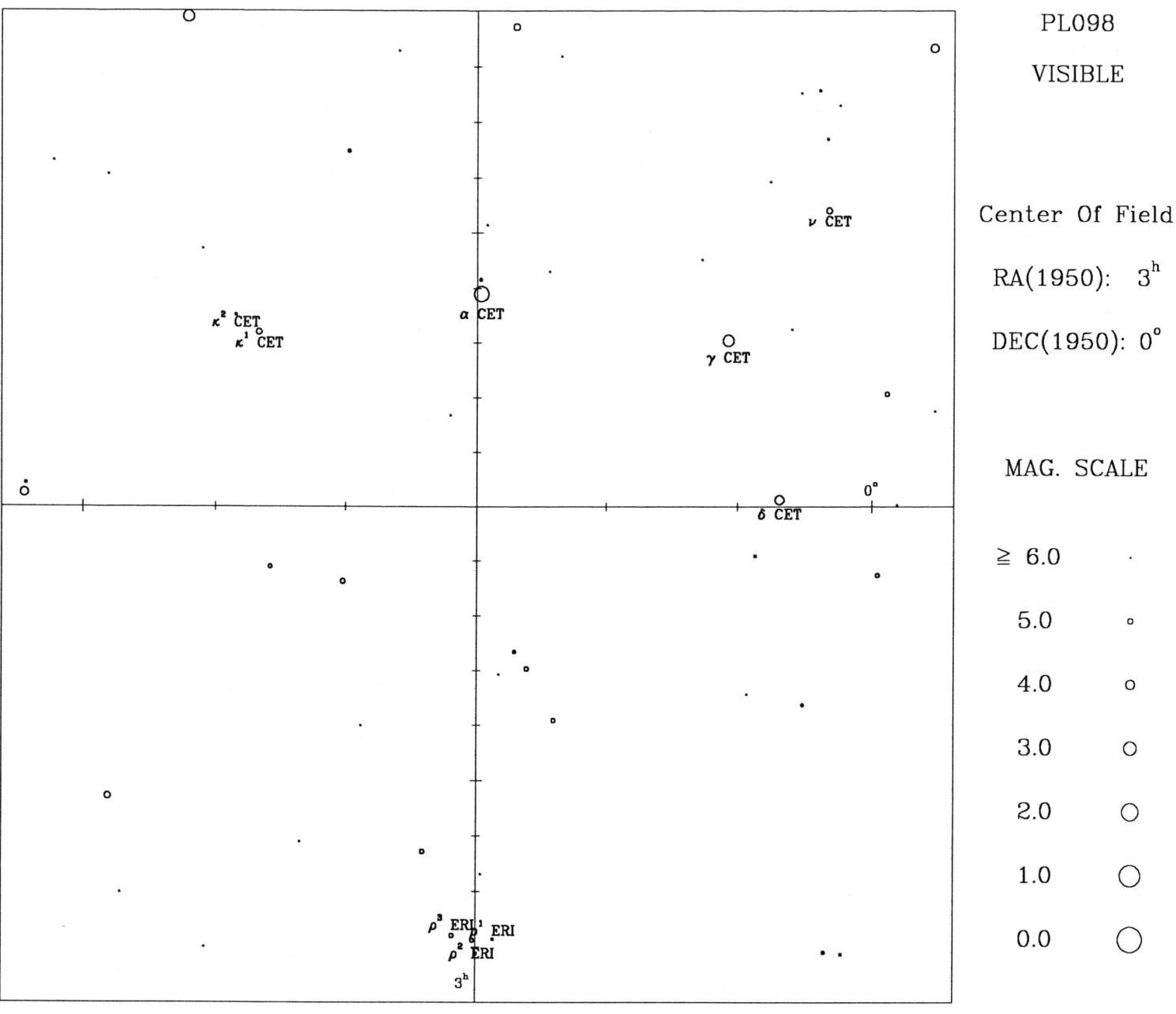

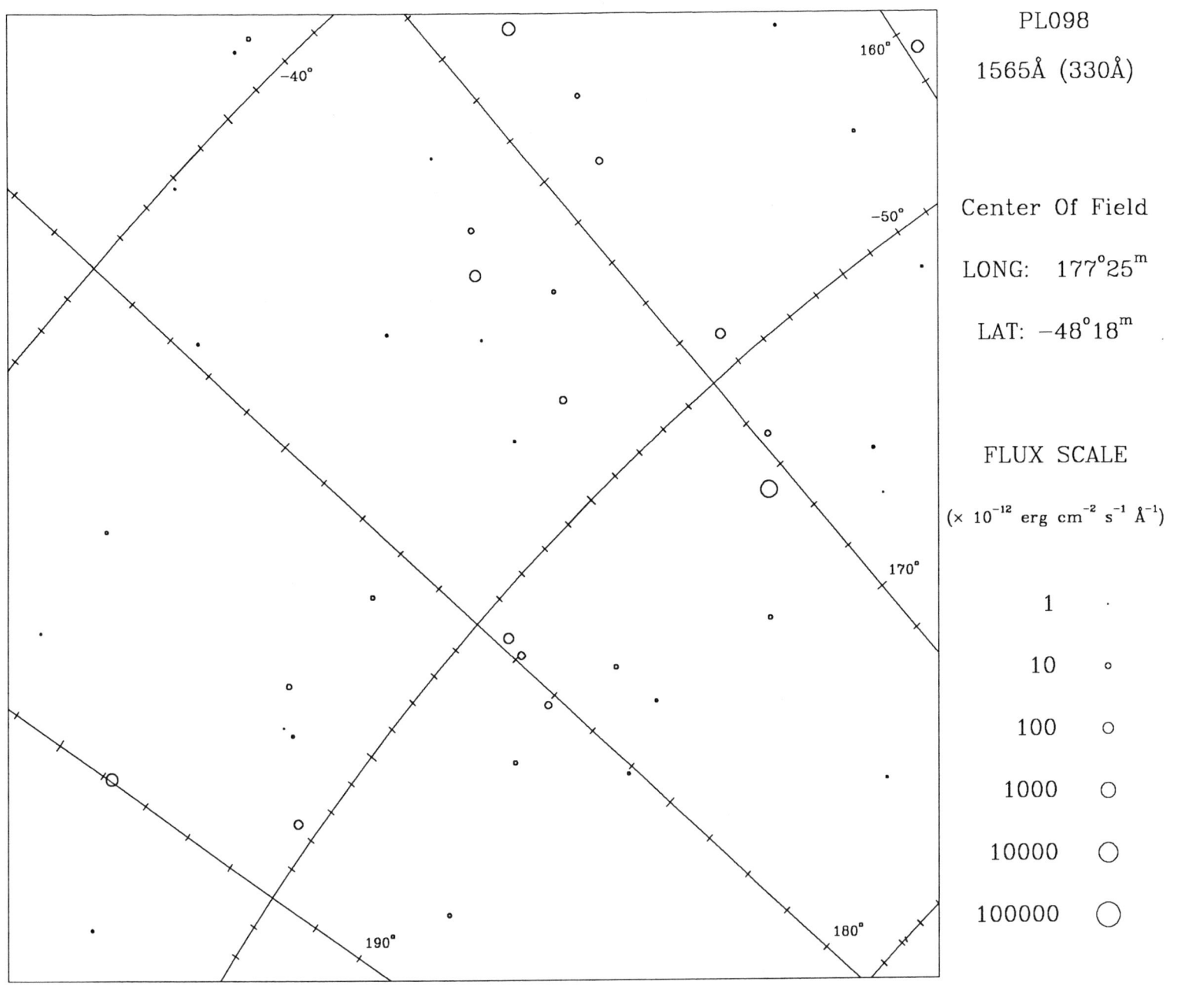

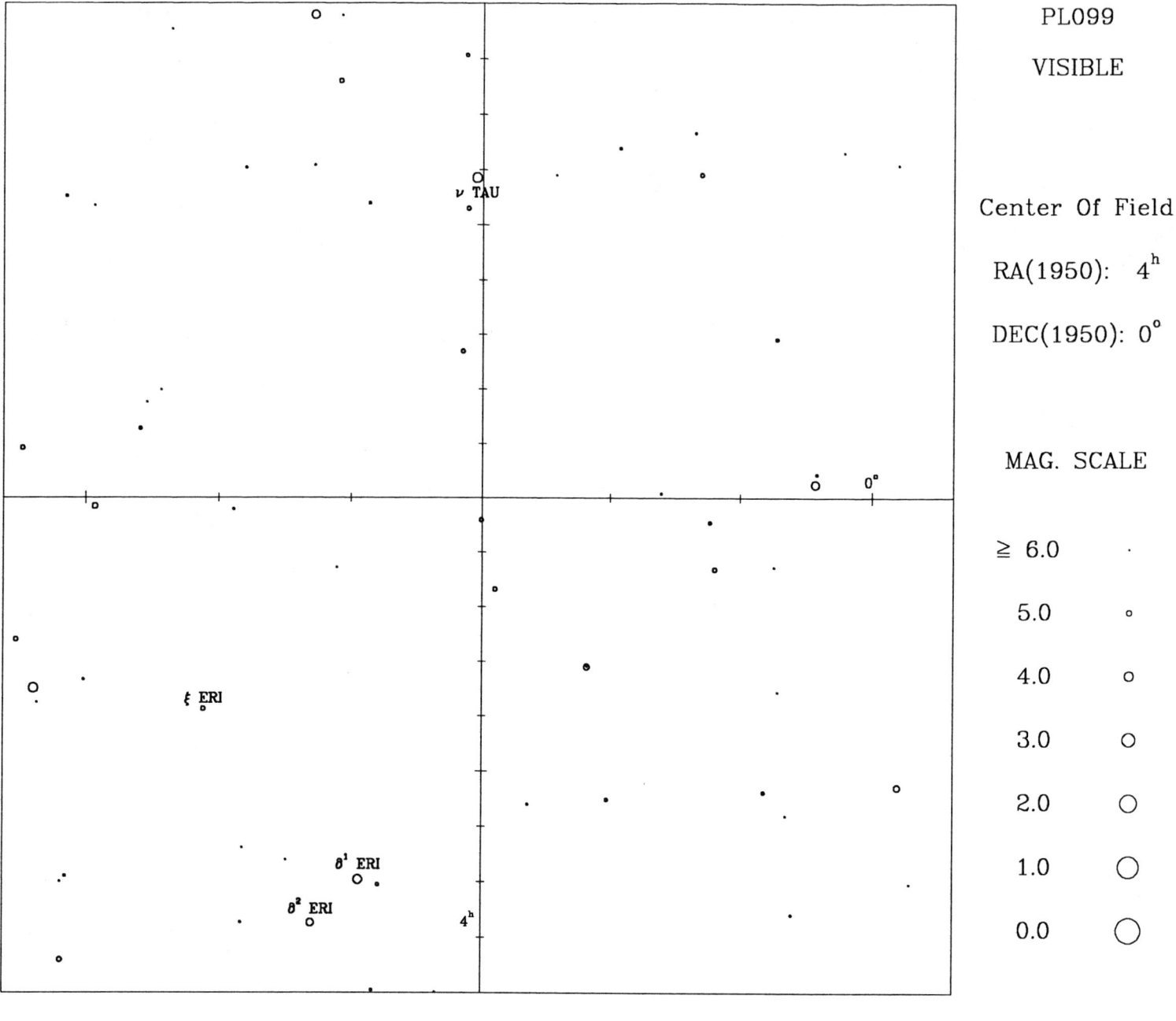

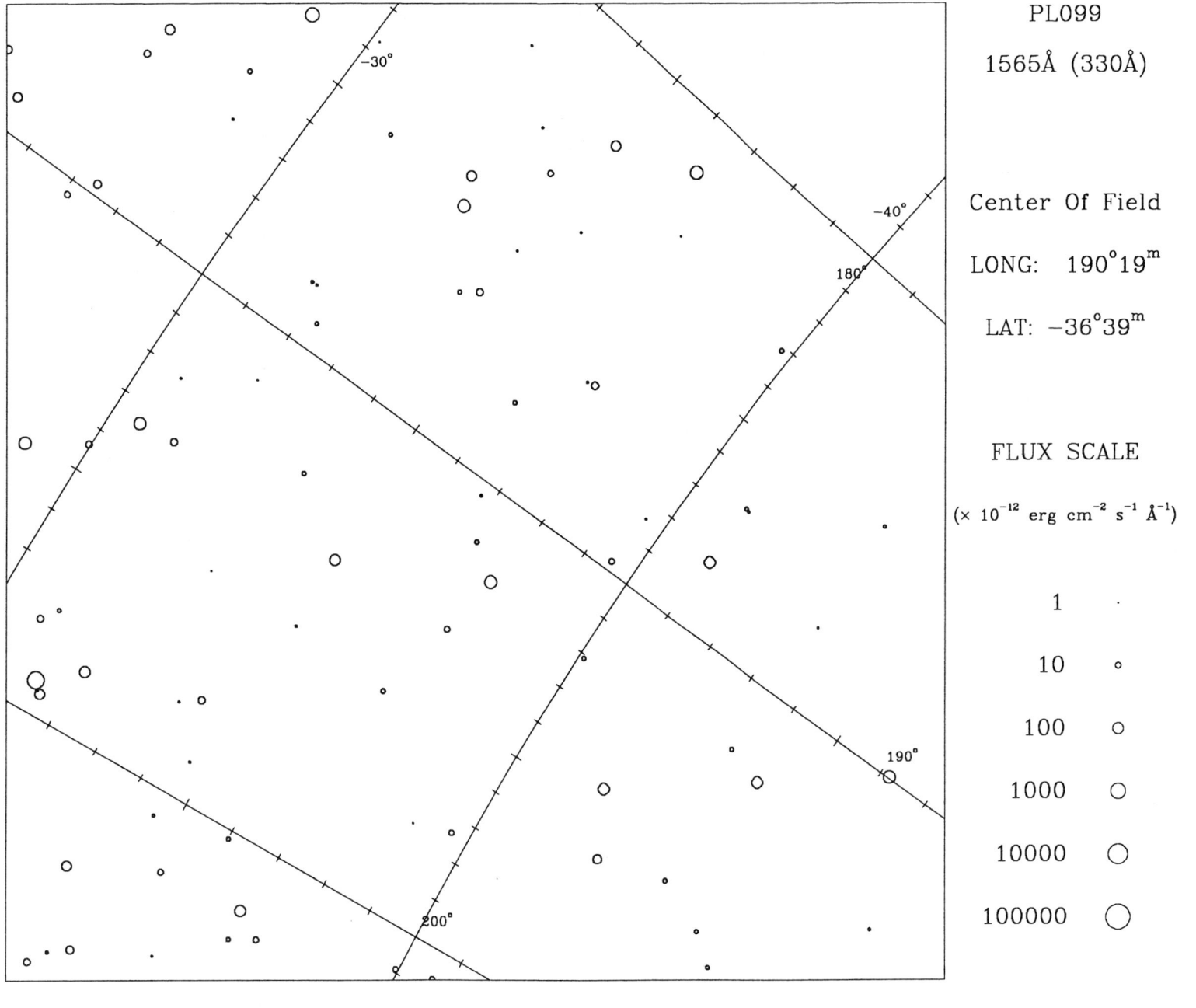

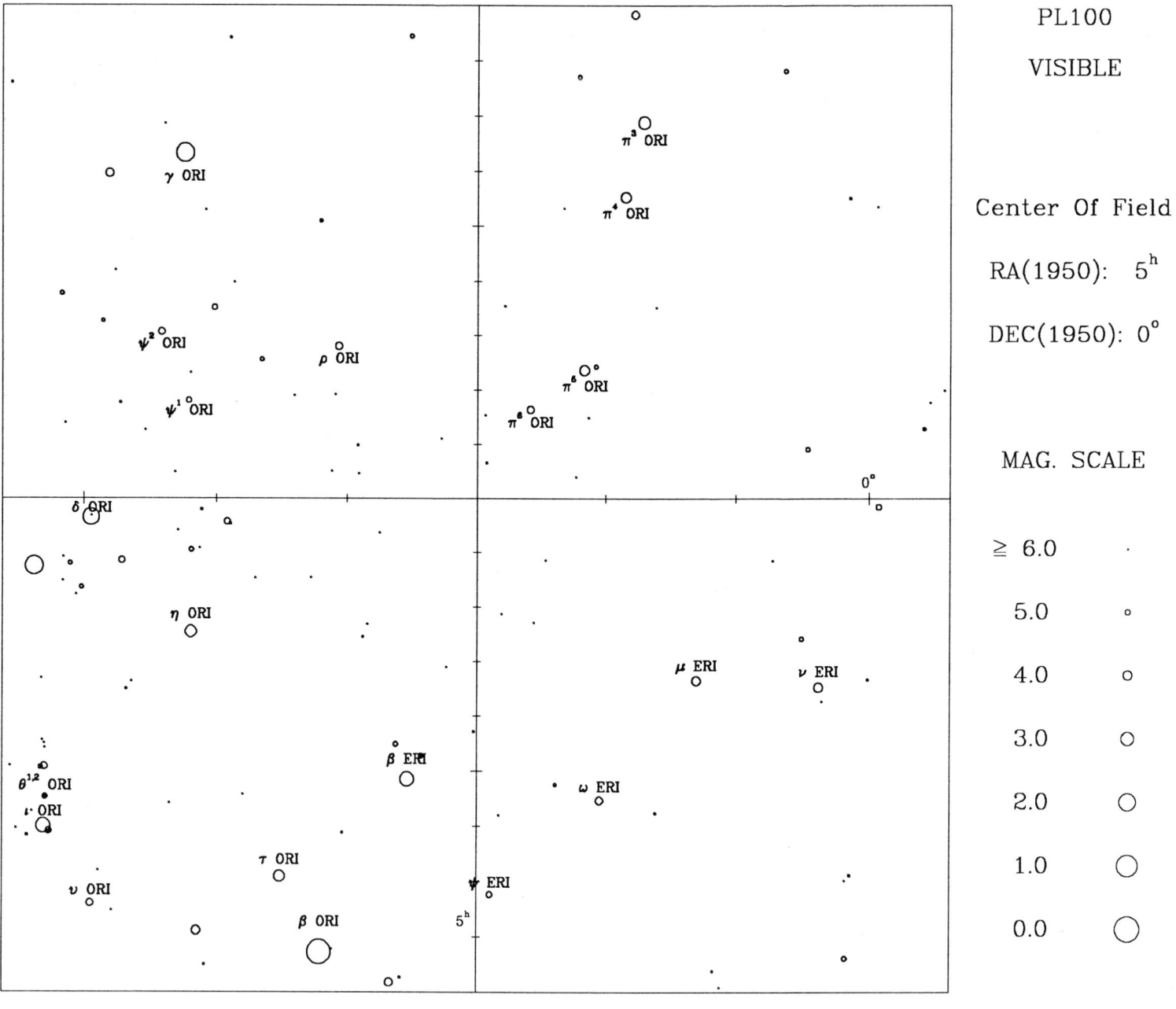

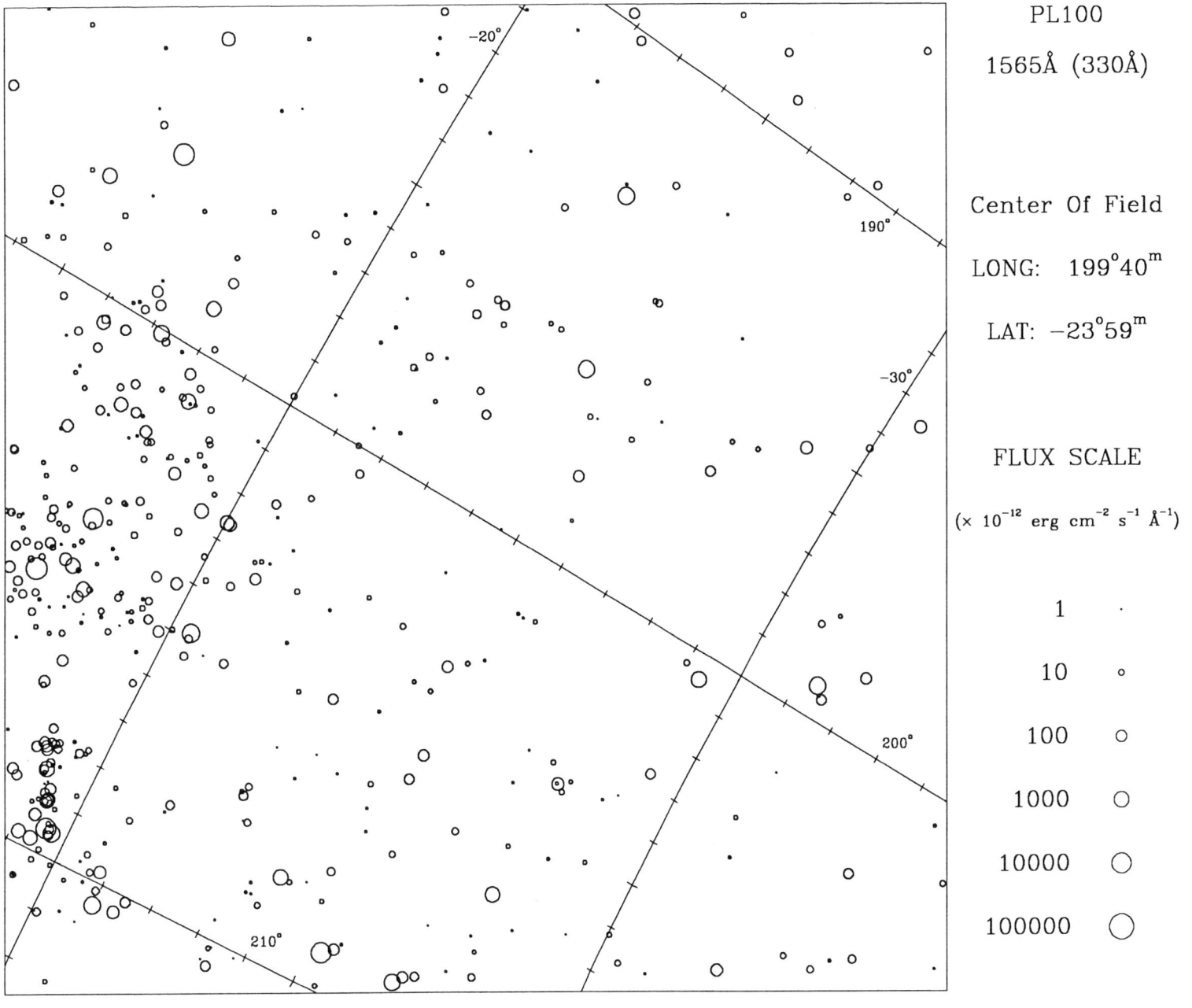

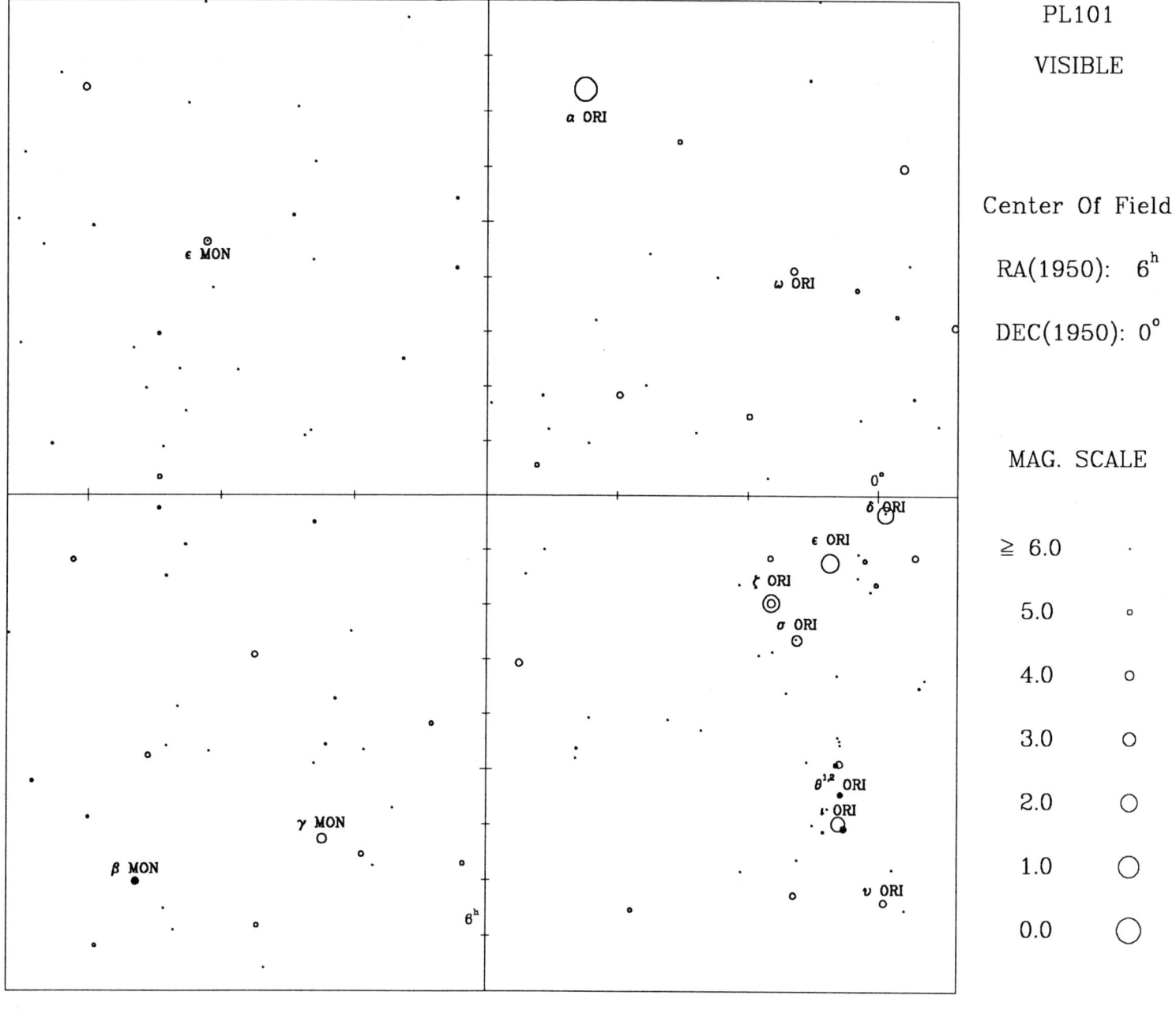

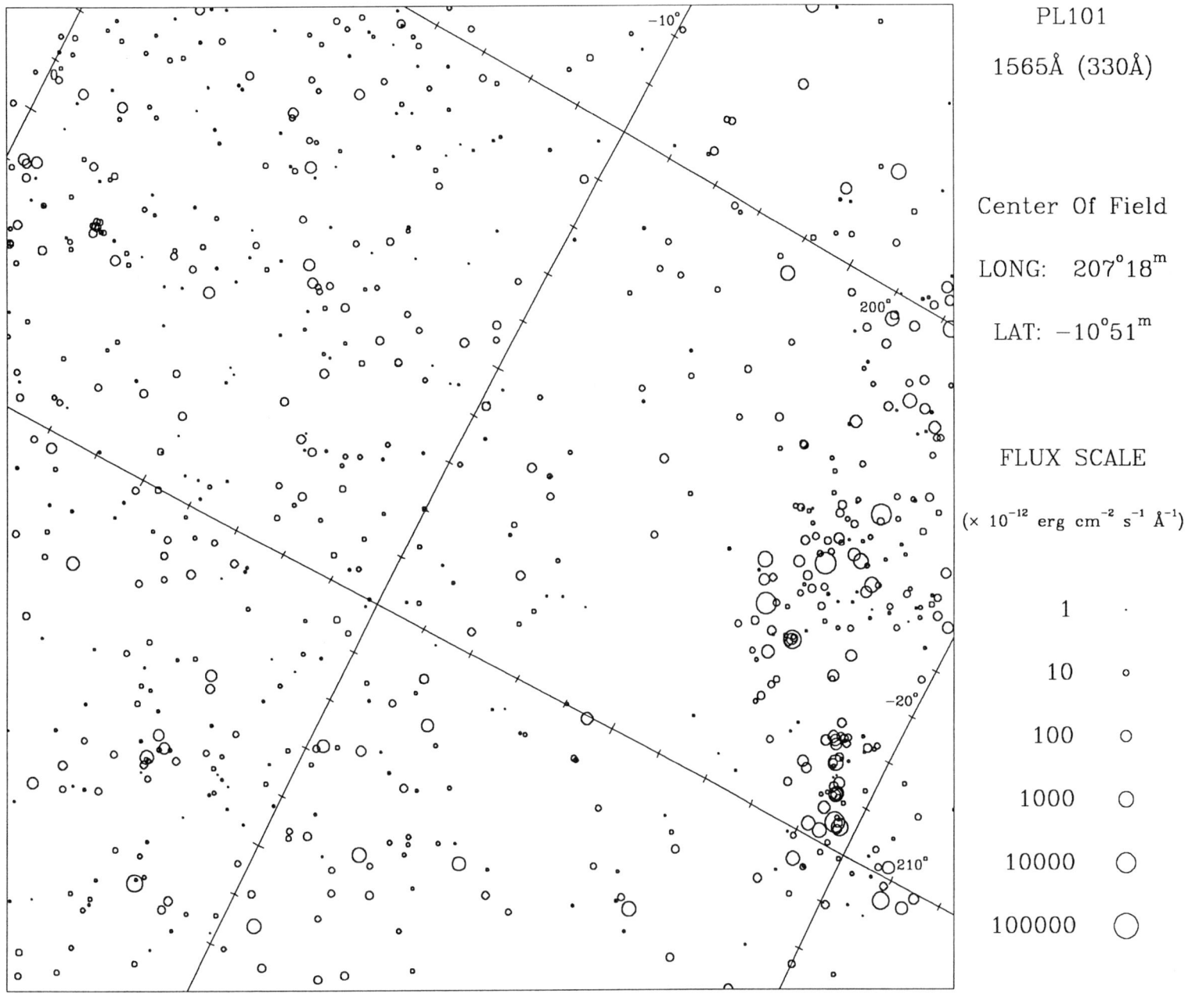

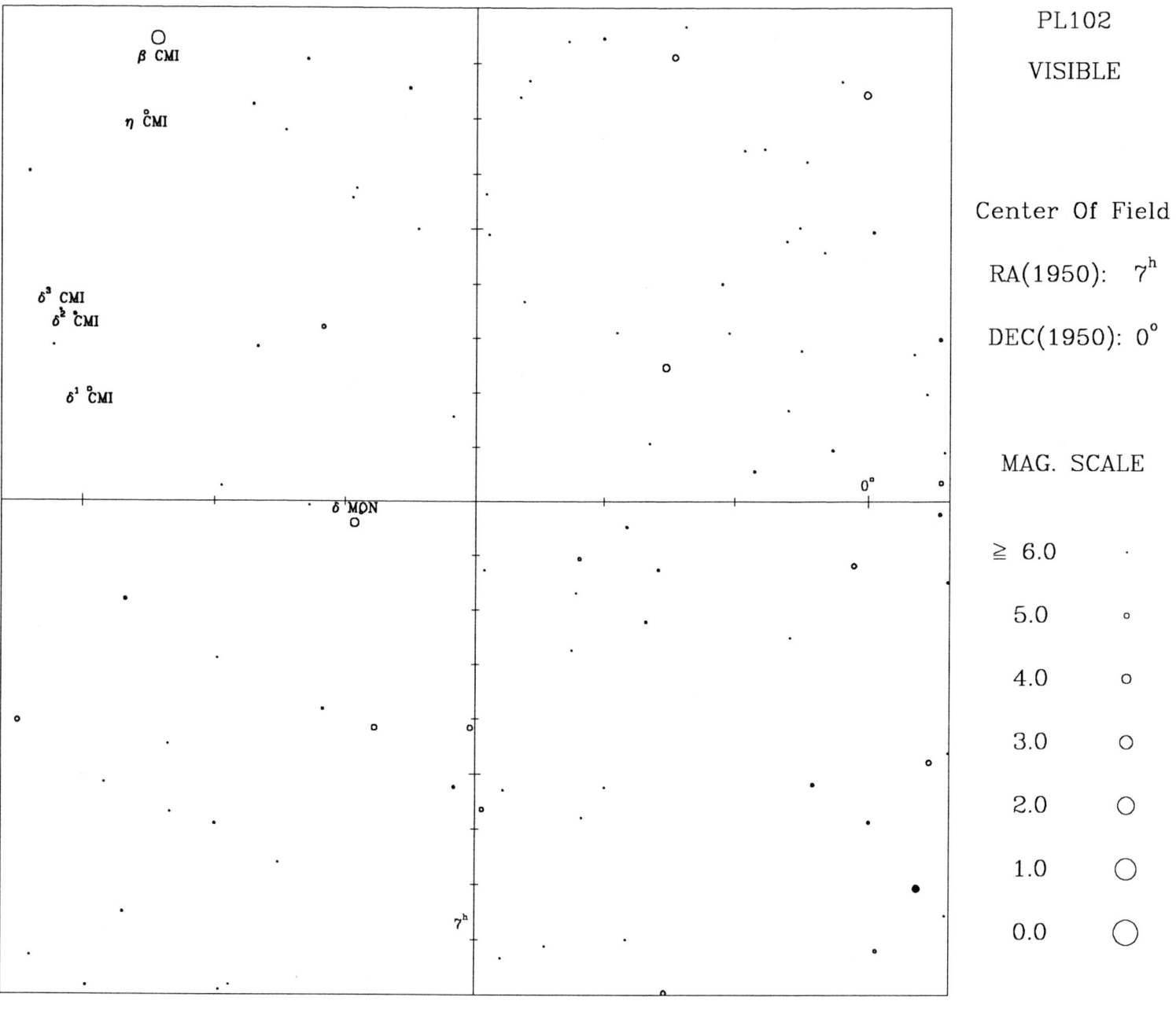

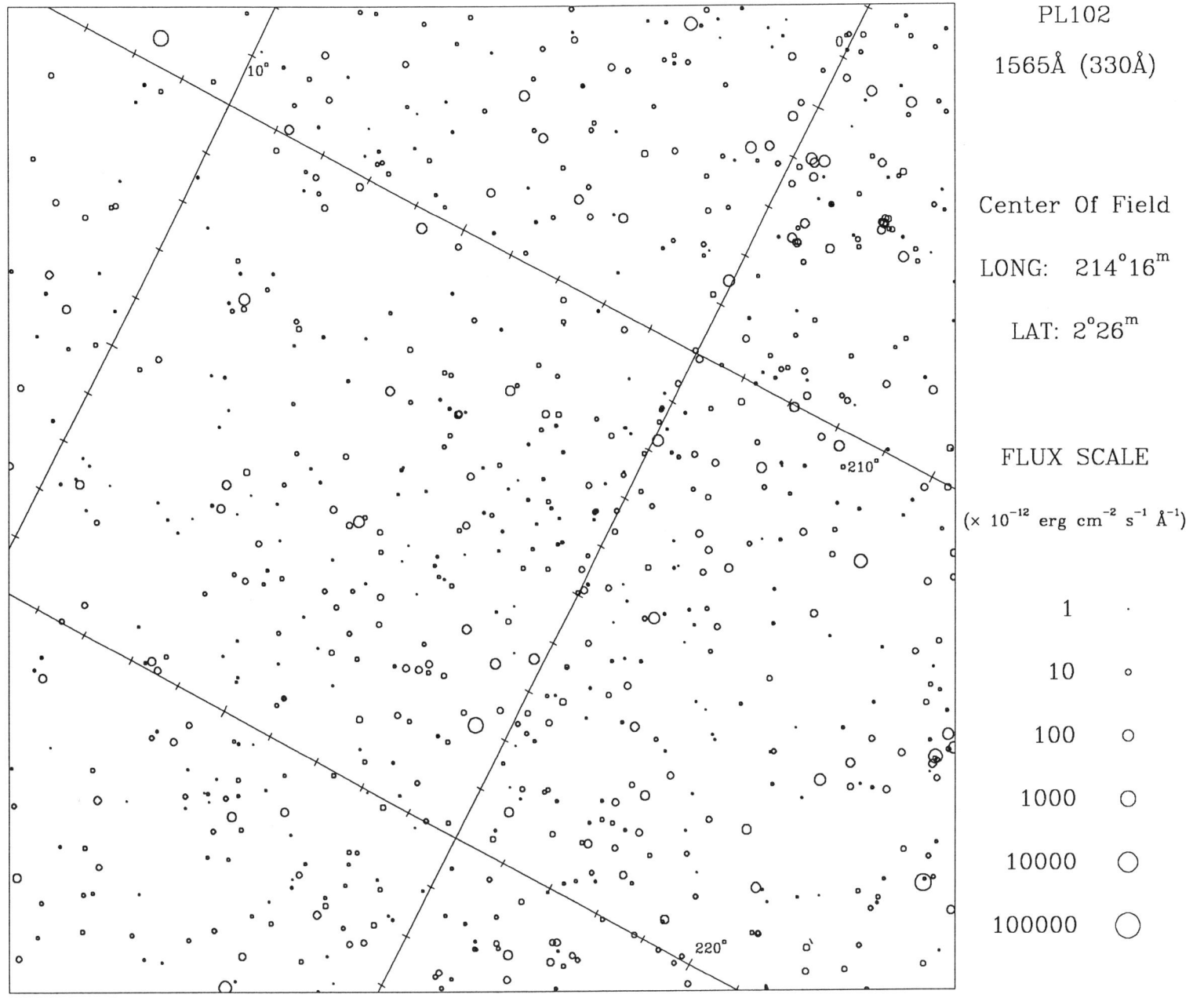

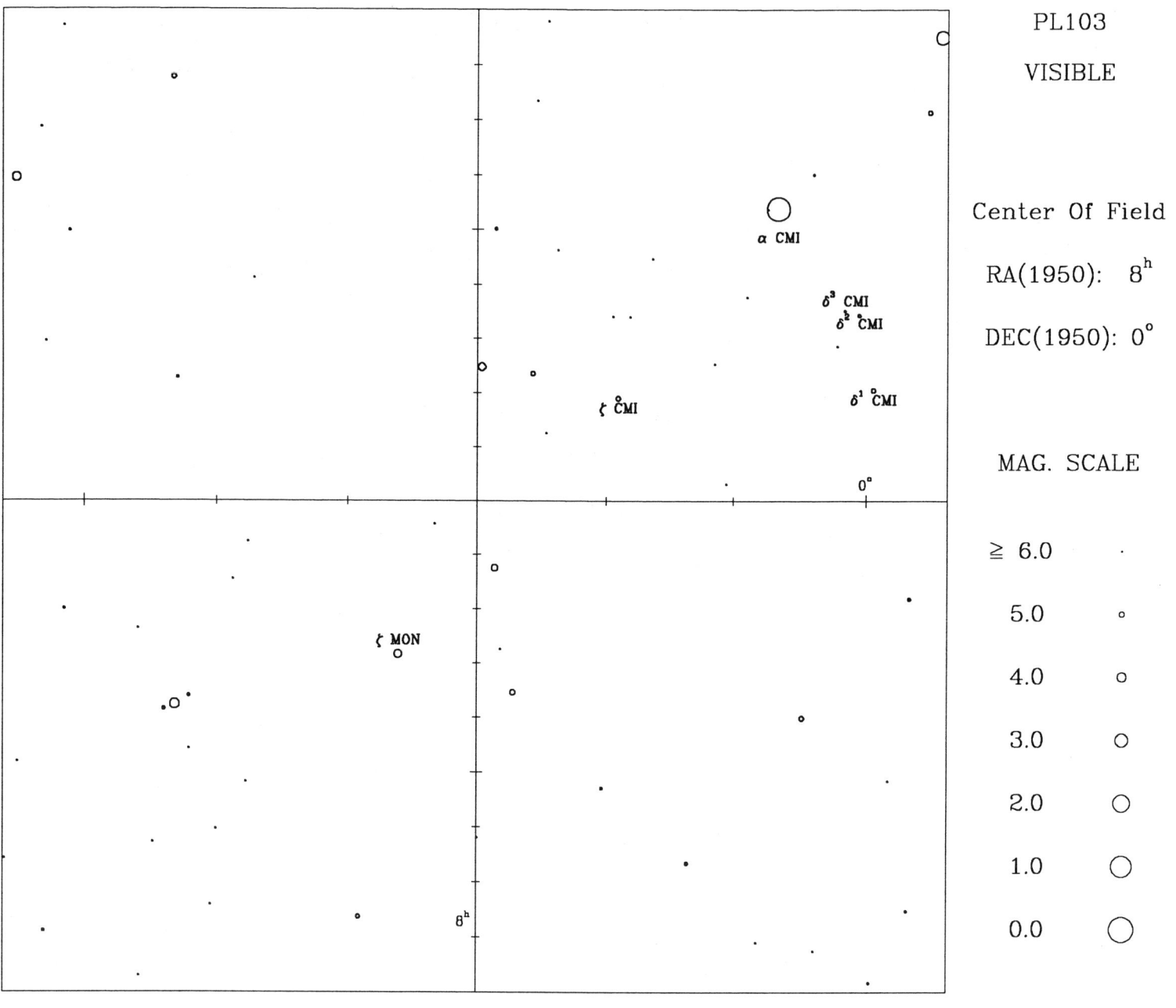

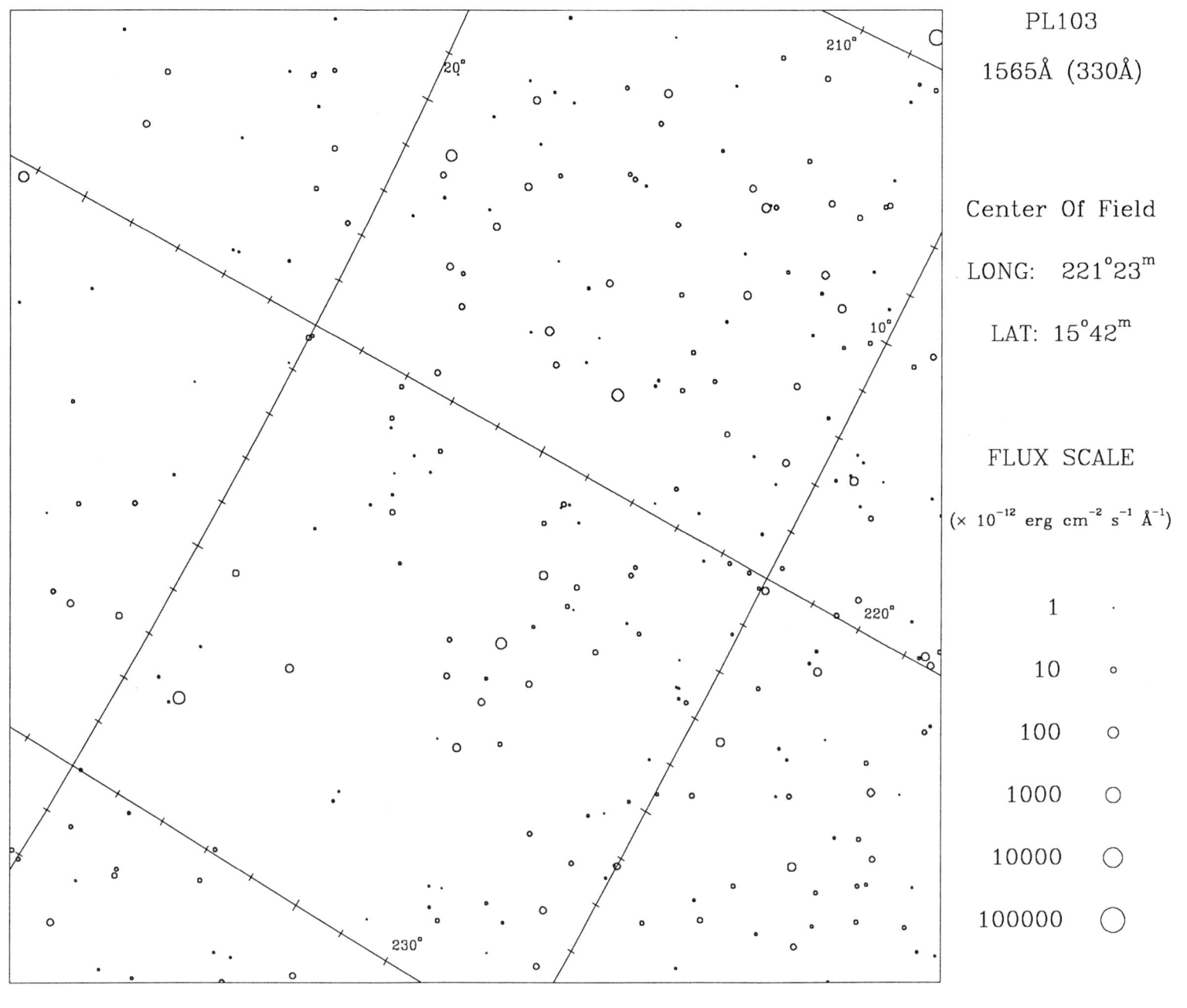

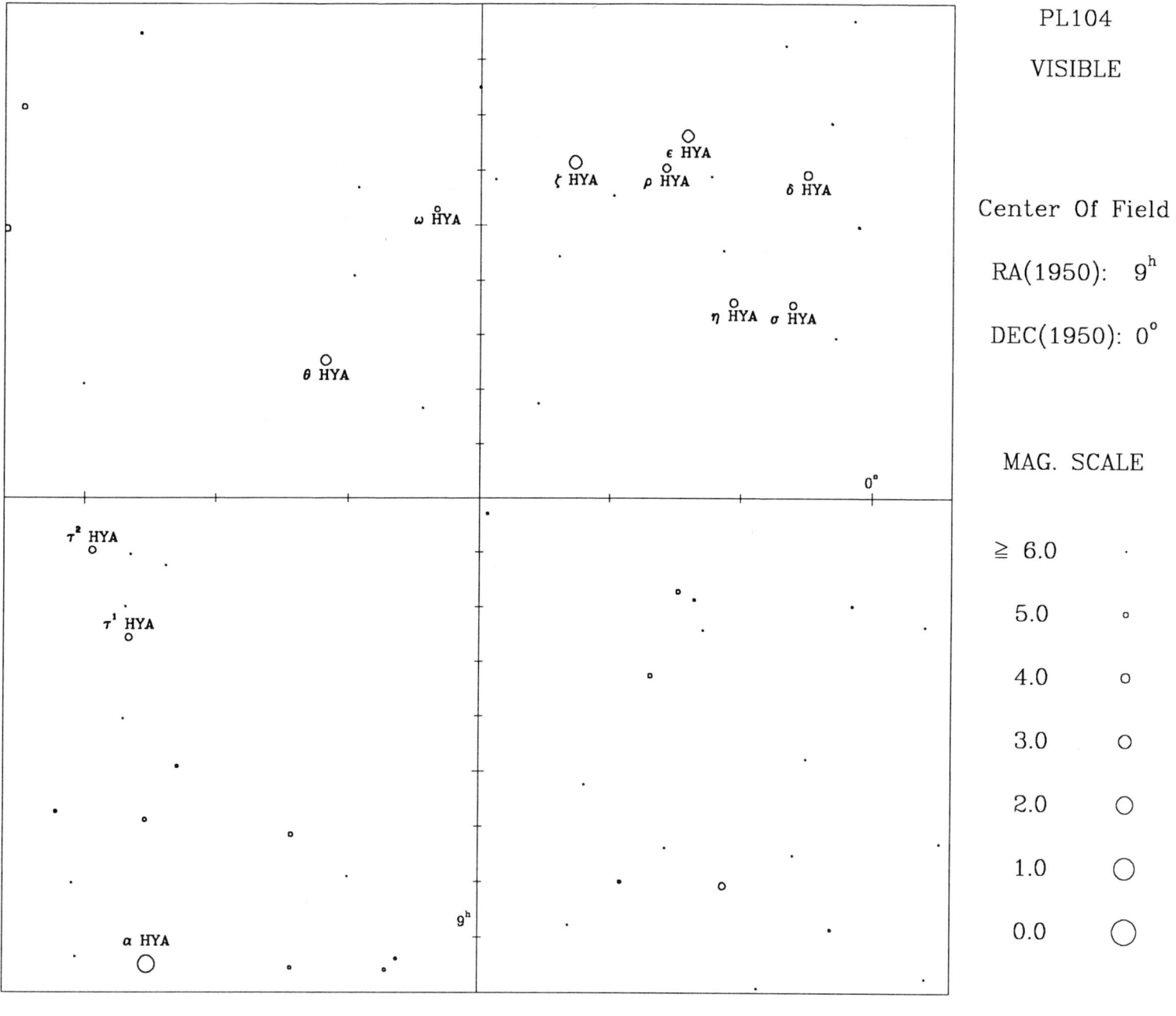

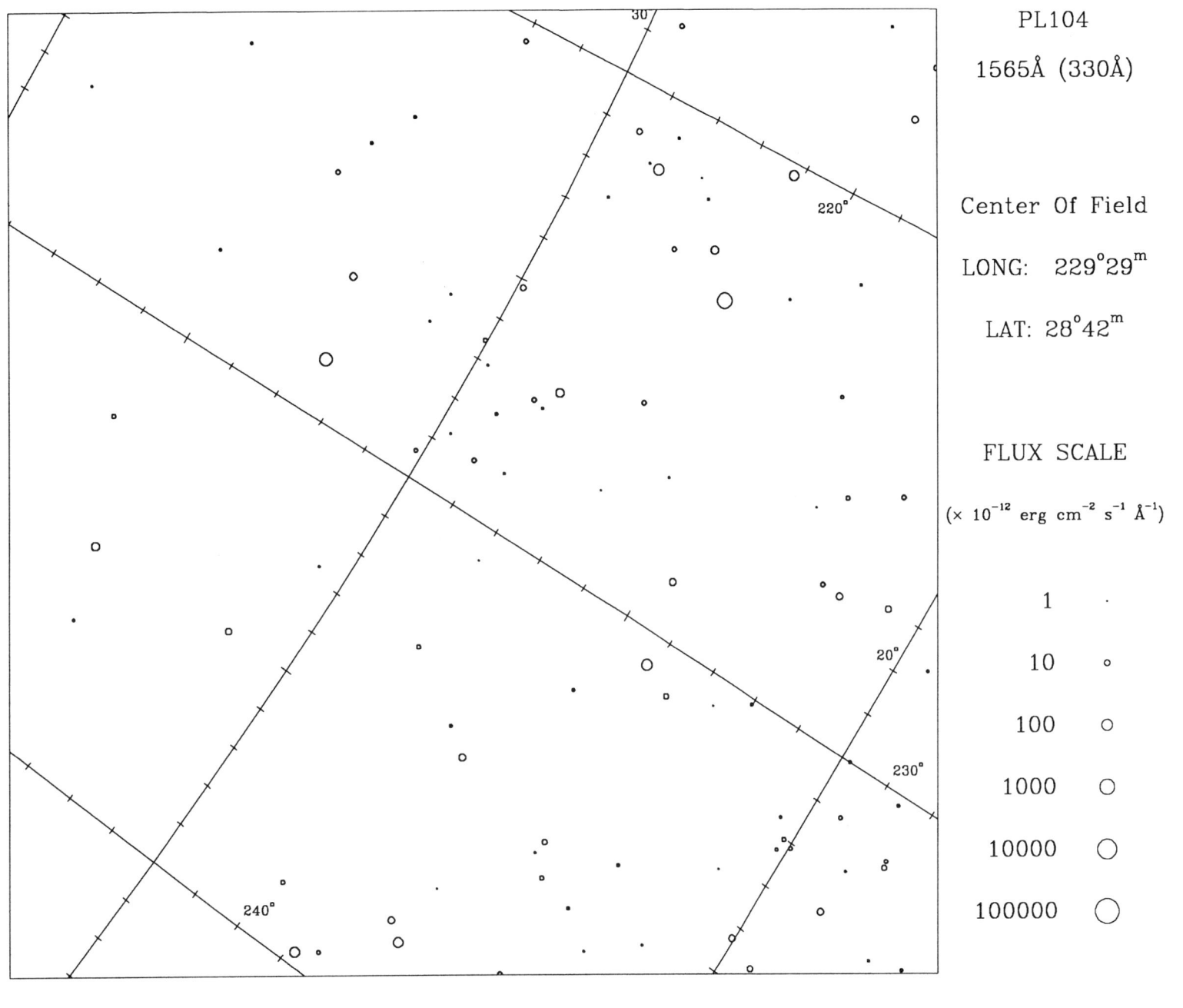

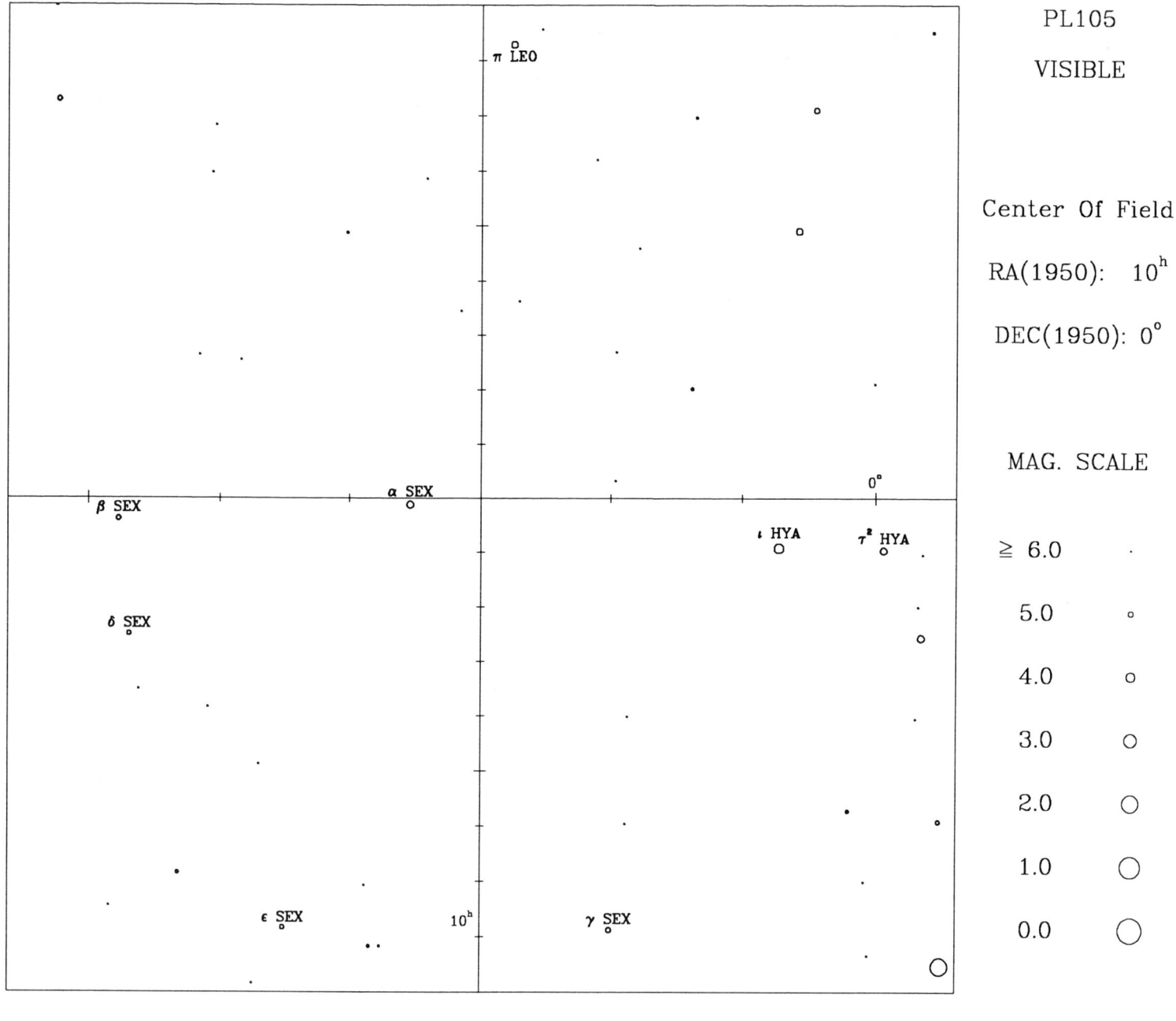

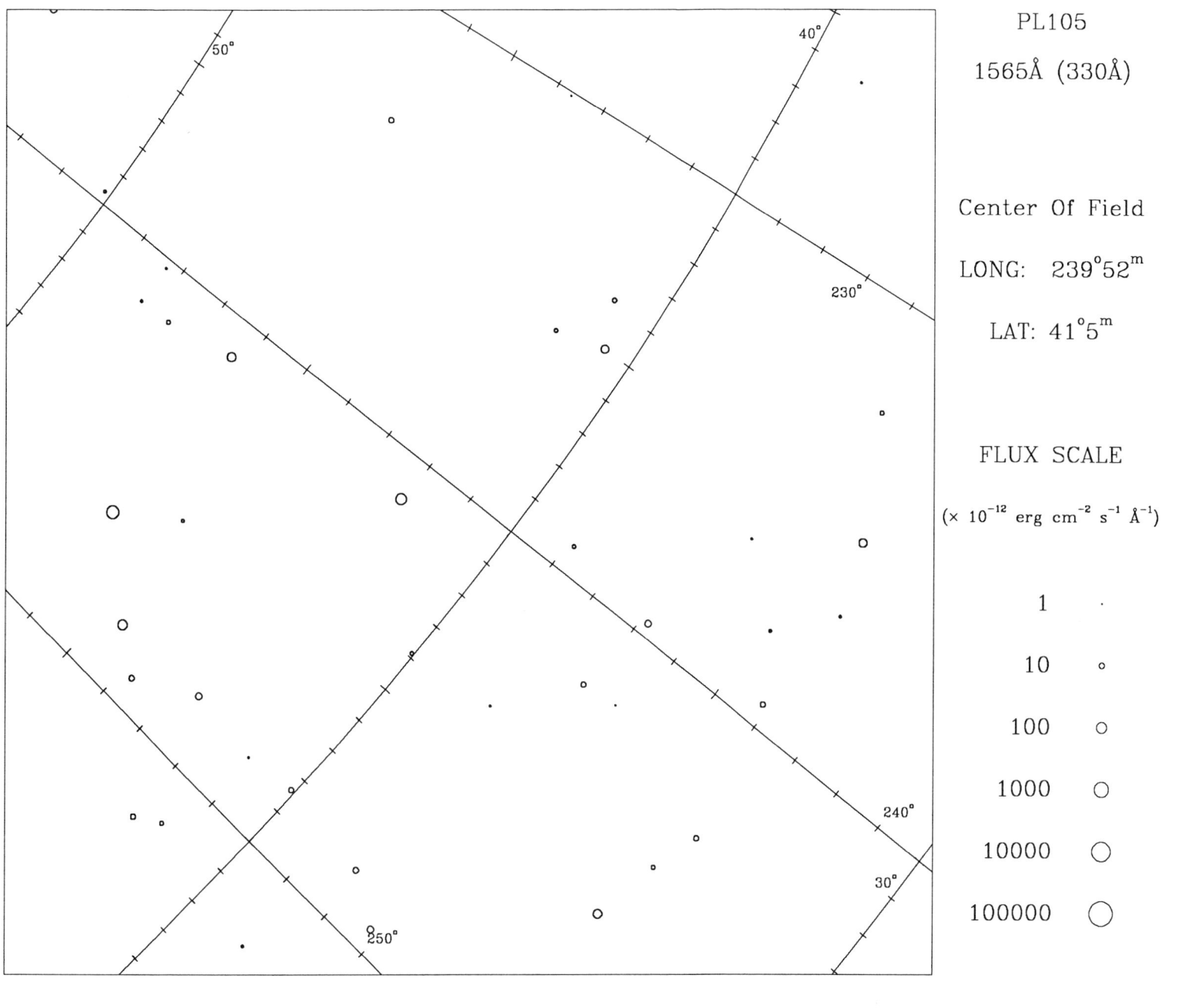

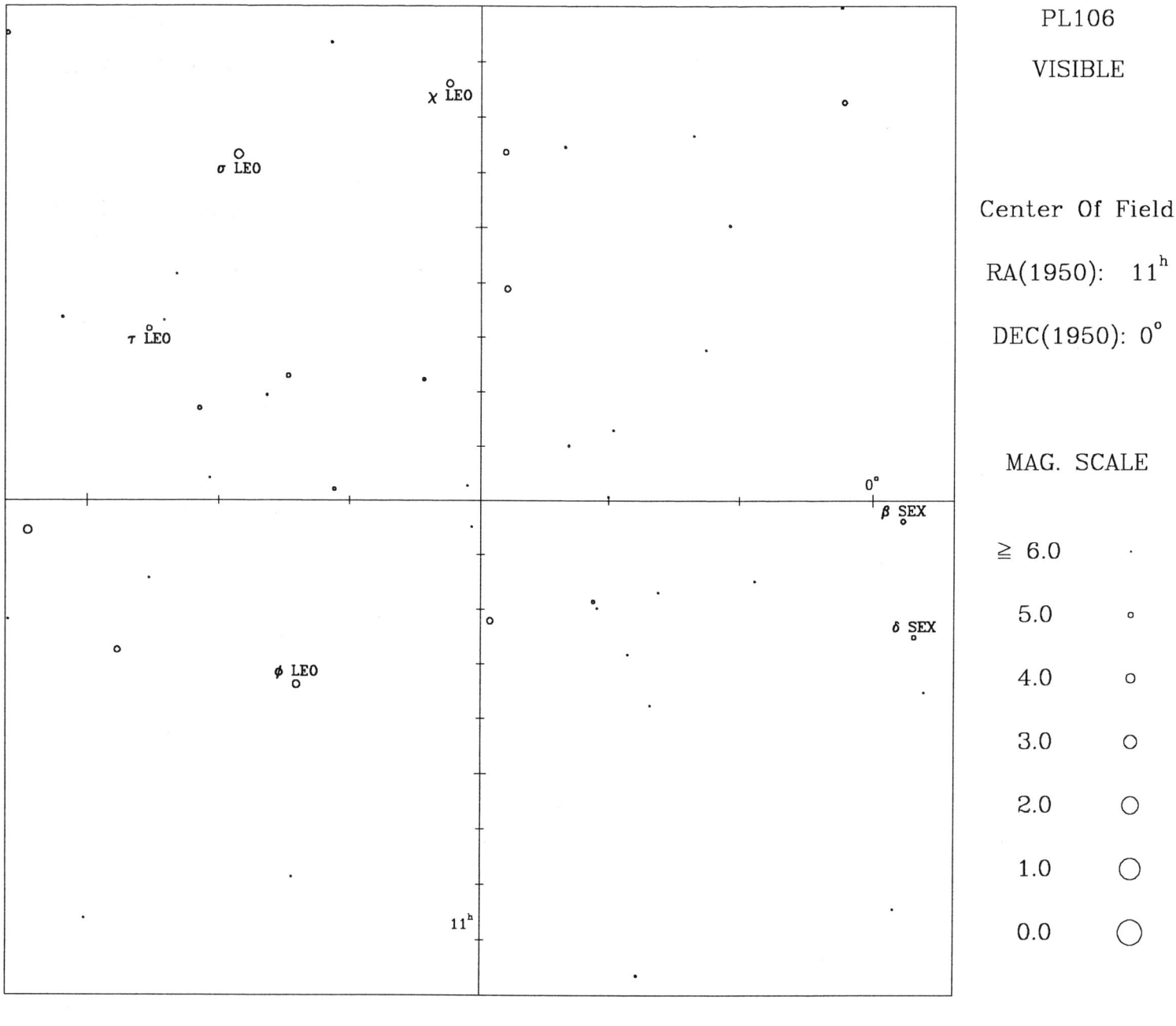

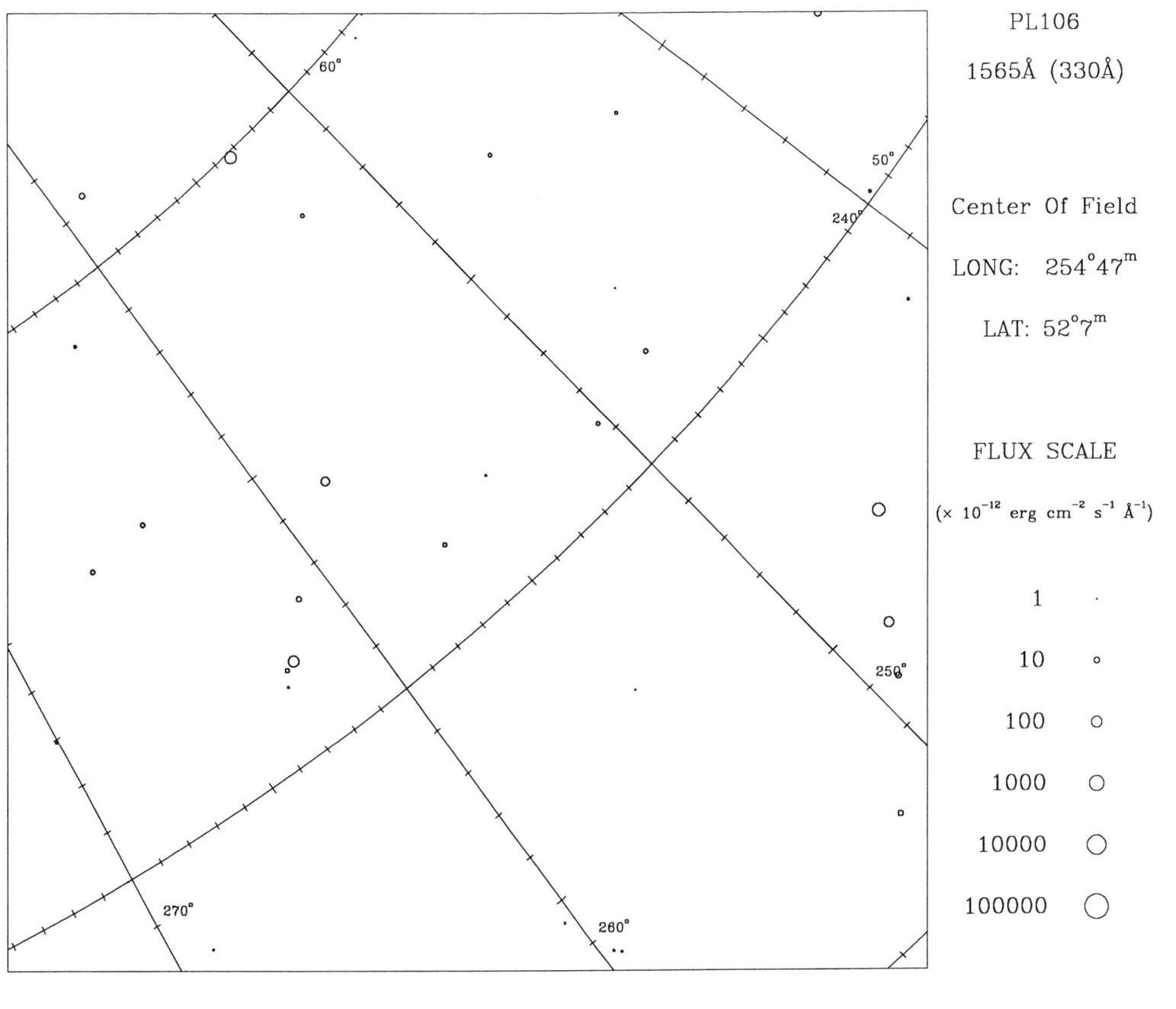

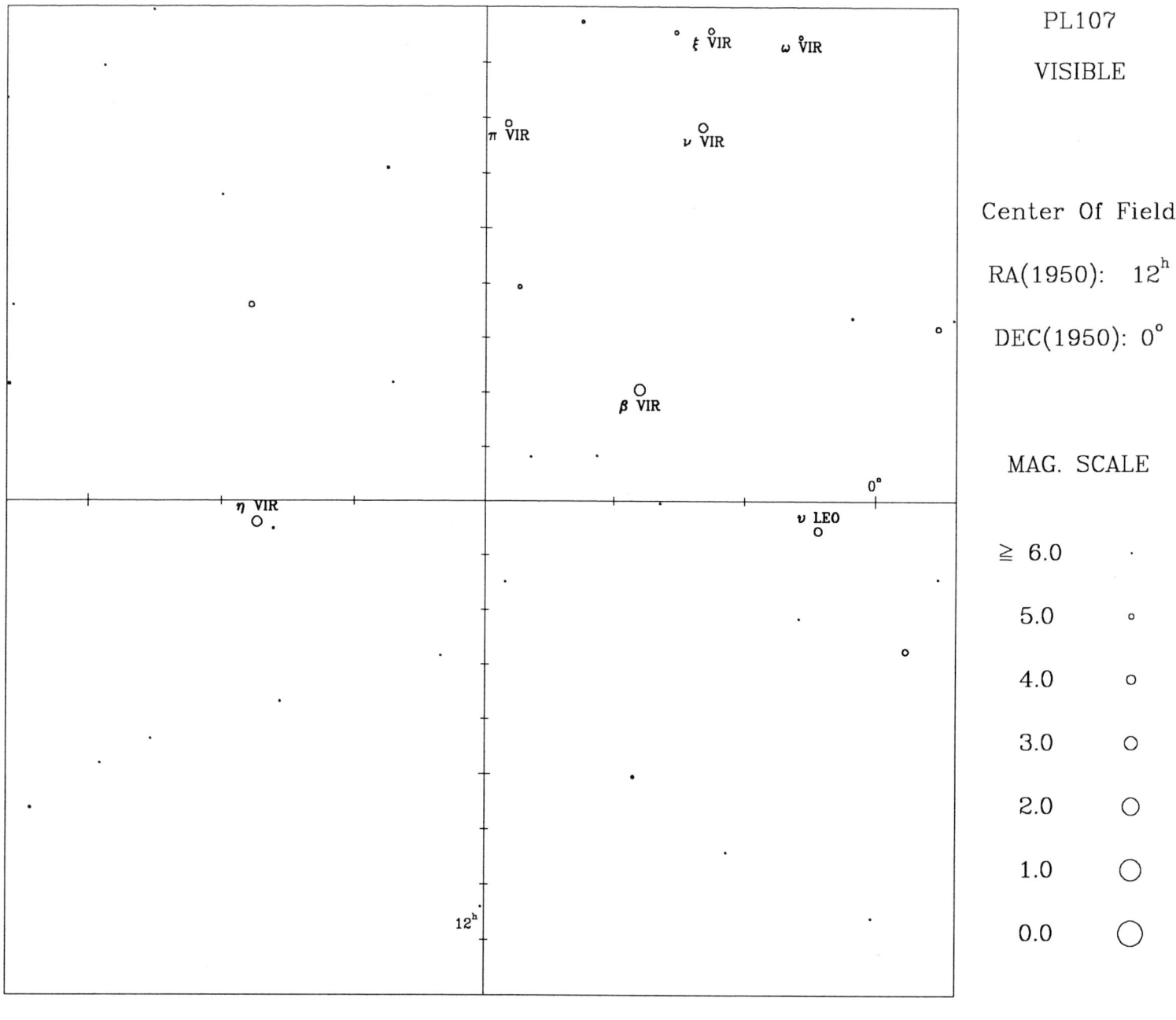

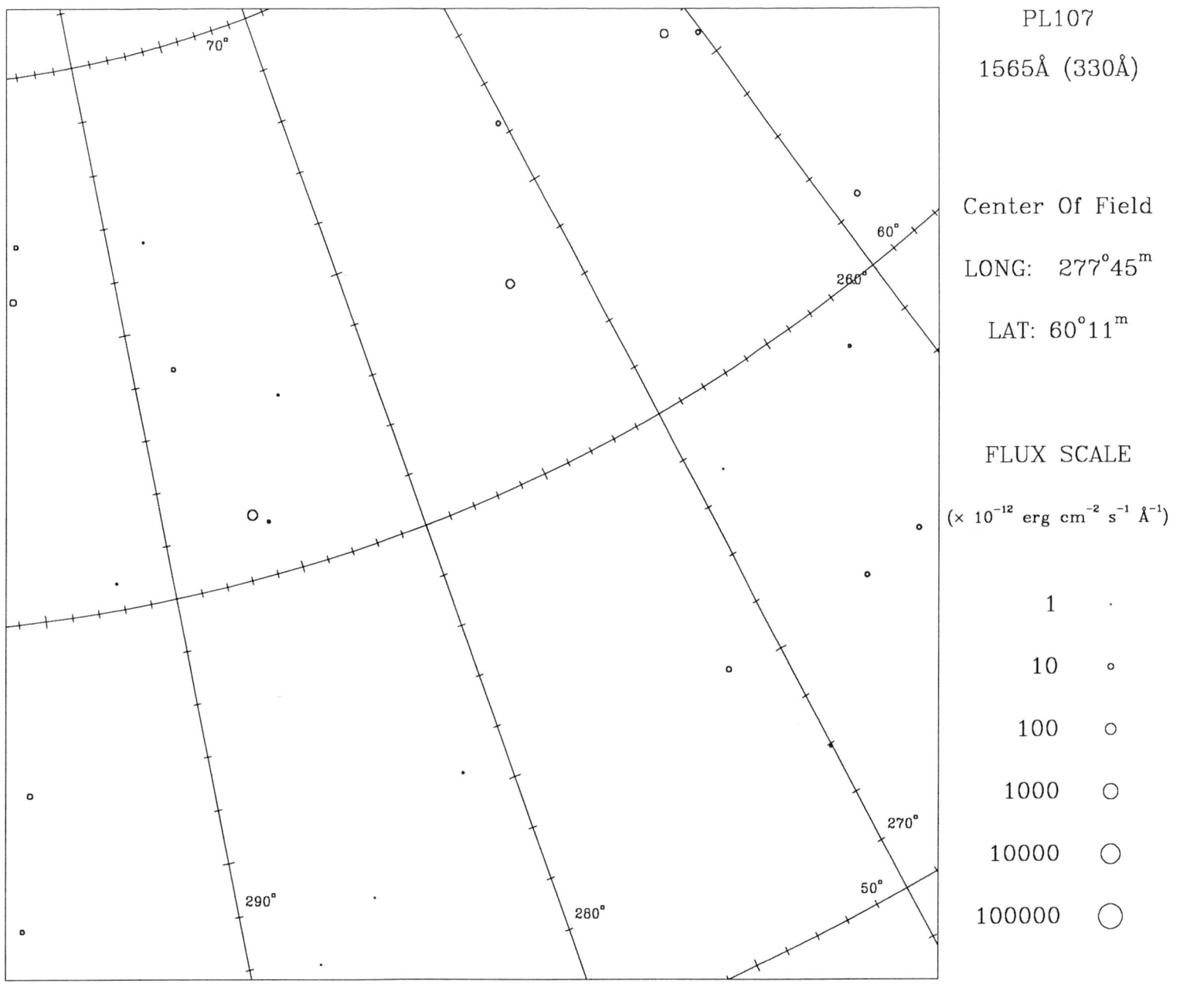

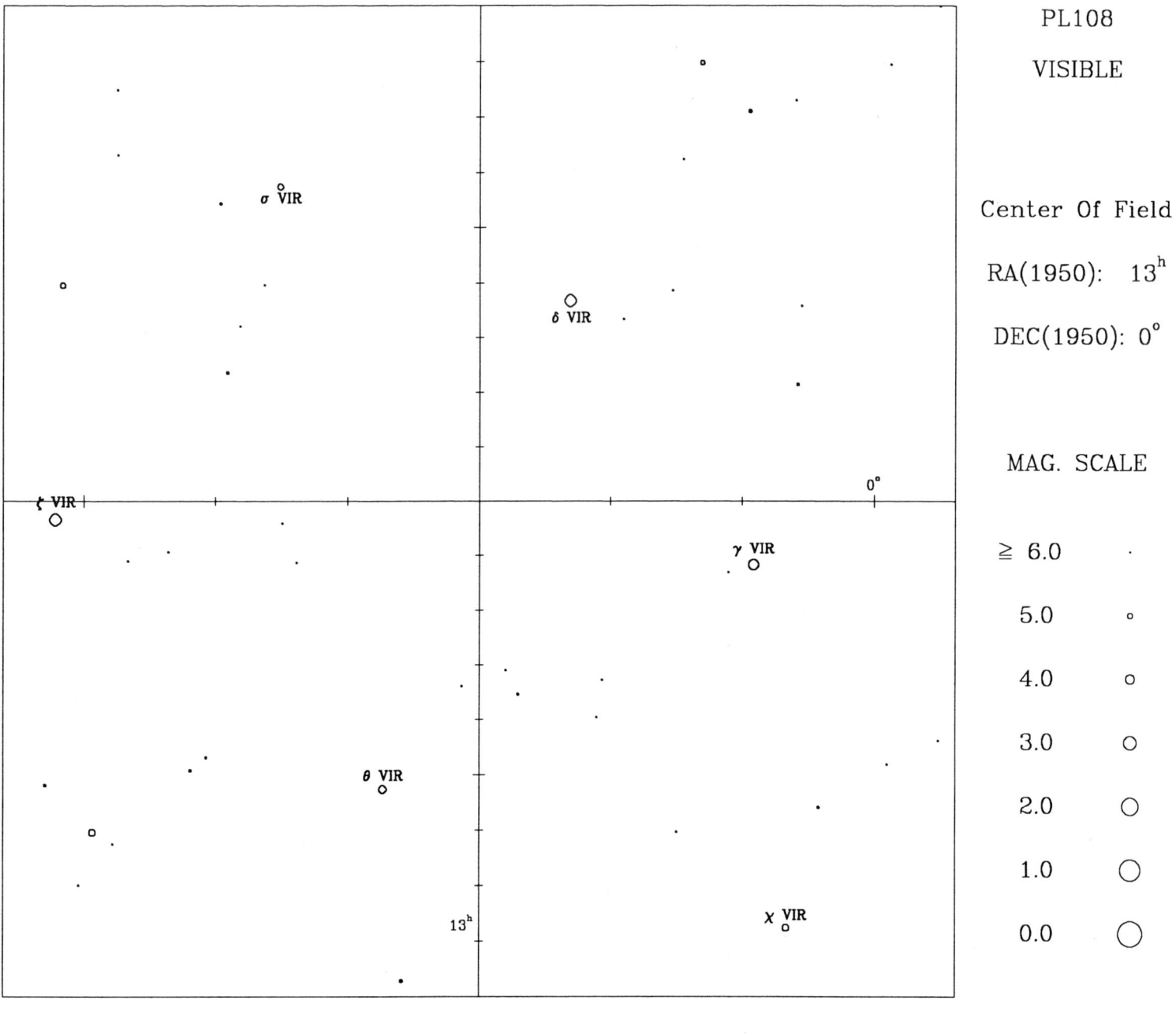

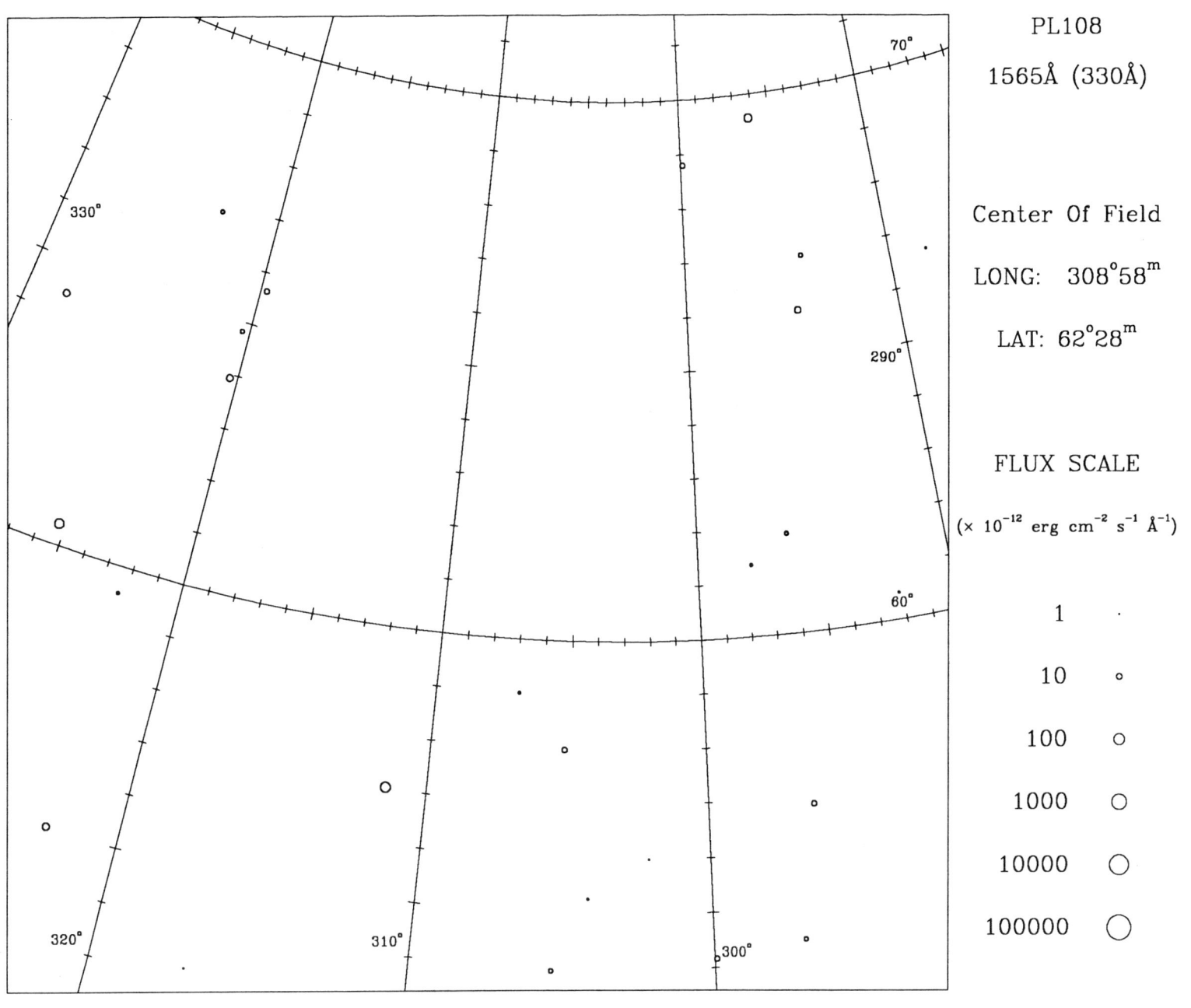

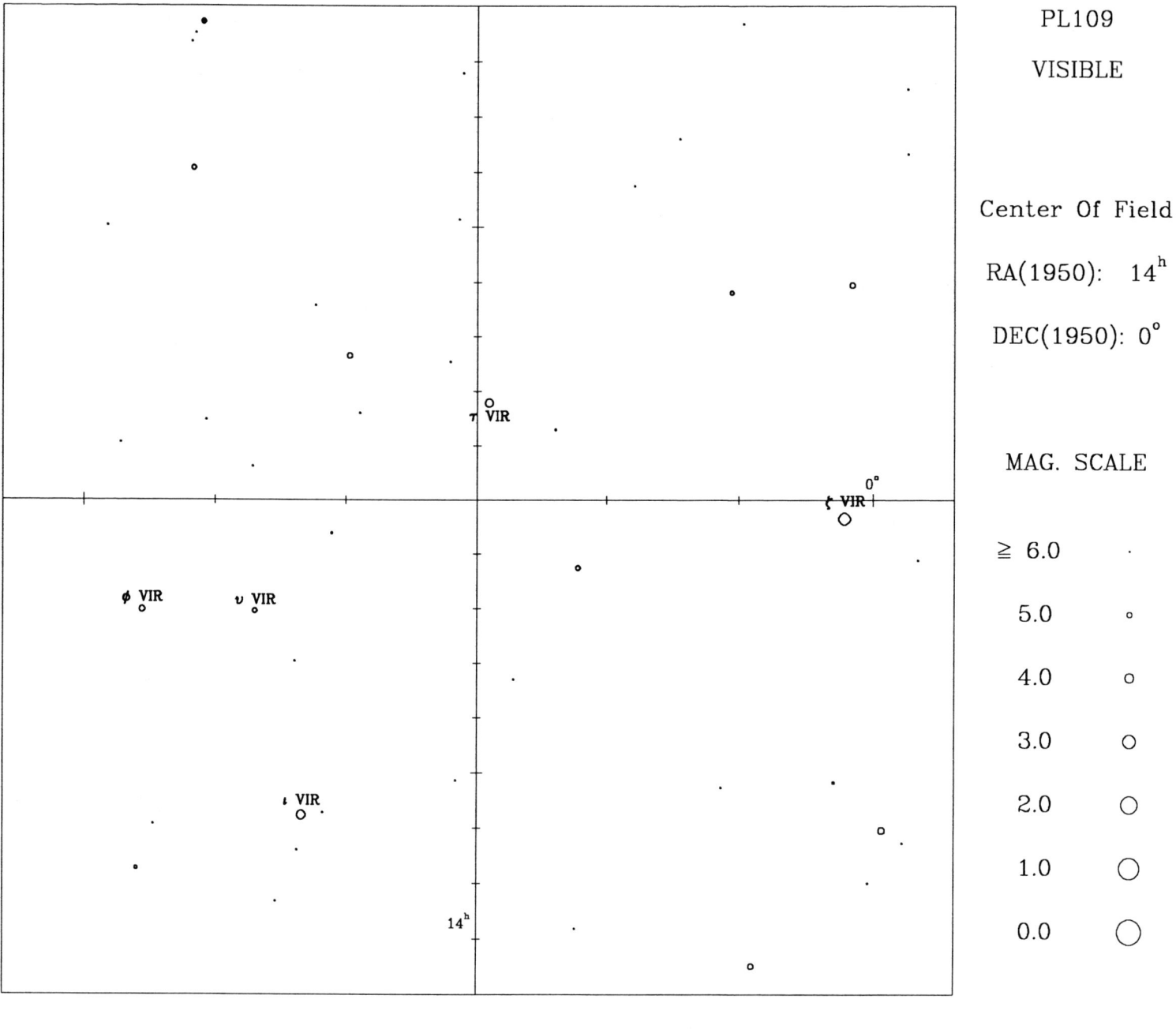

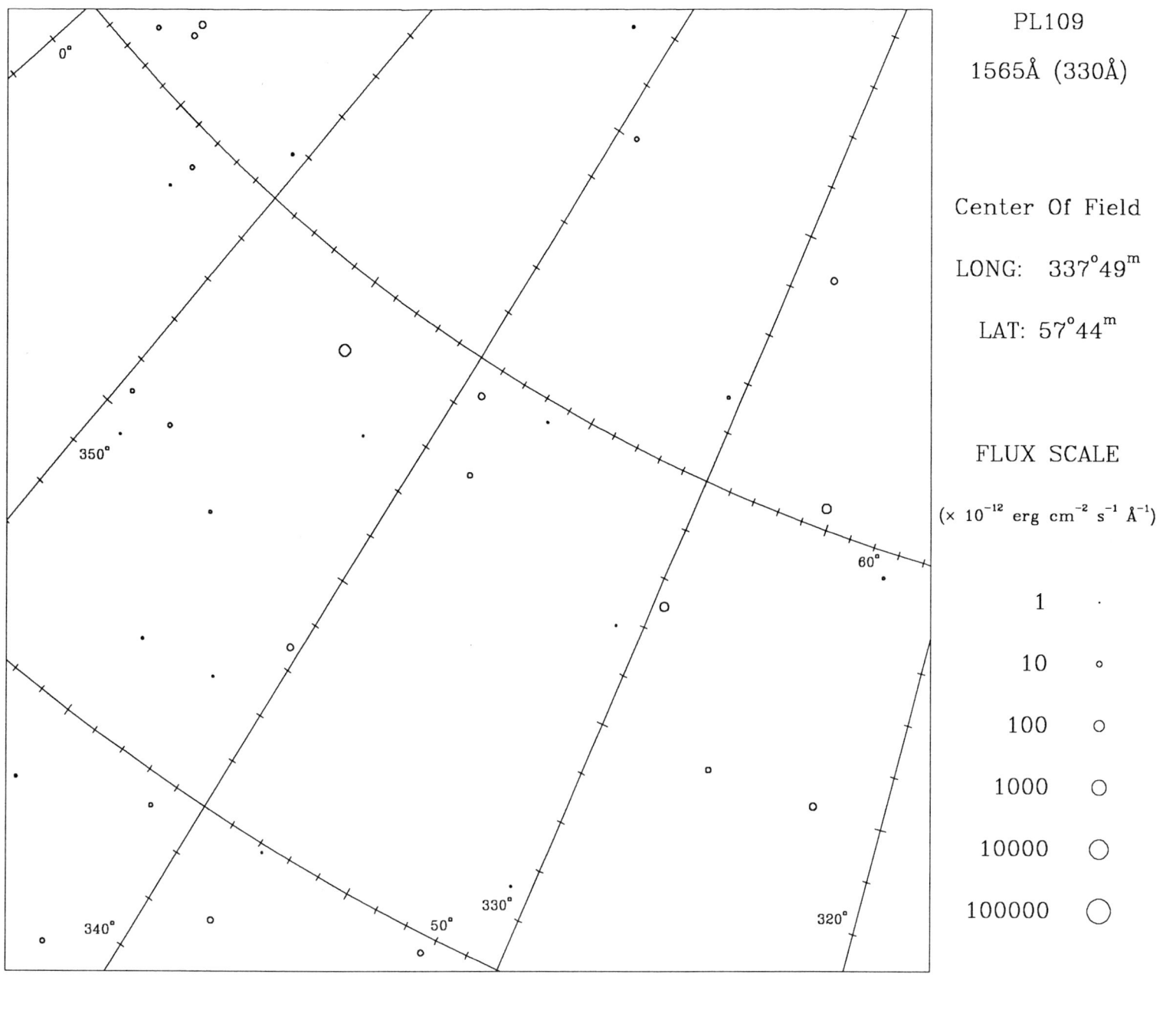

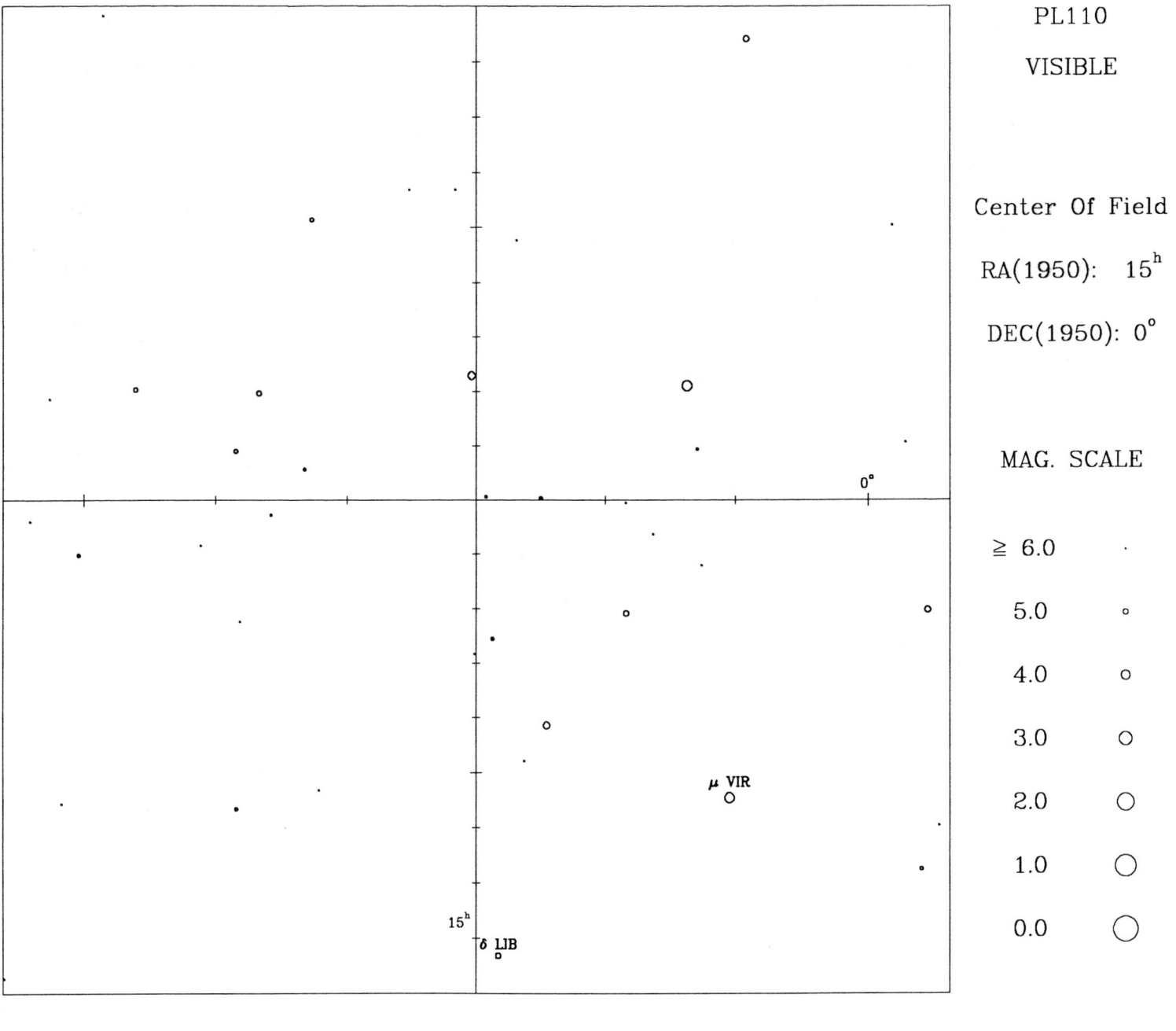

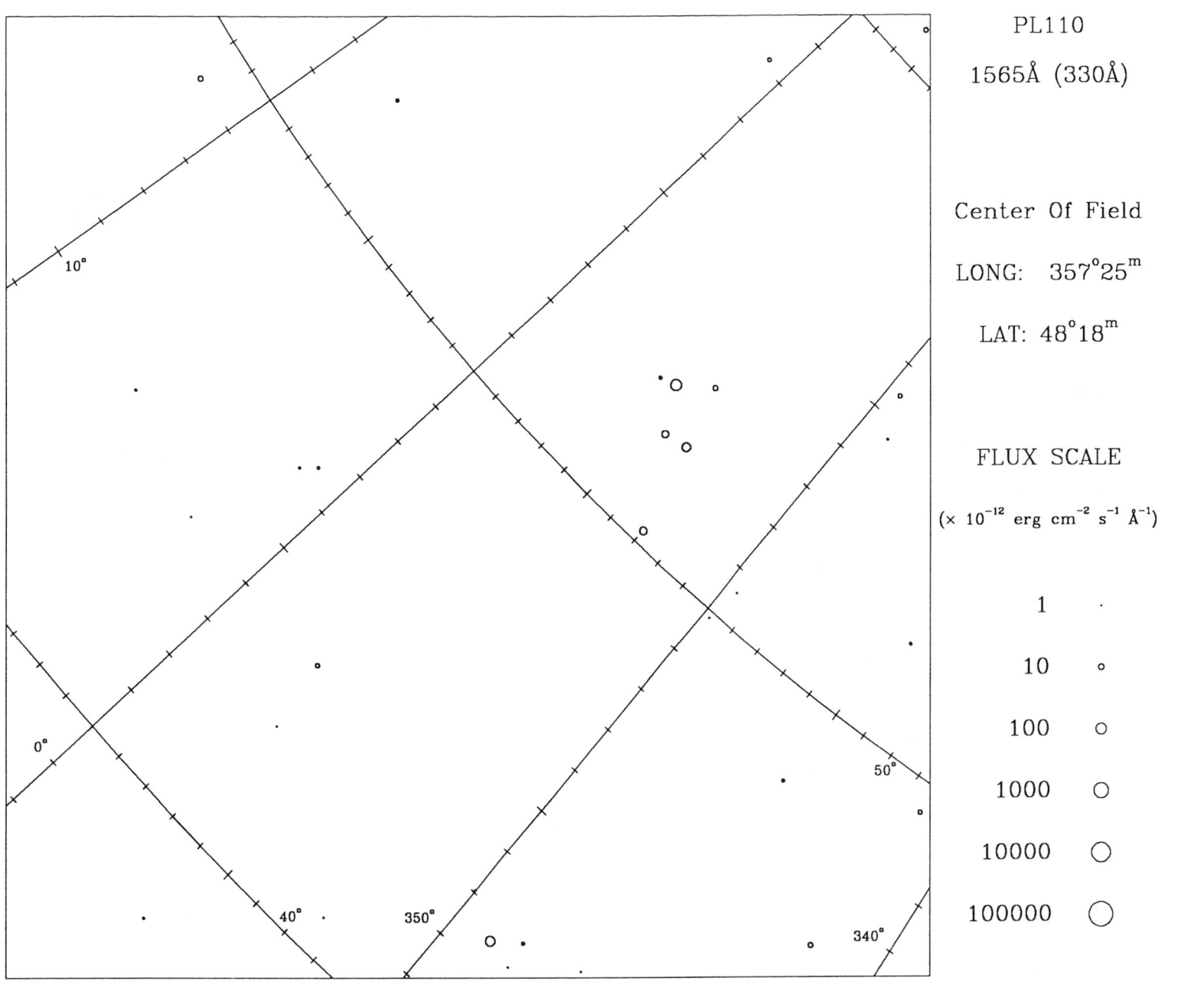

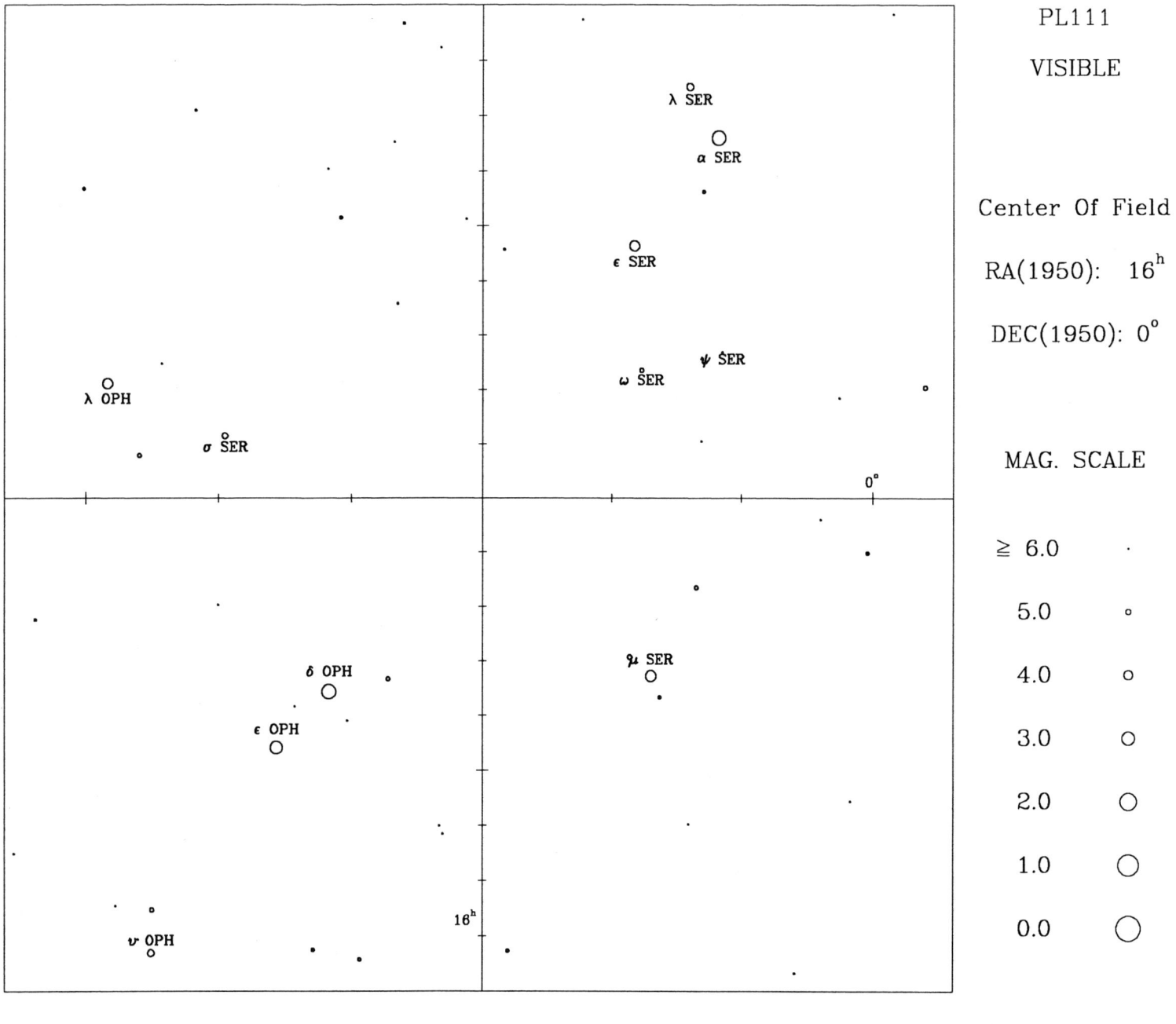

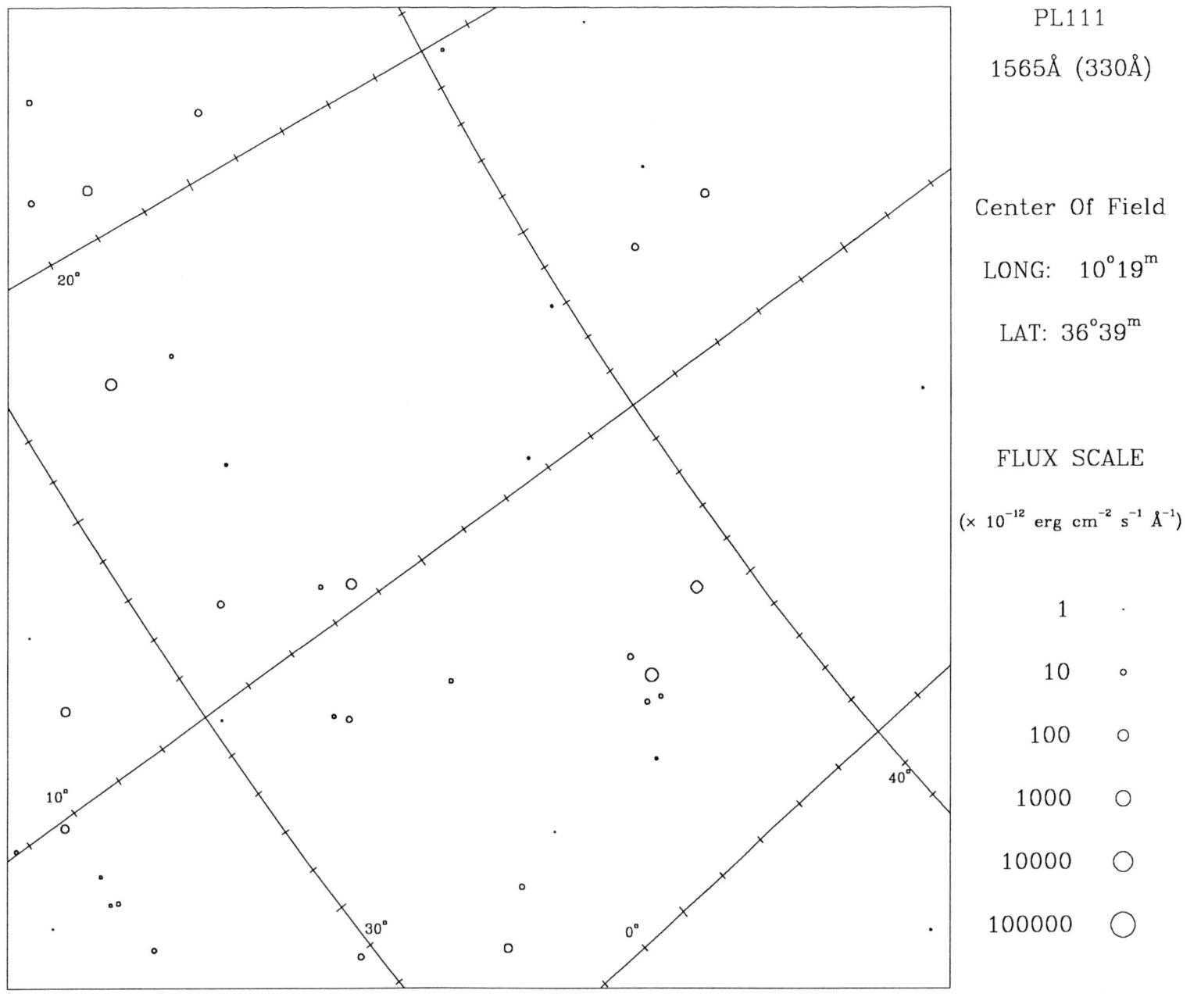

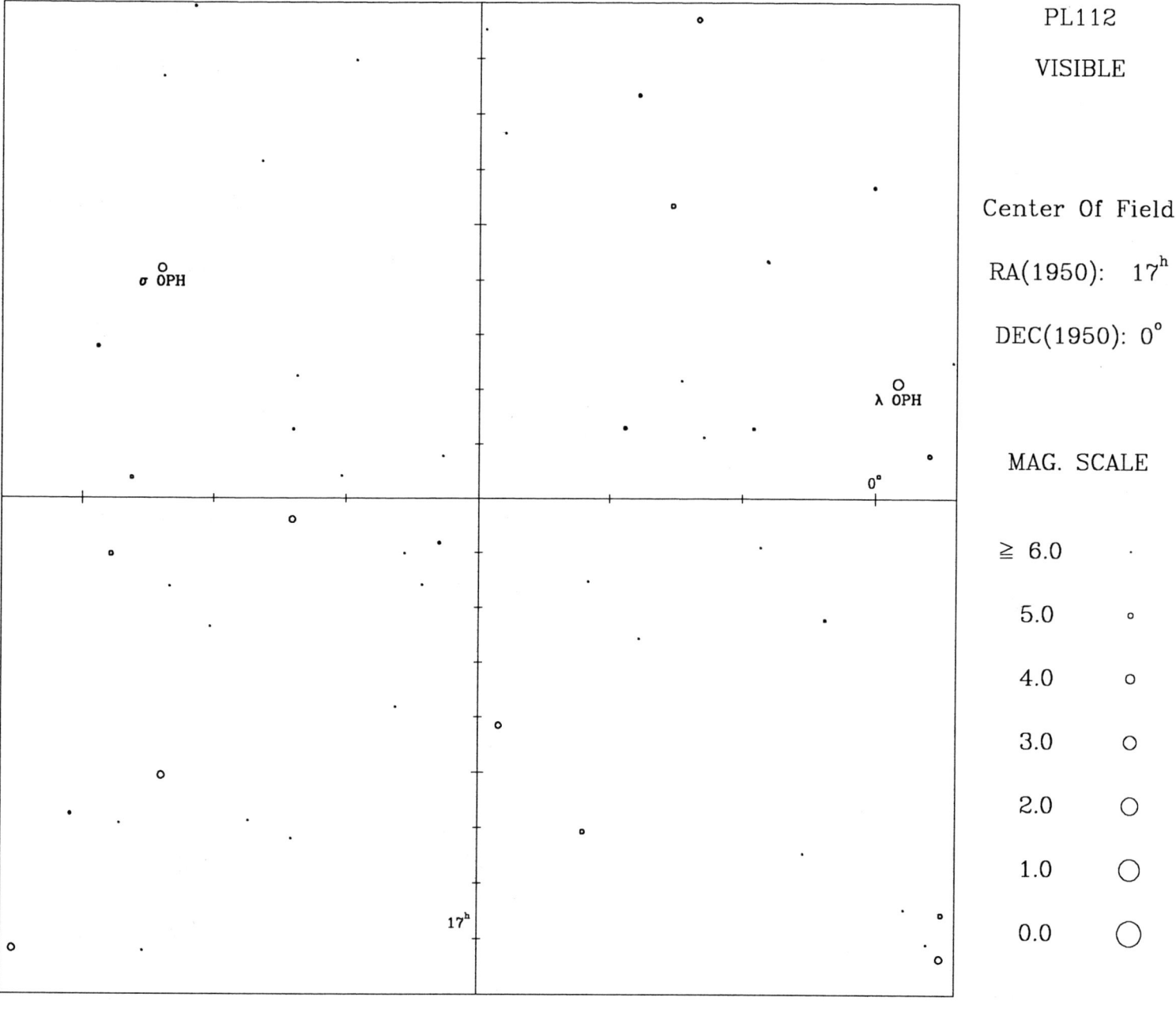

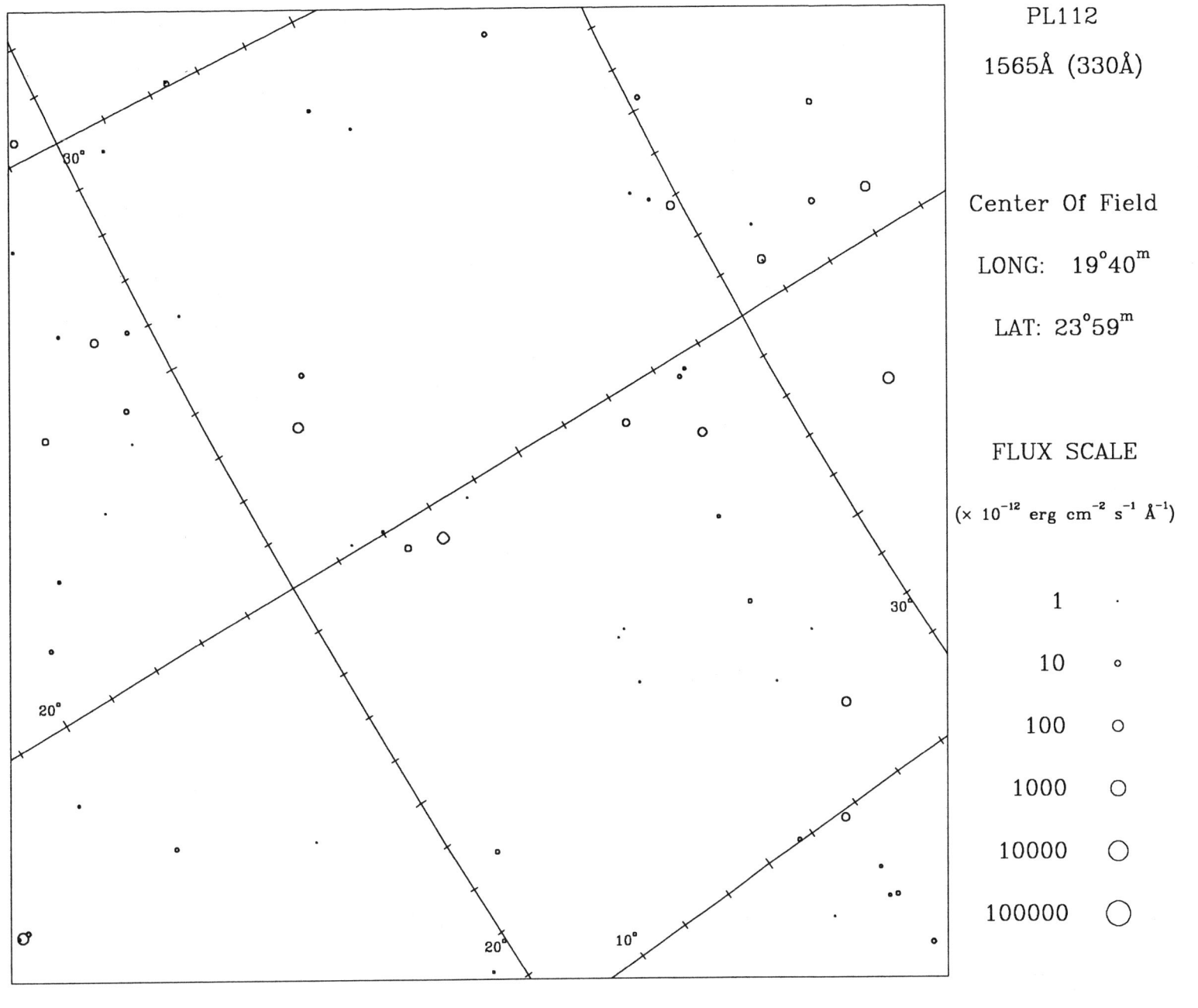

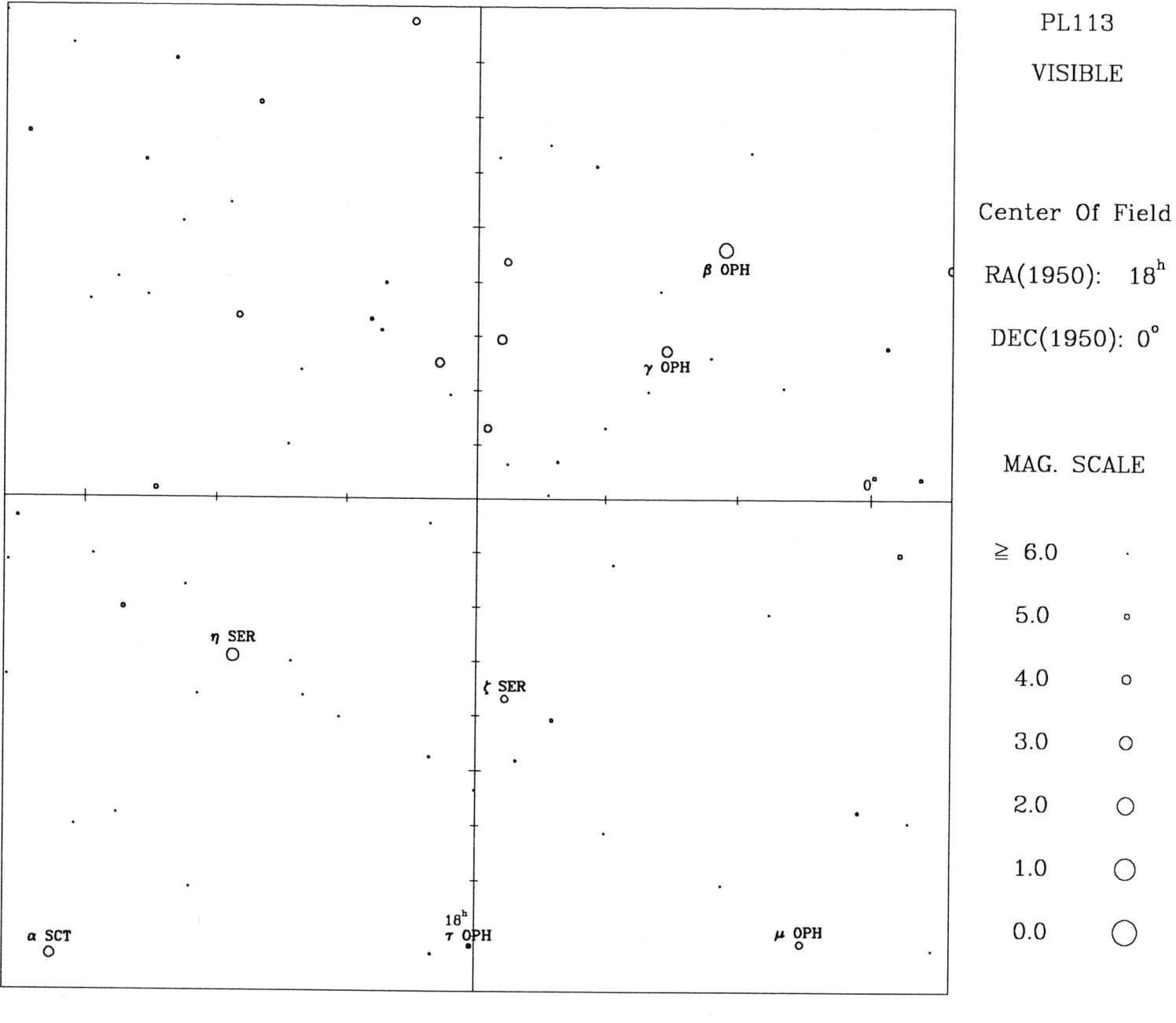

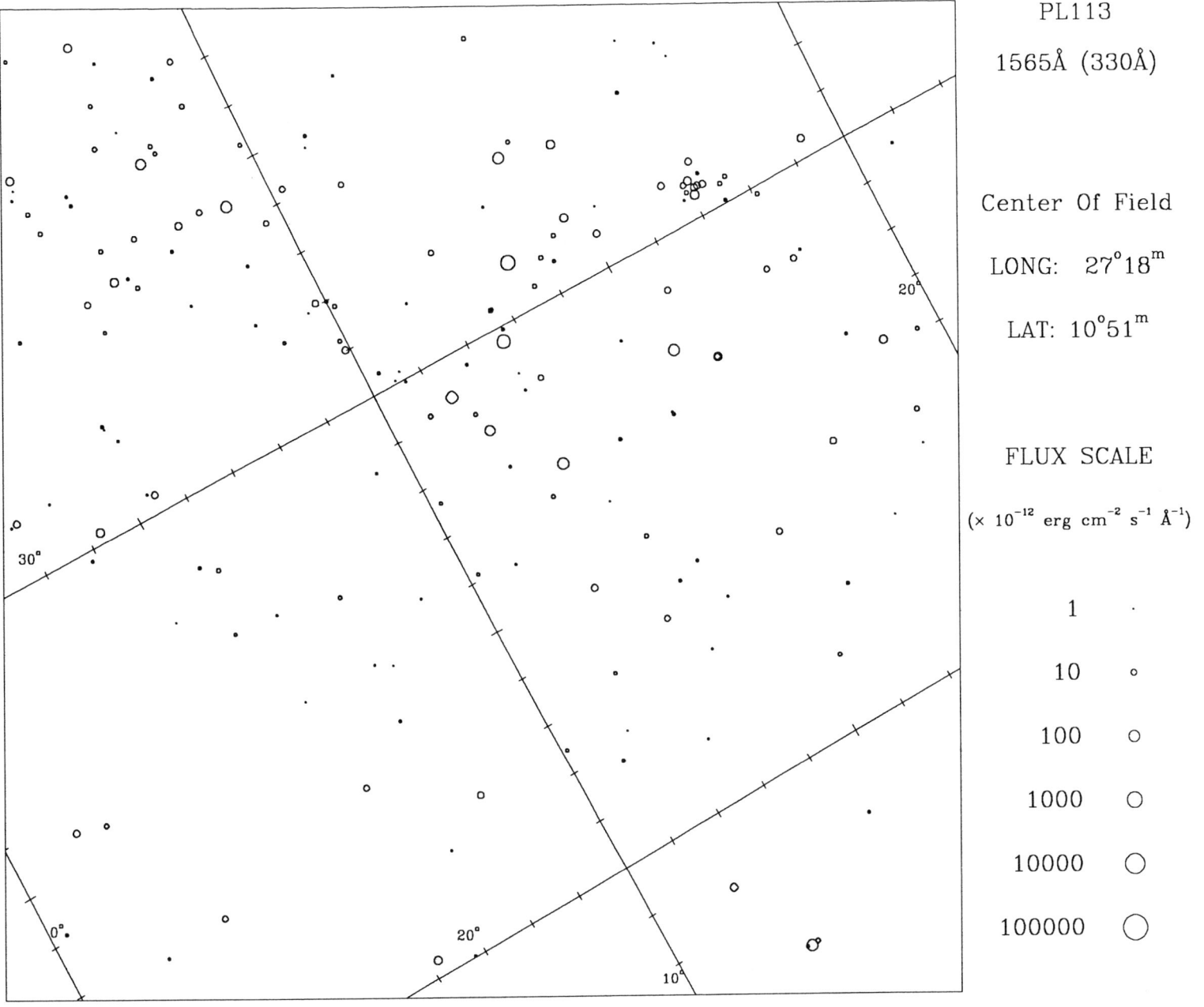

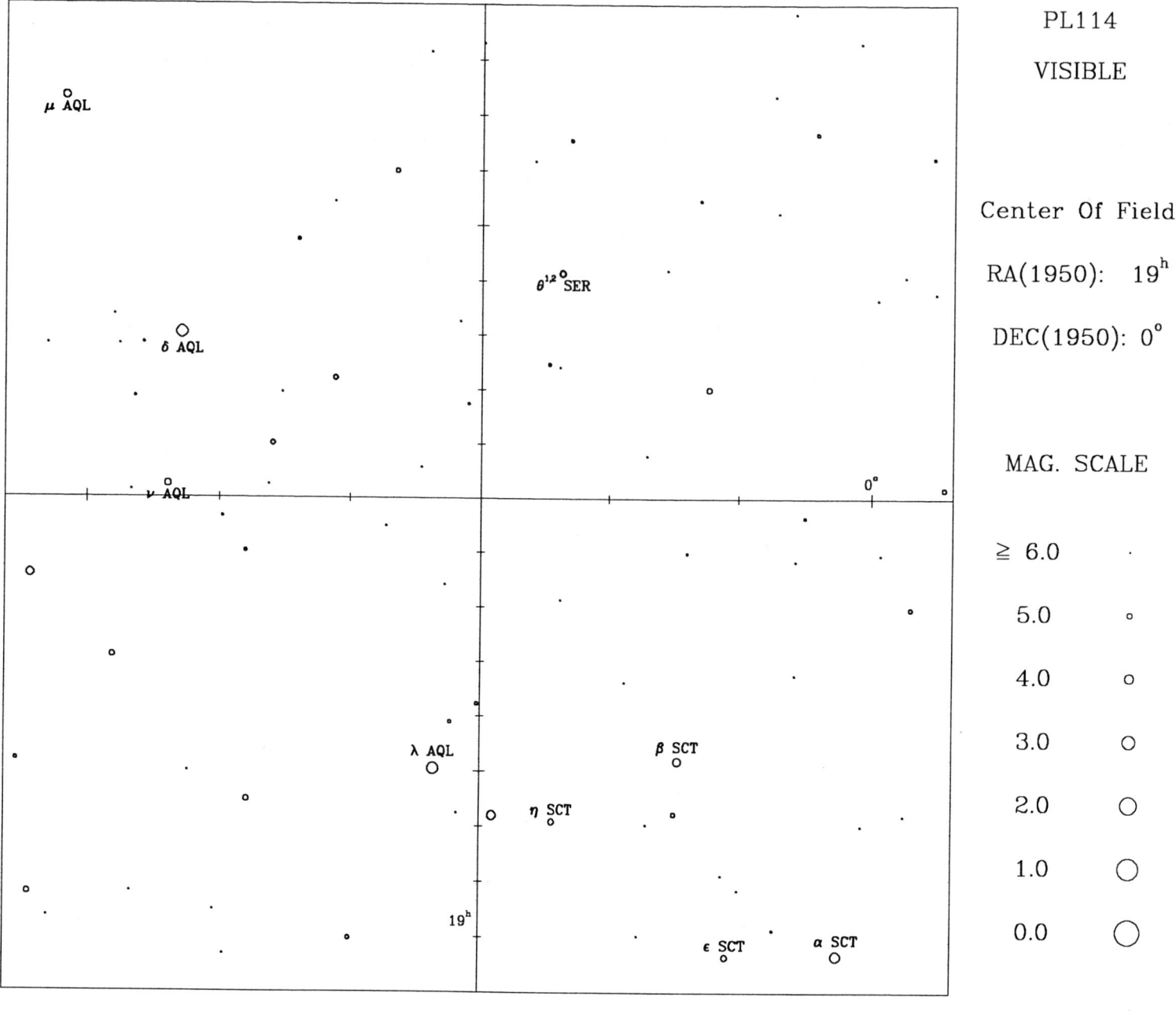

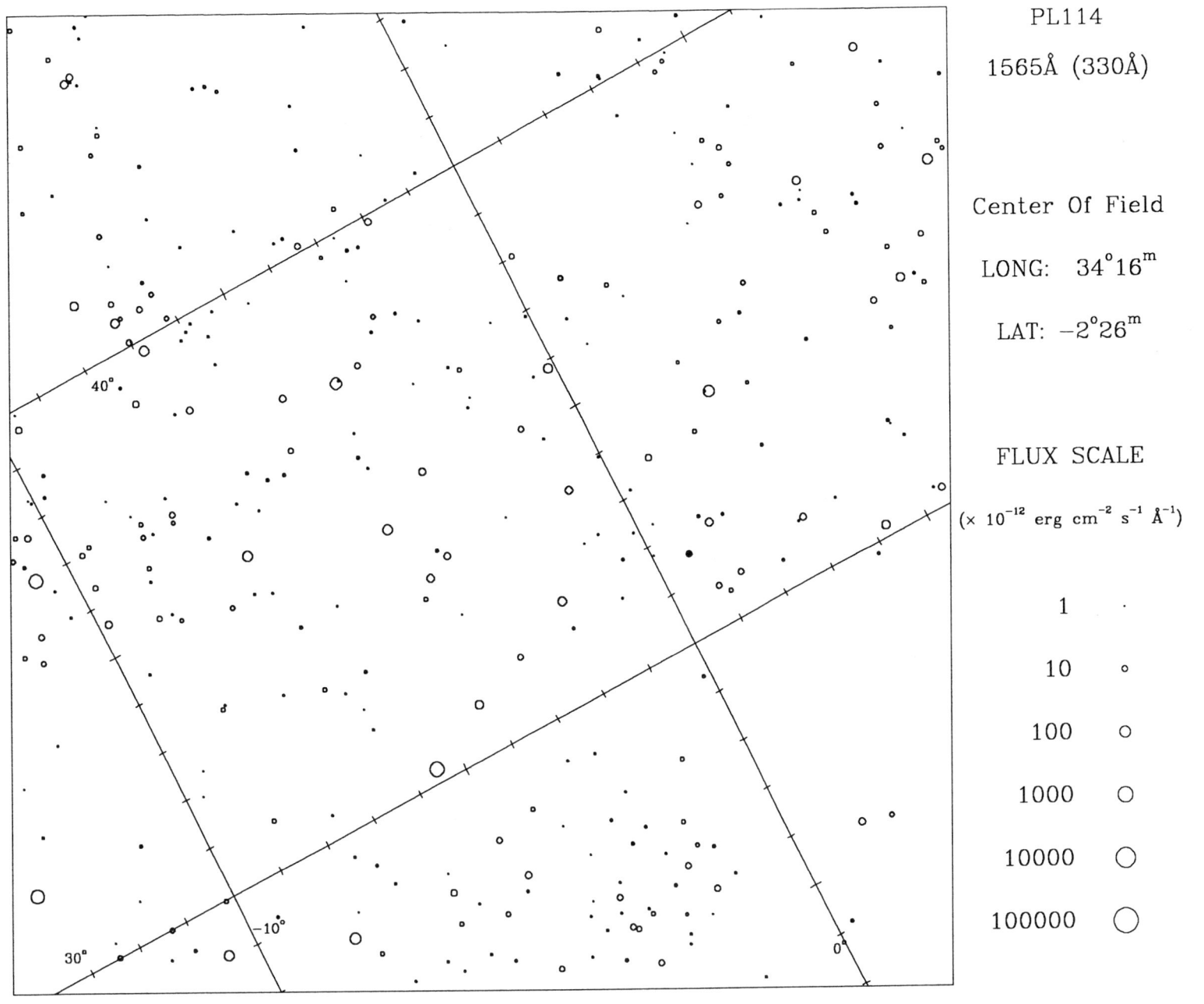

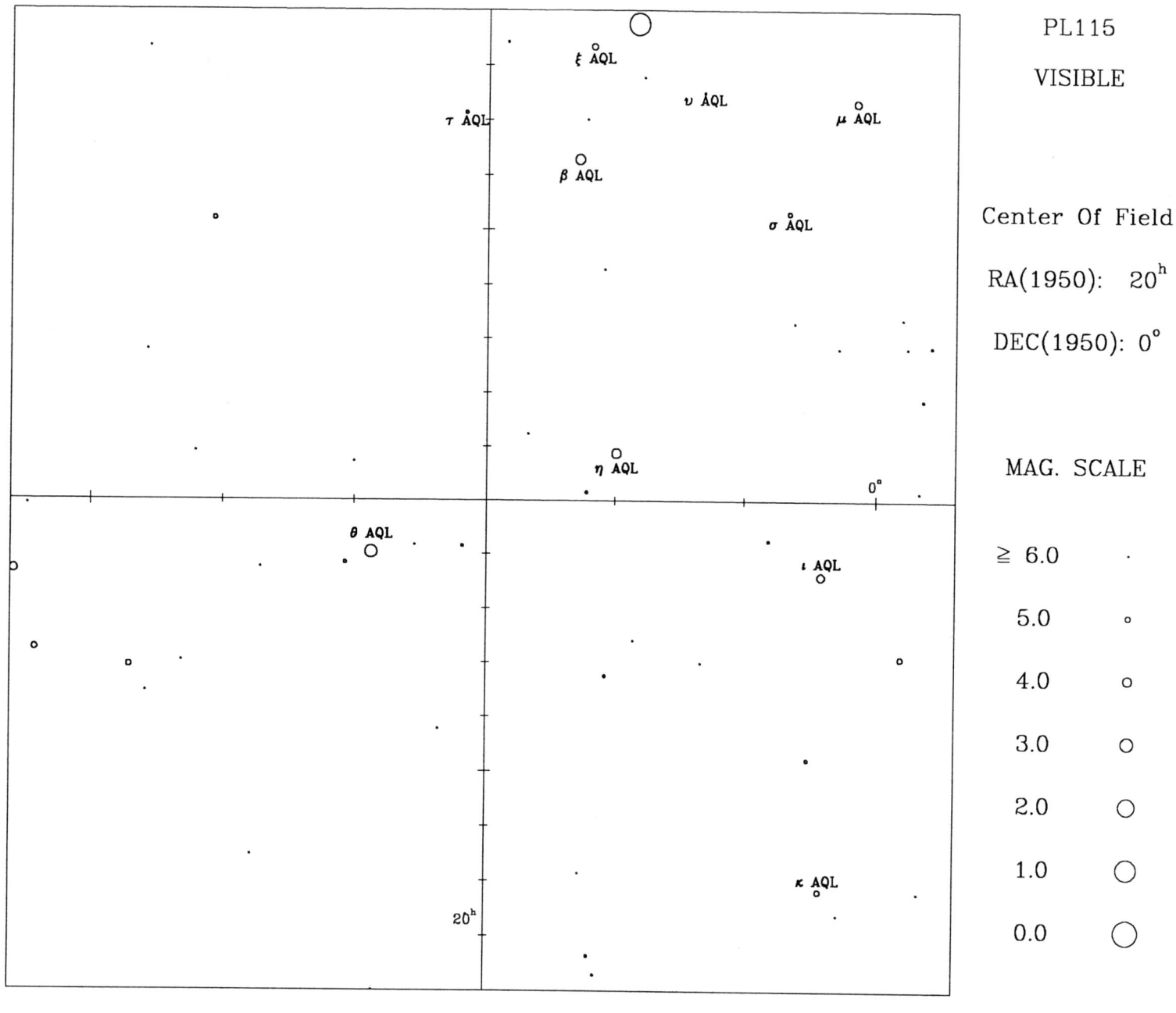

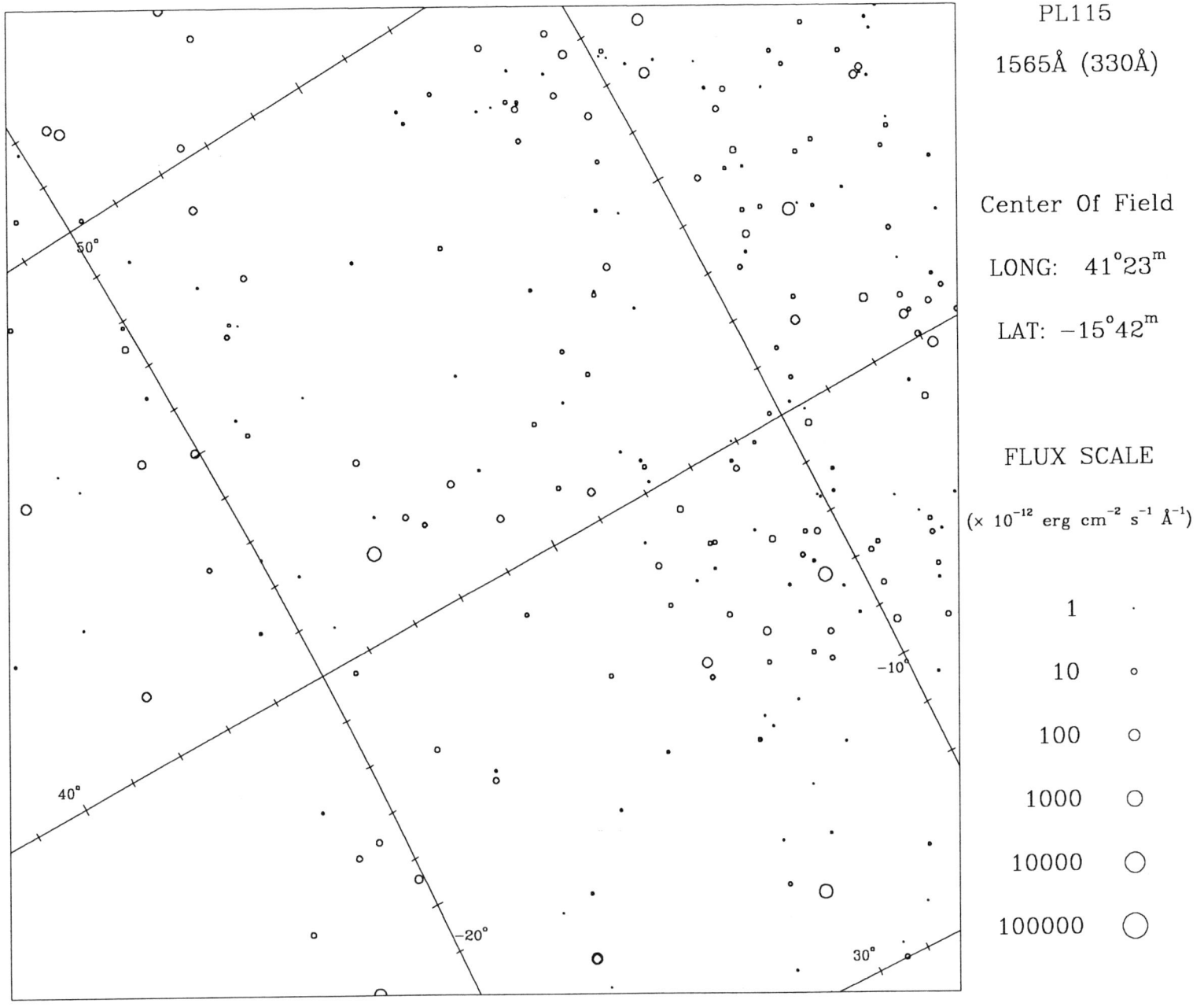

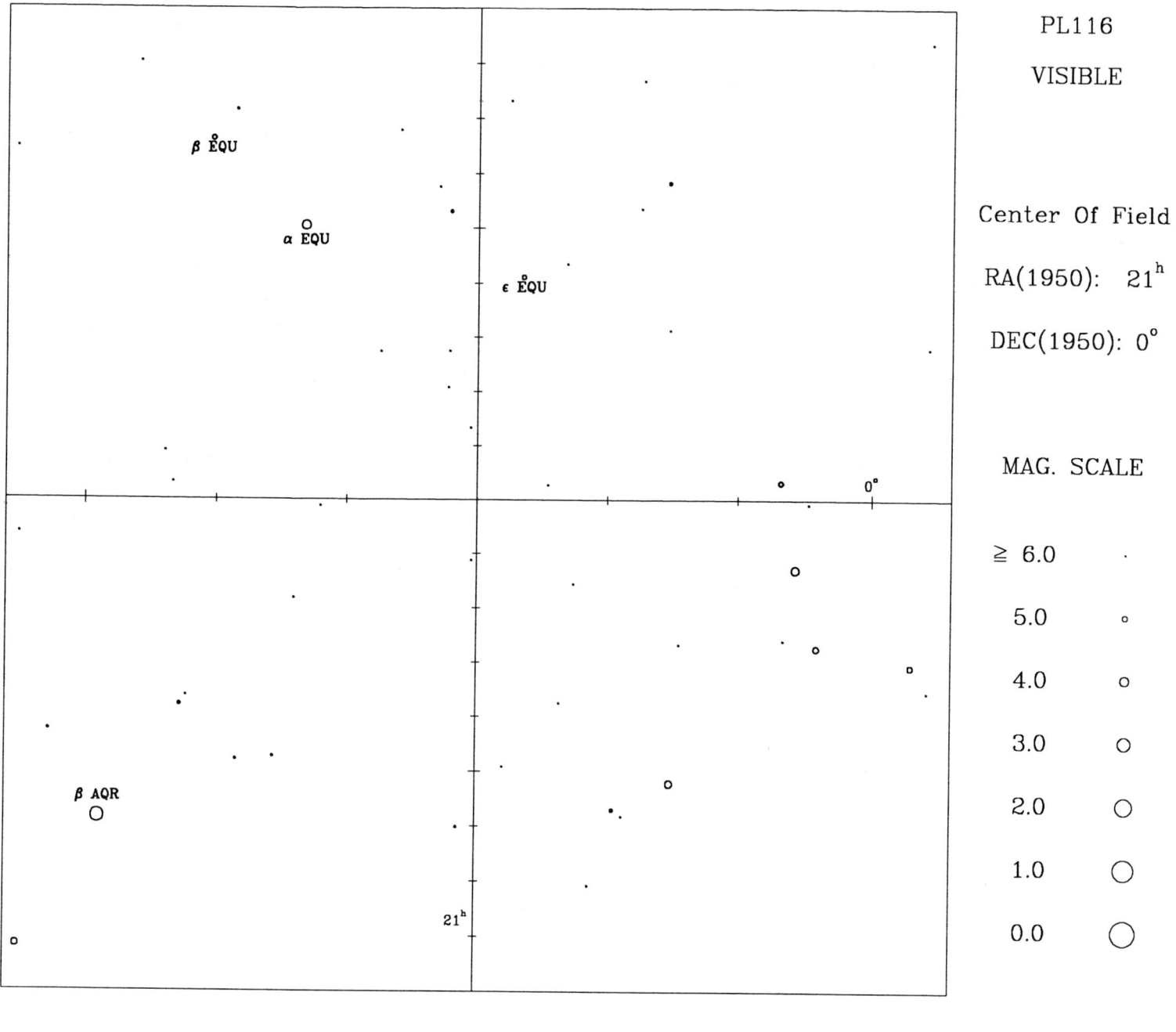

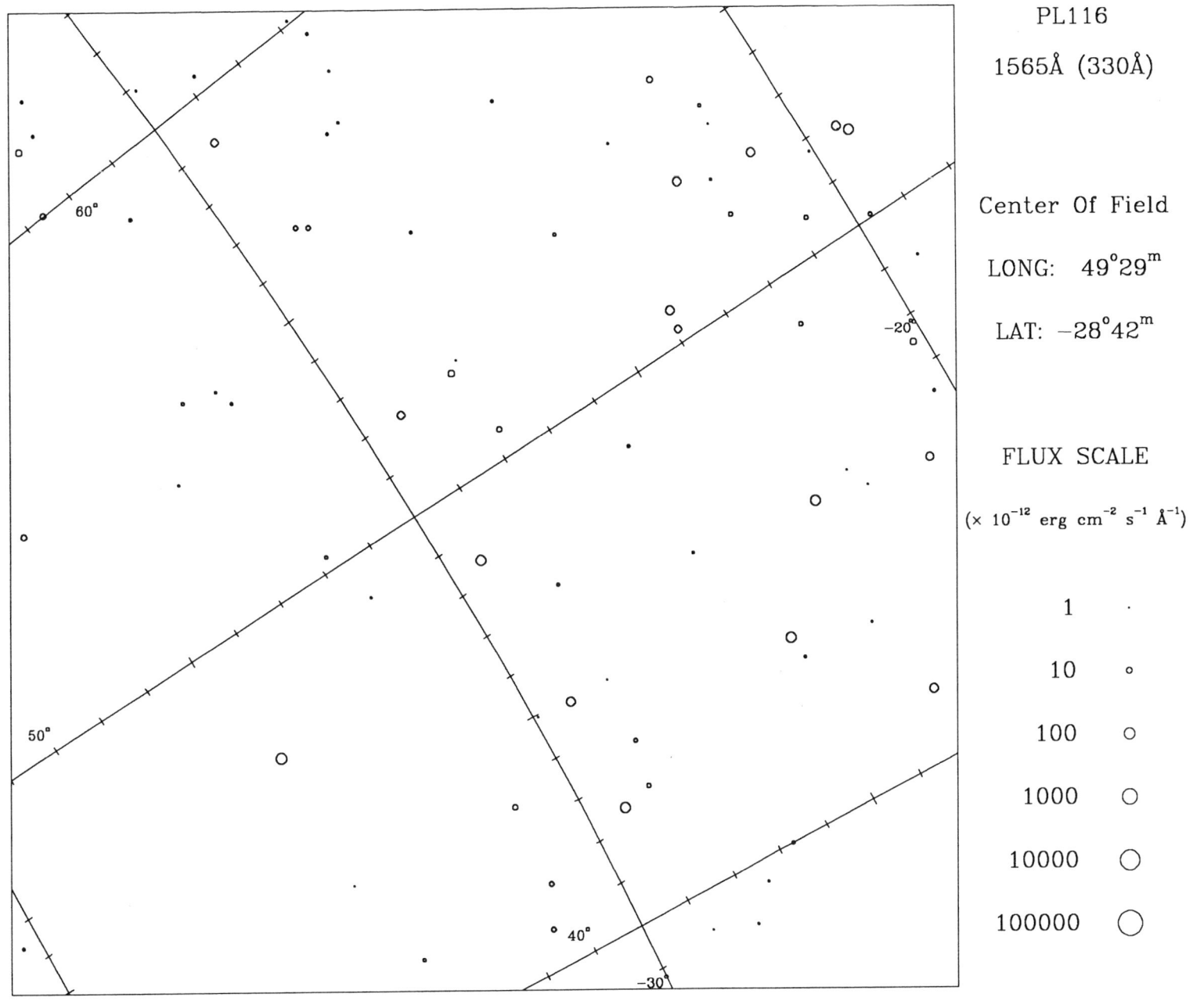

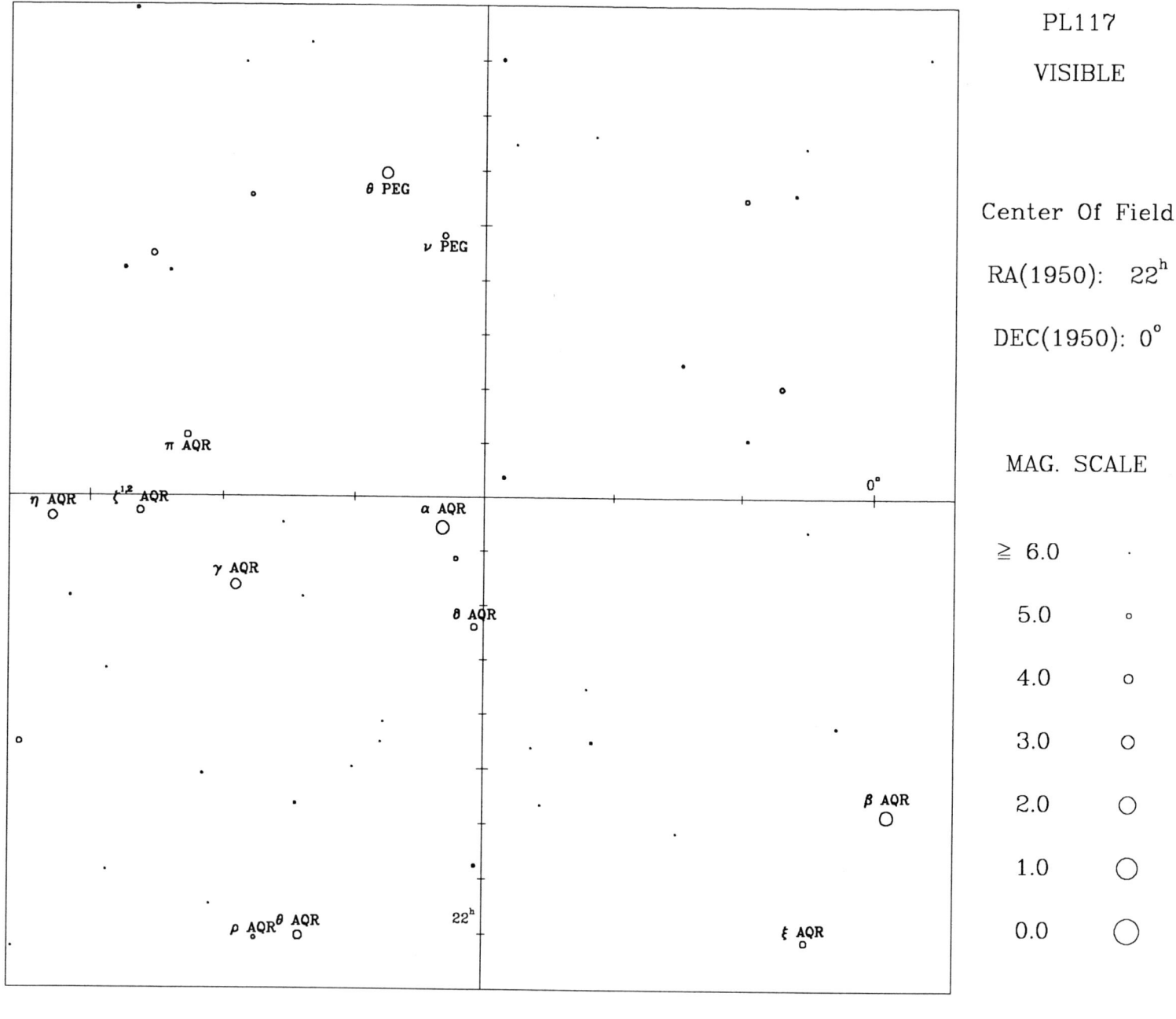

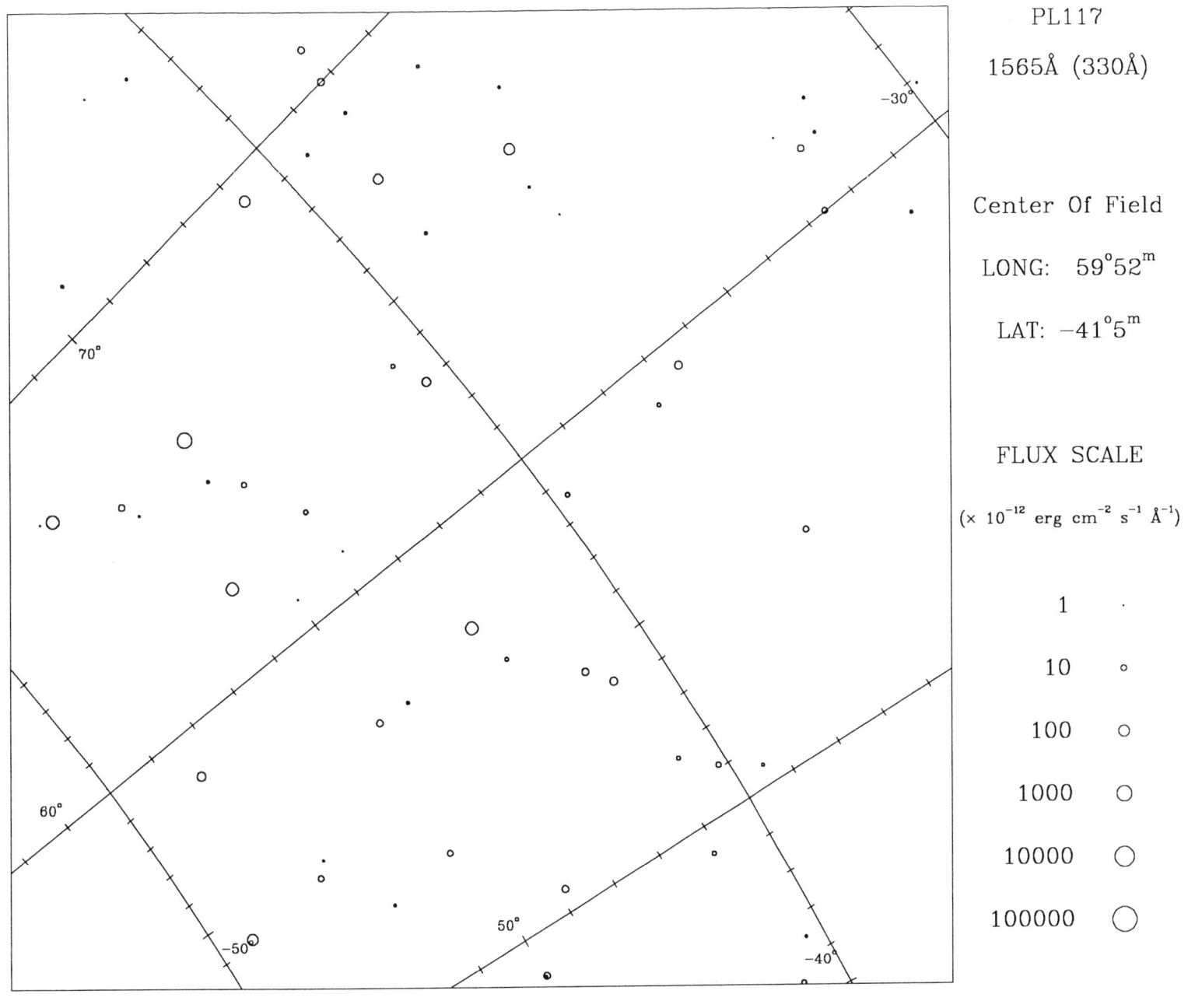

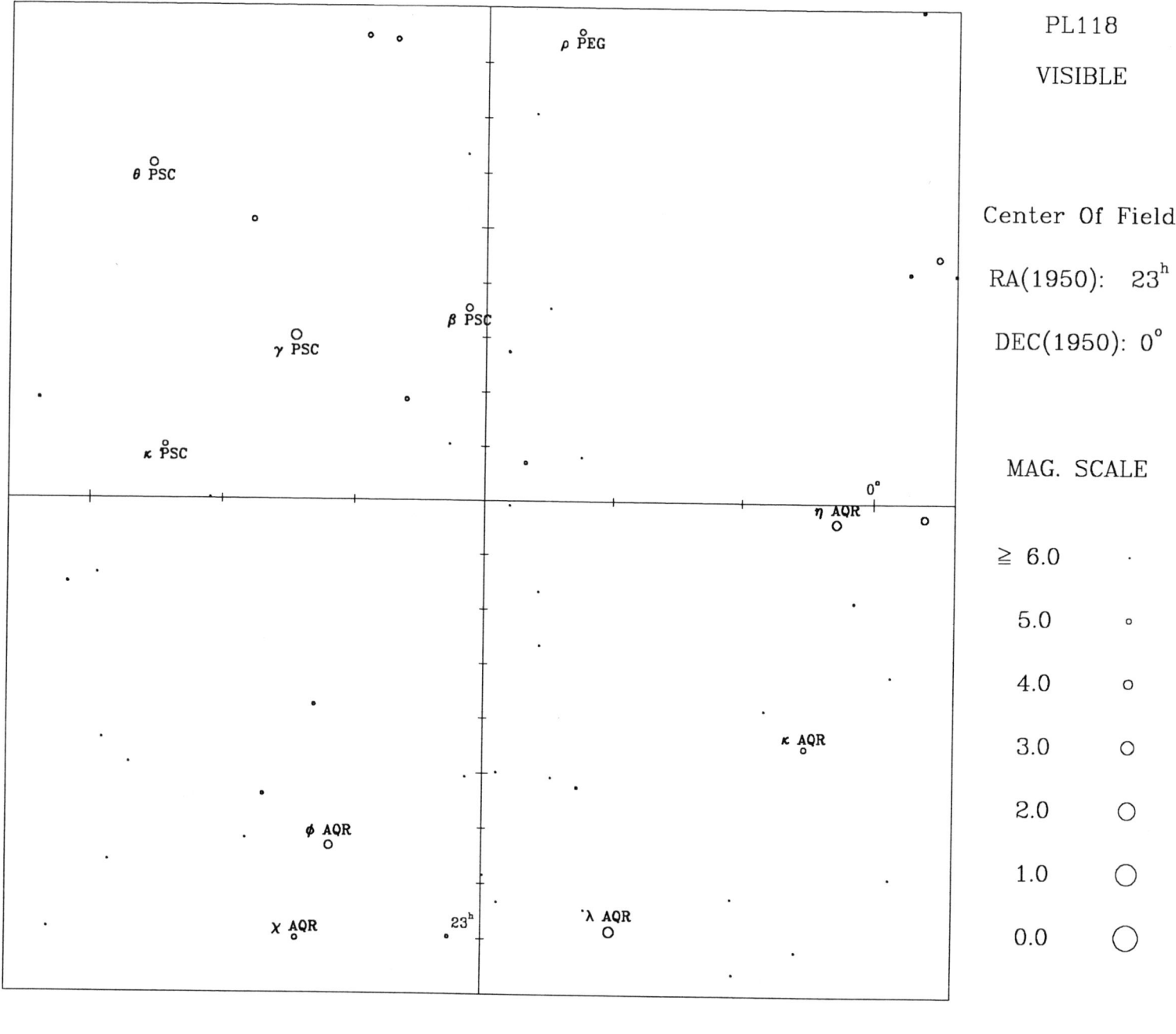

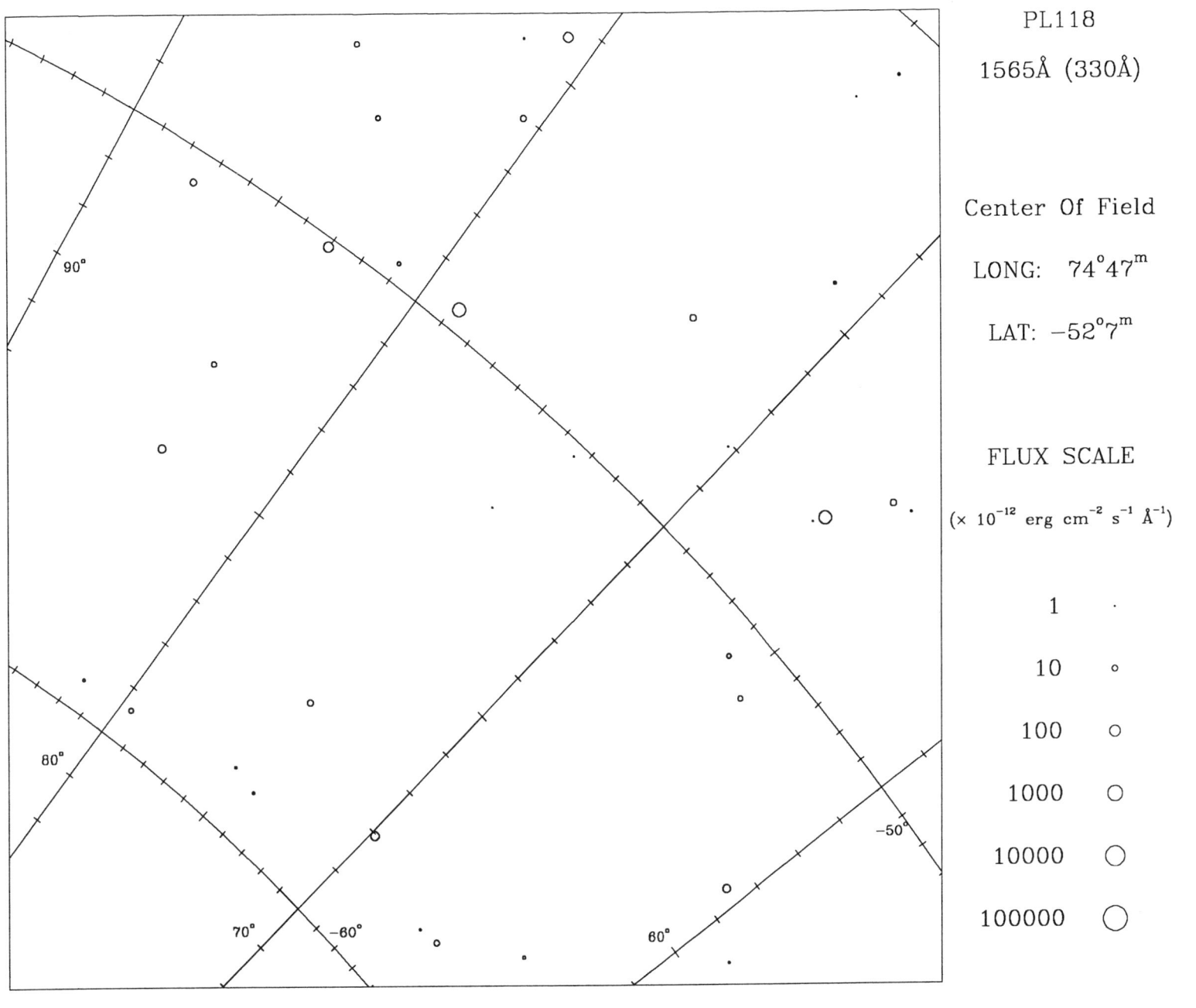

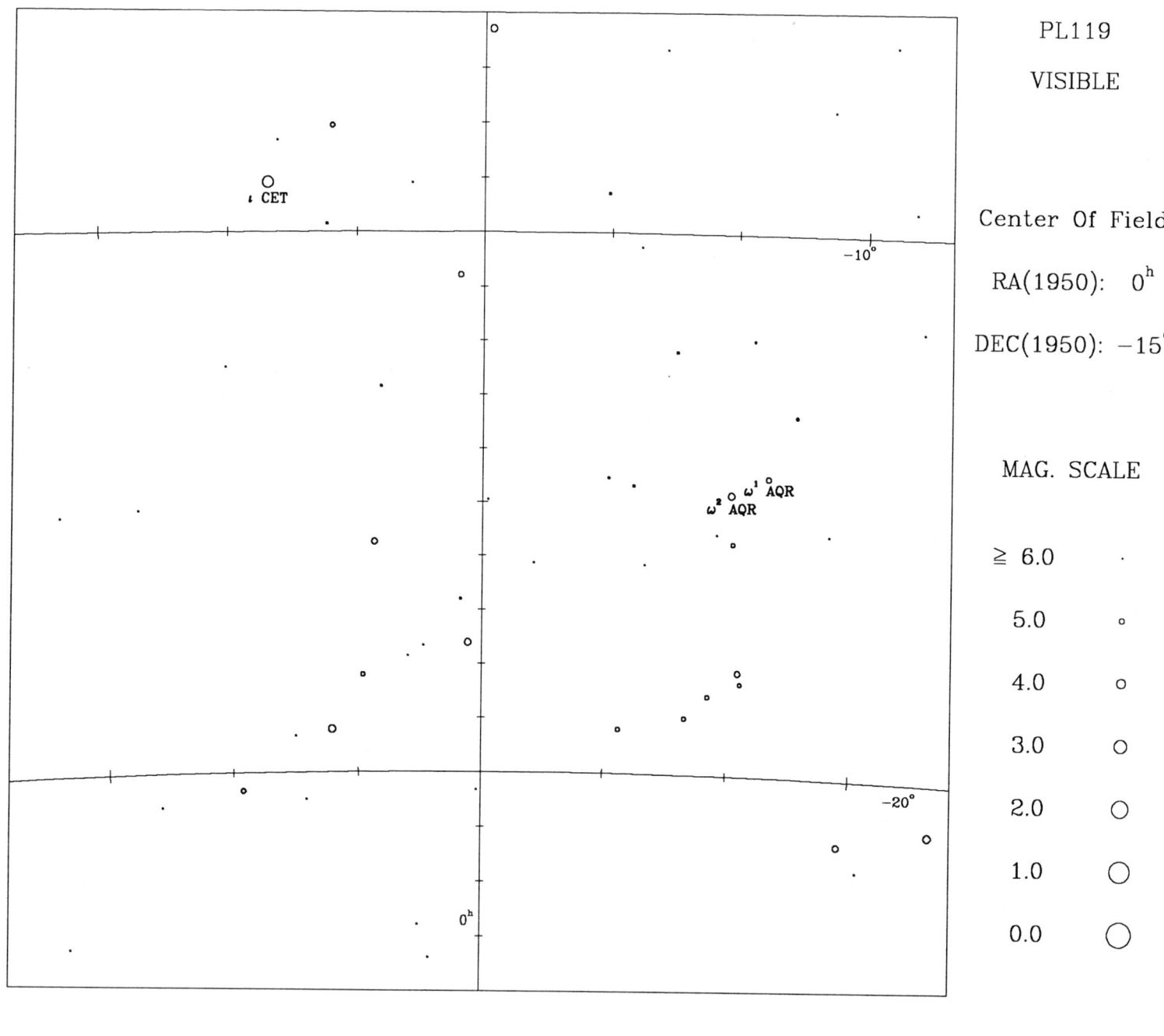

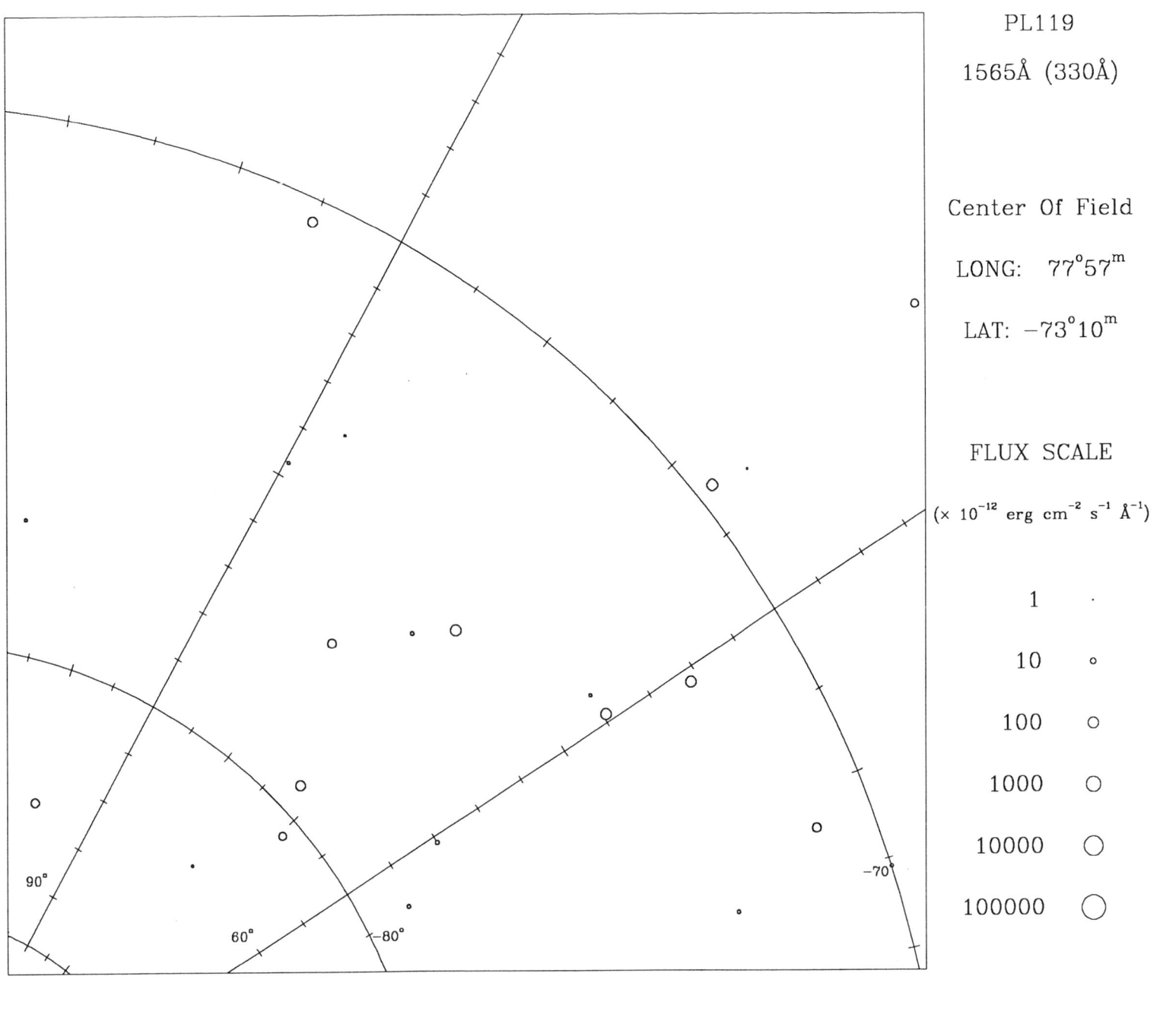

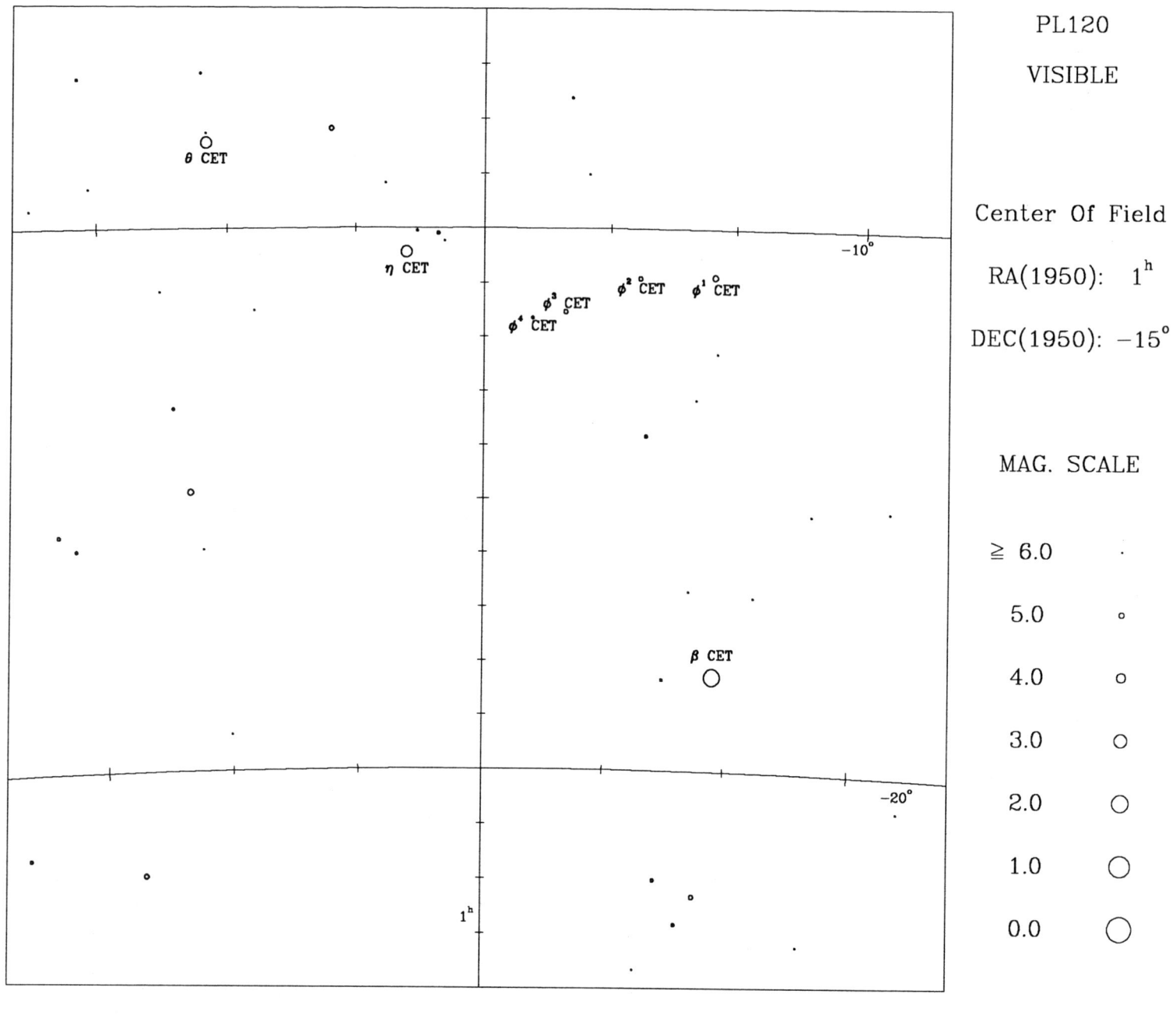

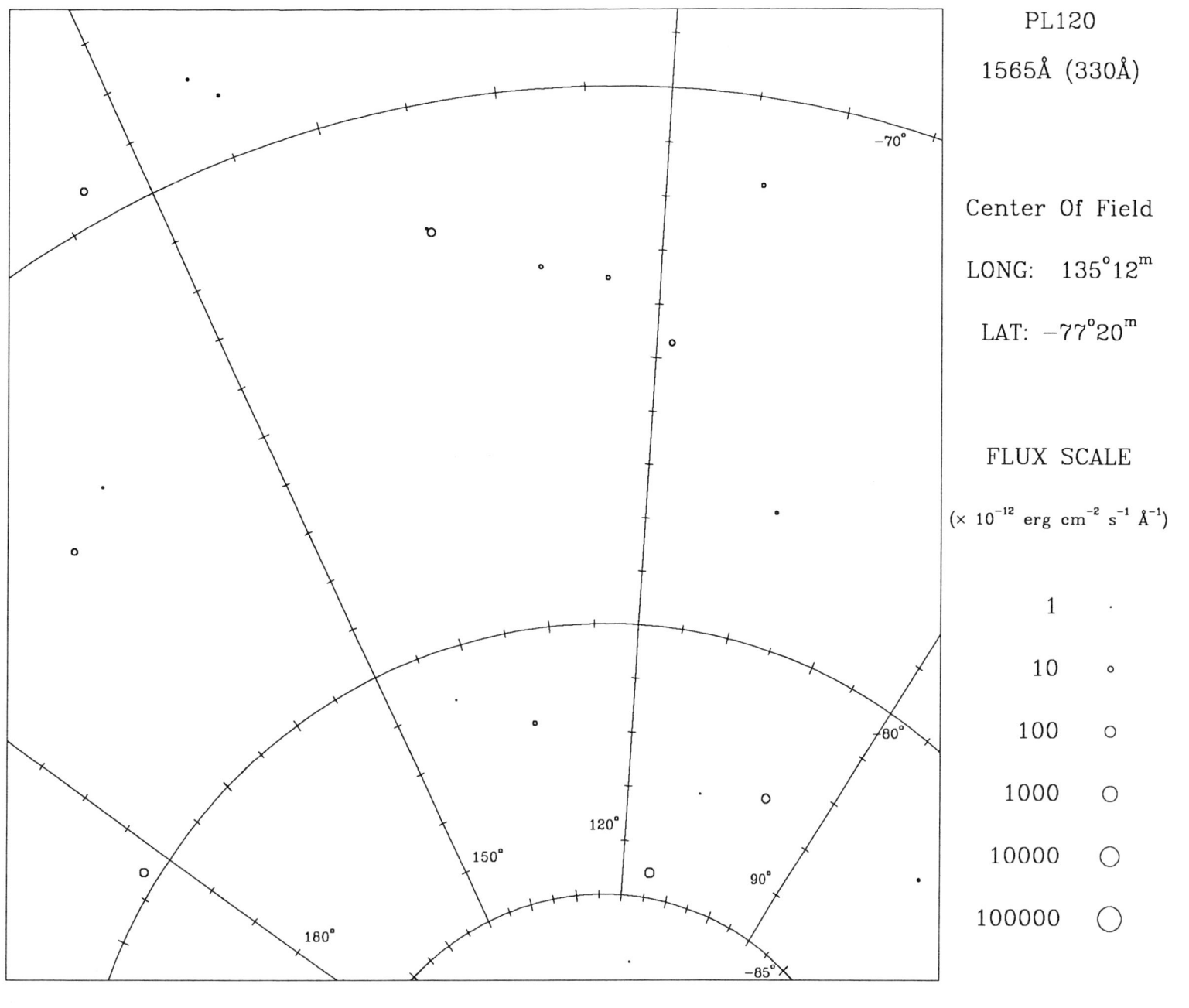

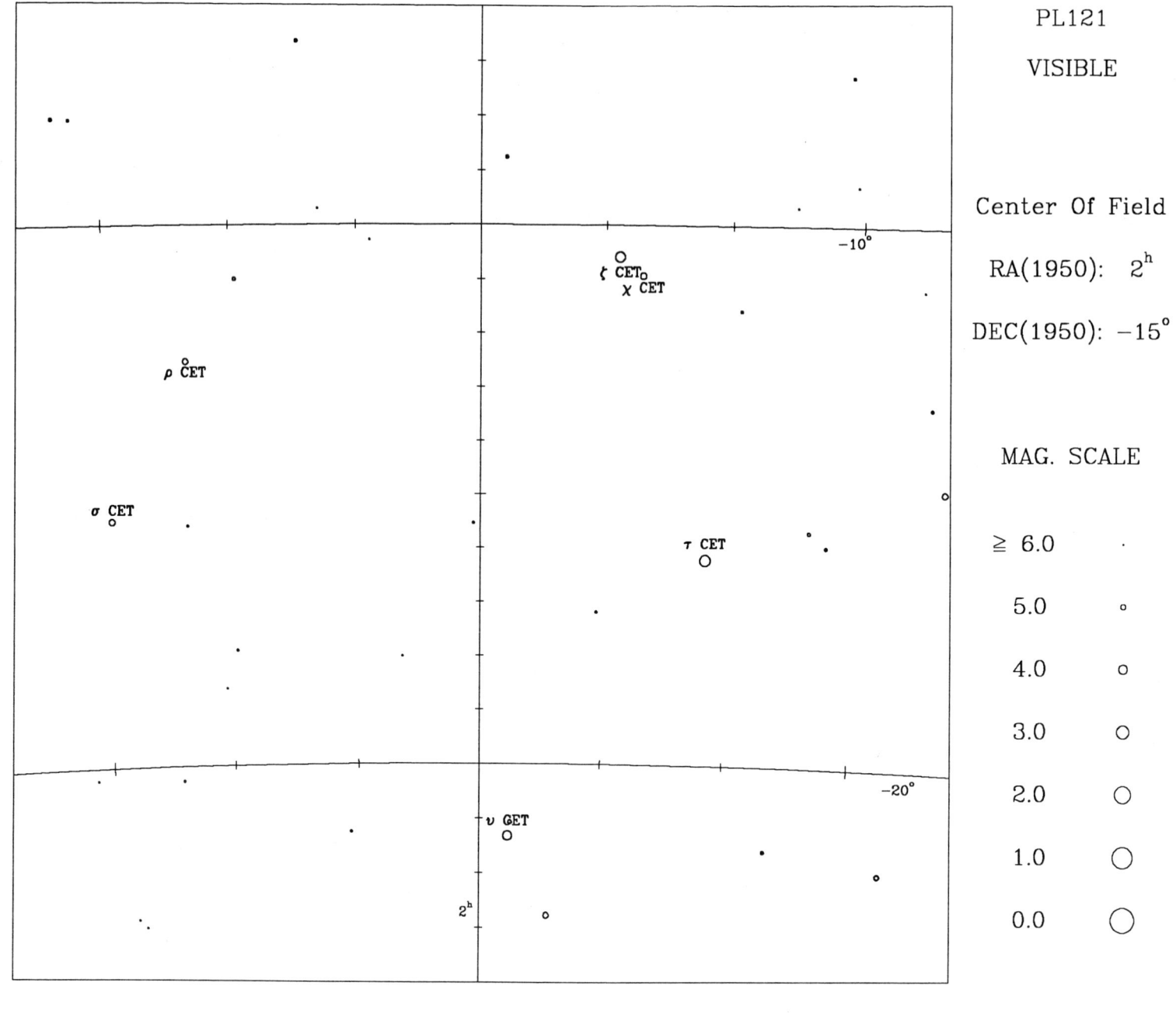

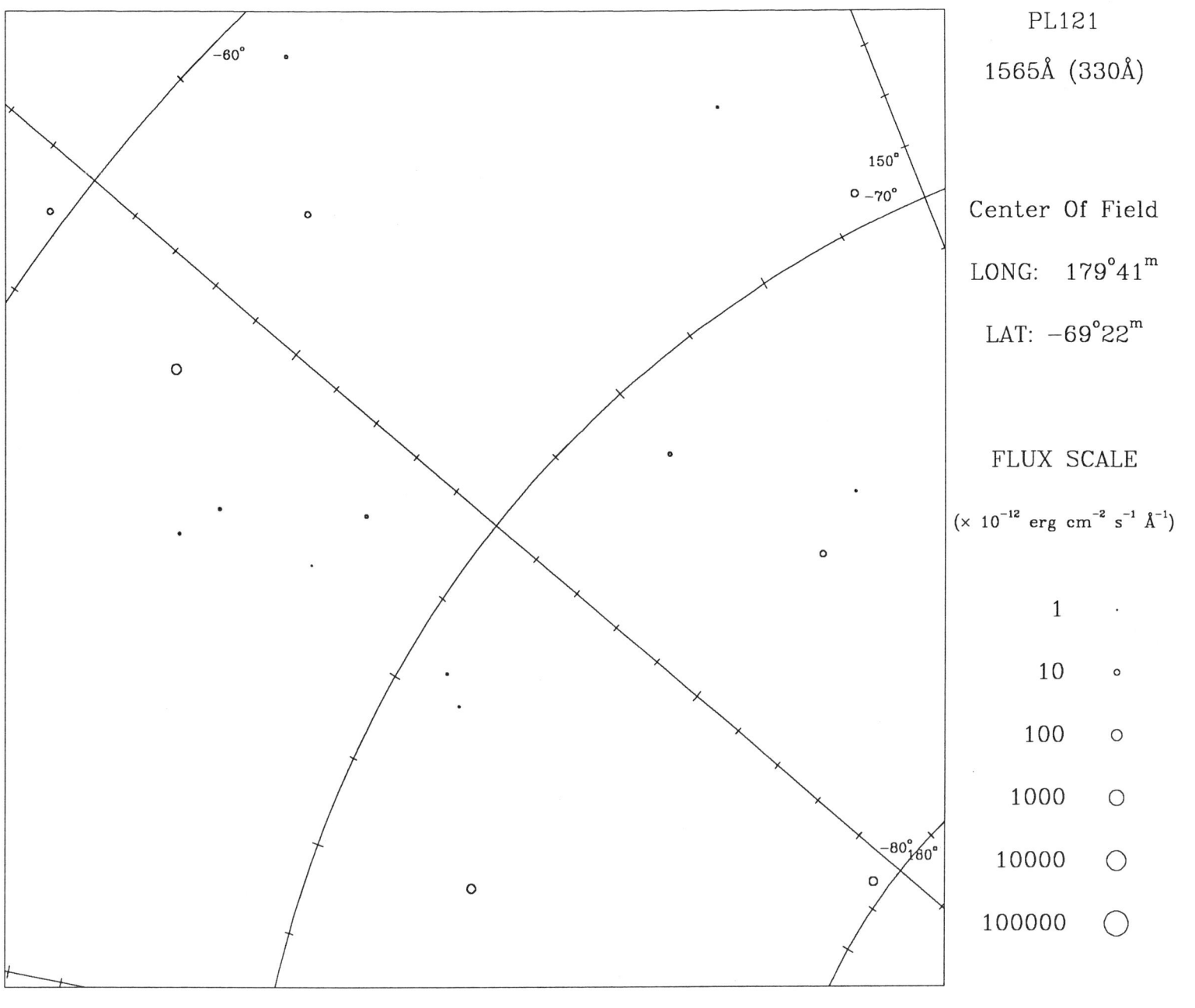

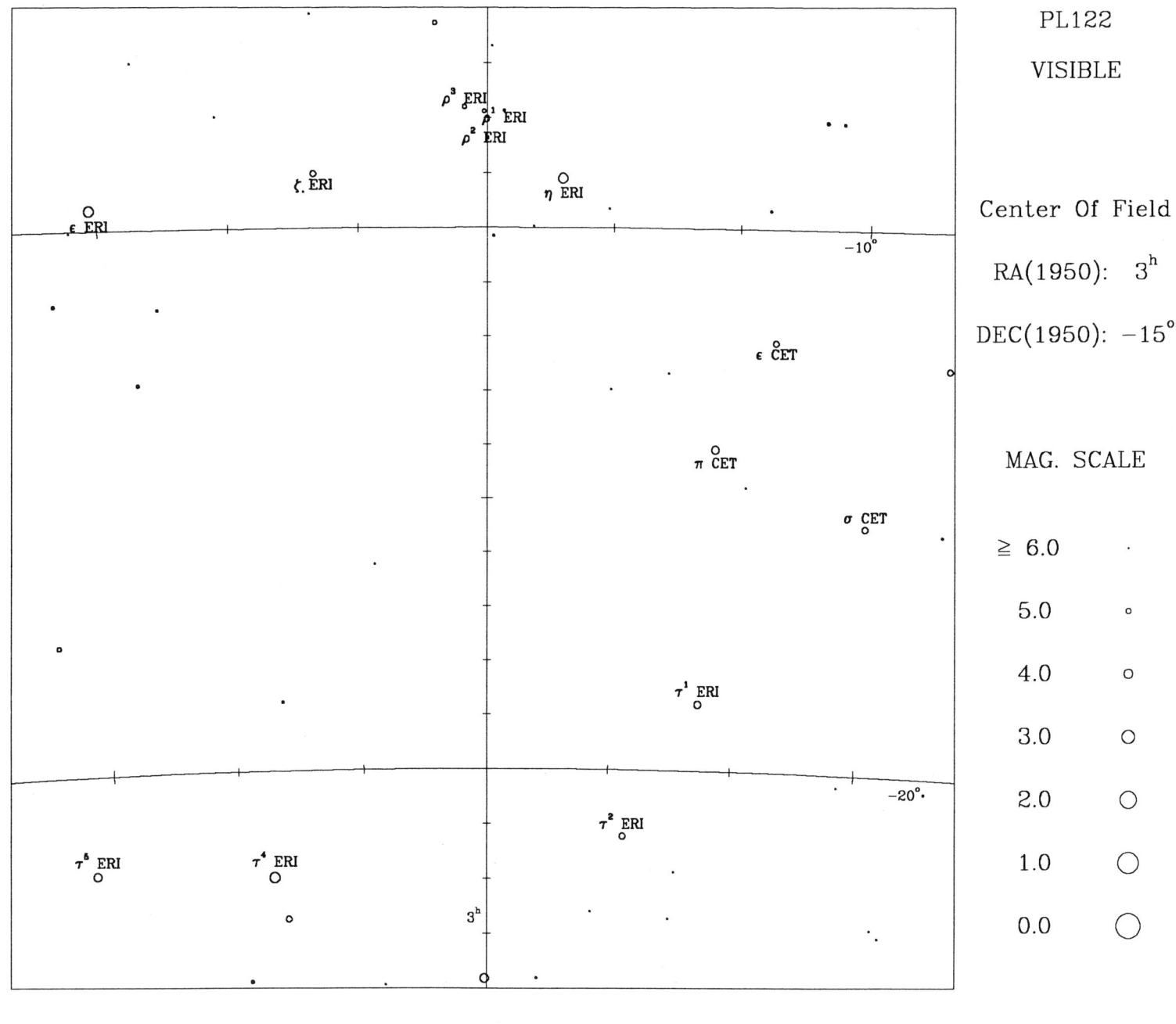

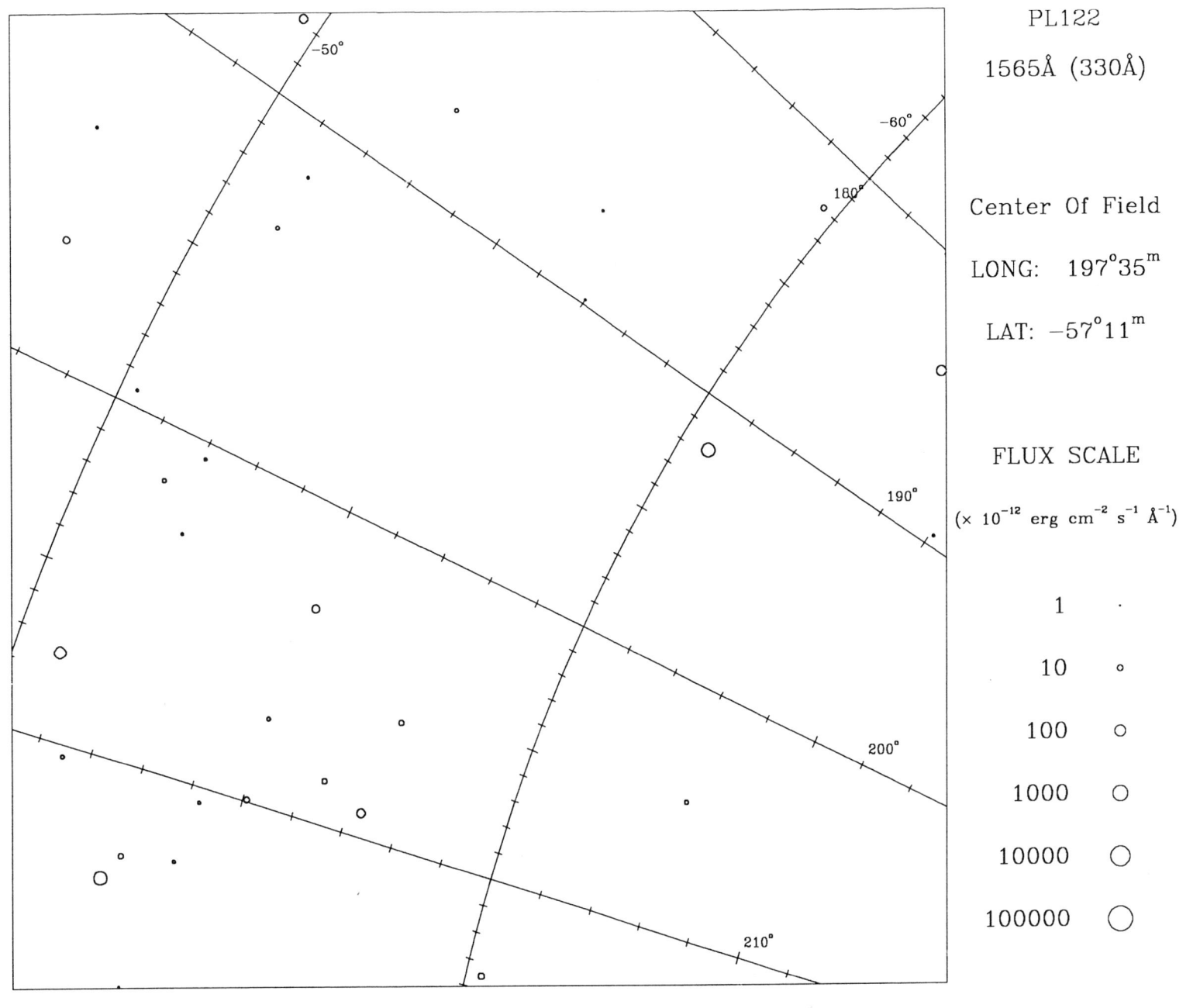

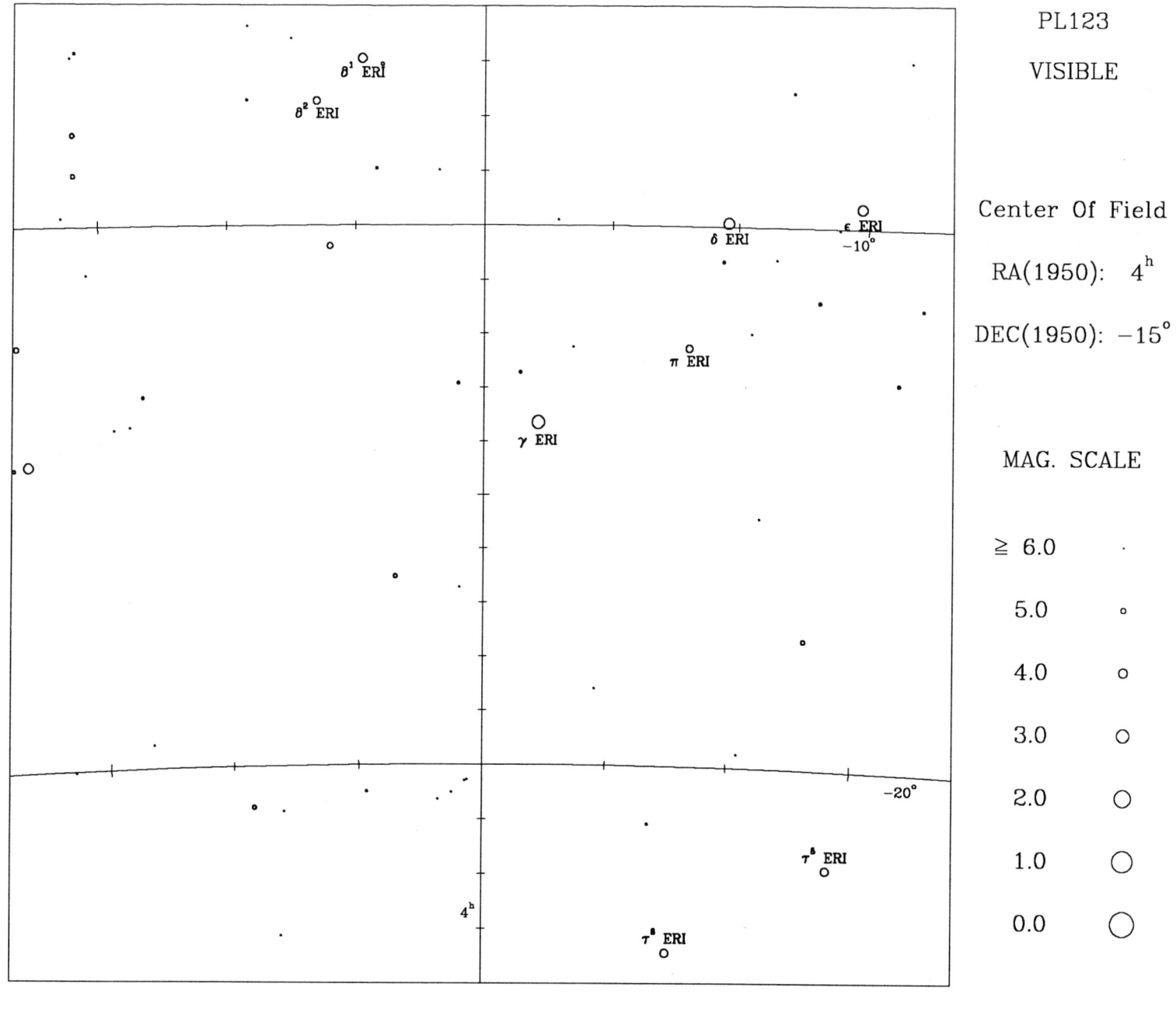

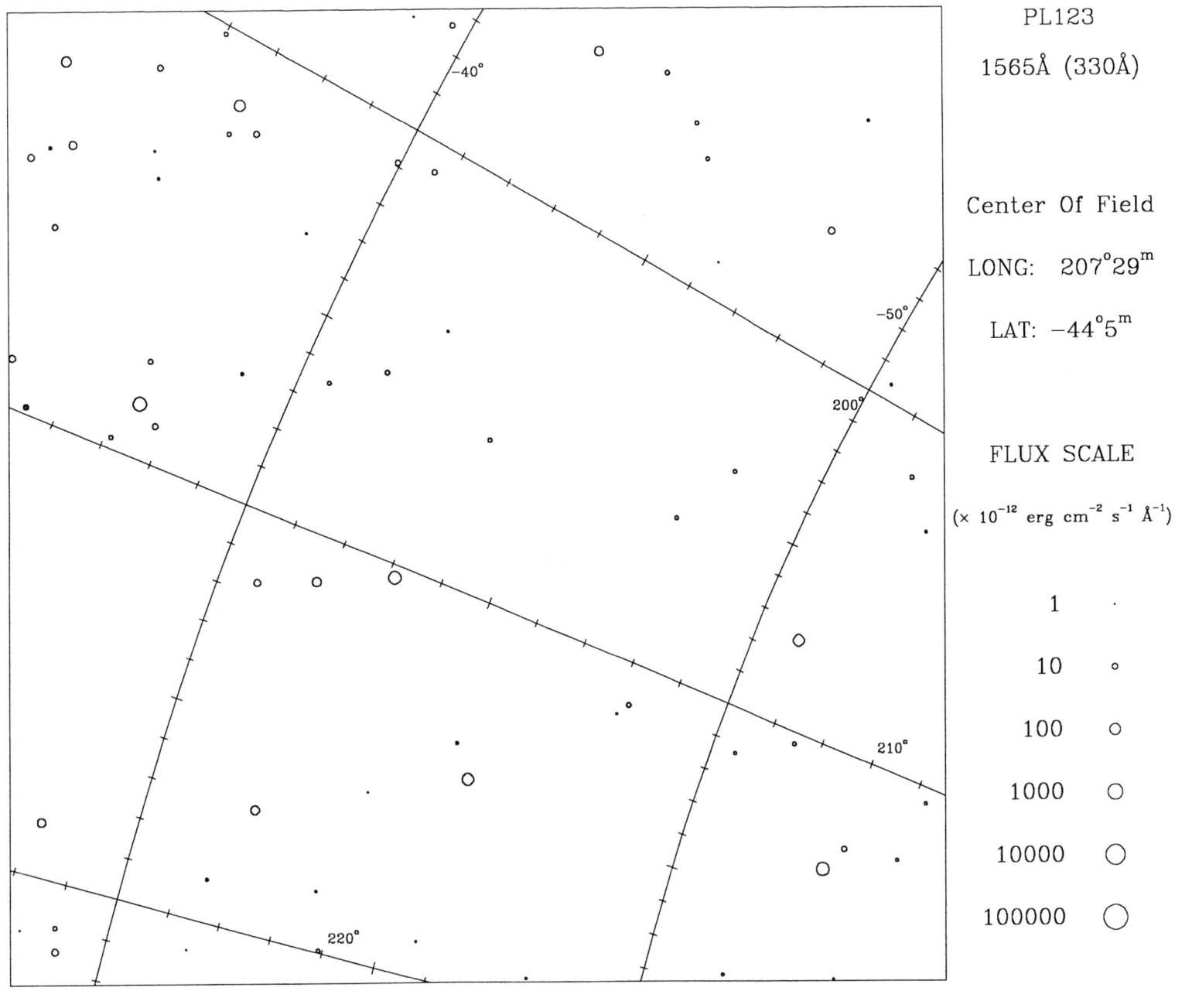

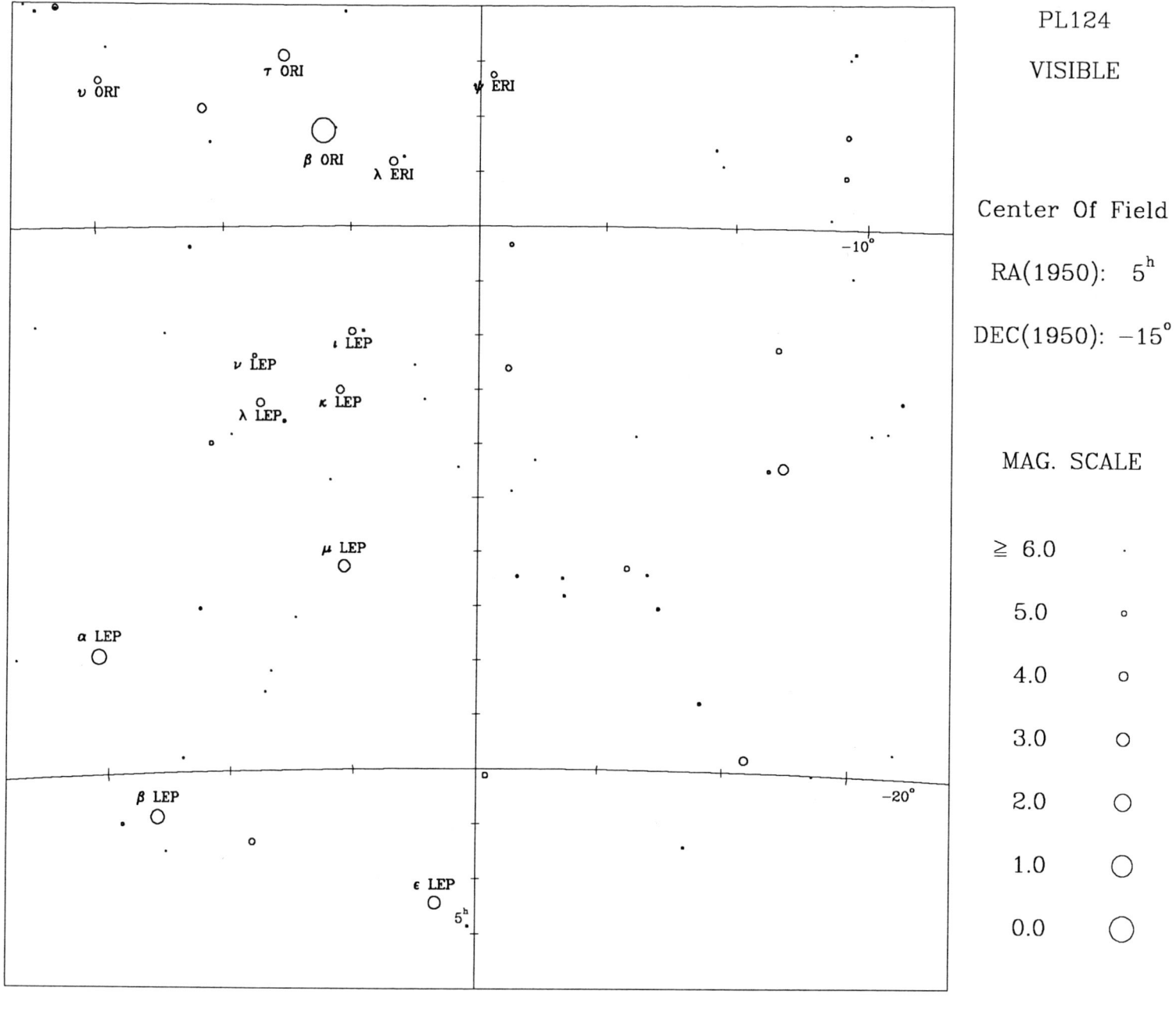

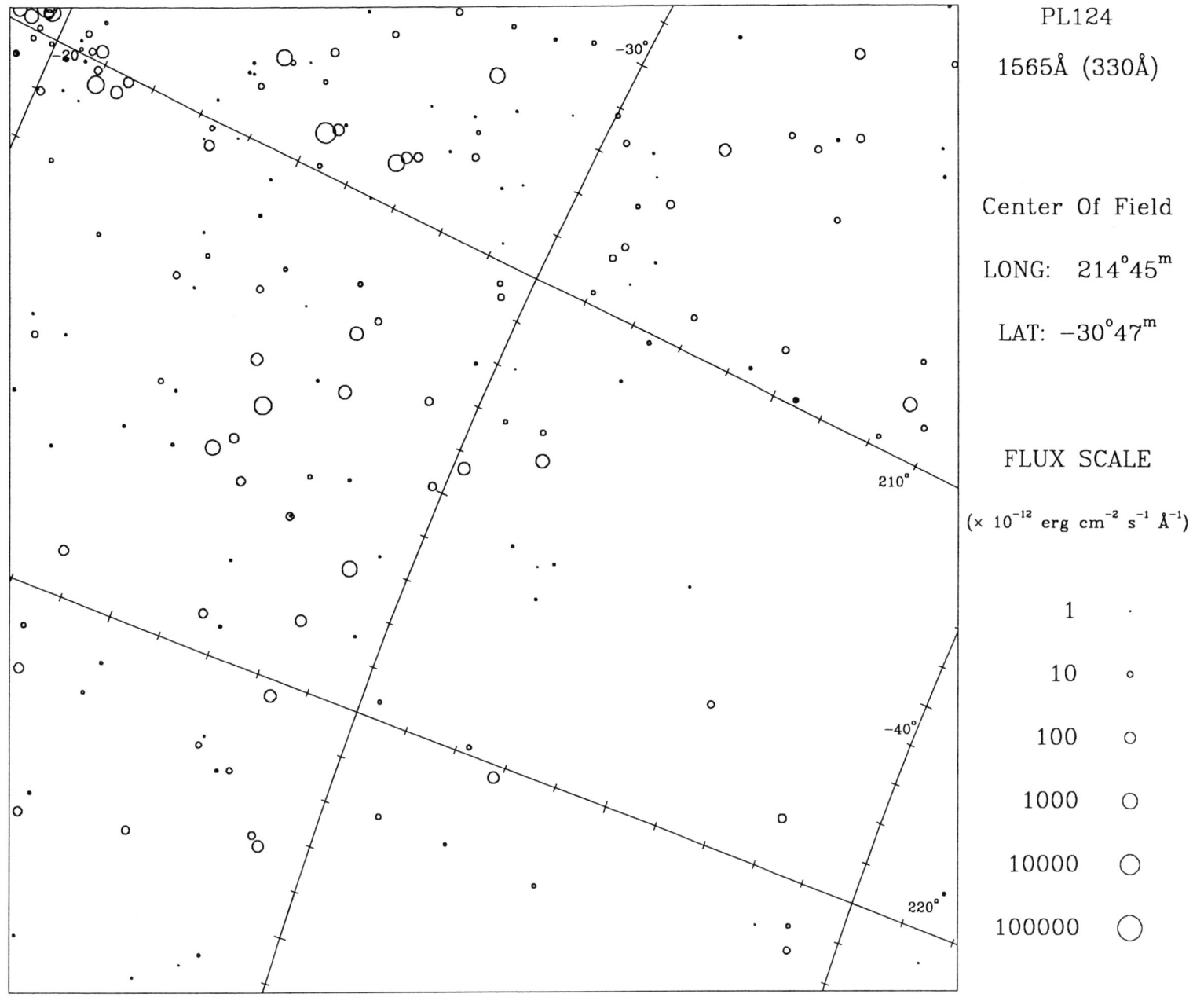

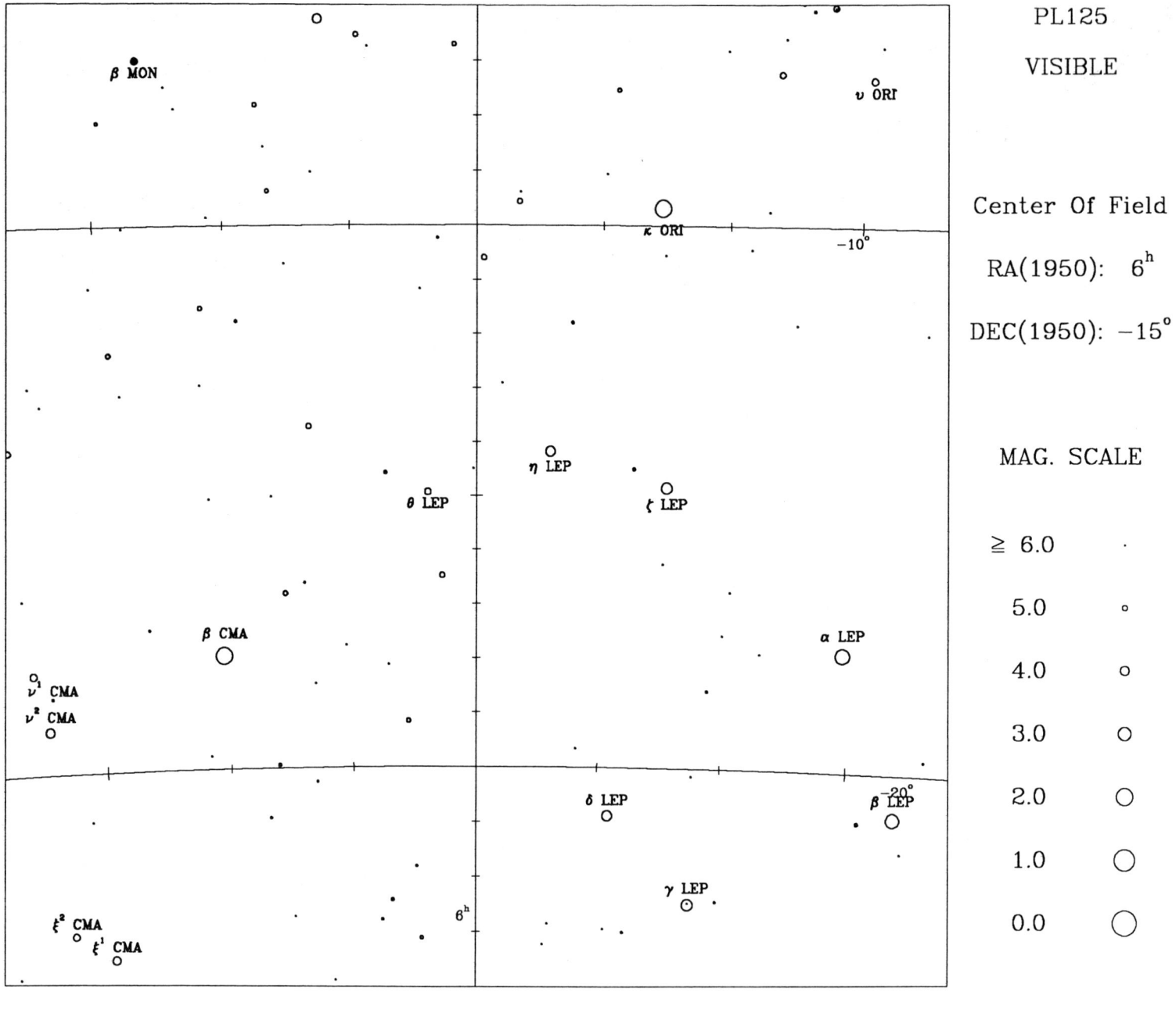

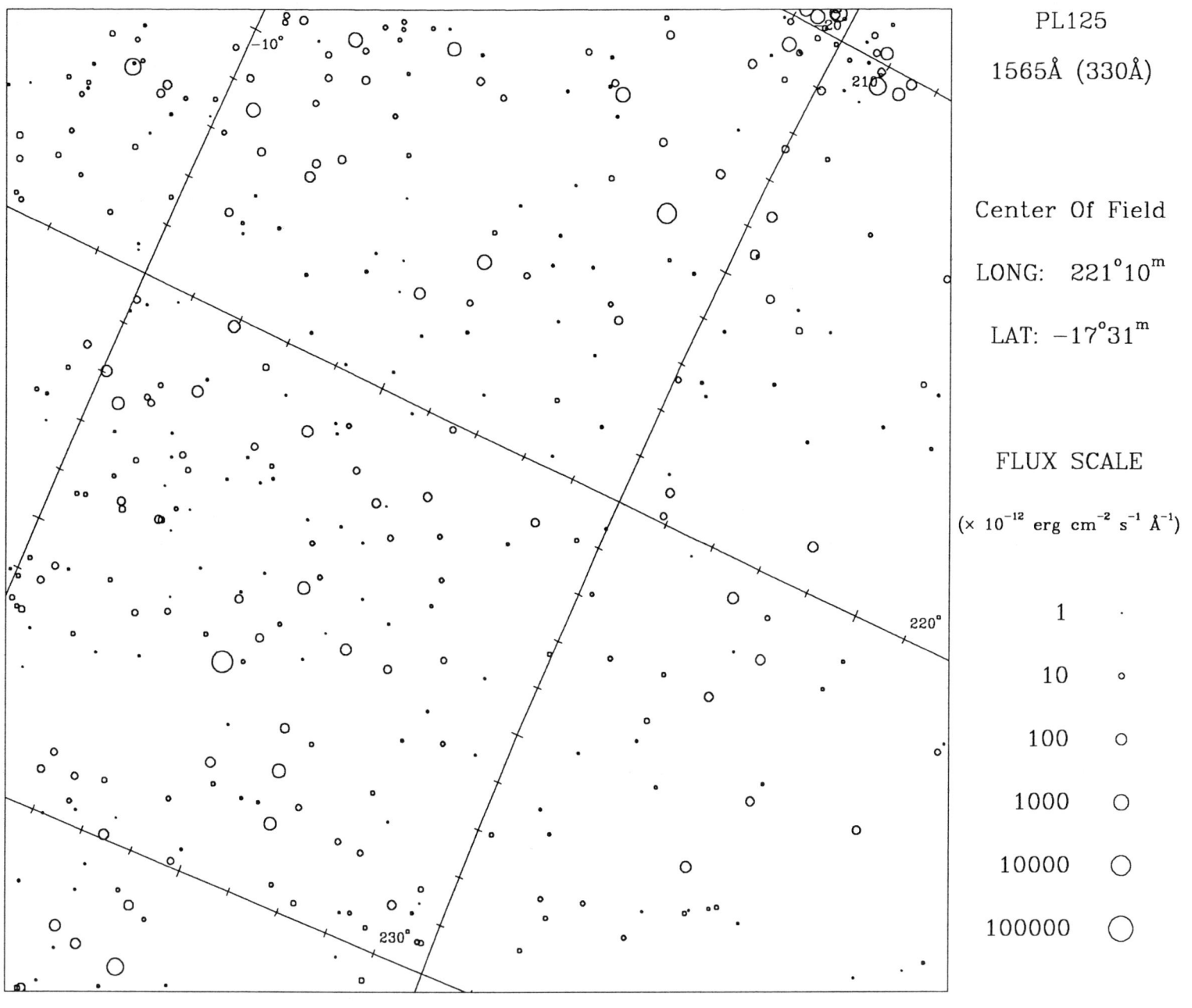

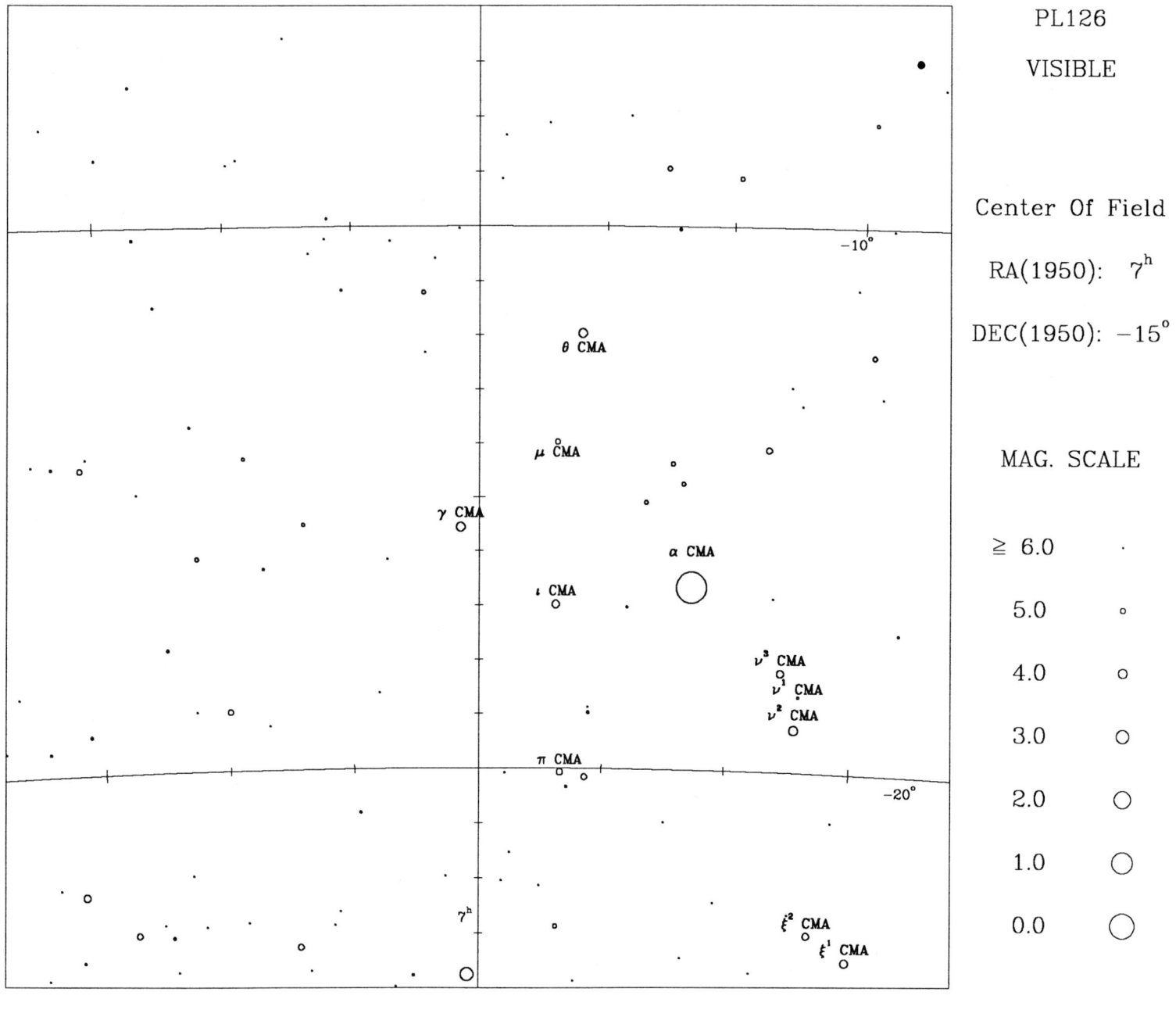

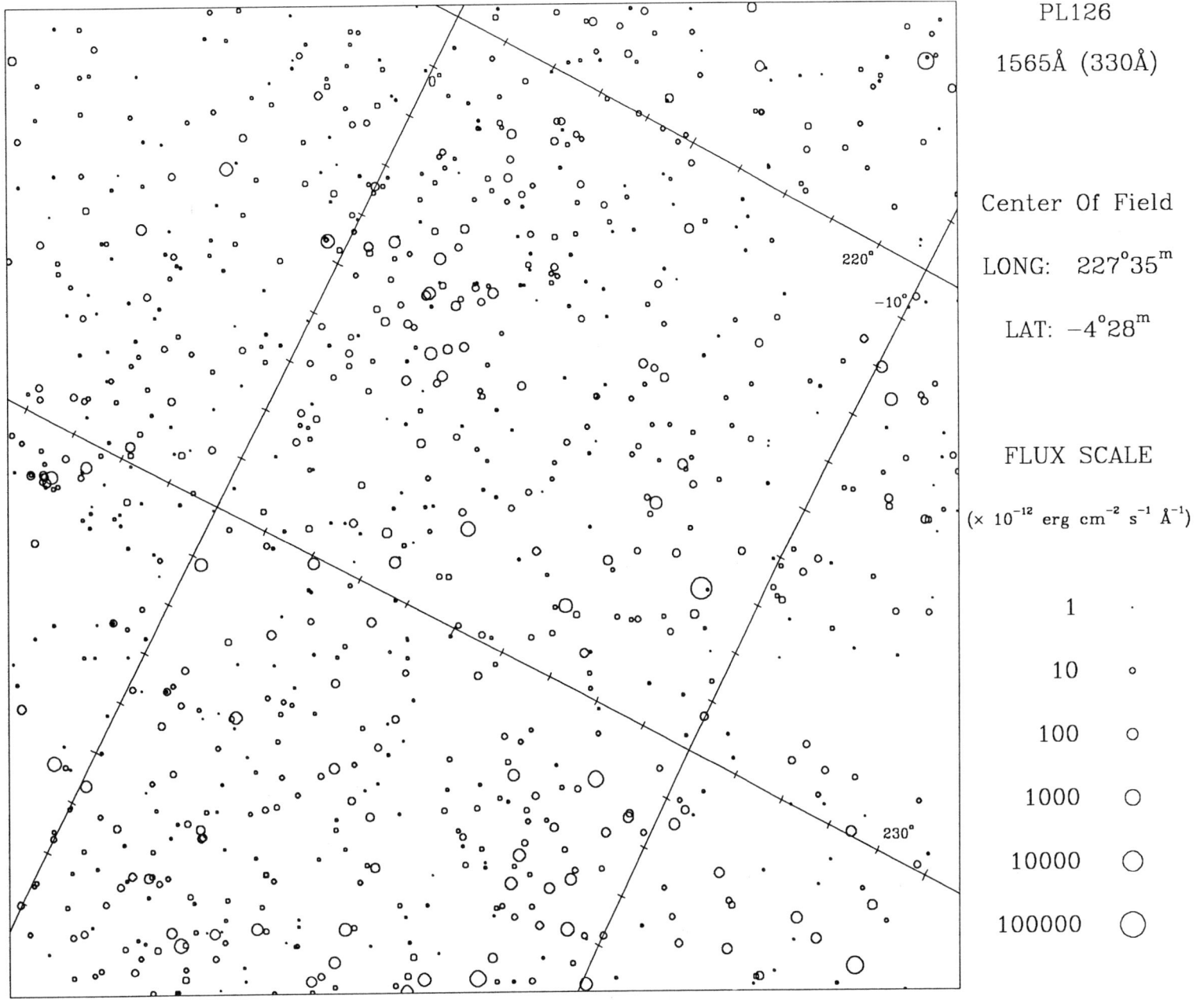

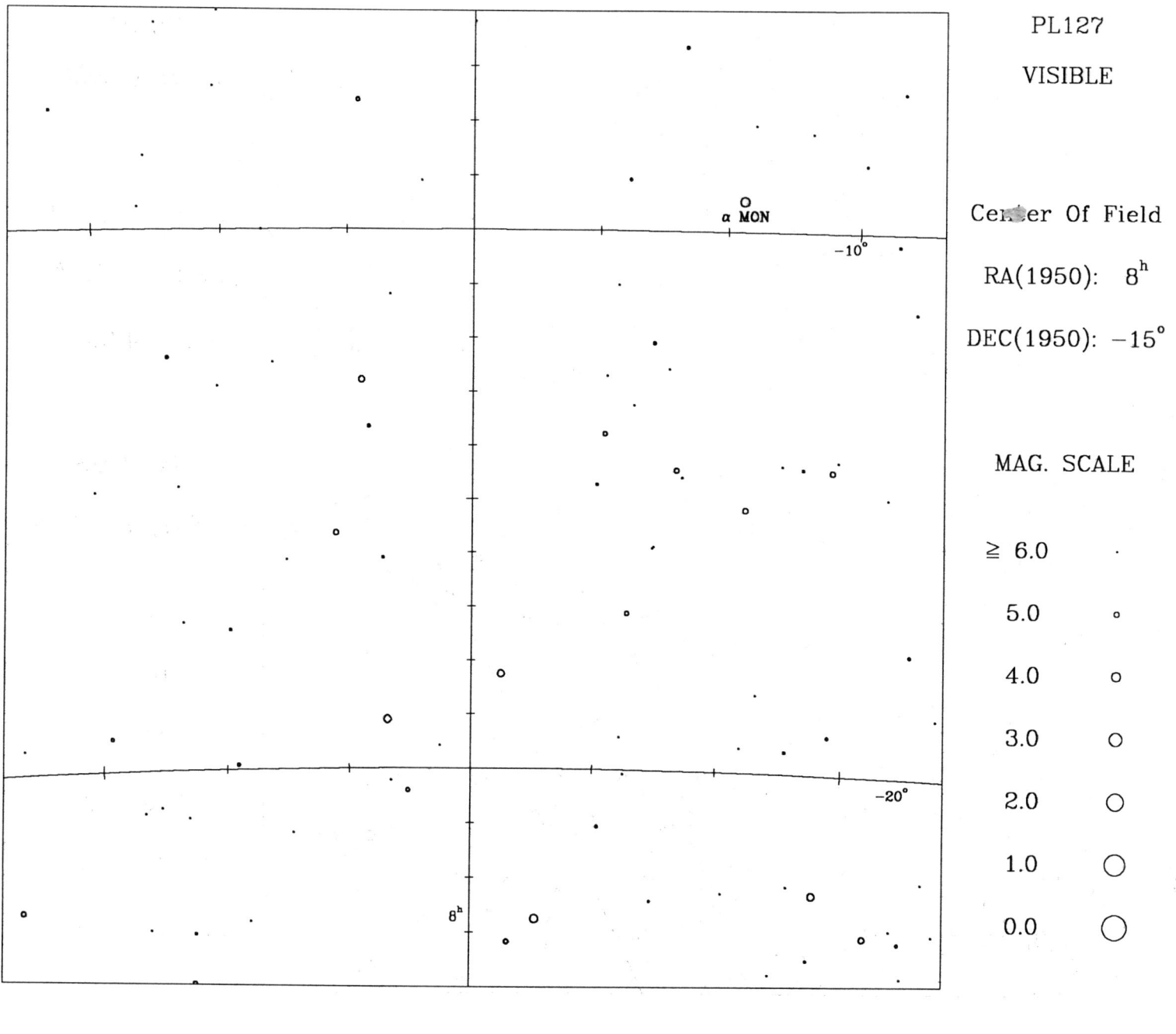

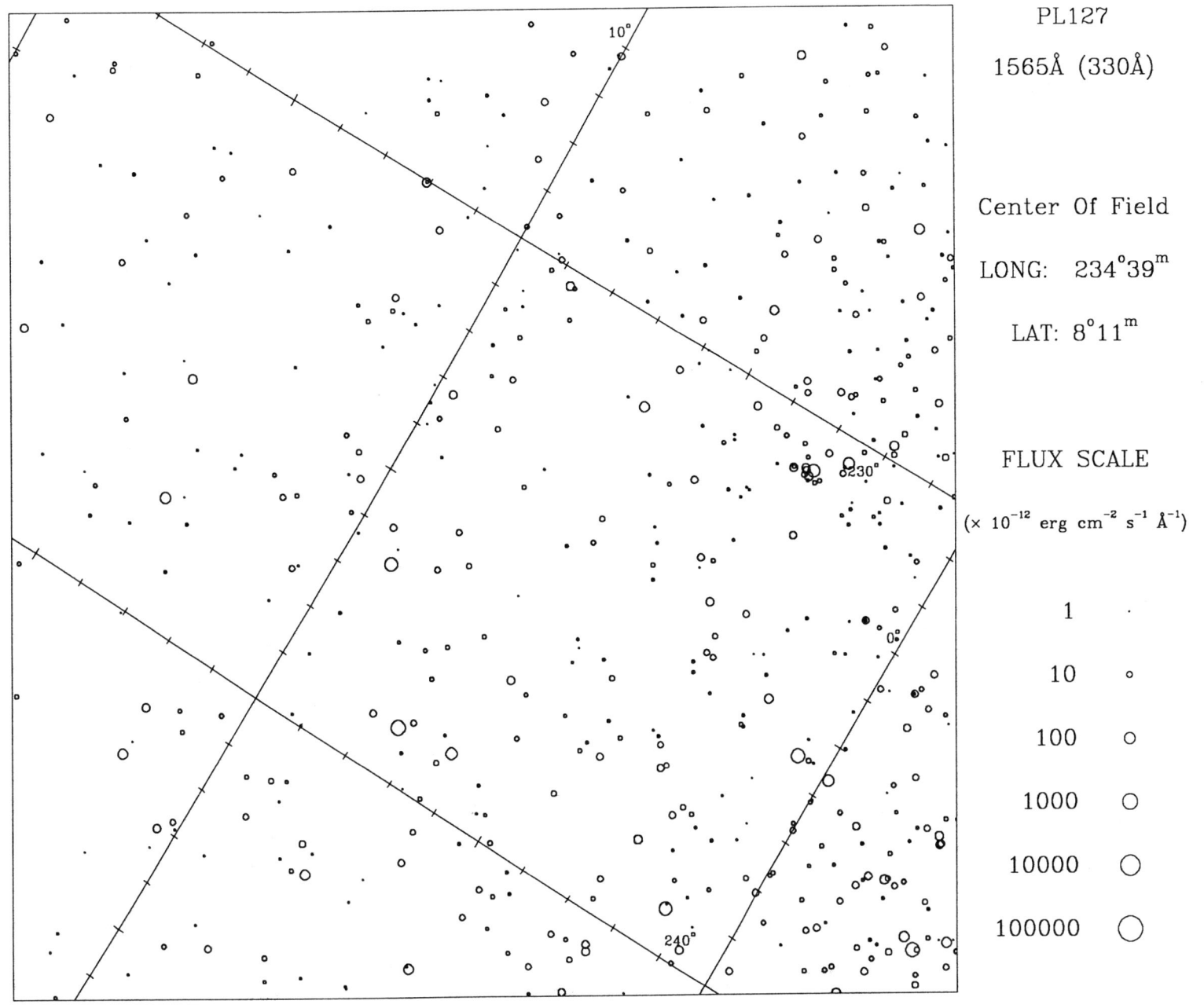

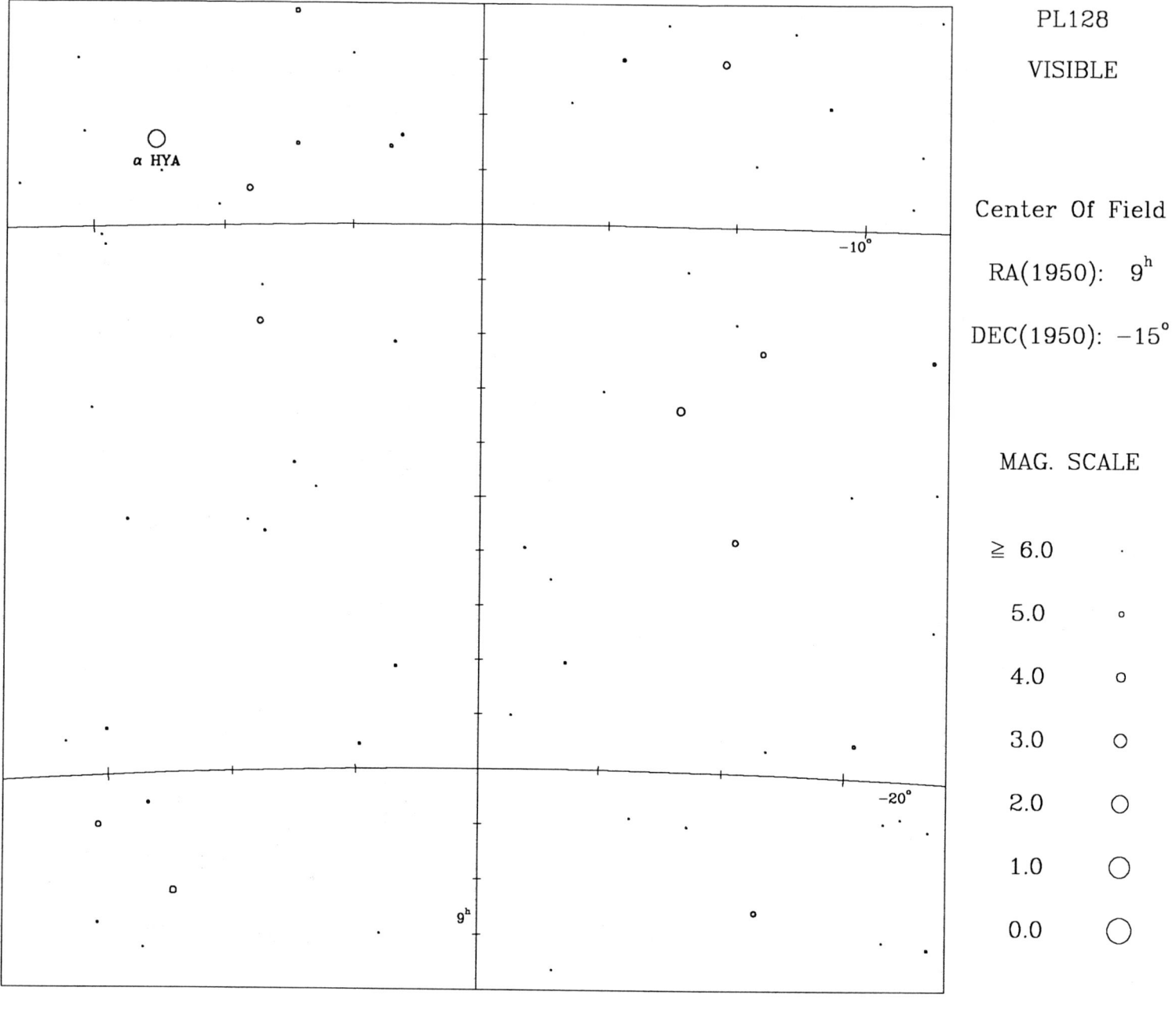

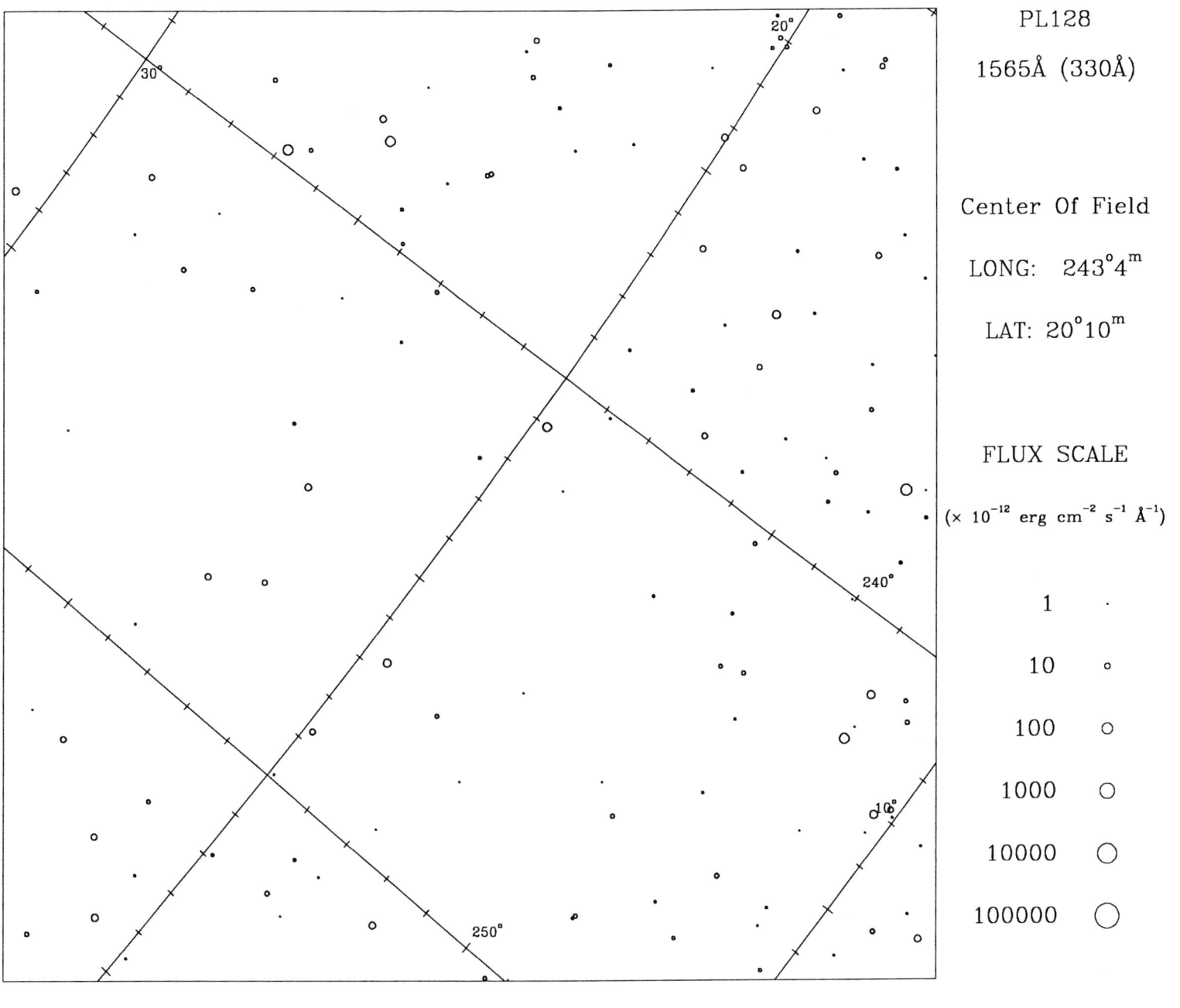

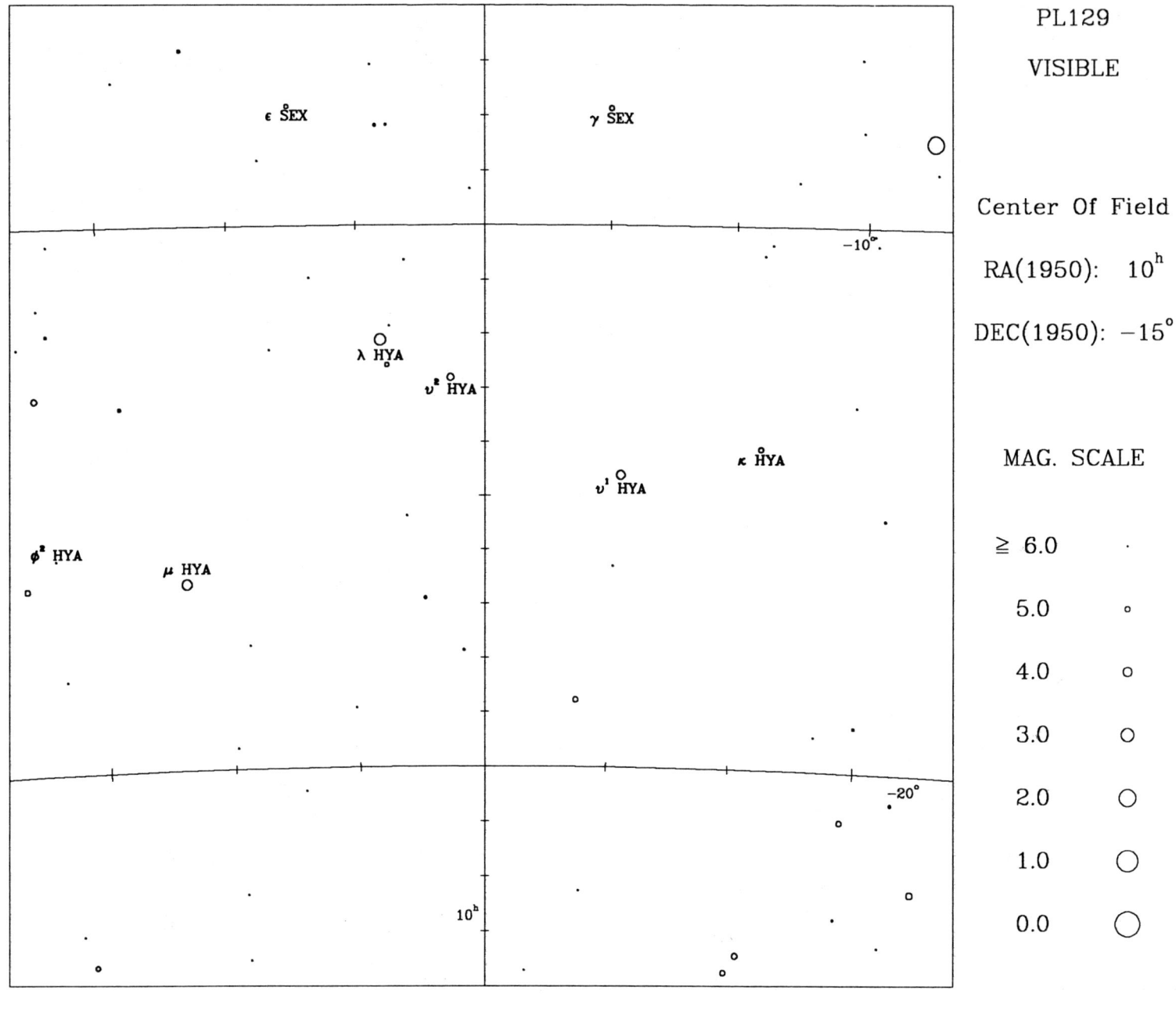

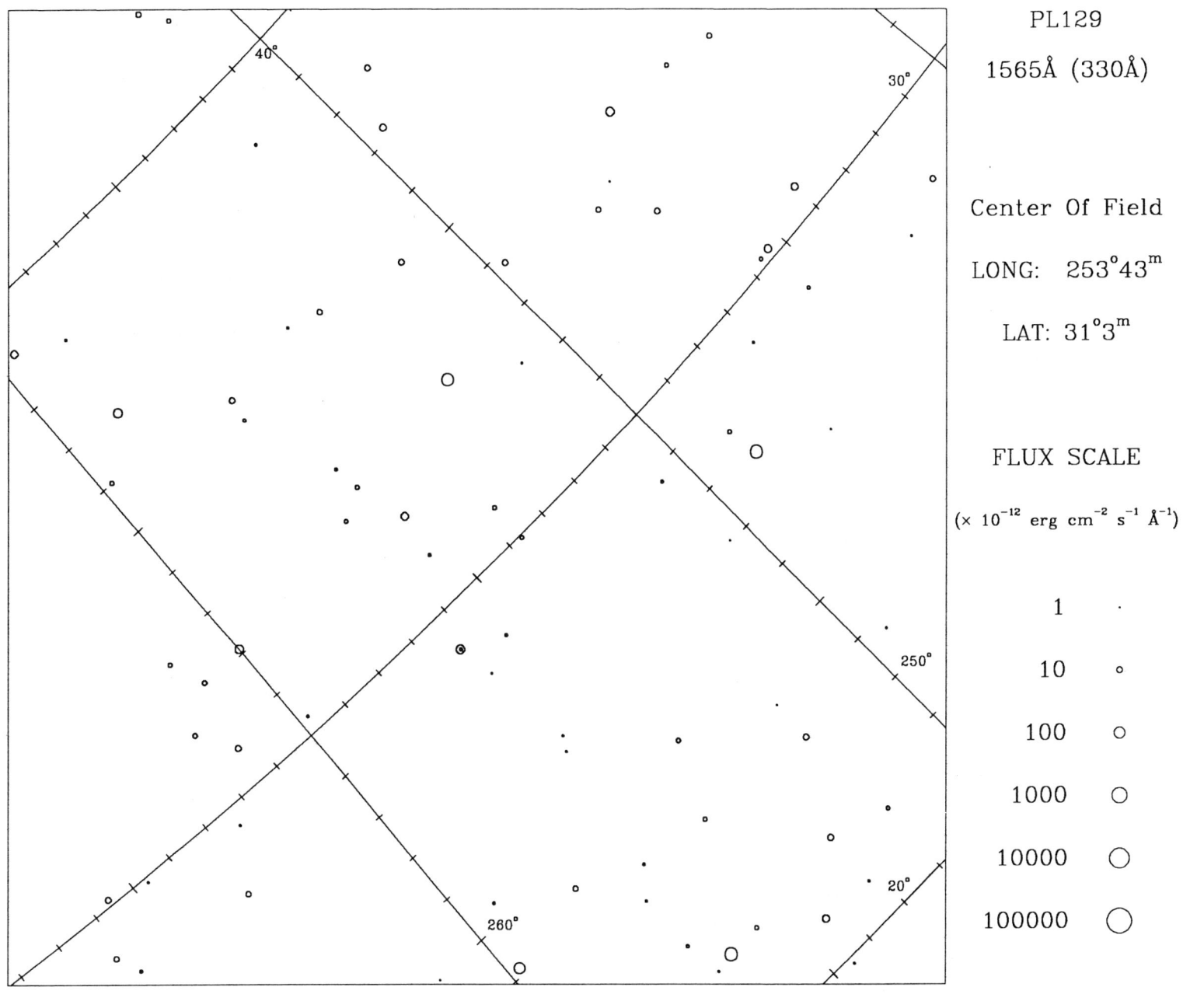

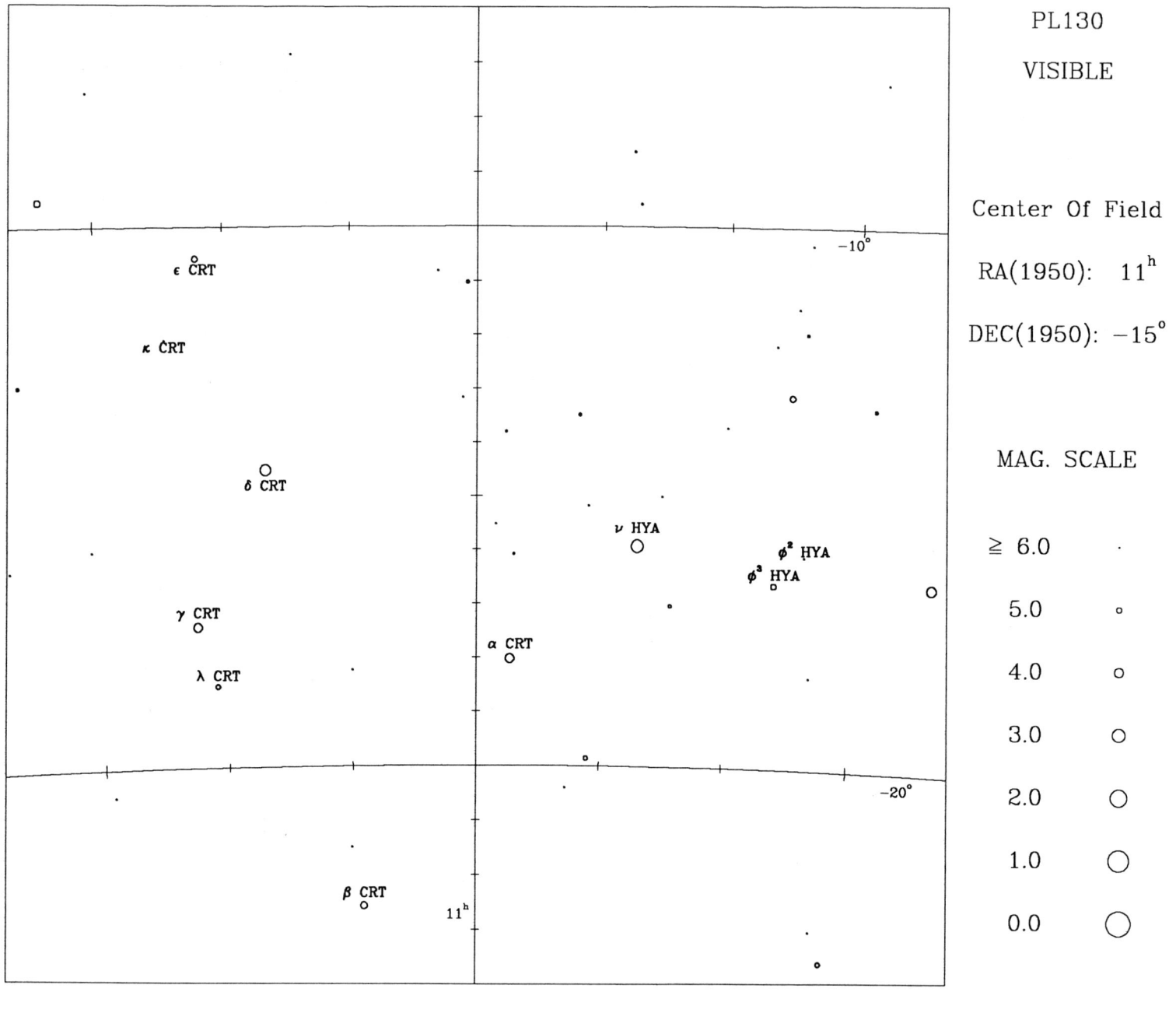

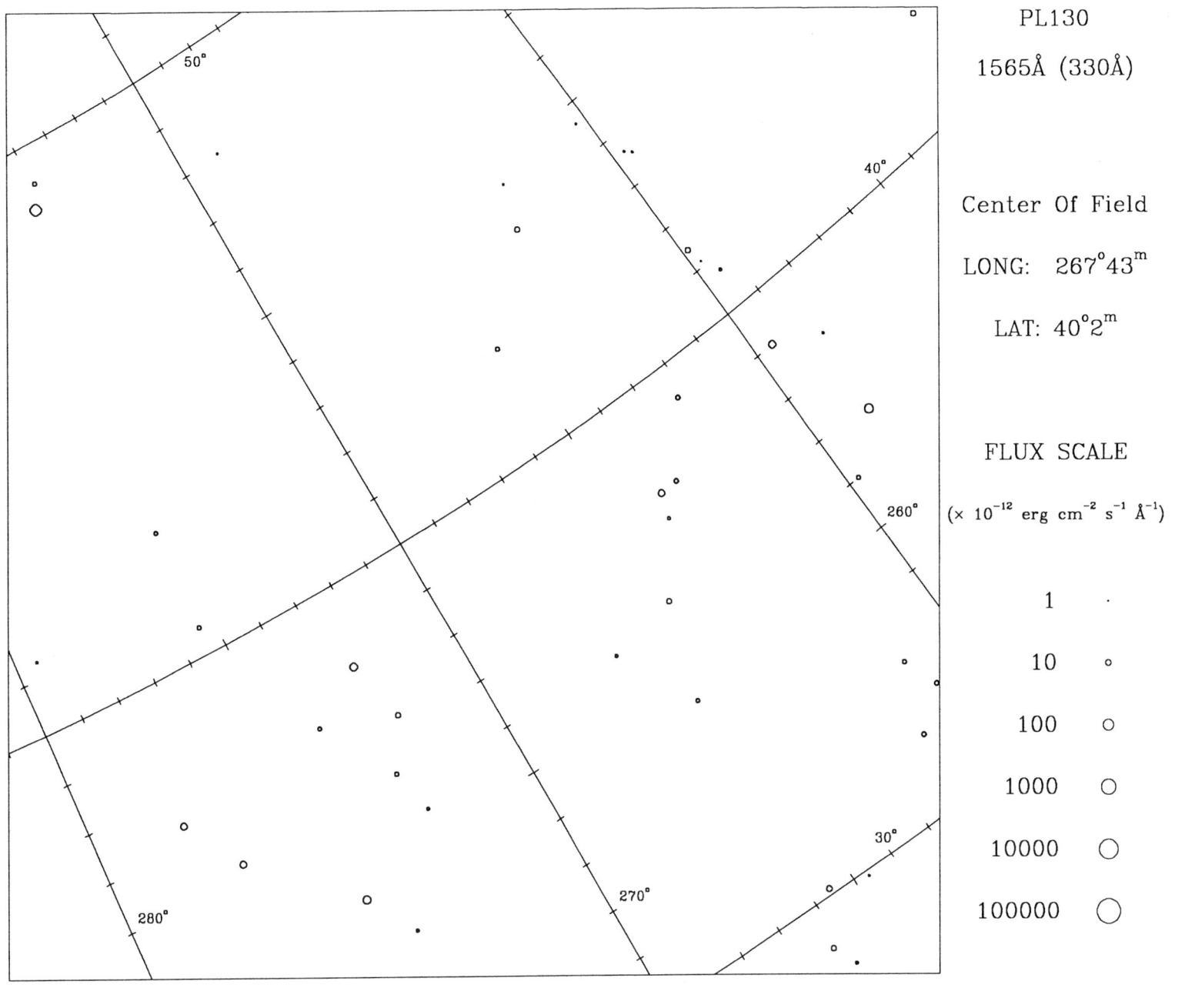

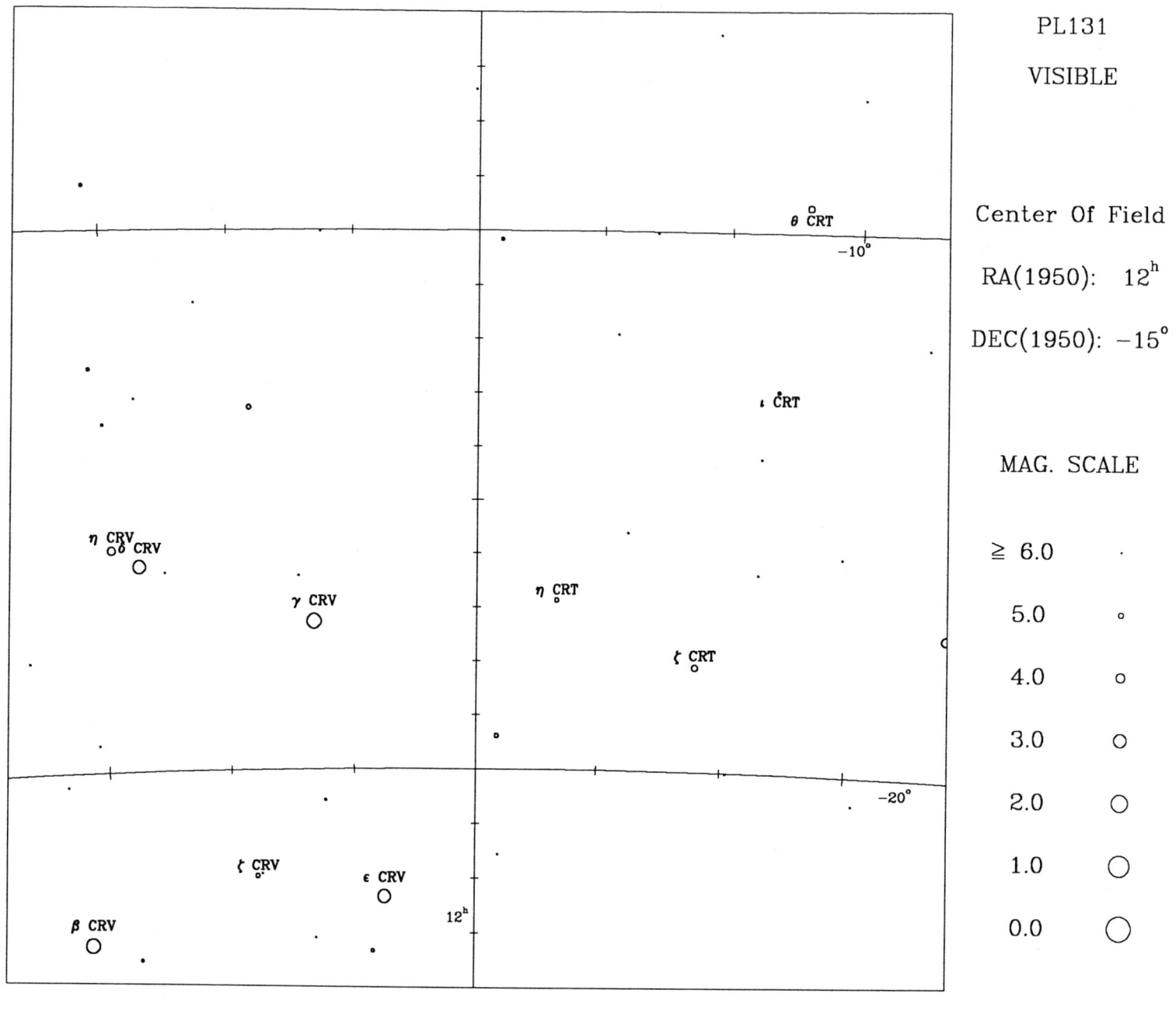

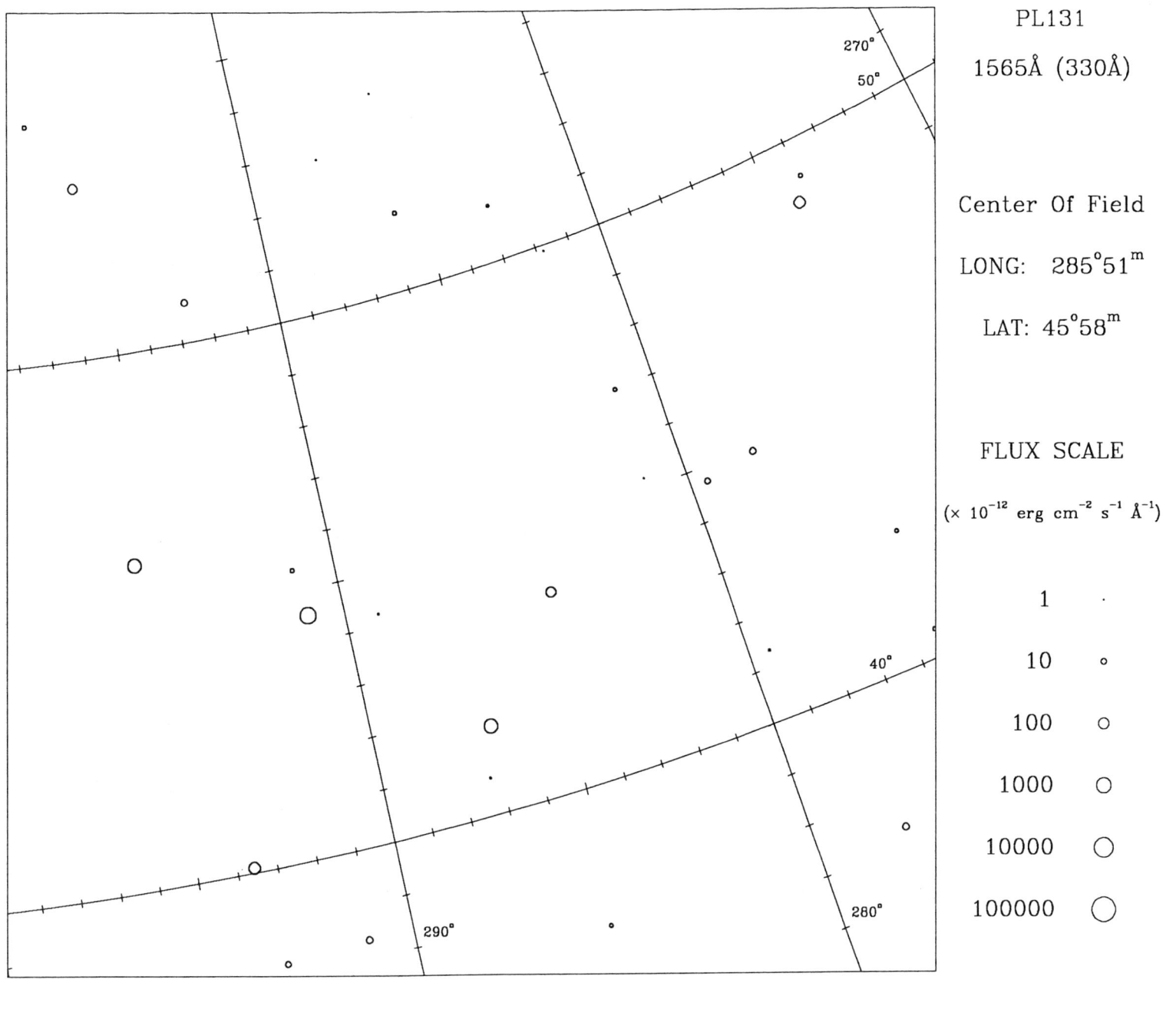

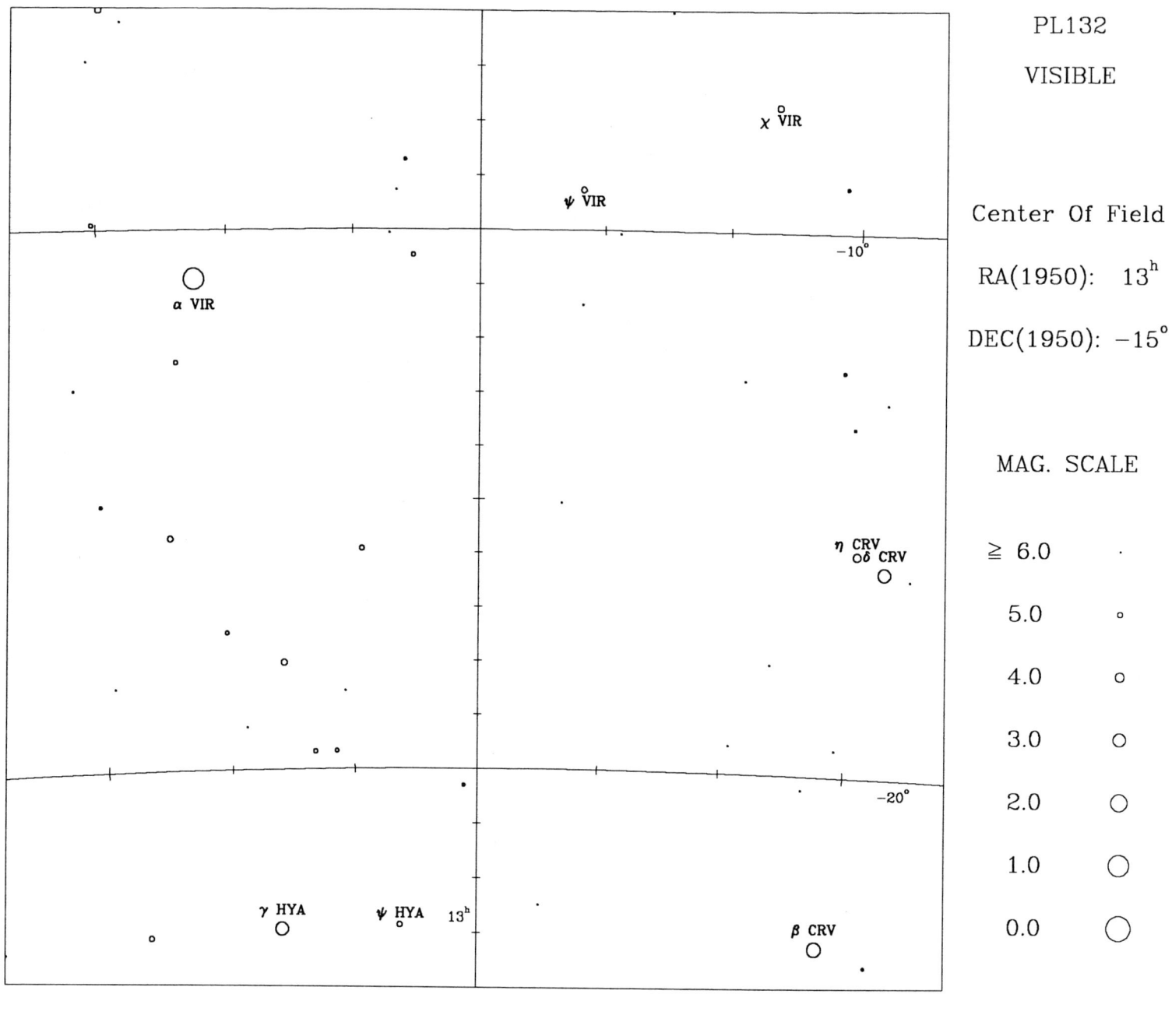

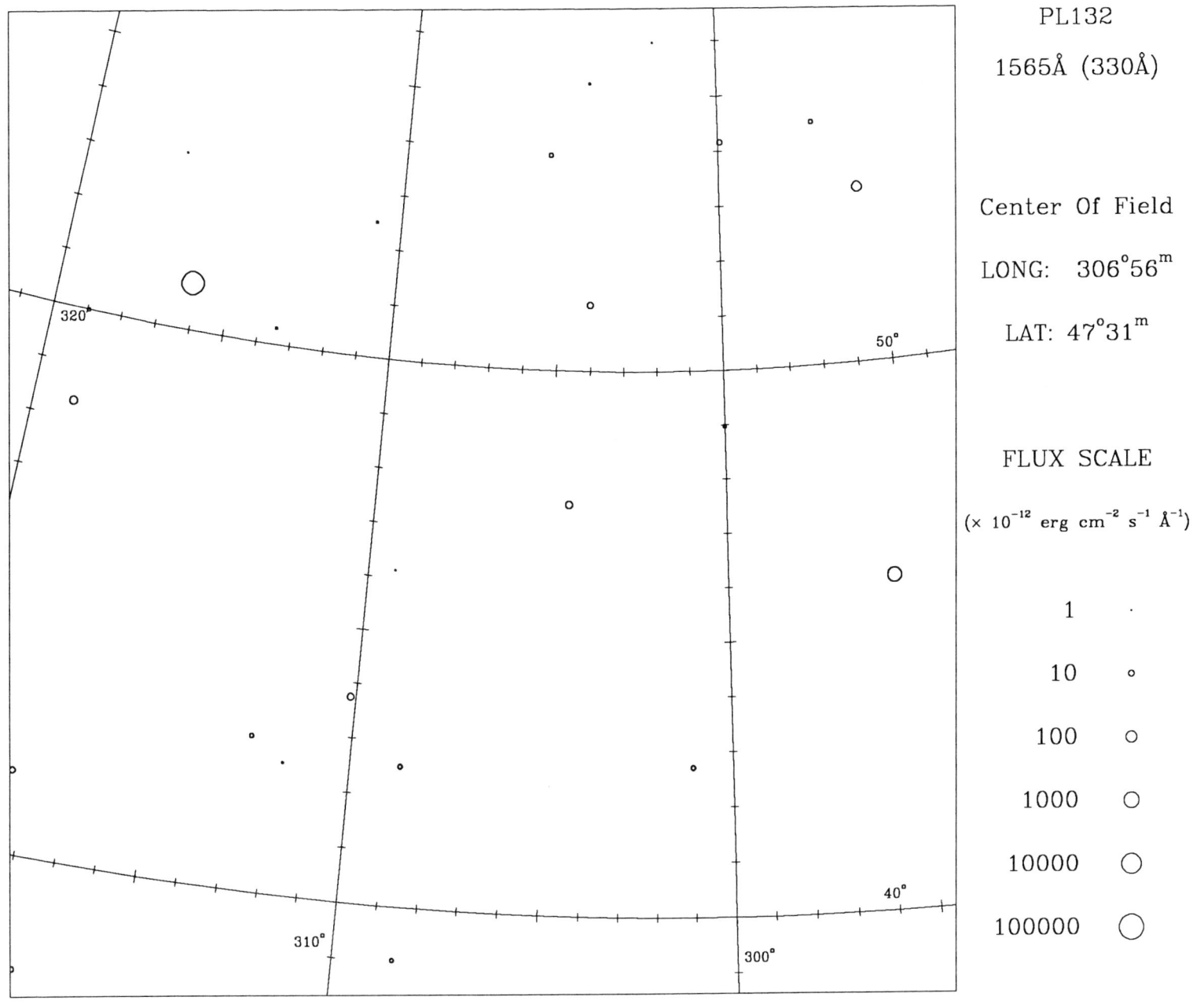

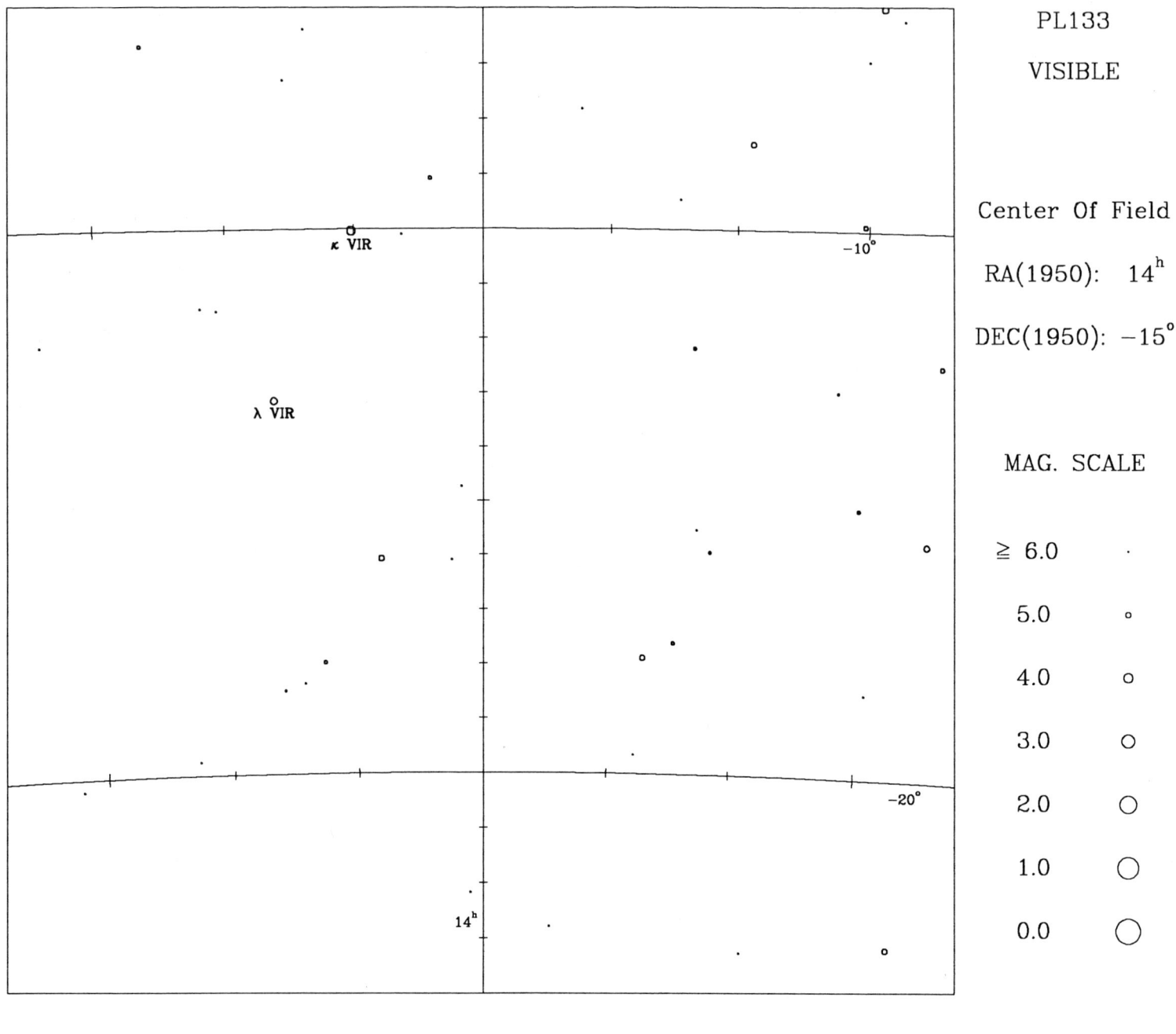

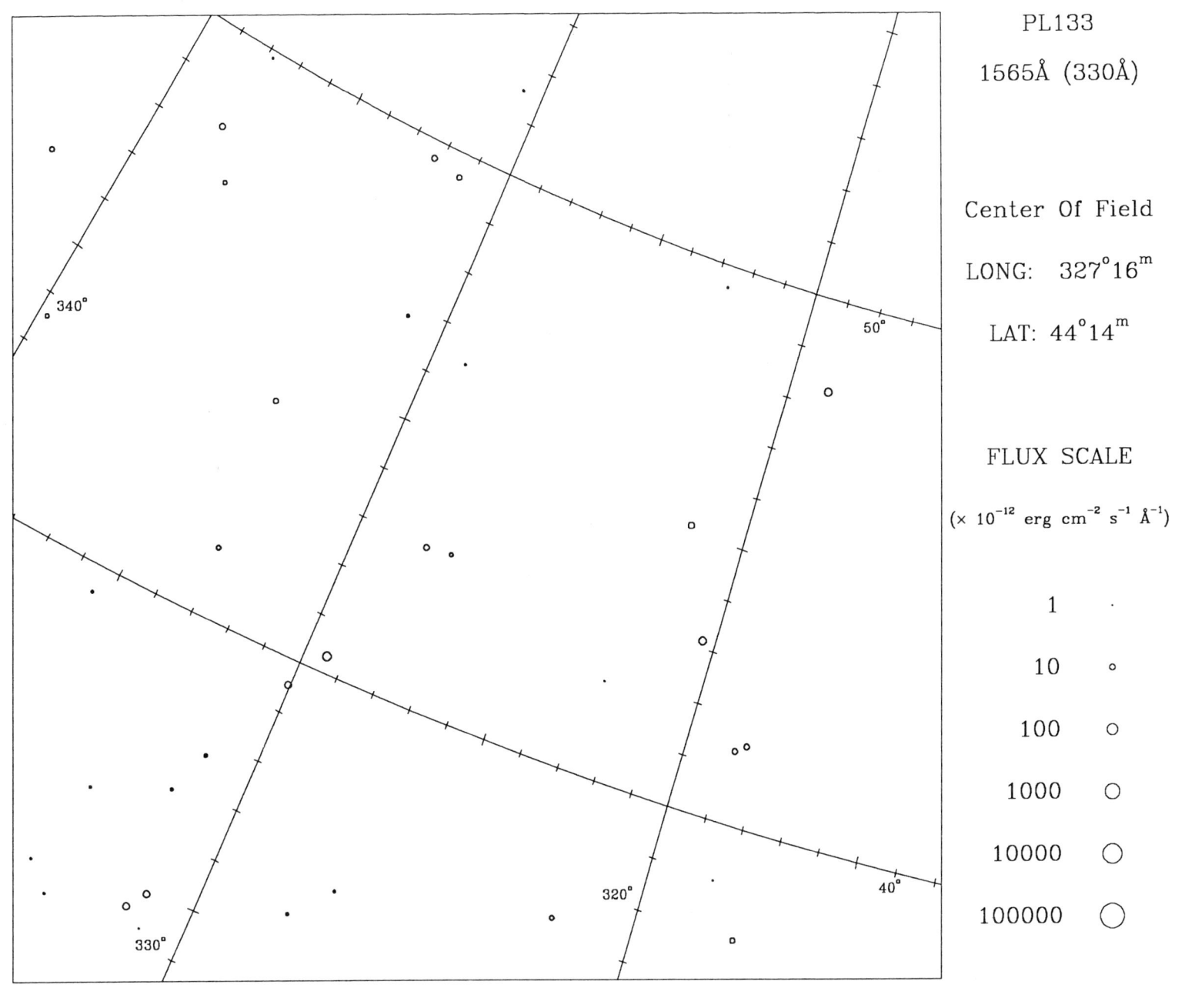

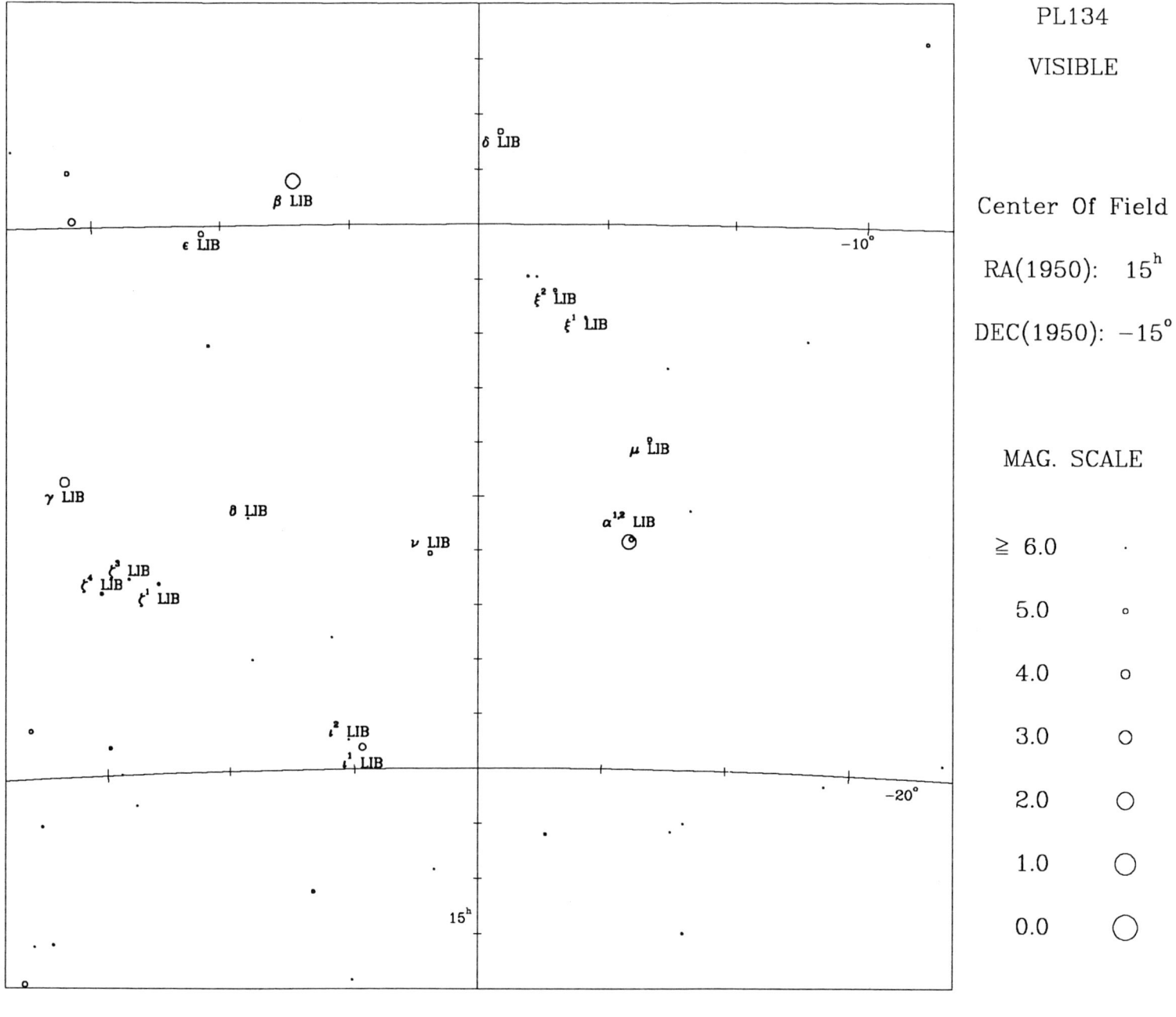

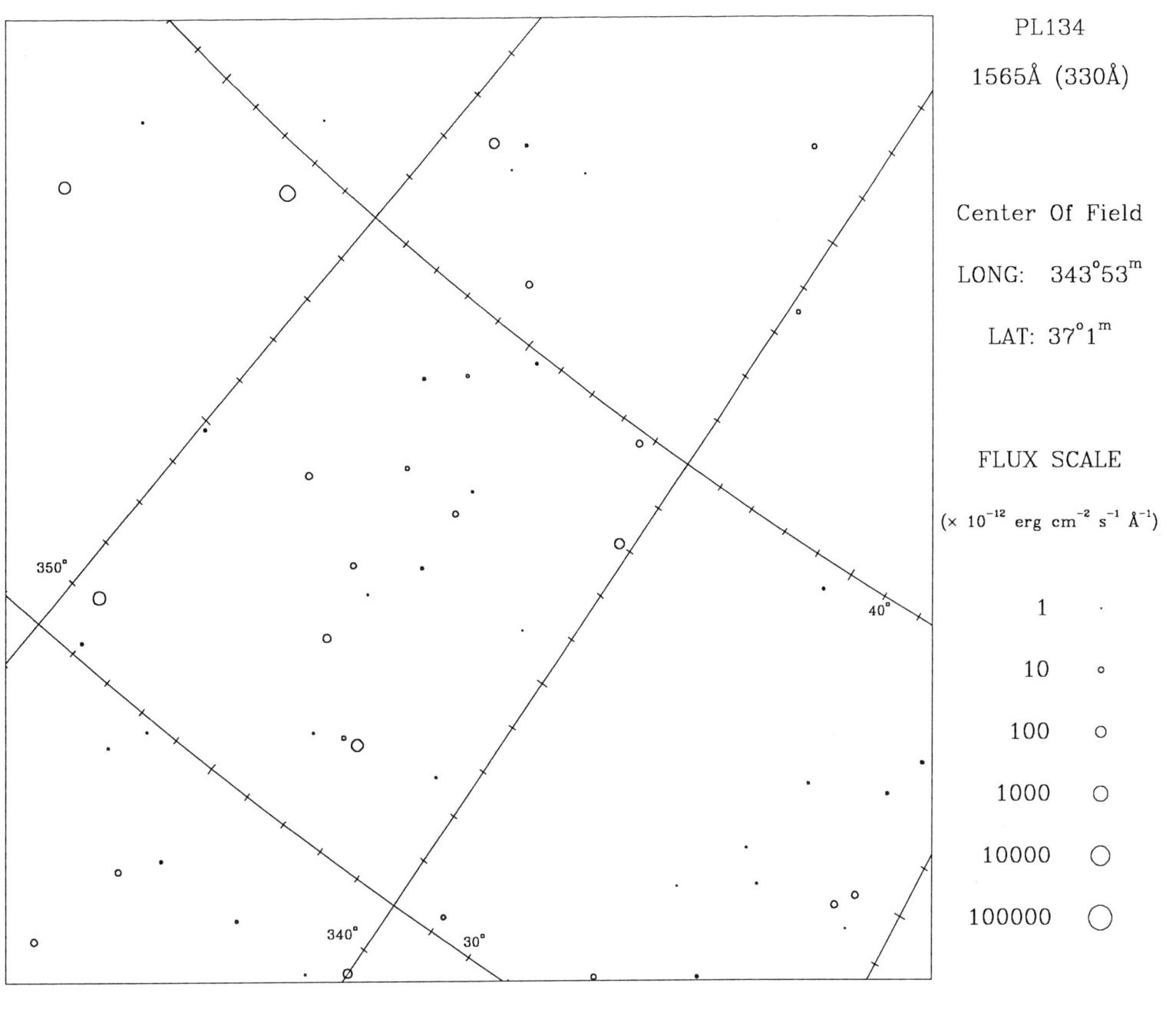

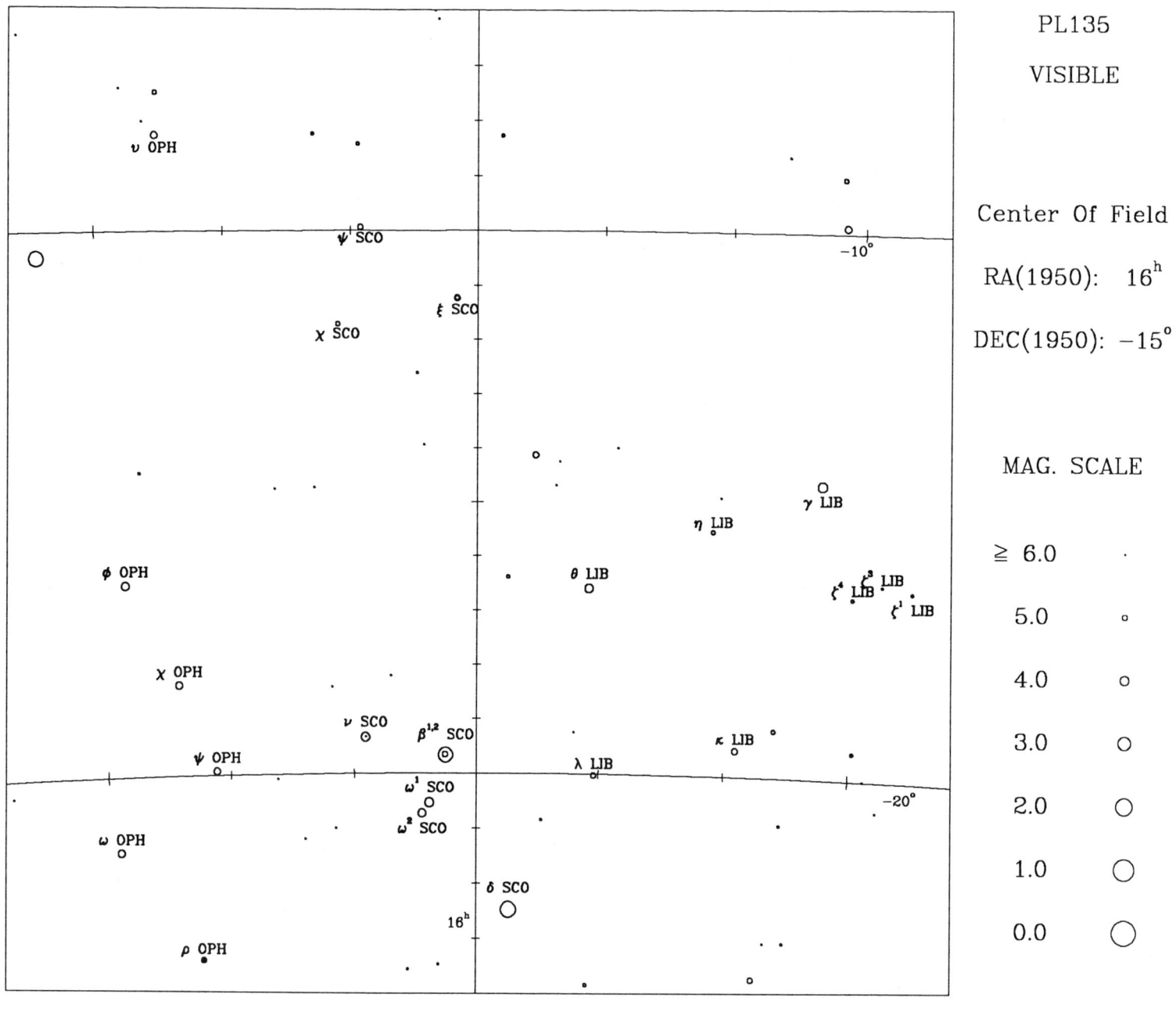

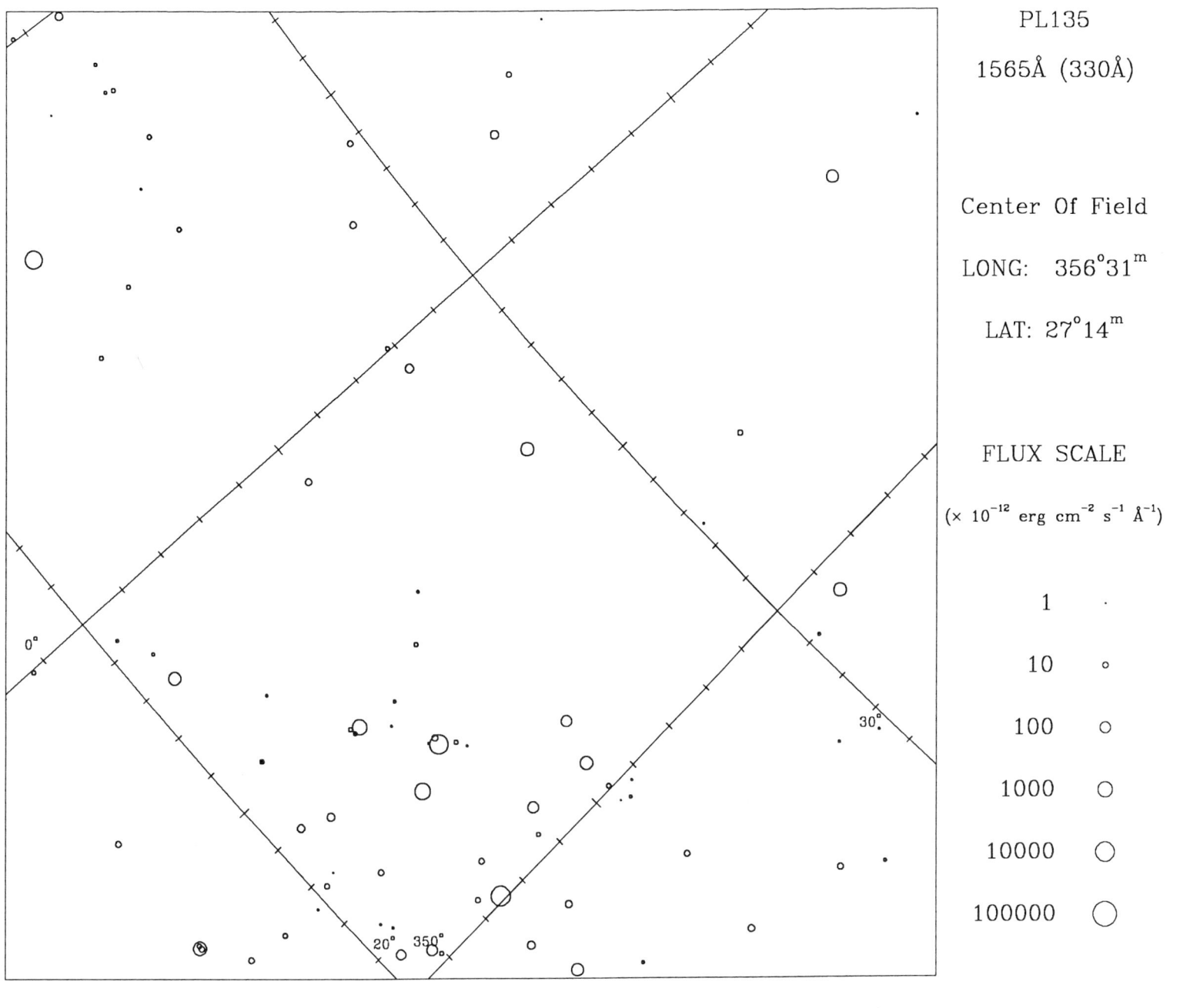

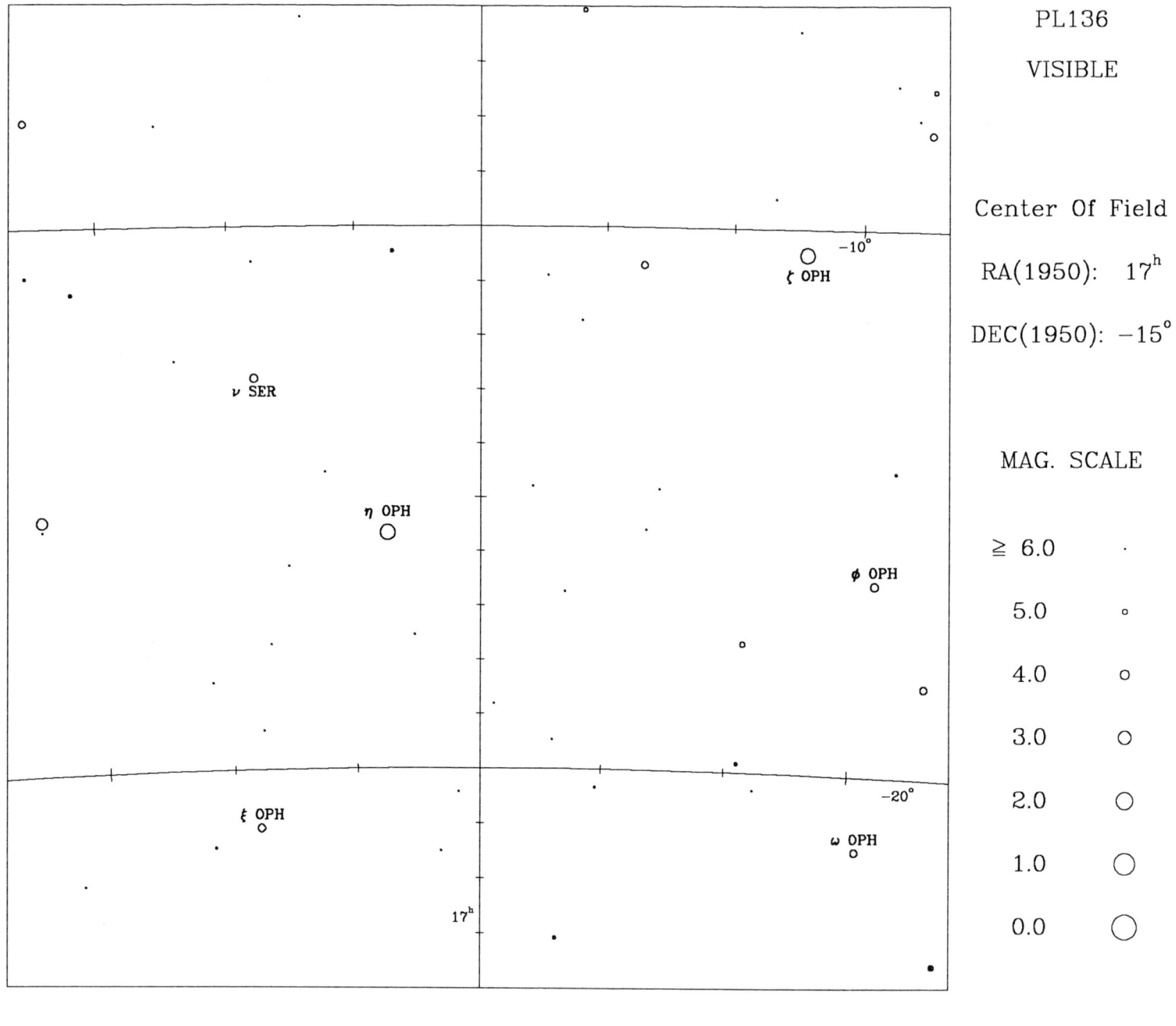

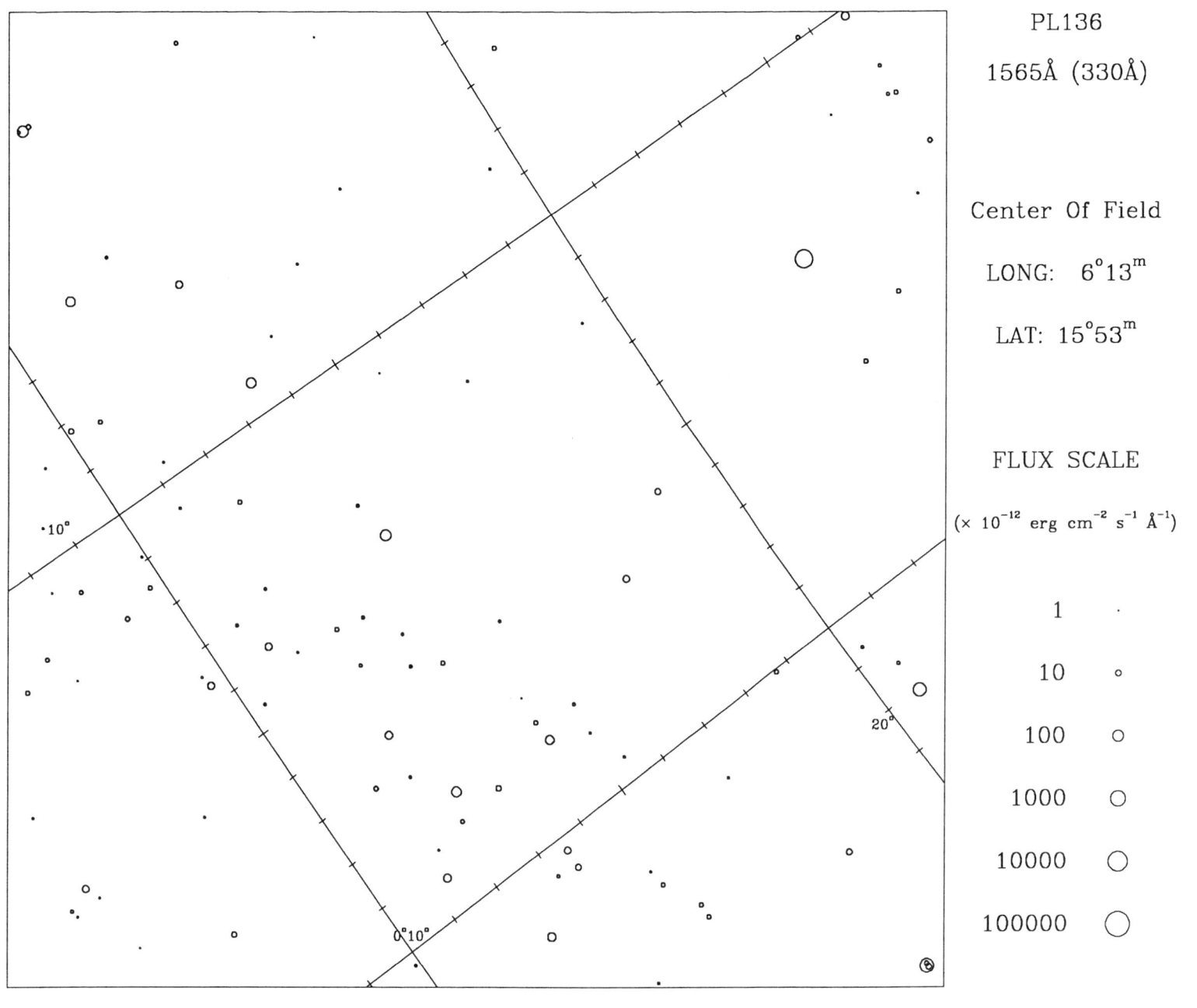

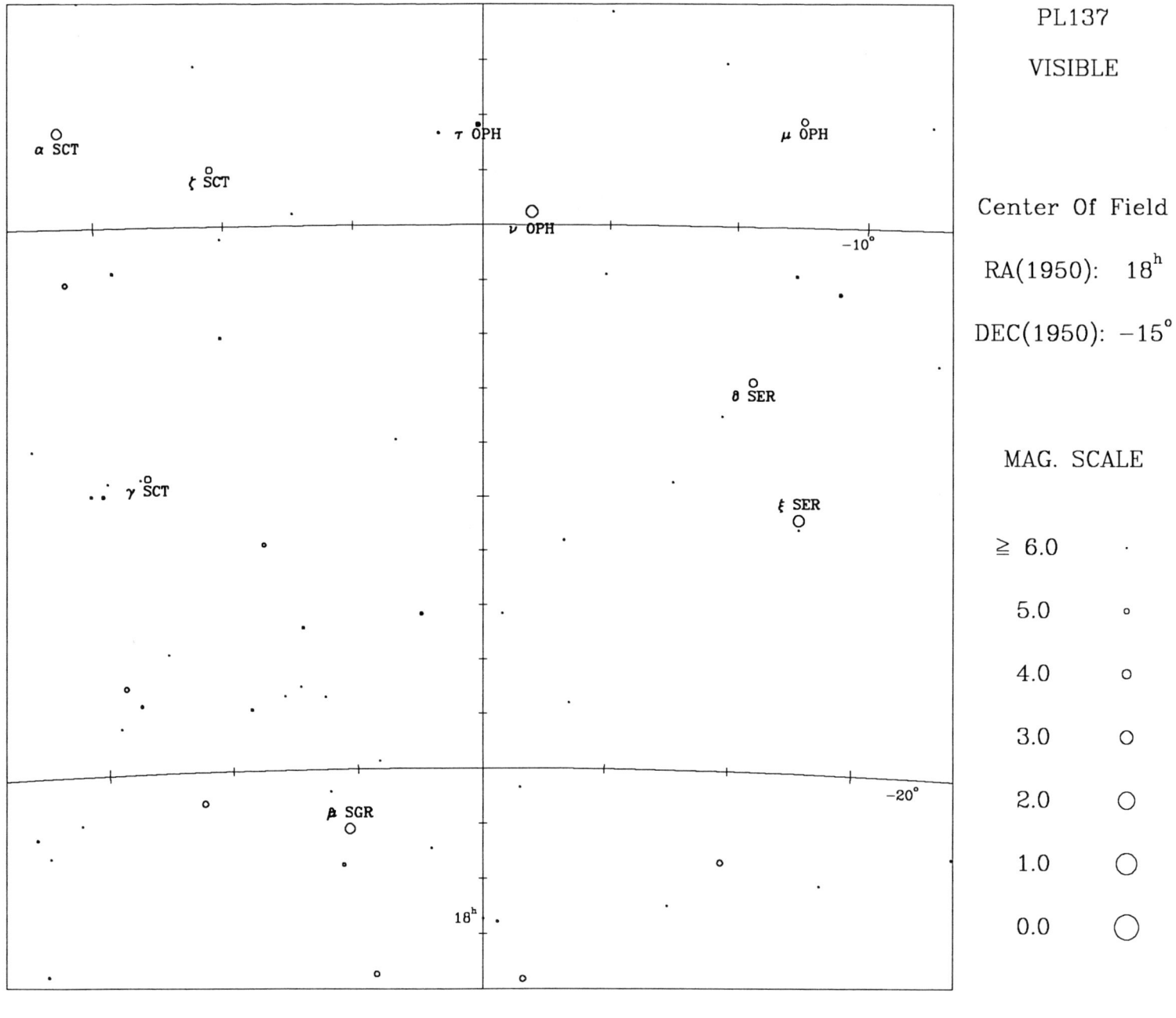

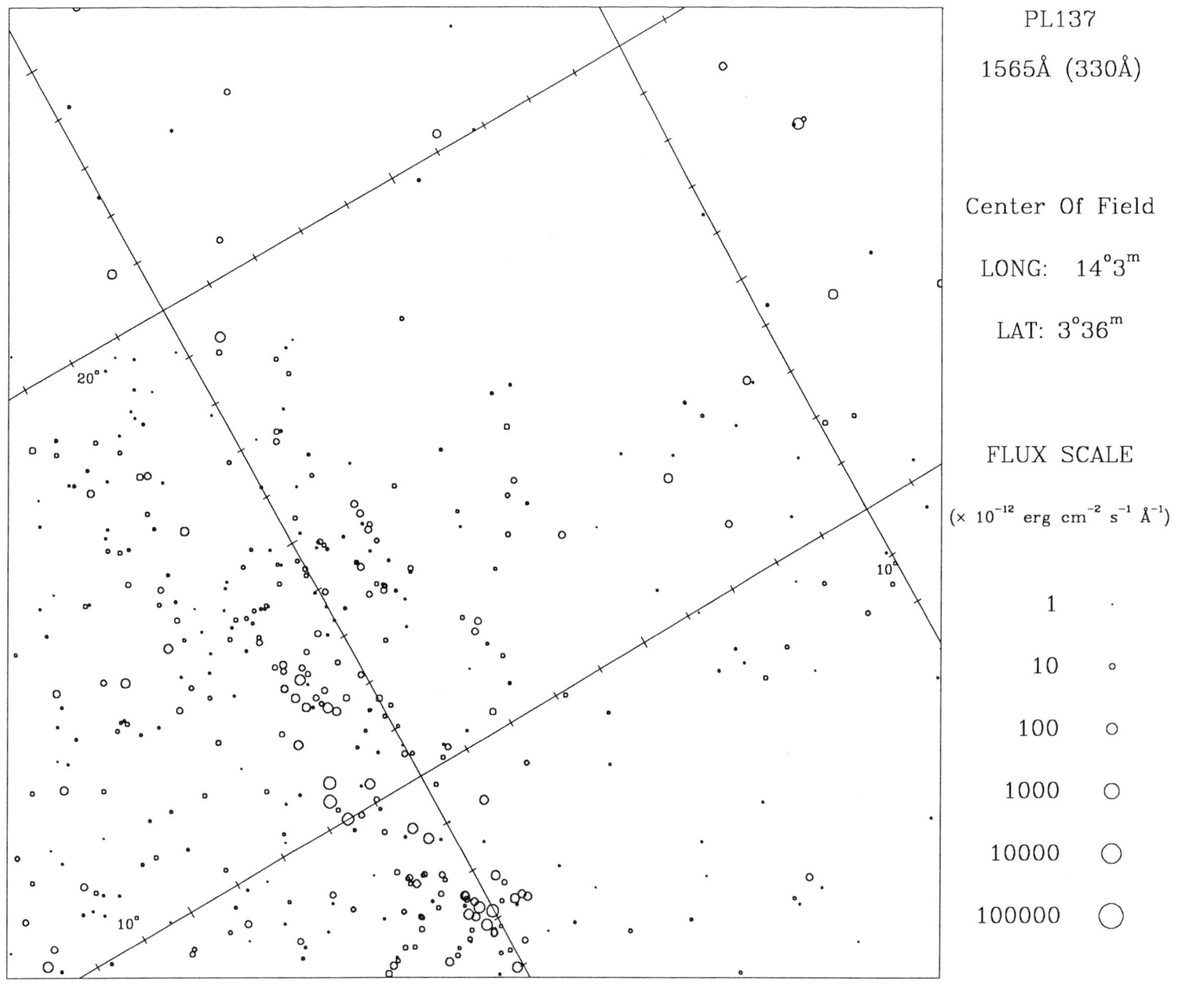

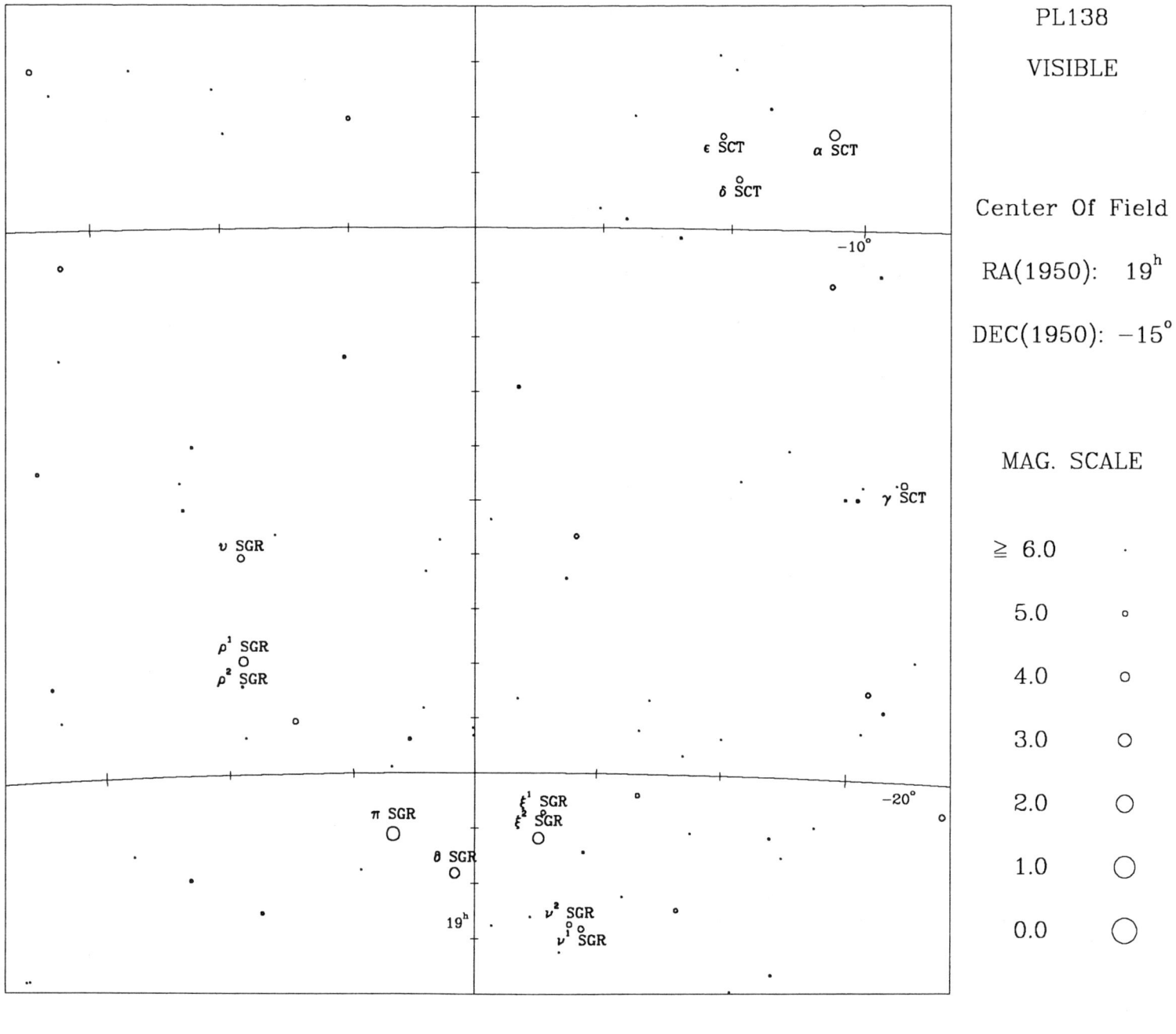

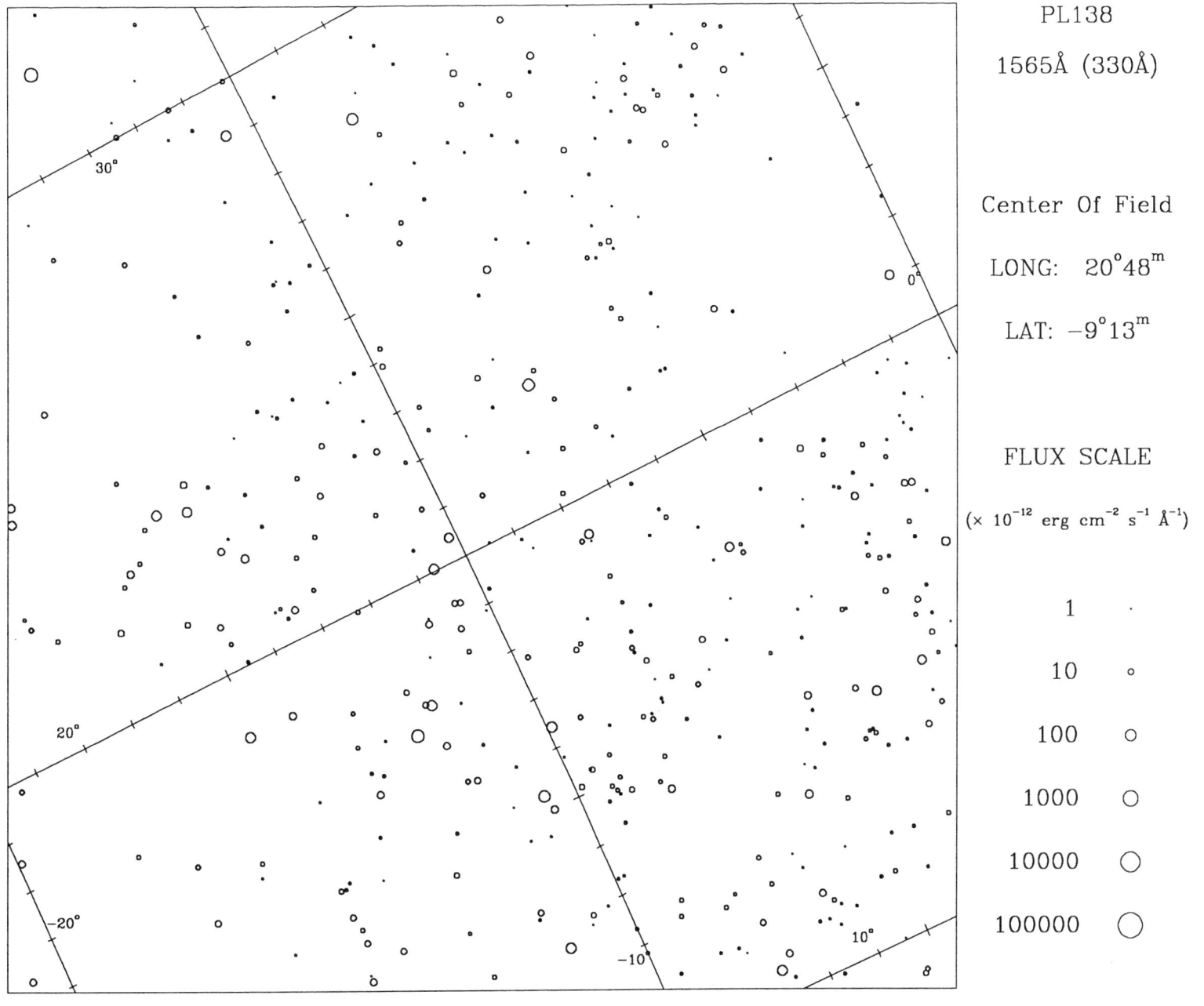

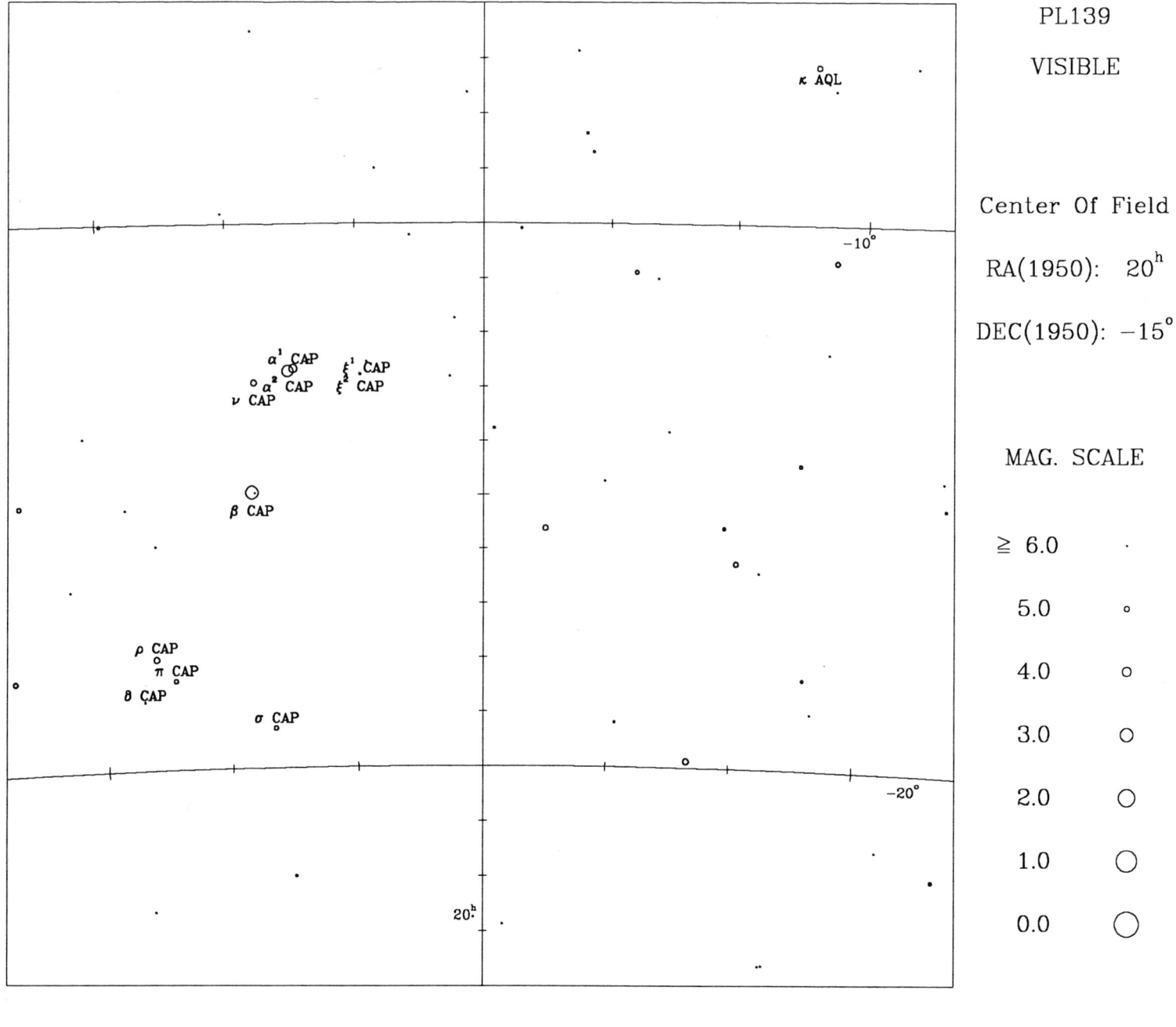

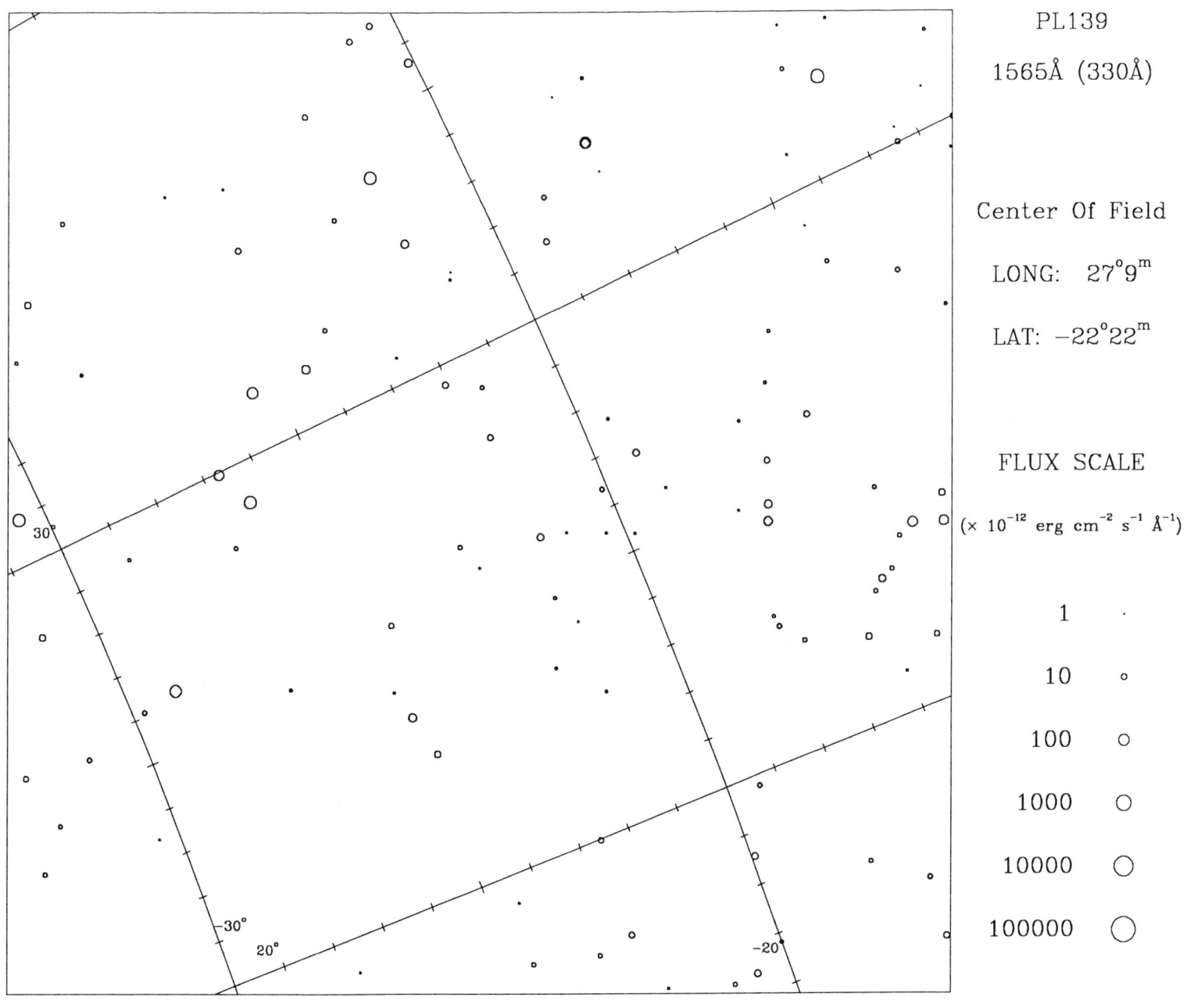

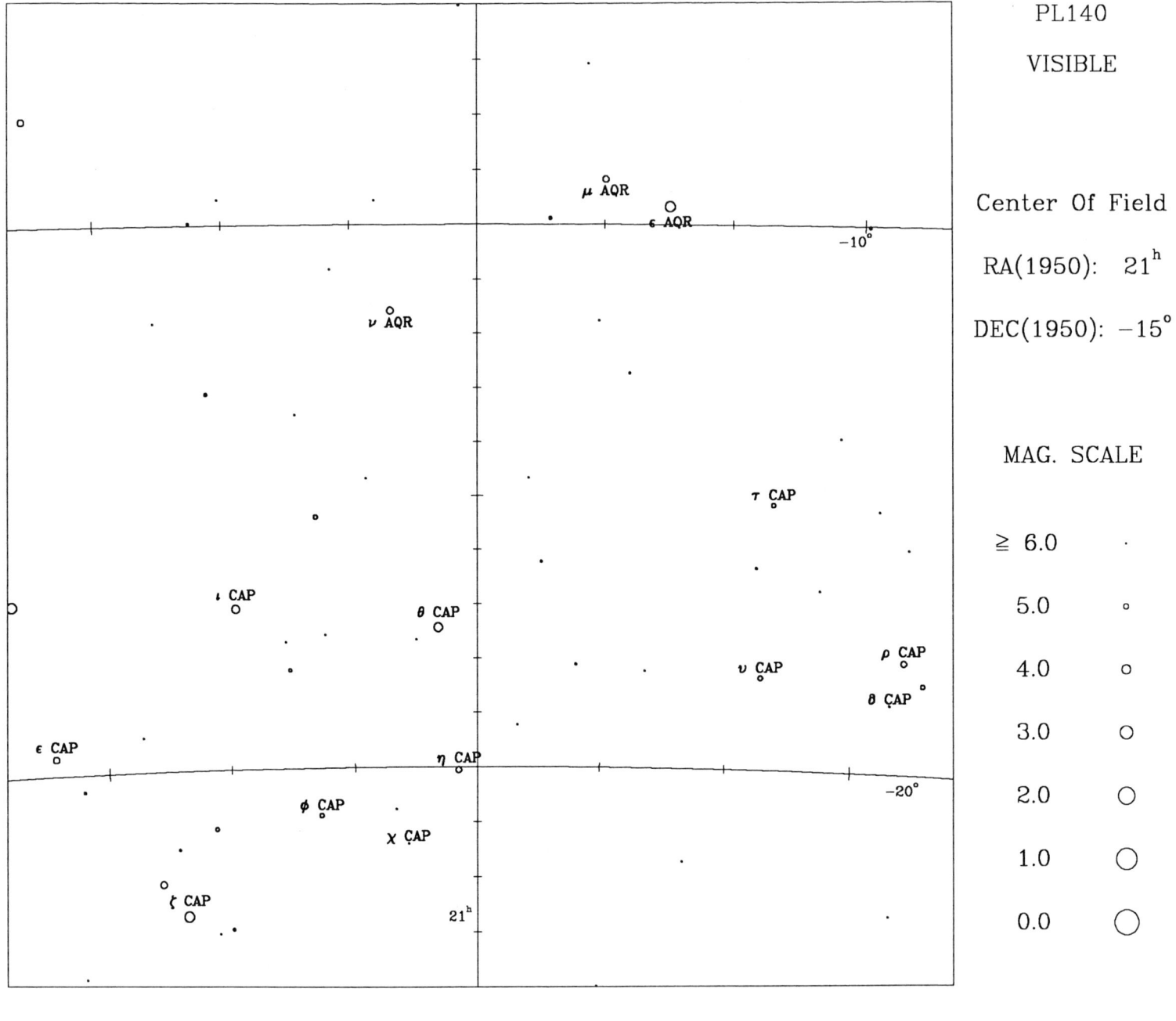

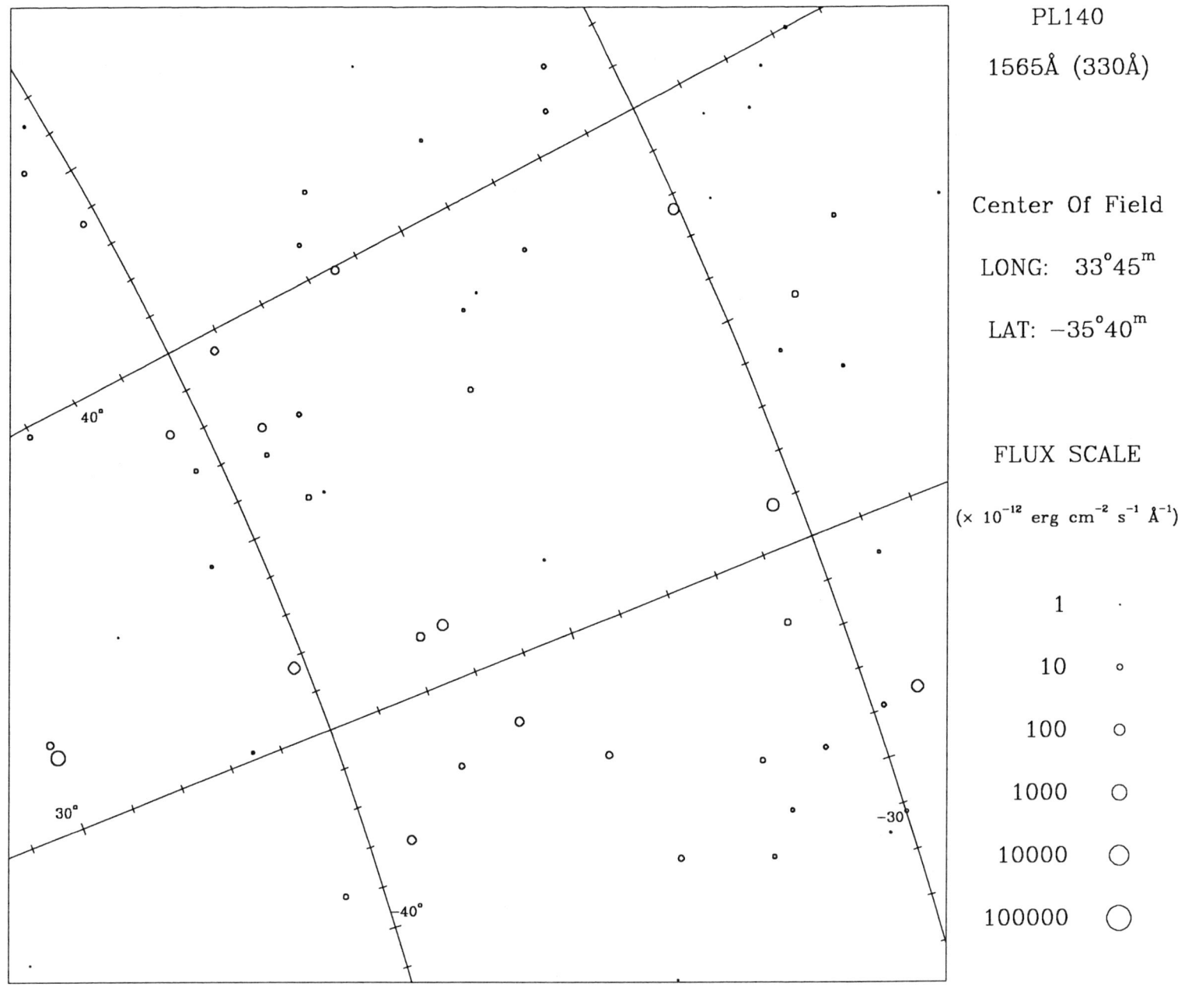

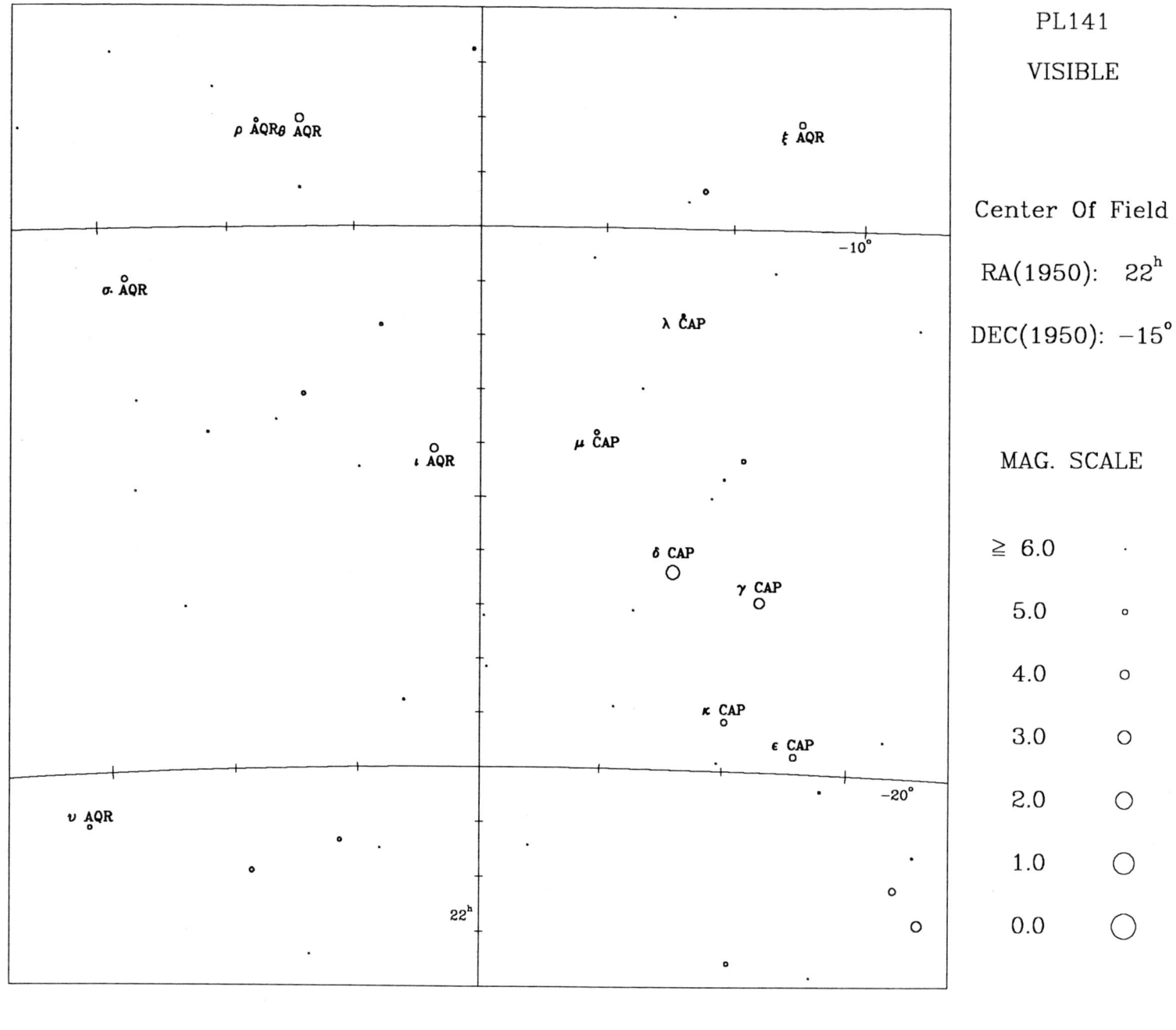

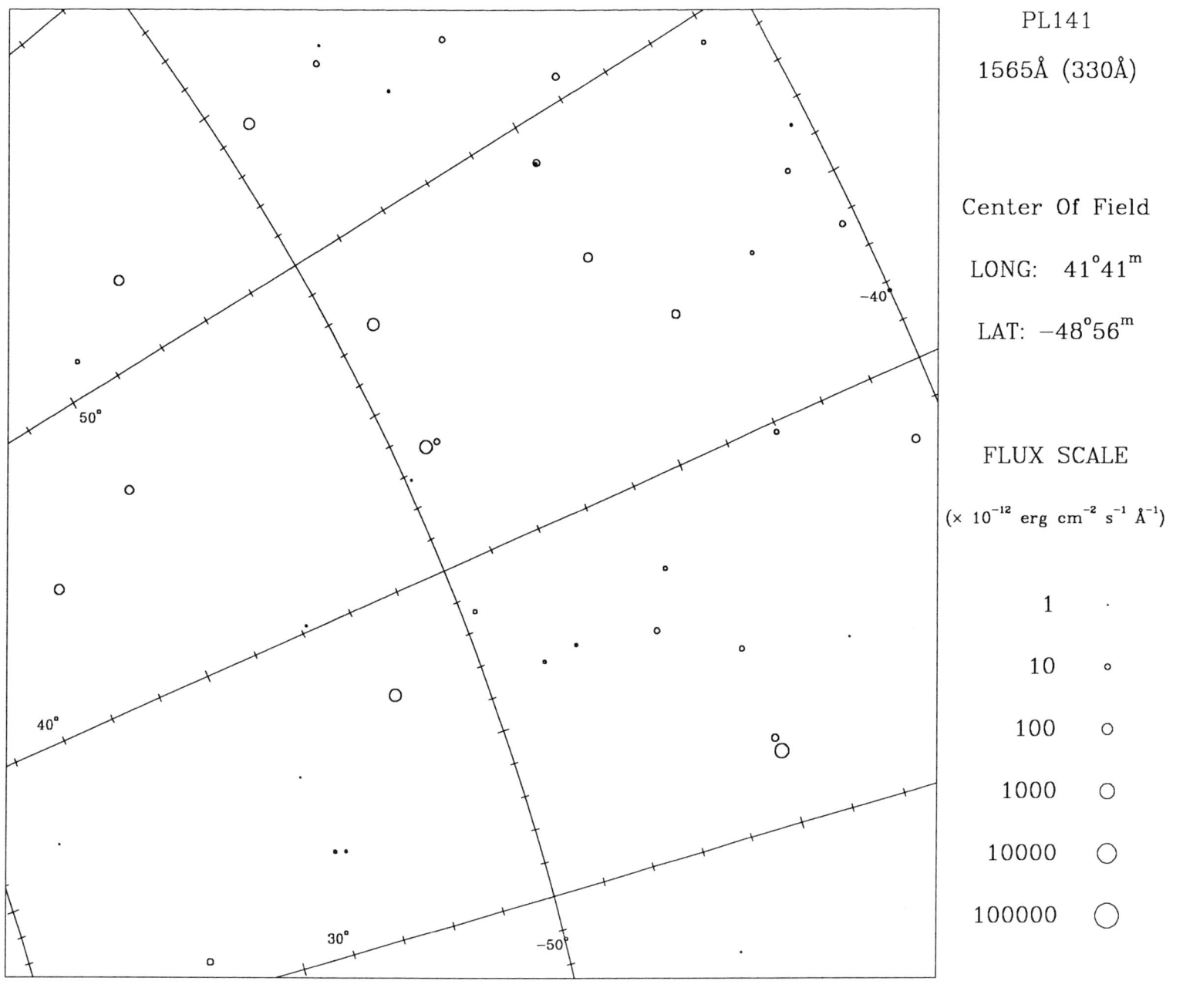

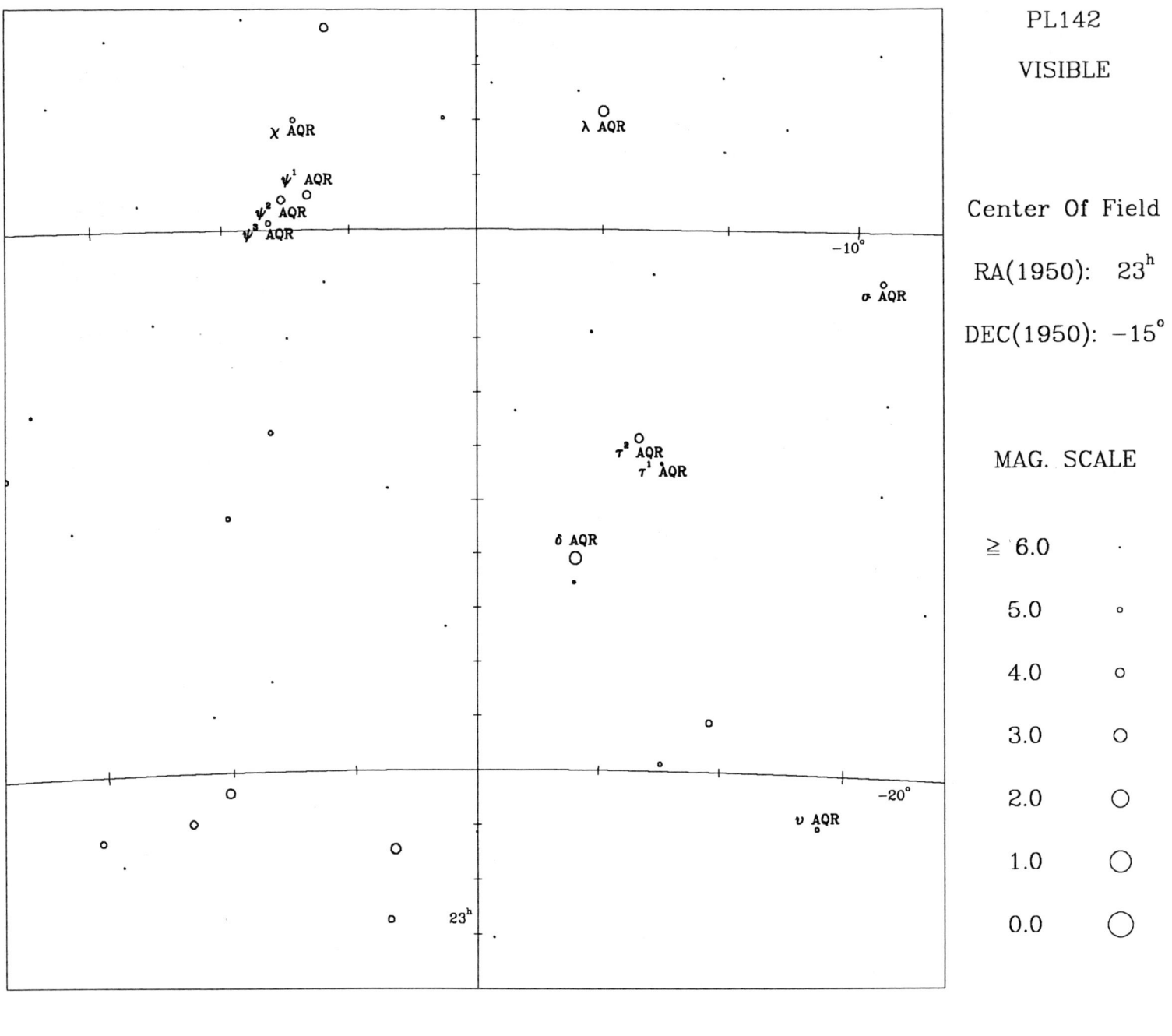

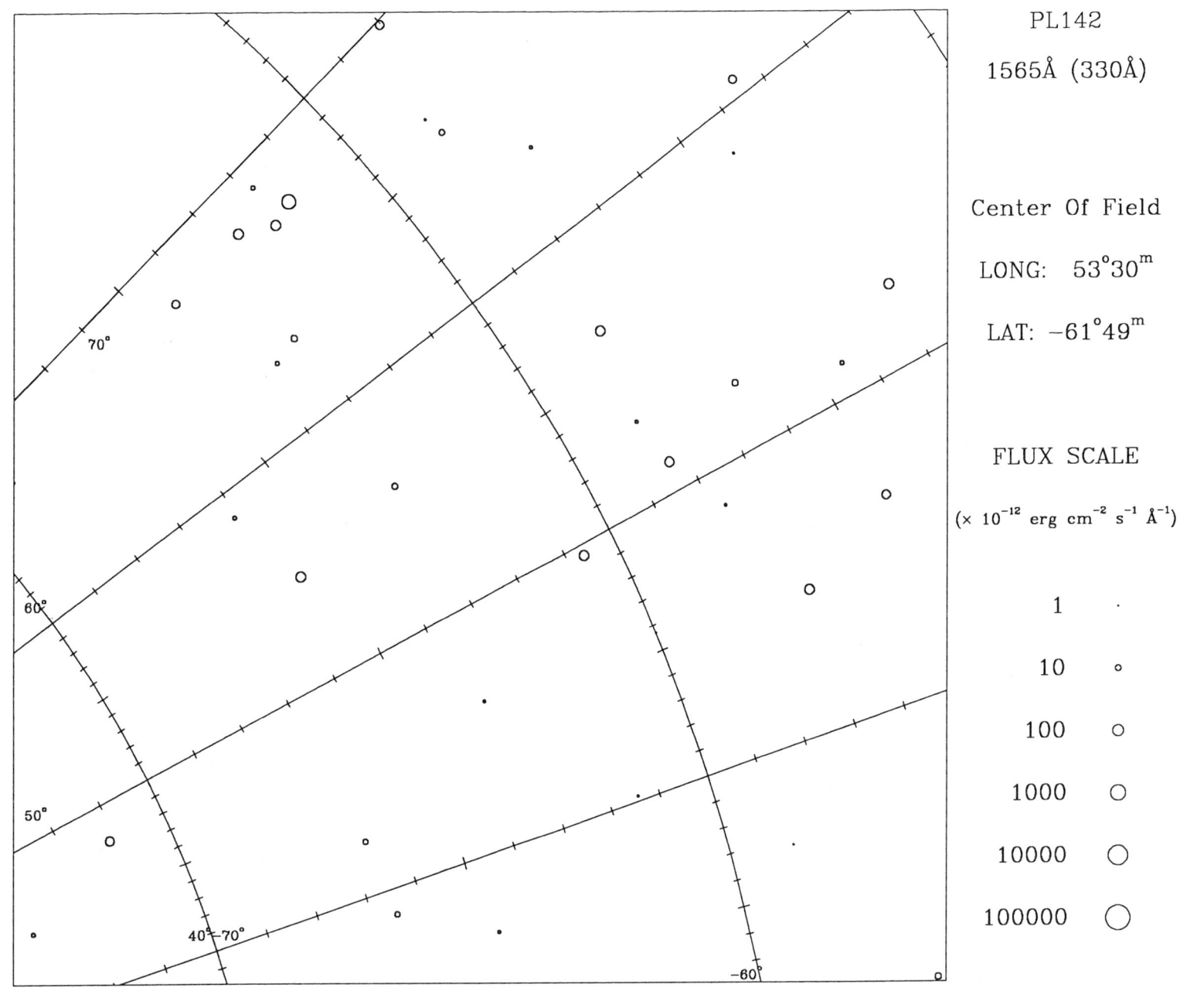

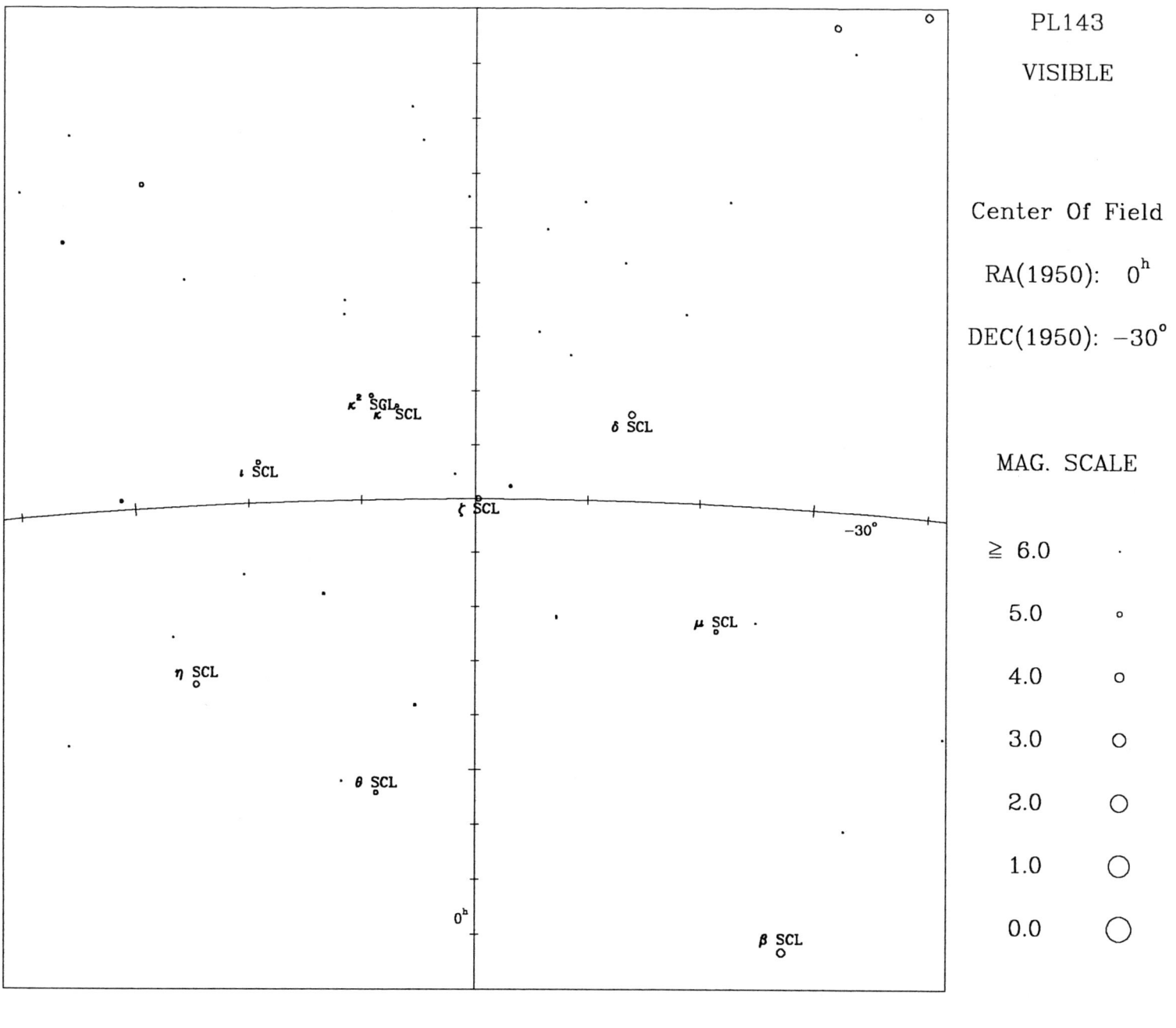

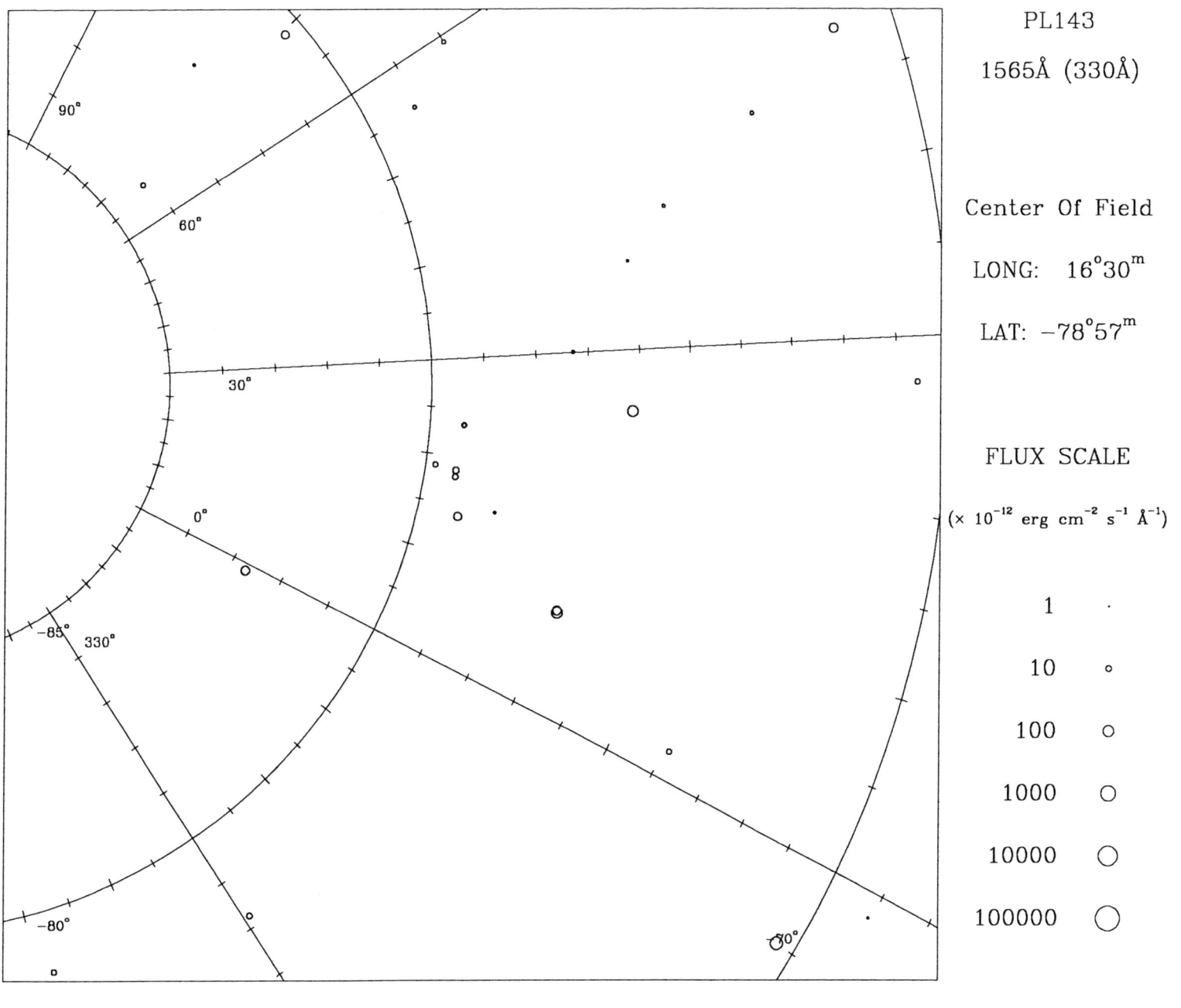

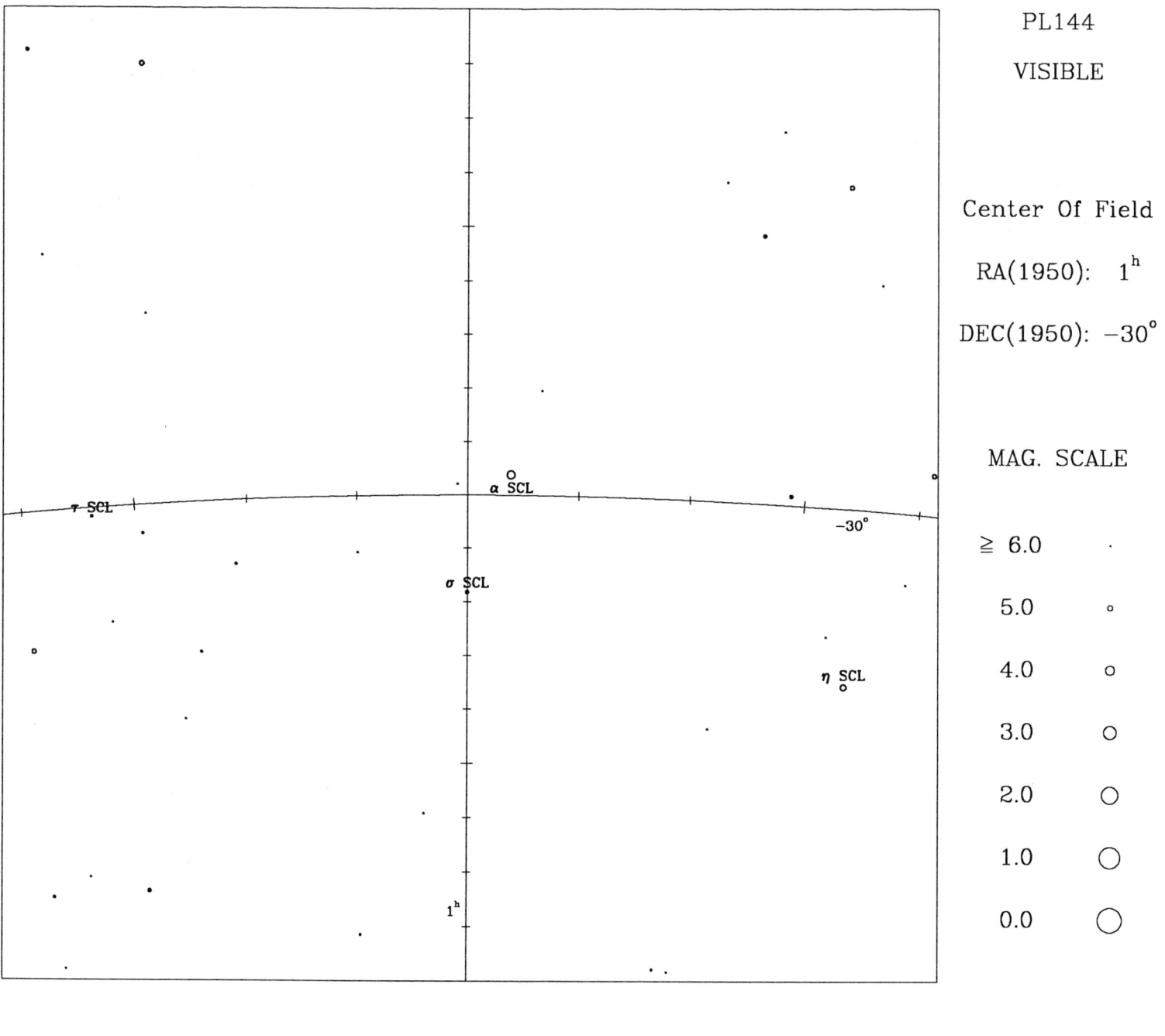

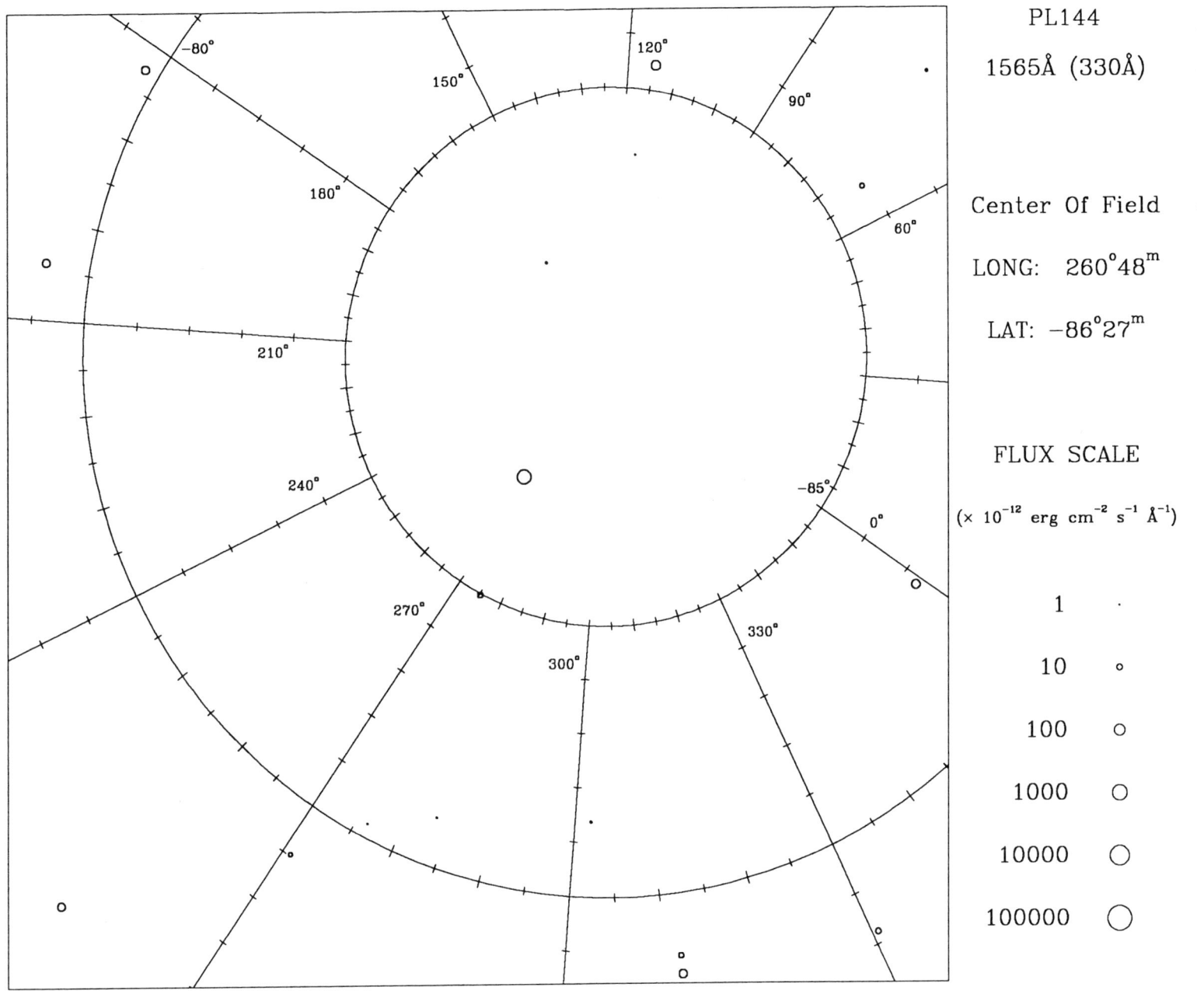

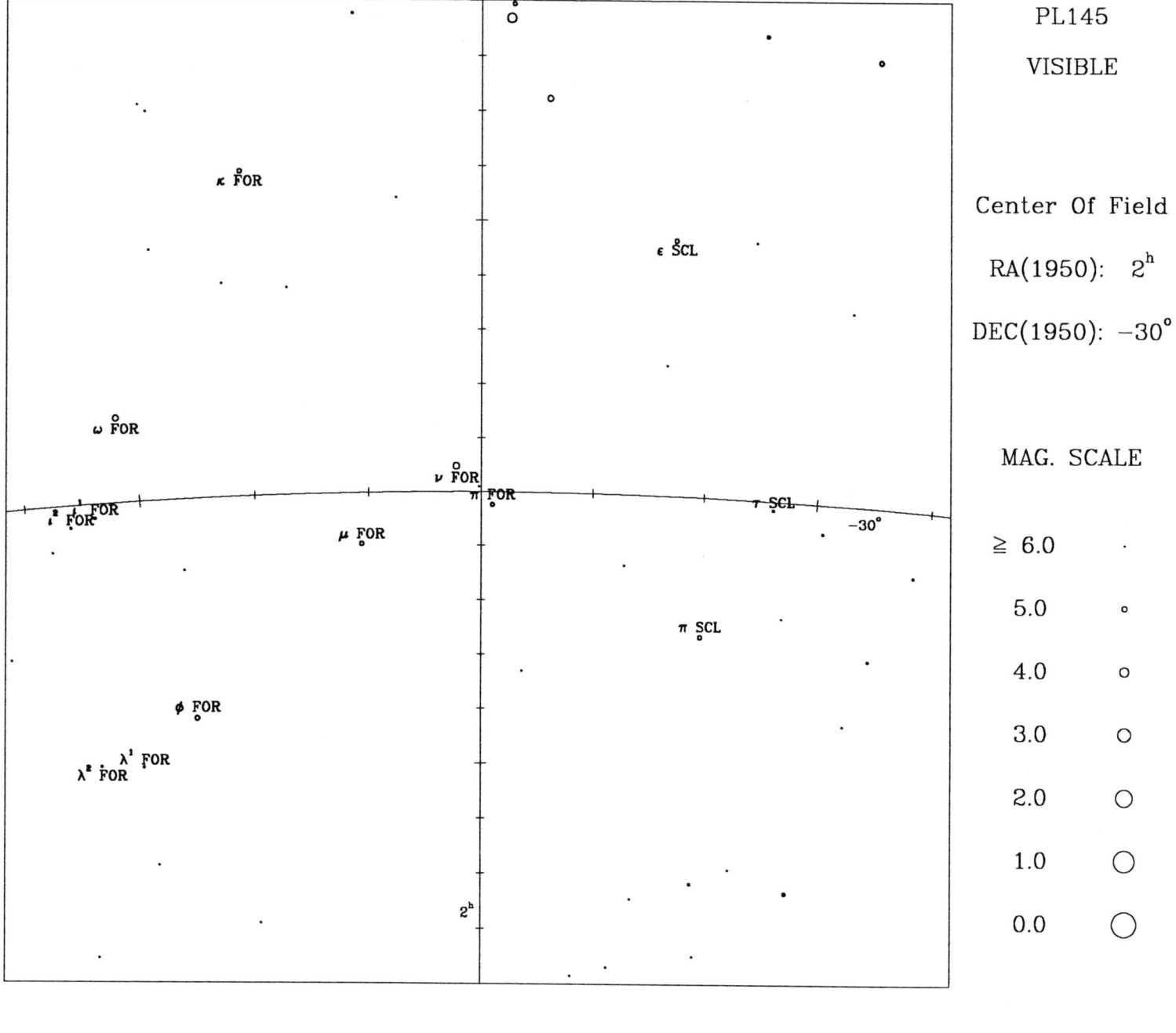

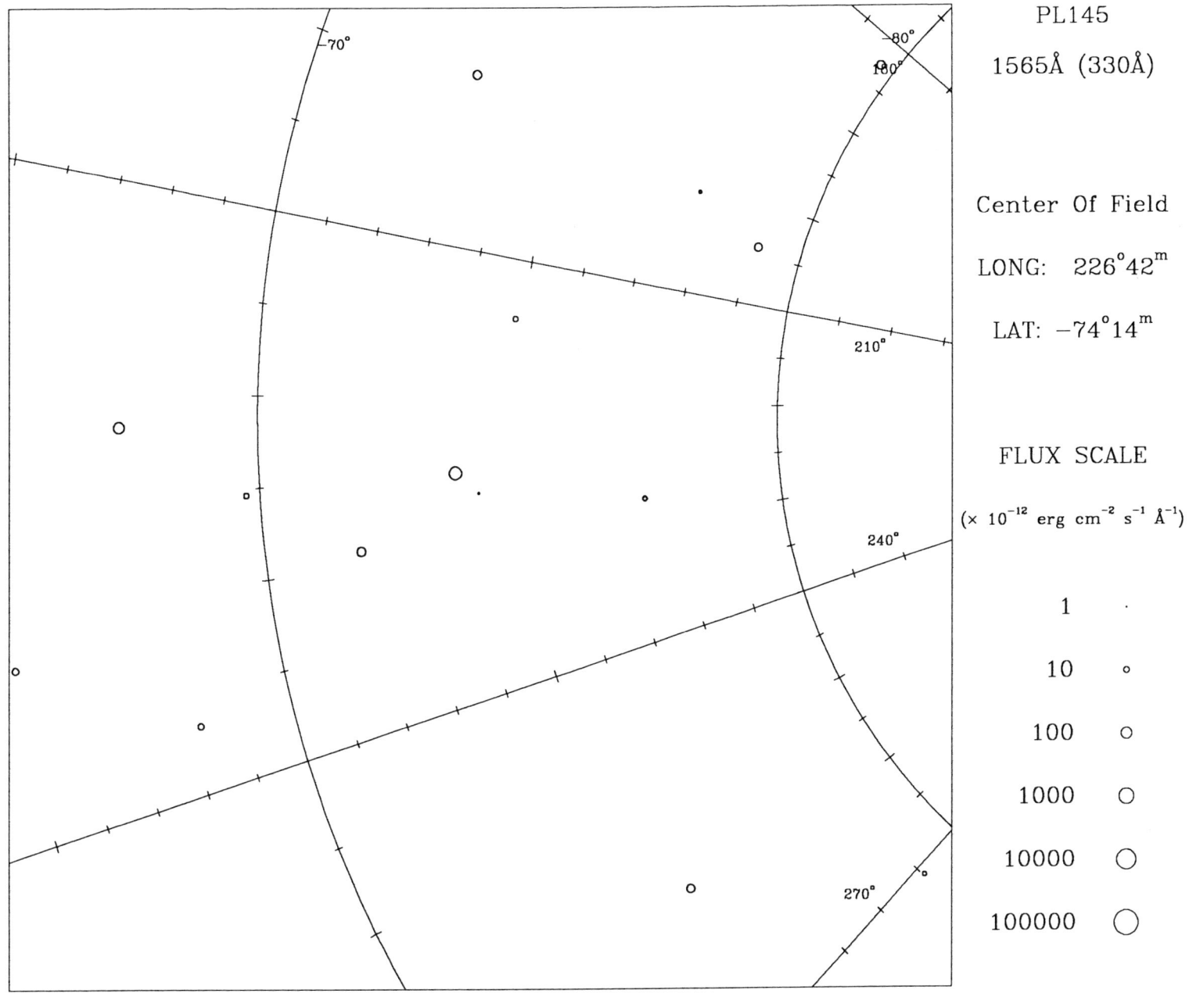

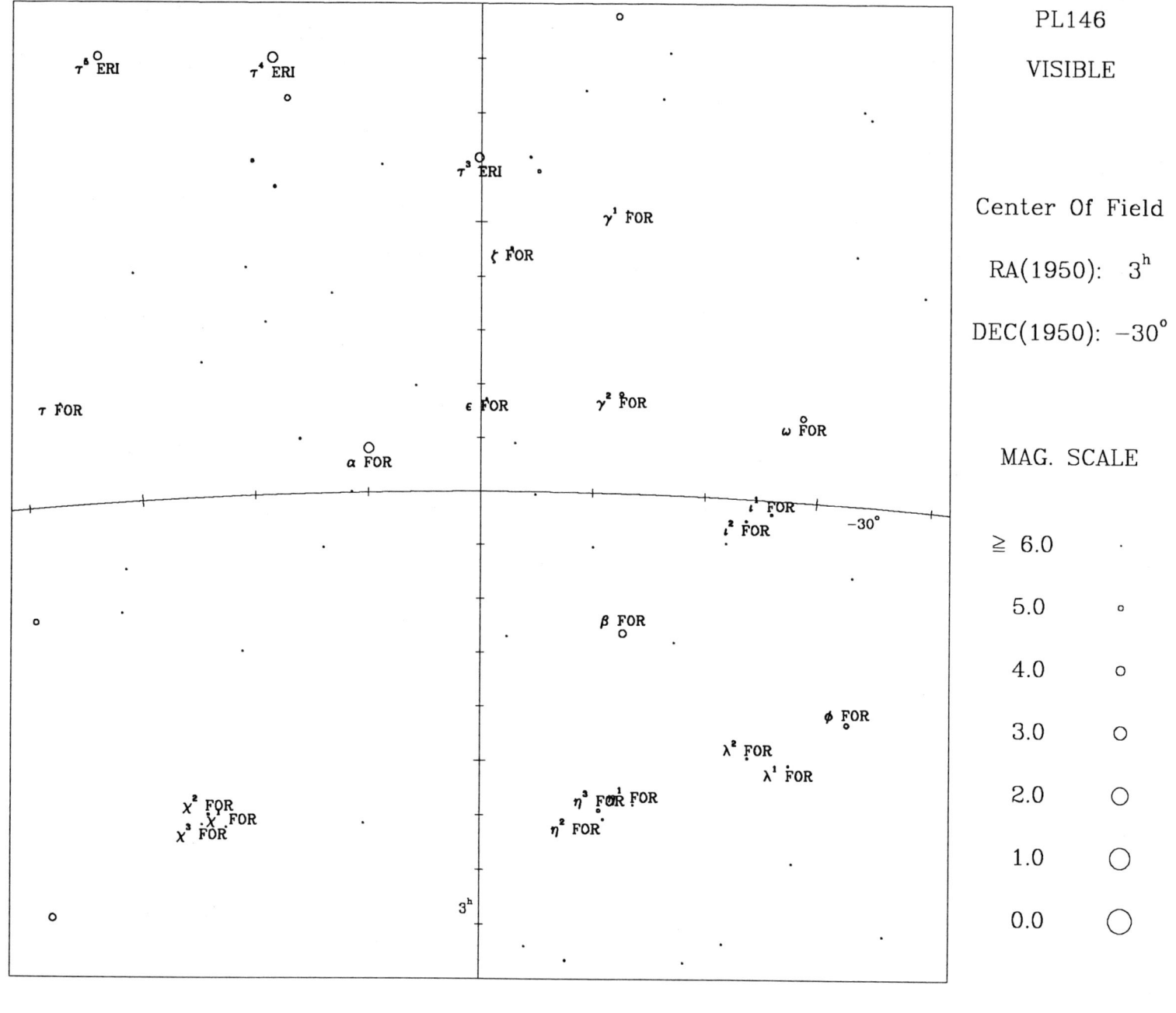

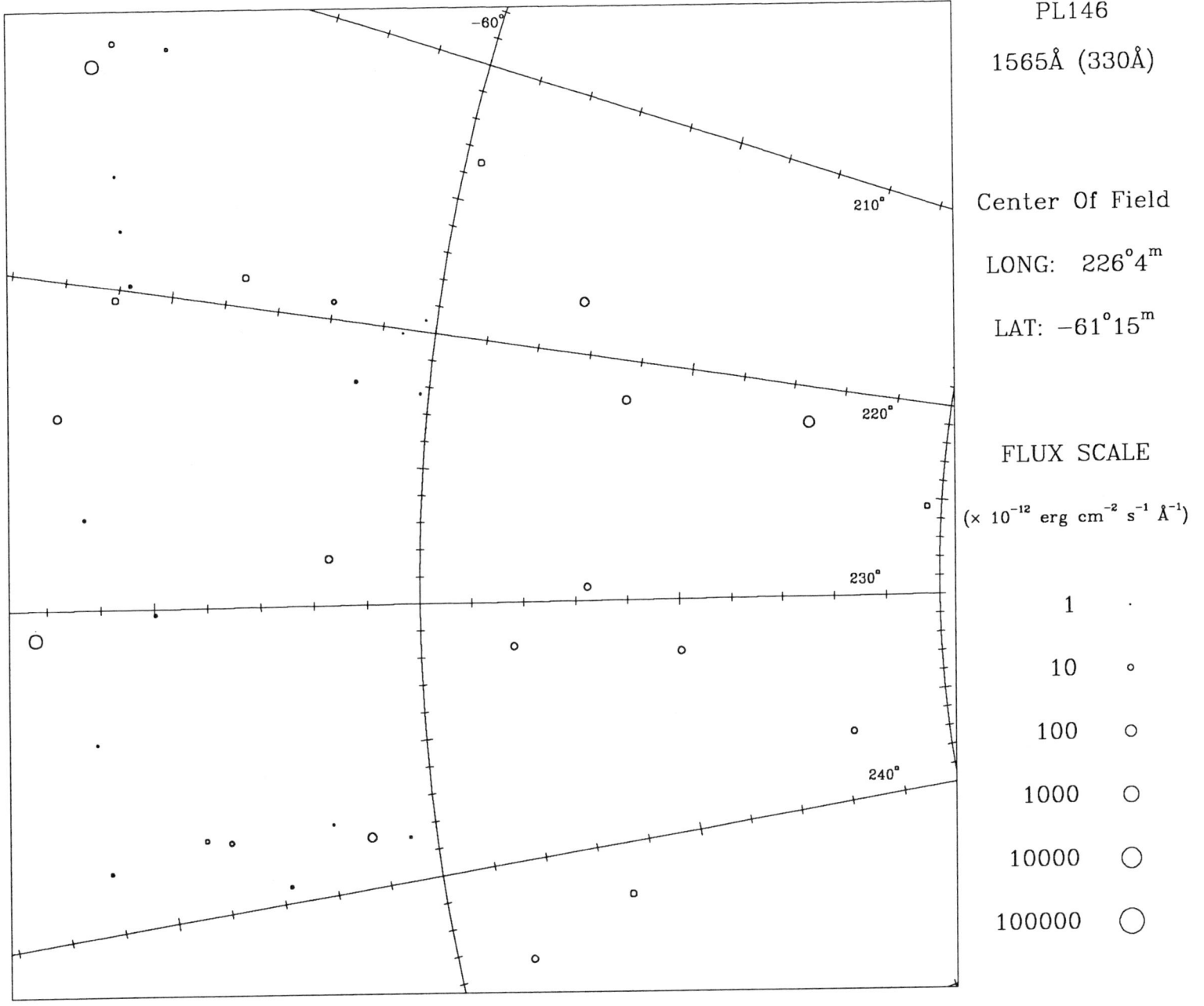

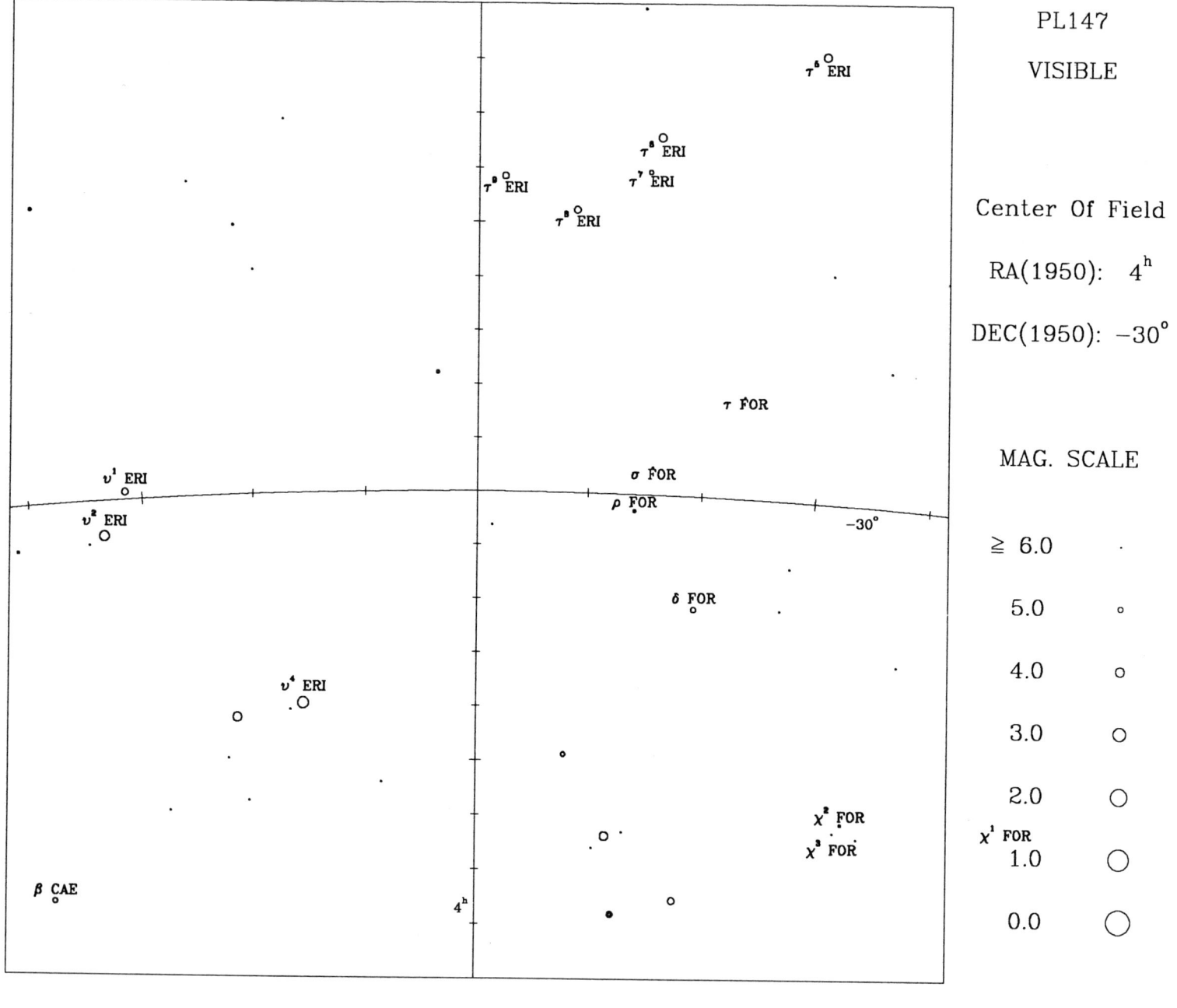

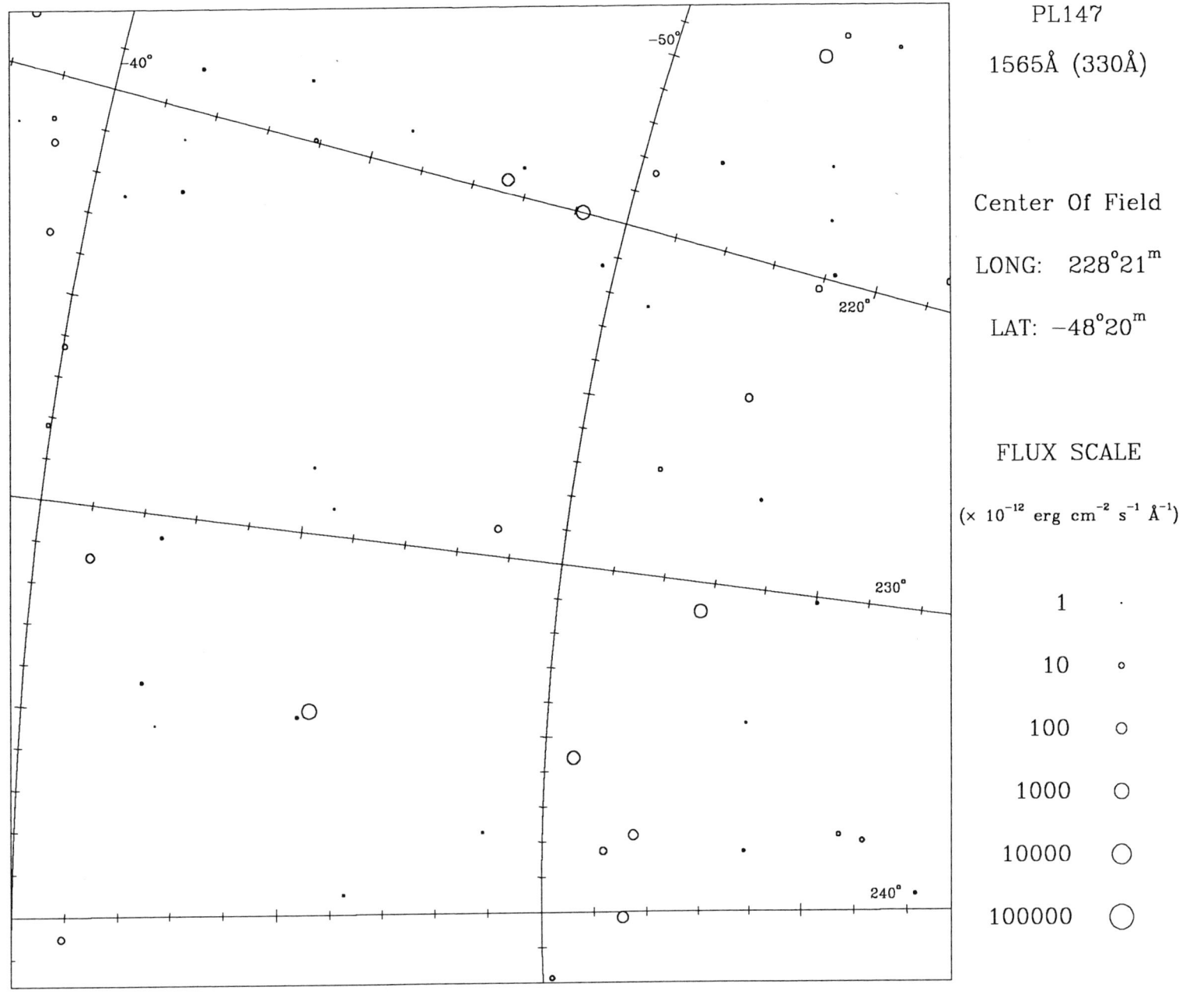

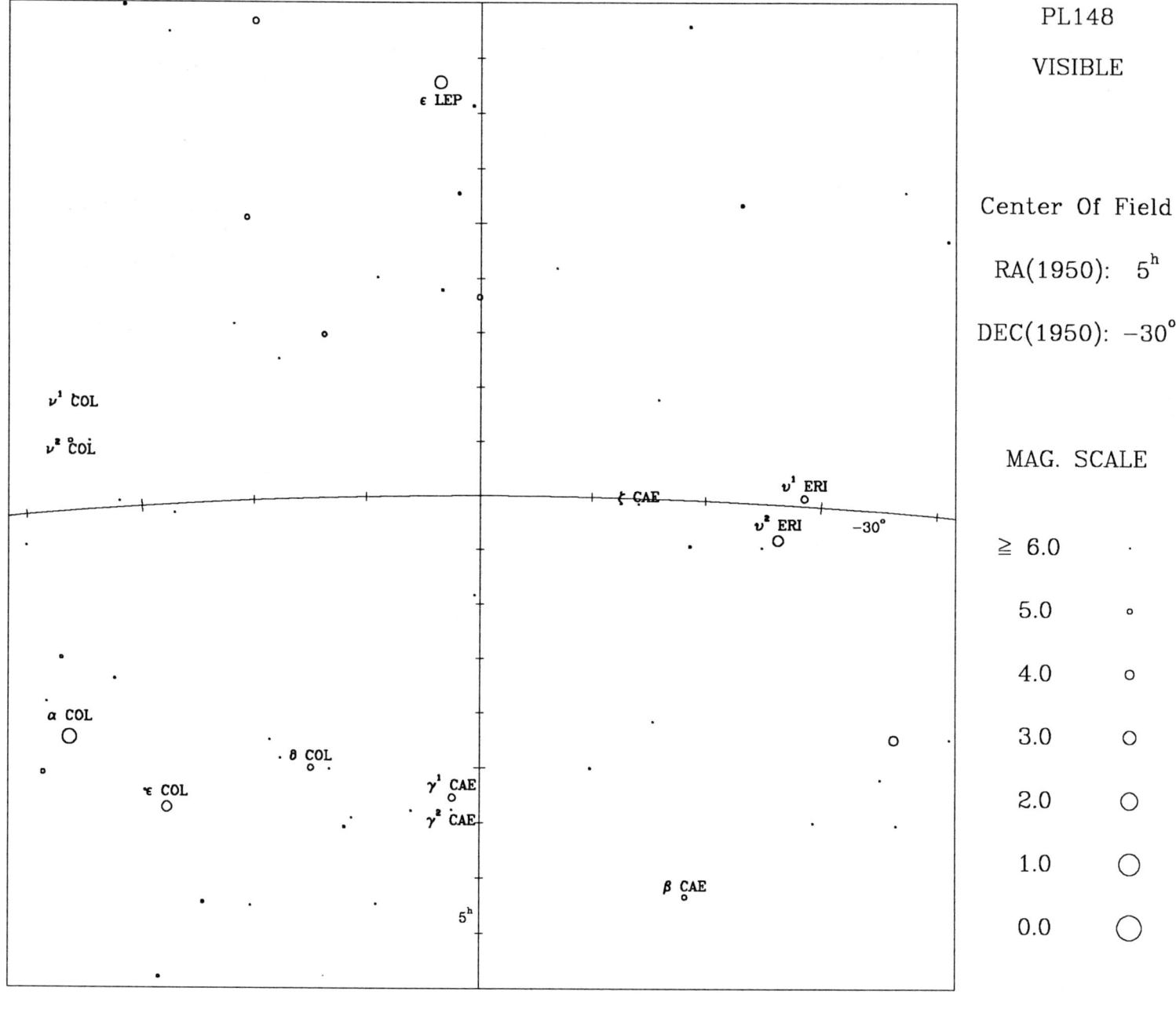

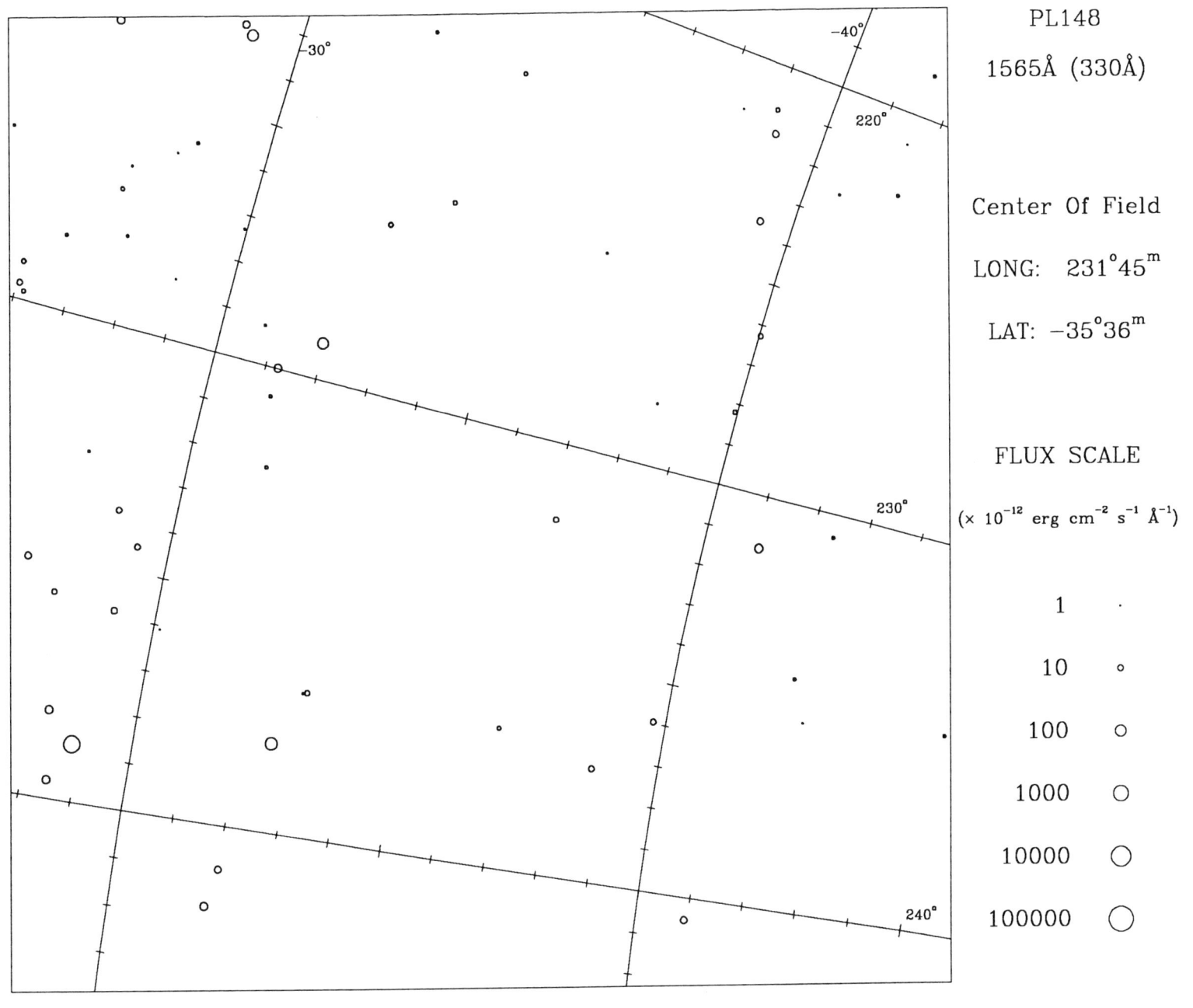

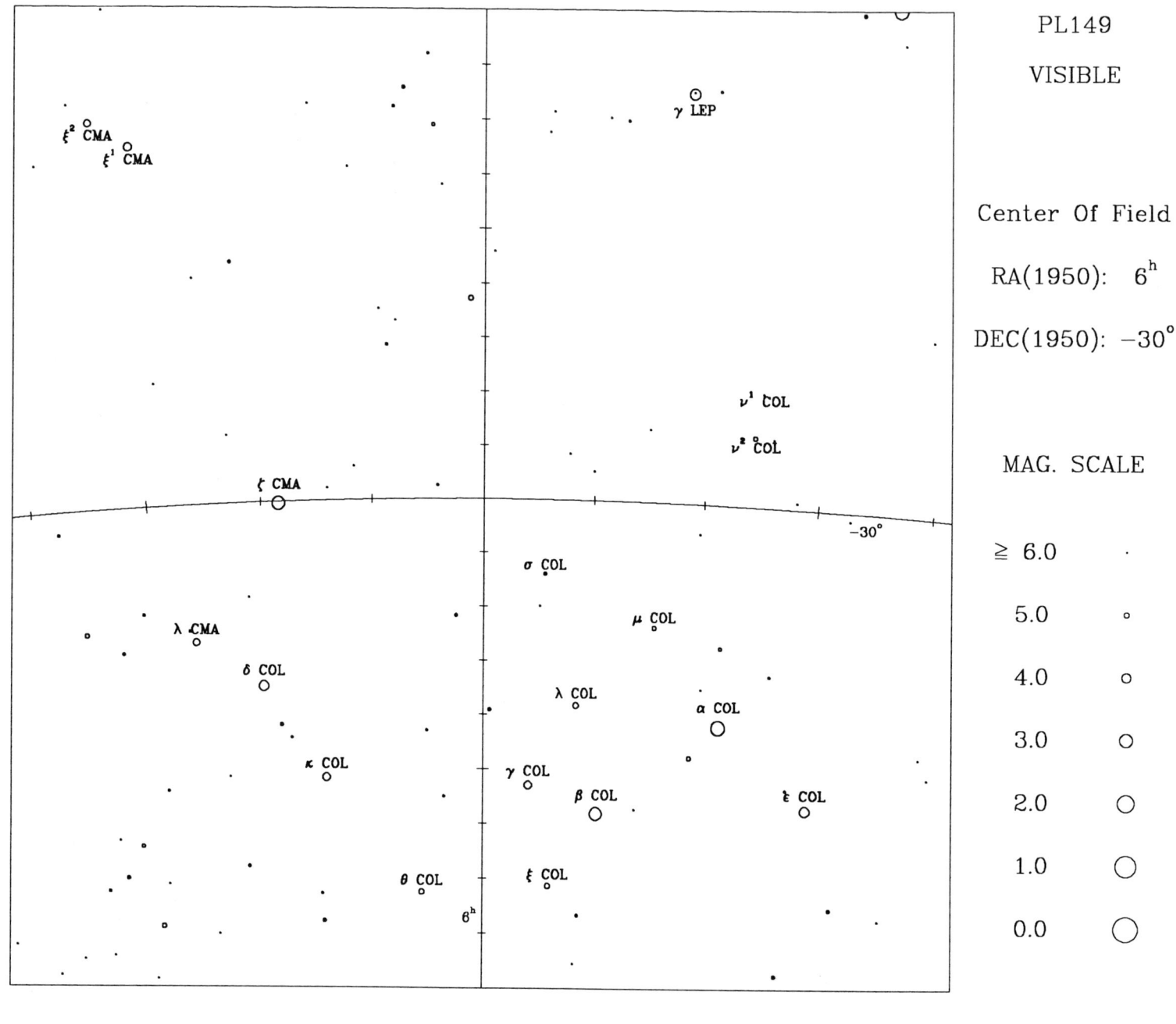

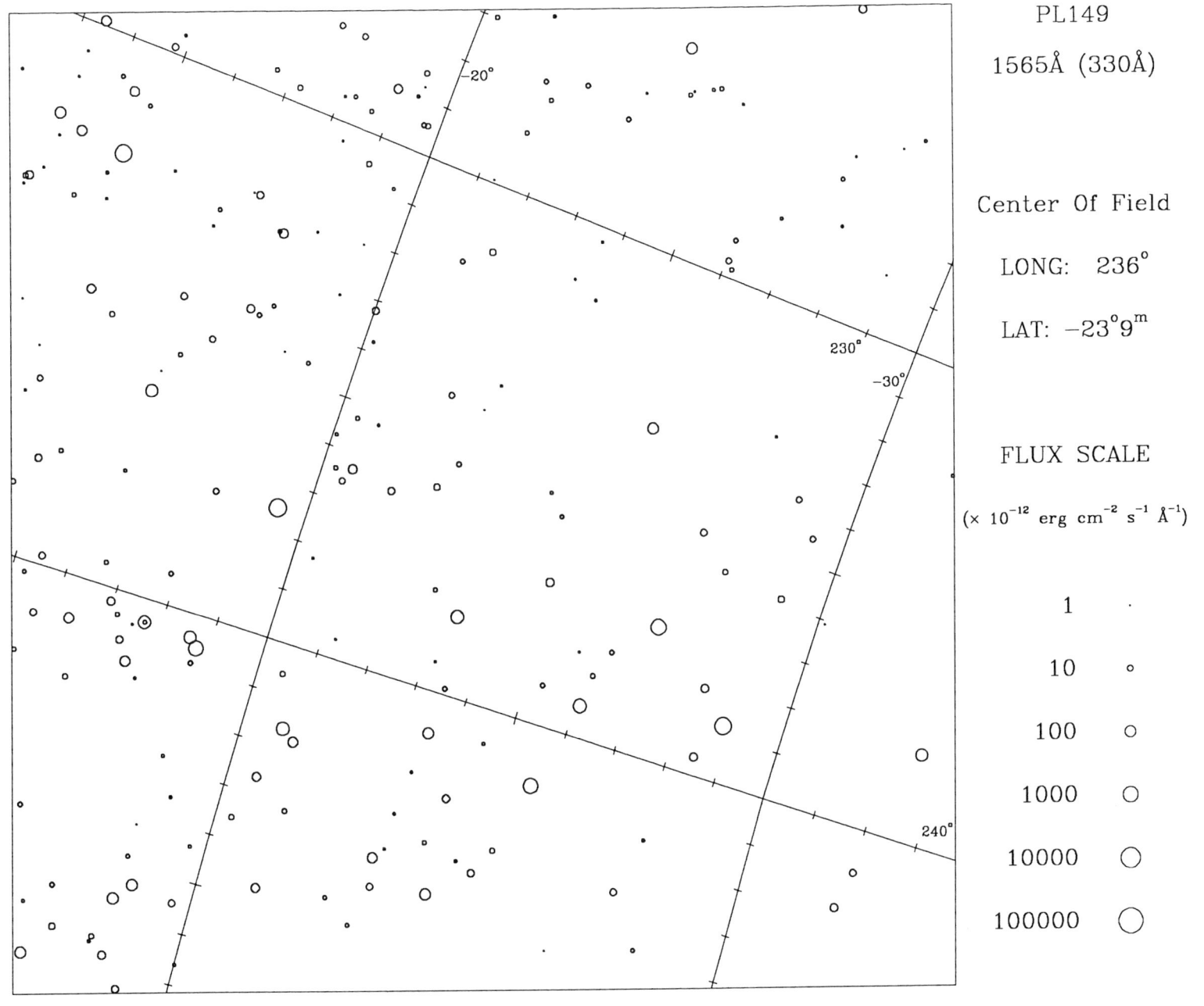

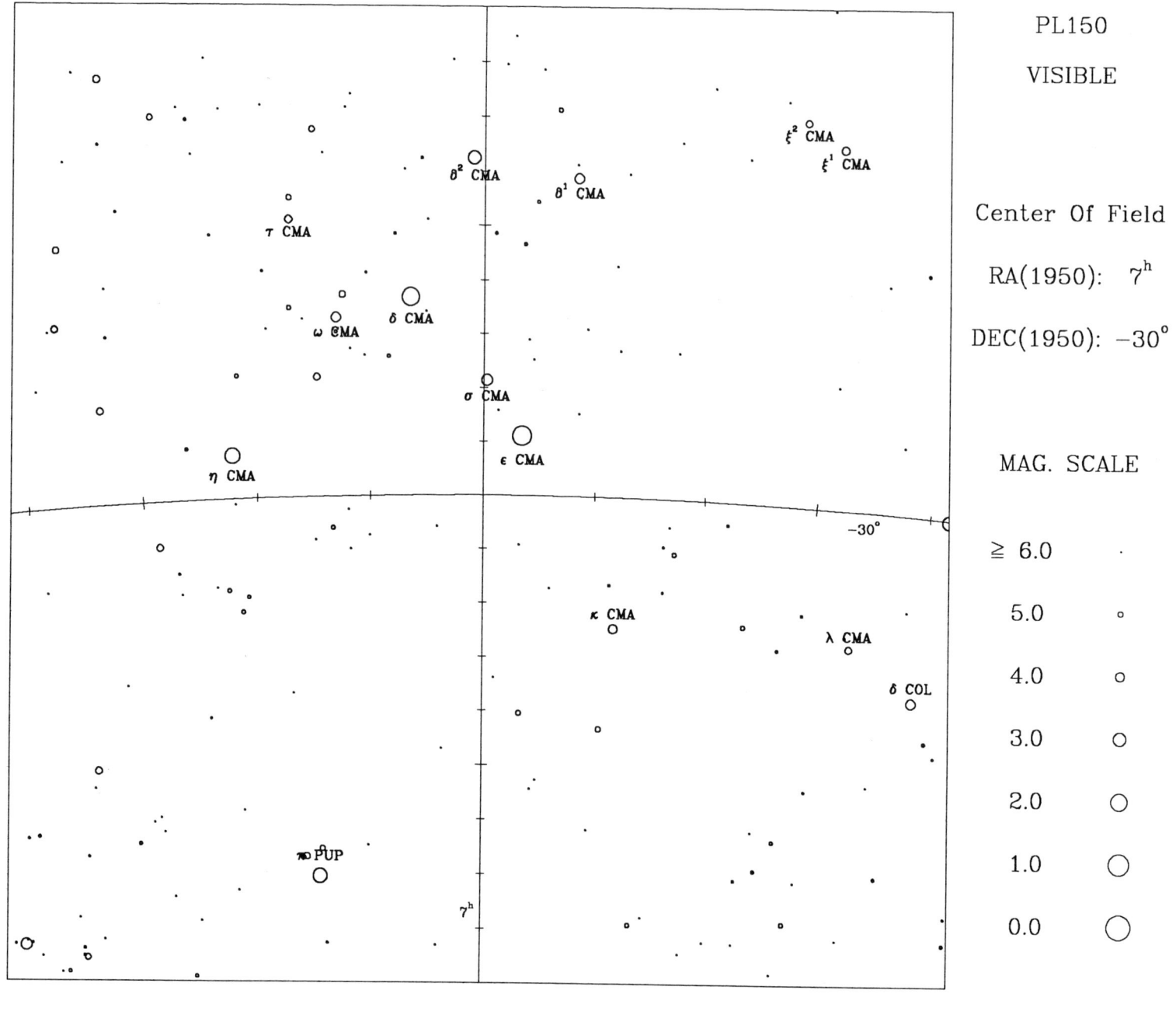

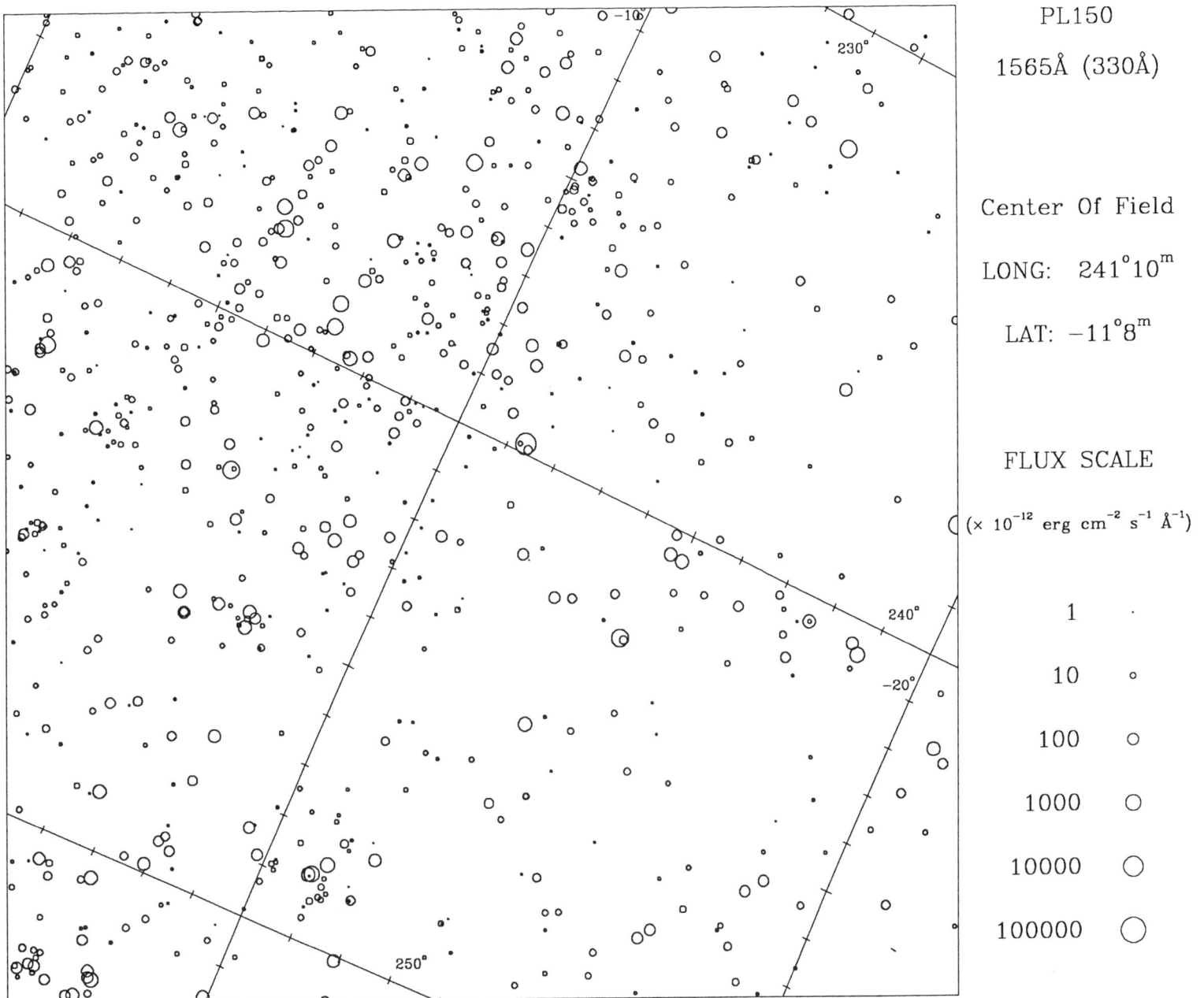

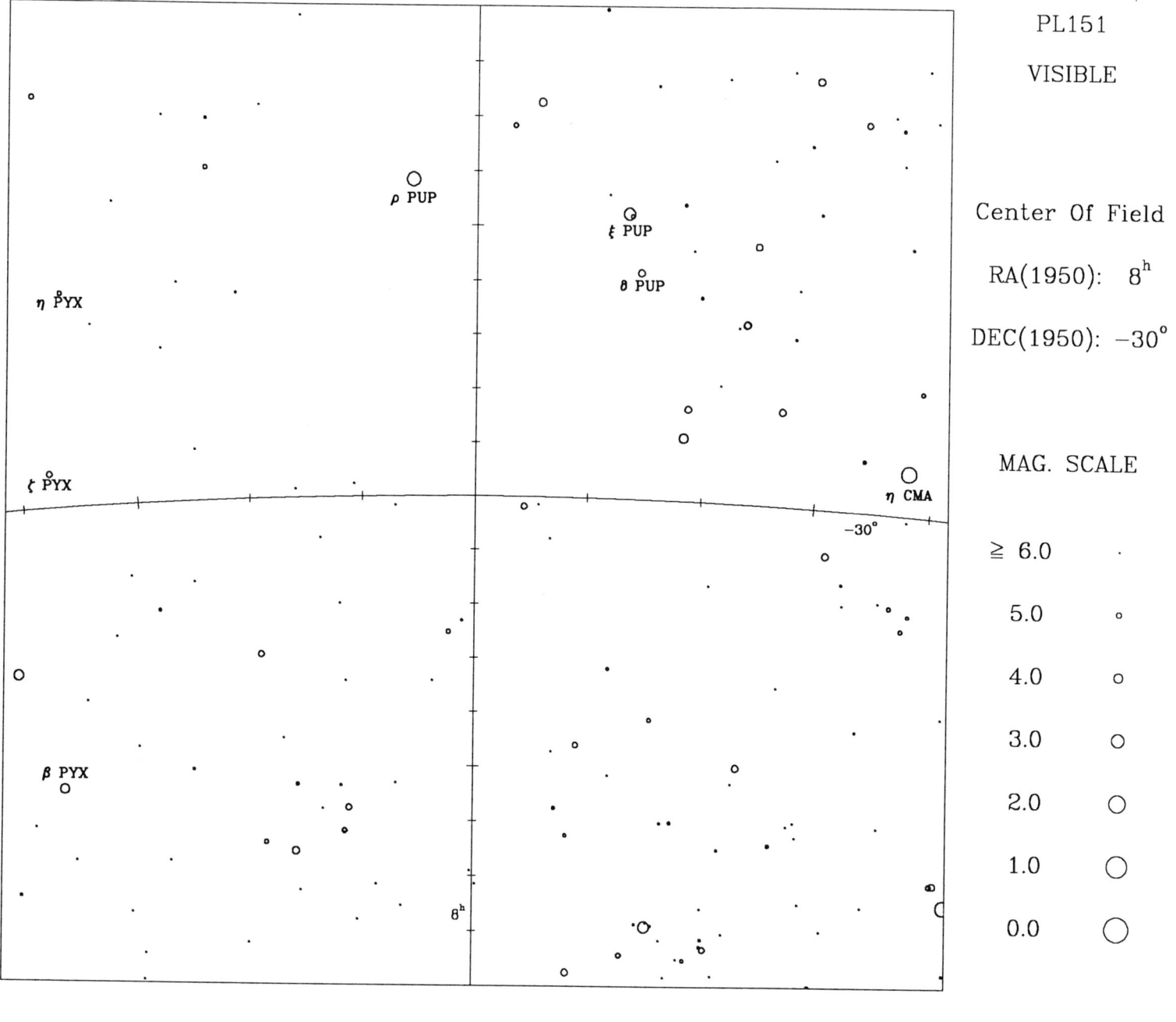

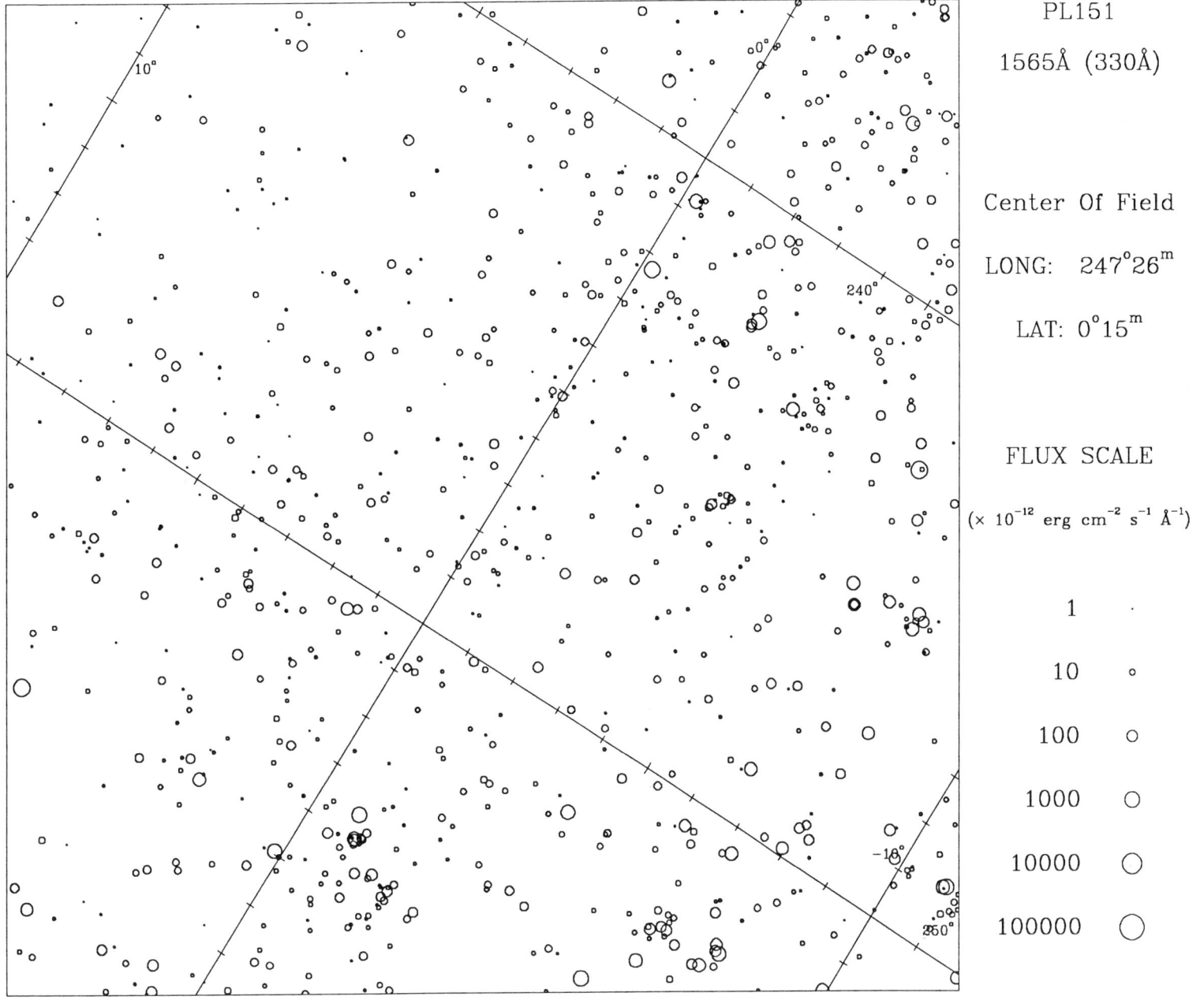

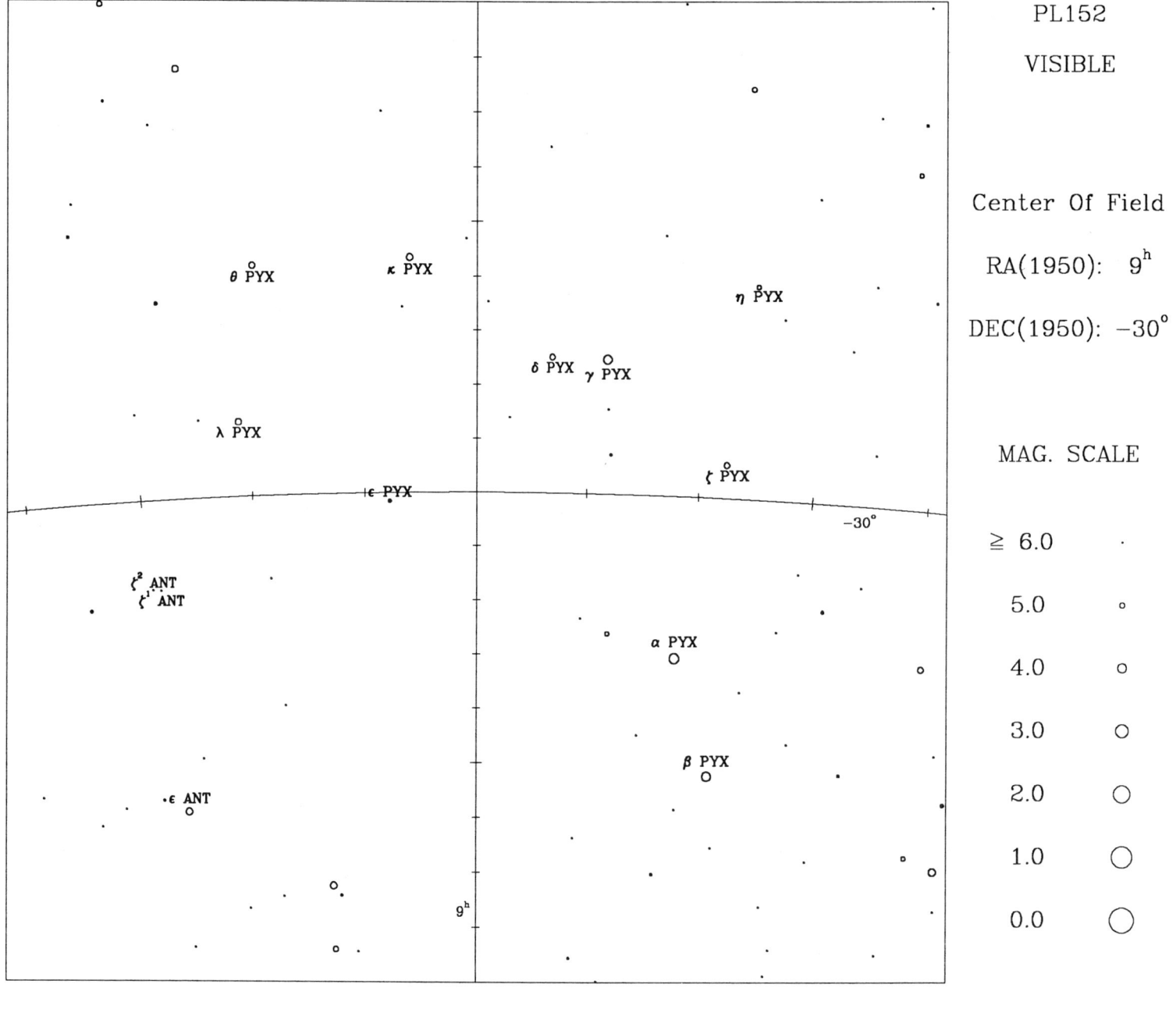

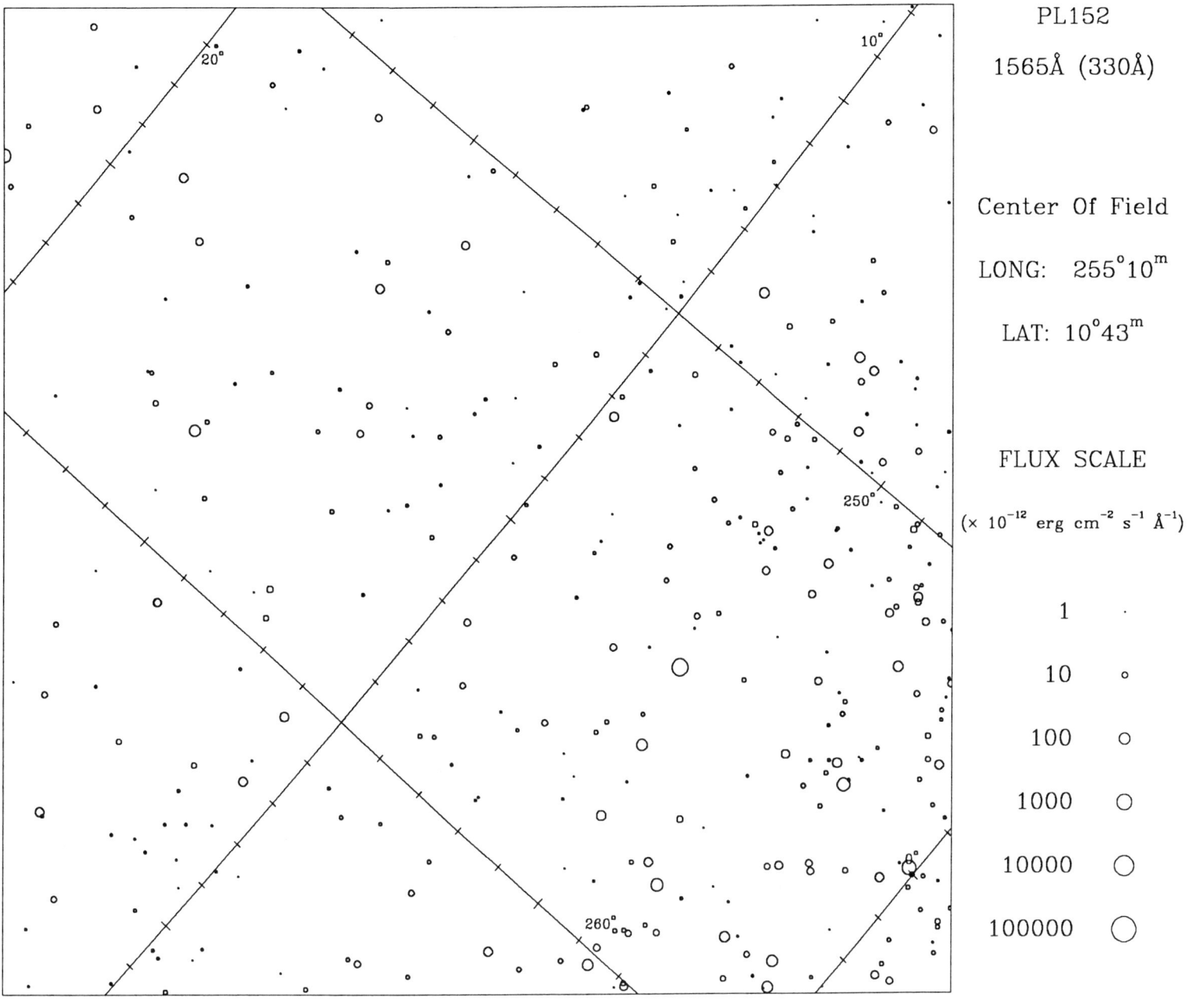

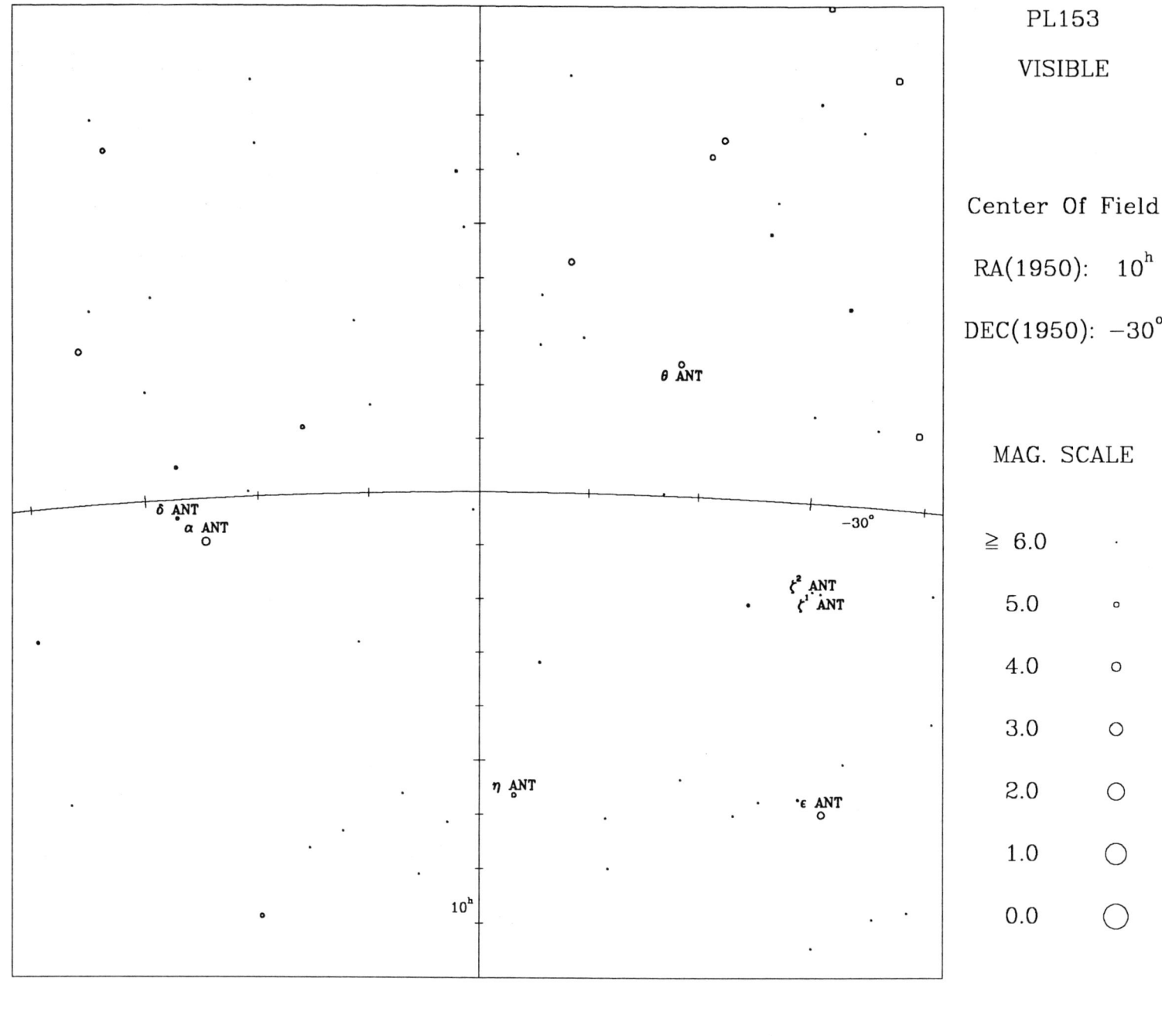

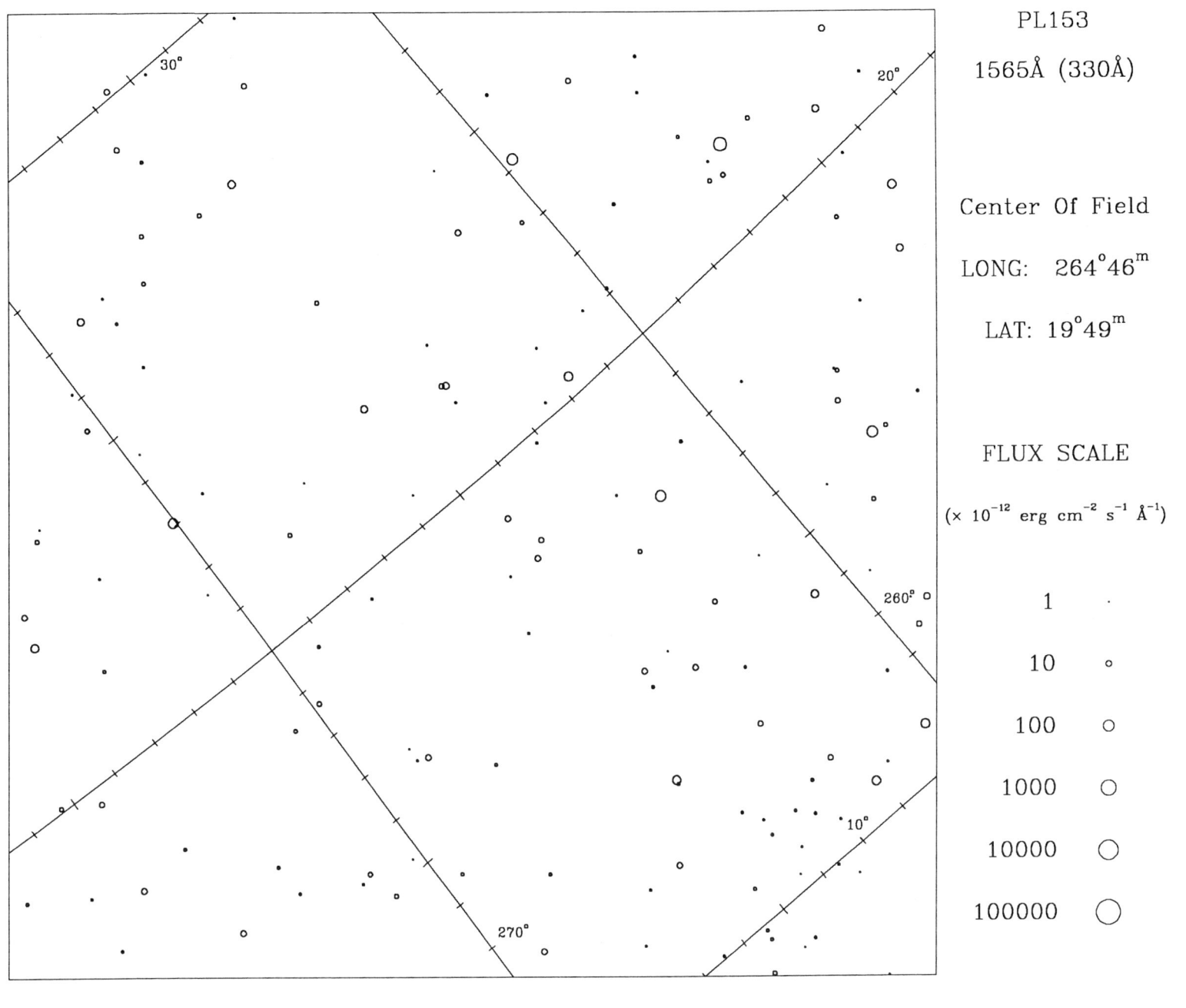

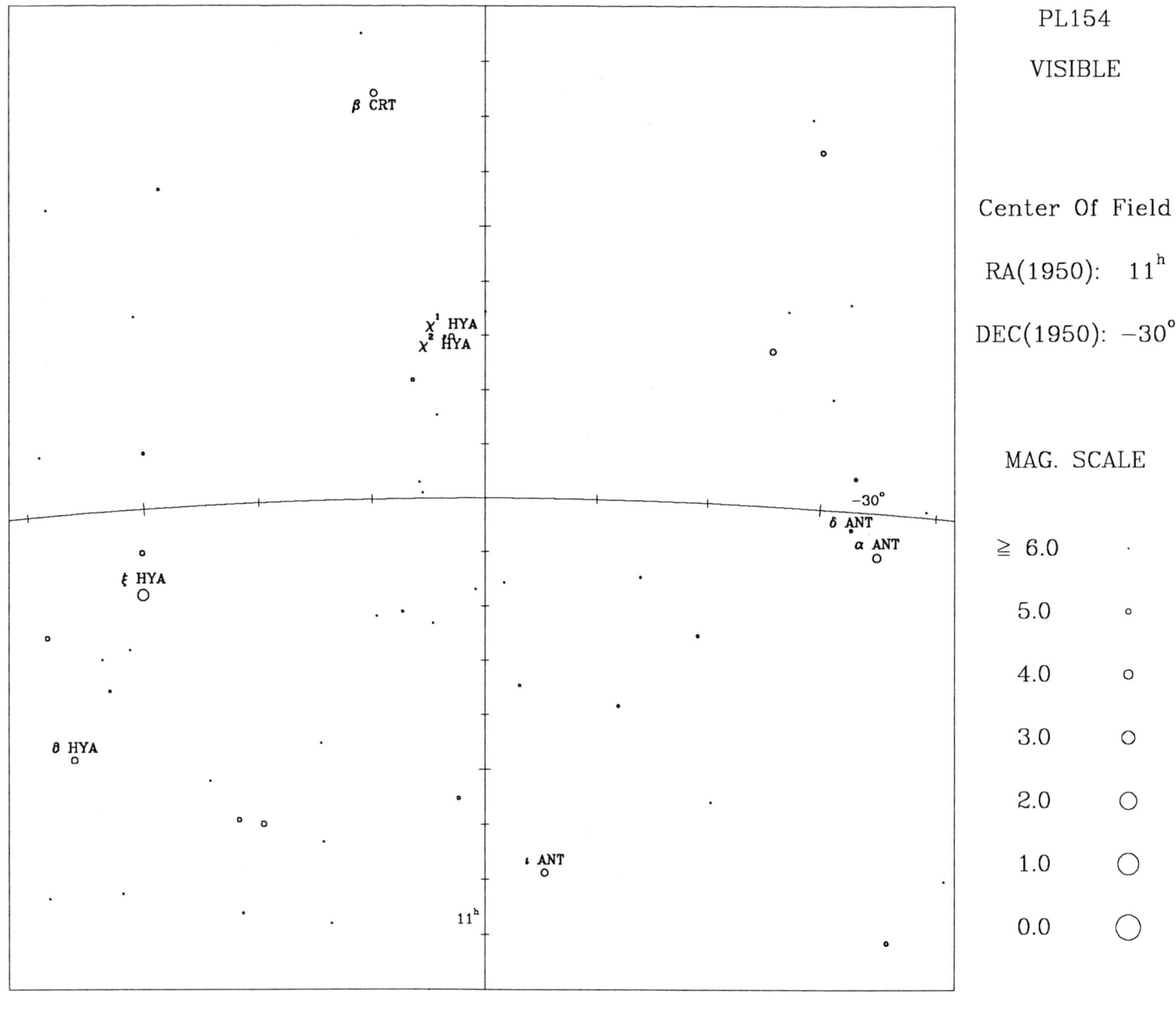

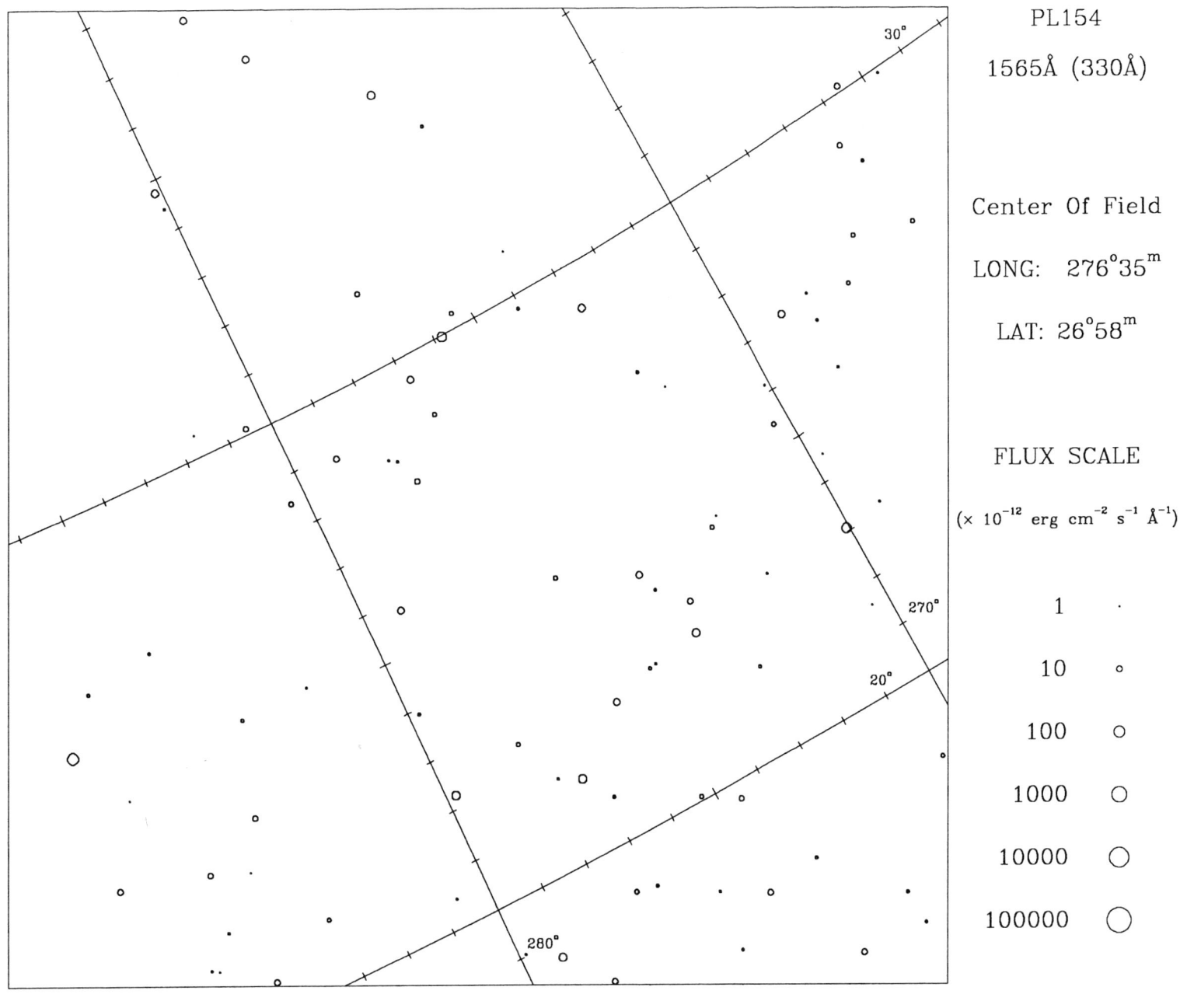

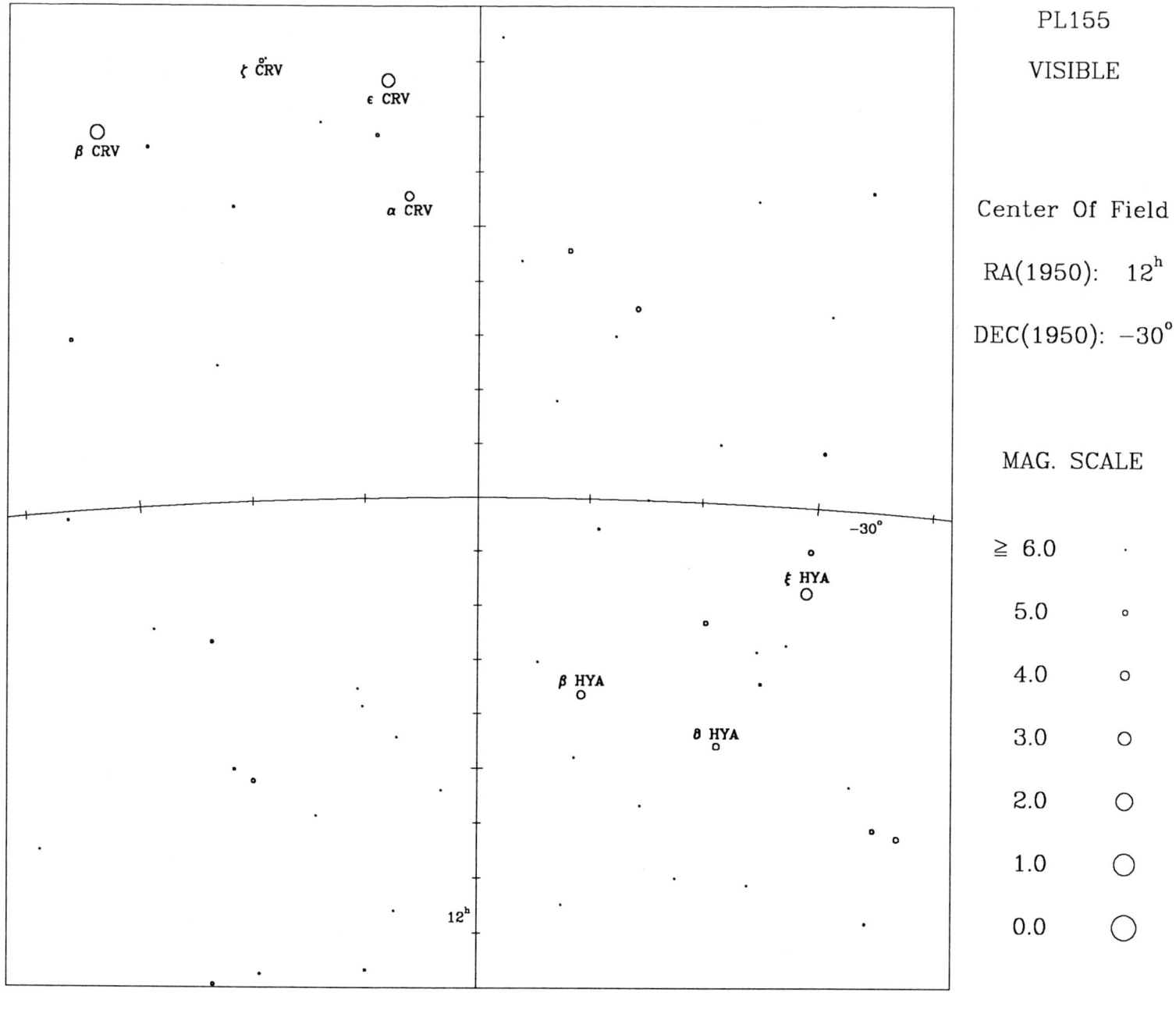

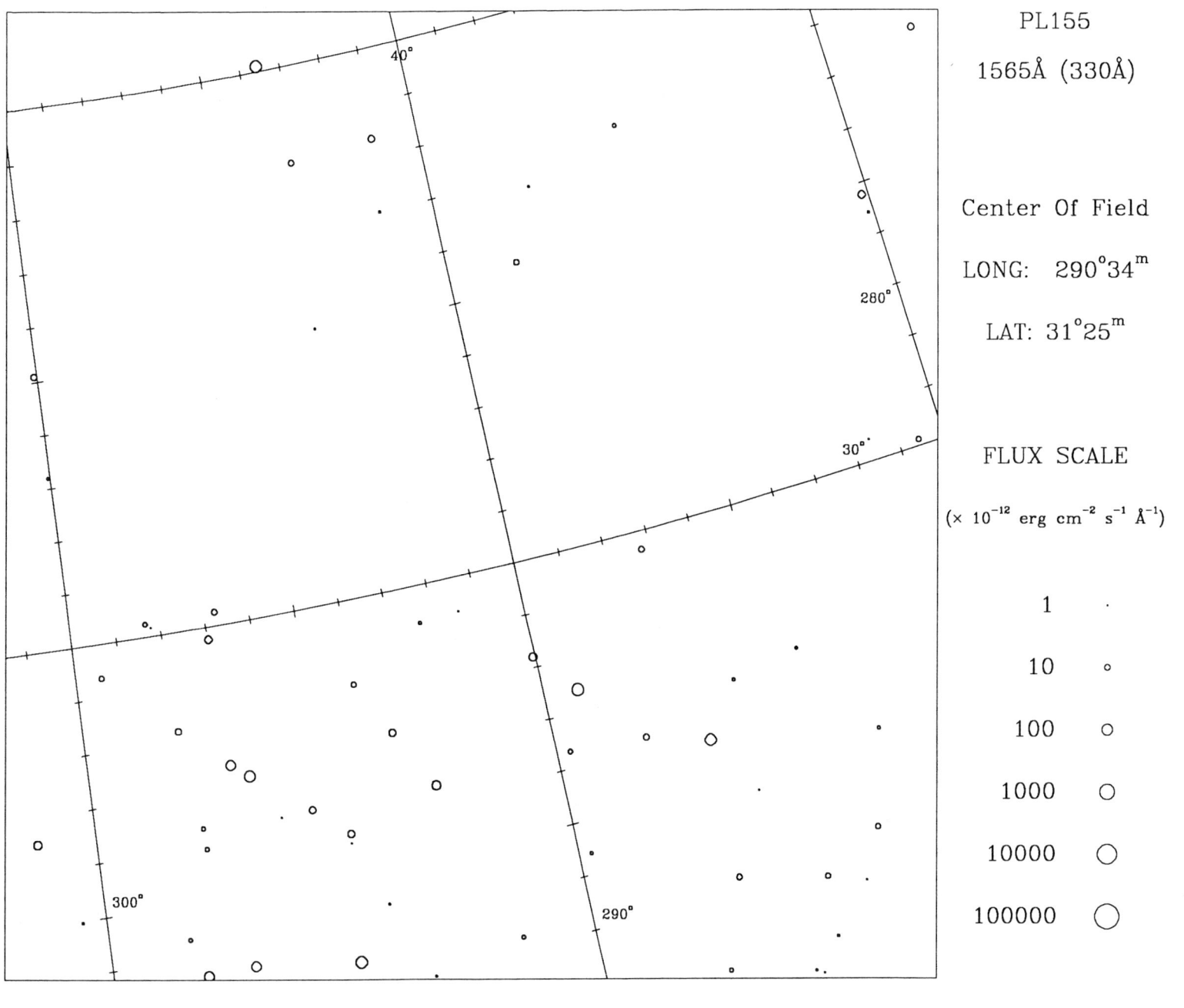

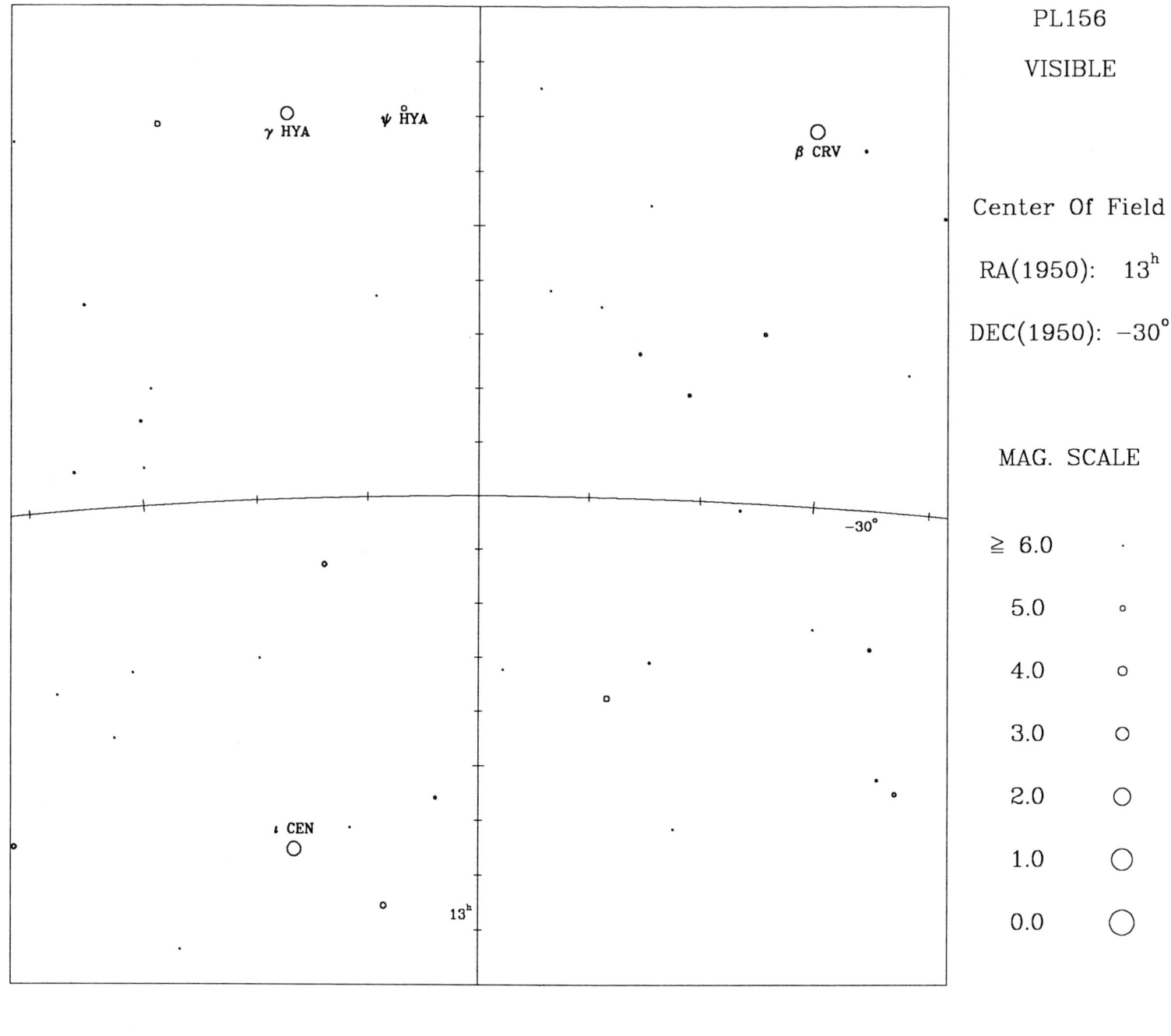

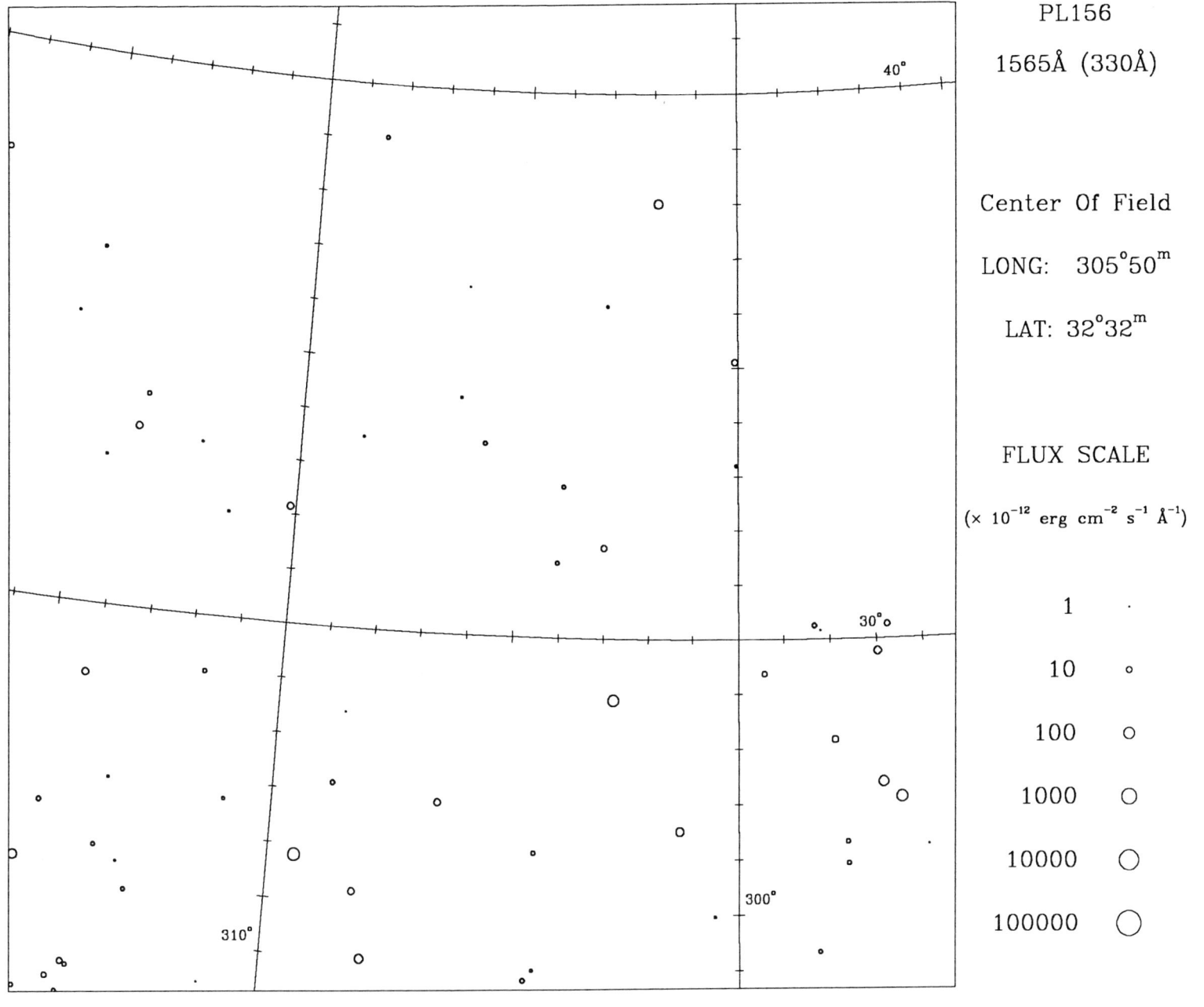

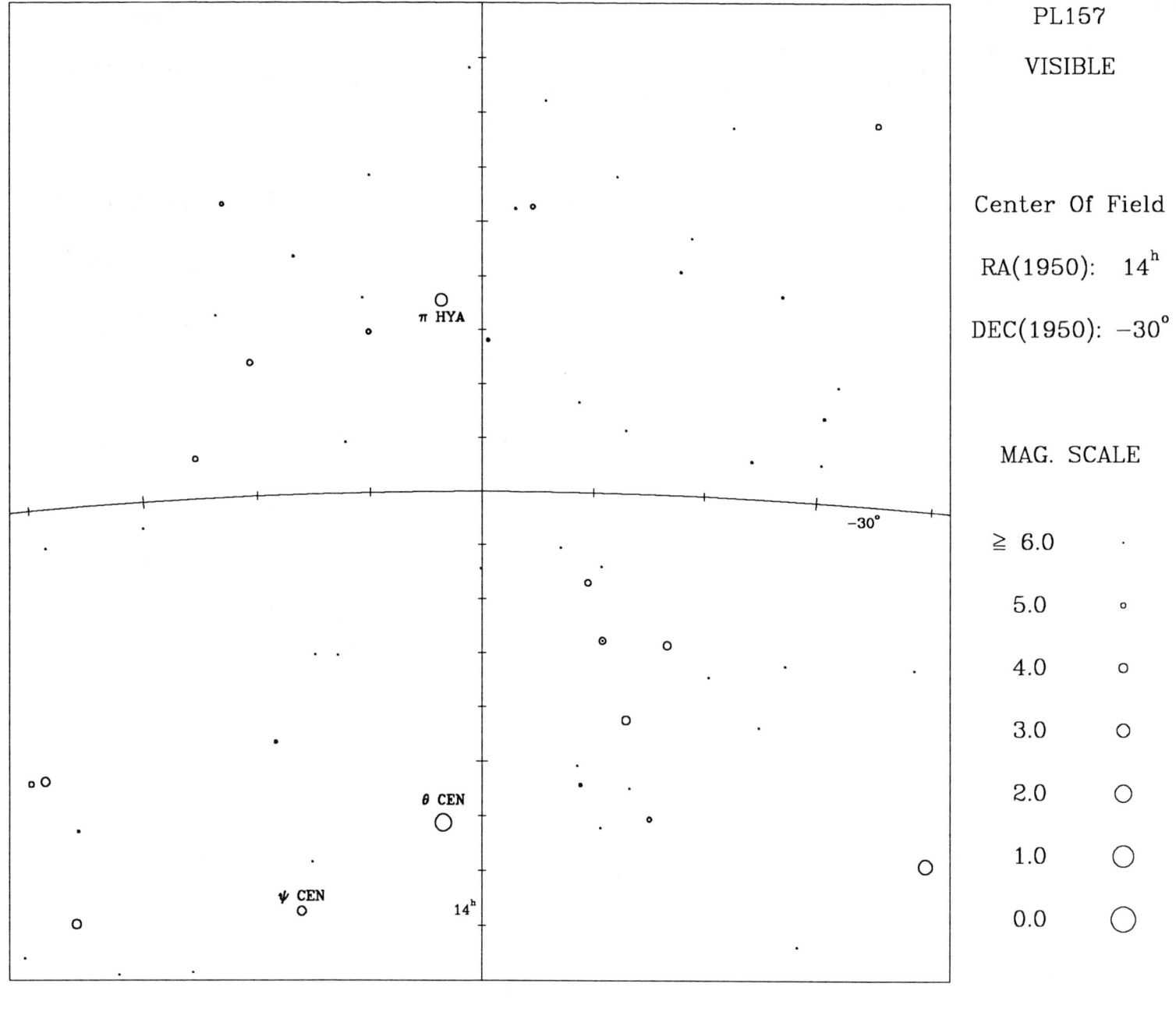

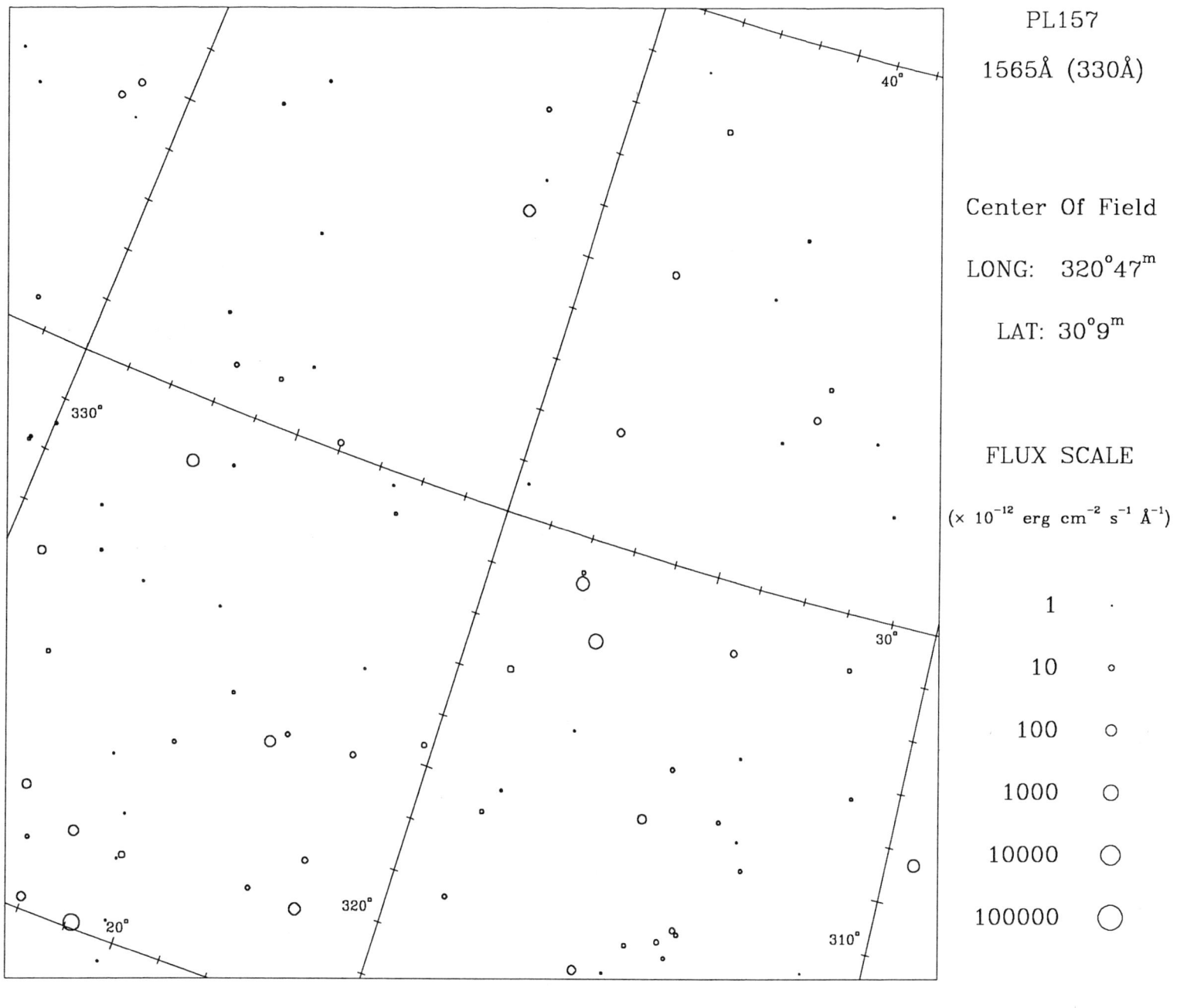

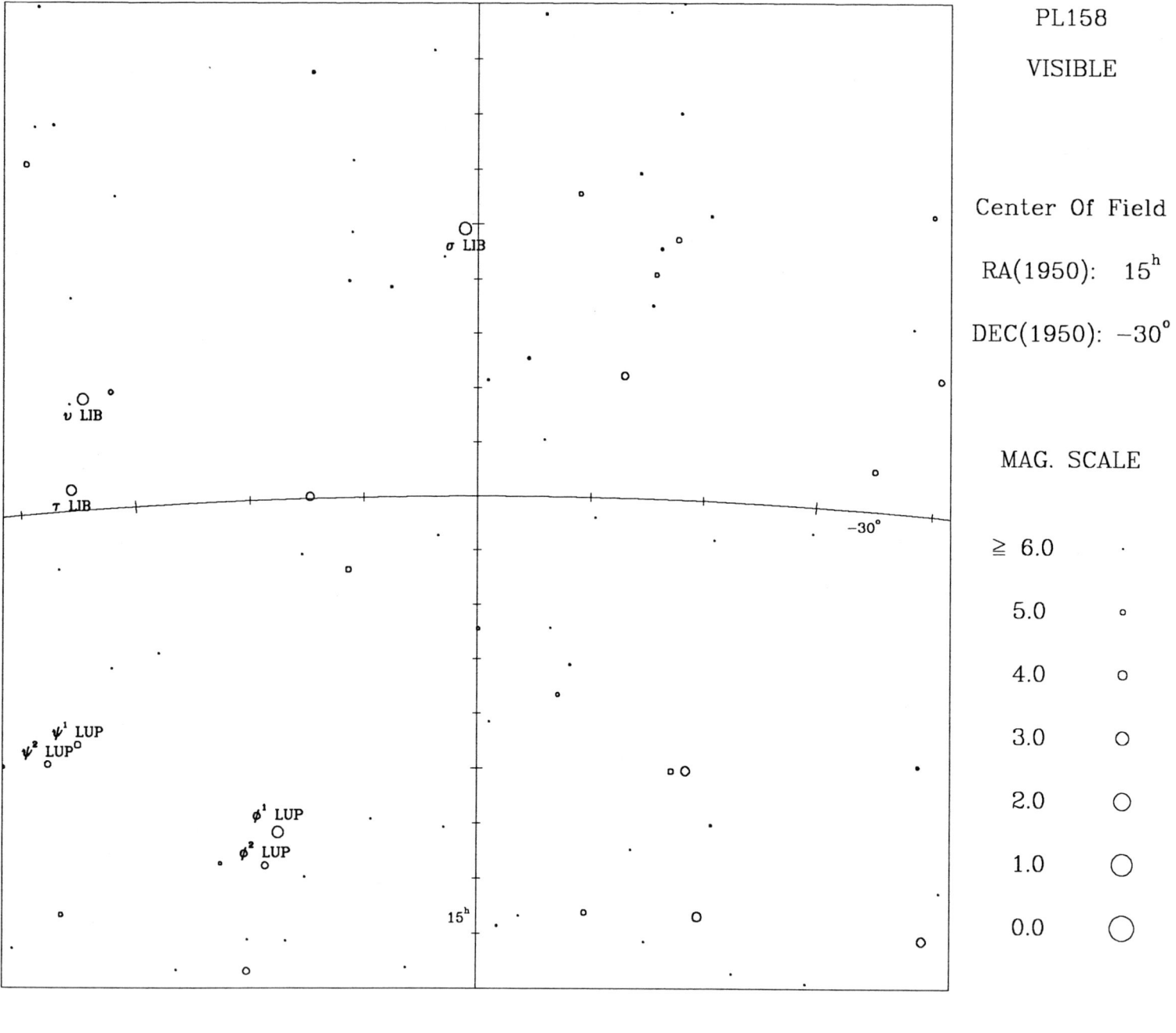

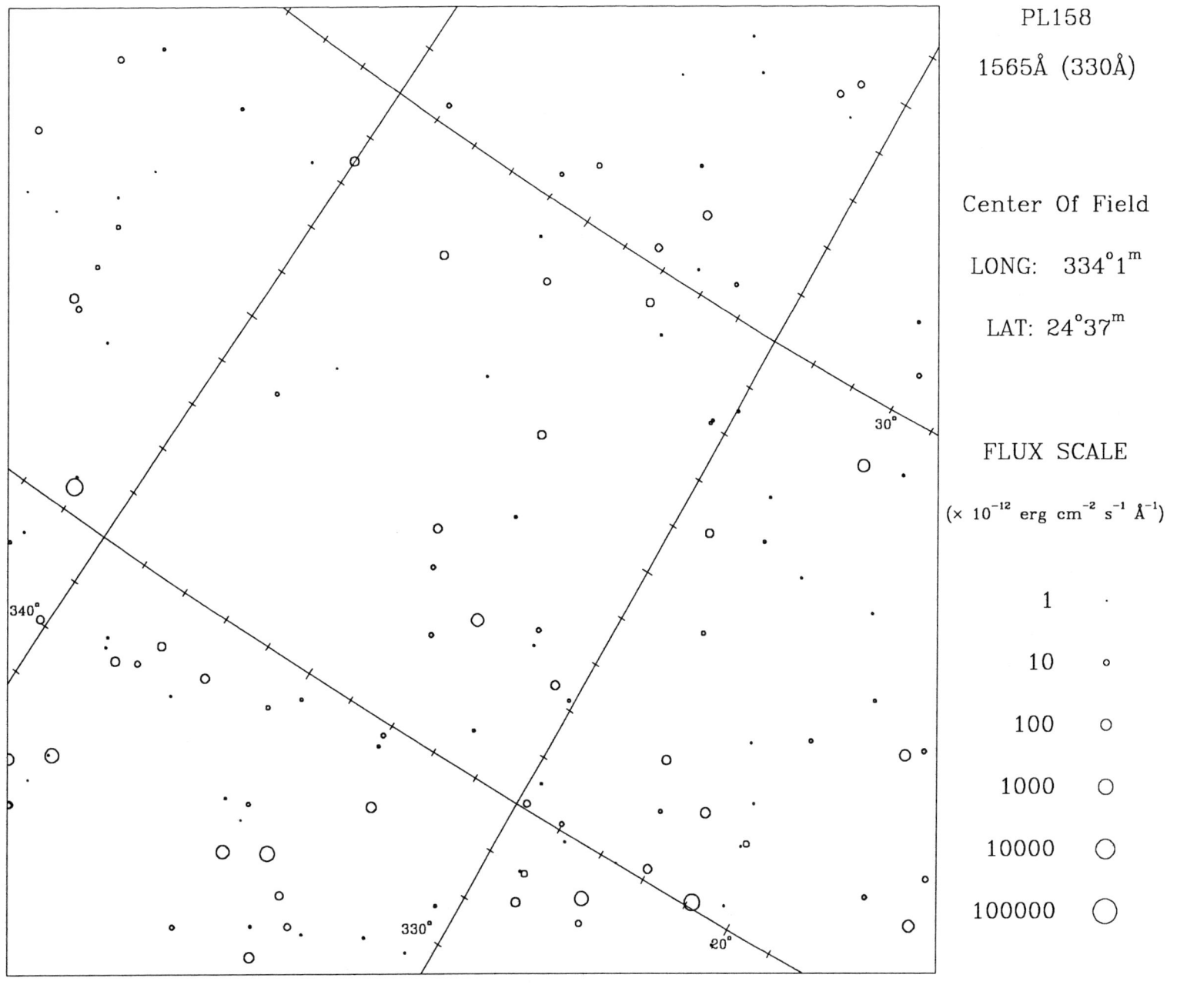

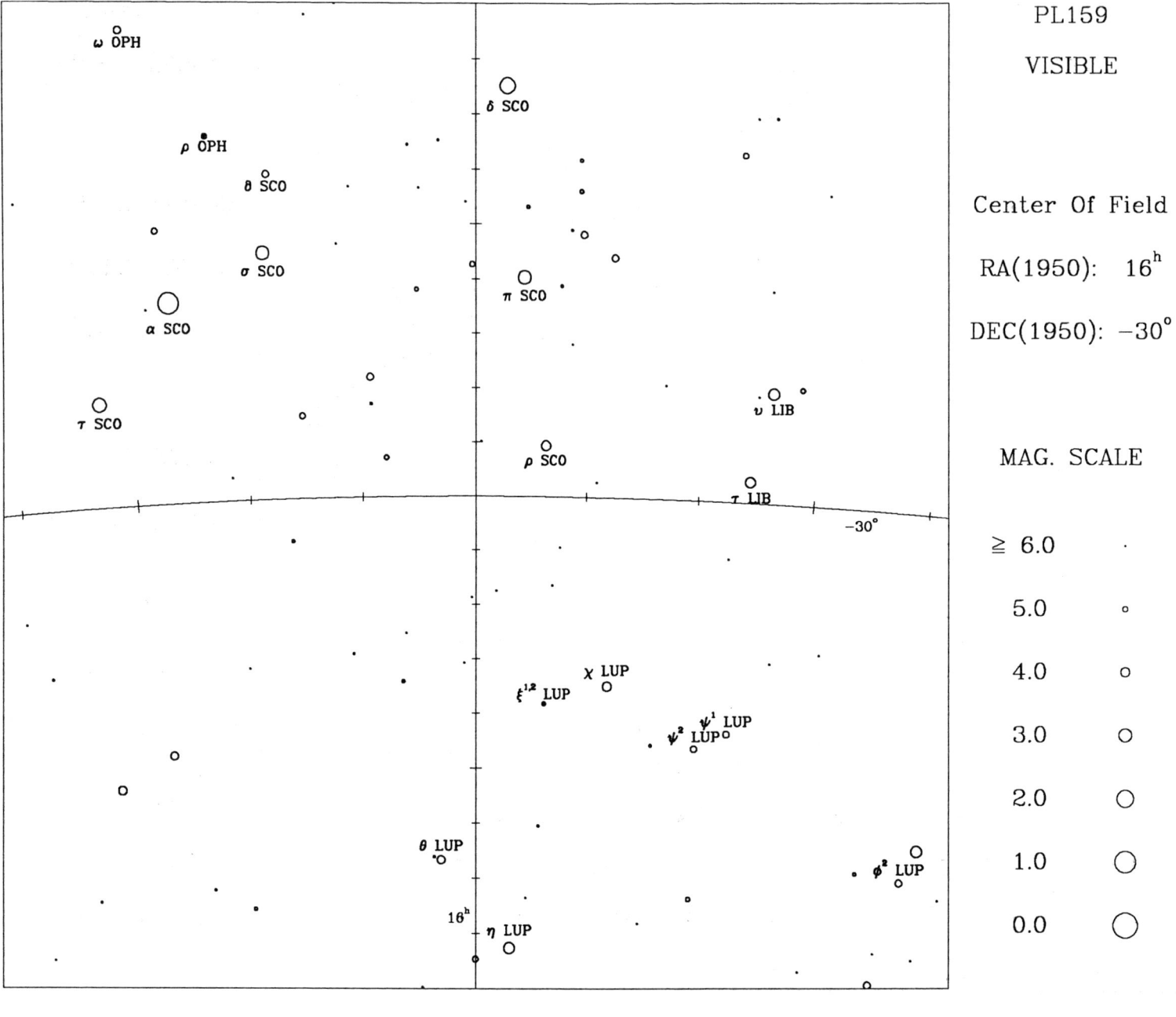

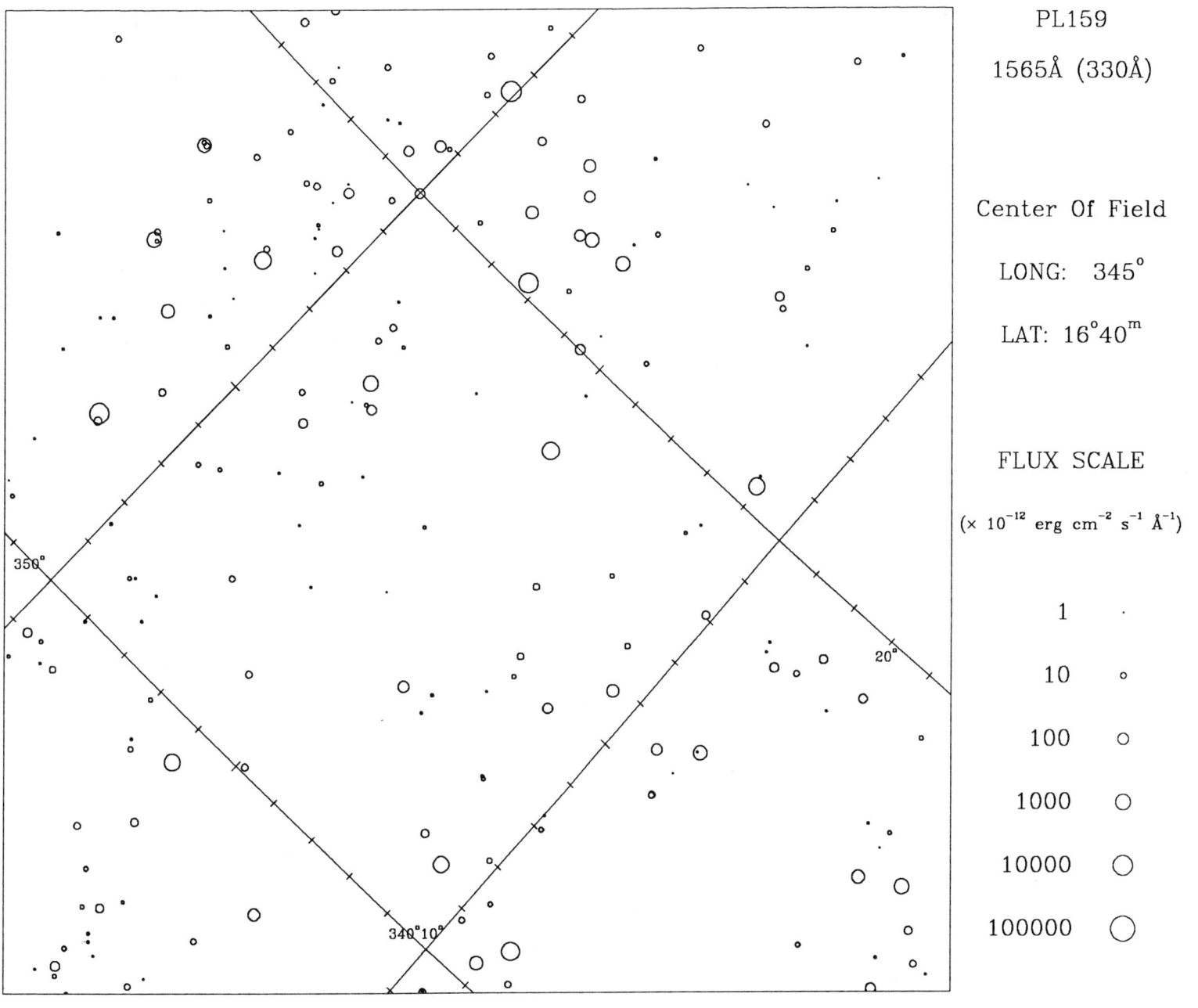

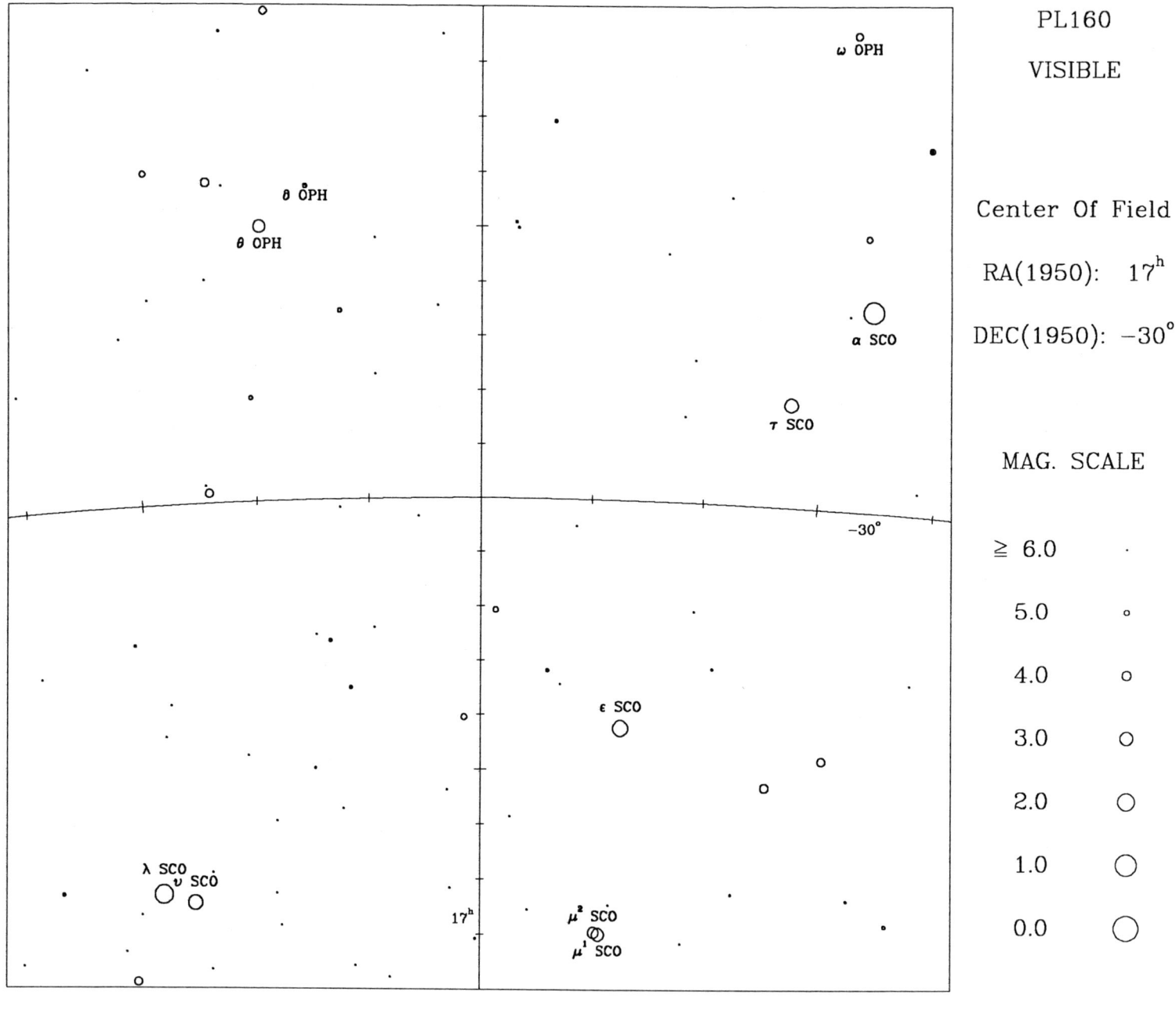

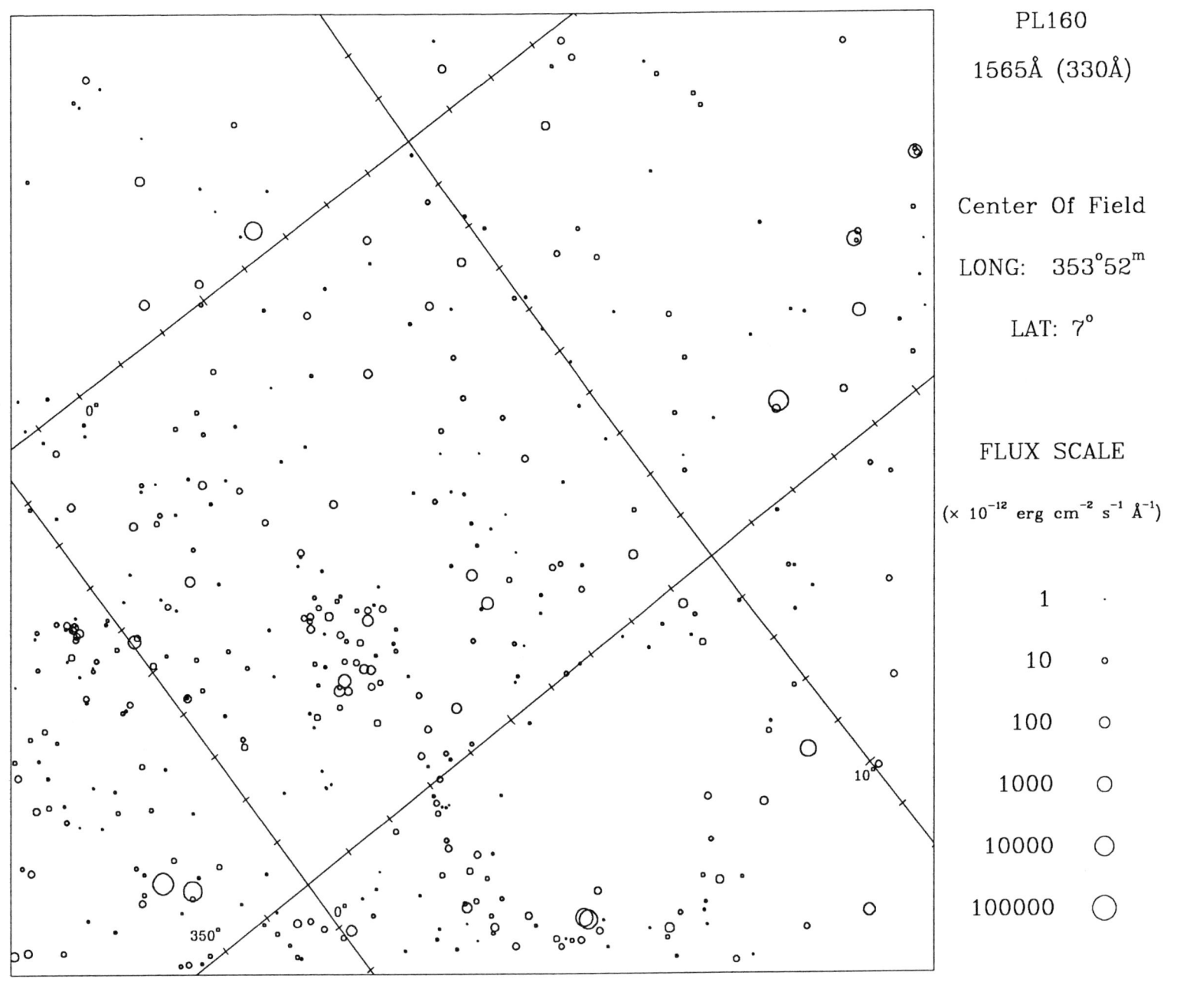

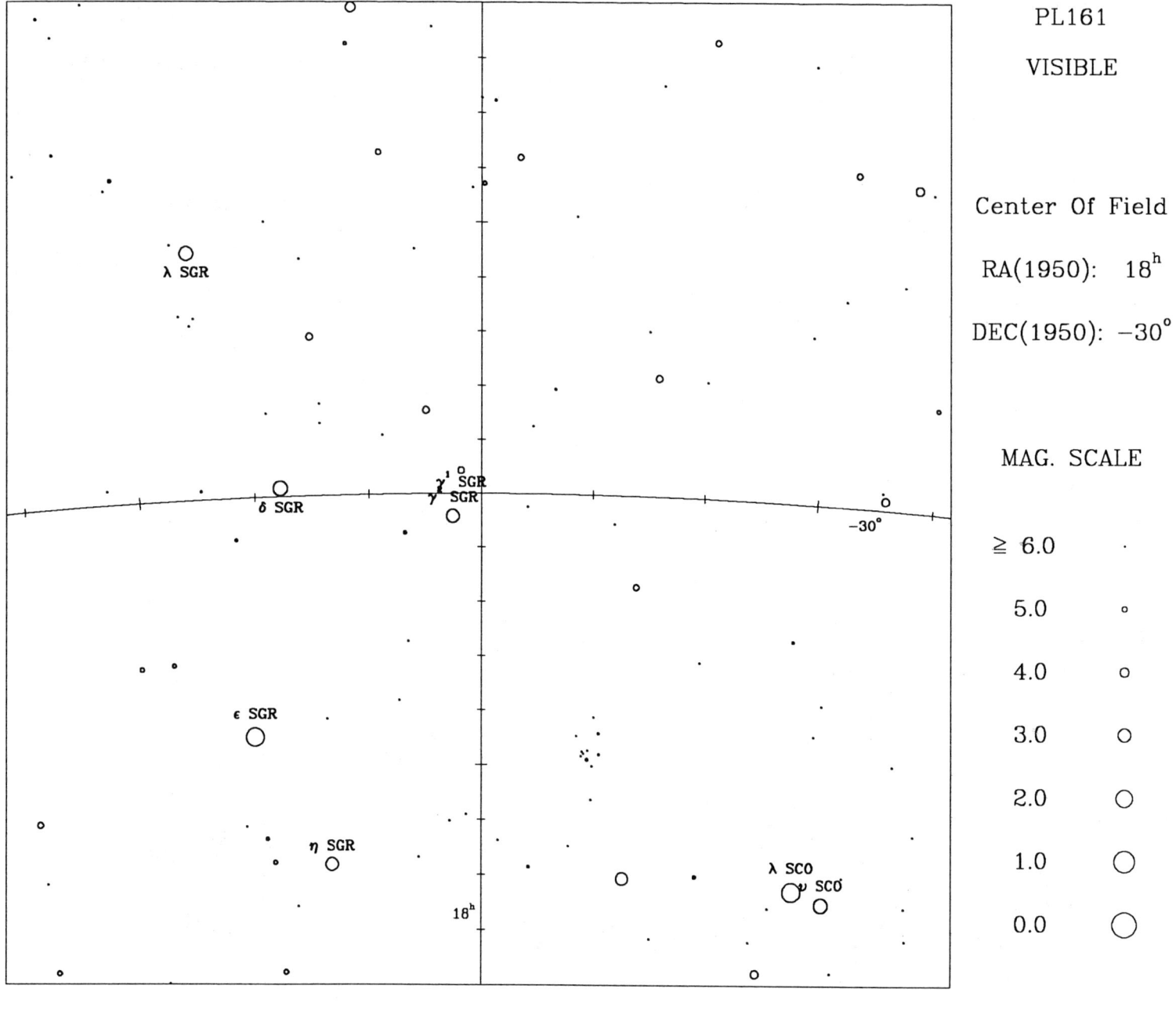

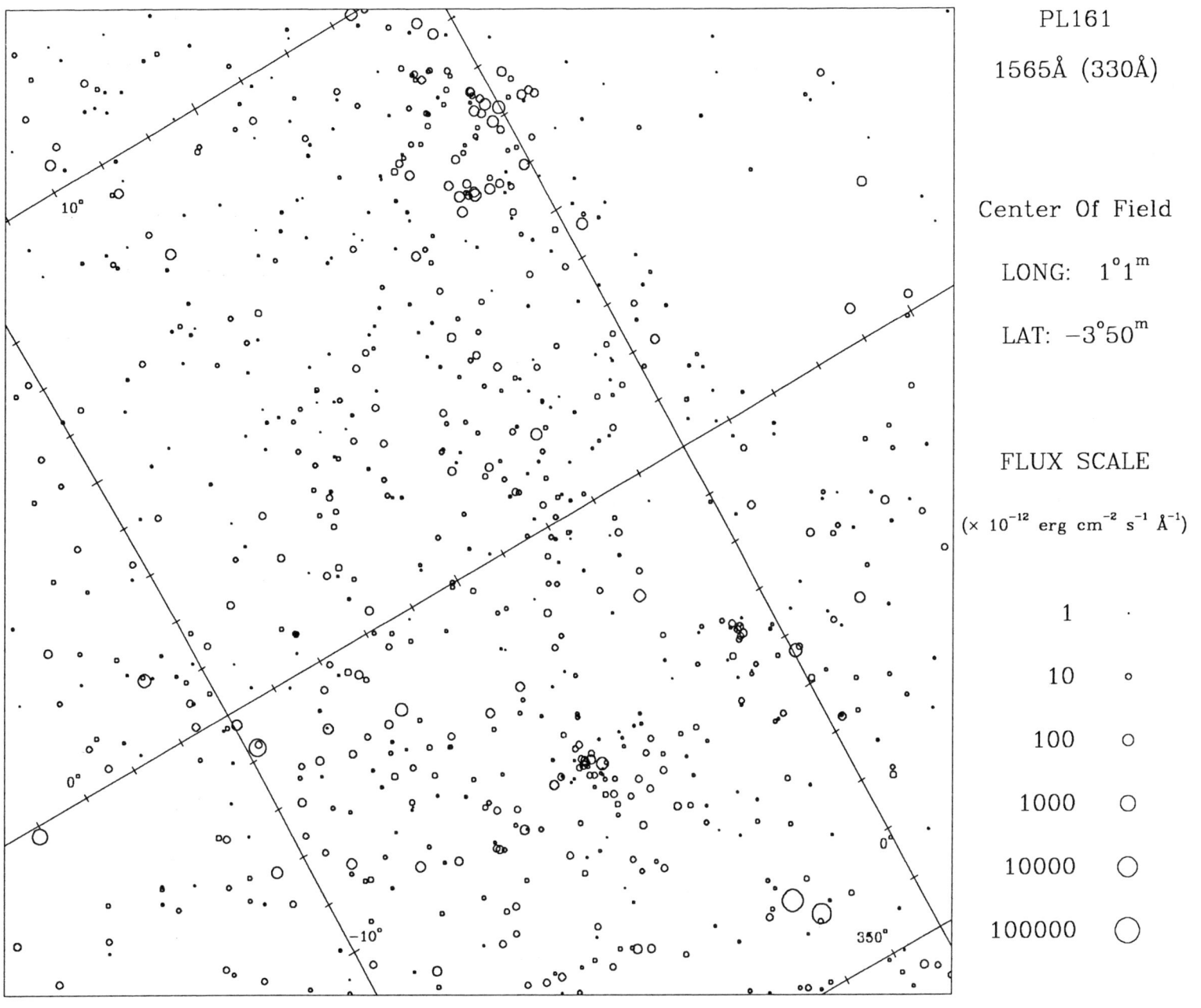

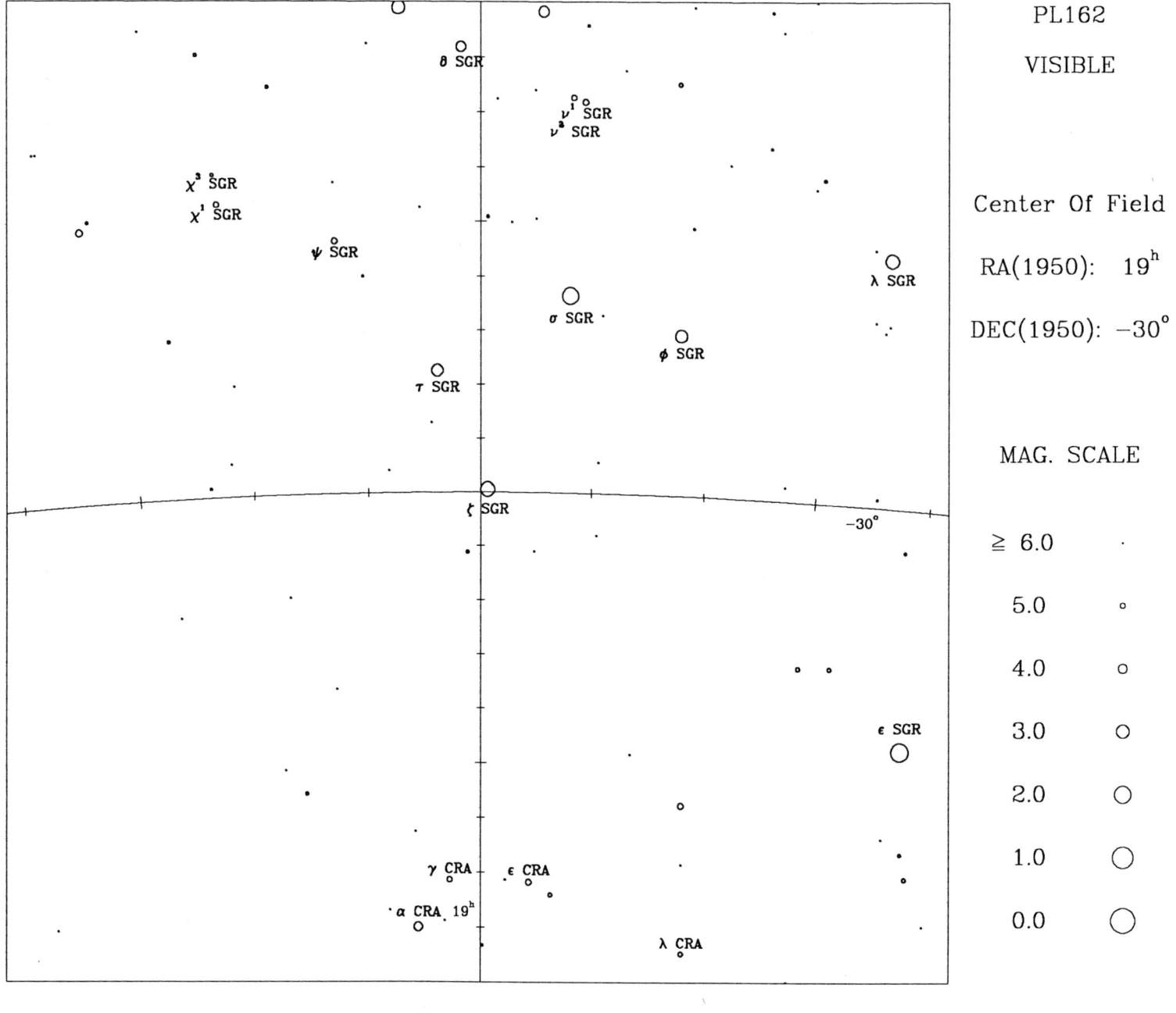

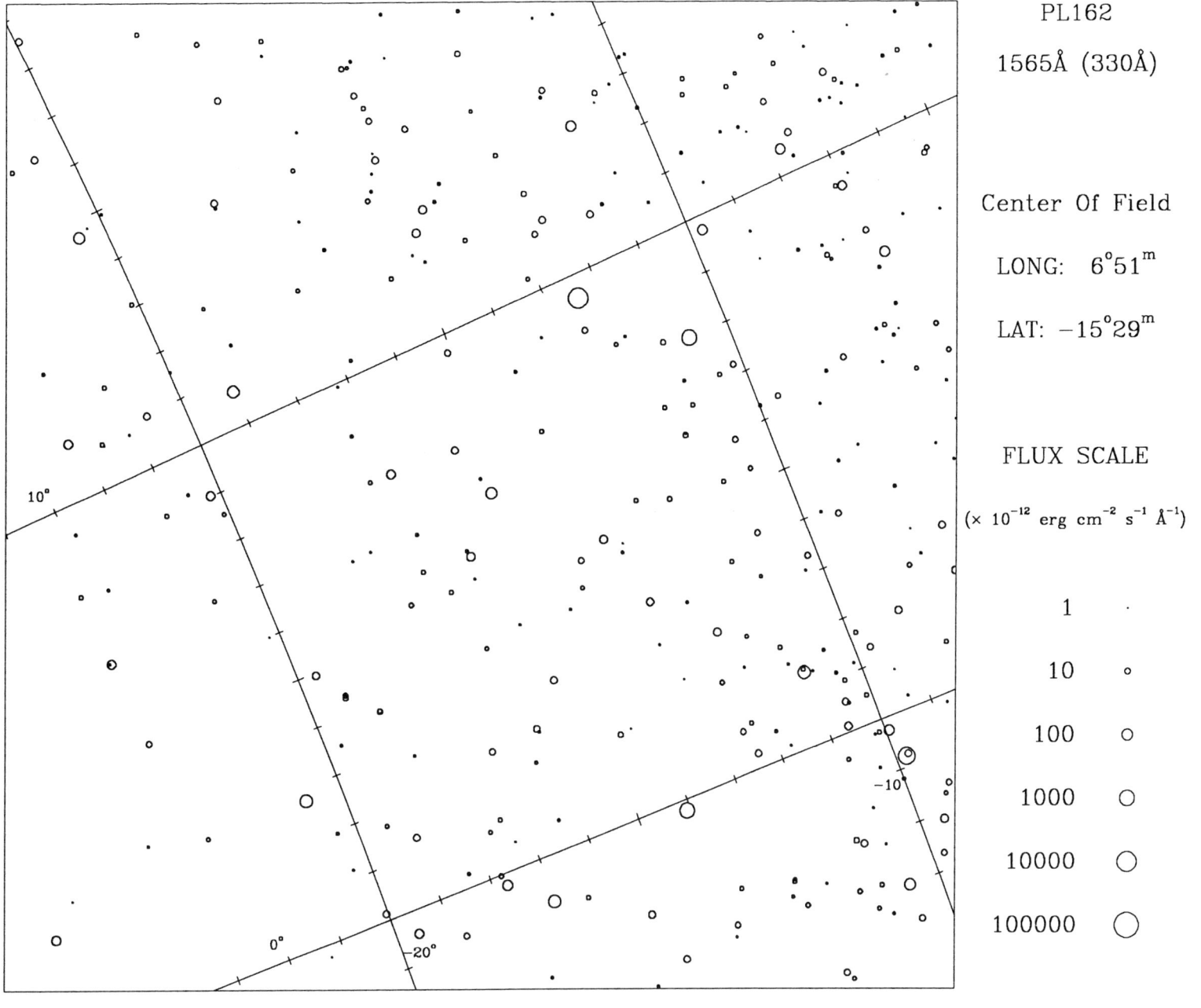

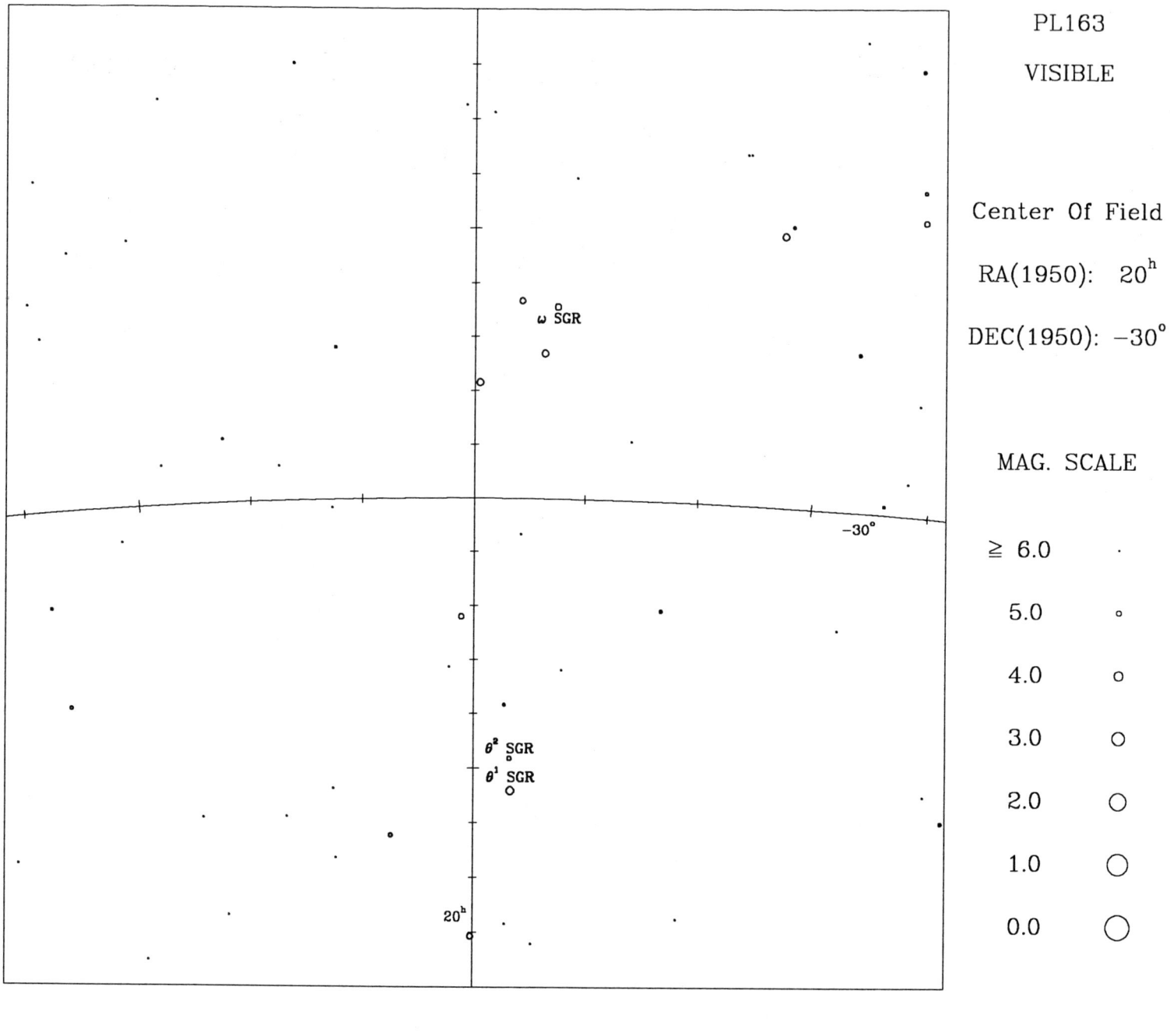

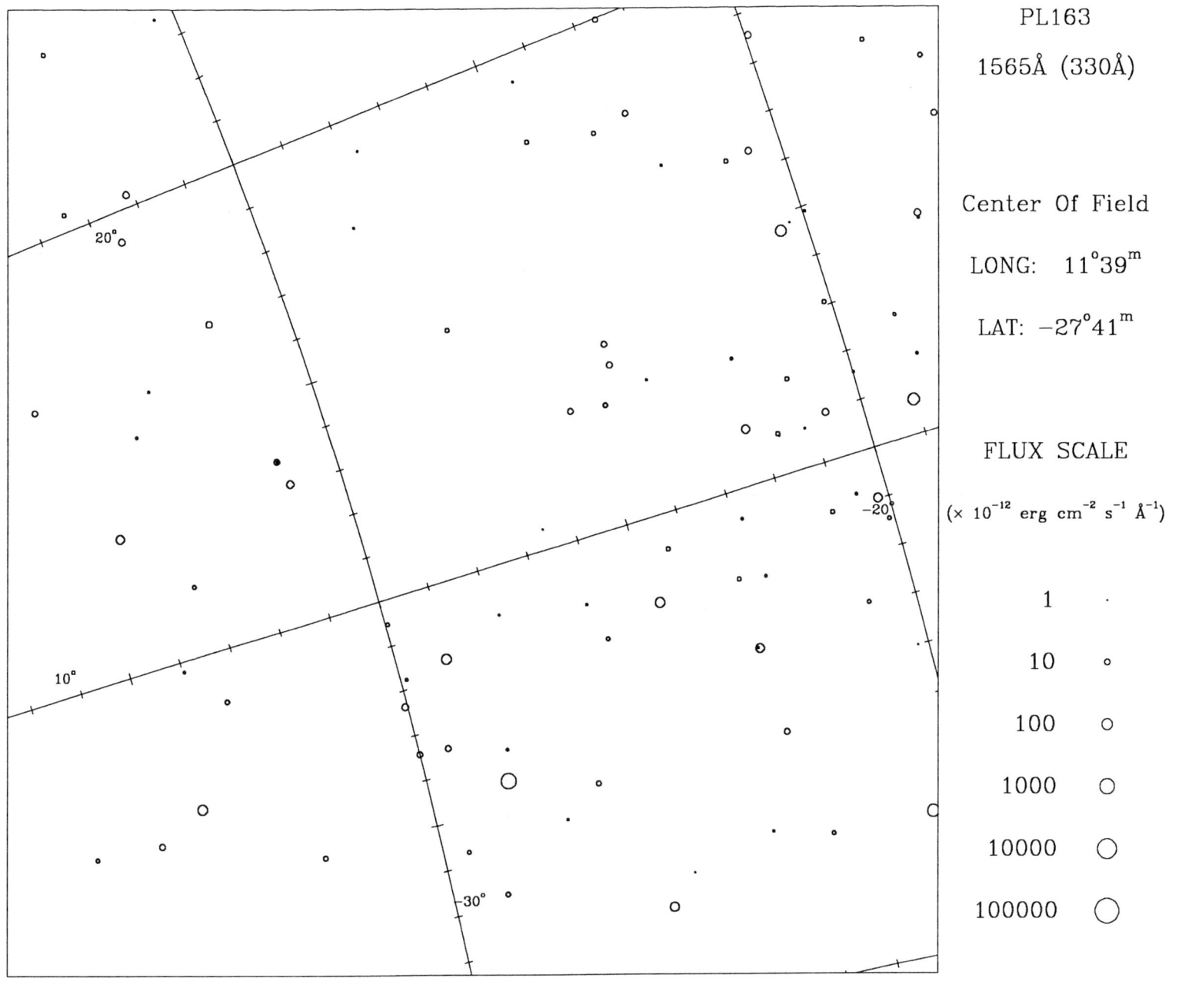

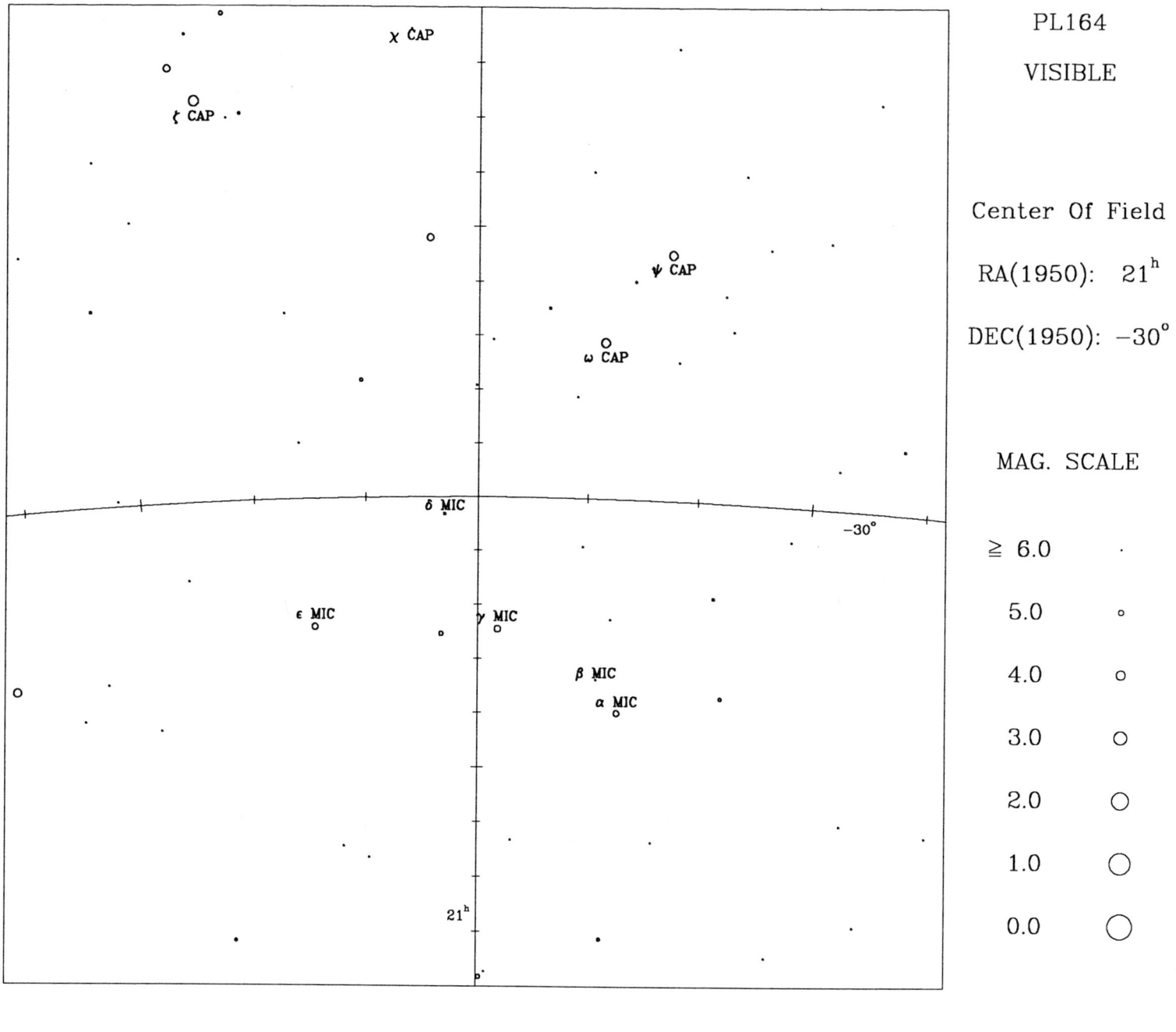

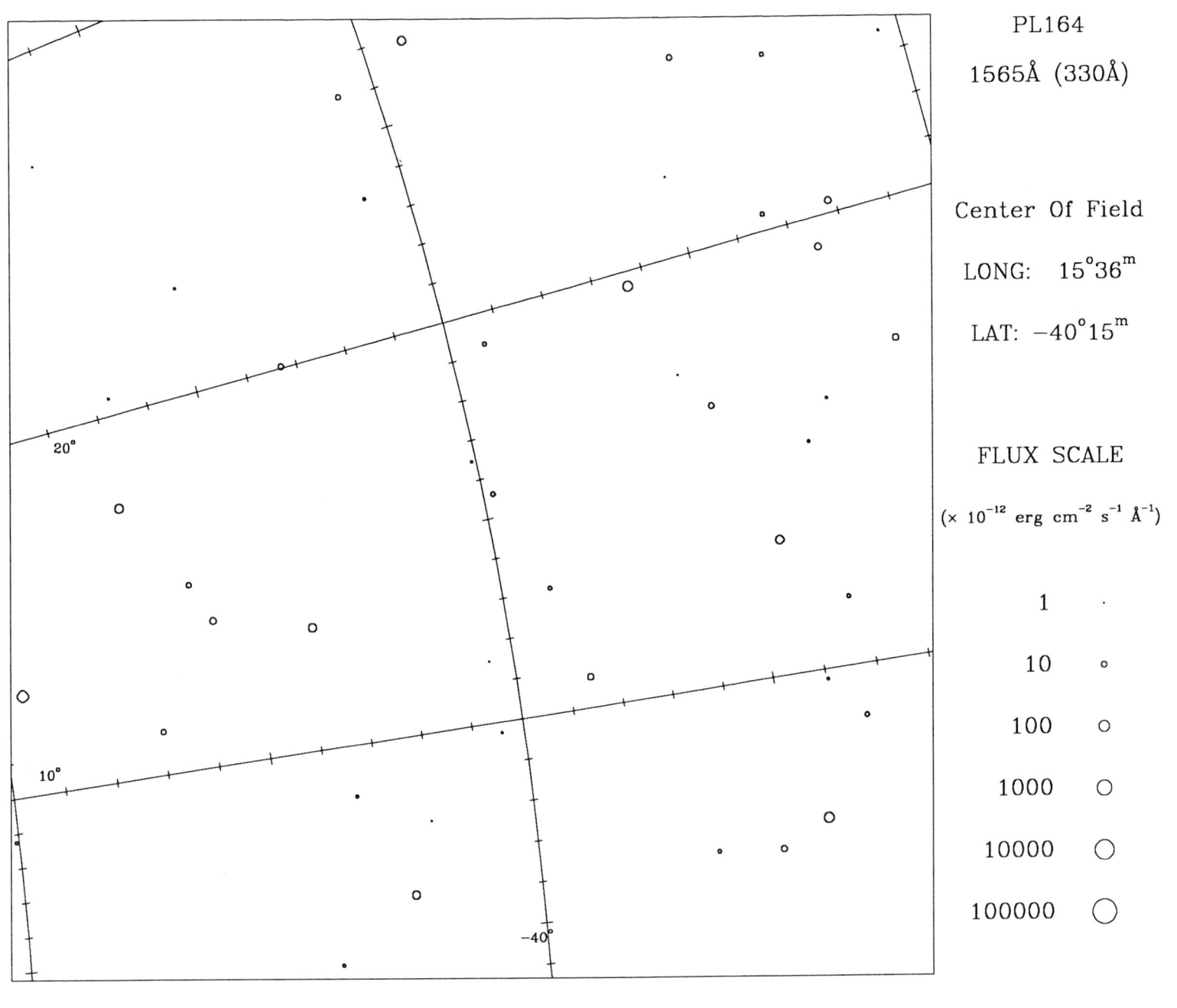

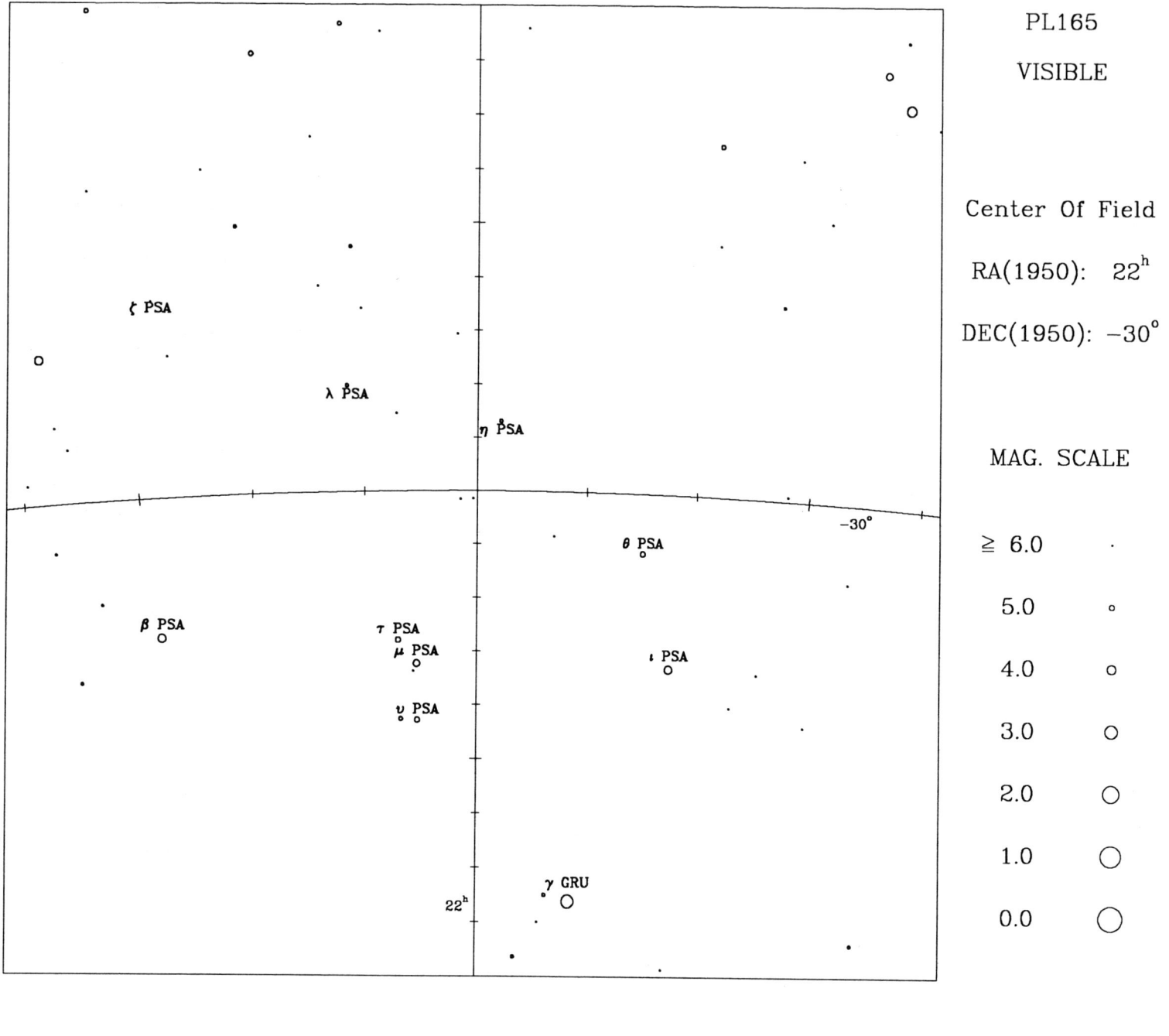

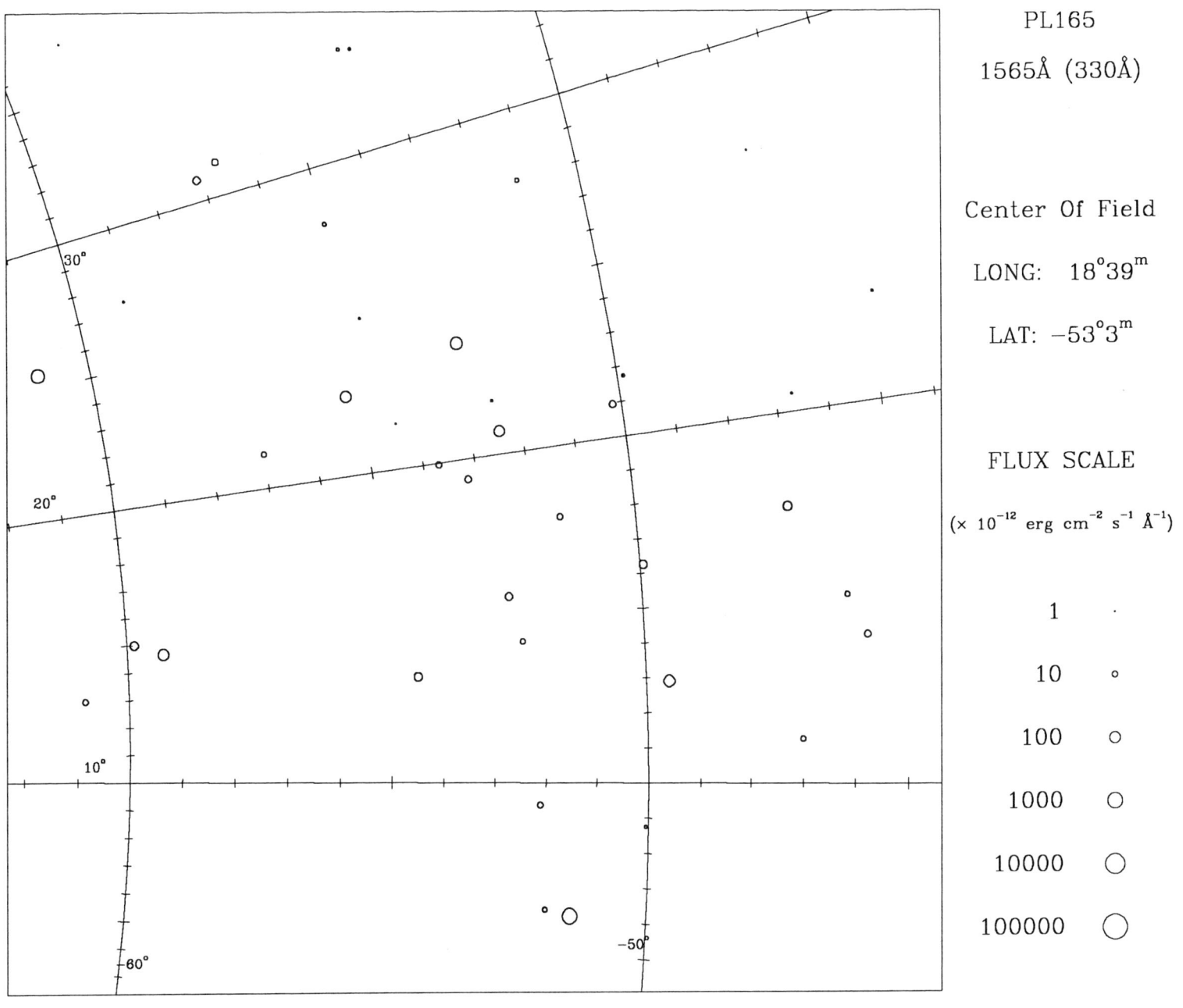

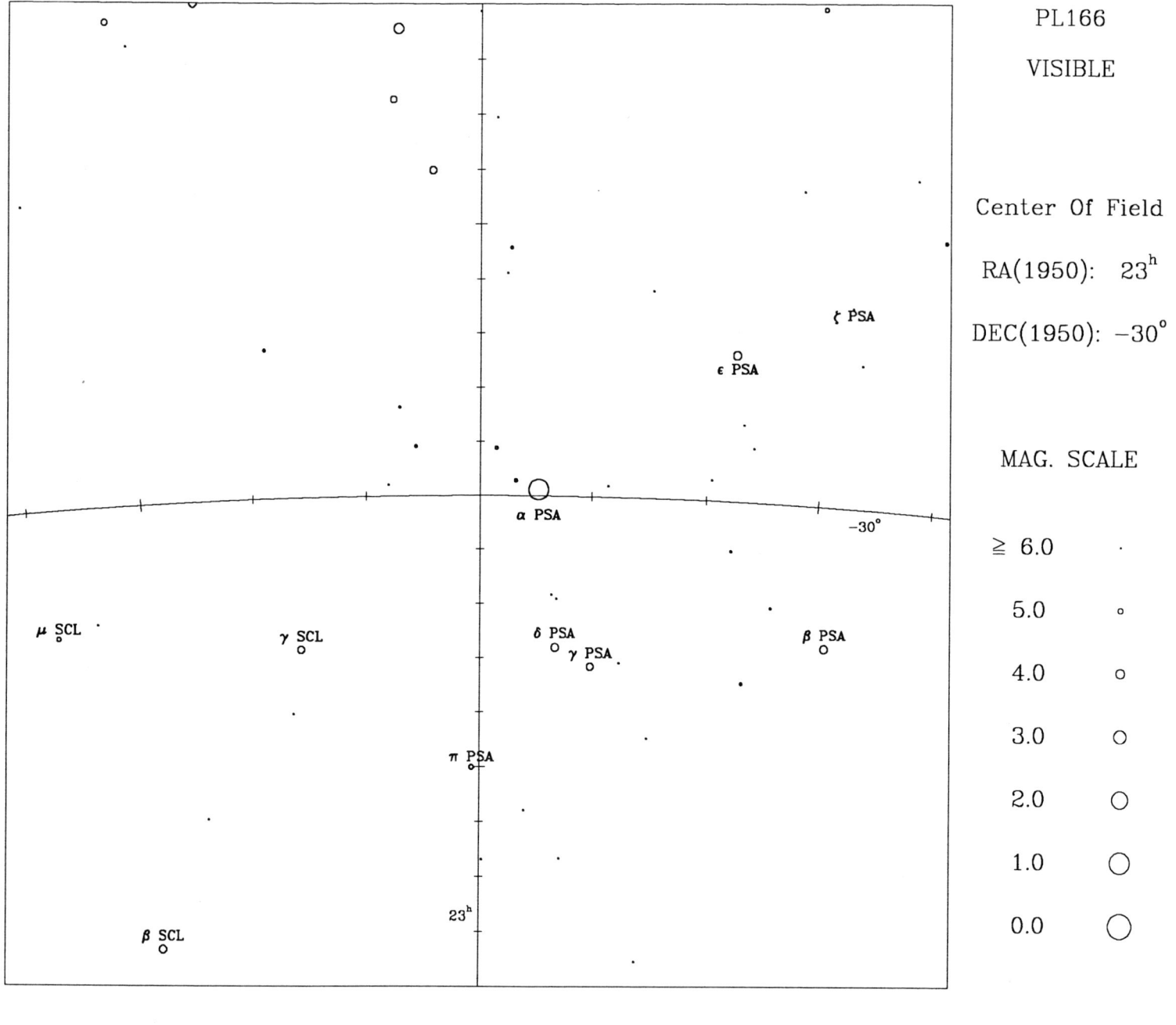

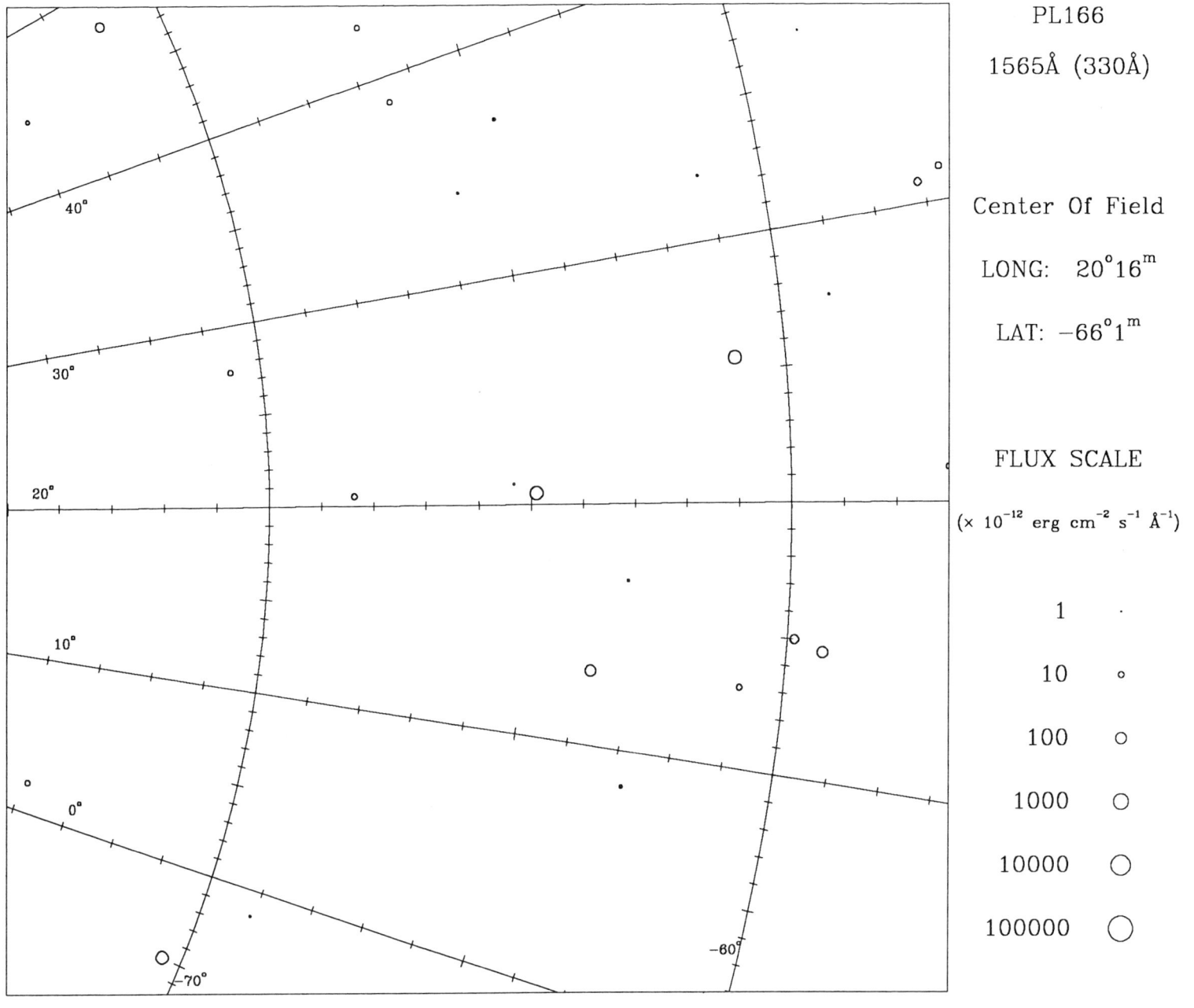

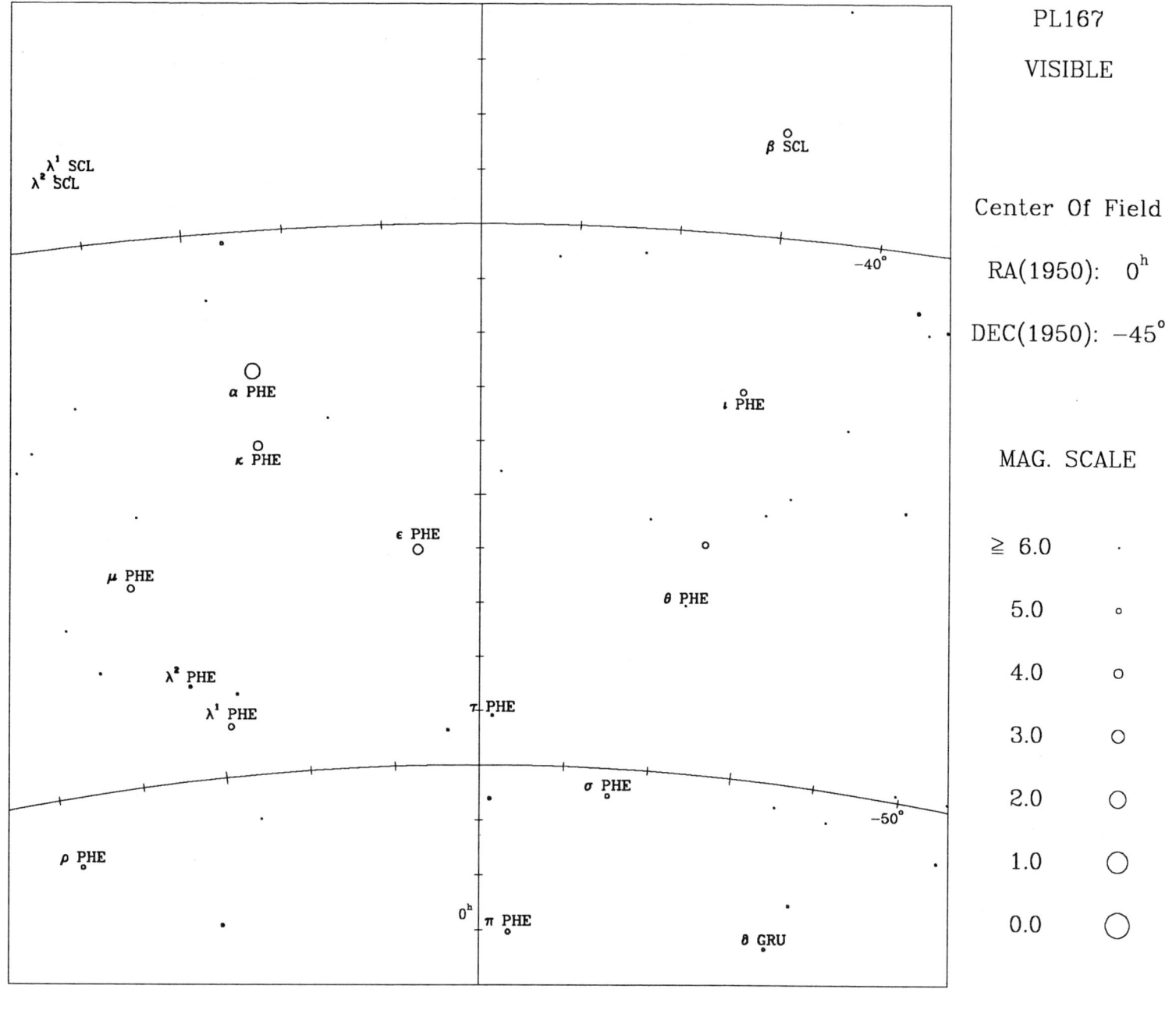

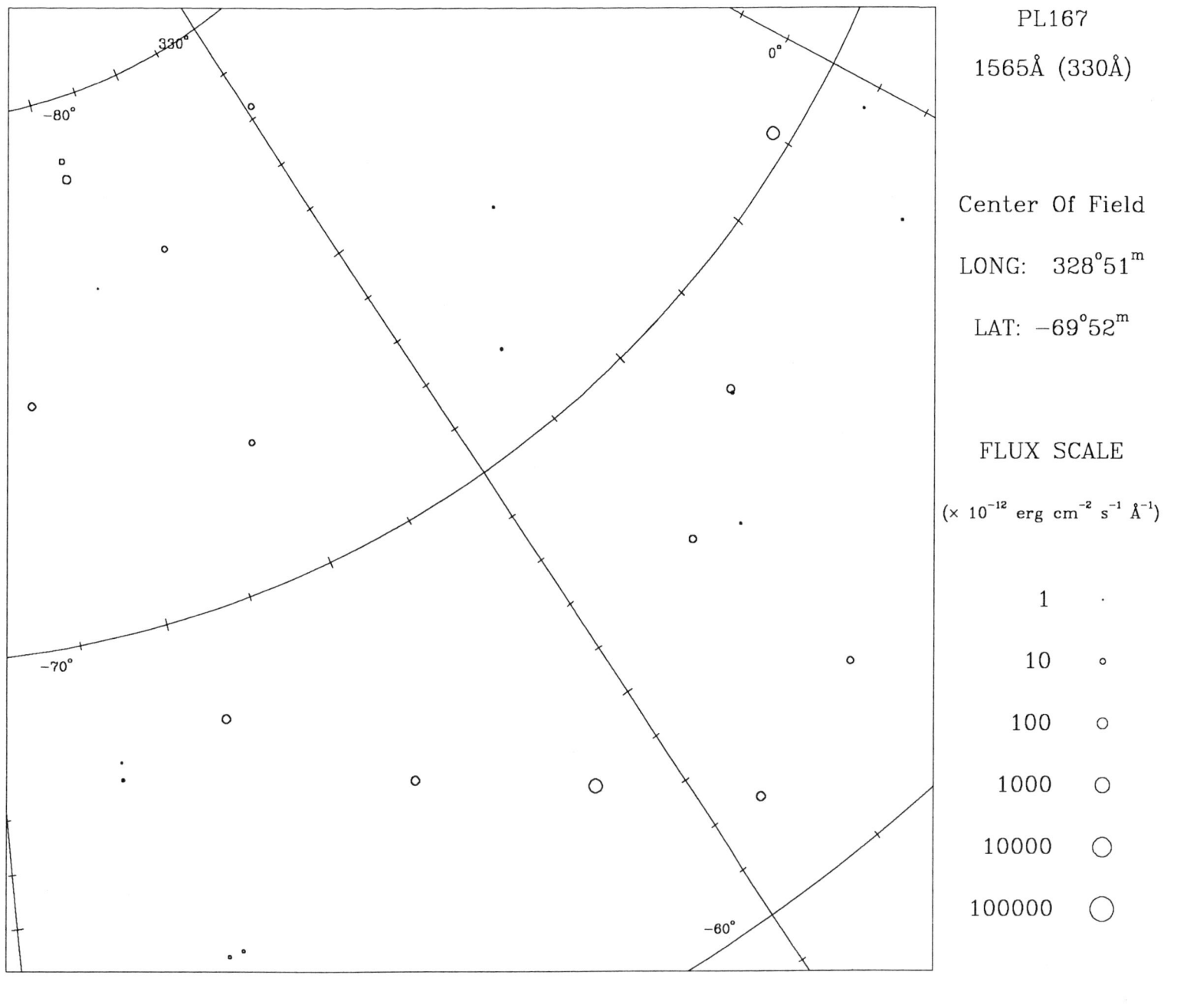

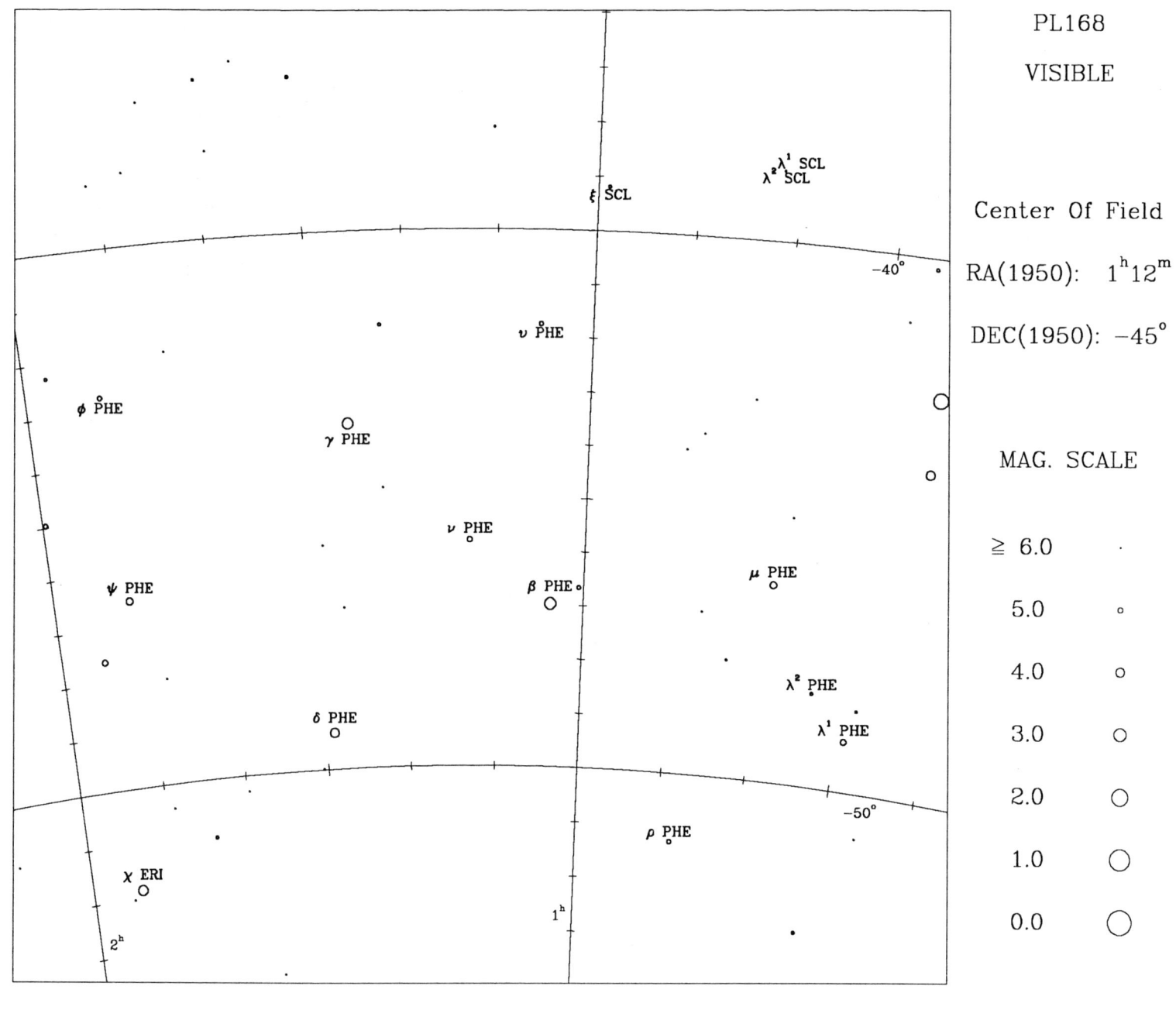

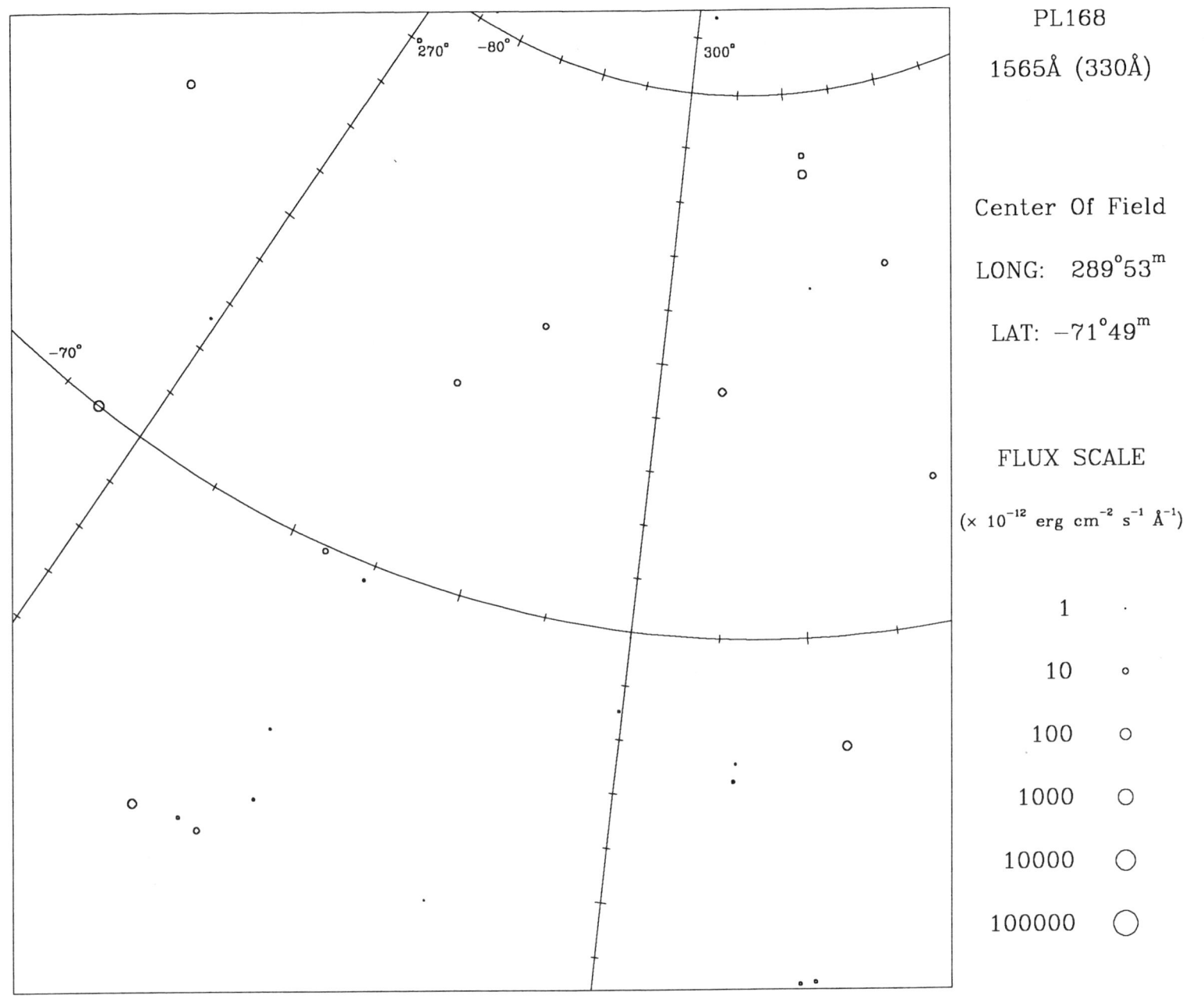

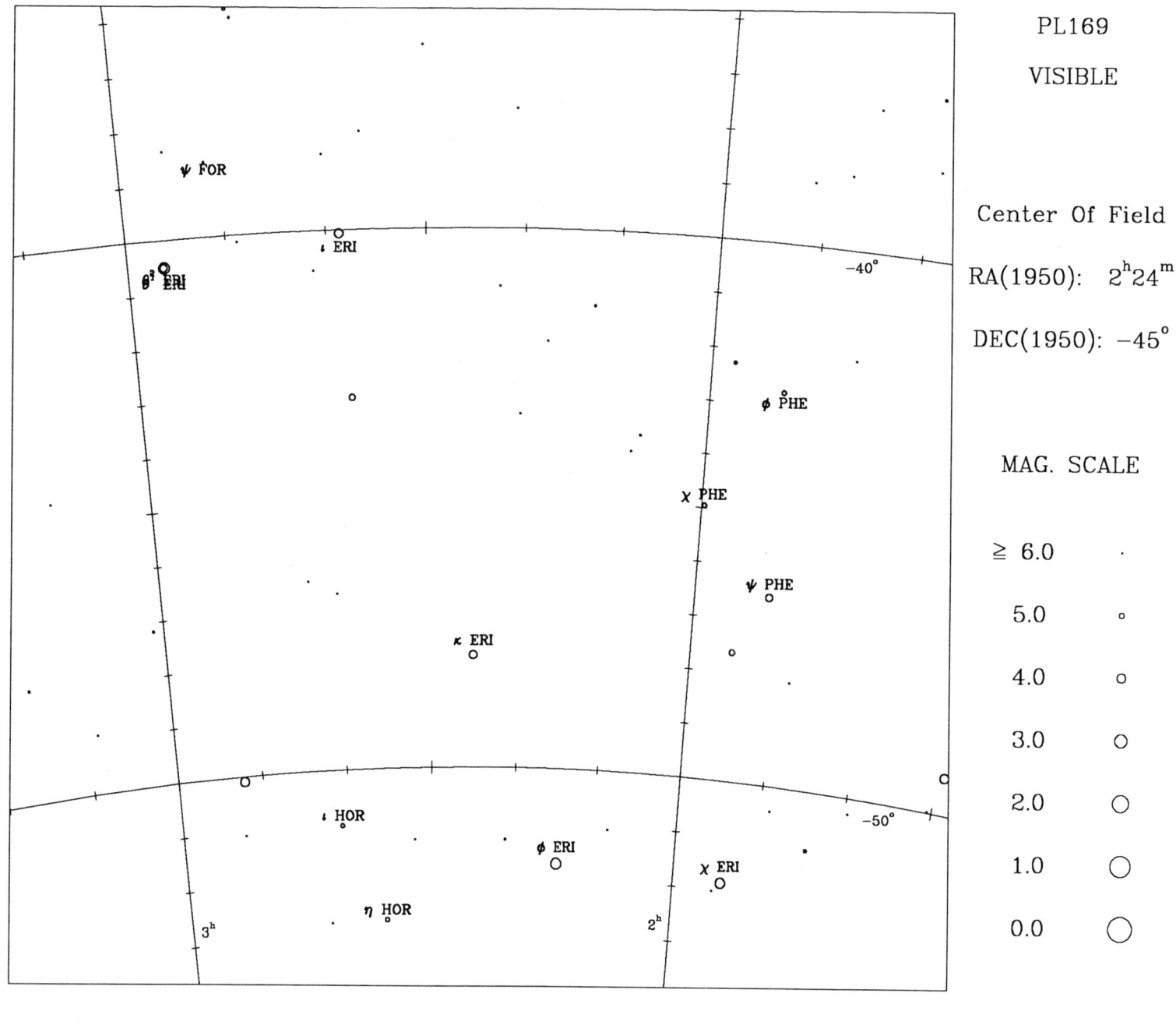

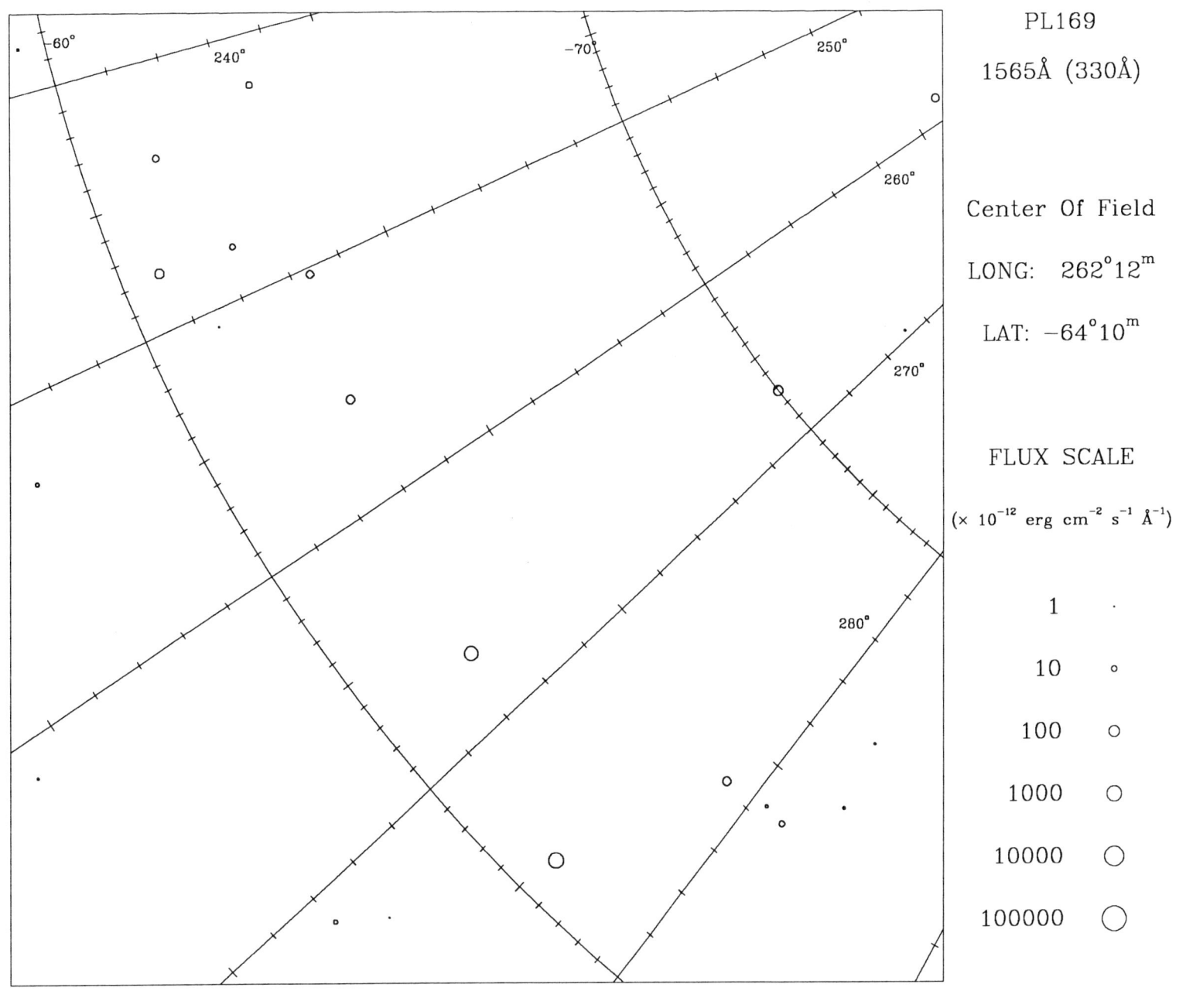

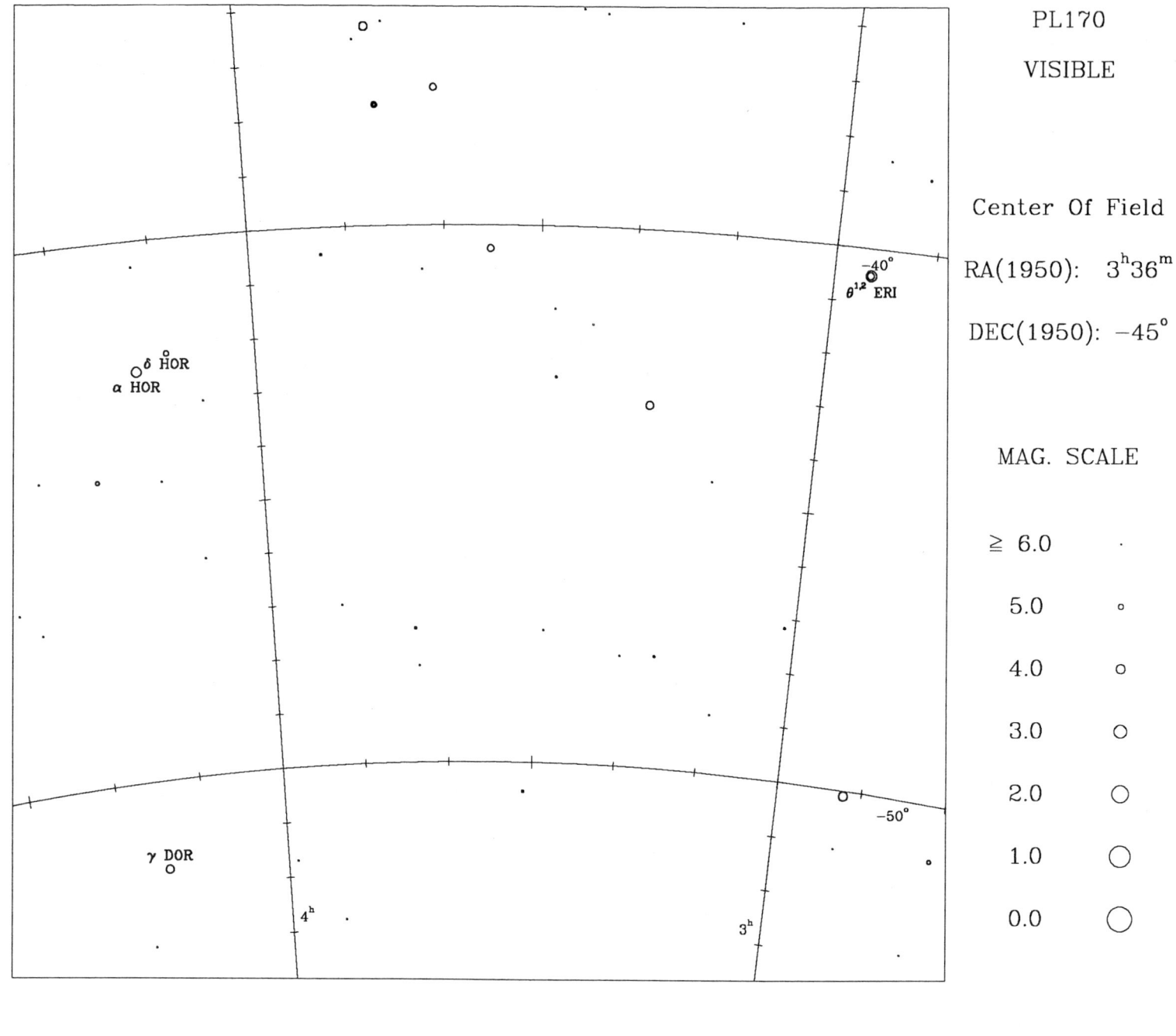

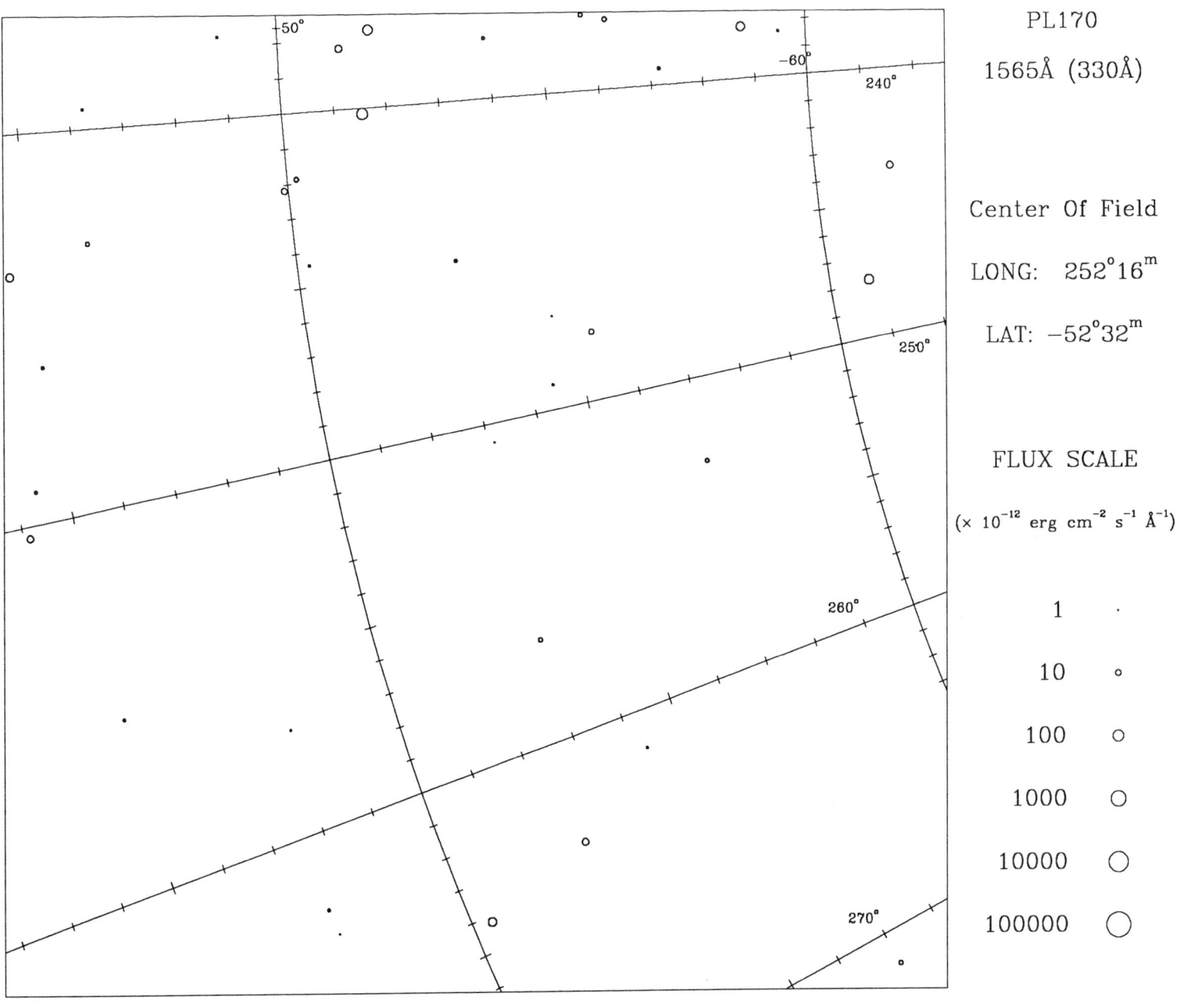

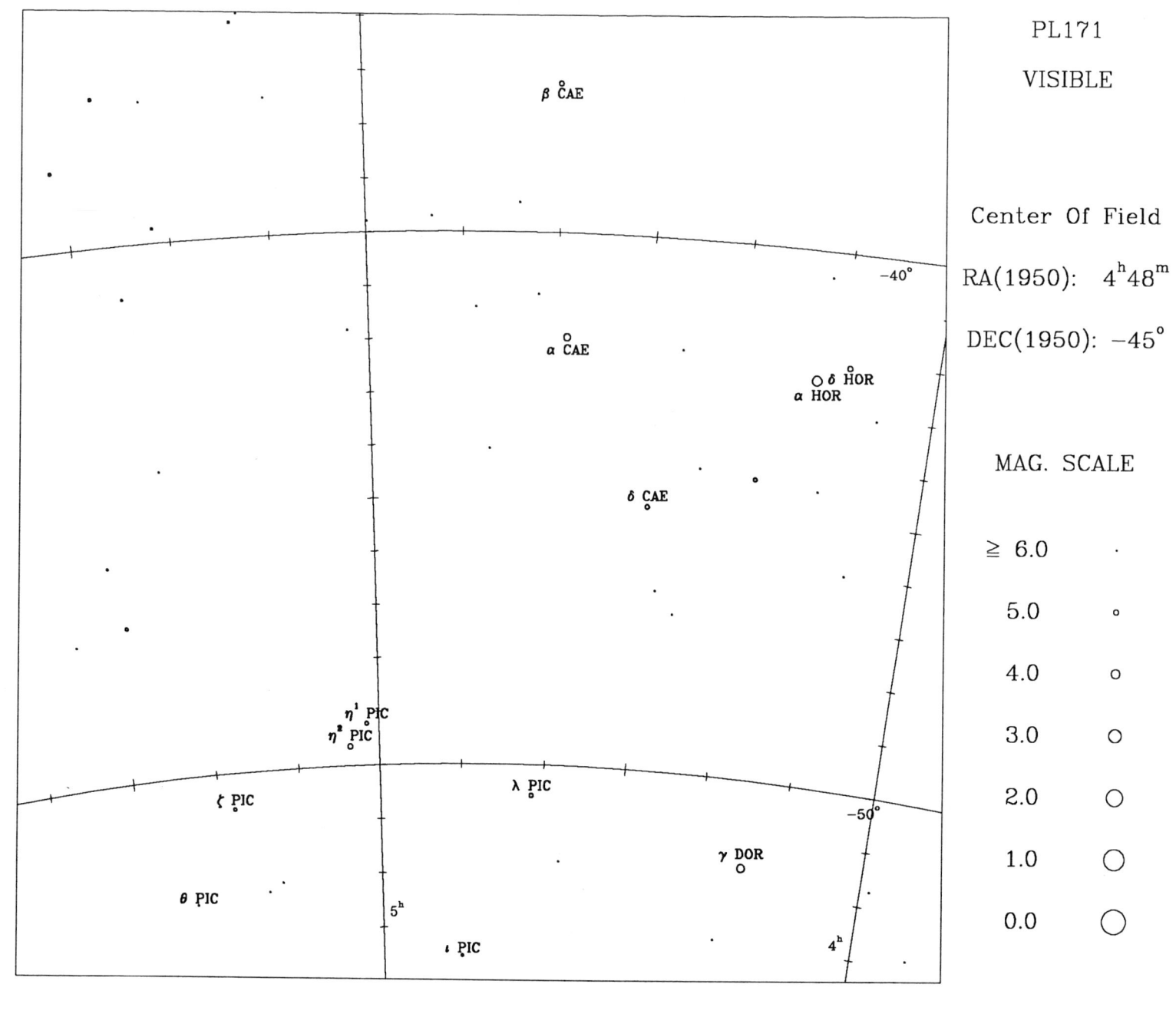

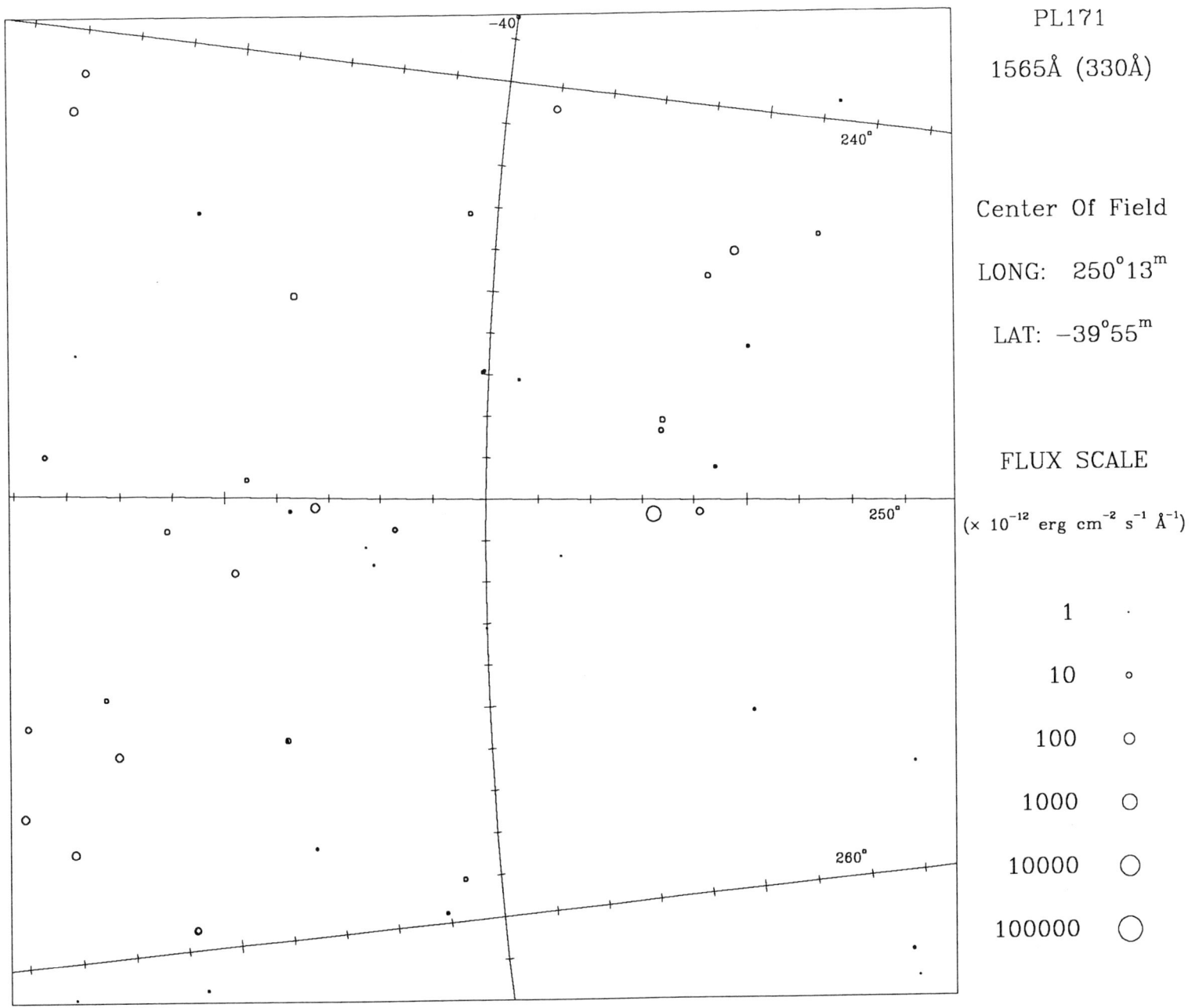

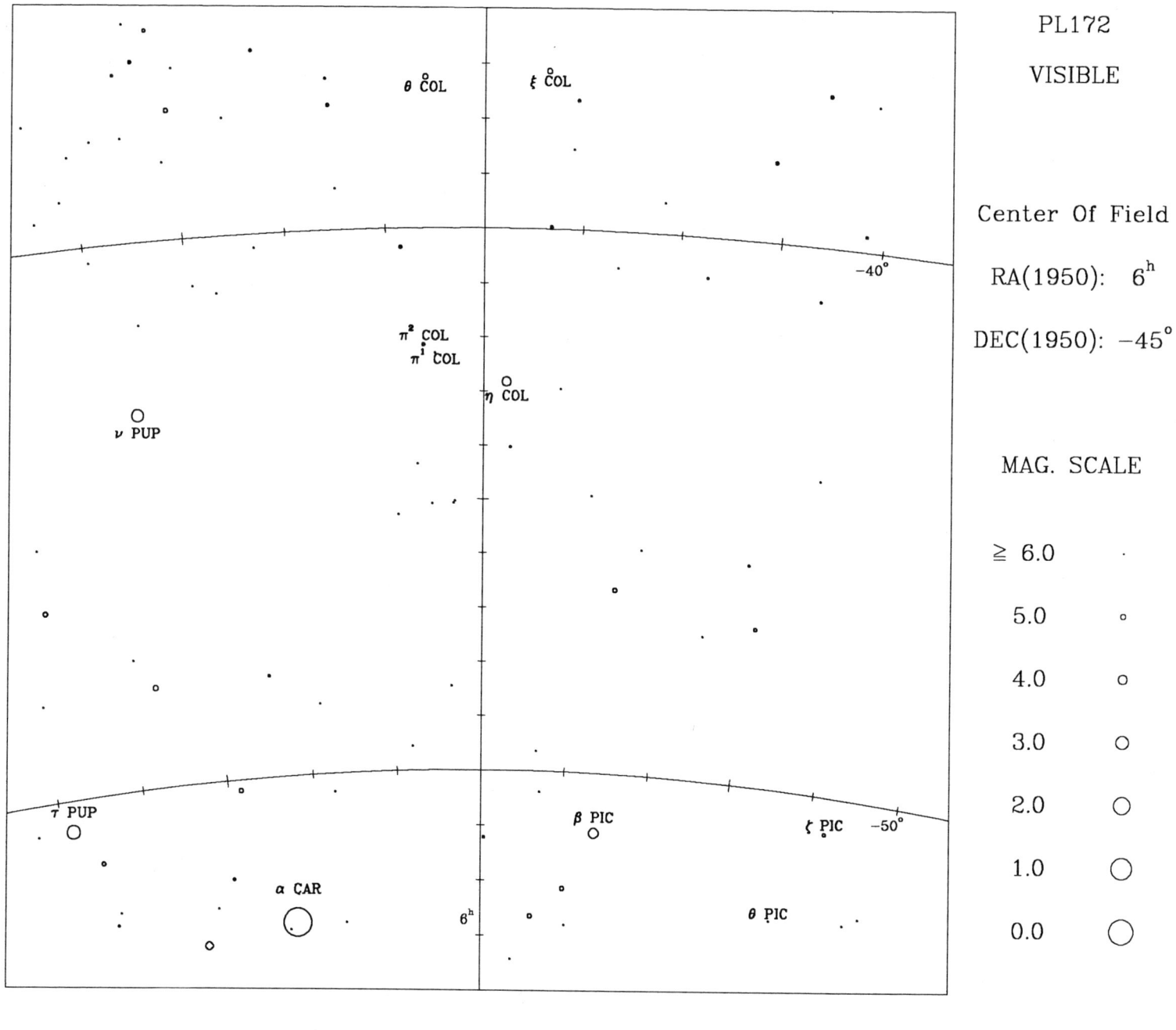

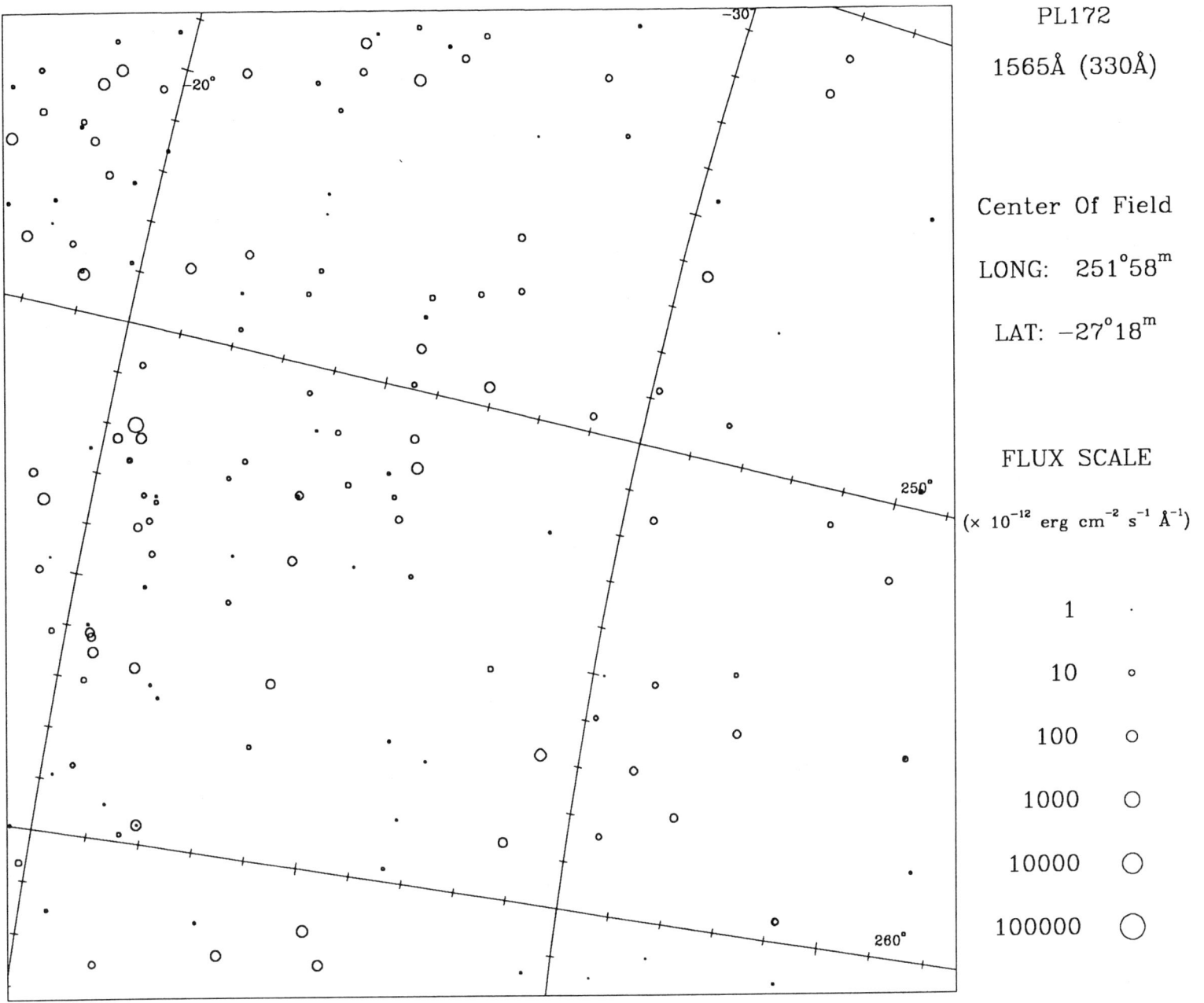

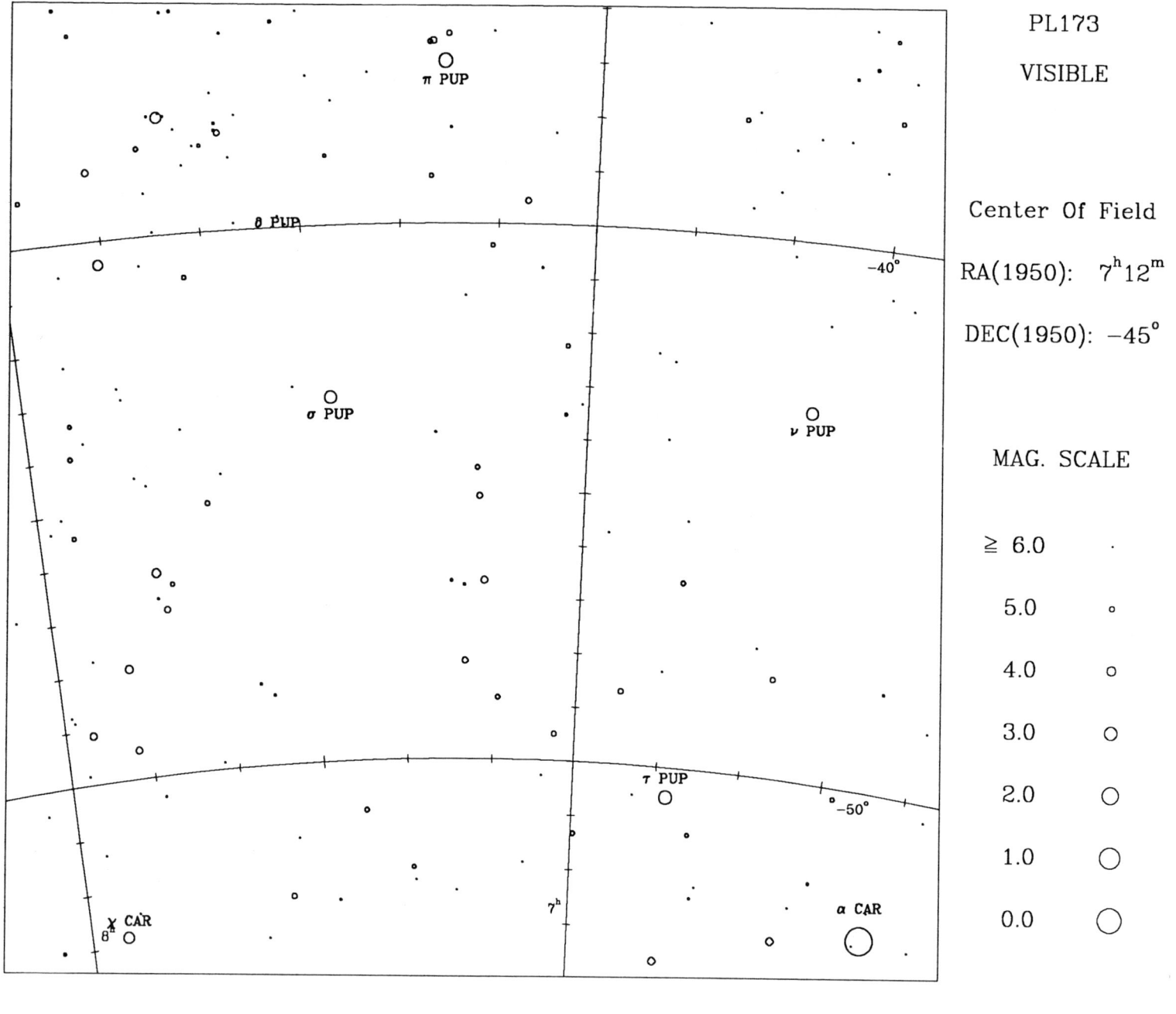

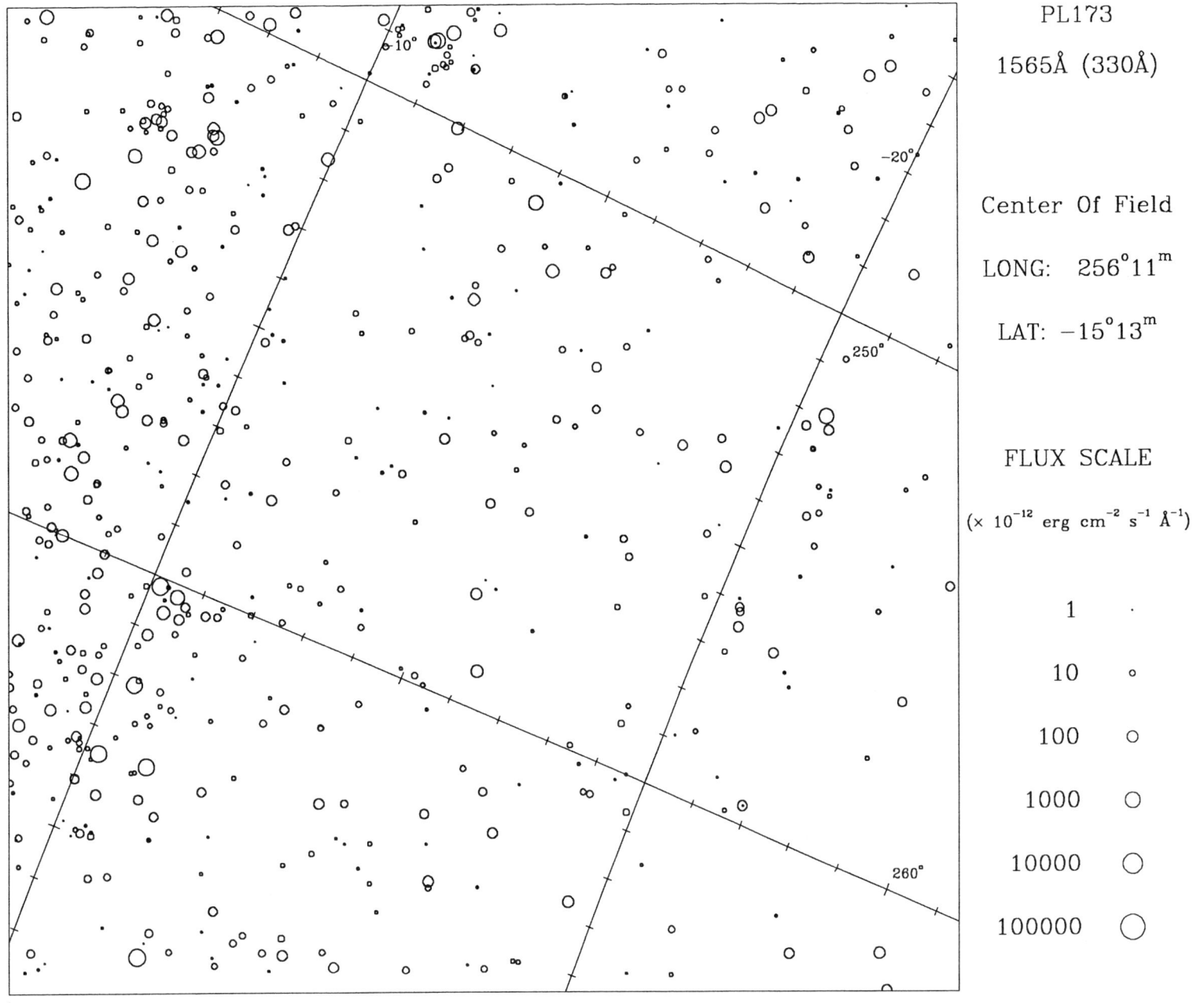

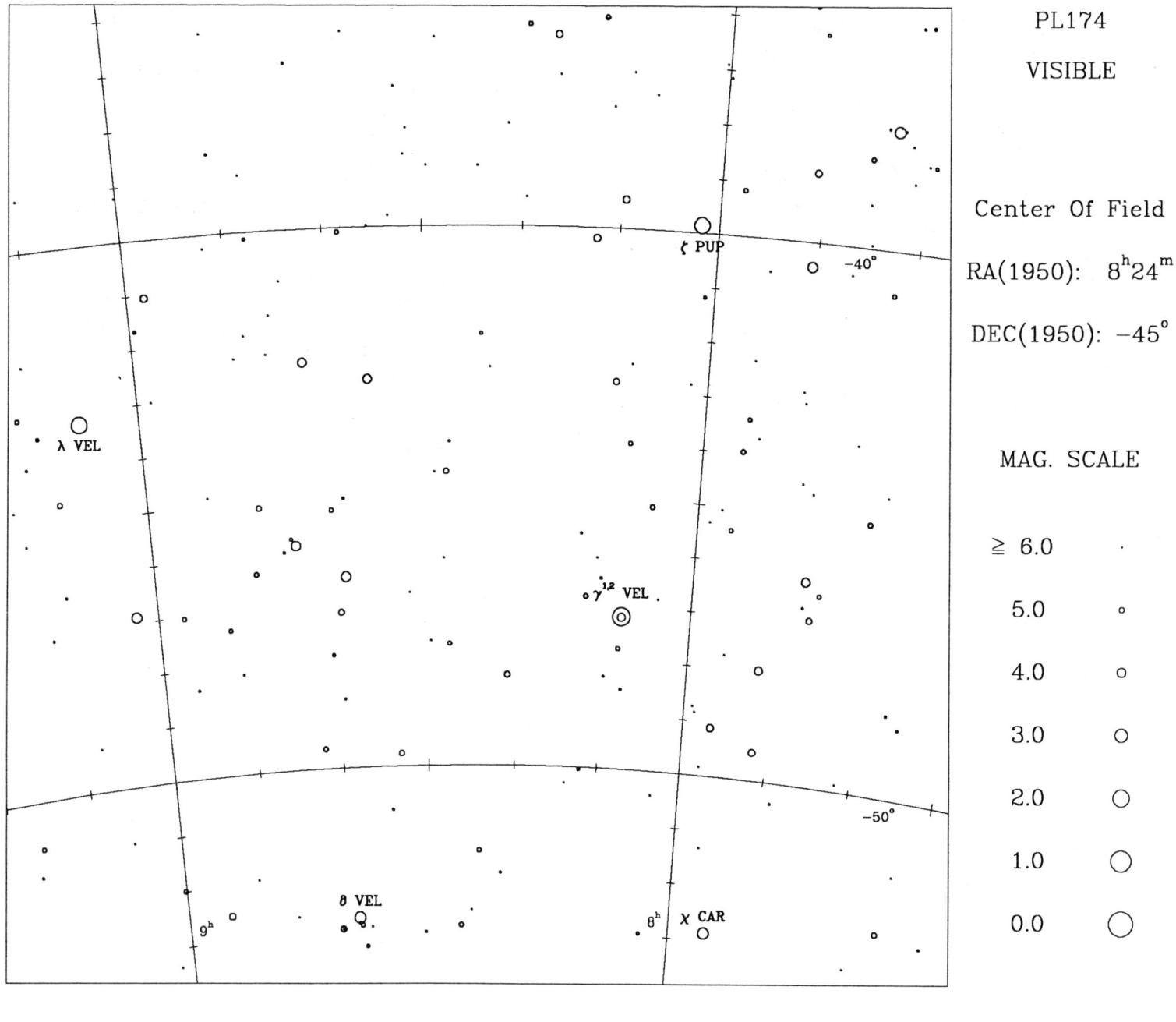

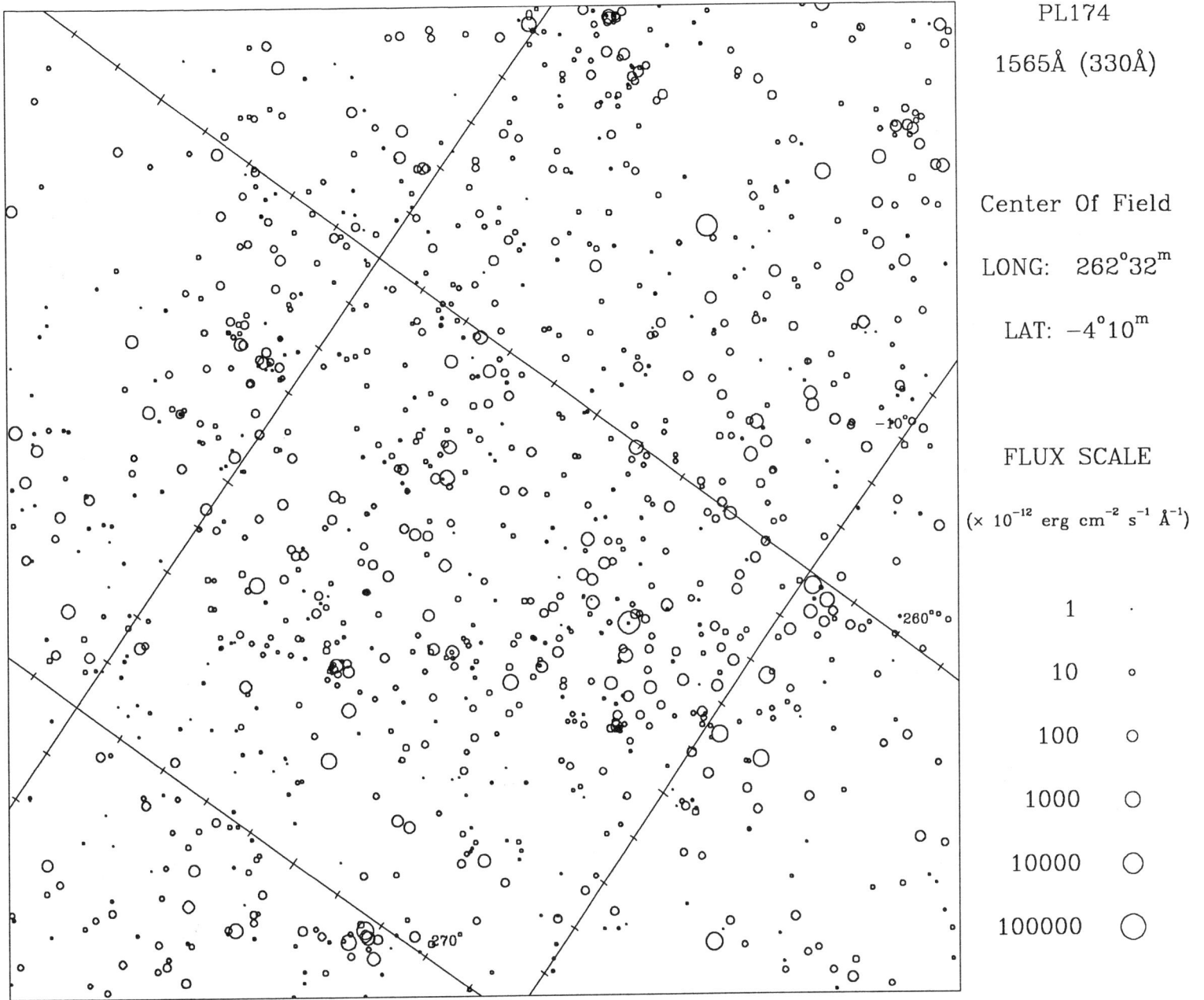

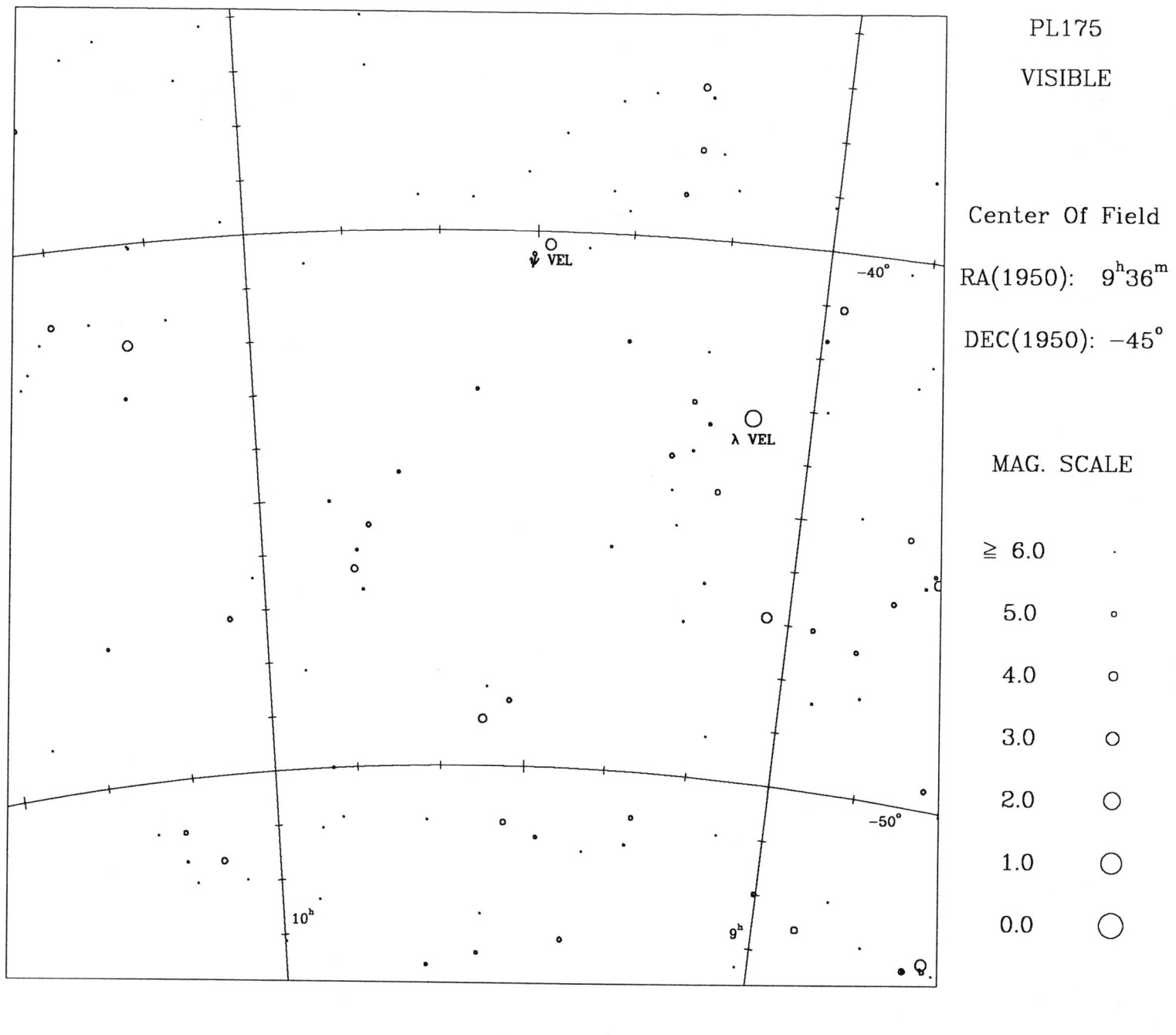

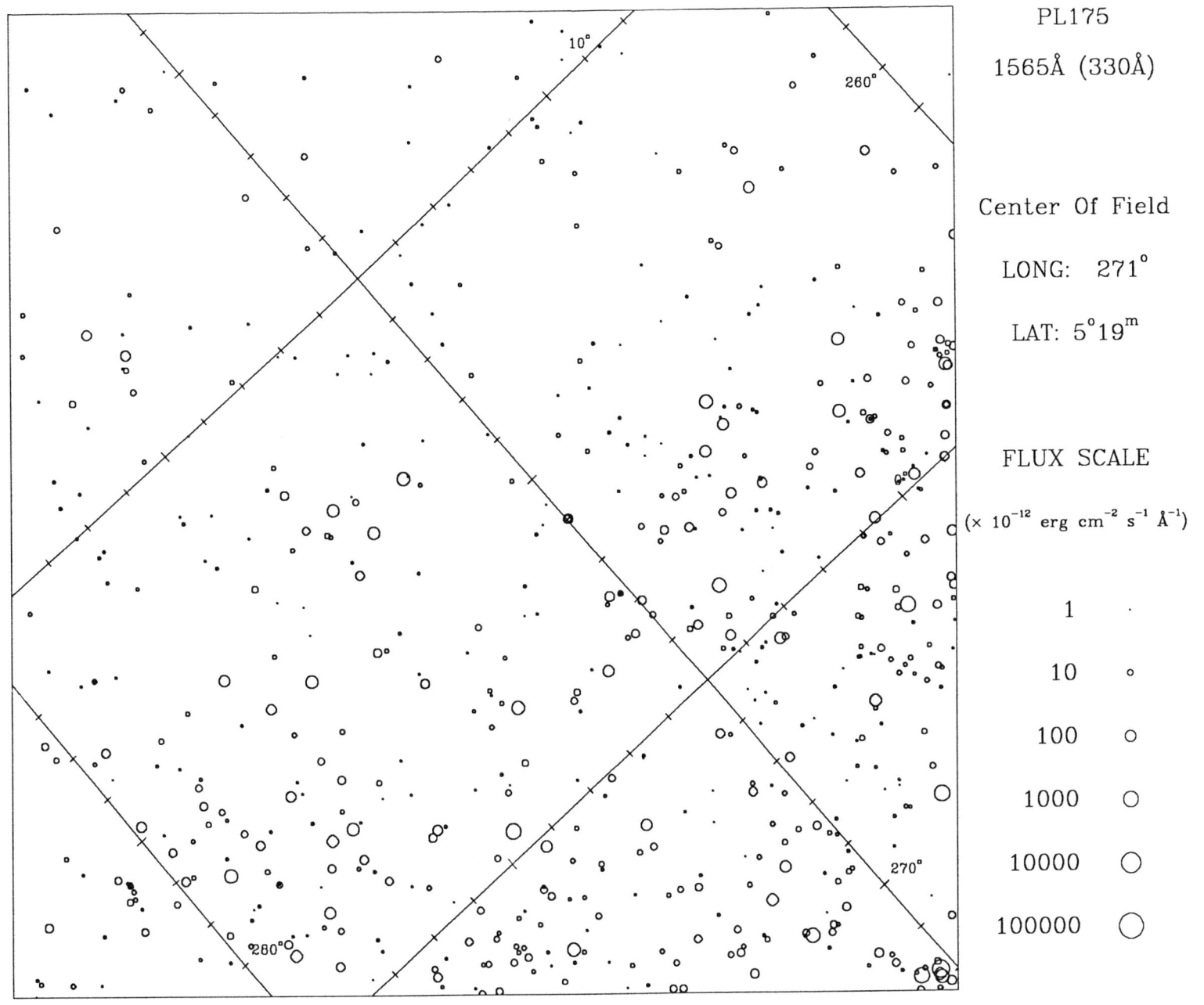

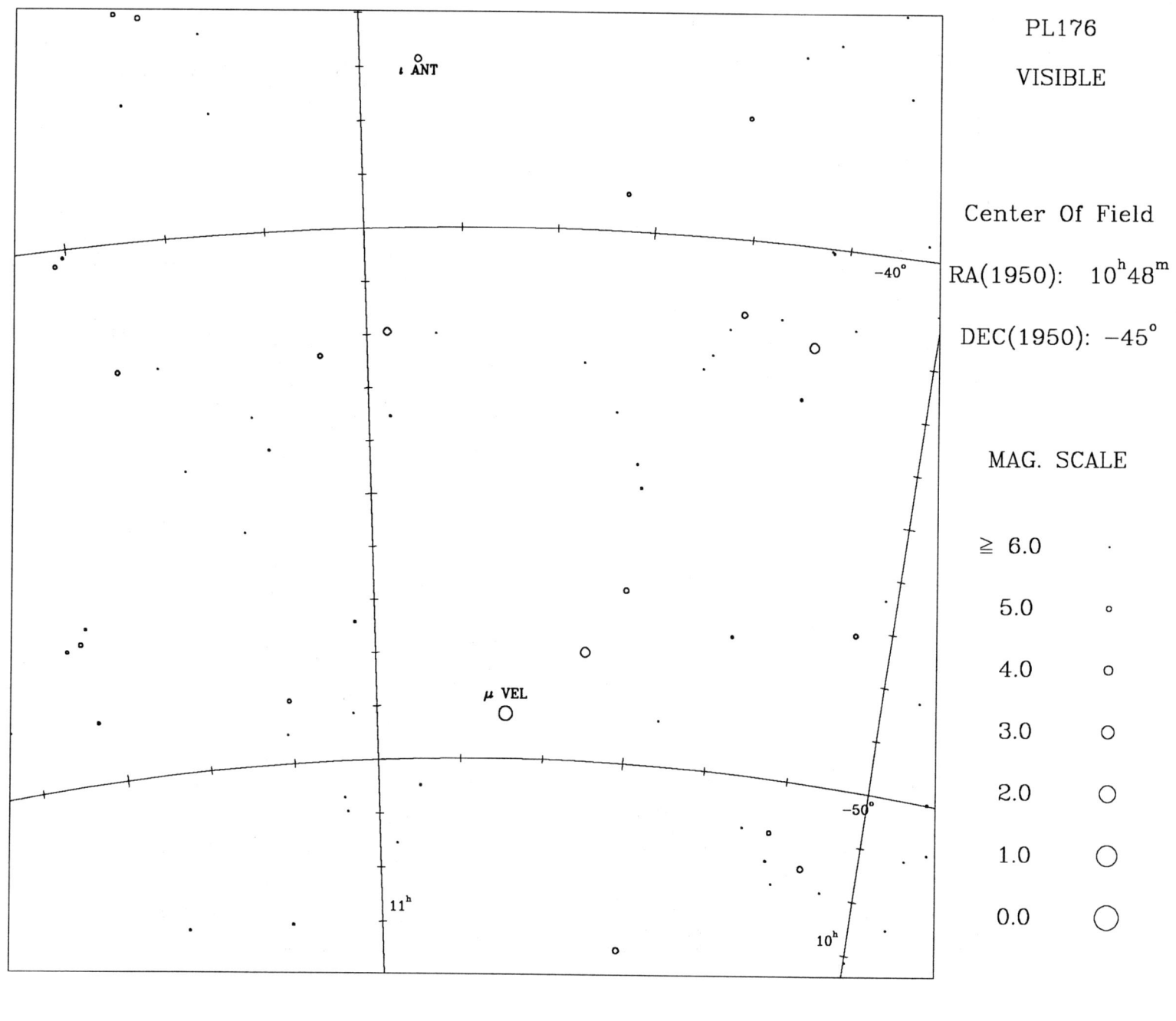

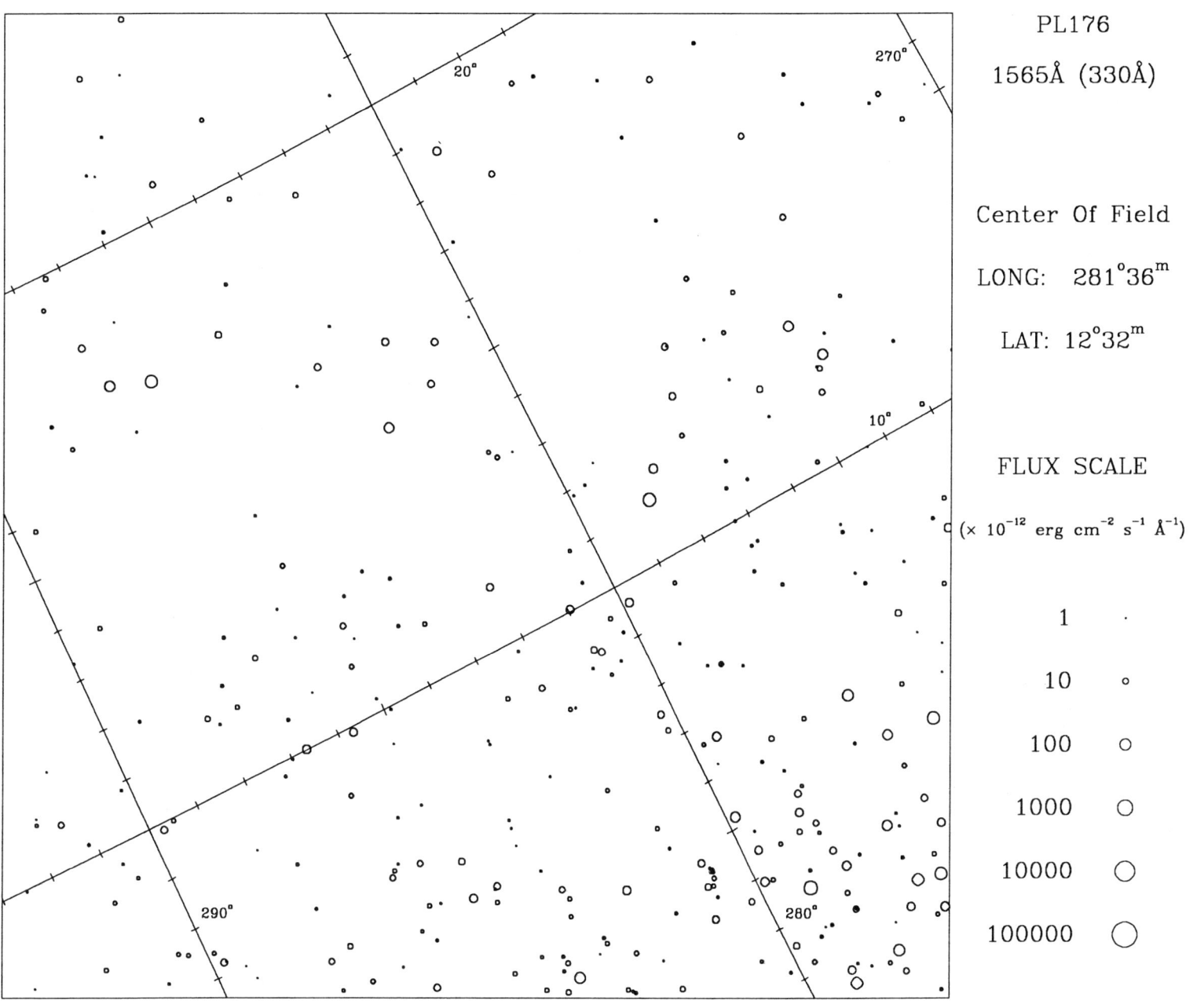

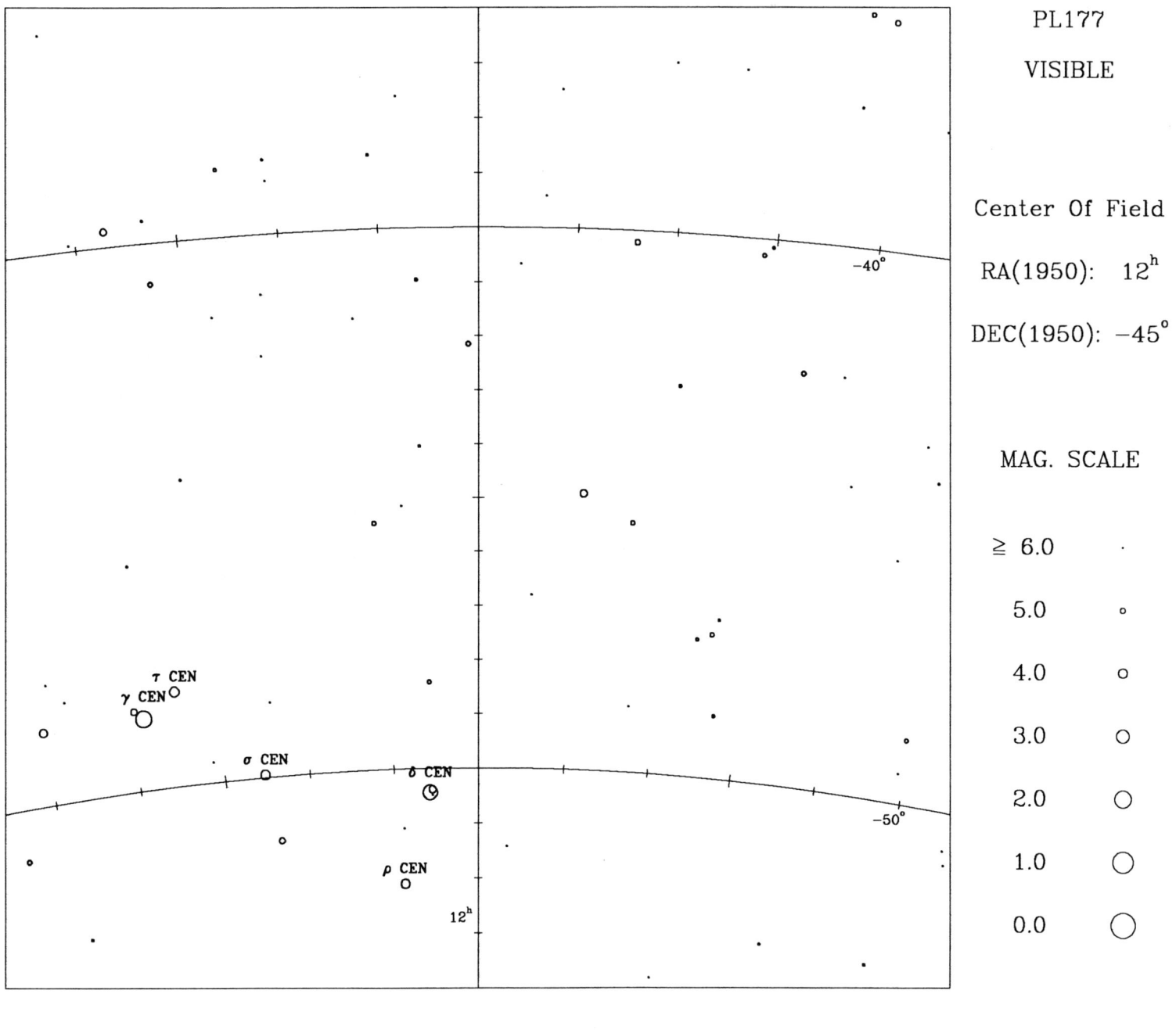

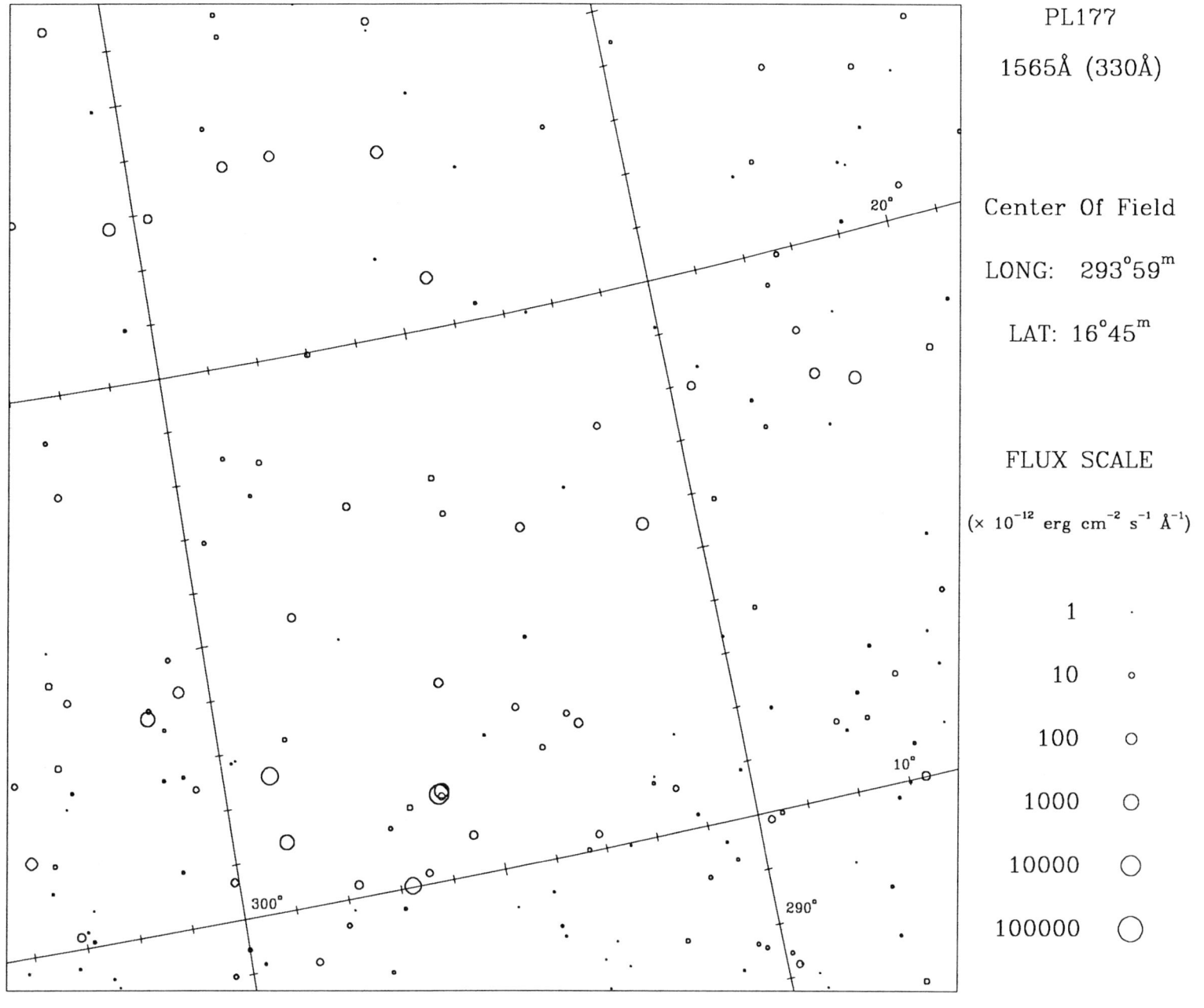

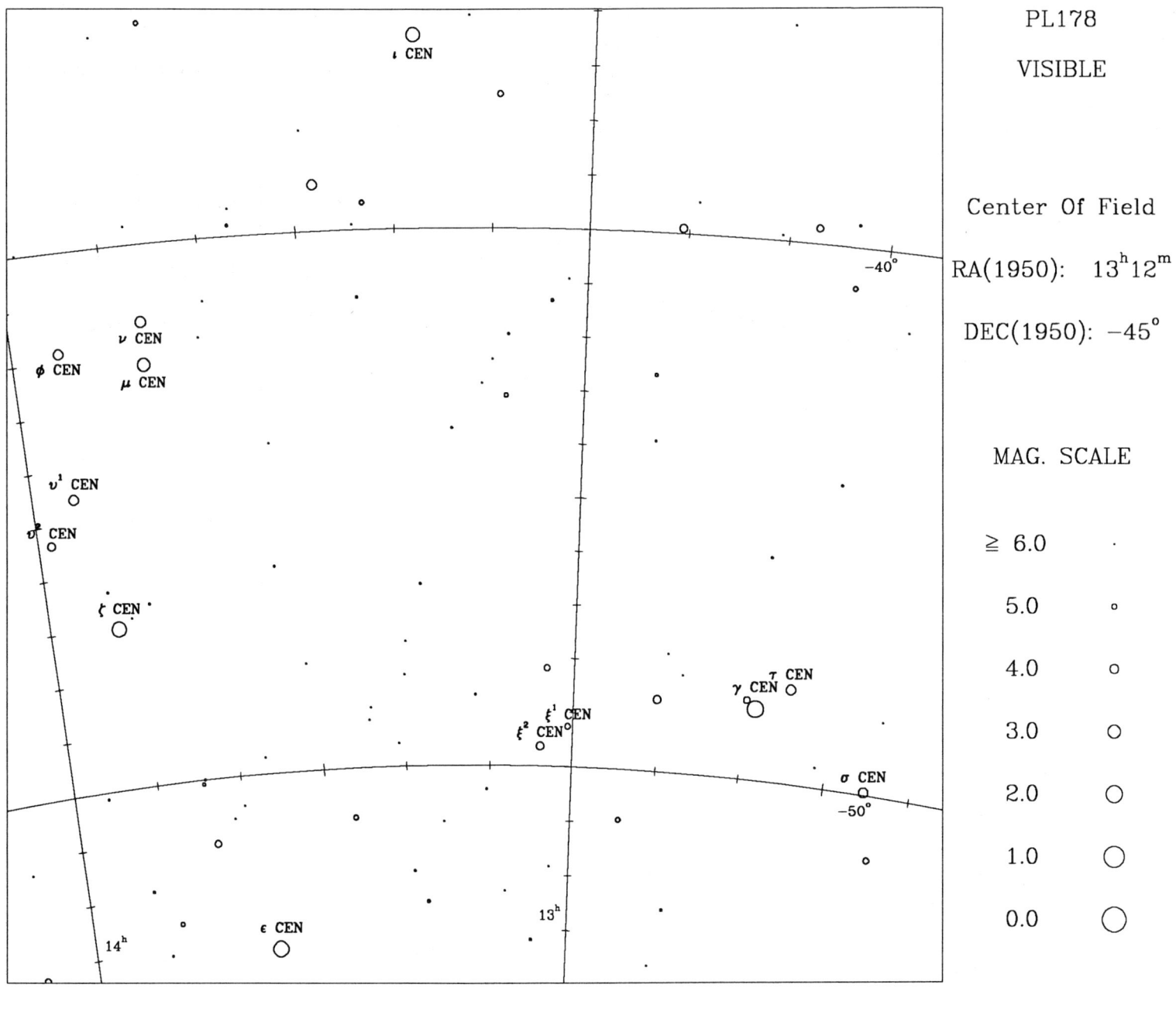

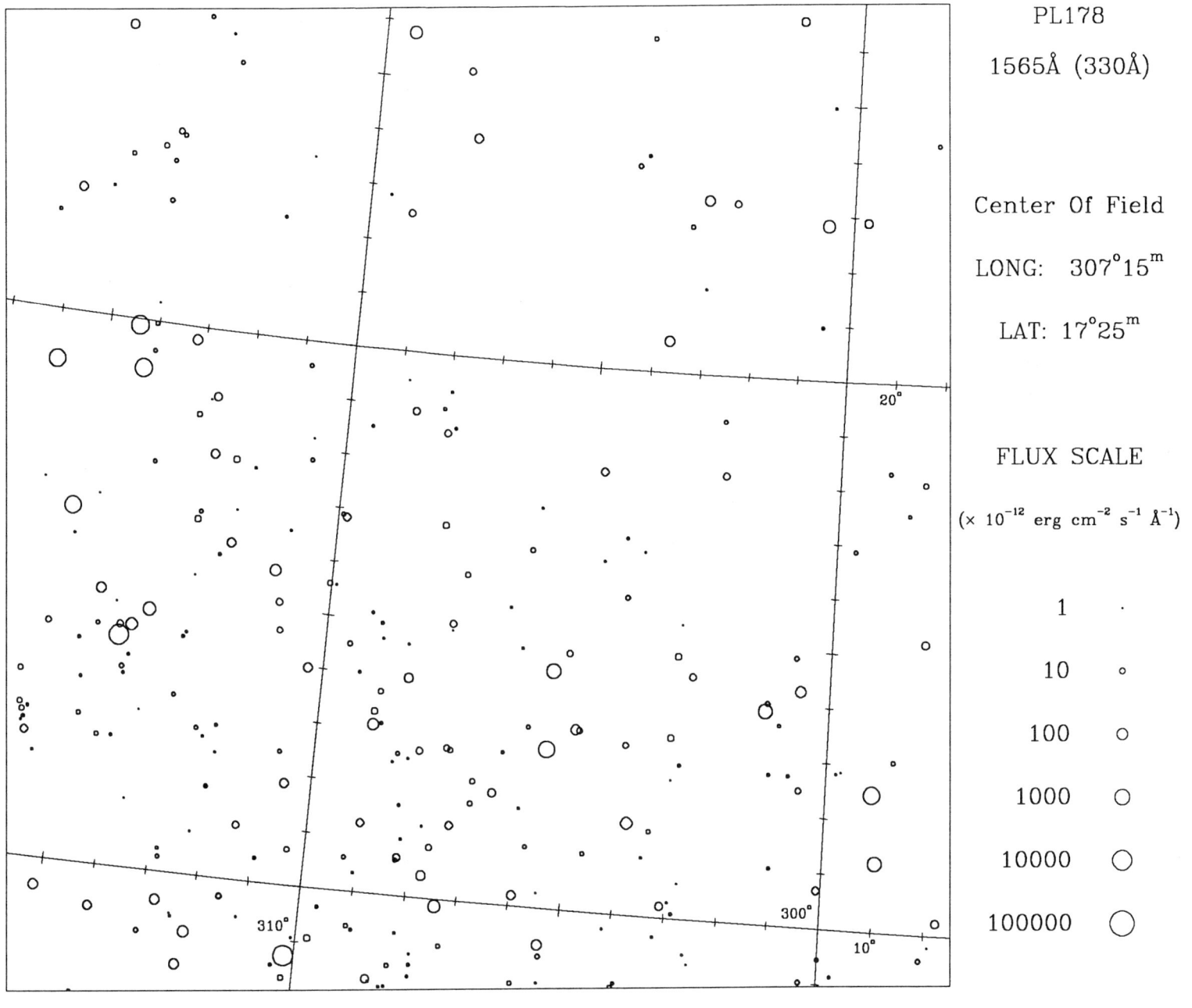

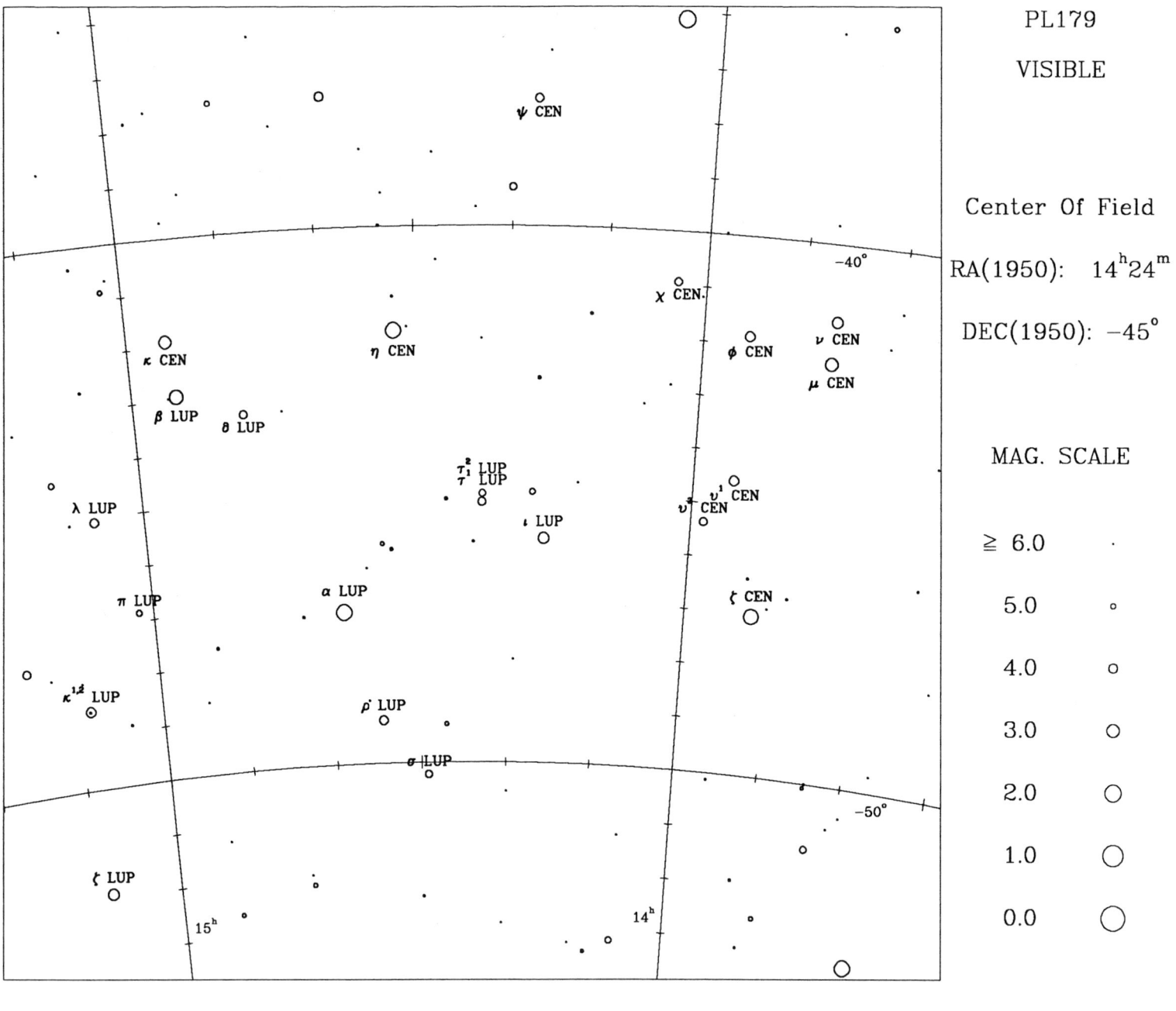

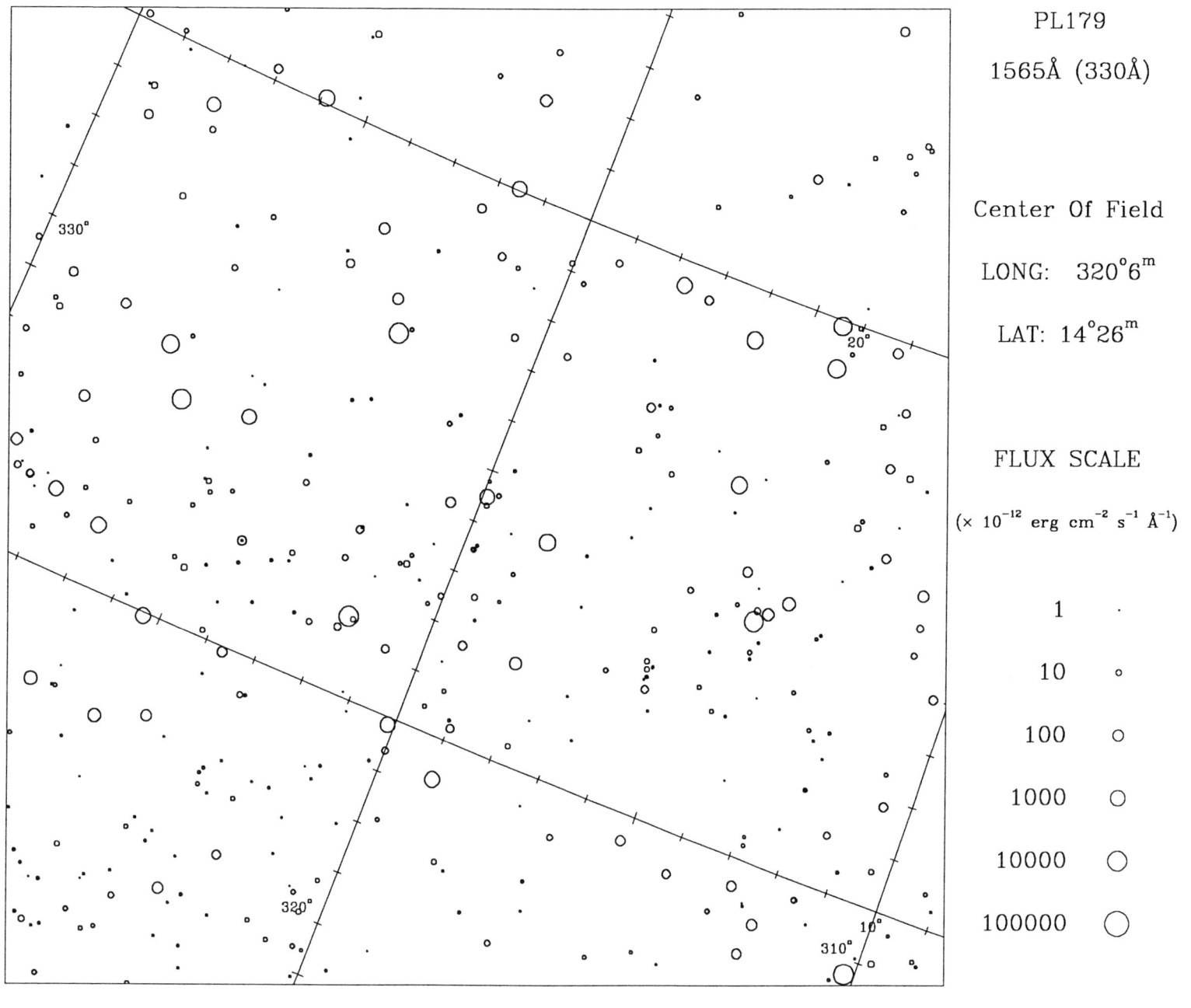

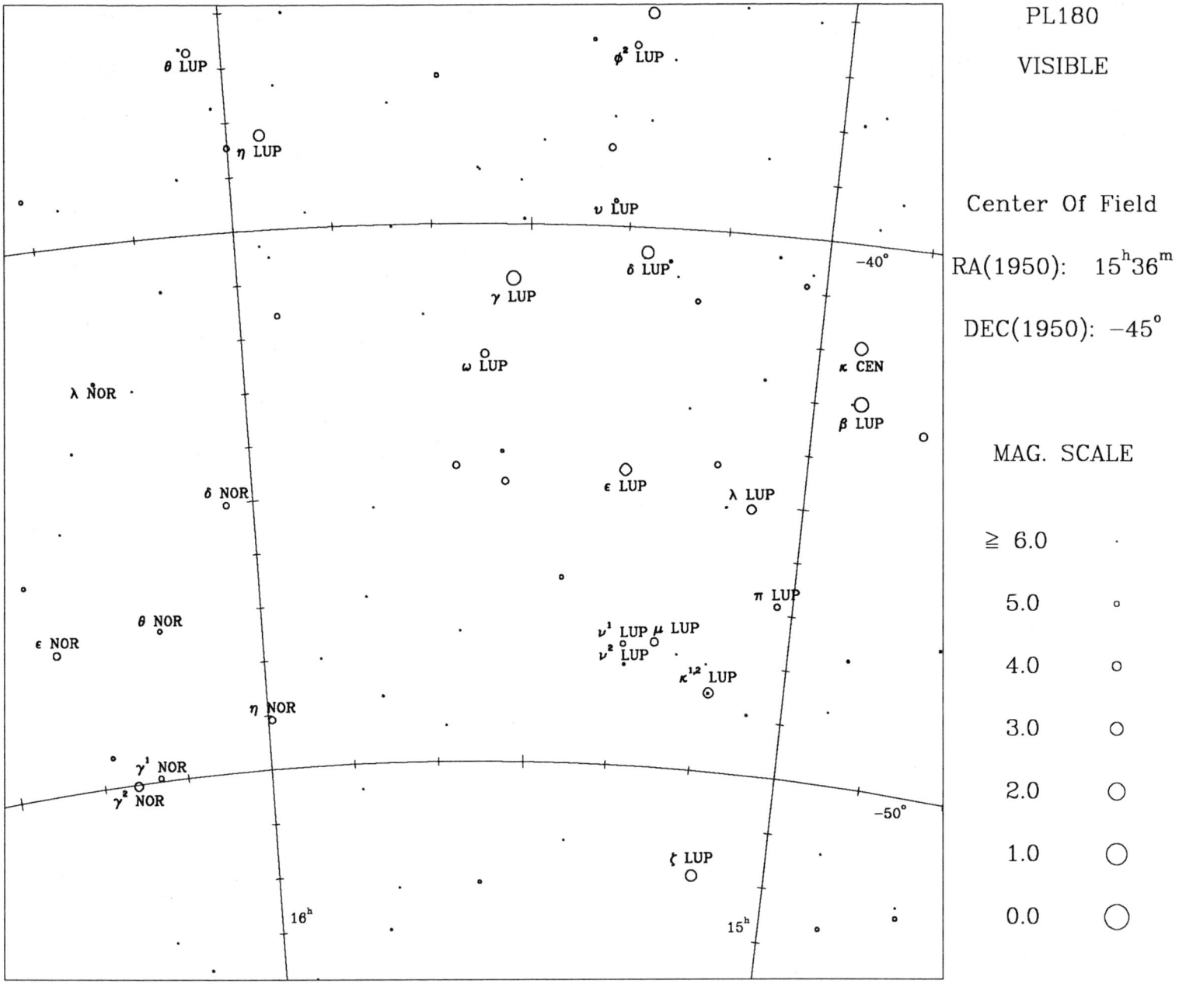

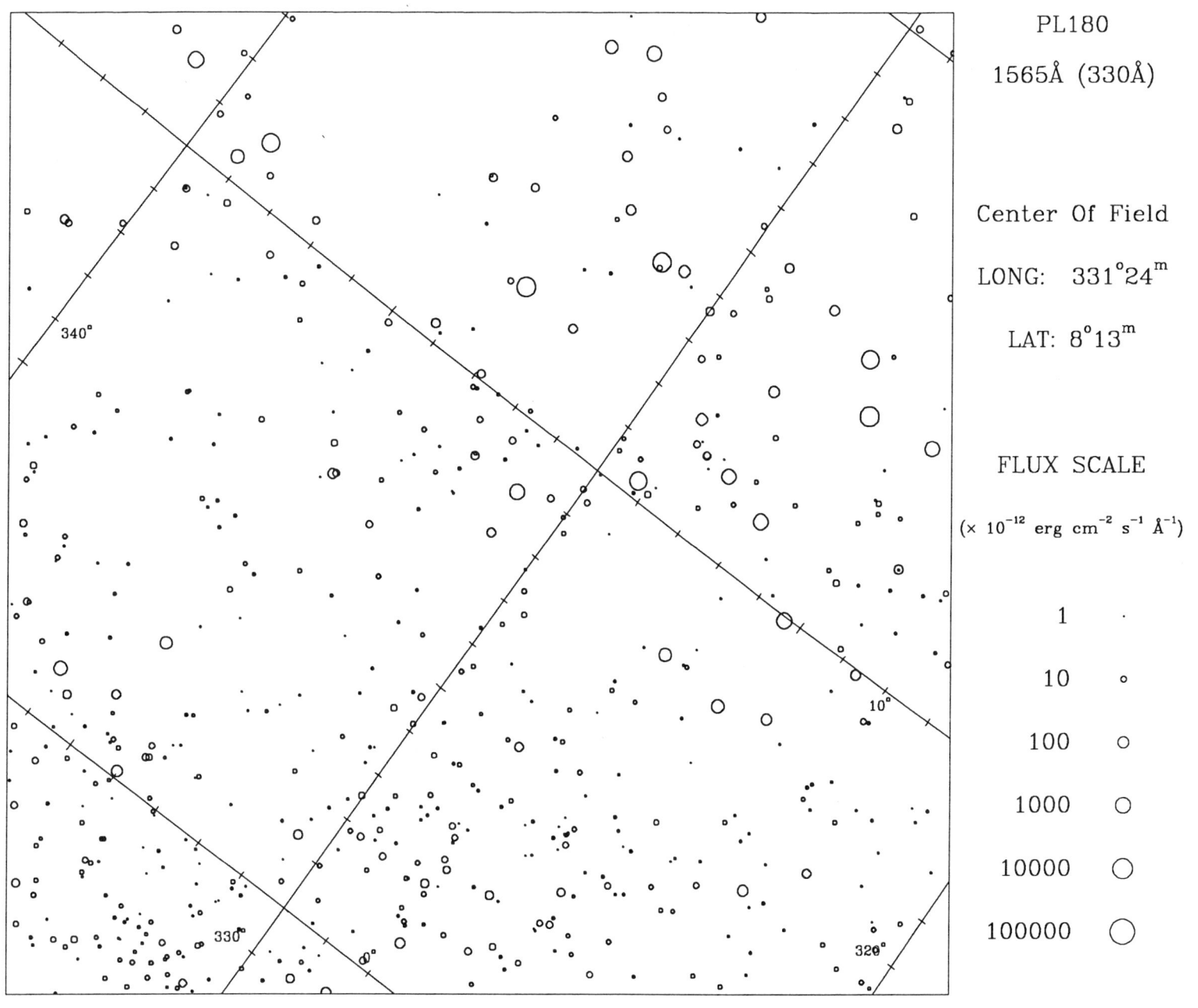

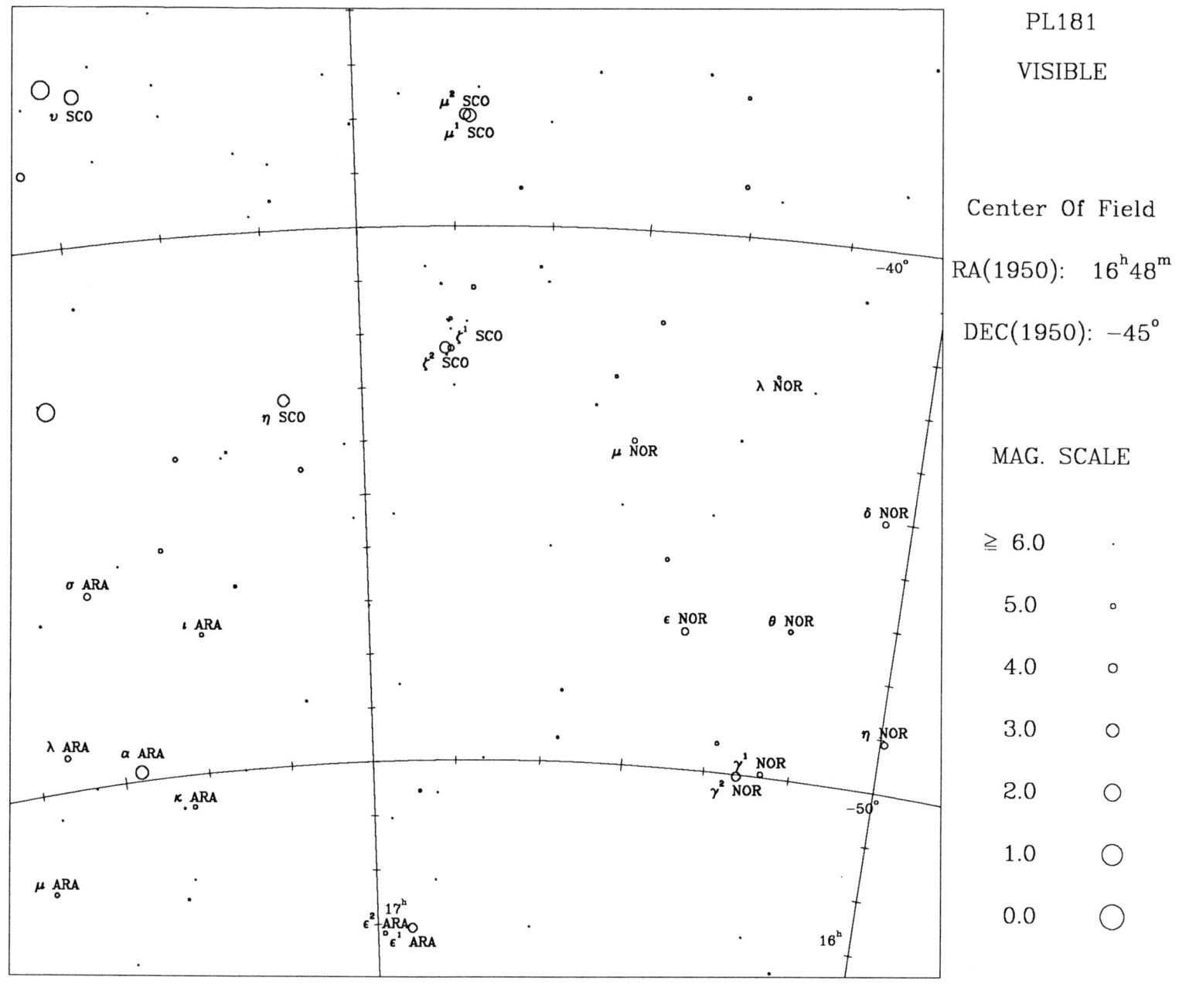

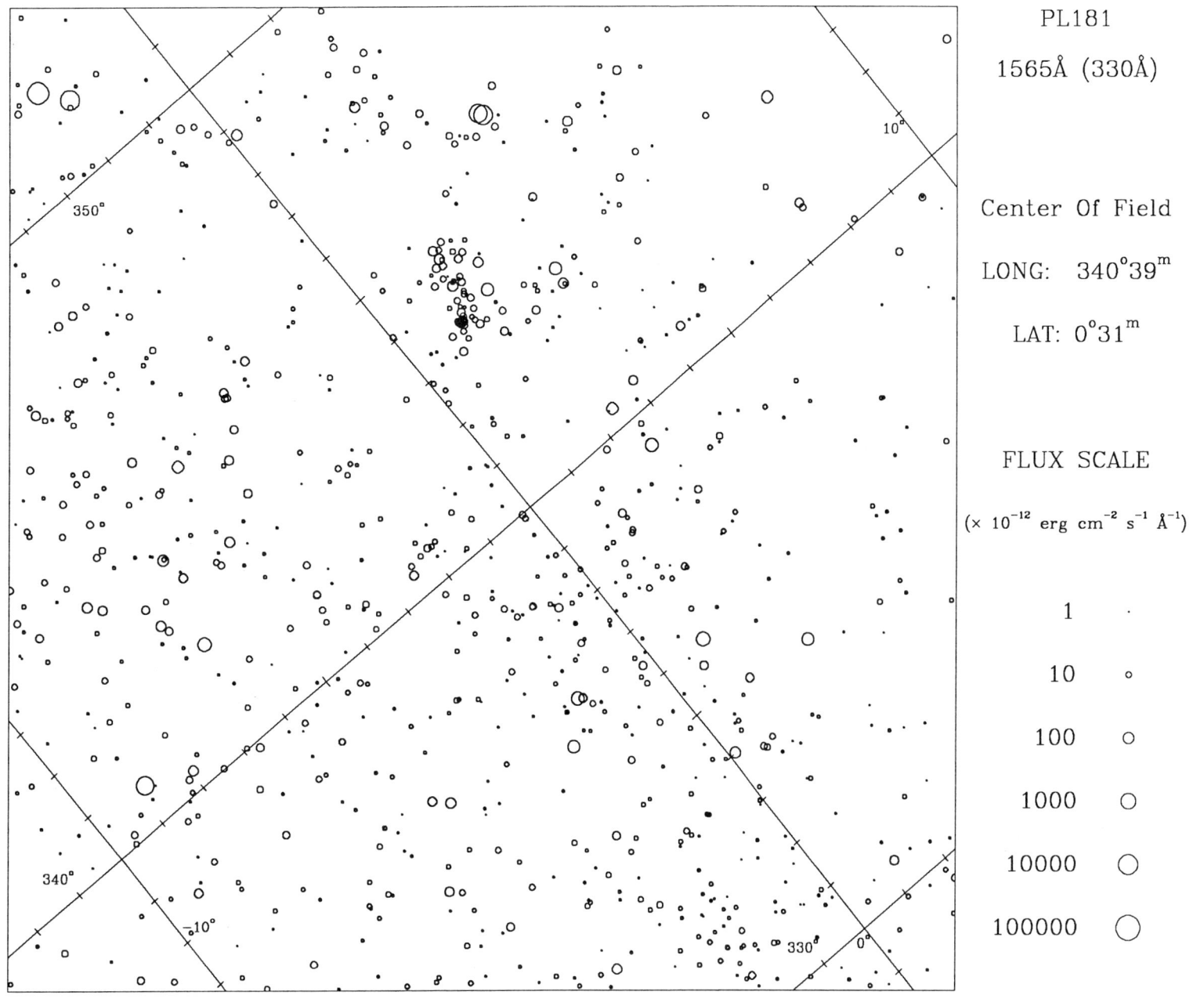

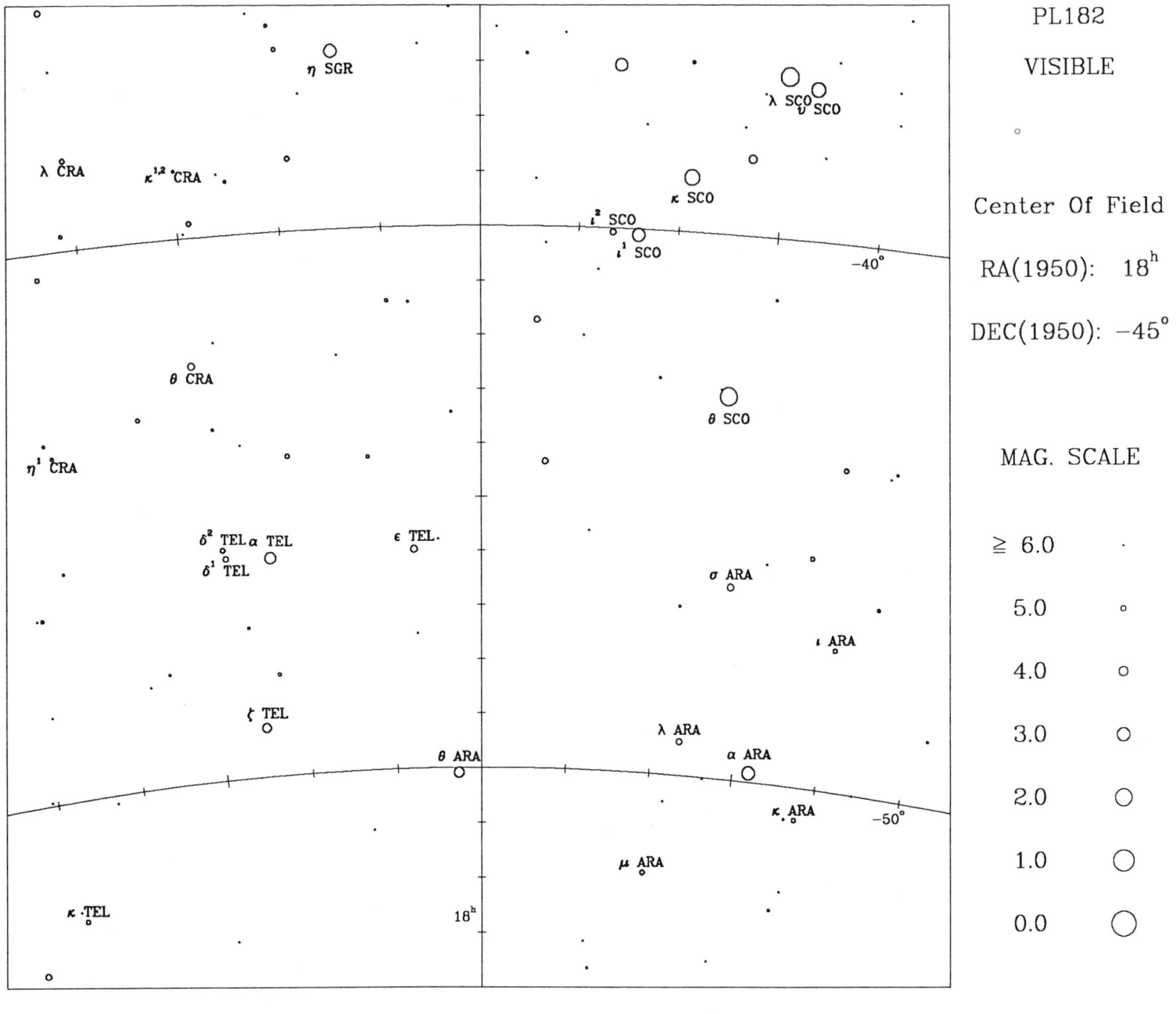

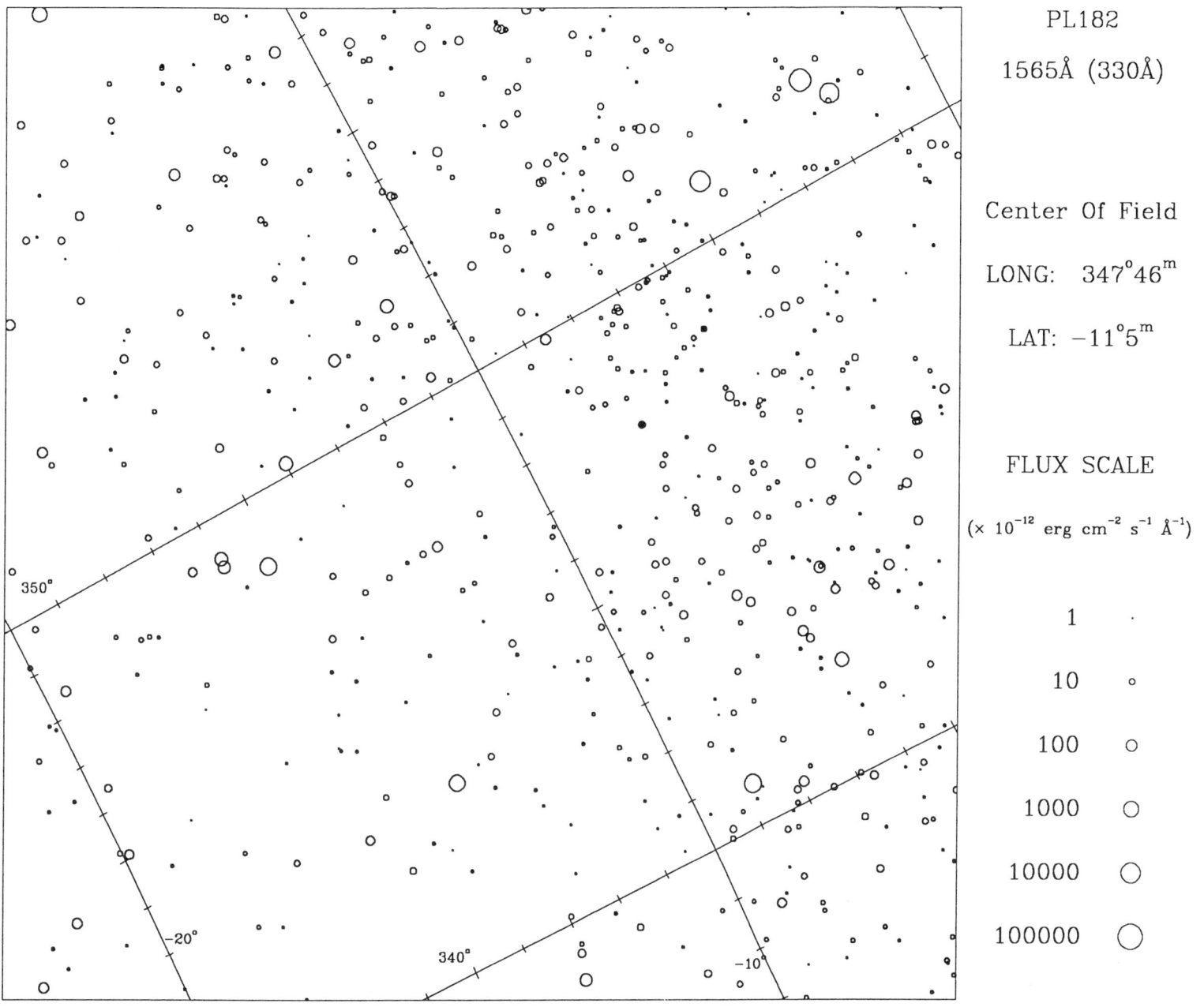

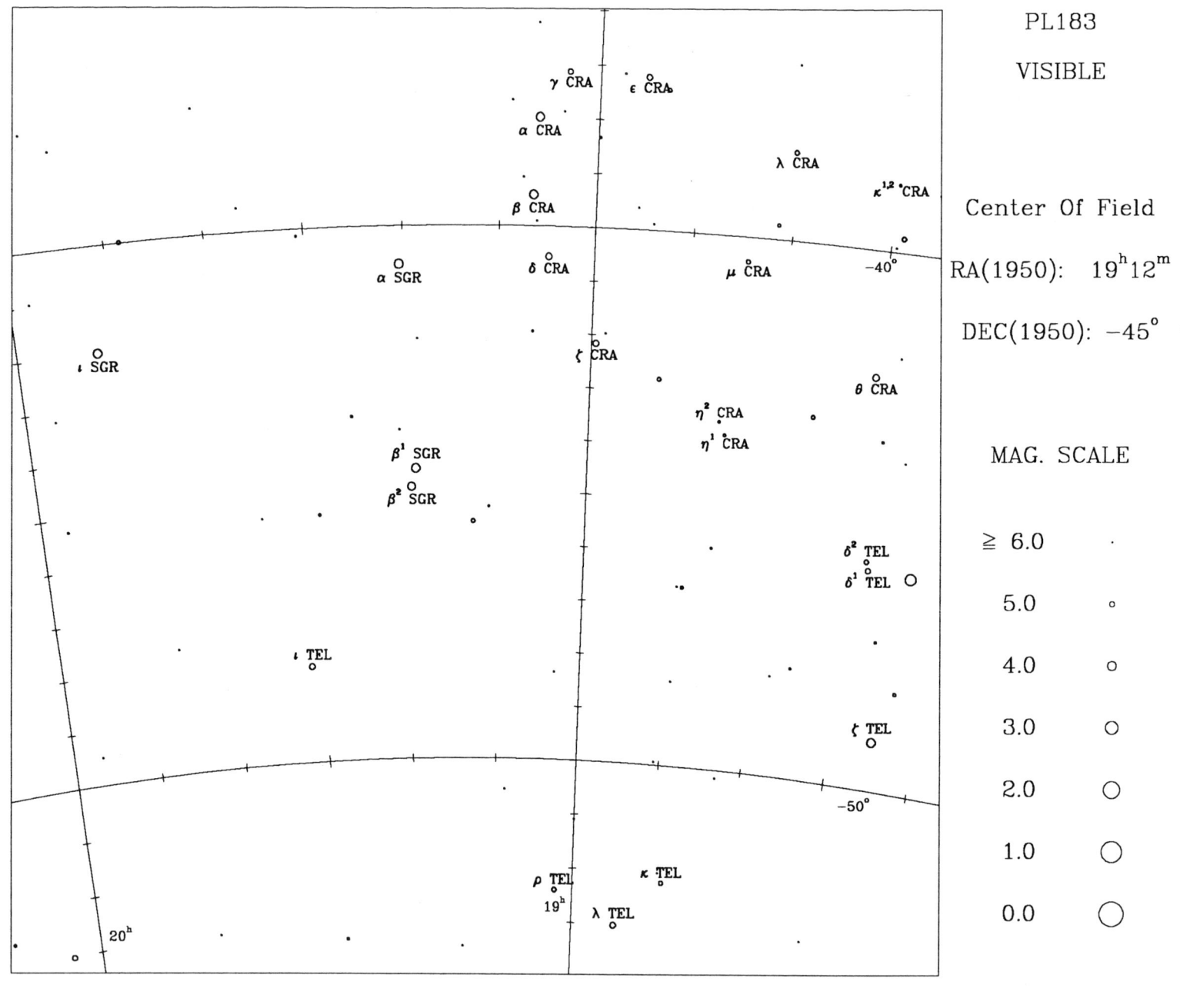

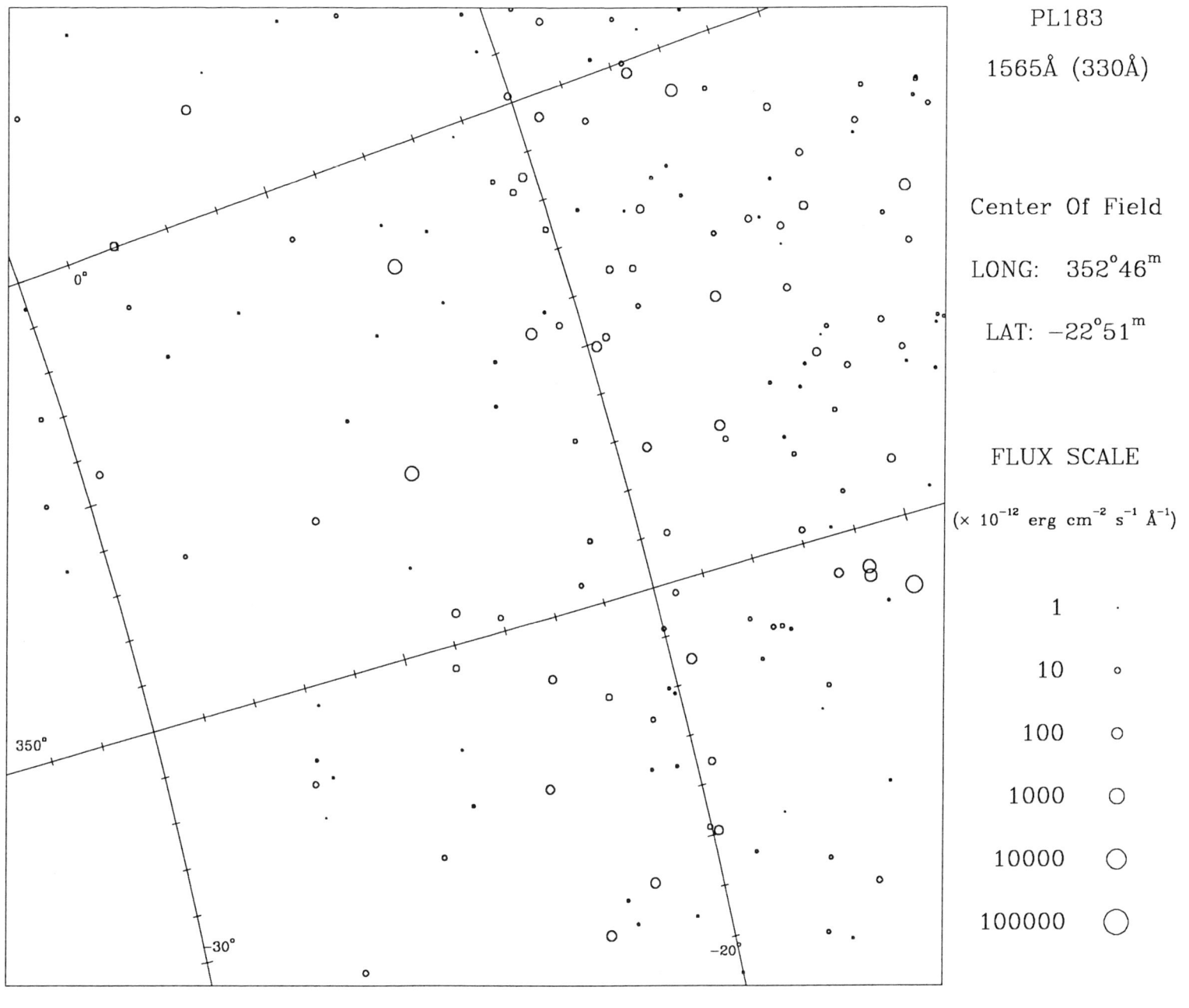

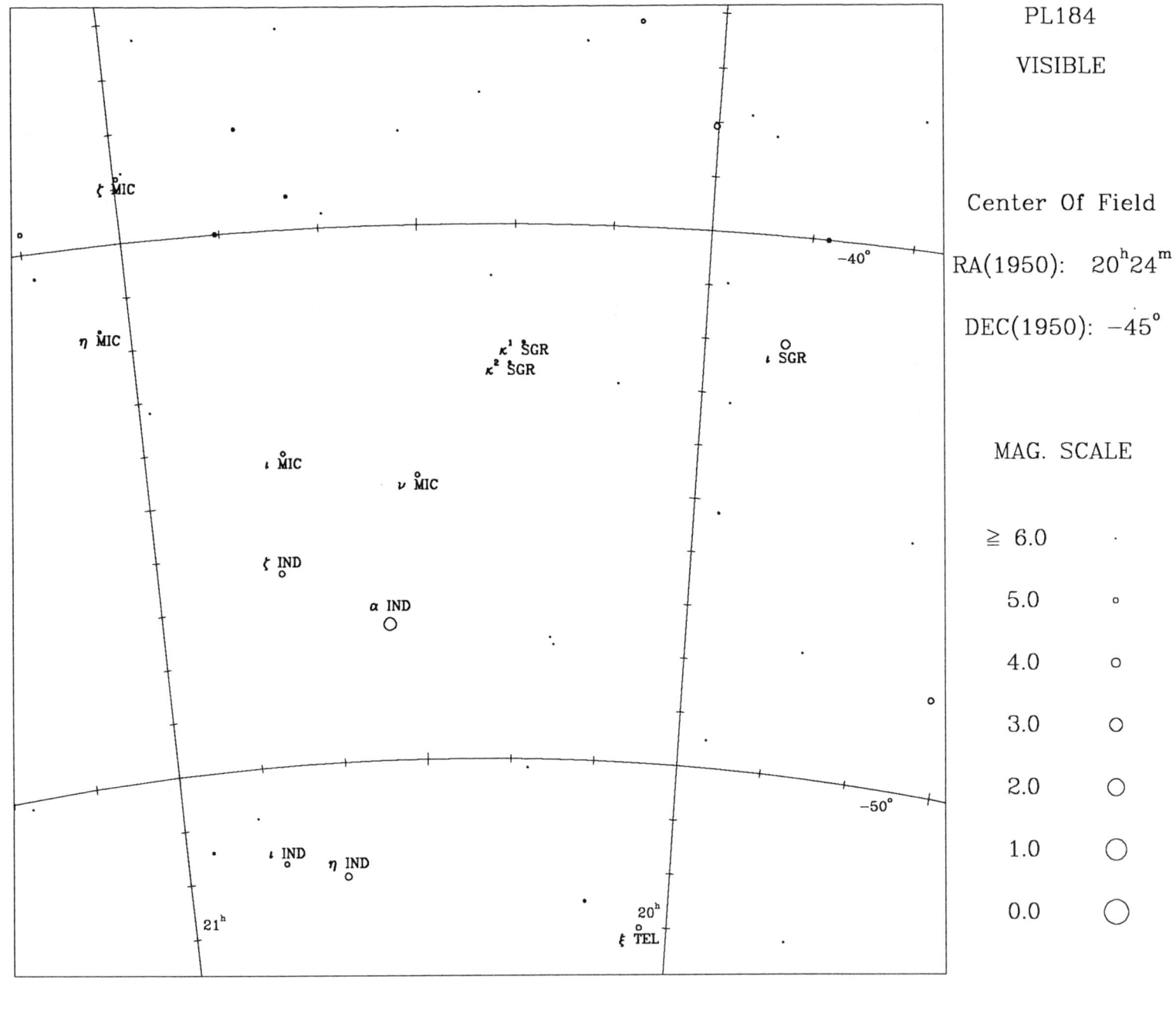

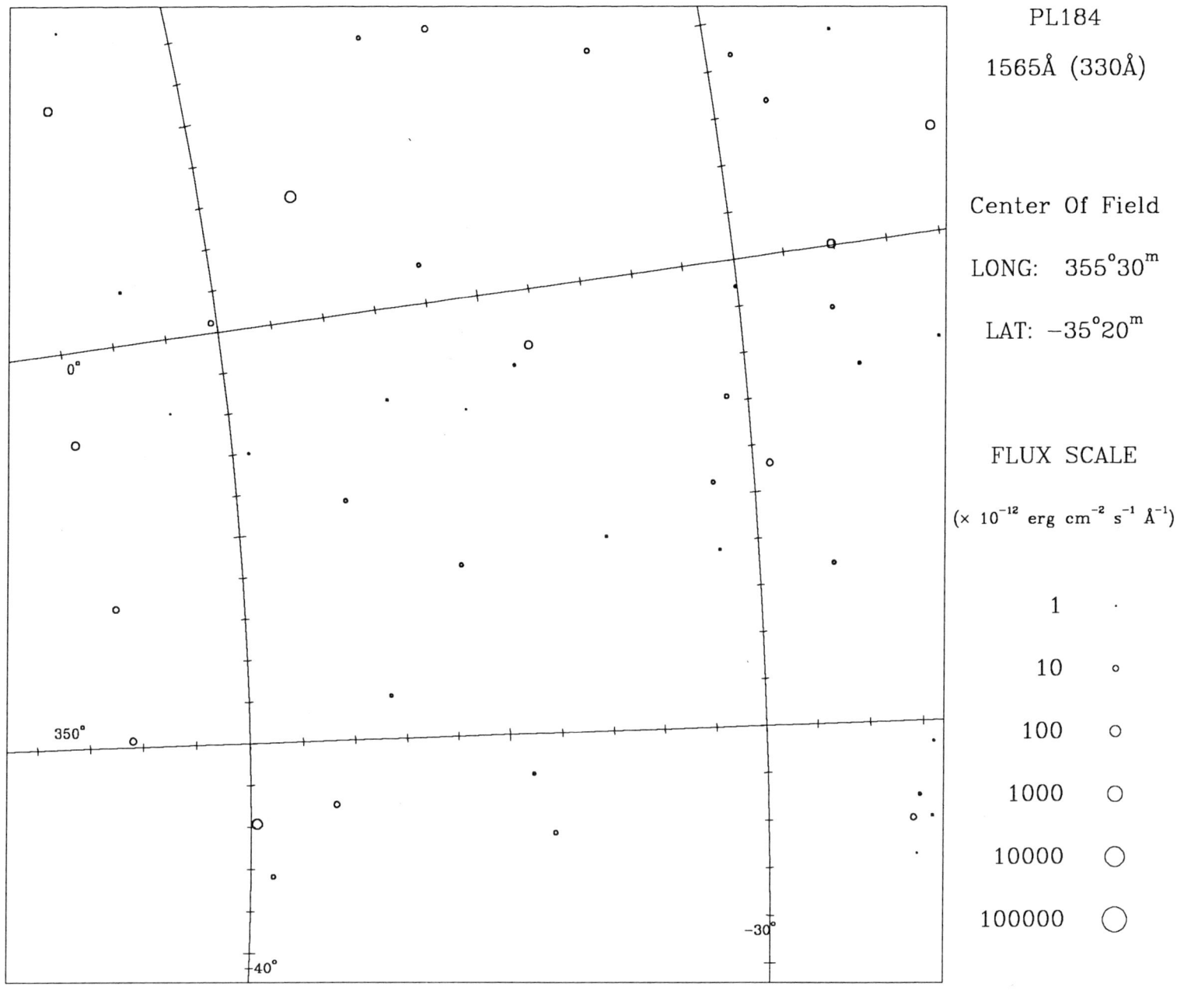

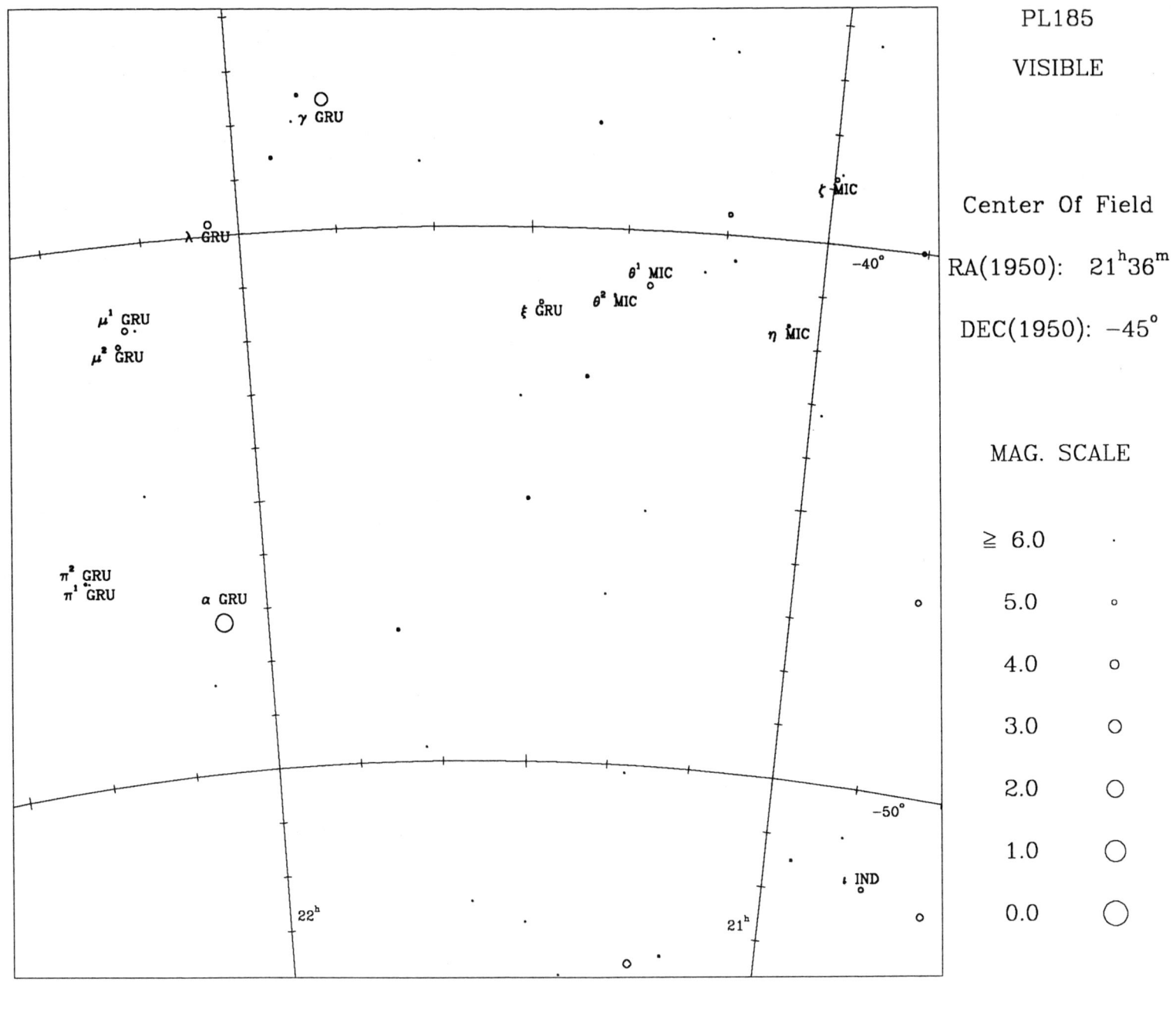

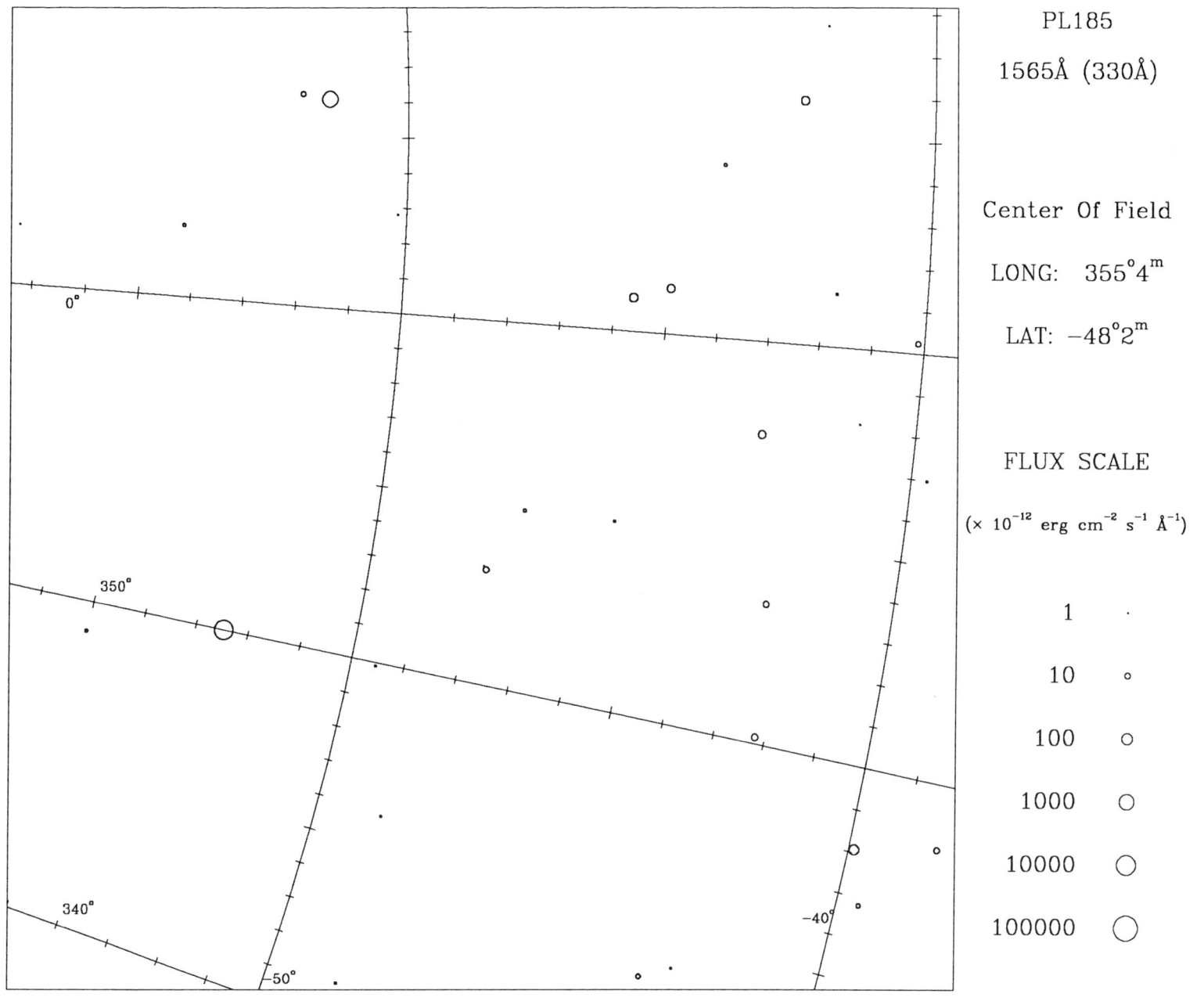

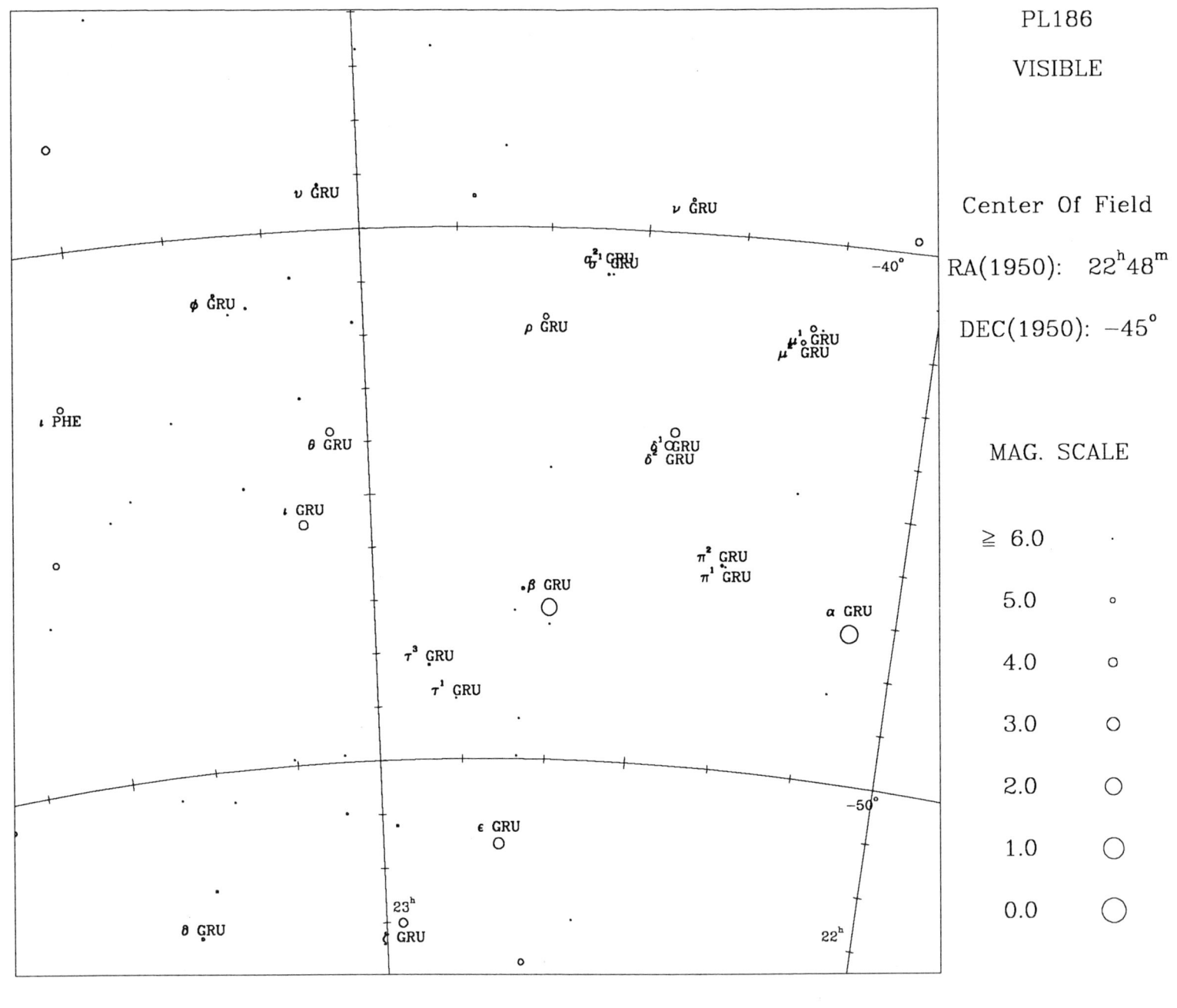

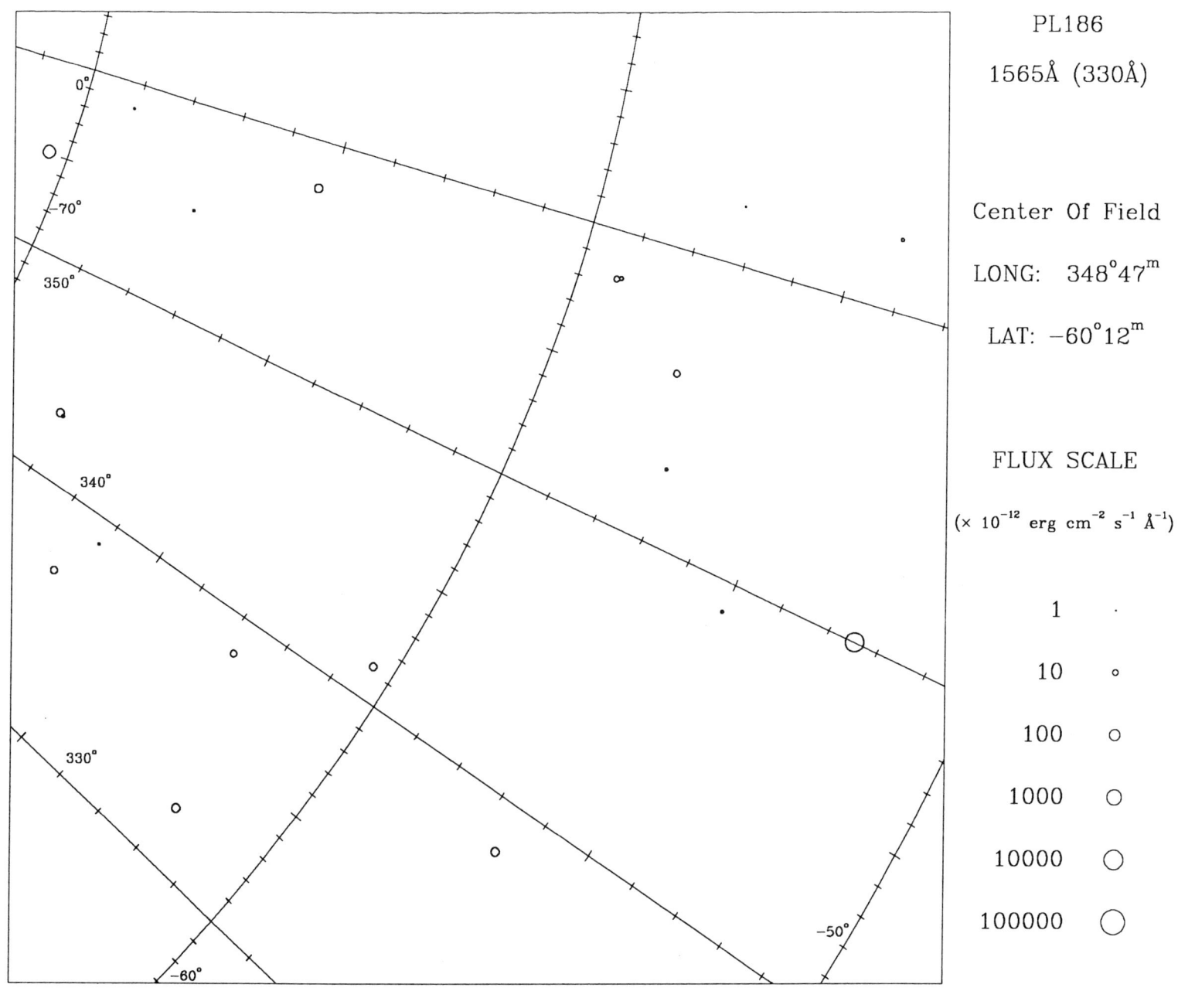

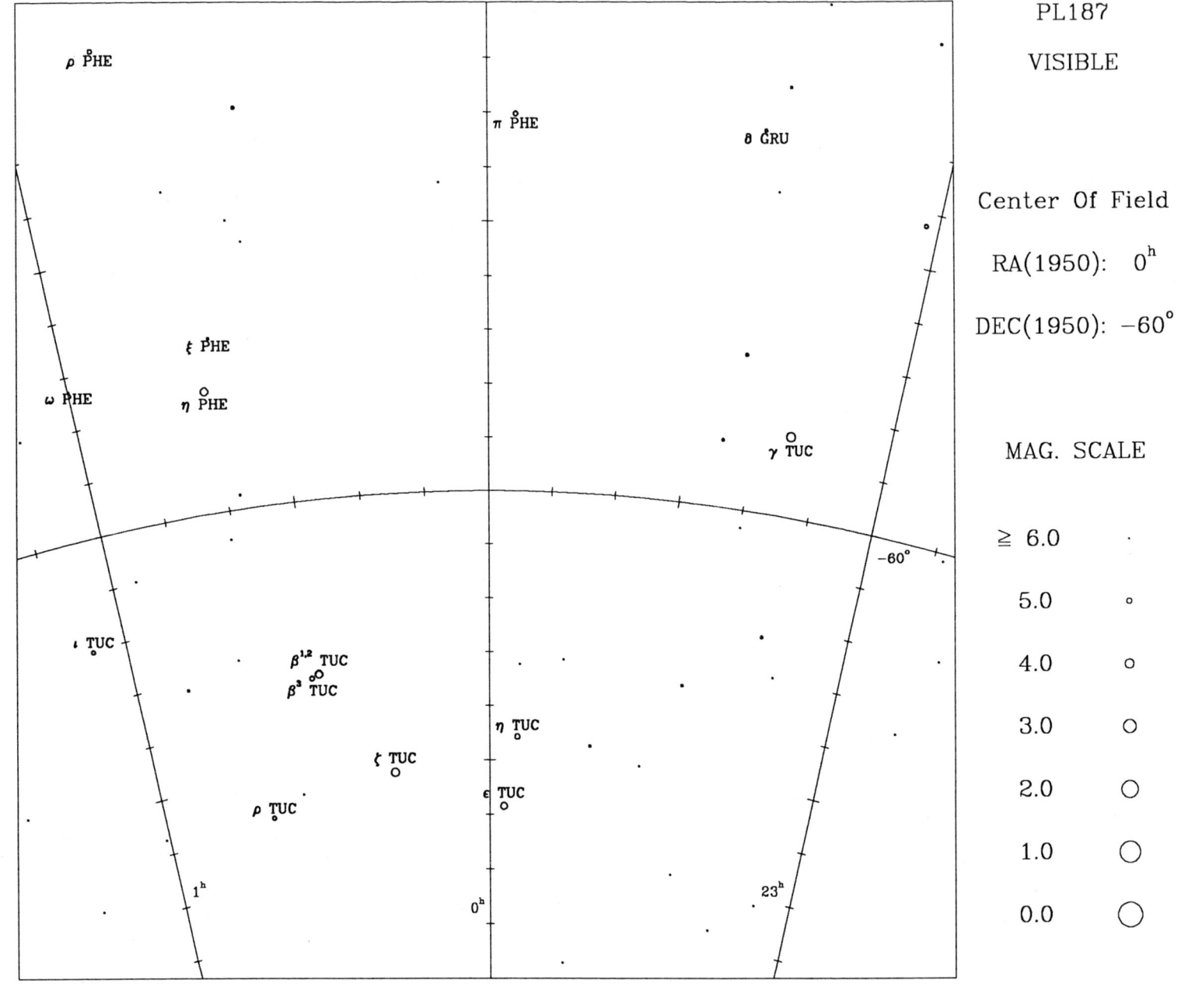

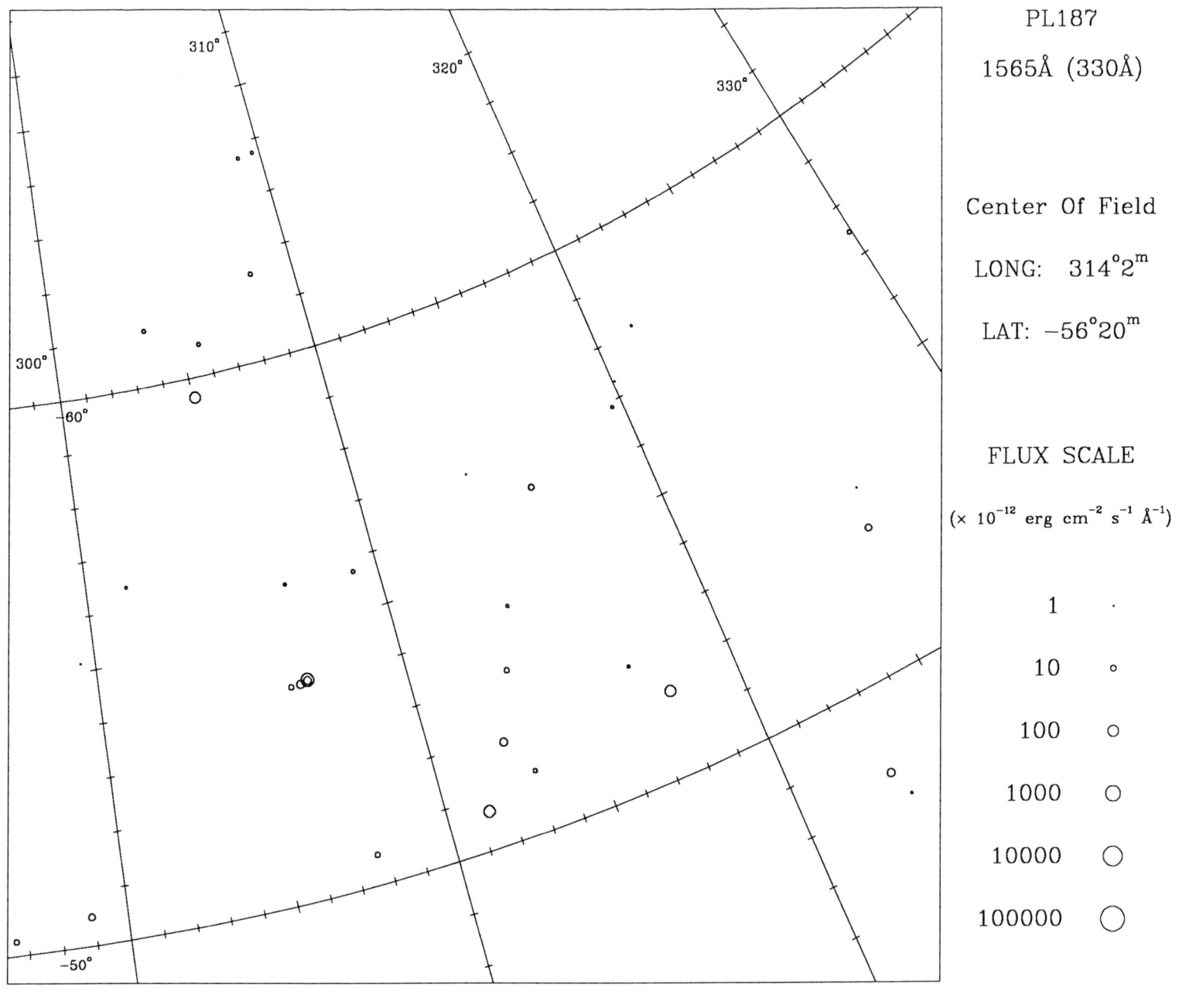

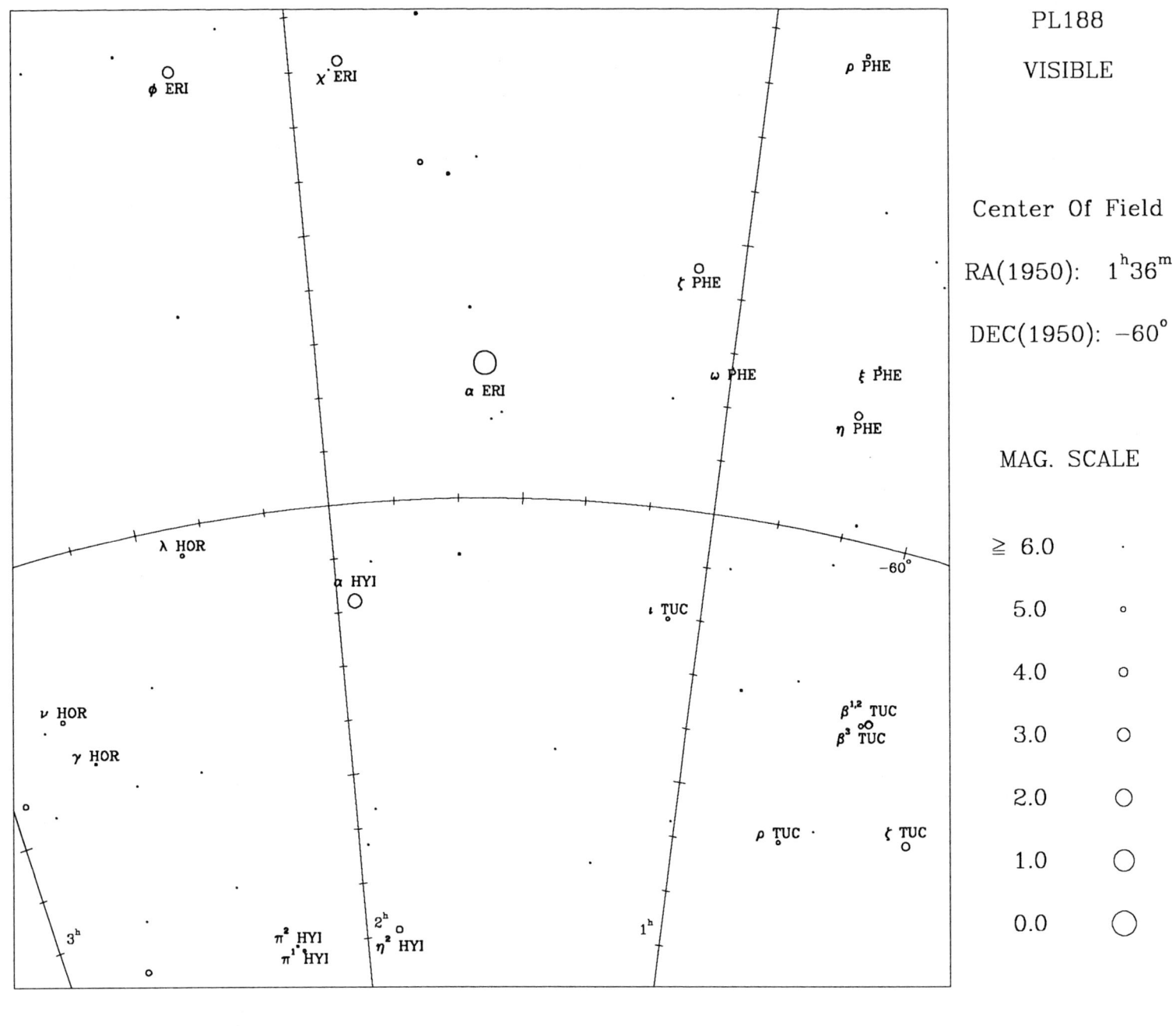

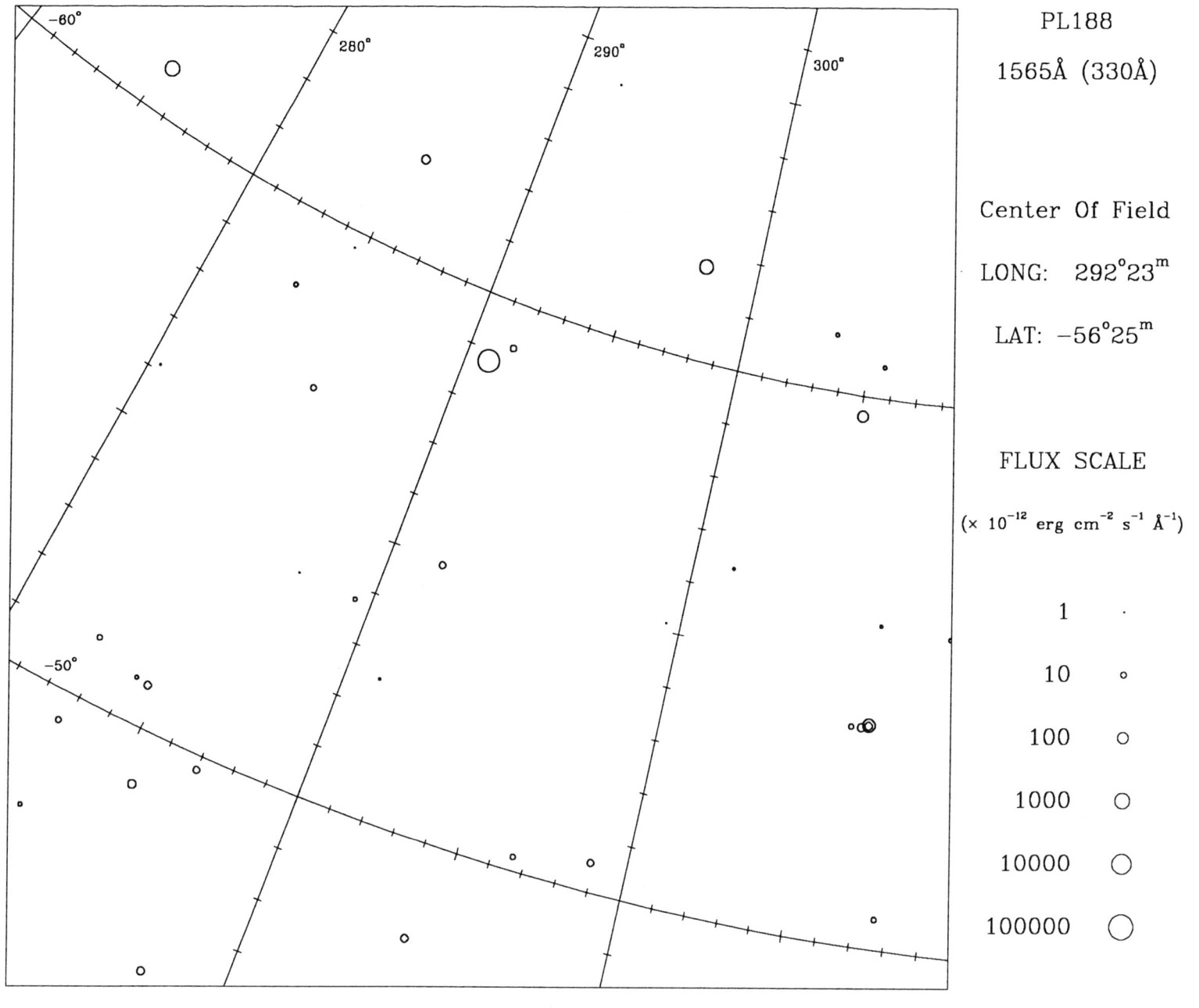

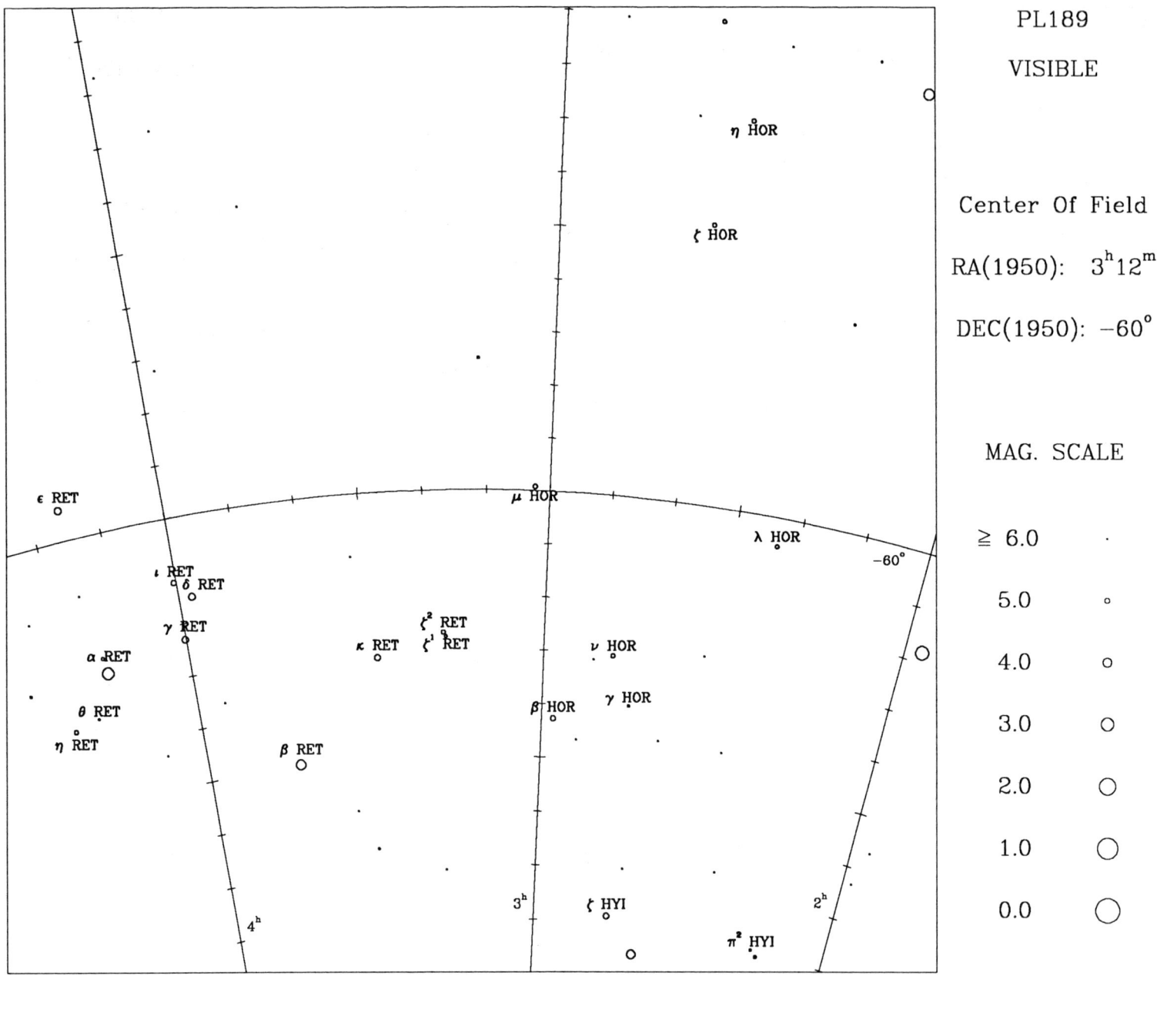

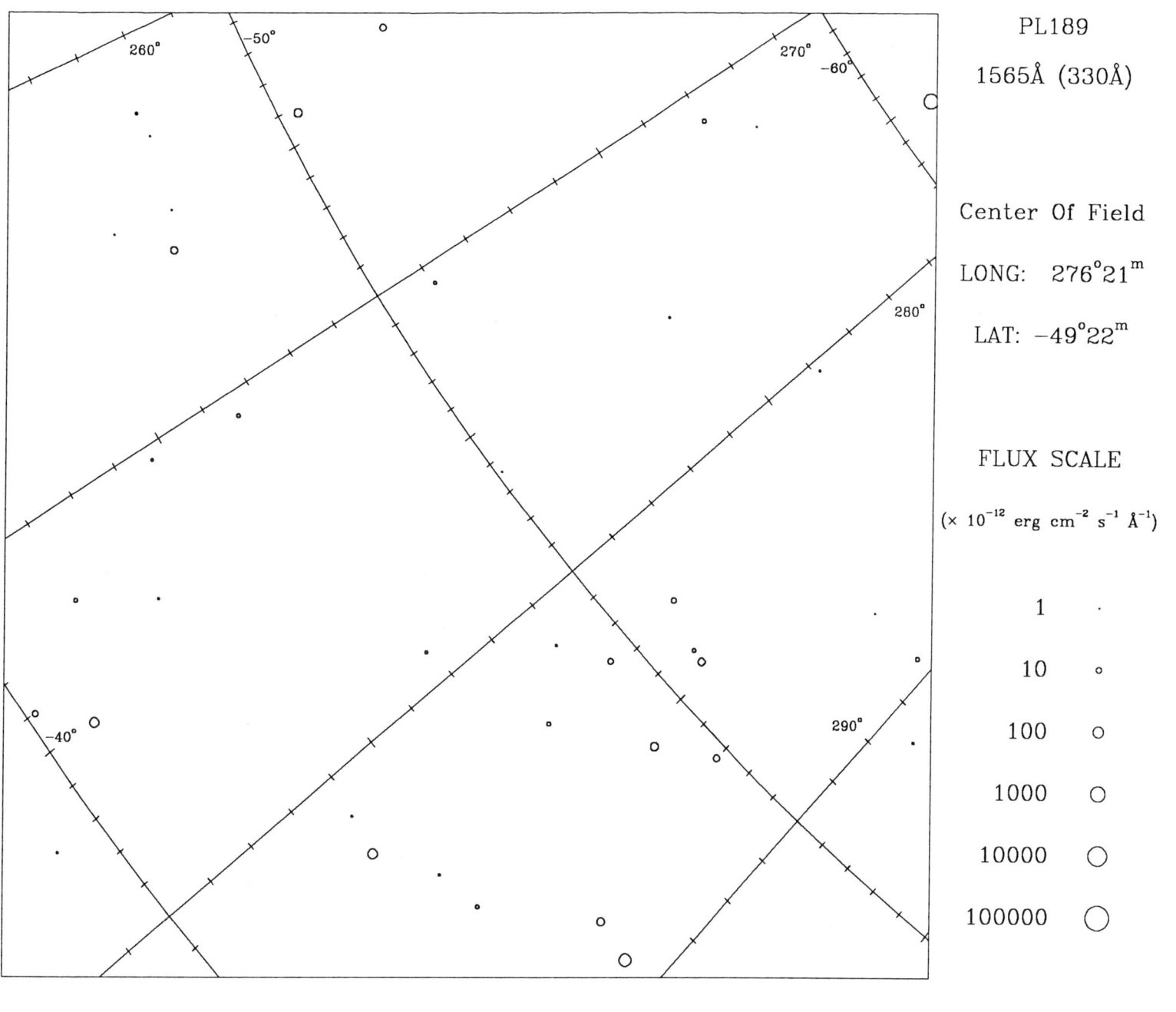

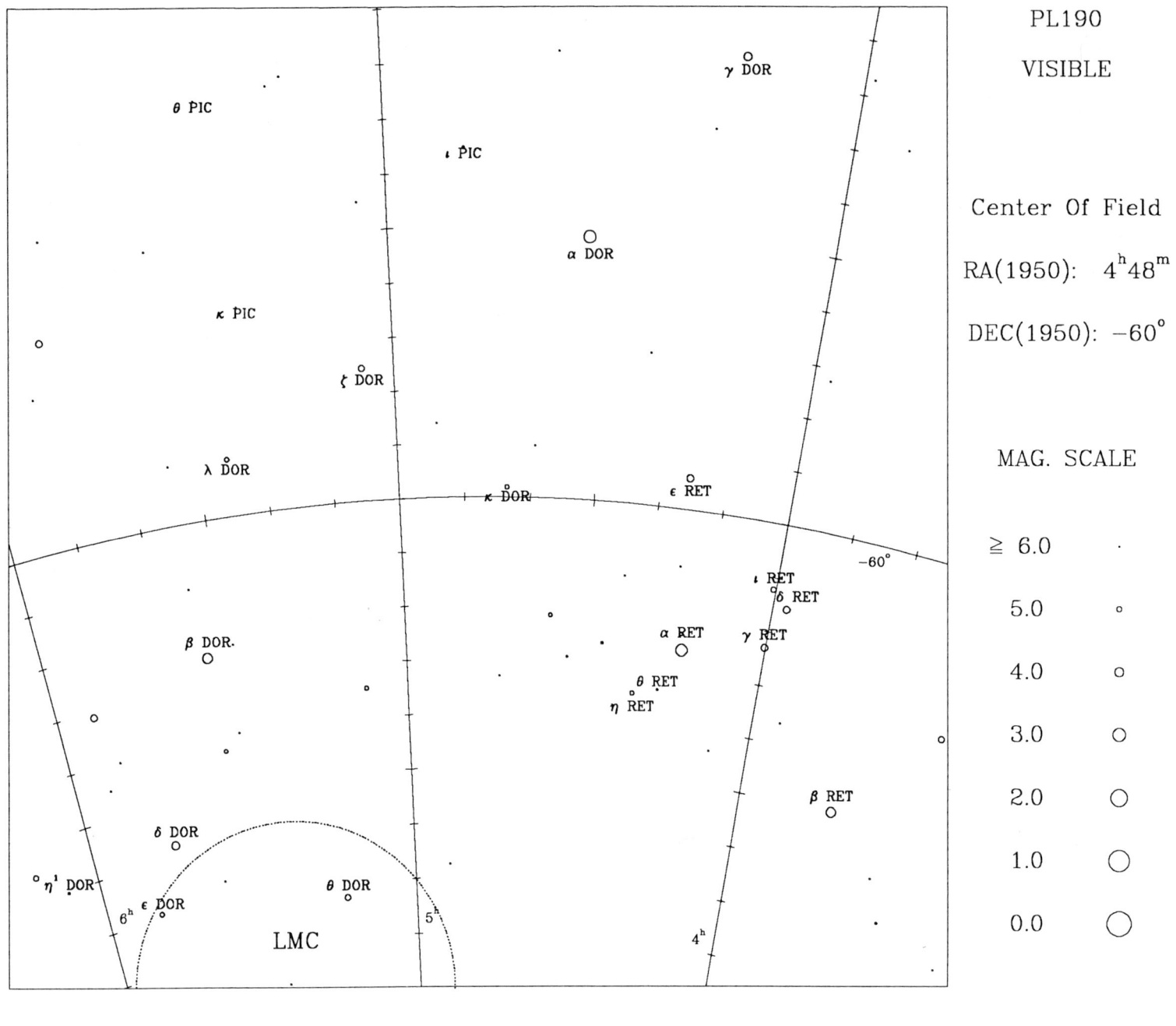

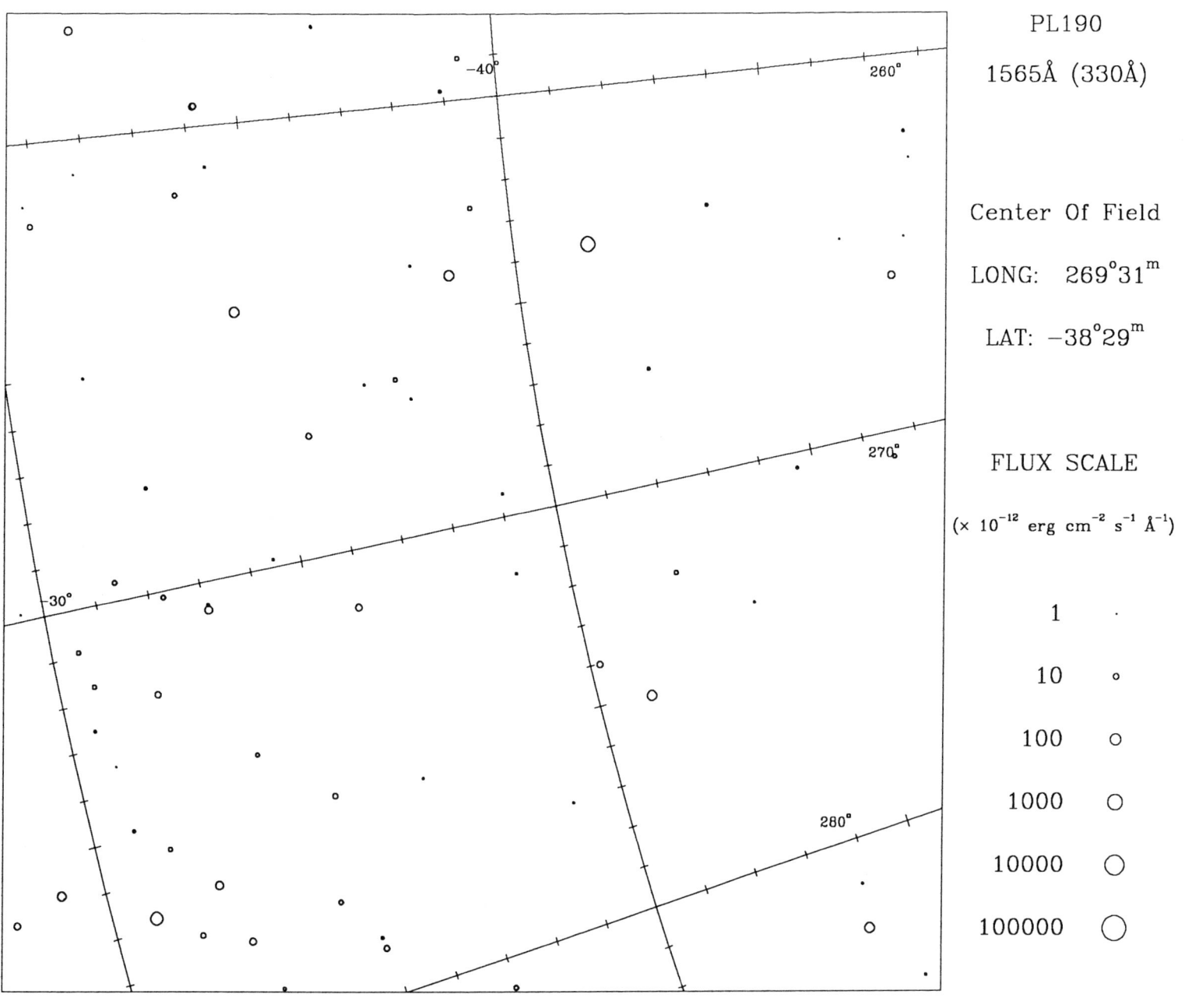

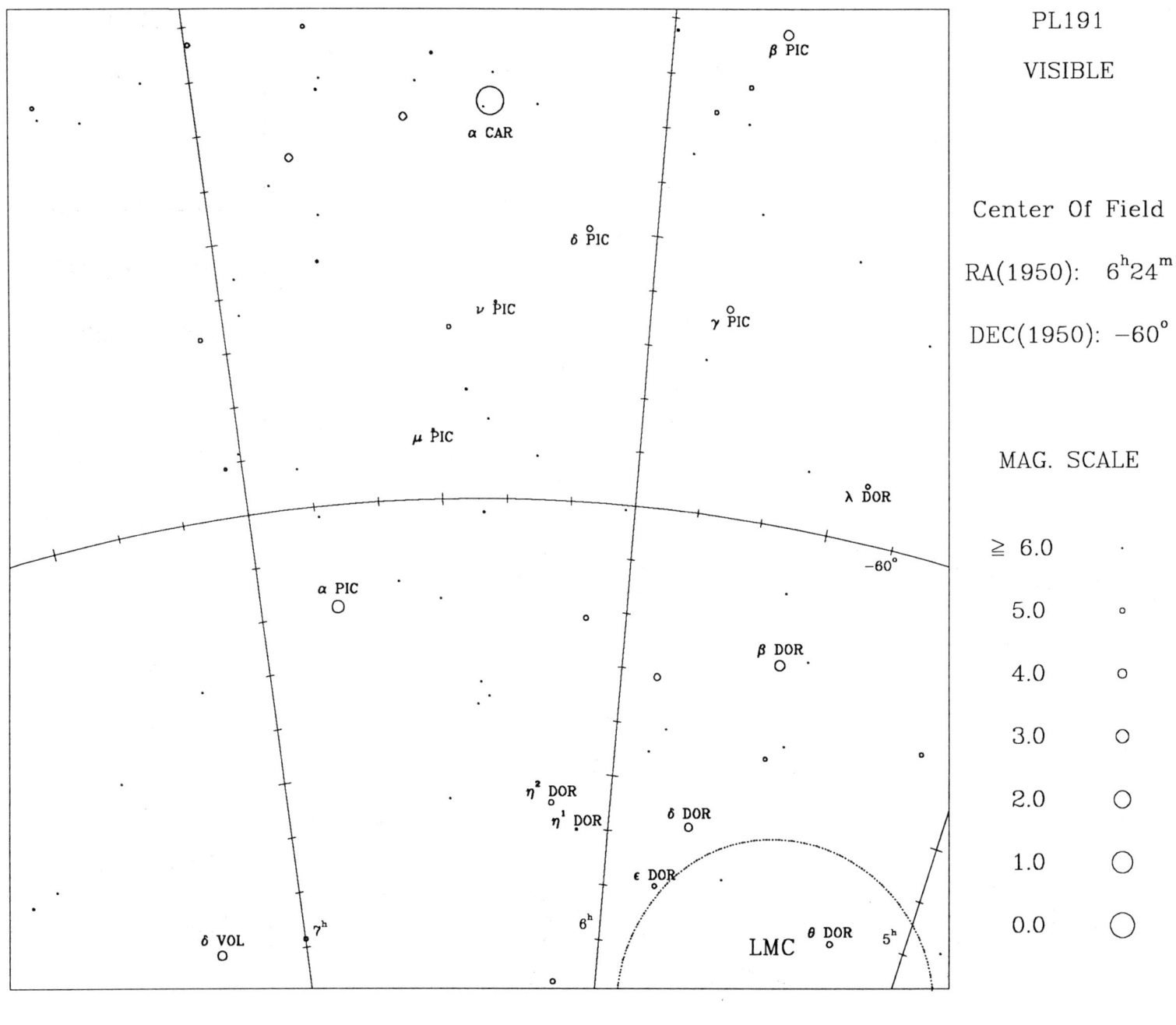

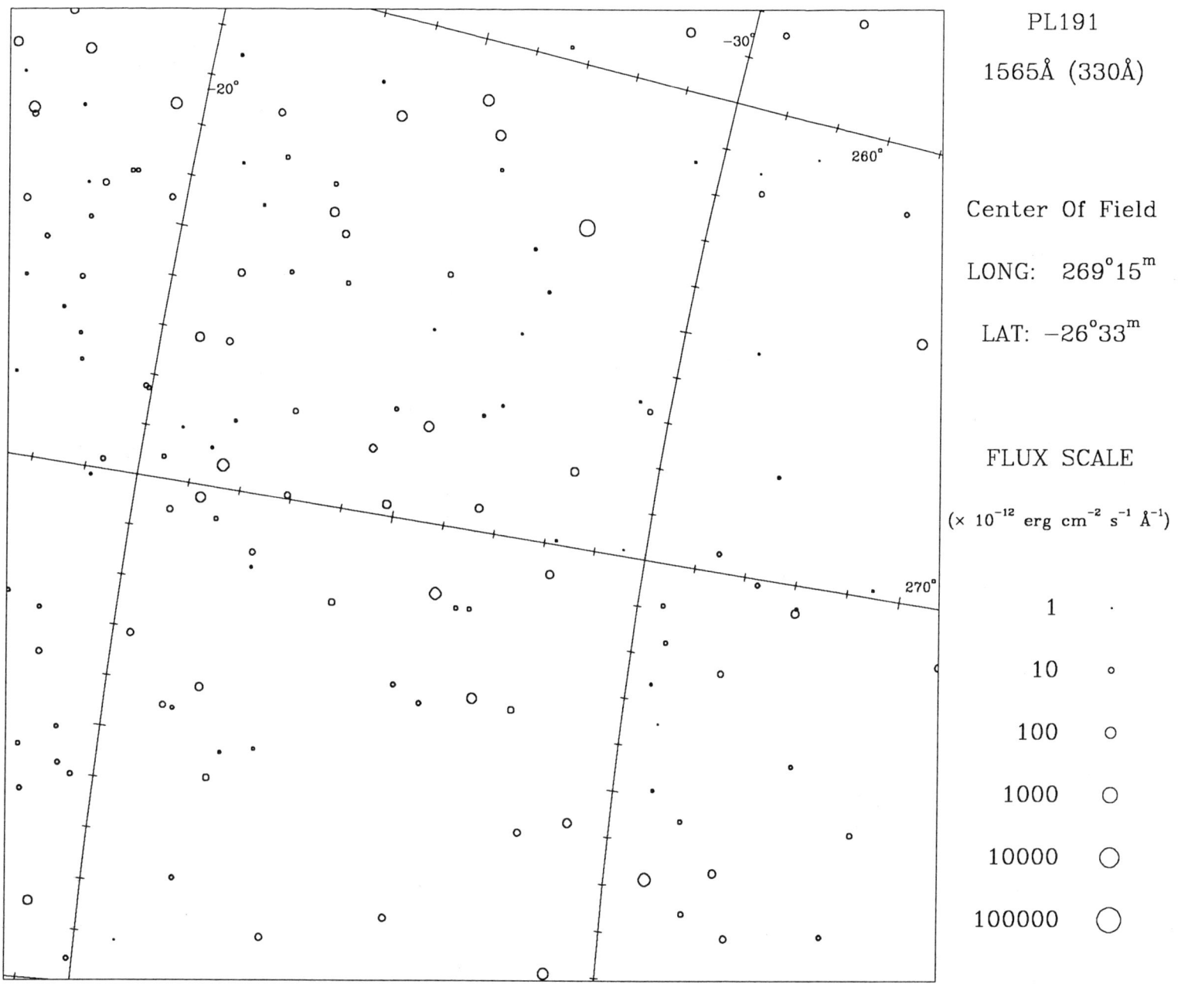

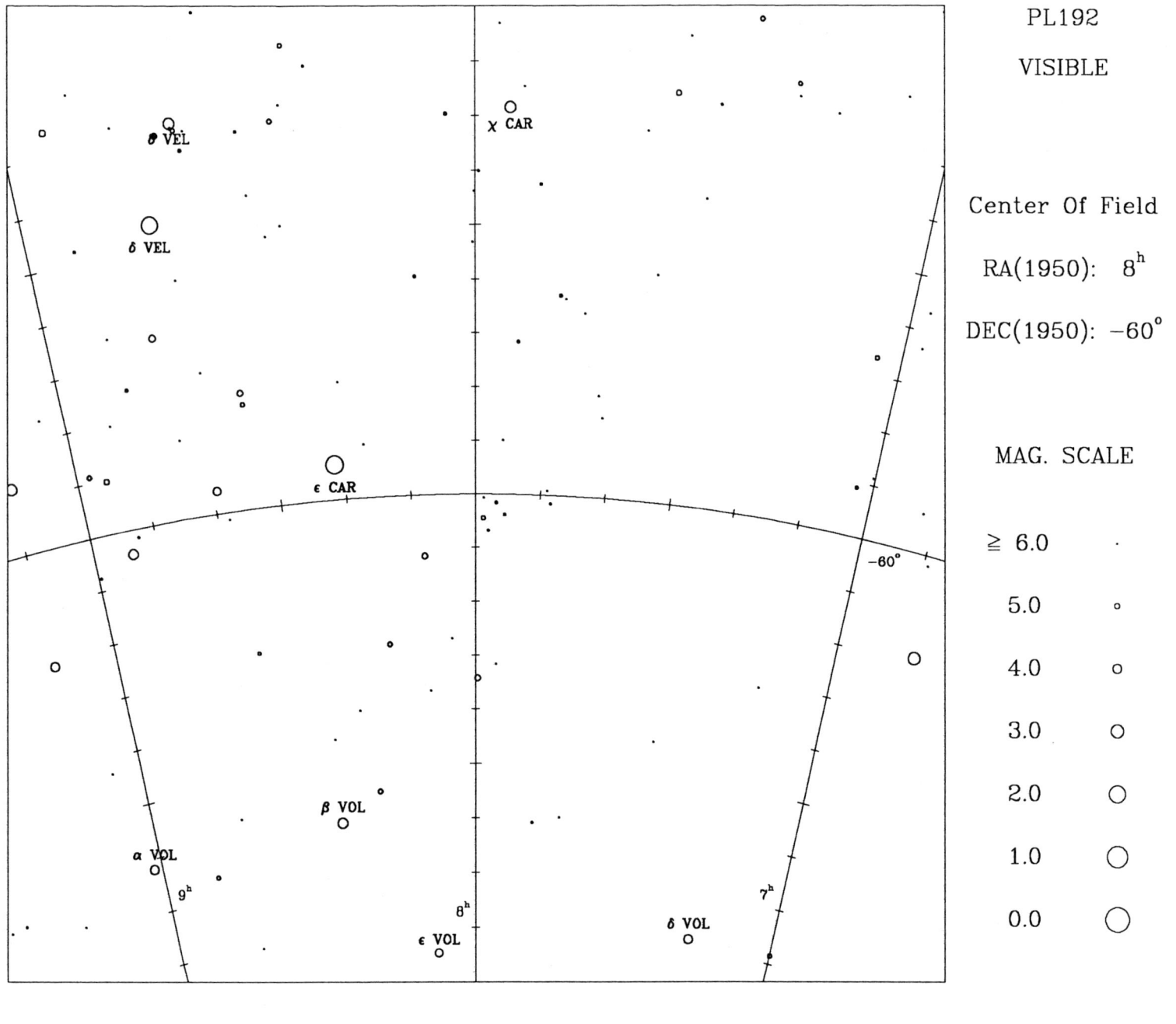

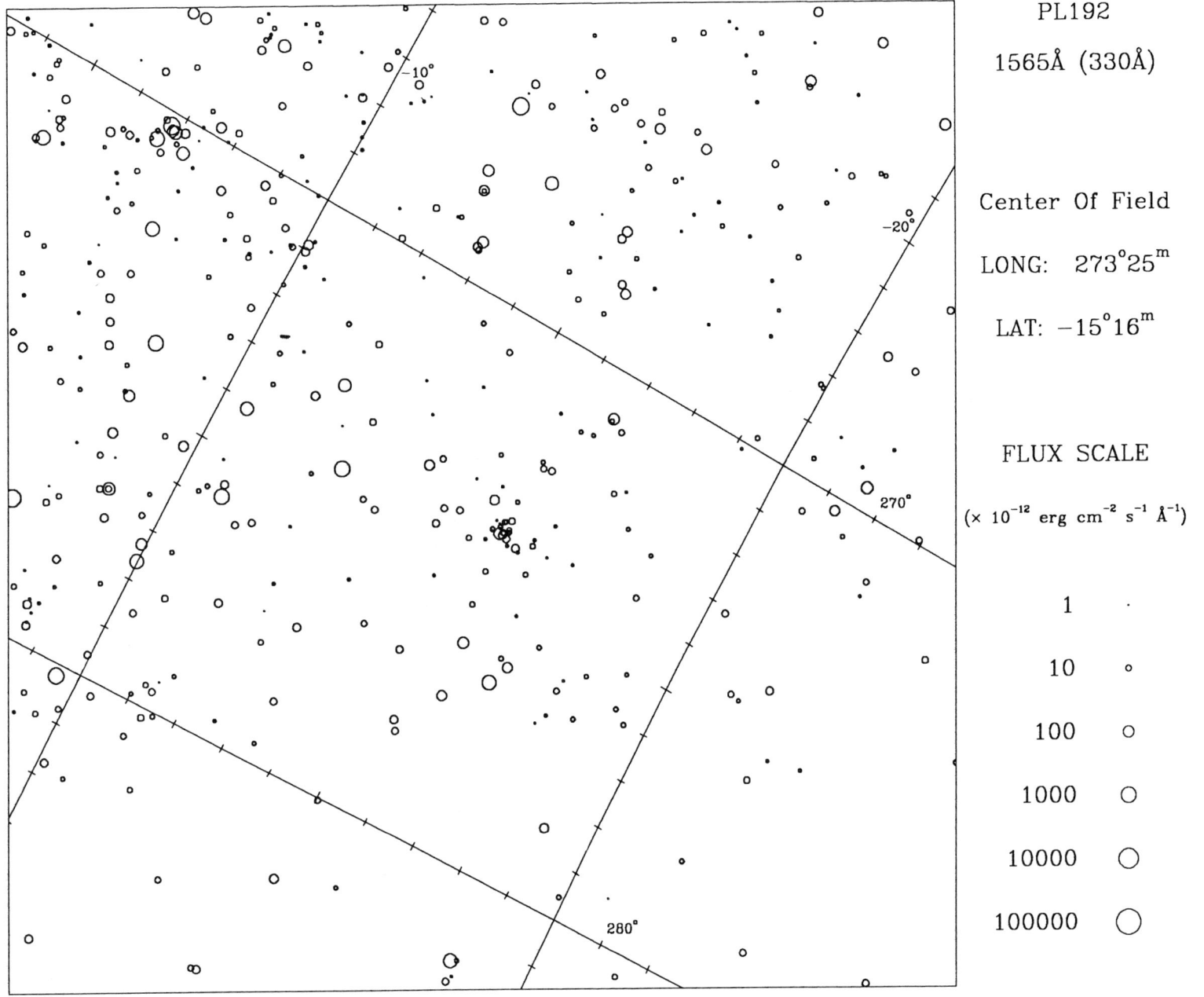

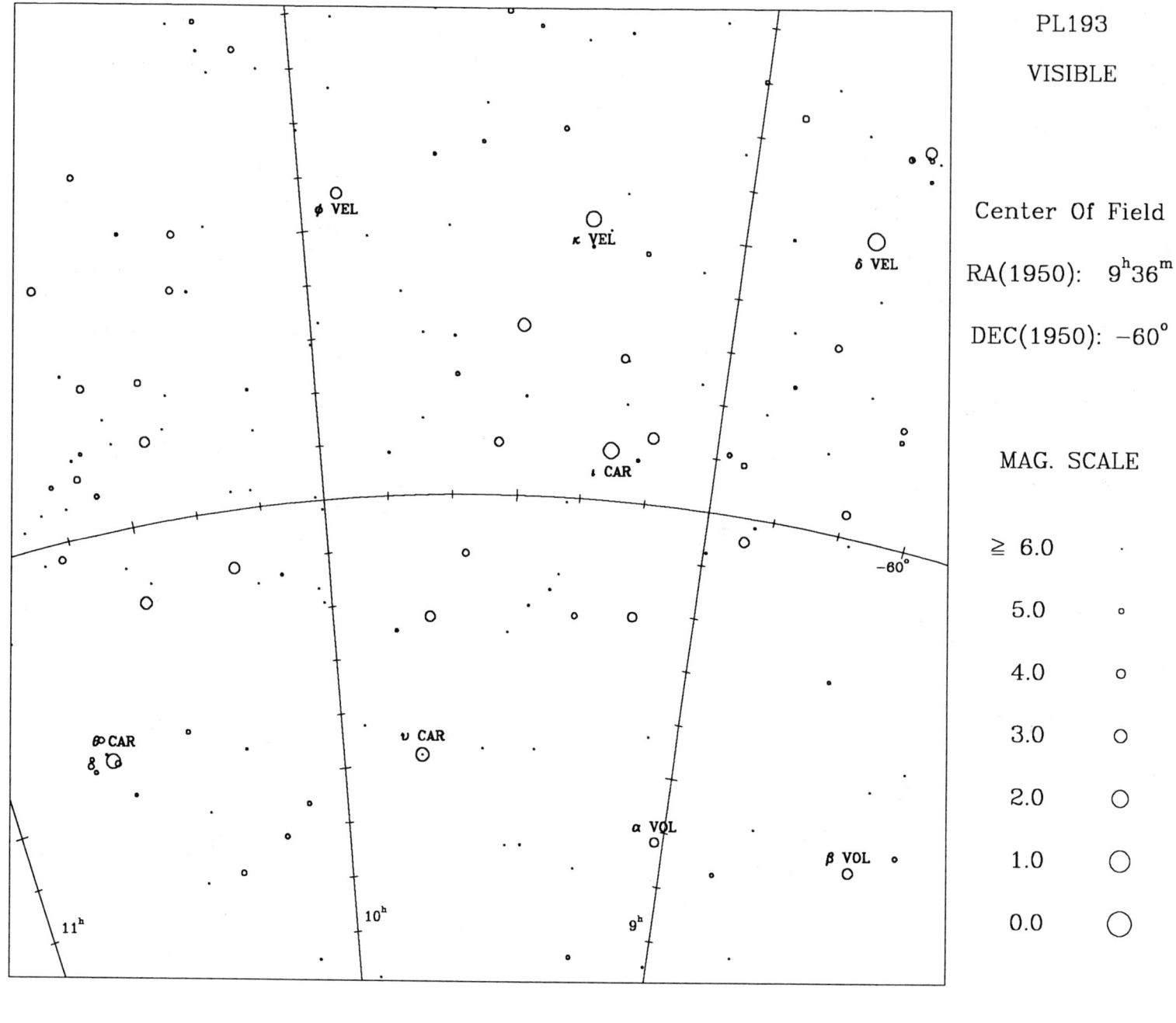

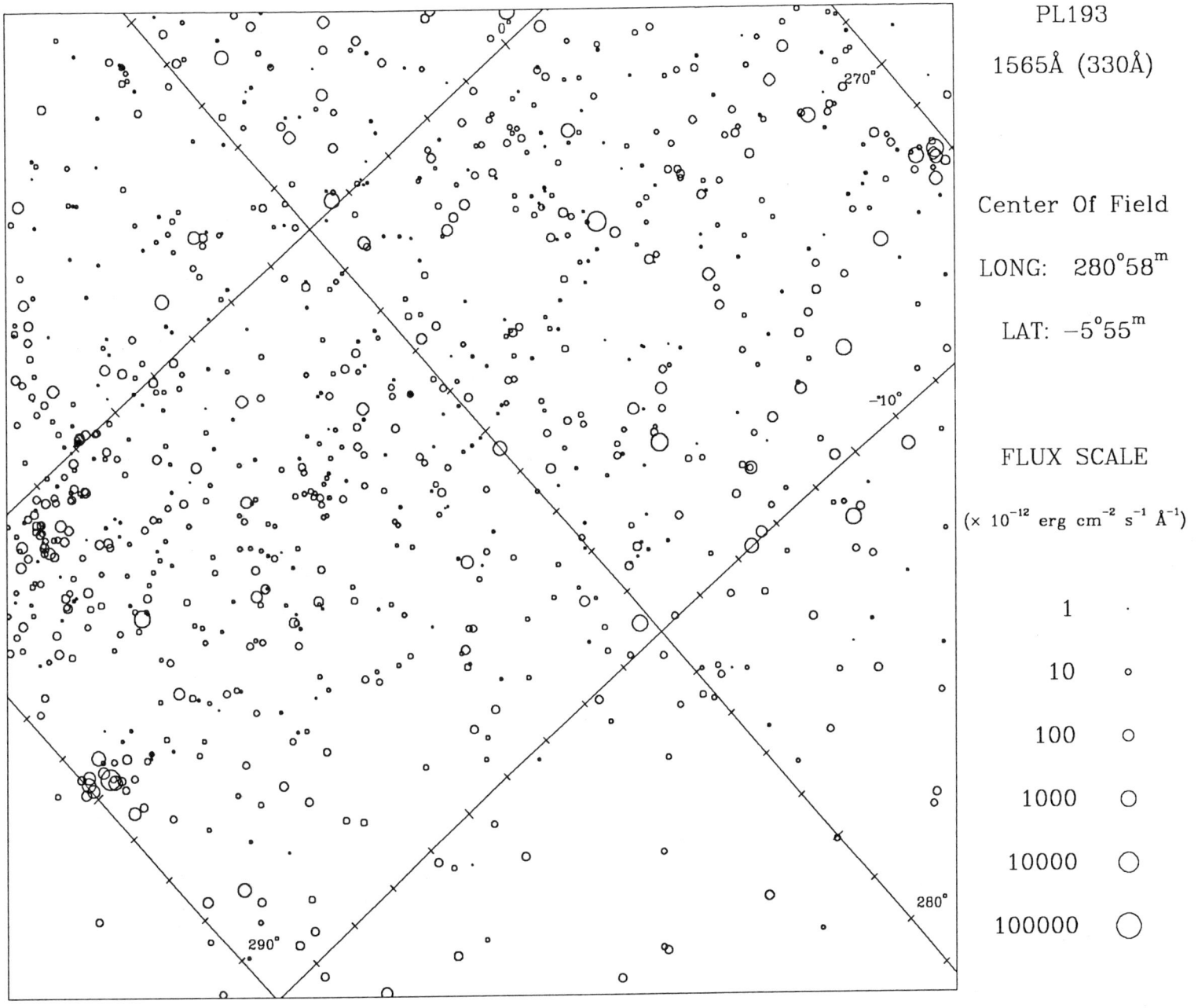

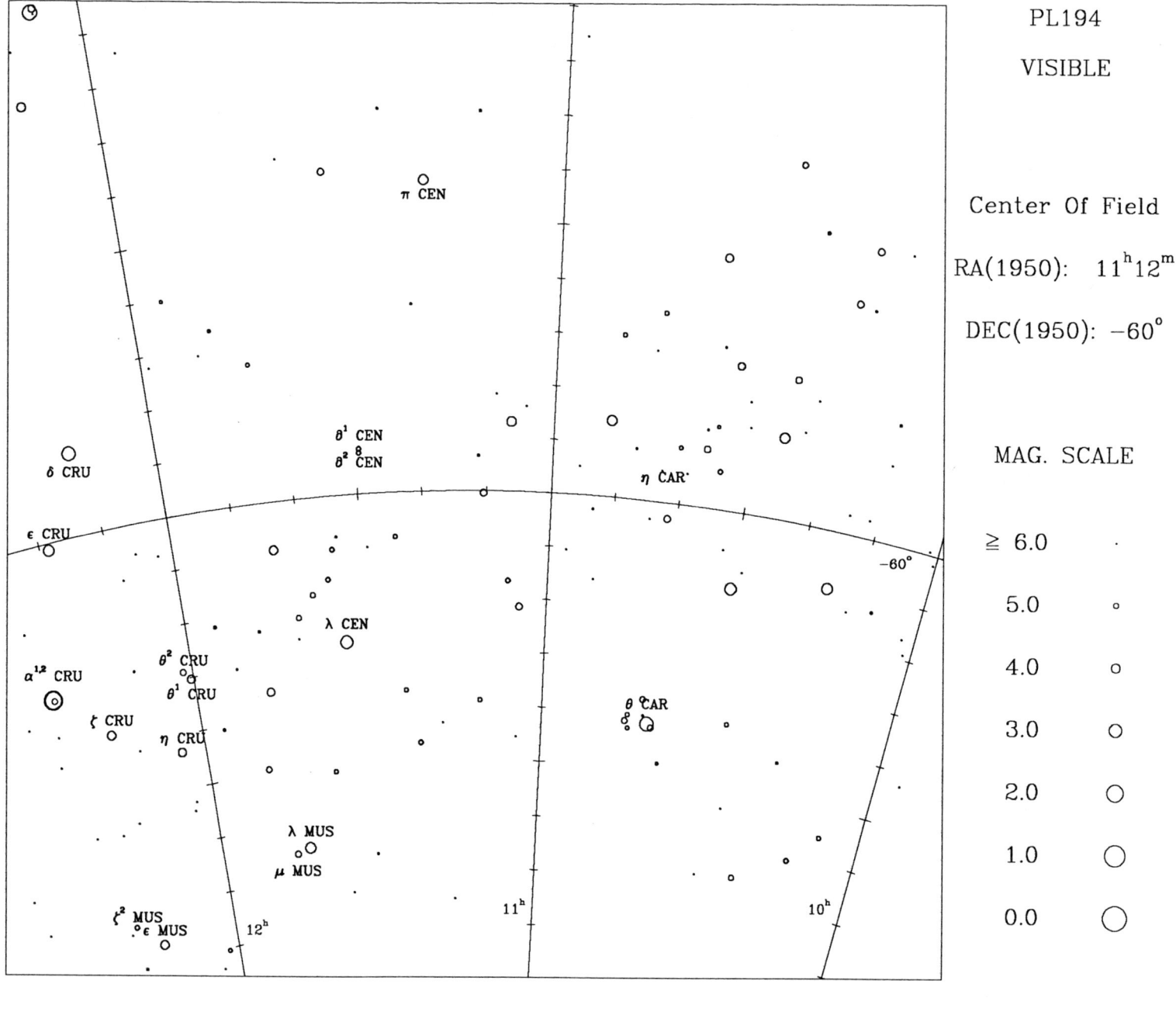

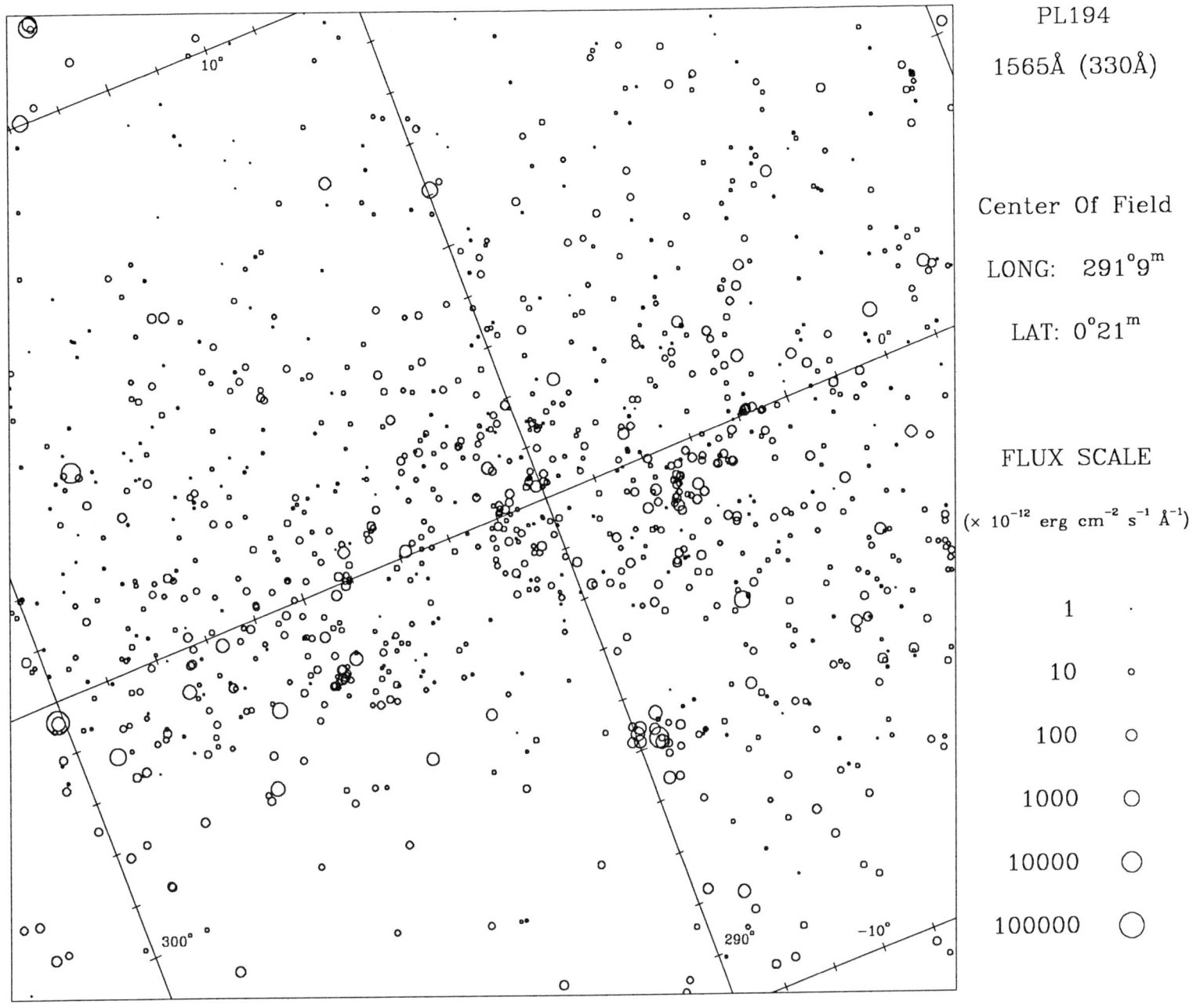

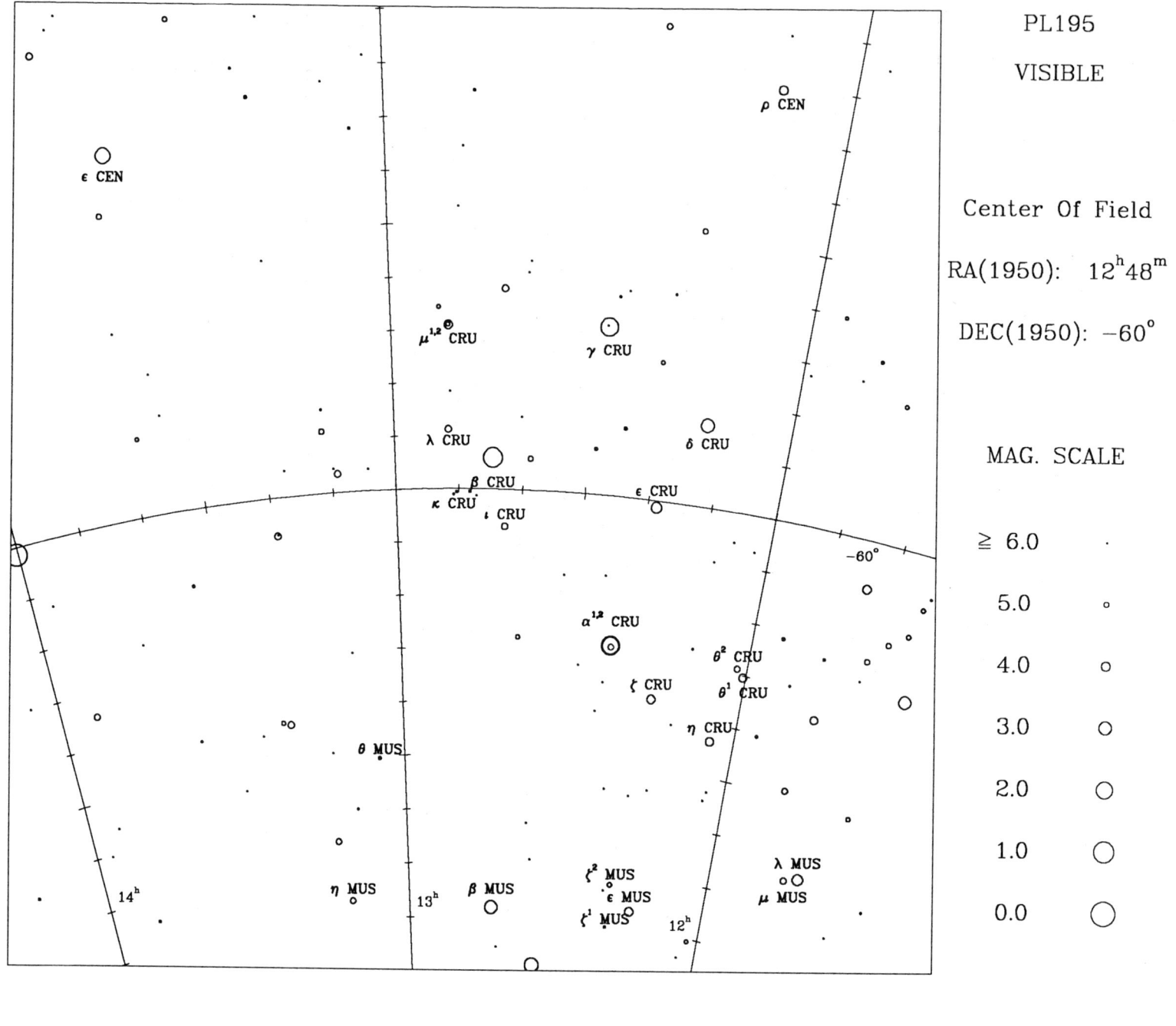

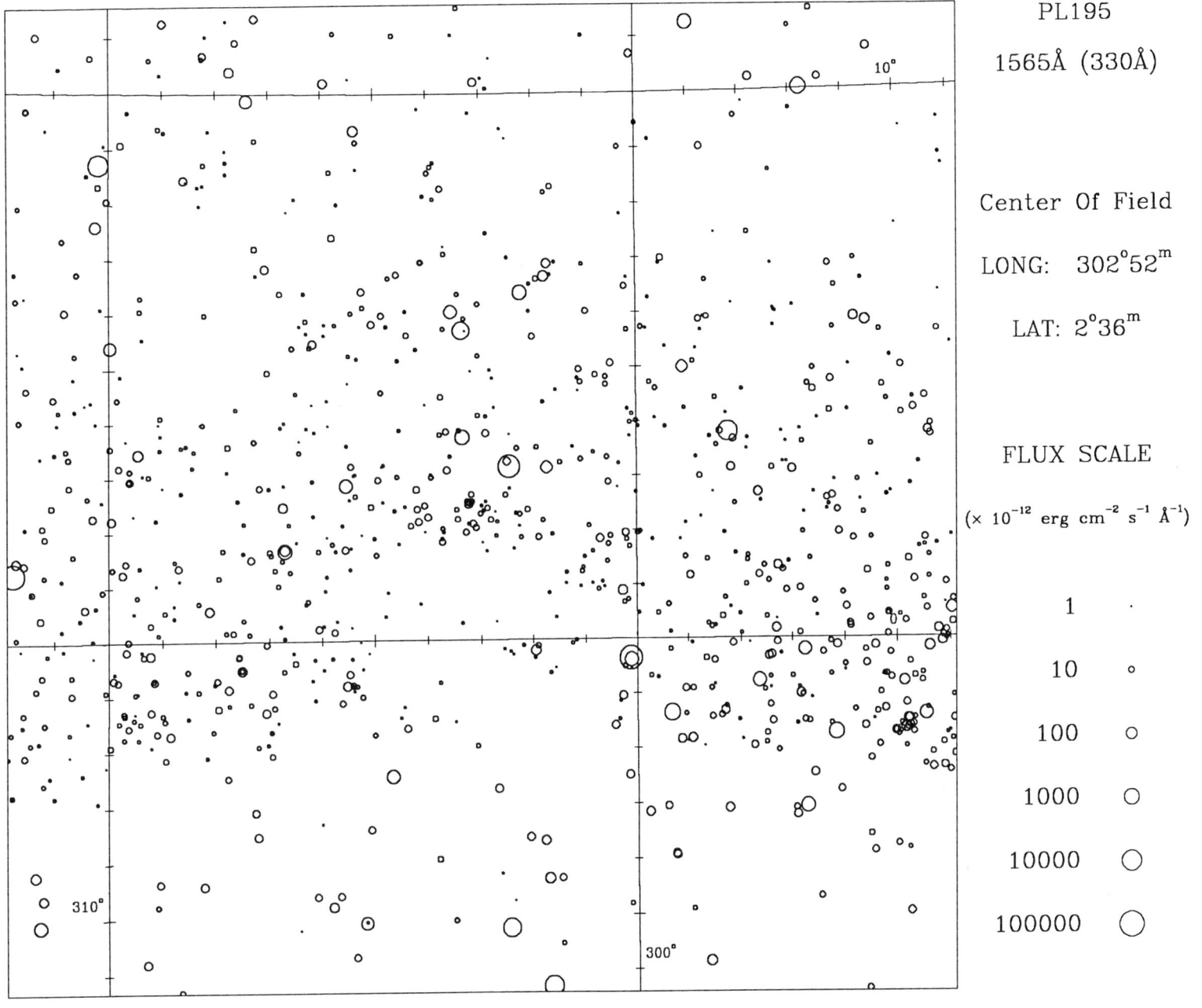

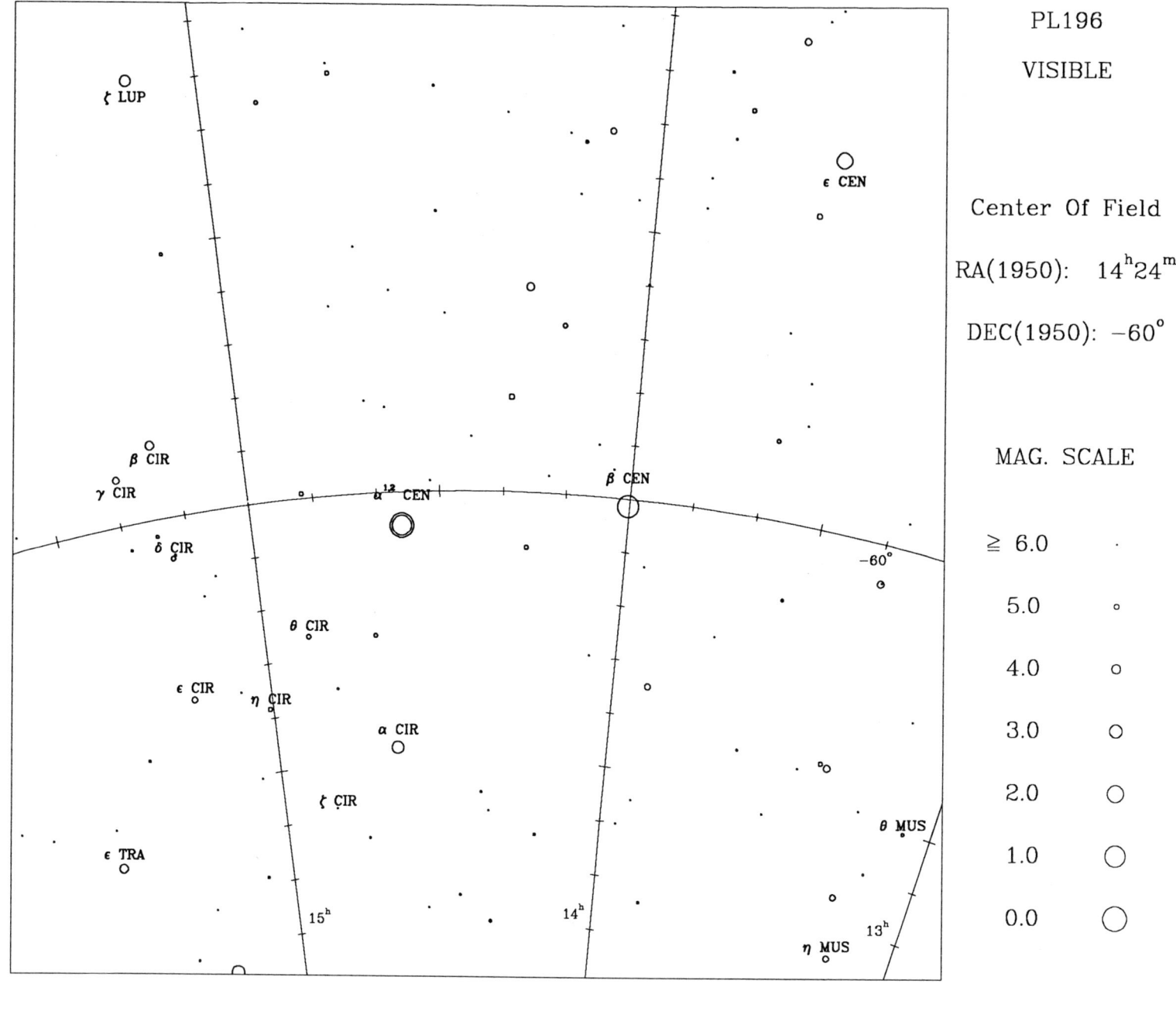

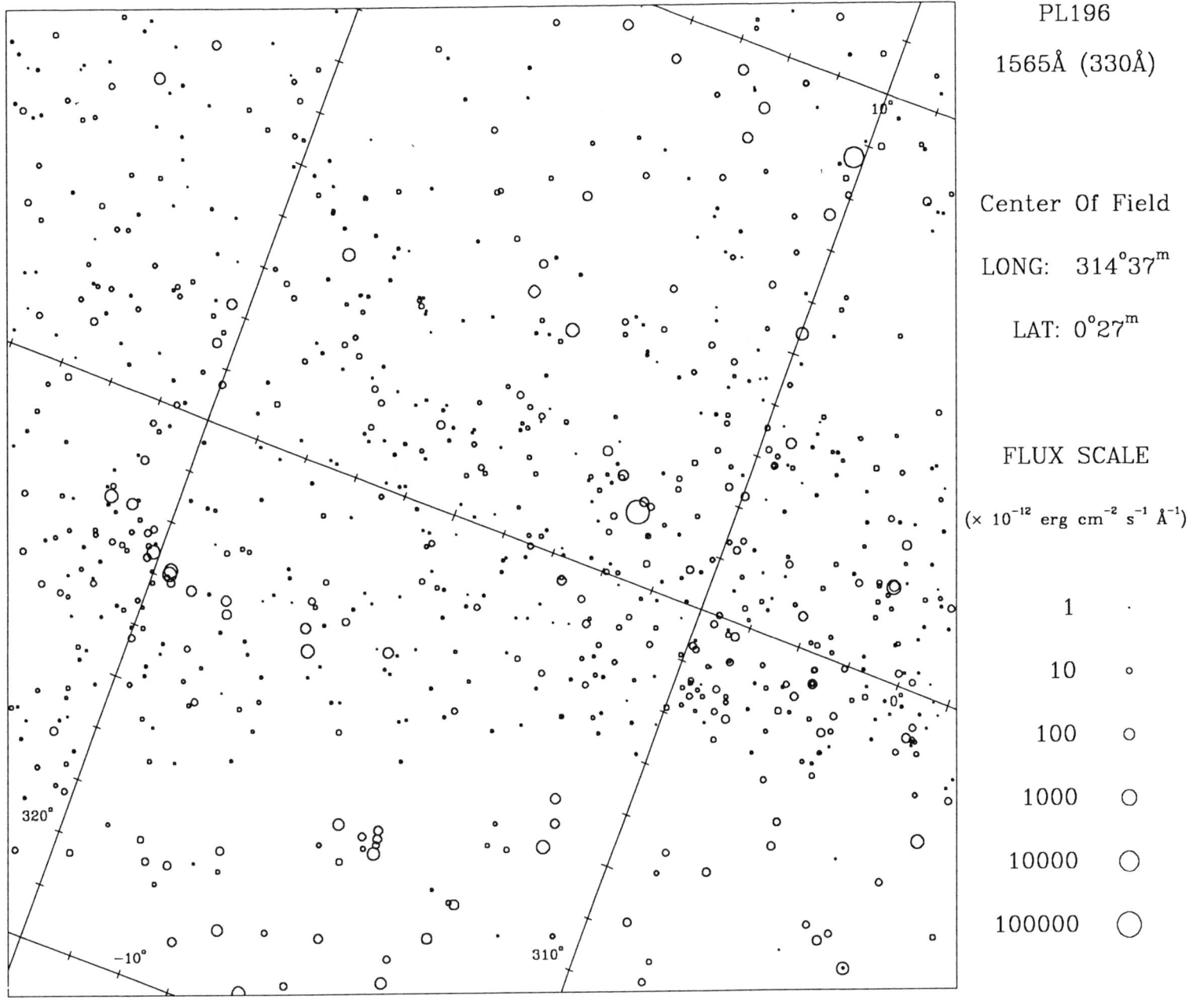

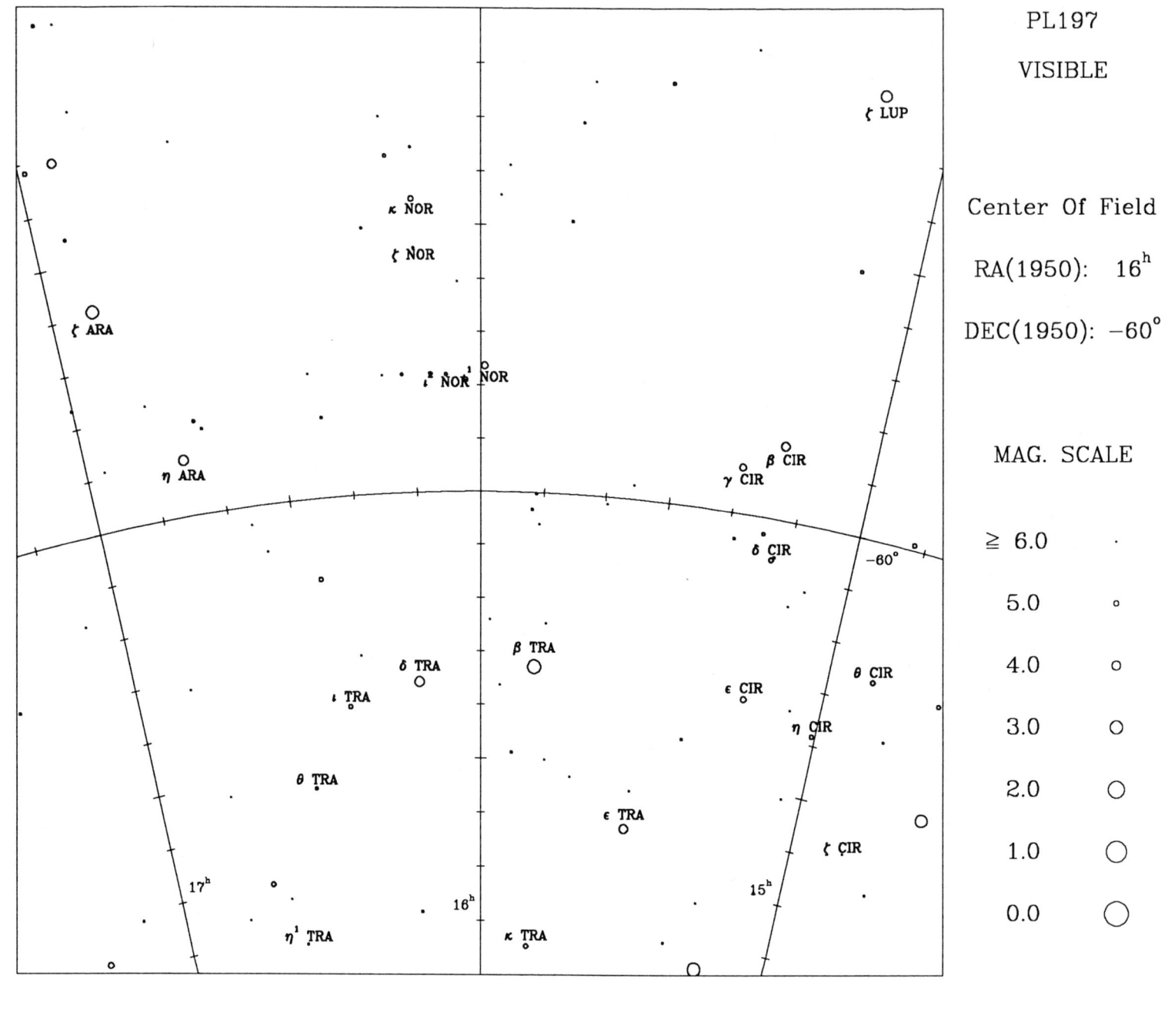

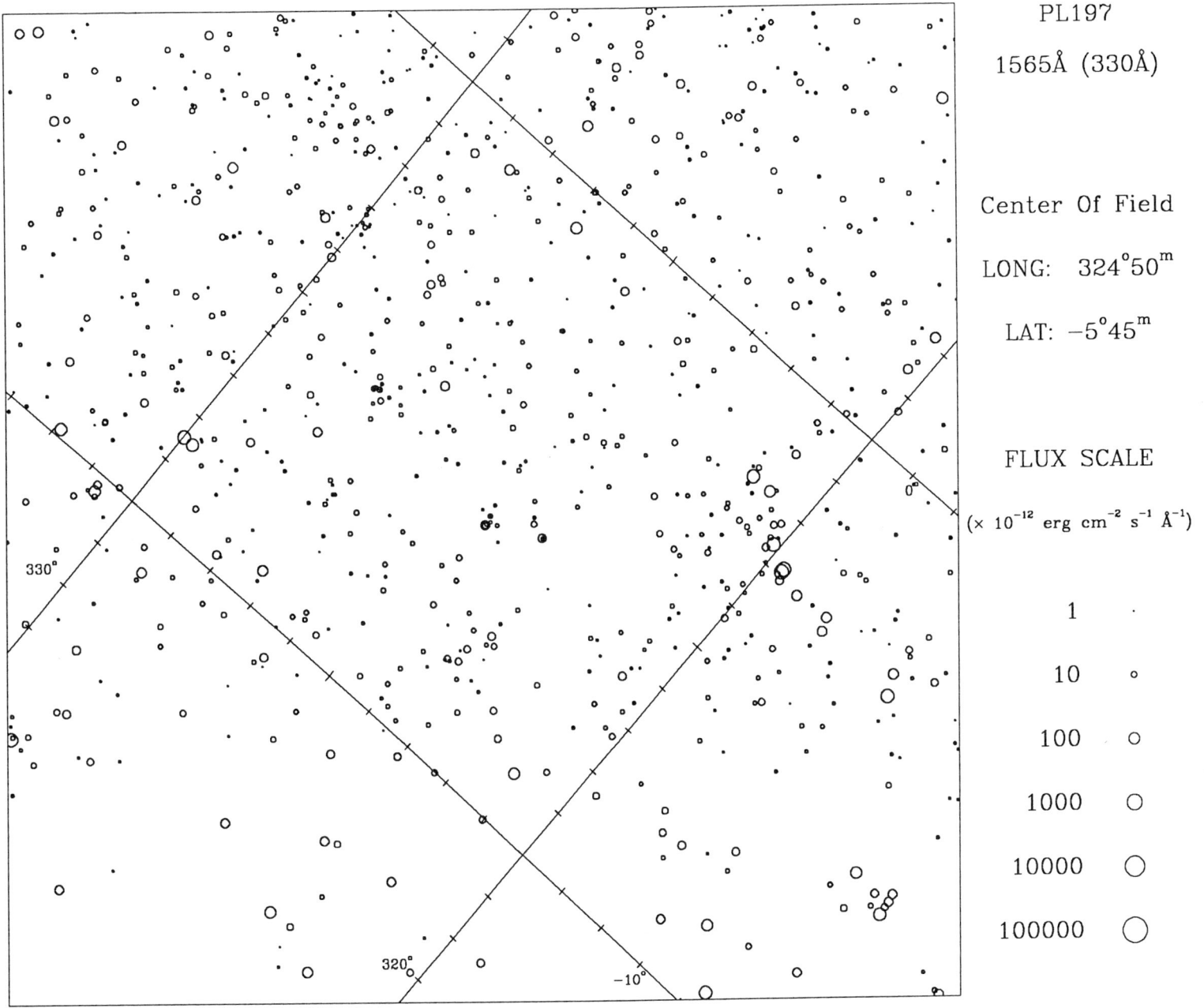

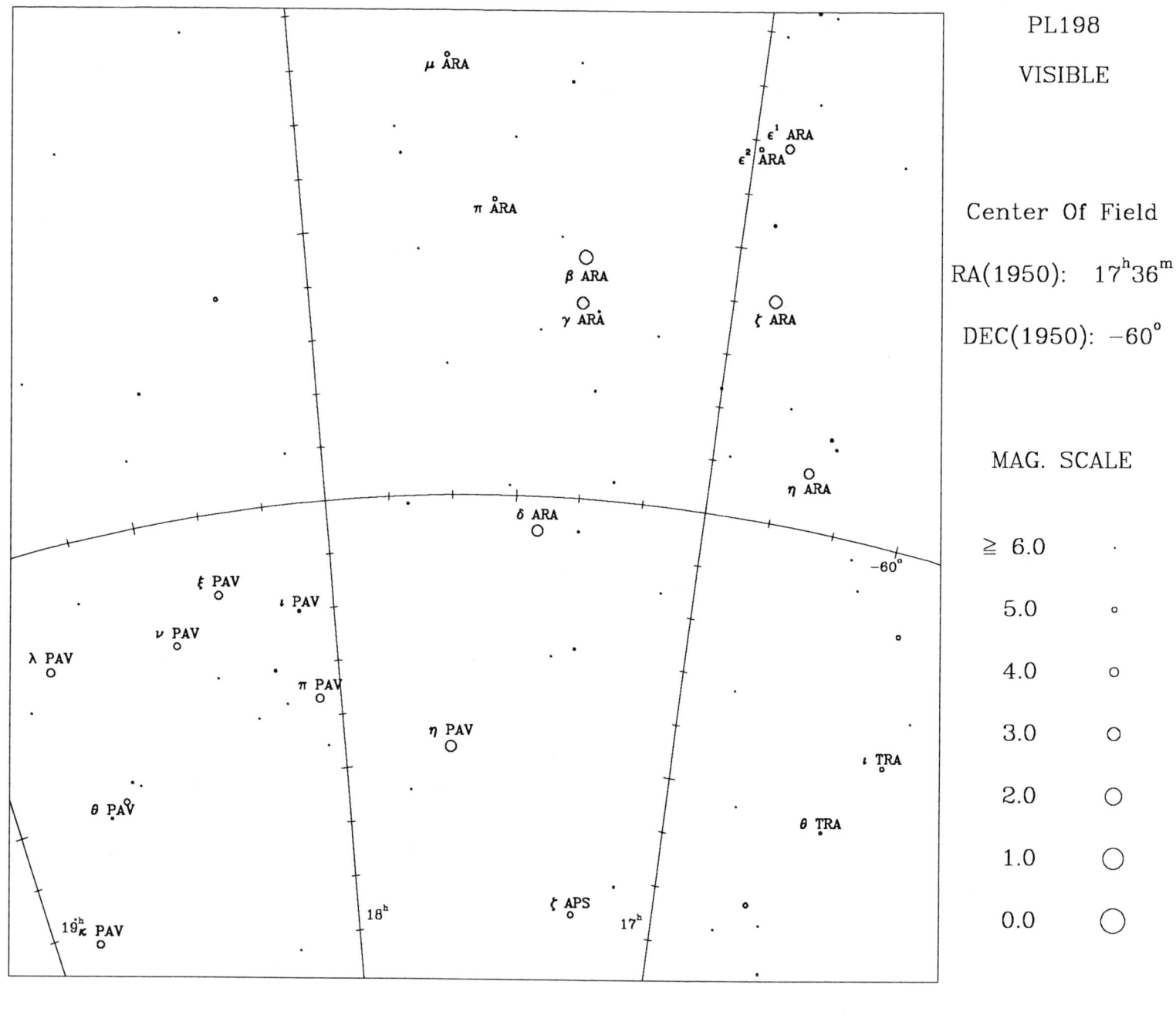

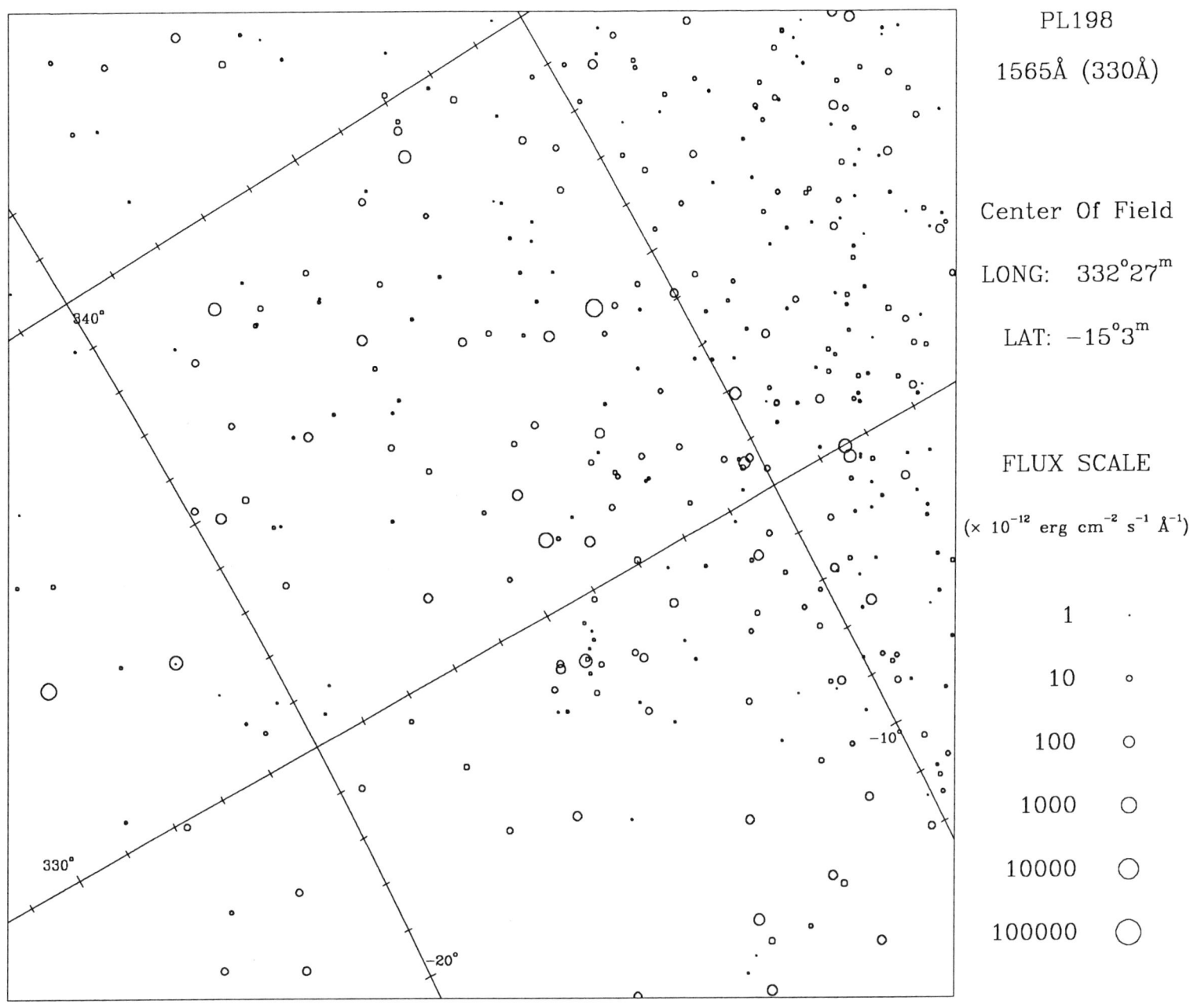

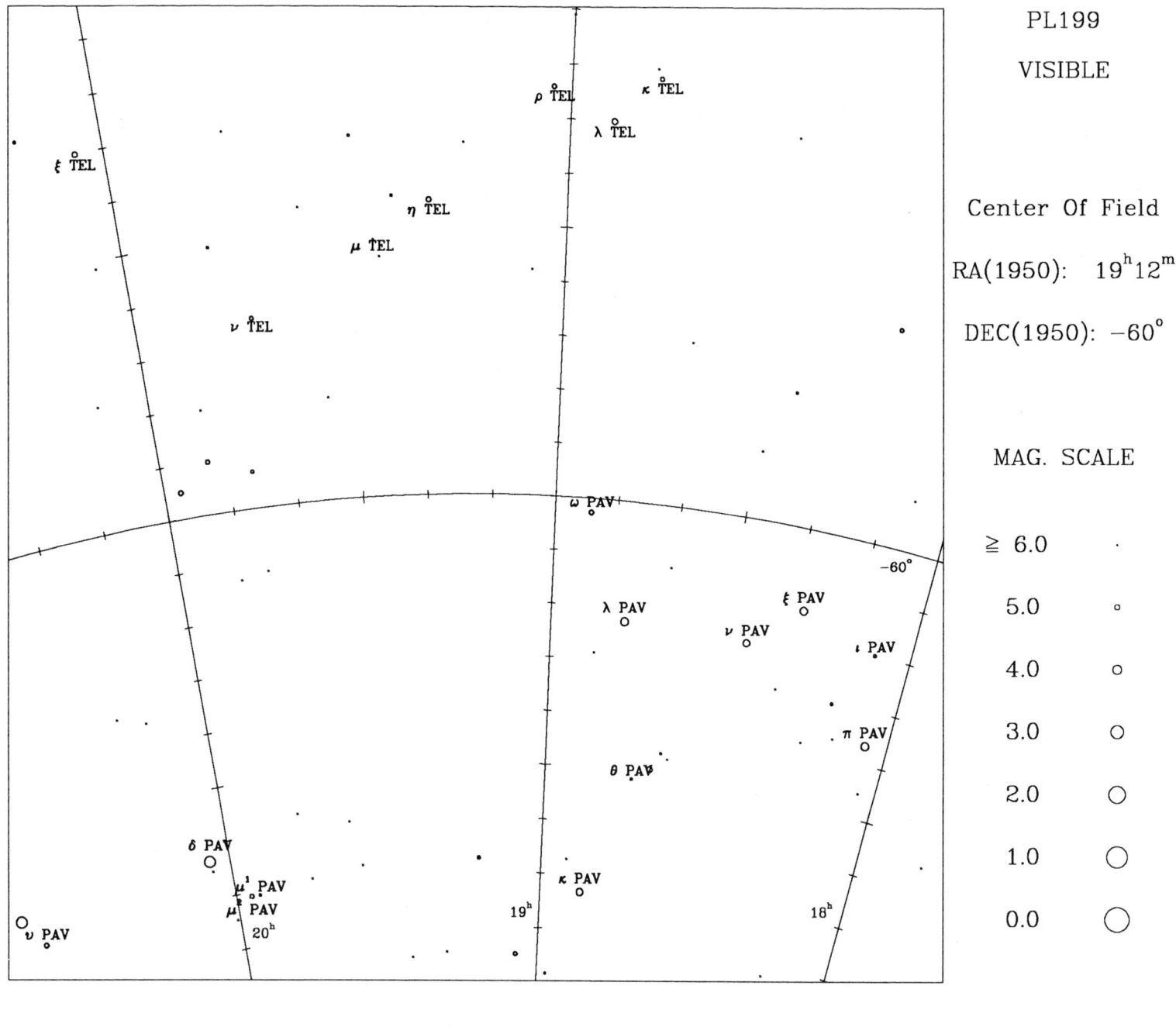

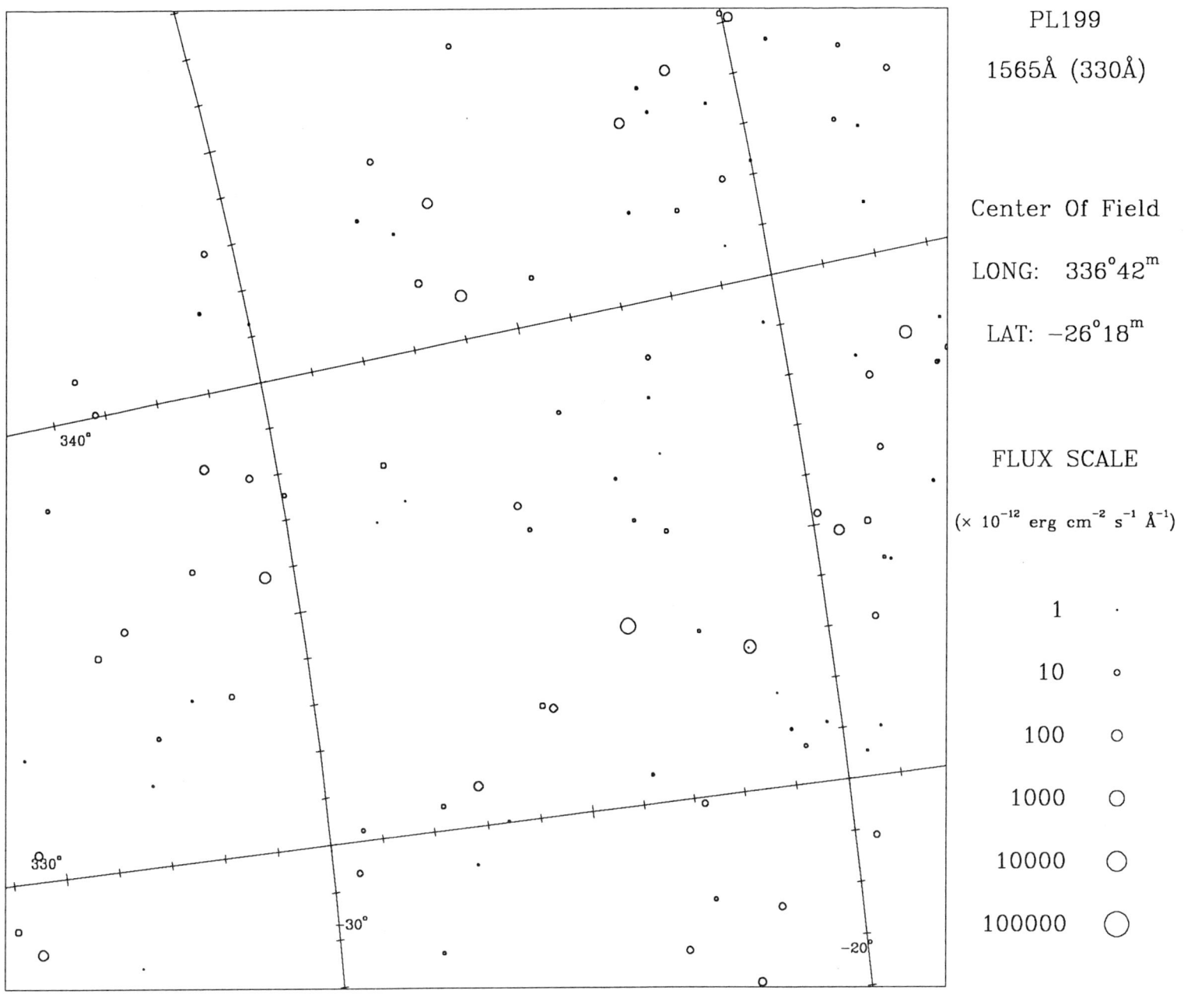

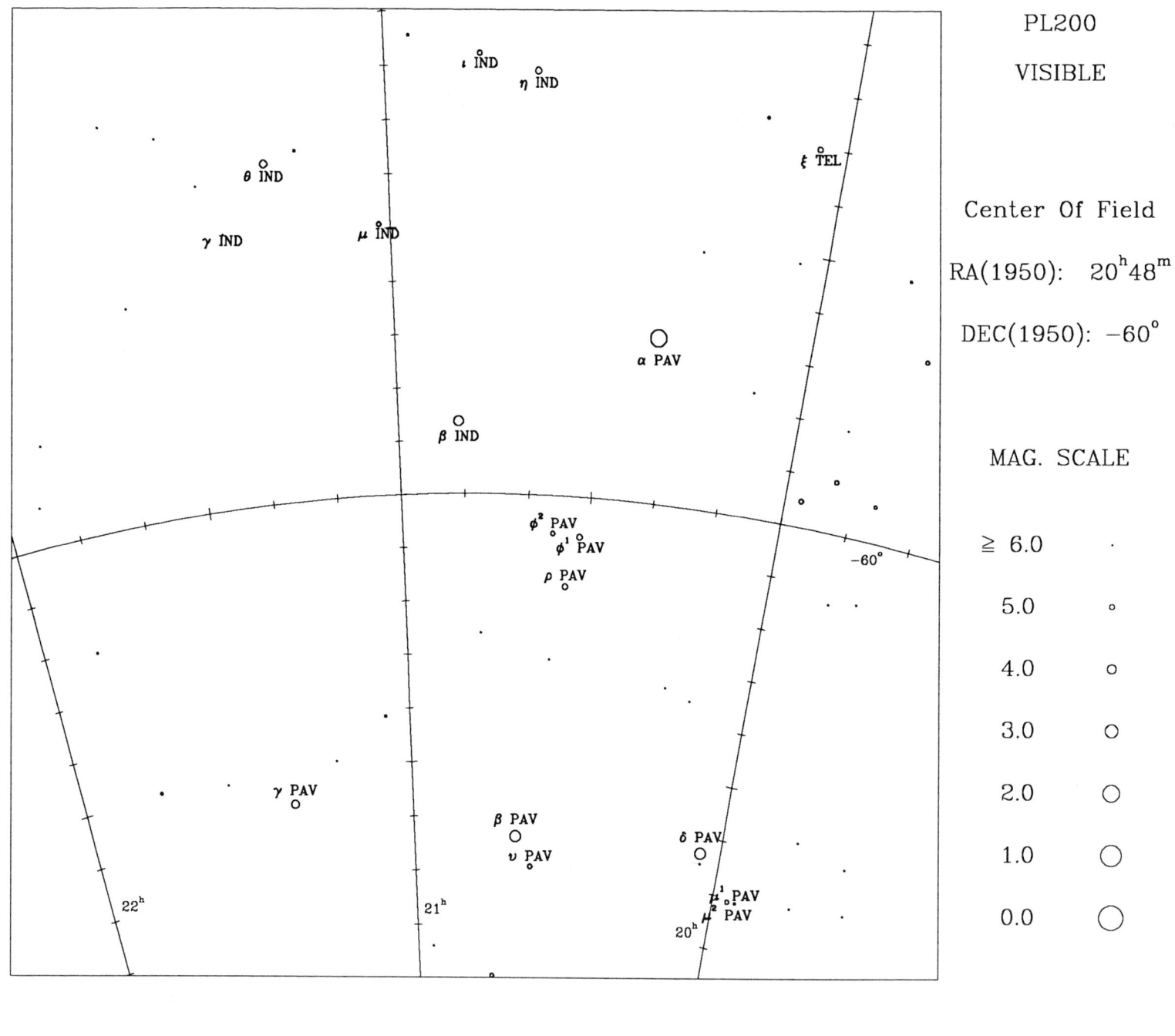

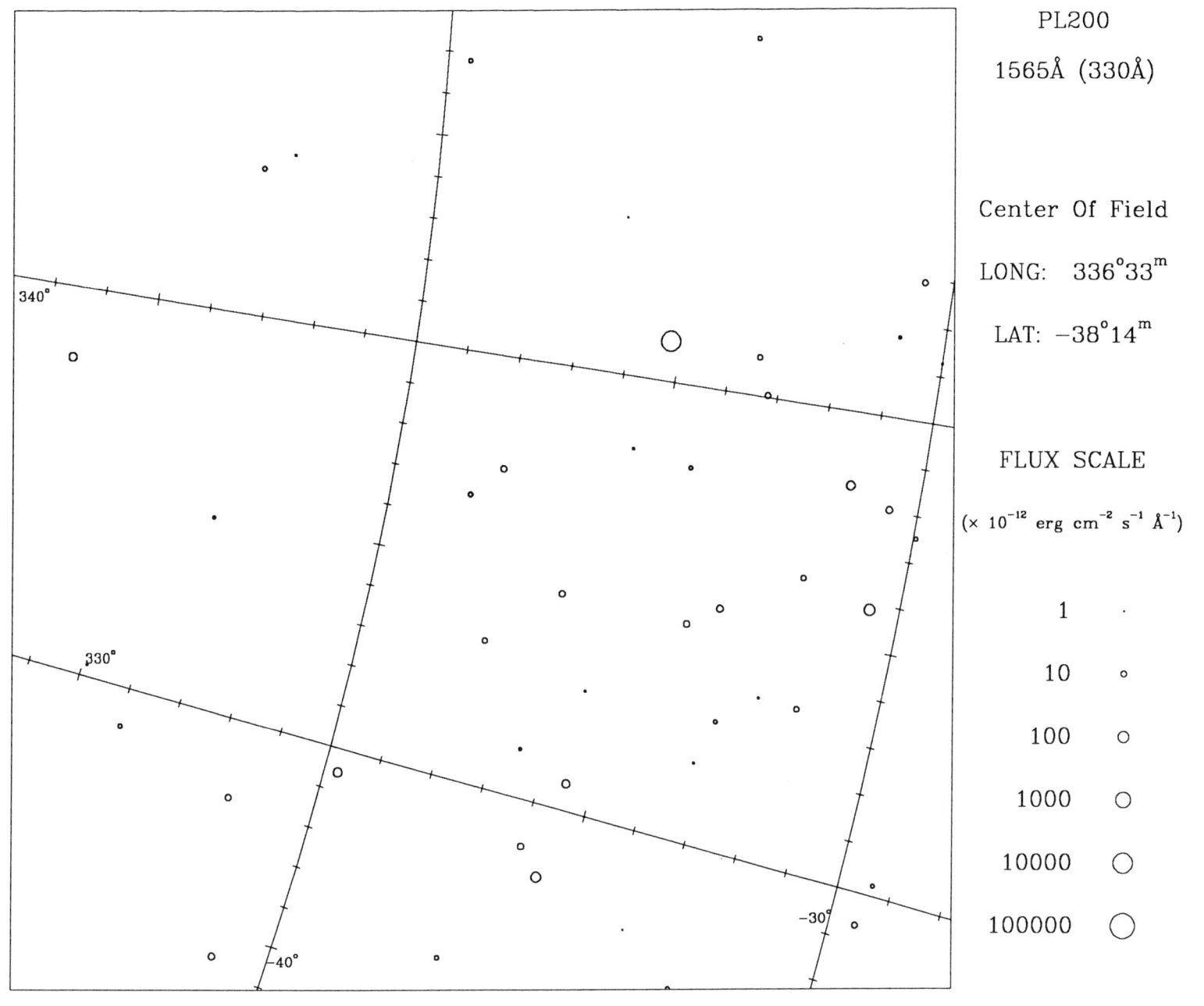

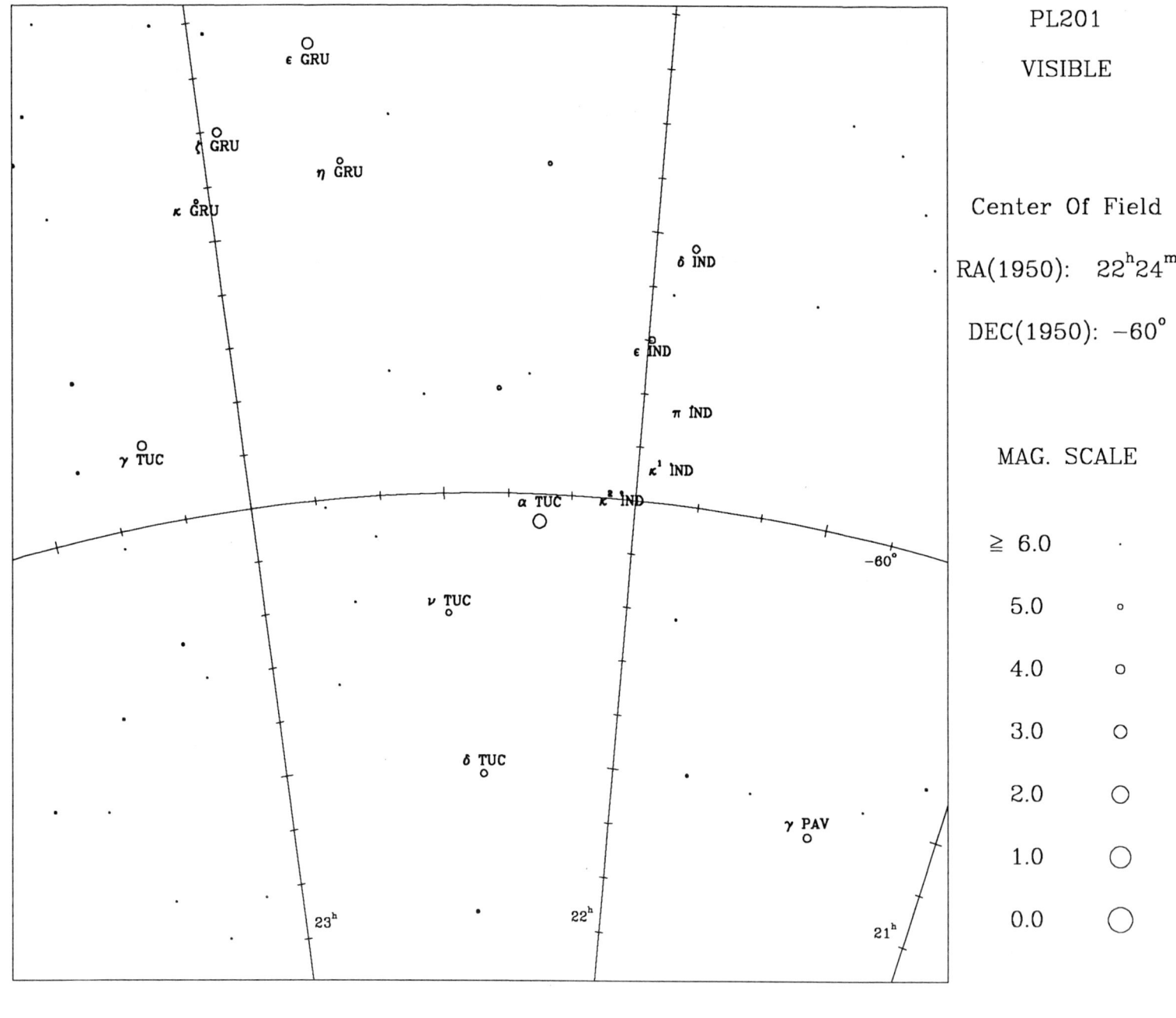

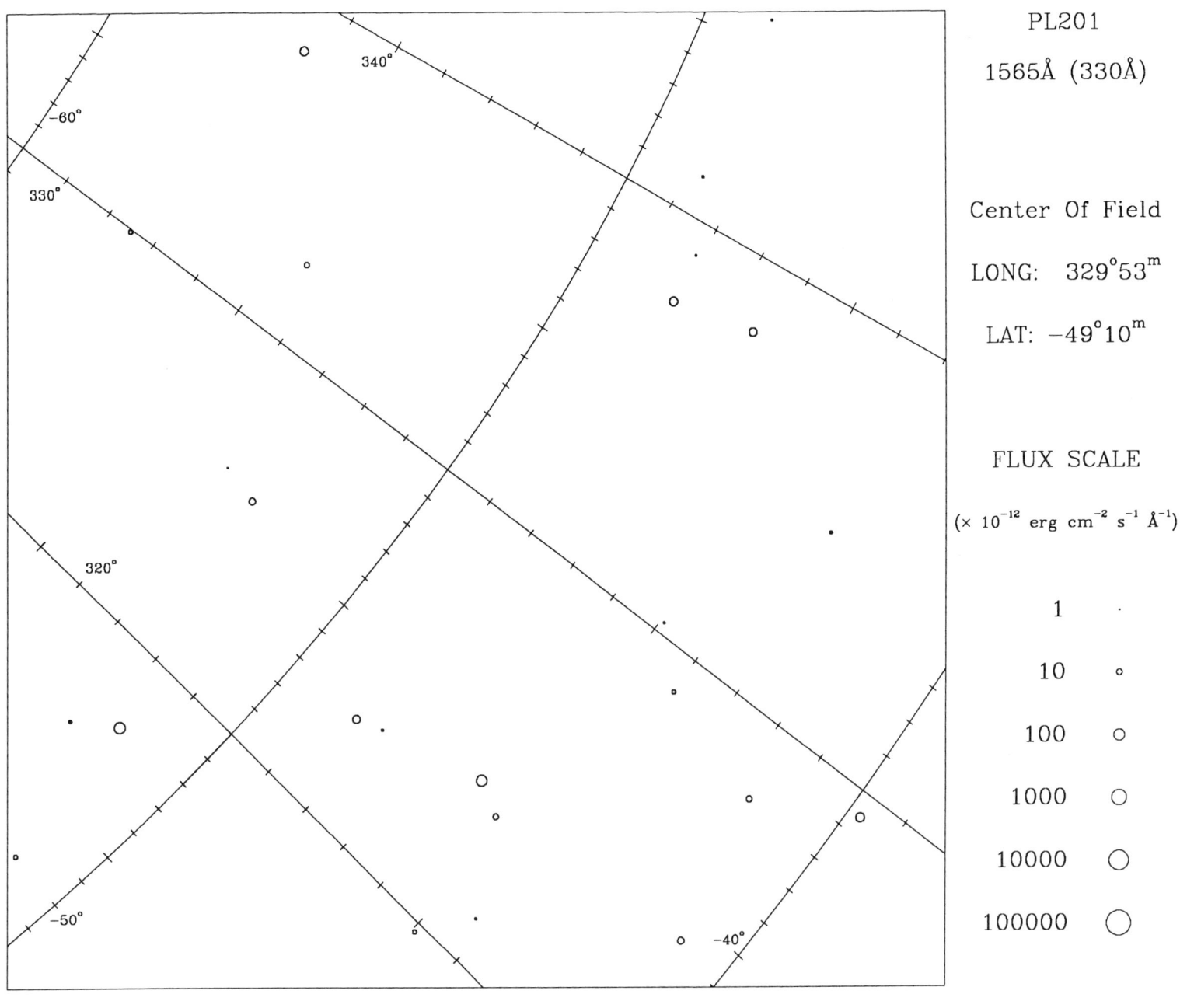

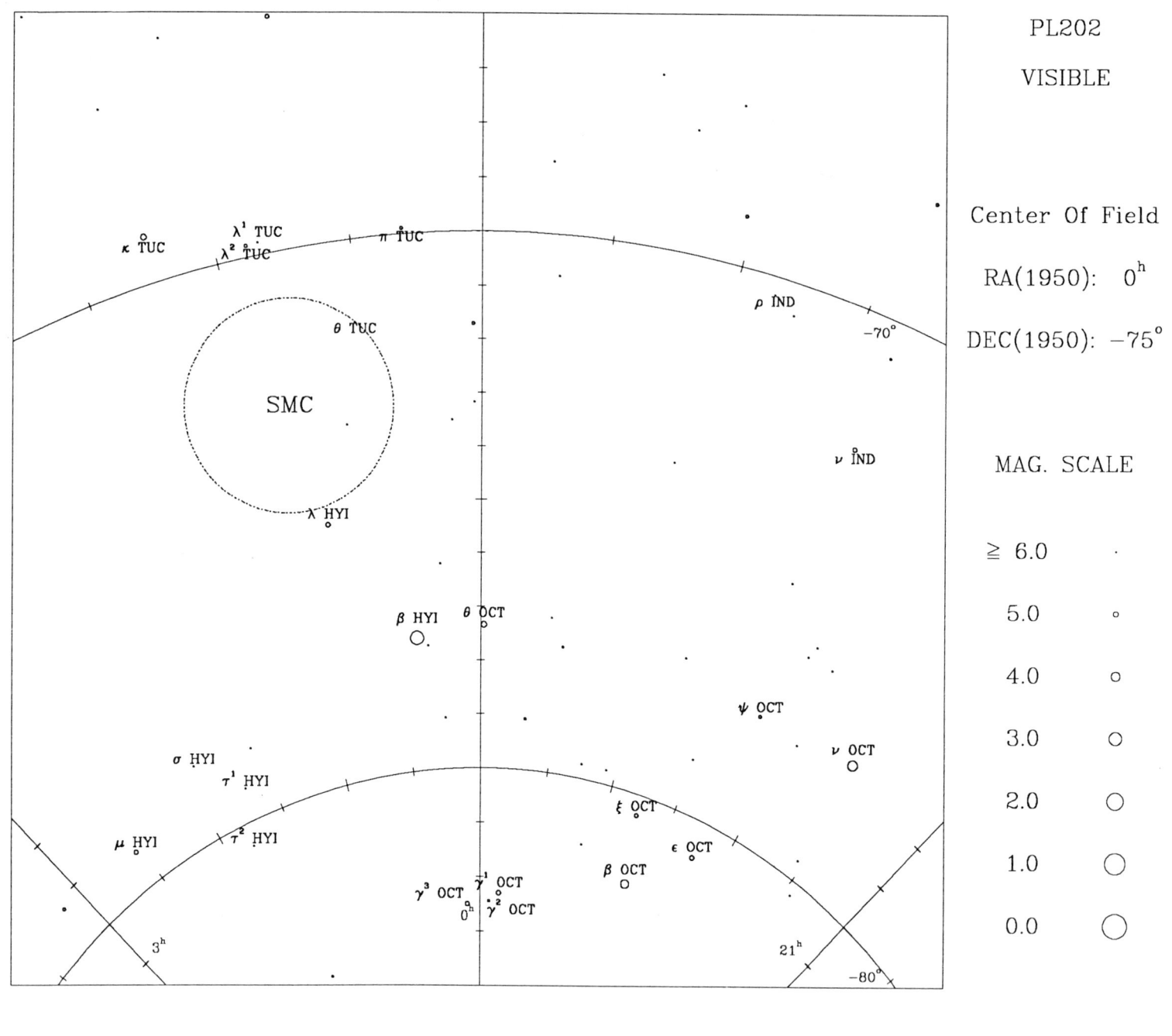

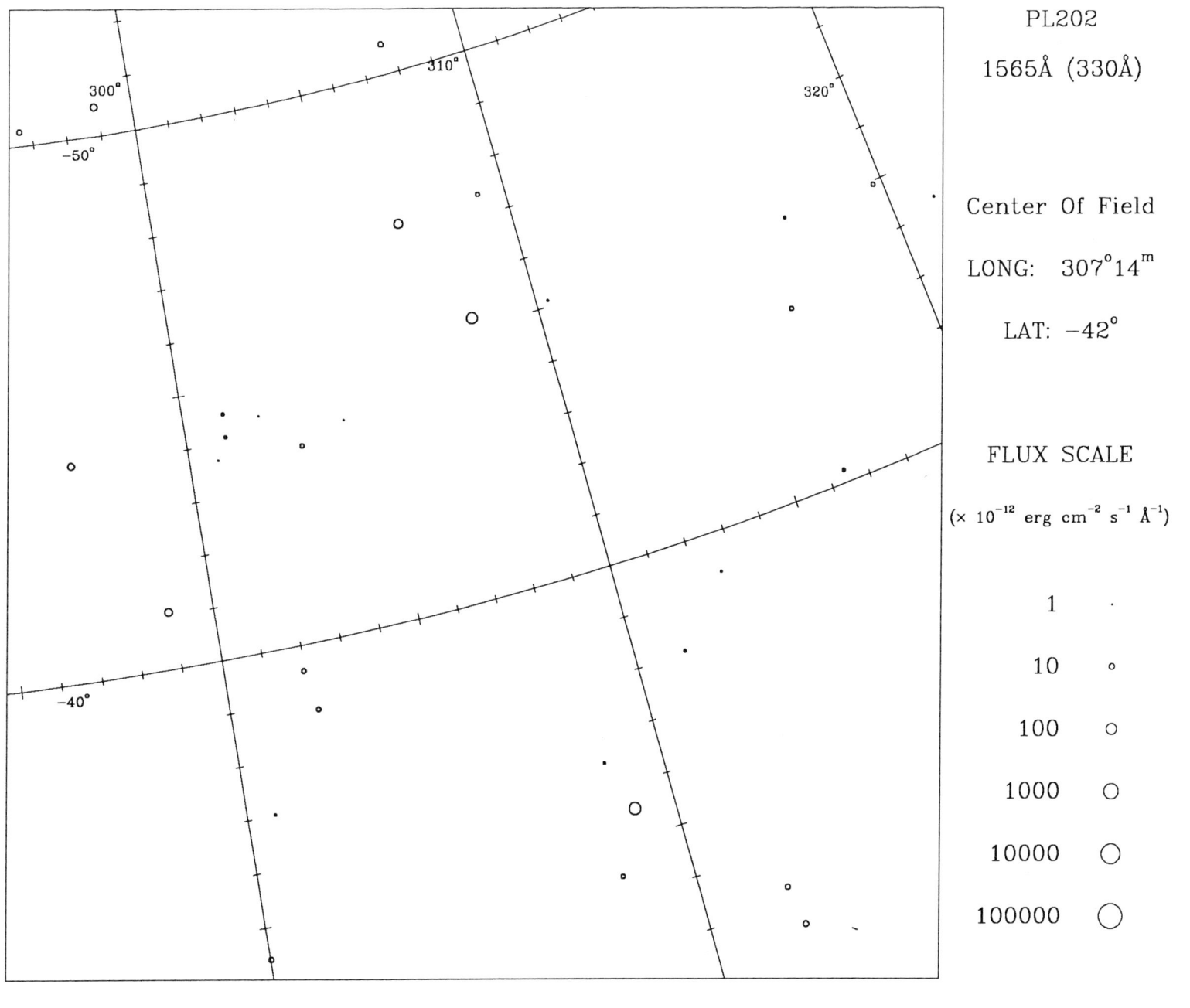

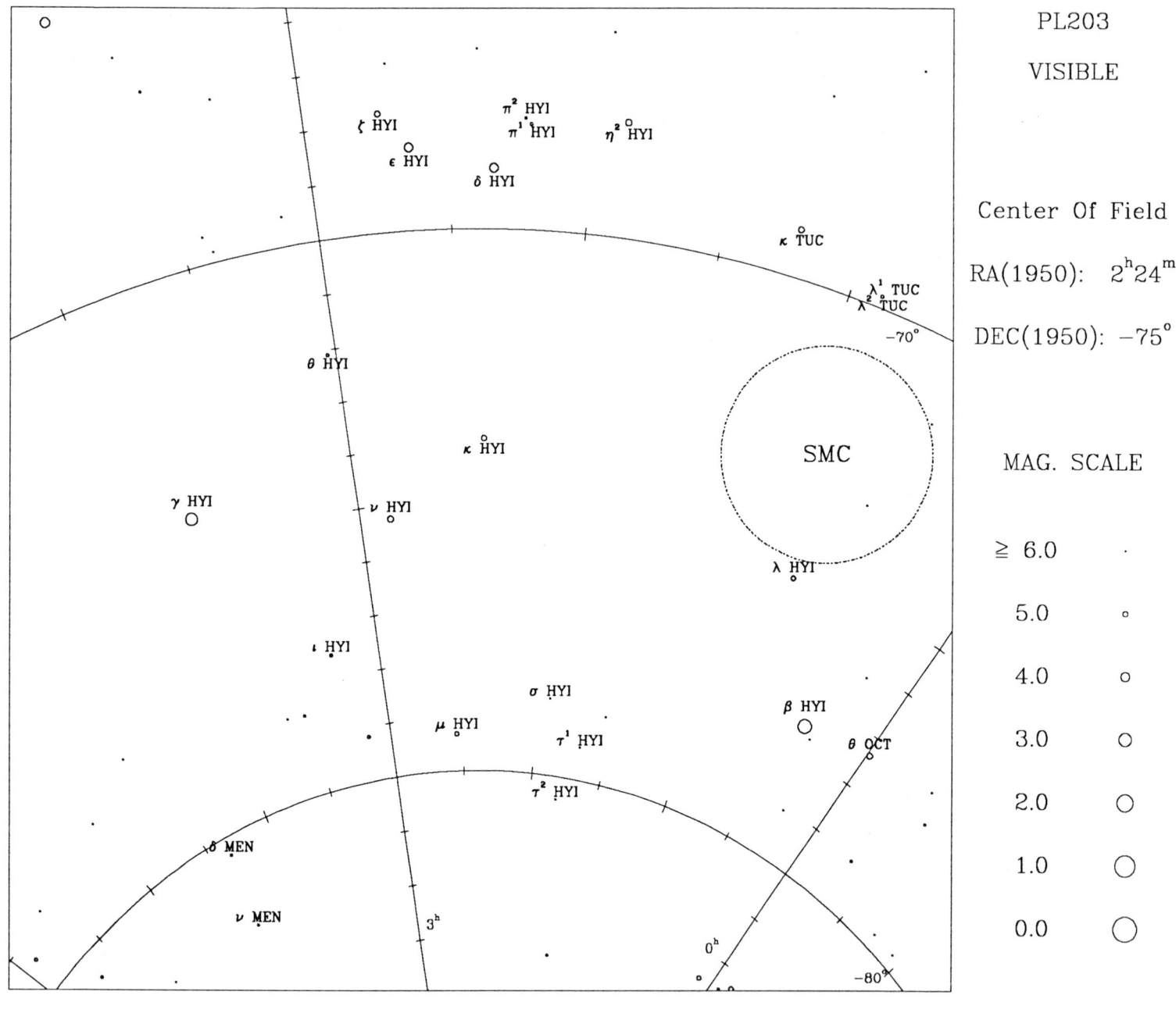

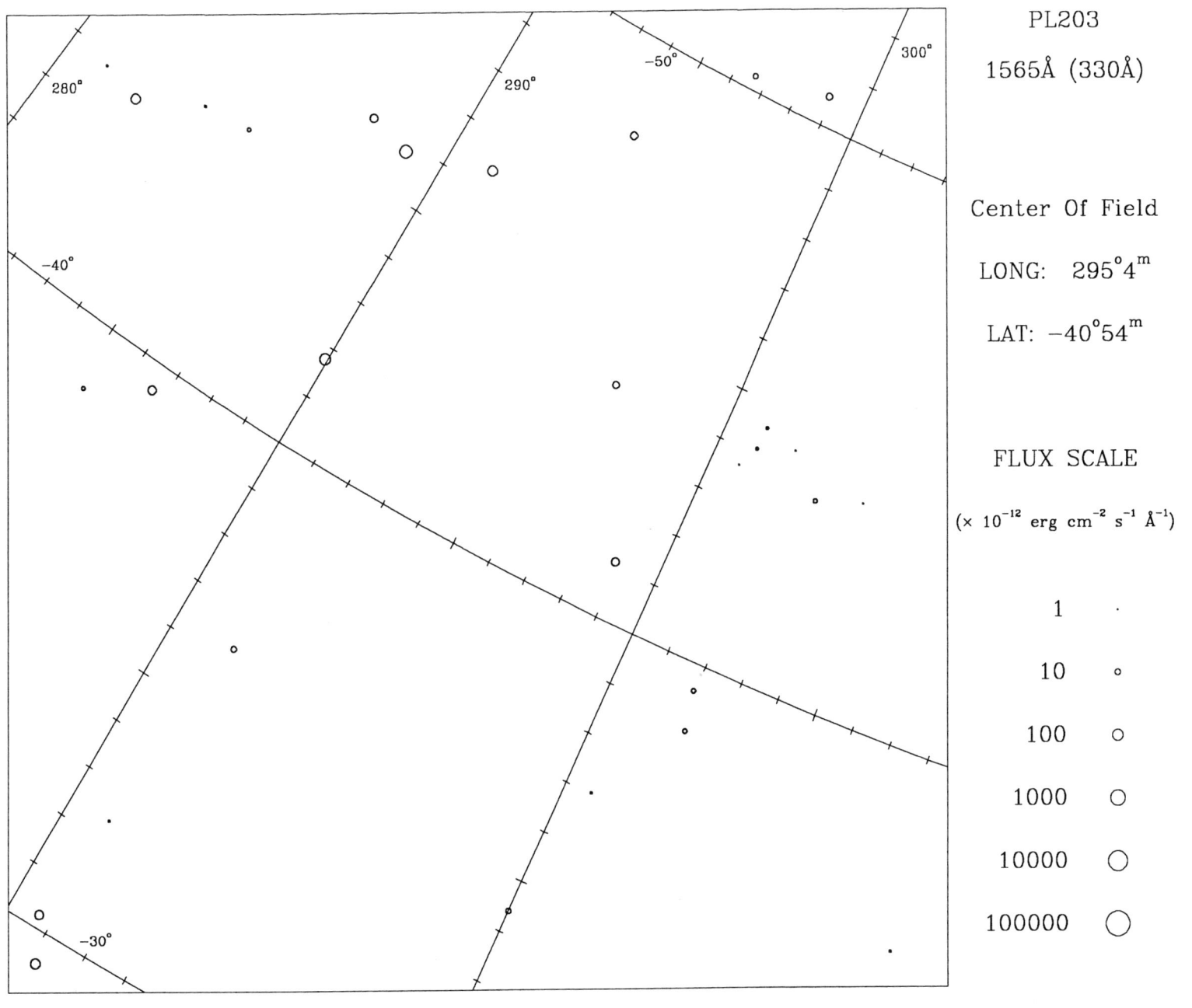

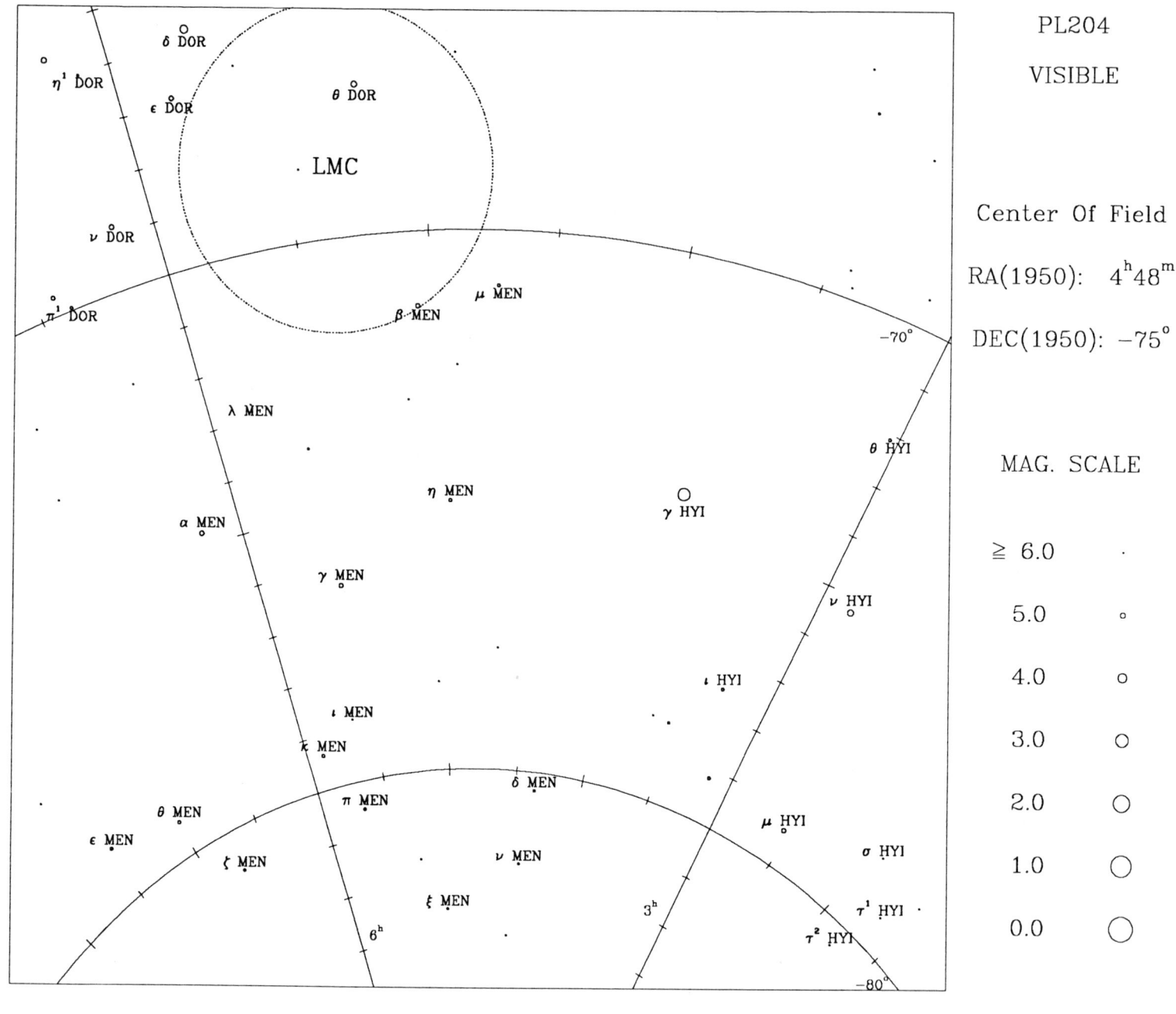

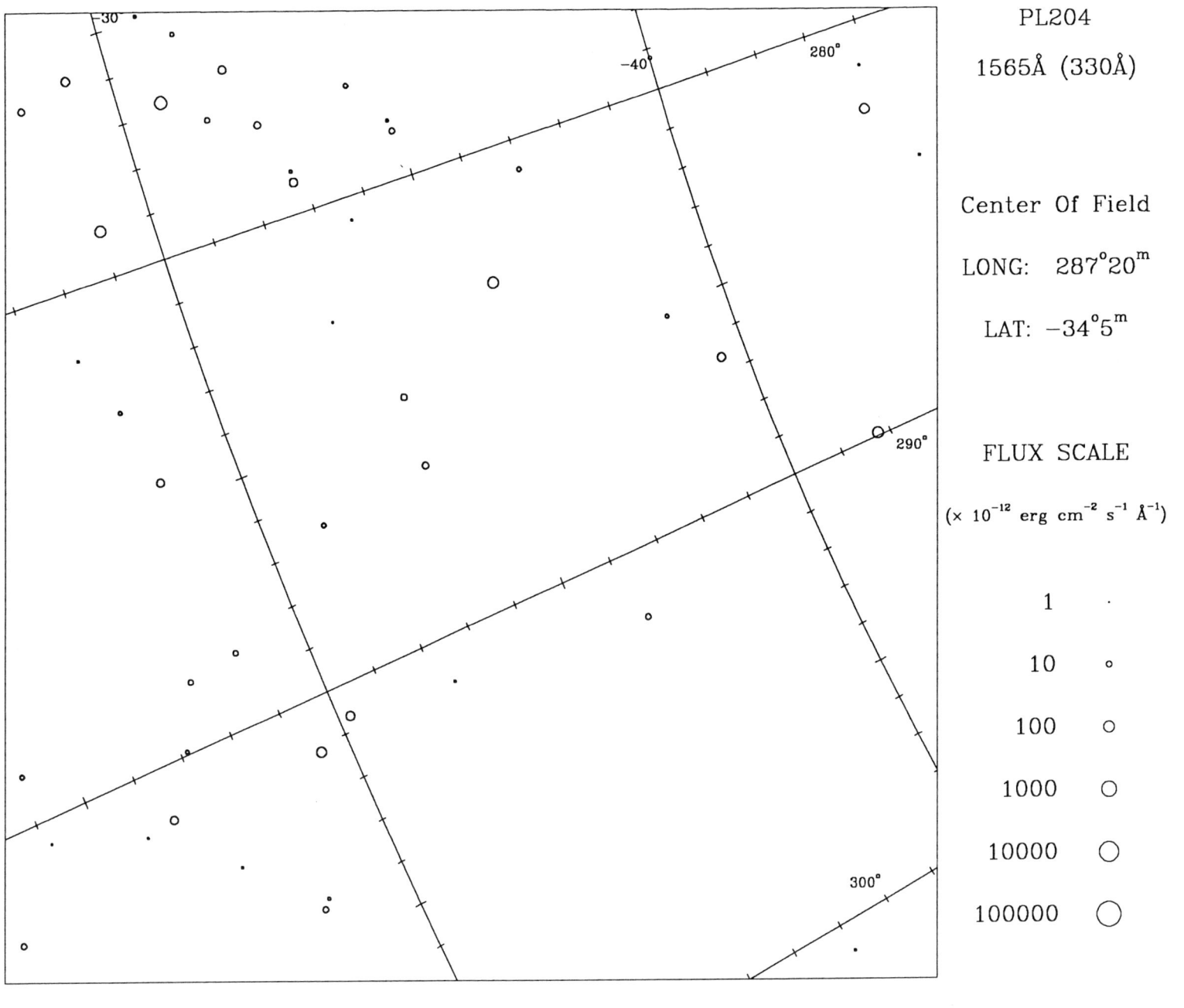

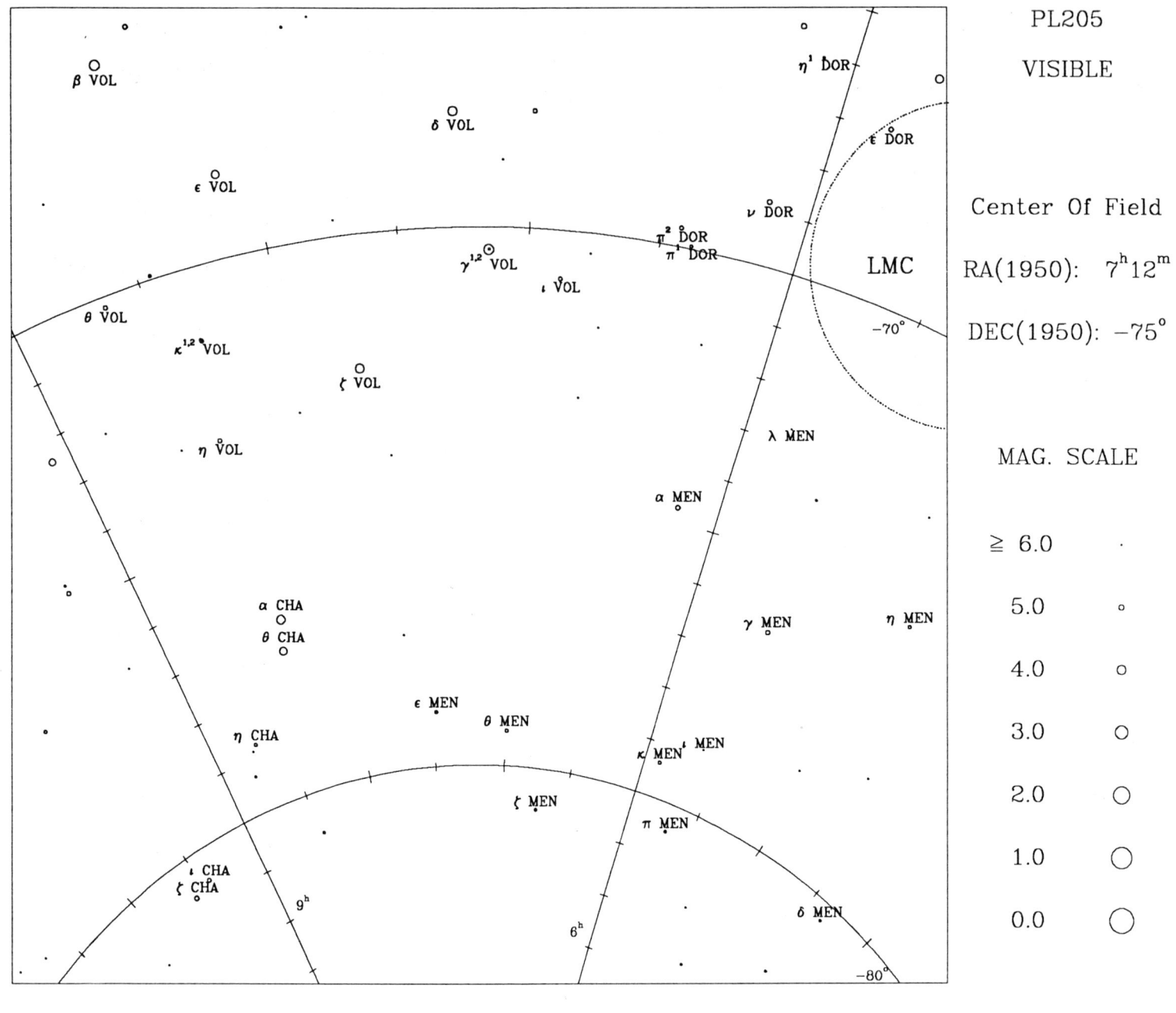

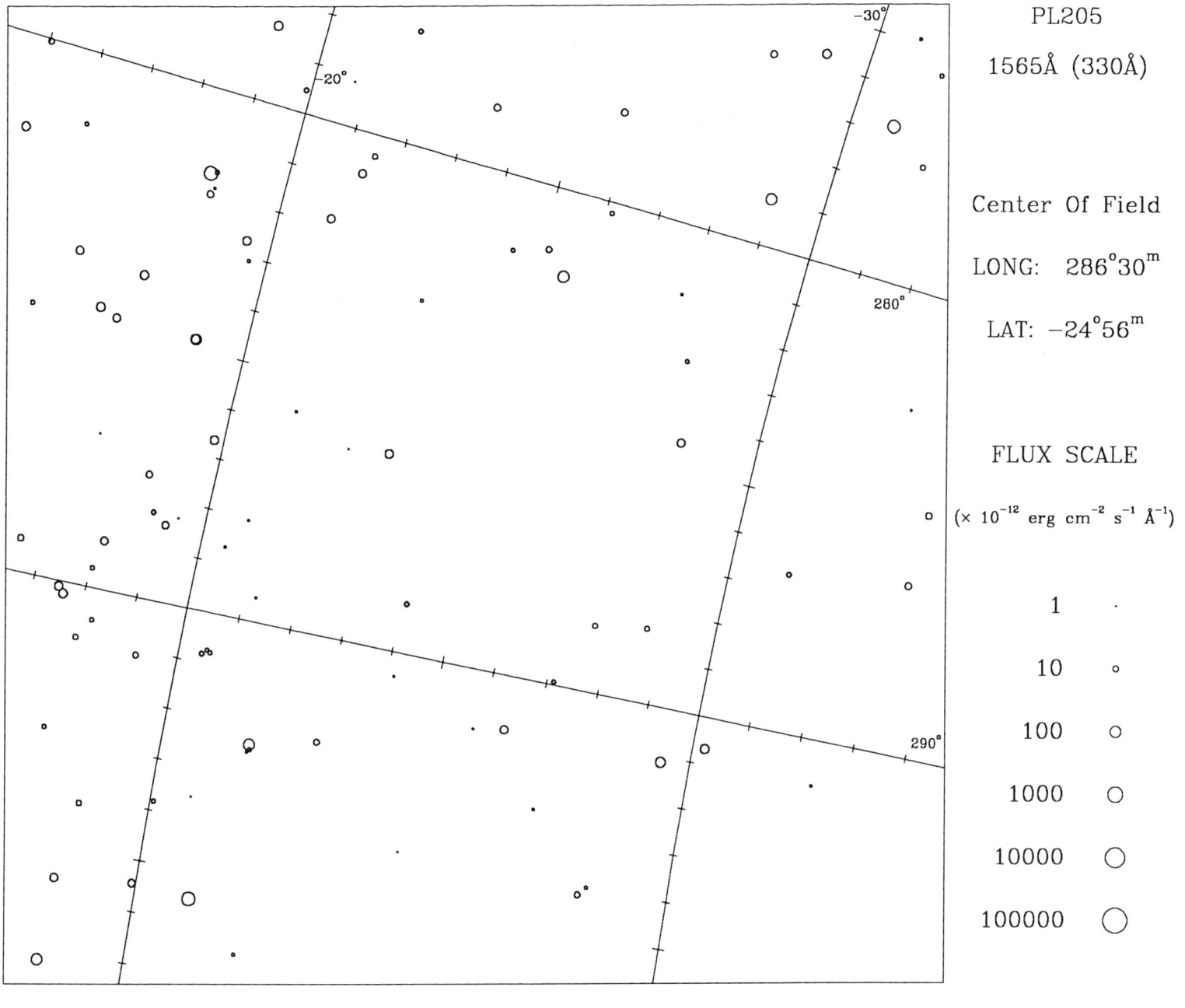

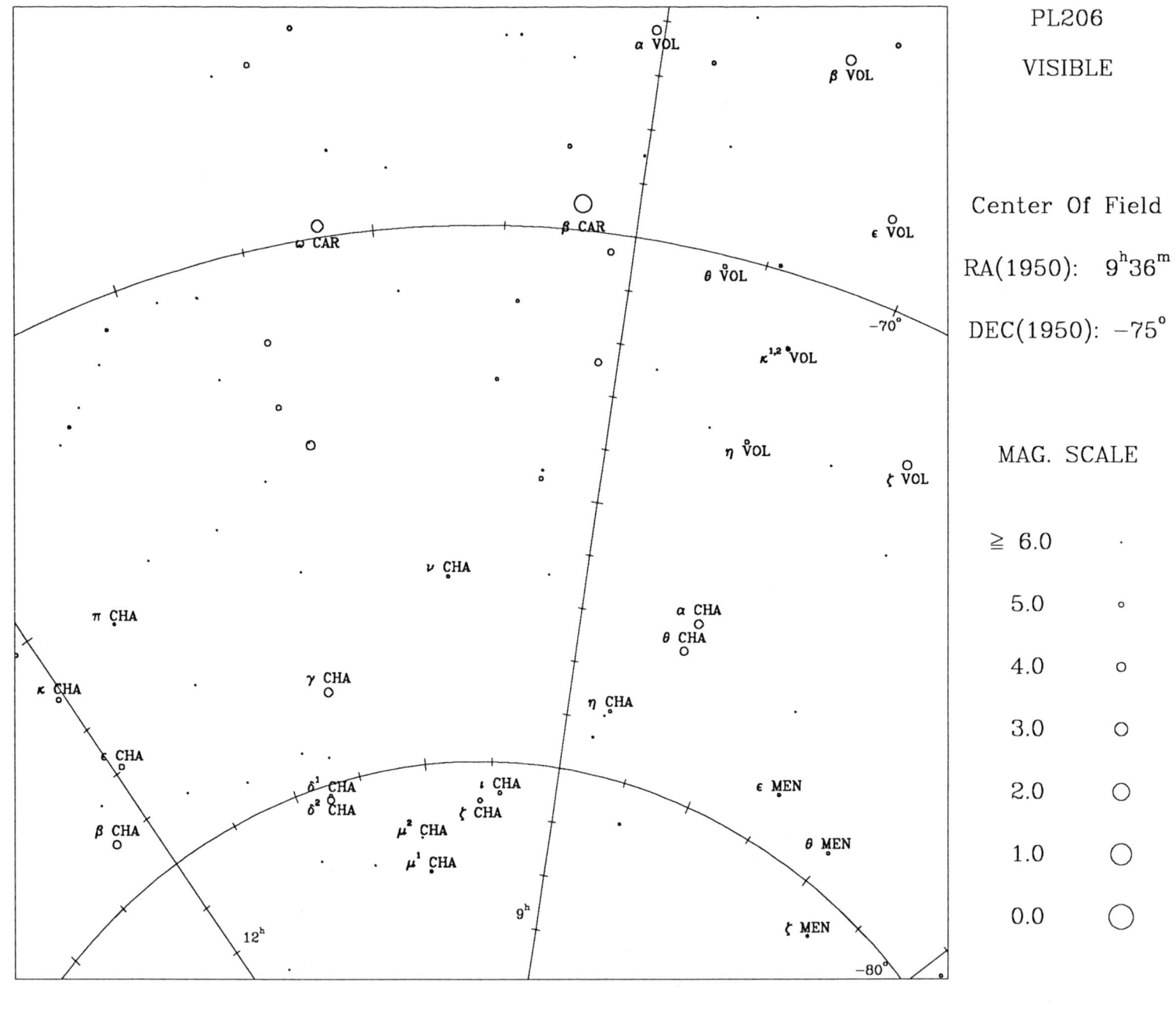

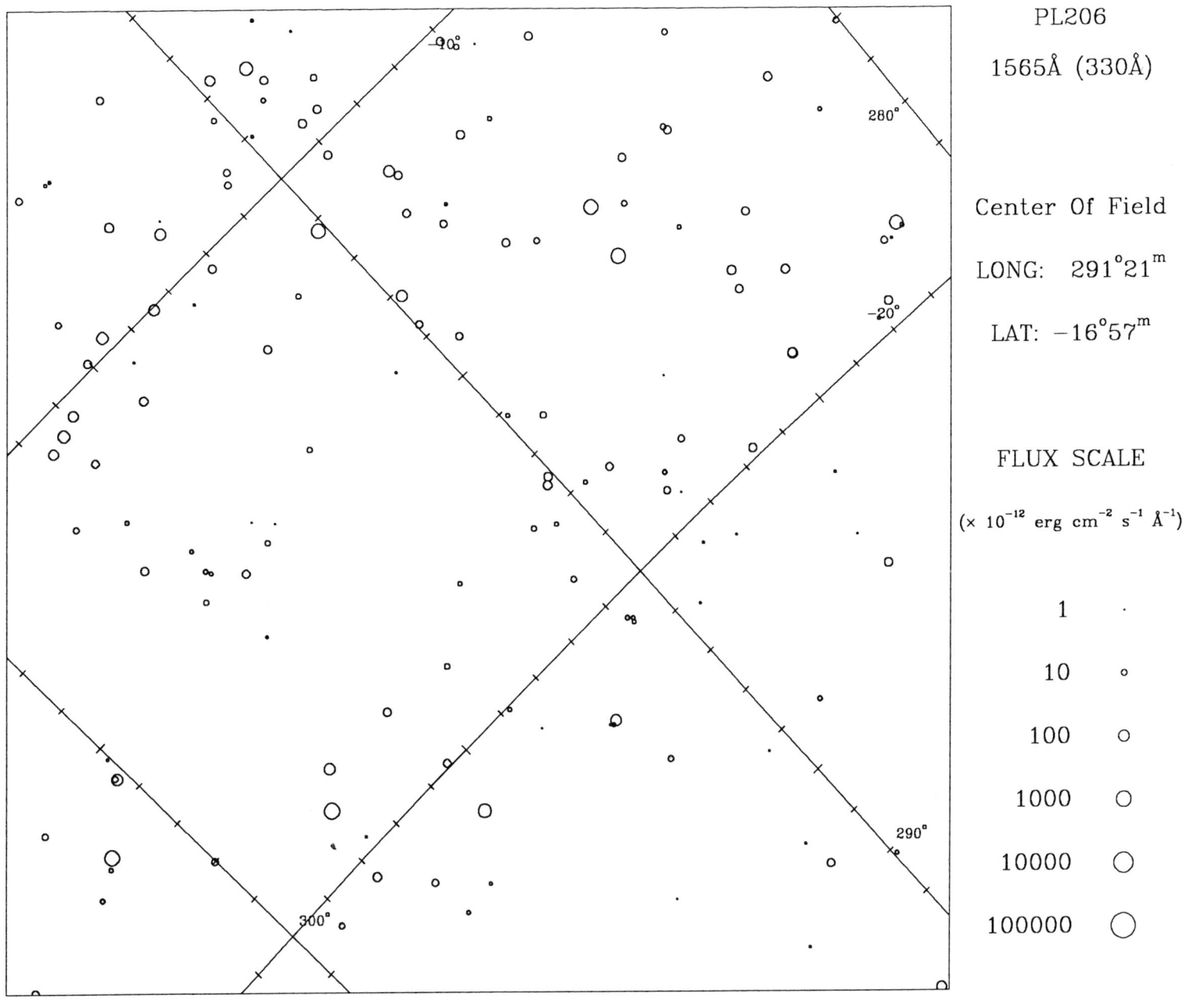

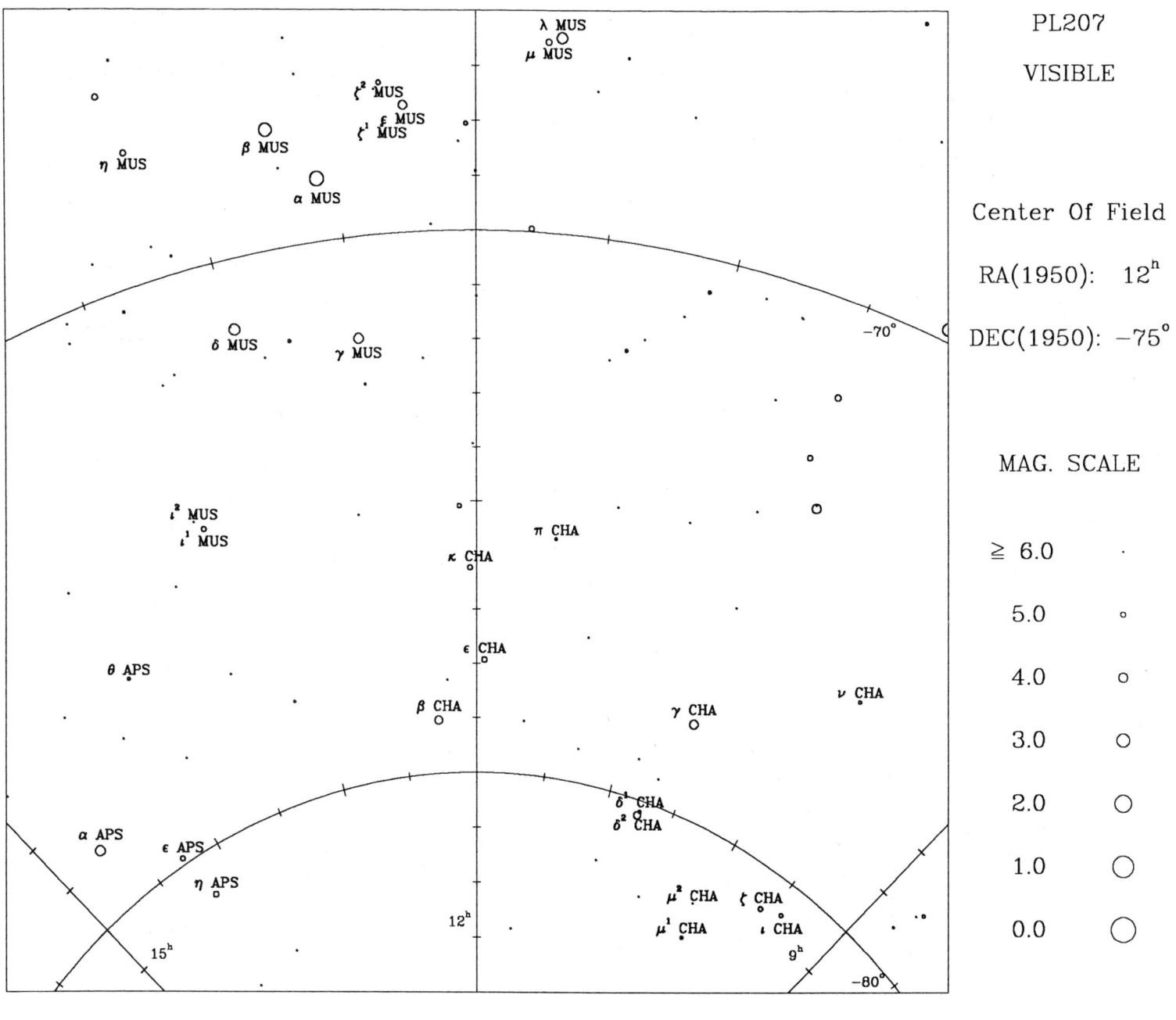

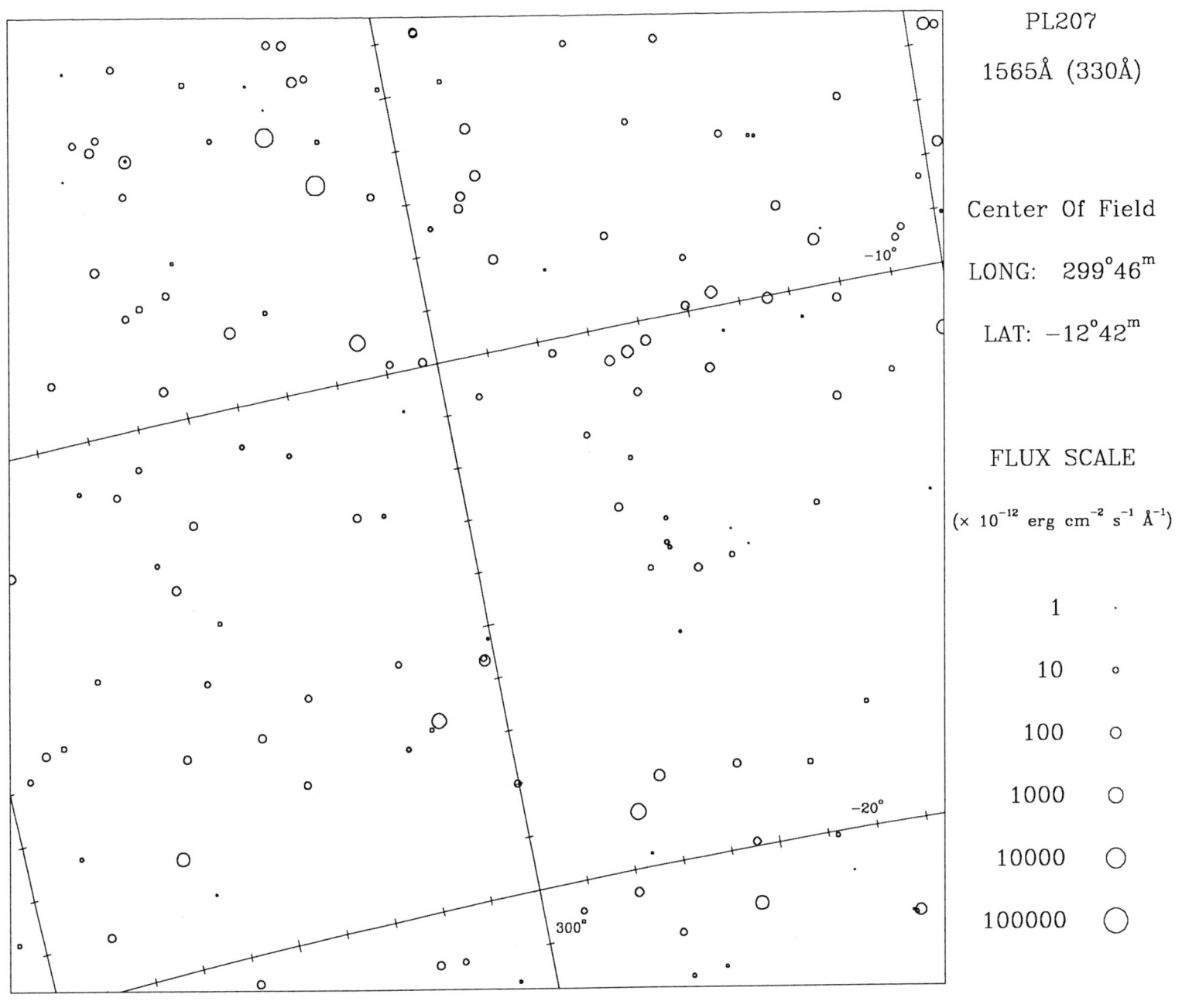

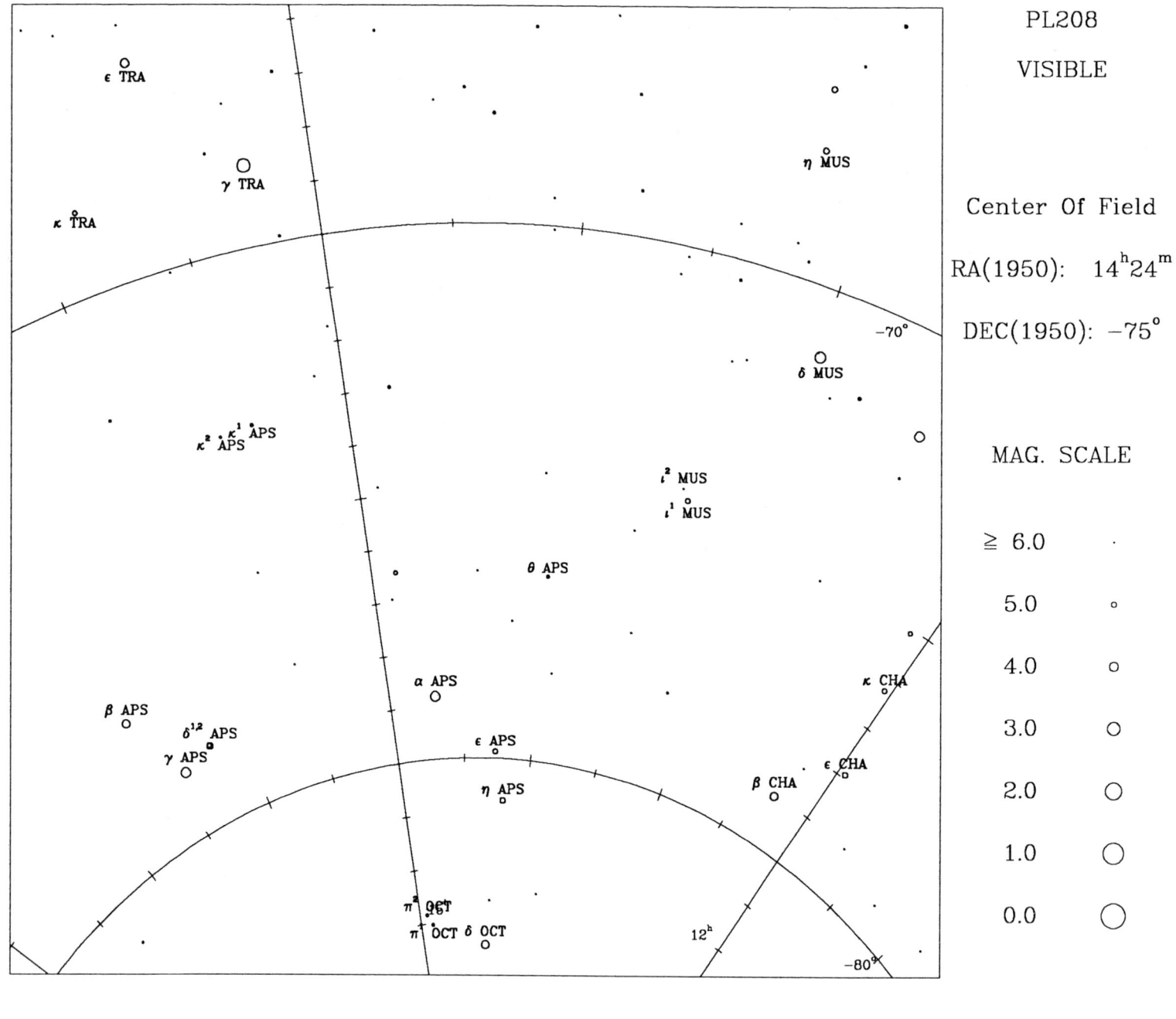

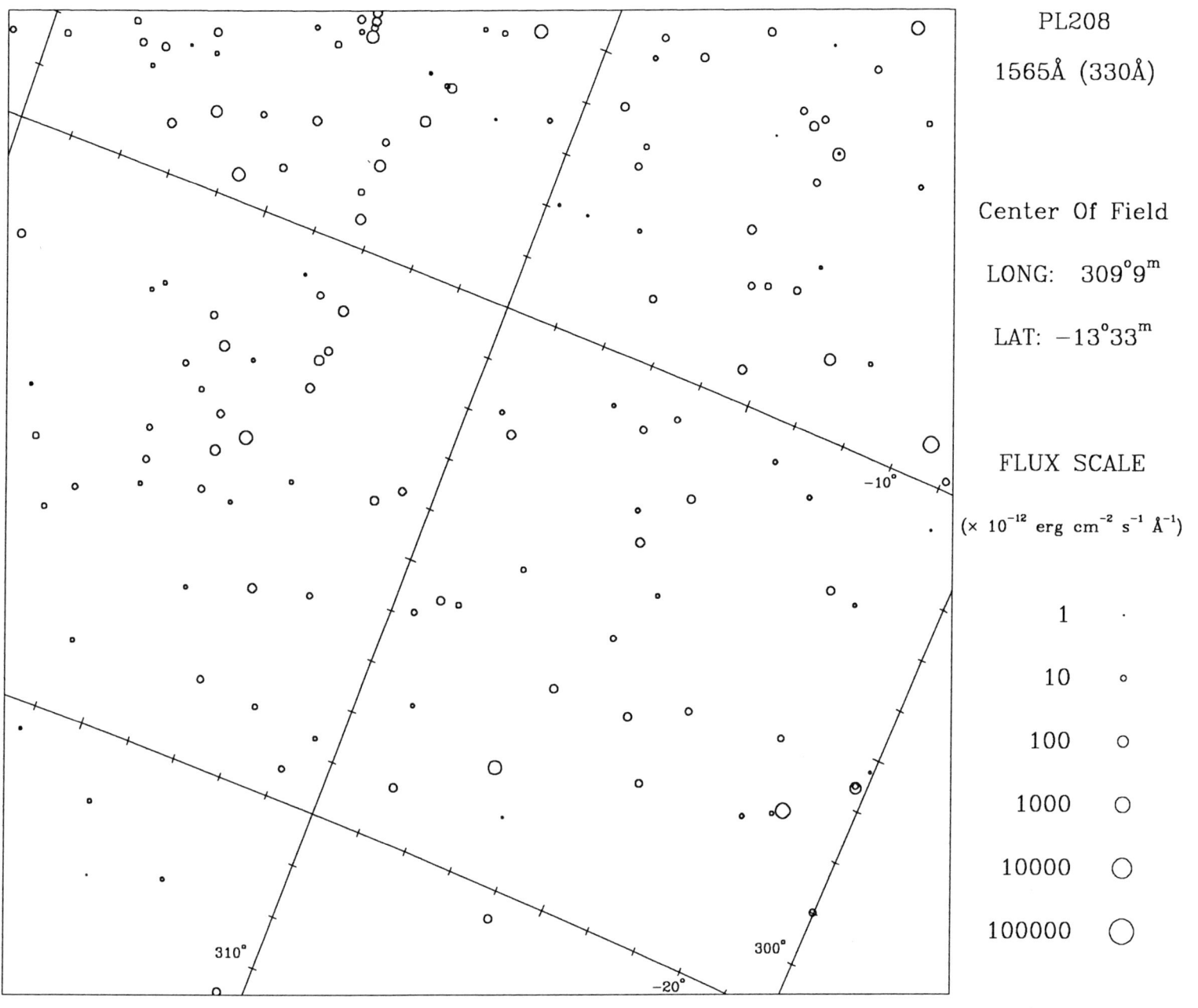

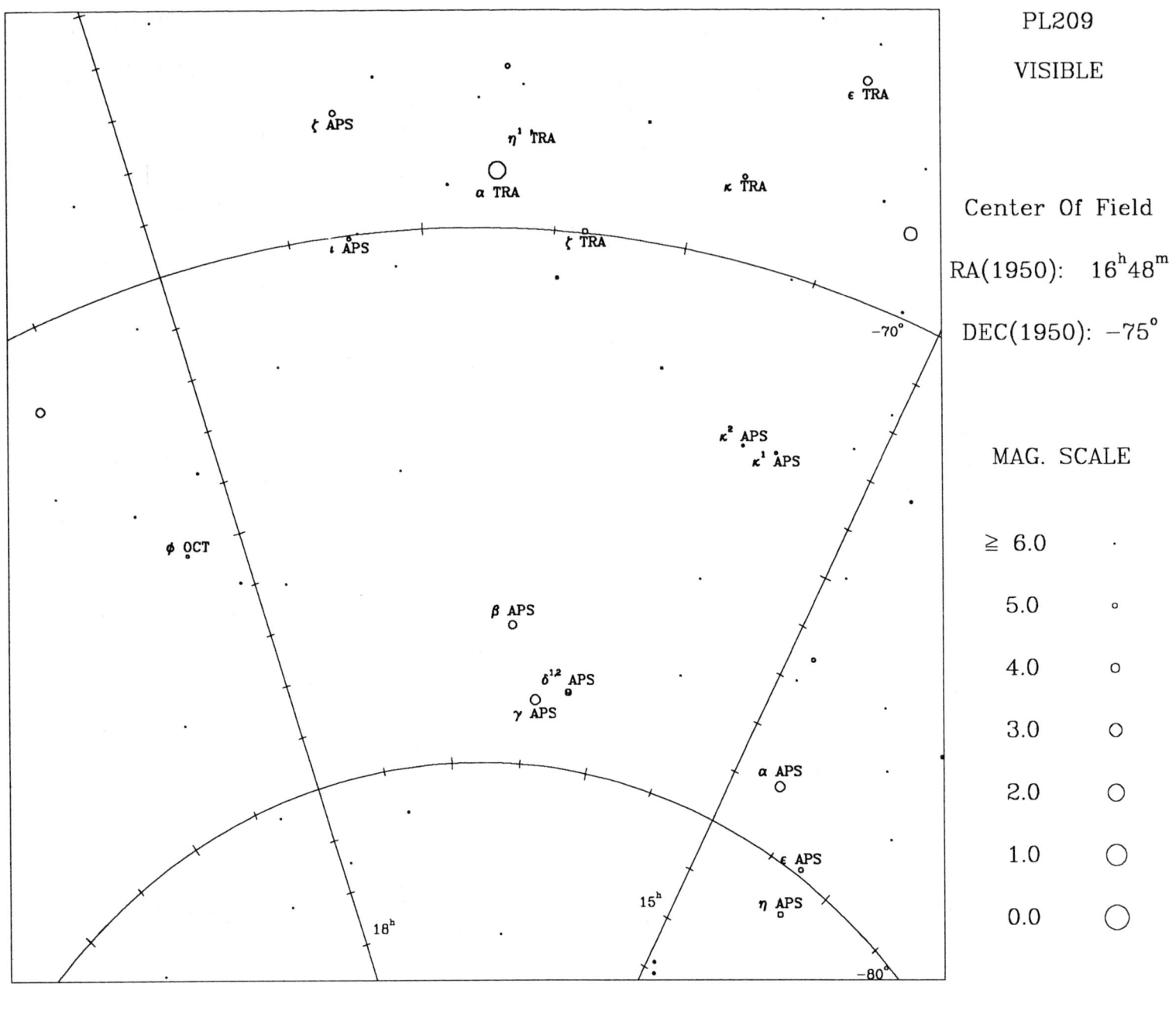

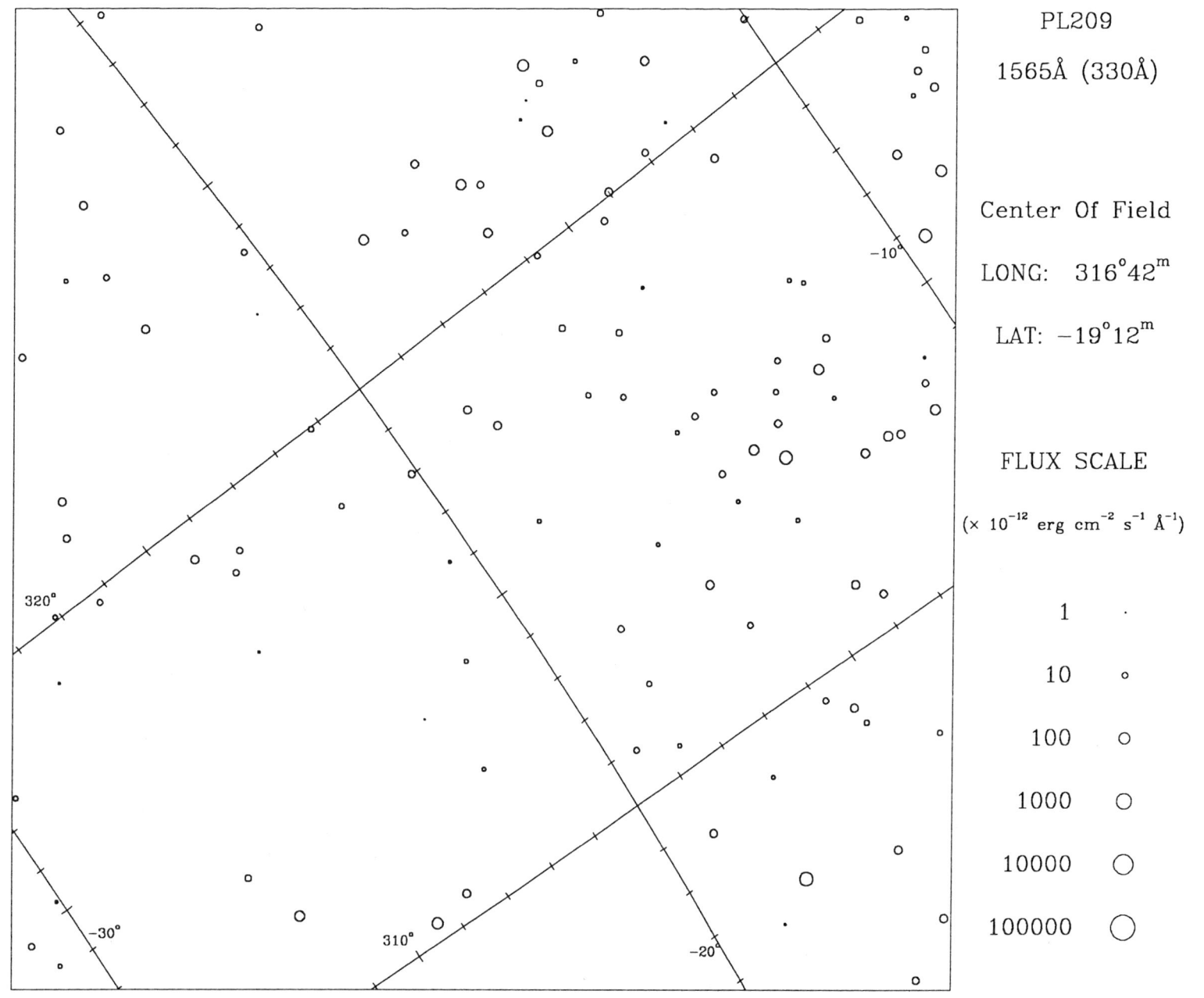

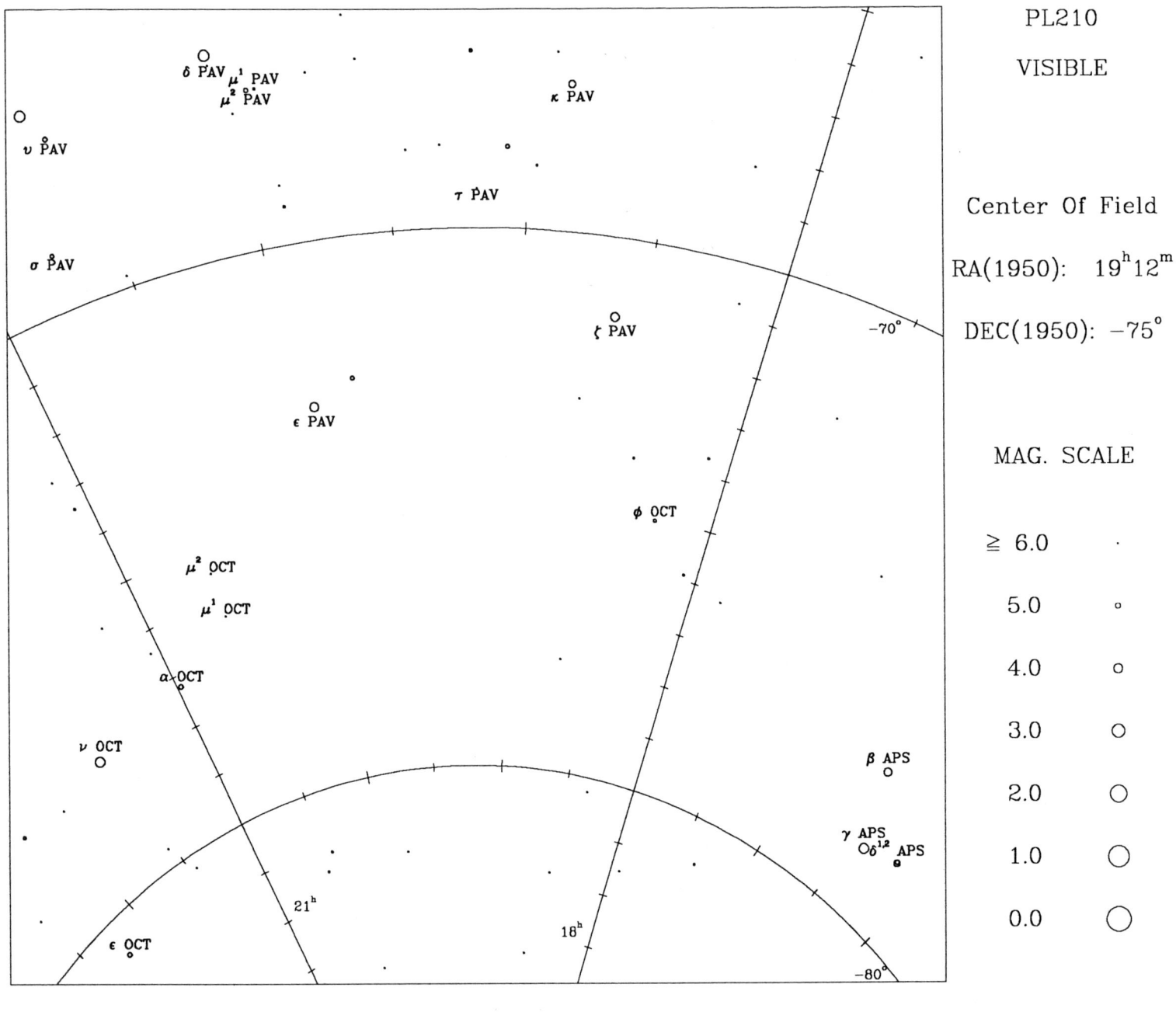

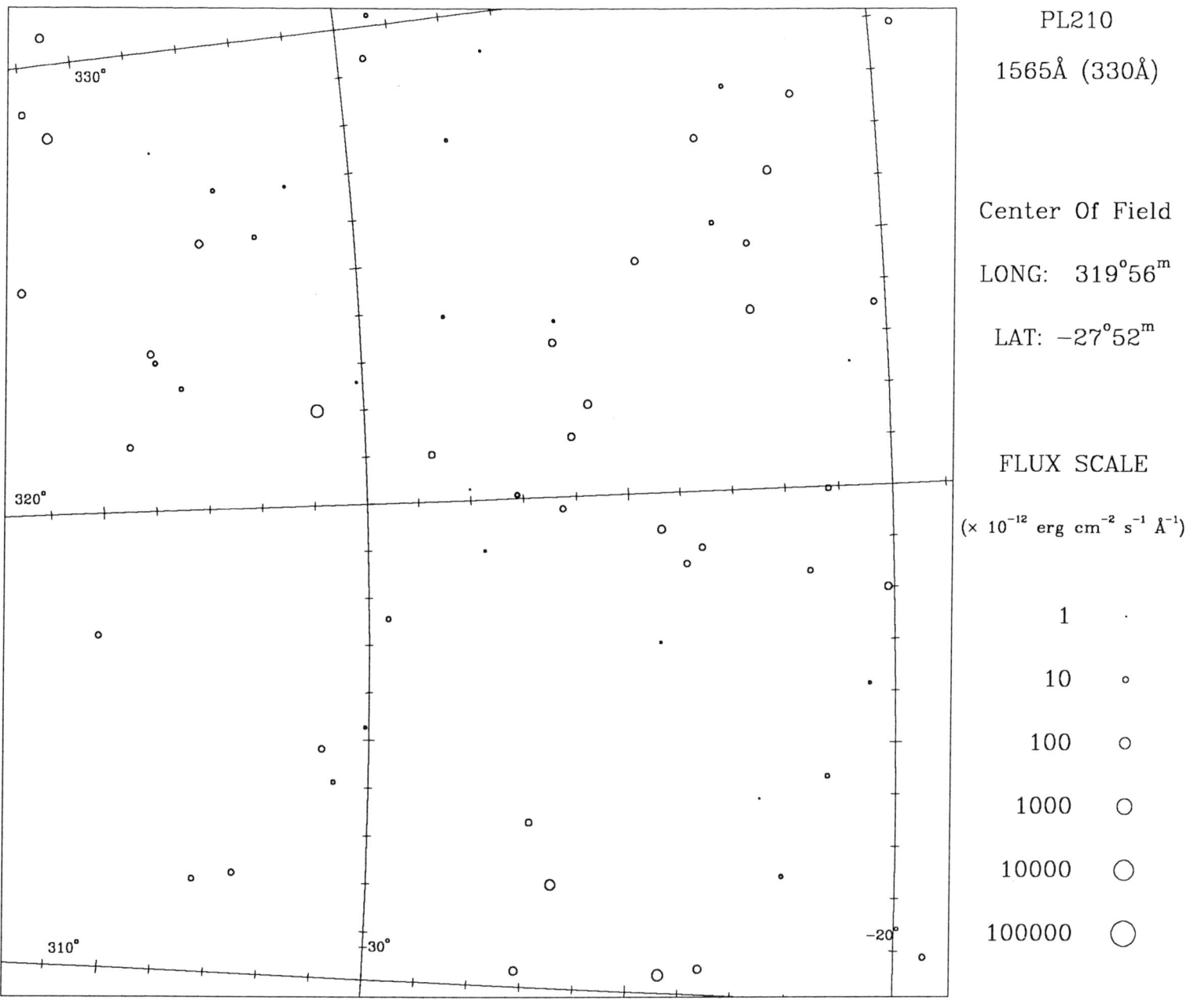

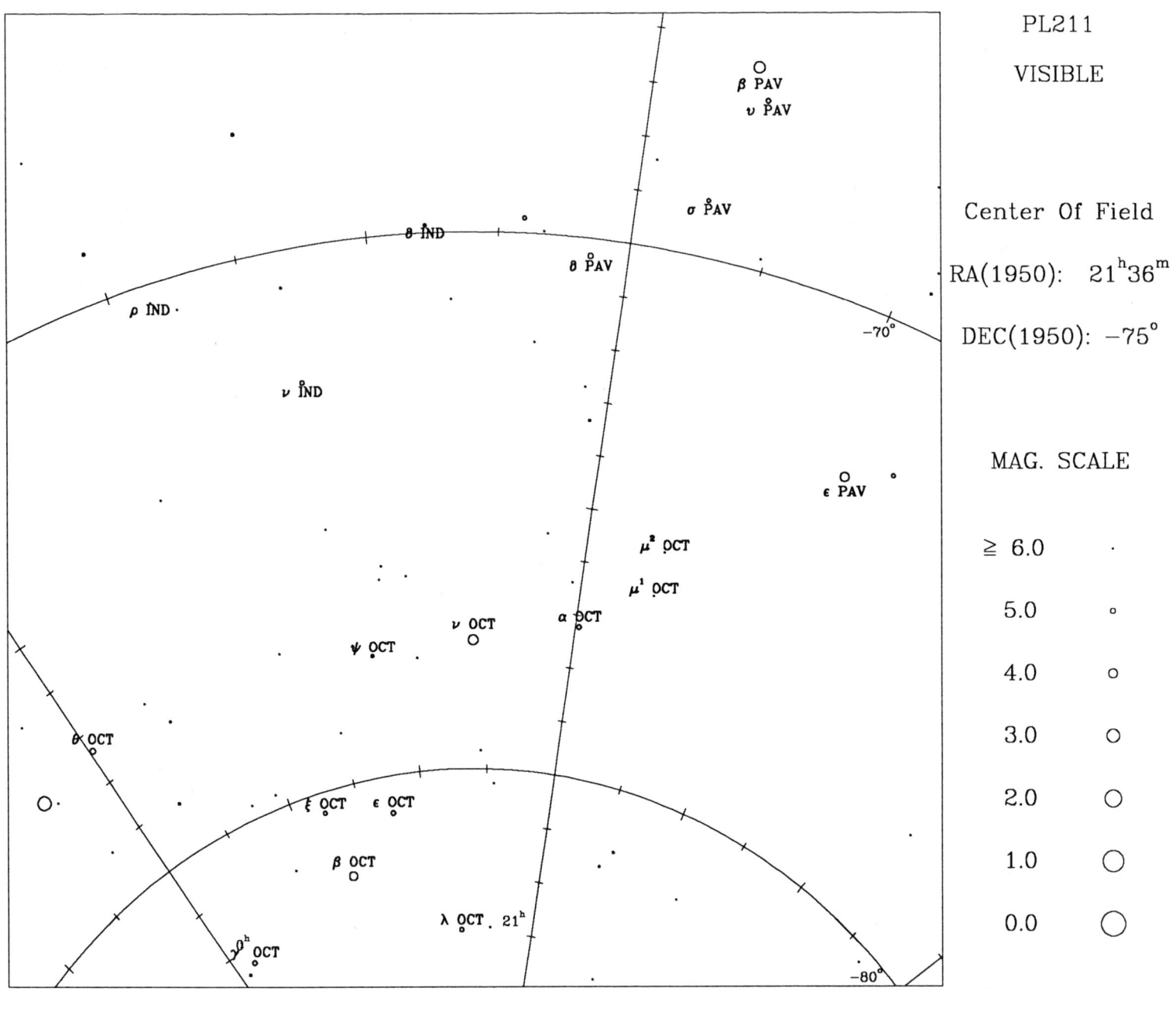

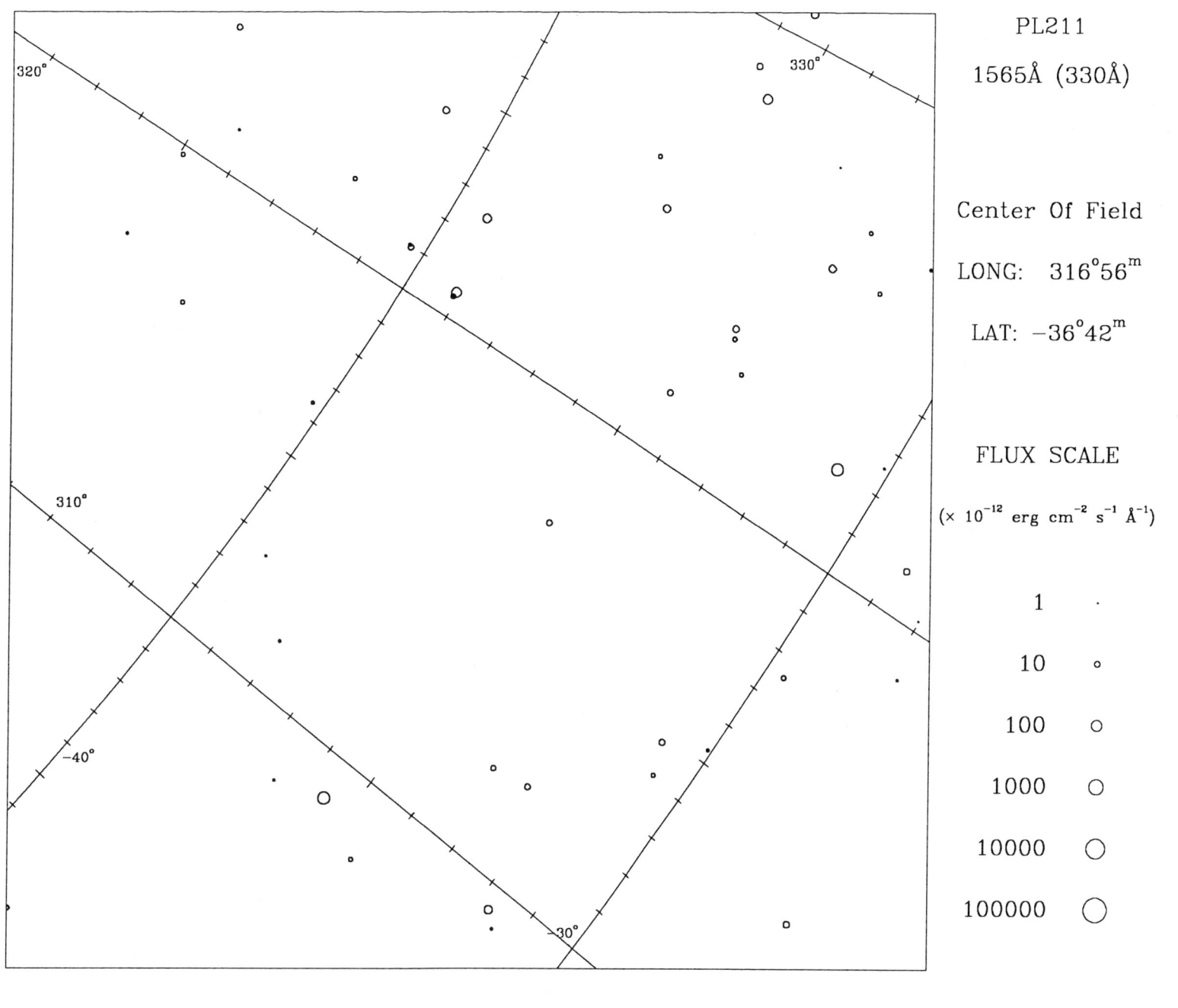

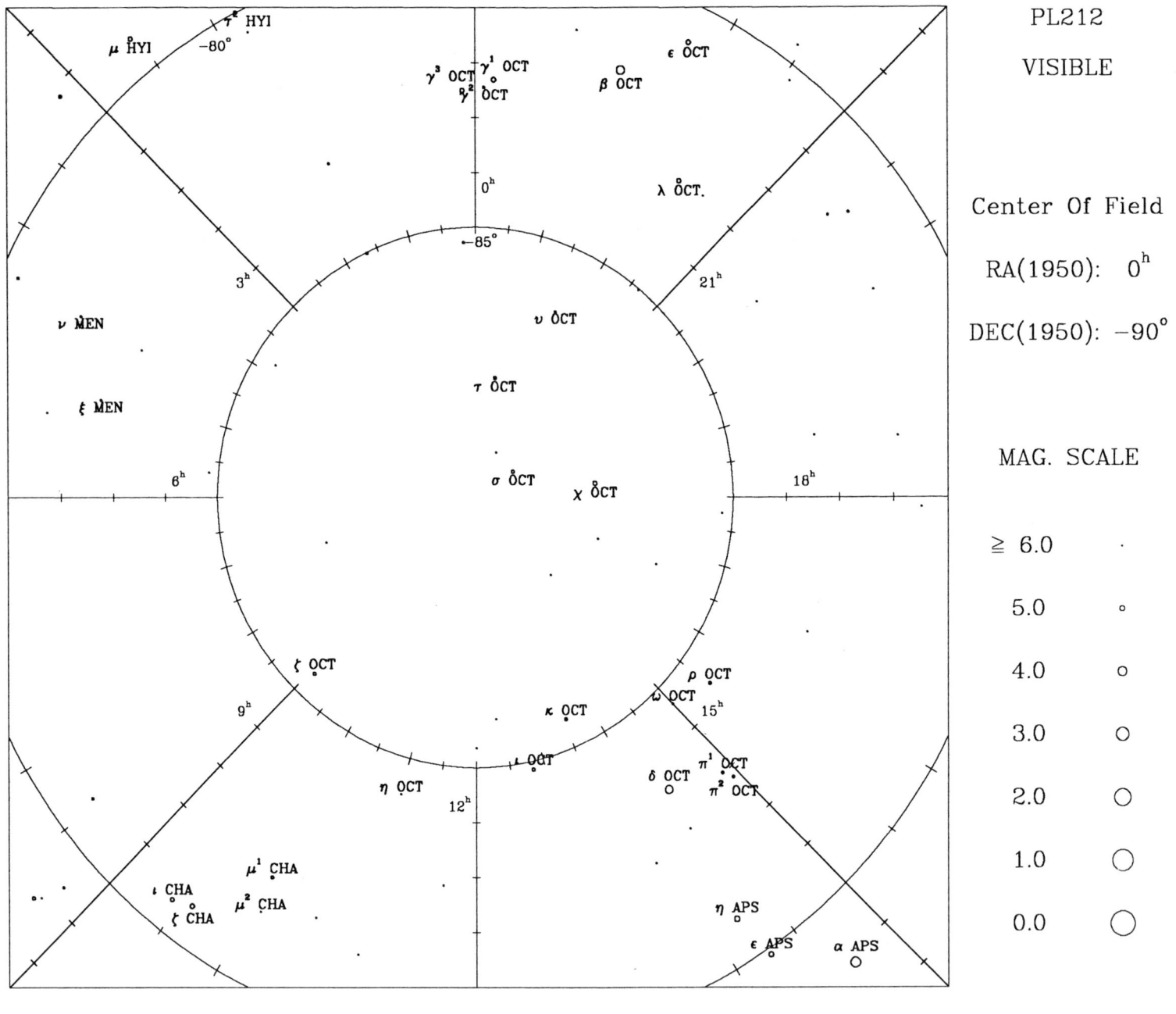

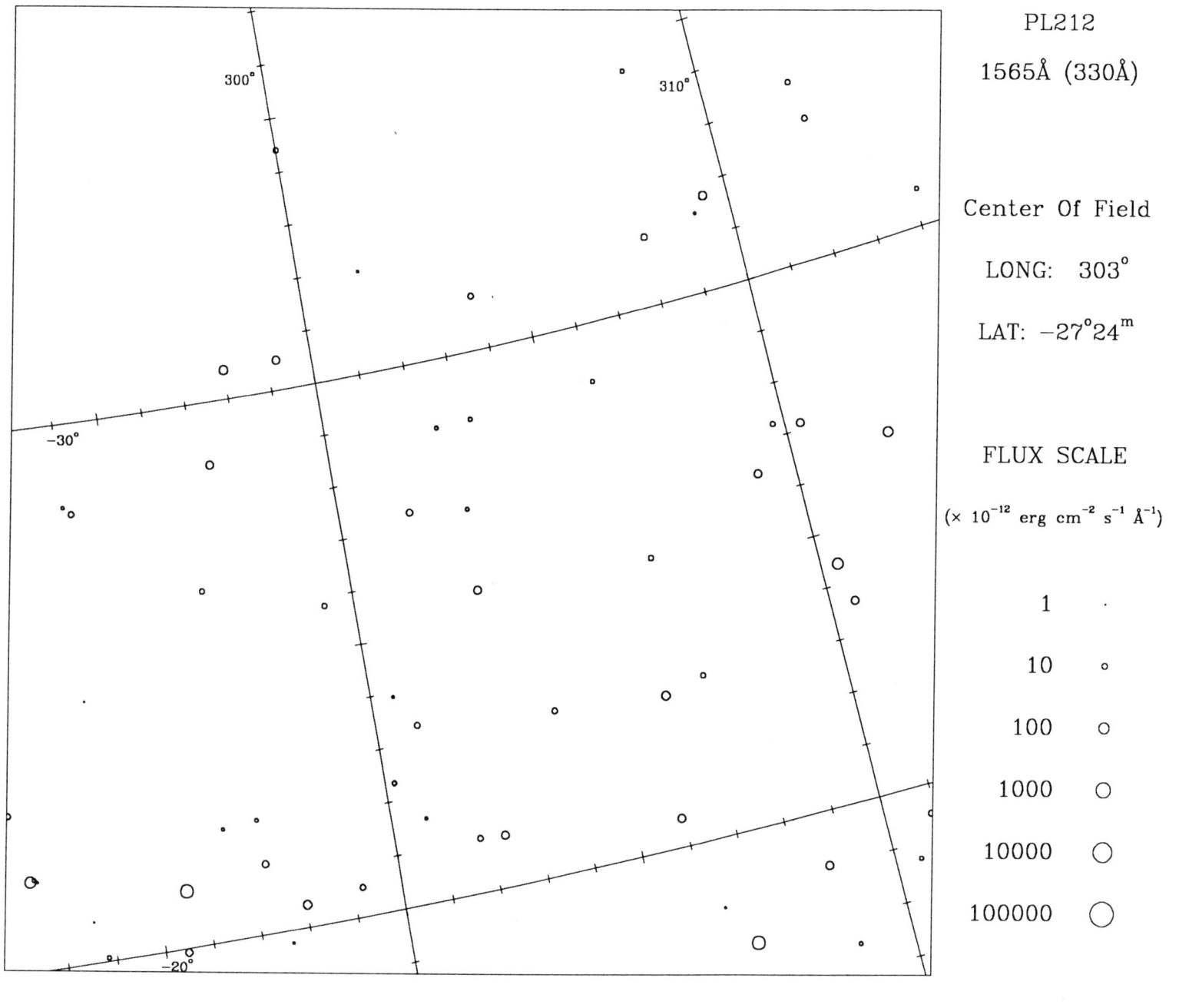

Index to Galactic Coordinates

PLATE	LONG. degrees	LAT. degrees	R.A. h m s	DEC. d m s
PL060	80.8	86.5	13 0 0.0	30 0 0.0
PL059	196.5	79.0	12 0 0.0	30 0 0.0
PL084	315.2	77.3	13 0 0.0	15 0 0.0
PL061	46.7	74.2	14 0 0.0	30 0 0.0
PL083	258.0	73.2	12 0 0.0	15 0 0.0
PL038	109.9	71.8	13 12 0.0	45 0 0.0
PL037	148.8	69.9	12 0 0.0	45 0 0.0
PL085	359.7	69.4	14 0 0.0	15 0 0.0
PL058	200.3	66.0	11 0 0.0	30 0 0.0
PL039	82.2	64.2	14 24 0.0	45 0 0.0
PL108	309.0	62.5	13 0 0.0	0 0 0.0
PL082	233.5	61.8	11 0 0.0	15 0 0.0
PL062	46.1	61.3	15 0 0.0	30 0 0.0
PL036	168.8	60.2	10 48 0.0	45 0 0.0
PL107	277.7	60.2	12 0 0.0	0 0 0.0
PL109	337.8	57.7	14 0 0.0	0 0 0.0
PL020	123.2	57.4	12 48 0.0	60 0 0.0
PL086	17.6	57.2	15 0 0.0	15 0 0.0
PL021	103.2	53.6	14 24 0.0	60 0 0.0
PL019	143.2	53.4	11 12 0.0	60 0 0.0
PL057	198.7	53.1	10 0 0.0	30 0 0.0
PL040	72.3	52.5	15 36 0.0	45 0 0.0
PL106	254.8	52.1	11 0 0.0	0 0 0.0
PL081	221.7	48.9	10 0 0.0	15 0 0.0
PL063	48.3	48.3	16 0 0.0	30 0 0.0
PL110	357.4	48.3	15 0 0.0	0 0 0.0
PL035	175.1	48.0	9 36 0.0	45 0 0.0
PL132	306.9	47.5	13 0 0.0	-15 0 0.0
PL131	285.9	46.0	12 0 0.0	-15 0 0.0
PL022	91.9	44.2	16 0 0.0	60 0 0.0
PL133	327.3	44.2	14 0 0.0	-15 0 0.0
PL087	27.5	44.1	16 0 0.0	15 0 0.0
PL018	154.2	44.0	9 36 0.0	60 0 0.0
PL007	127.2	42.0	12 0 0.0	75 0 0.0
PL105	239.9	41.1	10 0 0.0	0 0 0.0
PL008	115.1	40.9	14 24 0.0	75 0 0.0
PL056	195.6	40.2	9 0 0.0	30 0 0.0
PL130	267.7	40.0	11 0 0.0	-15 0 0.0
PL041	70.2	39.9	16 48 0.0	45 0 0.0
PL134	343.9	37.0	15 0 0.0	-15 0 0.0
PL006	136.9	36.7	9 36 0.0	75 0 0.0
PL111	10.3	36.7	16 0 0.0	0 0 0.0
PL080	213.8	35.7	9 0 0.0	15 0 0.0
PL064	51.8	35.6	17 0 0.0	30 0 0.0
PL034	175.5	35.3	8 24 0.0	45 0 0.0
PL009	107.3	34.1	16 48 0.0	75 0 0.0
PL023	88.7	32.5	17 36 0.0	60 0 0.0
PL156	305.8	32.5	13 0 0.0	-30 0 0.0
PL017	157.3	32.3	8 0 0.0	60 0 0.0
PL155	290.6	31.4	12 0 0.0	-30 0 0.0

Index to Galactic Coordinates

PLATE	LONG. degrees	LAT. degrees	R.A. h m s	DEC. d m s
PL129	253.7	31.0	10 0 0.0	-15 0 0.0
PL088	34.8	30.8	17 0 0.0	15 0 0.0
PL157	320.8	30.1	14 0 0.0	-30 0 0.0
PL104	229.5	28.7	9 0 0.0	0 0 0.0
PL005	139.9	27.9	7 12 0.0	75 0 0.0
PL055	191.6	27.7	8 0 0.0	30 0 0.0
PL001	123.0	27.4	0 0 0.0	90 0 0.0
PL042	72.0	27.3	18 0 0.0	45 0 0.0
PL135	356.5	27.2	16 0 0.0	-15 0 0.0
PL154	276.6	27.0	11 0 0.0	-30 0 0.0
PL010	106.5	24.9	19 12 0.0	75 0 0.0
PL158	334.0	24.6	15 0 0.0	-30 0 0.0
PL112	19.7	24.0	17 0 0.0	0 0 0.0
PL065	56.0	23.2	18 0 0.0	30 0 0.0
PL033	172.8	22.8	7 12 0.0	45 0 0.0
PL079	207.1	22.4	8 0 0.0	15 0 0.0
PL024	90.9	20.7	19 12 0.0	60 0 0.0
PL016	155.0	20.5	6 24 0.0	60 0 0.0
PL128	243.1	20.2	9 0 0.0	-15 0 0.0
PL153	264.8	19.8	10 0 0.0	-30 0 0.0
PL004	136.7	19.2	4 48 0.0	75 0 0.0
PL089	41.2	17.5	18 0 0.0	15 0 0.0
PL178	307.3	17.4	13 12 0.0	-45 0 0.0
PL011	111.4	16.9	21 36 0.0	75 0 0.0
PL159	345.0	16.7	16 0 0.0	-30 0 0.0
PL177	294.0	16.7	12 0 0.0	-45 0 0.0
PL136	6.2	15.9	17 0 0.0	-15 0 0.0
PL103	221.4	15.7	8 0 0.0	0 0 0.0
PL054	186.8	15.5	7 0 0.0	30 0 0.0
PL043	76.2	15.2	19 12 0.0	45 0 0.0
PL179	320.1	14.4	14 24 0.0	-45 0 0.0
PL003	129.2	13.5	2 24 0.0	75 0 0.0
PL002	119.8	12.7	0 0 0.0	75 0 0.0
PL176	281.6	12.5	10 48 0.0	-45 0 0.0
PL032	167.8	11.1	6 0 0.0	45 0 0.0
PL066	61.2	11.1	19 0 0.0	30 0 0.0
PL113	27.3	10.9	18 0 0.0	0 0 0.0
PL152	255.2	10.7	9 0 0.0	-30 0 0.0
PL025	96.8	10.3	20 48 0.0	60 0 0.0
PL015	149.0	10.1	4 48 0.0	60 0 0.0
PL078	200.8	9.2	7 0 0.0	15 0 0.0
PL127	234.7	8.2	8 0 0.0	-15 0 0.0
PL180	331.4	8.2	15 36 0.0	-45 0 0.0
PL160	353.9	7.0	17 0 0.0	-30 0 0.0
PL175	271.0	5.3	9 36 0.0	-45 0 0.0
PL090	47.6	4.5	19 0 0.0	15 0 0.0
PL044	82.5	4.2	20 24 0.0	45 0 0.0
PL053	181.0	3.8	6 0 0.0	30 0 0.0
PL137	14.1	3.6	18 0 0.0	-15 0 0.0
PL195	302.9	2.6	12 48 0.0	-60 0 0.0

Index to Galactic Coordinates

PLATE	LONG. degrees	LAT. degrees	R.A. h m s	DEC. d m s
PL102	214.3	2.4	7 0 0.0	0 0 0.0
PL026	105.8	2.3	22 24 0.0	60 0 0.0
PL014	140.0	2.2	3 12 0.0	60 0 0.0
PL031	160.7	0.5	4 48 0.0	45 0 0.0
PL194	291.1	0.4	11 12 0.0	-60 0 0.0
PL196	314.6	0.4	14 24 0.0	-60 0 0.0
PL151	247.4	0.2	8 0 0.0	-30 0 0.0
PL067	67.4	-0.2	20 0 0.0	30 0 0.0
PL181	340.7	-0.5	16 48 0.0	-45 0 0.0
PL012	116.9	-2.0	0 0 0.0	60 0 0.0
PL013	128.8	-2.1	1 36 0.0	60 0 0.0
PL114	34.3	-2.4	19 0 0.0	0 0 0.0
PL077	194.1	-3.6	6 0 0.0	15 0 0.0
PL161	1.0	-3.8	18 0 0.0	-30 0 0.0
PL174	262.5	-4.2	8 24 0.0	-45 0 0.0
PL126	227.6	-4.5	7 0 0.0	-15 0 0.0
PL045	91.0	-5.3	21 36 0.0	45 0 0.0
PL197	324.8	-5.7	16 0 0.0	-60 0 0.0
PL193	281.0	-5.9	9 36 0.0	-60 0 0.0
PL052	173.9	-7.0	5 0 0.0	30 0 0.0
PL030	151.4	-8.2	3 36 0.0	45 0 0.0
PL091	54.7	-8.2	20 0 0.0	15 0 0.0
PL138	20.8	-9.2	19 0 0.0	-15 0 0.0
PL068	75.2	-10.7	21 0 0.0	30 0 0.0
PL101	207.3	-10.9	6 0 0.0	0 0 0.0
PL150	241.2	-11.1	7 0 0.0	-30 0 0.0
PL182	347.8	-11.1	18 0 0.0	-45 0 0.0
PL046	101.6	-12.5	22 48 0.0	45 0 0.0
PL207	299.8	-12.7	12 0 0.0	-75 0 0.0
PL208	309.2	-13.5	14 24 0.0	-75 0 0.0
PL029	140.1	-14.4	2 24 0.0	45 0 0.0
PL198	332.5	-15.0	17 36 0.0	-60 0 0.0
PL173	256.2	-15.2	7 12 0.0	-45 0 0.0
PL192	273.4	-15.3	8 0 0.0	-60 0 0.0
PL162	6.8	-15.5	19 0 0.0	-30 0 0.0
PL115	41.4	-15.7	20 0 0.0	0 0 0.0
PL076	186.2	-15.9	5 0 0.0	15 0 0.0
PL027	114.0	-16.7	0 0 0.0	45 0 0.0
PL051	165.0	-16.7	4 0 0.0	30 0 0.0
PL206	291.4	-16.9	9 36 0.0	-75 0 0.0
PL028	127.3	-17.4	1 12 0.0	45 0 0.0
PL125	221.2	-17.5	6 0 0.0	-15 0 0.0
PL209	316.7	-19.2	16 48 0.0	-75 0 0.0
PL069	84.8	-19.8	22 0 0.0	30 0 0.0
PL092	63.1	-20.2	21 0 0.0	15 0 0.0
PL139	27.1	-22.4	20 0 0.0	-15 0 0.0
PL183	352.8	-22.8	19 12 0.0	-45 0 0.0
PL149	236.0	-23.2	6 0 0.0	-30 0 0.0
PL100	199.7	-24.0	5 0 0.0	0 0 0.0
PL050	154.0	-24.6	3 0 0.0	30 0 0.0

Index to Galactic Coordinates

PLATE	LONG. degrees	LAT. degrees	R.A. h m s	DEC. d m s
PL205	286.5	-24.9	7 12 0.0	-75 0 0.0
PL199	336.7	-26.3	19 12 0.0	-60 0 0.0
PL191	269.2	-26.5	6 24 0.0	-60 0 0.0
PL070	96.6	-27.0	23 0 0.0	30 0 0.0
PL075	176.5	-27.2	4 0 0.0	15 0 0.0
PL172	252.0	-27.3	6 0 0.0	-45 0 0.0
PL212	303.0	-27.4	0 0 0.0	-90 0 0.0
PL163	11.6	-27.7	20 0 0.0	-30 0 0.0
PL210	319.9	-27.9	19 12 0.0	-75 0 0.0
PL116	49.5	-28.7	21 0 0.0	0 0 0.0
PL049	140.8	-30.1	2 0 0.0	30 0 0.0
PL124	214.8	-30.8	5 0 0.0	-15 0 0.0
PL093	73.7	-31.0	22 0 0.0	15 0 0.0
PL047	110.6	-31.4	0 0 0.0	30 0 0.0
PL048	125.8	-32.5	1 0 0.0	30 0 0.0
PL204	287.3	-34.1	4 48 0.0	-75 0 0.0
PL184	355.5	-35.3	20 24 0.0	-45 0 0.0
PL148	231.8	-35.6	5 0 0.0	-30 0 0.0
PL140	33.8	-35.7	21 0 0.0	-15 0 0.0
PL099	190.3	-36.7	4 0 0.0	0 0 0.0
PL211	316.9	-36.7	21 36 0.0	-75 0 0.0
PL074	163.9	-37.0	3 0 0.0	15 0 0.0
PL200	336.5	-38.2	20 48 0.0	-60 0 0.0
PL190	269.5	-38.5	4 48 0.0	-60 0 0.0
PL171	250.2	-39.9	4 48 0.0	-45 0 0.0
PL094	87.7	-40.0	23 0 0.0	15 0 0.0
PL164	15.6	-40.2	21 0 0.0	-30 0 0.0
PL203	295.1	-40.9	2 24 0.0	-75 0 0.0
PL117	59.9	-41.1	22 0 0.0	0 0 0.0
PL202	307.2	-42.0	0 0 0.0	-75 0 0.0
PL123	207.5	-44.1	4 0 0.0	-15 0 0.0
PL073	147.3	-44.2	2 0 0.0	15 0 0.0
PL071	105.9	-46.0	0 0 0.0	15 0 0.0
PL072	126.9	-47.5	1 0 0.0	15 0 0.0
PL185	355.1	-48.0	21 36 0.0	-45 0 0.0
PL098	177.4	-48.3	3 0 0.0	0 0 0.0
PL147	228.3	-48.3	4 0 0.0	-30 0 0.0
PL141	41.7	-48.9	22 0 0.0	-15 0 0.0
PL201	329.9	-49.2	22 24 0.0	-60 0 0.0
PL189	276.4	-49.4	3 12 0.0	-60 0 0.0
PL118	74.8	-52.1	23 0 0.0	0 0 0.0
PL170	252.3	-52.5	3 36 0.0	-45 0 0.0
PL165	18.7	-53.1	22 0 0.0	-30 0 0.0
PL187	314.0	-56.3	0 0 0.0	-60 0 0.0
PL188	292.4	-56.4	1 36 0.0	-60 0 0.0
PL122	197.6	-57.2	3 0 0.0	-15 0 0.0
PL097	157.8	-57.7	2 0 0.0	0 0 0.0
PL095	97.7	-60.2	0 0 0.0	0 0 0.0
PL186	348.8	-60.2	22 48 0.0	-45 0 0.0
PL146	226.1	-61.3	3 0 0.0	-30 0 0.0

Index to Galactic Coordinates

PLATE	LONG. degrees	LAT. degrees	R.A. h m s	DEC. d m s
PL142	53.5	-61.8	23 0 0.0	-15 0 0.0
PL096	129.0	-62.5	1 0 0.0	0 0 0.0
PL169	262.2	-64.2	2 24 0.0	-45 0 0.0
PL166	20.3	-66.0	23 0 0.0	-30 0 0.0
PL121	179.7	-69.4	2 0 0.0	-15 0 0.0
PL167	328.8	-69.9	0 0 0.0	-45 0 0.0
PL168	289.9	-71.8	1 12 0.0	-45 0 0.0
PL119	78.0	-73.2	0 0 0.0	-15 0 0.0
PL145	226.7	-74.2	2 0 0.0	-30 0 0.0
PL120	135.2	-77.3	1 0 0.0	-15 0 0.0
PL143	16.5	-79.0	0 0 0.0	-30 0 0.0
PL144	260.8	-86.5	1 0 0.0	-30 0 0.0

OCT 2 1 1988